4th EDITION

Nutrition Now

Judith E. Brown
University of Minnesota

THOMSON
WADSWORTH

Australia • Canada • Mexico • Singapore • Spain •
United Kingdom • United States

Publisher: Peter Marshall
Development Editor: Elizabeth Howe
Assistant Editor: Elesha Feldman
Editorial Assistant: Lisa Michel
Technology Project Manager: Travis Metz
Marketing Manager: Jennifer Somerville
Marketing Assistant: Melanie Banfield
Advertising Project Manager: Shemika Britt
Project Manager, Editorial Production: Sandra Craig
Creative Director: Rob Hugel
Print/Media Buyer: Barbara Britton
Permissions Editor: Kiely Sexton

Production: Ann Borman
Text Designer: Lisa Buckley
Photo Researcher: Anne Sheroff
Copy Editor: Chris Thillen
Illustrators: Ann Borman, Stan Maddock
Cover Designer: Larry Didona
Cover Images: Watermelon: © John Burwell/Getty Images;
group of friends: © Ariel Skelley/CORBIS
Cover Printer: RR Donnelley
Compositor: Parkwood Composition Service
Printer: RR Donnelley

For more information about our products, contact us at:
Thomson Learning Academic Resource Center
1-800-423-0563
For permission to use material from this text or product,
submit a request online at
http://www.thomsonrights.com.
Any additional questions about permissions can be submitted by
email to thomsonrights@thomson.com.

All Unit opening photos: Photo Disc

Library of Congress Control Number: 2004106297

Student Edition: ISBN 0-534-62325-5

Instructor's Edition: ISBN 0-534-62306-9

Thomson Wadsworth
10 Davis Drive
Belmont, CA 94002-3098
USA

Asia
Thomson Learning
5 Shenton Way #01-01
UIC Building
Singapore 068808

Australia/New Zealand
Thomson Learning
102 Dodds Street
Southbank, Victoria 3006
Australia

Canada
Nelson
1120 Birchmount Road
Toronto, Ontario M1K 5G4
Canada

Europe/Middle East/Africa
Thomson Learning
High Holborn House
50/51 Bedford Row
London WC1R 4LR
United Kingdom

Latin America
Thomson Learning
Seneca, 53
Colonia Polanco
11560 Mexico D.F.
Mexico

Spain/Portugal
Paraninfo
Calle Magallanes, 25
28015 Madrid, Spain

Recommended Dietary Allowances (RDA) and Adequate Intakes (AI) for Vitamins

Age (yr)	Thiamin RDA (mg/day)	Riboflavin RDA (mg/day)	Niacin RDA (mg/day)[a]	Biotin AI (µg/day)	Pantothenic acid AI (mg/day)	Vitamin B$_6$ RDA (mg/day)	Folate RDA (µg/day)[b]	Vitamin B$_{12}$ RDA (µg/day)	Choline AI (mg/day)	Vitamin C RDA (mg/day)	Vitamin A RDA (µg/day)[c]	Vitamin D AI (µg/day)[d]	Vitamin E RDA (mg/day)[e]	Vitamin K AI (µg/day)
Infants														
0–0.5	0.2	0.3	2	5	1.7	0.1	65	0.4	125	40	400	5	4	2.0
0.5–1	0.3	0.4	4	6	1.8	0.3	80	0.5	150	50	500	5	5	2.5
Children														
1–3	0.5	0.5	6	8	2	0.5	150	0.9	200	15	300	5	6	30
4–8	0.6	0.6	8	12	3	0.6	200	1.2	250	25	400	5	7	55
Males														
9–13	0.9	0.9	12	20	4	1.0	300	1.8	375	45	600	5	11	60
14–18	1.2	1.3	16	25	5	1.3	400	2.4	550	75	900	5	15	75
19–30	1.2	1.3	16	30	5	1.3	400	2.4	550	90	900	5	15	120
31–50	1.2	1.3	16	30	5	1.3	400	2.4	550	90	900	5	15	120
51–70	1.2	1.3	16	30	5	1.7	400	2.4	550	90	900	10	15	120
>70	1.2	1.3	16	30	5	1.7	400	2.4	550	90	900	15	15	120
Females														
9–13	0.9	0.9	12	20	4	1.0	300	1.8	375	45	600	5	11	60
14–18	1.0	1.0	14	25	5	1.2	400	2.4	400	65	700	5	15	75
19–30	1.1	1.1	14	30	5	1.3	400	2.4	425	75	700	5	15	90
31–50	1.1	1.1	14	30	5	1.3	400	2.4	425	75	700	5	15	90
51–70	1.1	1.1	14	30	5	1.5	400	2.4	425	75	700	10	15	90
>70	1.1	1.1	14	30	5	1.5	400	2.4	425	75	700	15	15	90
Pregnancy														
≤18	1.4	1.4	18	30	6	1.9	600	2.6	450	80	750	5	15	75
19–30	1.4	1.4	18	30	6	1.9	600	2.6	450	85	770	5	15	90
31–50	1.4	1.4	18	30	6	1.9	600	2.6	450	85	770	5	15	90
Lactation														
≤18	1.4	1.6	17	35	7	2.0	500	2.8	550	115	1200	5	19	75
19–30	1.4	1.6	17	35	7	2.0	500	2.8	550	120	1300	5	19	90
31–50	1.4	1.6	17	35	7	2.0	500	2.8	550	120	1300	5	19	90

NOTE: For all nutrients, values for infants are AI.
[a] Niacin recommendations are expressed as niacin equivalents (NE), except for recommendations for infants younger than 6 months, which are expressed as preformed niacin.
[b] Folate recommendations are expressed as dietary folate equivalents (DFE).
[c] Vitamin A recommendations are expressed as retinol activity equivalents (RAE).
[d] Vitamin D recommendations are expressed as cholecalciferol and assume an absence of adequate exposure to sunlight.
[e] Vitamin E recommendations are expressed as α-tocopherol.

Recommended Dietary Allowances (RDA) and Adequate Intakes (AI) for Minerals

Age (yr)	Sodium AI (mg/day)	Chloride AI (mg/day)	Potassium AI (mg/day)	Calcium AI (mg/day)	Phosphorus RDA (mg/day)	Magnesium RDA (mg/day)	Iron RDA (mg/day)	Zinc RDA (mg/day)	Iodine RDA (µg/day)	Selenium RDA (µg/day)	Copper RDA (µg/day)	Manganese AI (mg/day)	Fluoride AI (mg/day)	Chromium AI (µg/day)	Molybdenum RDA (µg/day)
Infants															
0–0.5	120	180	400	210	100	30	0.27	2	110	15	200	0.003	0.01	0.2	2
0.5–1	370	570	700	270	275	75	11	3	130	20	220	0.6	0.5	5.5	3
Children															
1–3	1000	1500	3000	500	460	80	7	3	90	20	340	1.2	0.7	11	17
4–8	1200	1900	3800	800	500	130	10	5	90	30	440	1.5	1.0	15	22
Males															
9–13	1500	2300	4500	1300	1250	240	8	8	120	40	700	1.9	2	25	34
14–18	1500	2300	4700	1300	1250	410	11	11	150	55	890	2.2	3	35	43
19–30	1500	2300	4700	1000	700	400	8	11	150	55	900	2.3	4	35	45
31–50	1500	2300	4700	1000	700	420	8	11	150	55	900	2.3	4	35	45
51–70	1300	2000	4700	1200	700	420	8	11	150	55	900	2.3	4	30	45
>70	1200	1800	4700	1200	700	420	8	11	150	55	900	2.3	4	30	45
Females															
9–13	1500	2300	4500	1300	1250	240	8	8	120	40	700	1.6	2	21	34
14–18	1500	2300	4700	1300	1250	360	15	9	150	55	890	1.6	3	24	43
19–30	1500	2300	4700	1000	700	310	18	8	150	55	900	1.8	3	25	45
31–50	1500	2300	4700	1000	700	320	18	8	150	55	900	1.8	3	25	45
51–70	1300	2000	4700	1200	700	320	8	8	150	55	900	1.8	3	20	45
>70	1200	1800	4700	1200	700	320	8	8	150	55	900	1.8	3	20	45
Pregnancy															
≤18	1500	2300	4700	1300	1250	400	27	12	220	60	1000	2.0	3	29	50
19–30	1500	2300	4700	1000	700	350	27	11	220	60	1000	2.0	3	30	50
31–50	1500	2300	4700	1000	700	360	27	11	220	60	1000	2.0	3	30	50
Lactation															
≤18	1500	2300	5100	1300	1250	360	10	14	290	70	1300	2.6	3	44	50
19–30	1500	2300	5100	1000	700	310	9	12	290	70	1300	2.6	3	45	50
31–50	1500	2300	5100	1000	700	320	9	12	290	70	1300	2.6	3	45	50

Tolerable Upper Intake Levels (UL) for Vitamins*

Age (yr)	Niacin (mg/day)[a]	Vitamin B6 (mg/day)	Folate (μg/day)[a]	Choline (mg/day)	Vitamin C (mg/day)	Vitamin A (μg/day)[b]	Vitamin D (μg/day)	Vitamin E (mg/day)[c]
Infants								
0–0.5	—	—	—	—	—	600	25	—
0.5–1	—	—	—	—	—	600	25	—
Children								
1–3	10	30	300	1000	400	600	50	200
4–8	15	40	400	1000	650	900	50	300
9–13	20	60	600	2000	1200	1700	50	600
Adolescents								
14–18	30	80	800	3000	1800	2800	50	800
Adults								
19–70	35	100	1000	3500	2000	3000	50	1000
>70	35	100	1000	3500	2000	3000	50	1000
Pregnancy								
≤18	30	80	800	3000	1800	2800	50	800
19–50	35	100	1000	3500	2000	3000	50	1000
Lactation								
≤18	30	80	800	3000	1800	2800	50	800
19–50	35	100	1000	3500	2000	3000	50	1000

* UL = The maximum level of daily nutrient intake that is likely to pose no risk of adverse effects. Unless otherwise specified, the UL represents total intake from food, water, and supplements. Due to lack of suitable data, ULs could not be established for vitamin K, thiamin, riboflavin. vitamin B12, pantothenic acid, biotin, or carotenoids. In the absence of ULs, extra caution may be warranted in consuming levels above recommended intakes.

[a] The UL for niacin and folate apply to synthetic forms obtained from supplements, fortified foods, or a combination of the two.
[b] The UL for vitamin A applies to preformed vitamin A only.
[c] The UL for vitamin E applies to any form of supplemental α-tocopherol, fortified foods, or a combination of the two.

Tolerable Upper Intake Levels (UL) for Minerals*

Age (yr)	Sodium (mg/day)	Chloride (mg/day)	Calcium (mg/day)	Phosphorus (mg/day)	Magnesium (mg/day)[d]	Iron (mg/day)[b]
Infants						
0–0.5	—[e]	—[e]	—	—	—	40
0.5–1	—[e]	—[e]	—	—	—	40
Children						
1–3	1500	2300	2500	3000	65	40
4–8	1900	2900	2500	3000	110	40
9–13	2200	3400	2500	4000	350	40
Adolescents						
14–18	2300	3600	2500	4000	350	45
Adults						
19–70	2300	3600	2500	4000	350	45
>70	2300	3600	2500	3000	350	45
Pregnancy						
≤18	2300	3600	2500	3500	350	45
19–50	2300	3600	2500	3500	350	45
Lactation						
≤18	2300	3600	2500	4000	350	45
19–50	2300	3600	2500	4000	350	45

* UL = The maximum level of daily nutrient intake that is likely to pose no risk of adverse effects. Unless otherwise specified, the UL represents total intake from food, water, and supplements. Due to lack of suitable data, ULs could not be established for arsenic, chromium, and silicon. In the absence of ULs, extra caution may be warranted in consuming levels above recommended intakes.

[d] The UL for magnesium applies to synthetic forms obtained from supplements or drugs only.
[e] Source of intake should be from human milk (or formula) and food only.

About the Author

Judith E. Brown has been a professor of nutrition and public health at the University of Minnesota. She received her Ph.D. in human nutrition from Florida State University and her M.P.H. in public health nutrition from the University of Michigan. She has received competitively funded research grants from the National Institutes of Health, the Centers for Disease Control and Prevention, and the Maternal and Child Health Bureau and has over 100 publications in the scientific literature including the *New England Journal of Medicine,* the *Journal of the American Medical Association,* and the *Journal of the American Dietetic Association.* A recipient of the Agnes Higgins Award in Maternal Nutrition from the March of Dimes, Dr. Brown is a registered dietitian and the successful author of *Everywoman's Guide to Nutrition* and *Nutrition for Pregnancy.*

Contents in Brief

Contents

Nutrition Timeline →

1621

First Thanksgiving feast at Plymouth colony.

Photo Disc

1702

First coffeehouse in America opens in Philadelphia

Photo Disc

1734

Scurvy recognized

1744

First record of ice cream in America at Maryland colony

Photo Disc

Nutrition Timeline

1747

Lind publishes "Treatise on Scurvy," citrus identified as cure.

1750

Ojibway and Sioux war over control of wild rice stands.

1762

Sandwich invented by the Earl of Sandwich.

1771

Potato heralded as famine food.

1774

Americans drink more coffee in protest over Britain's tea tax.

Stefano Bianchetti/CORBIS

1775

Lavoisier ("the father of the science of nutrition") discovers the energy producing property of food.

1816

Protein and amino
acids identified
followed by
carbohydrates and
fats in the mid 1800s

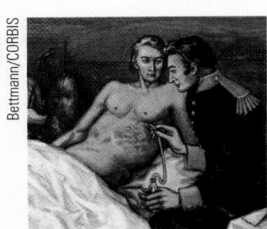

Bettmann/CORBIS

1833

Beaumont's
experiments on a
wounded man's
stomach greatly
expands knowledge
about digestion

1871

Proteins, carbohydrates,
and fats determined
to be insufficient
to support life; that
there are other
"essential" components.

Bettmann/CORBIS

1895

First milk station
providing children with
uncontaminated milk
opens in New York City.

Nutrition Timeline →

1896
Atwater publishes
Proximate Composition
of Food Materials.

1906
Pure Food and Drug Act passed by
President Theodore Roosevelt
to protect consumers against
contaminated foods.

Bettmann/CORBIS

1910
Pasteurized milk
introduced.

Photo Disc

1912
Funk suggested scurvy,
beriberi, and pellagra
caused by deficiency of
"vitamines" in the diet.

Photo Disc

1913
First vitamin discovered
(vitamin A).

1914

Goldberger identifies the cause of pellagra (niacin deficiency) in poor children to be a missing component of the diet rather than a germ as others believed.

1916

First dietary guidance material produced for the public released. It was titled "Food for Young Children."

1917

First food groups published
The Five Food Groups:
Milk and Meat;
Vegetables and Fruits;
Cereals; Fats and Fat Foods;
Sugars and Sugary Foods

Morton Salt Co.

1921

First fortified food produced: Iodized salt. It was needed to prevent widespread iodine deficiency goiter in many parts of the U.S.

Nutrition Timeline →

1928
American Society for Nutritional Sciences and the Journal of Nutrition founded.

1929
Essential fatty acids identified.

Photo Disc

1930s
Vitamin C identified in 1932, followed by pantothenic acid and riboflavin in 1933, and vitamin K in 1934.

Photo Disc

1937
Pellagra found to be due to a deficiency of niacin.

1941
First refined grain enrichment standards developed.

1941

FDR Library

First Recommended Dietary Allowances (RDAs) announced by President Franklin Roosevelt on radio.

1946

National School Lunch Act passed.

Photo Disc

1947

Vitamin B12 identified.

1953

Photo Disc

Double helix structure of DNA discovered.

Photo Disc

1965

Food Stamp Act passed, Food Stamp program established.

Nutrition Timeline

1966
Child Nutrition Act added school breakfast to the National School Lunch Program.

Photo Disc

1968
First national nutrition survey in U.S. launched (The Ten State Nutrition Survey.)

Photo Disc

1970
First Canadian national nutrition survey launched (Nutrition Canada National Survey).

1972
Special Supplemental Food and Nutrition Program for Women, Infants, and Children (WIC) established.

1977
Dietary Goals for the U.S. issued.

1978
First Health Objectives for the Nation released.

1989

First national scientific consensus report on diet and chronic disease published.

Photo Disc

1992

The Food Guide pyramid is released by the USDA.

1997

RDAs expanded to Dietary Reference Intakes (DRIs).

1998

Folic acid fortification of refined grain products begins.

Photo Disc

2003

Sequencing of DNA in the human genome completed. Marks beginning of new era of research in nutrient-gene interactions.

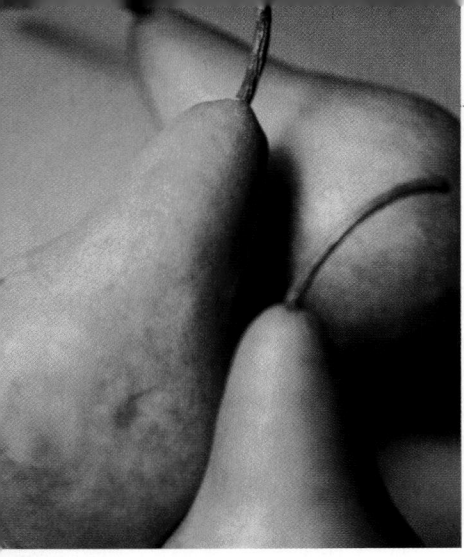

Preface

It is with great pleasure that we present to you the fourth edition of *Nutrition Now*. You will find that this edition differs from the third in many ways. The few years since the last edition have been busy ones in terms of advances in knowledge about the science of nutrition. New DRIs have been developed for macronutrients and electrolytes. As a group, nutritionists and consumers are coming to terms with the practical implications of changes in recommendations for fat intake. Trans fatty acids are finding their way out of processed foods. News about low-carb diets and products, functional foods, fast foods, and the obesity and diabetes epidemics frequently make headlines. New calls for the regulation of herbs as drugs can be heard; ephedra has been banned for weight loss. Supersizing is losing favor and fast food and other restaurants are beginning to compete for health-, carb-, and calorie-conscious consumers. We are coming closer to understanding the depth and breadth of effects of nutrient-gene interactions on health and disease. As the fourth edition goes to press, committees charged with revising the Food Guide Pyramid and the Dietary Guidelines for Americans are at work. Official proposals related to changing reference levels of nutrients used in nutrition labeling have surfaced, and will lead to an updating of % Daily Value figures listed in Nutrition Facts panels in the future. These past few years have been marked by many changes, and the pace of developments in the field of nutrition is escalating. This edition of *Nutrition Now* captures these changes and attempts to point students in the directions the field may be headed next.

Updates to the fourth edition of *Nutrition Now* are sufficiently extensive to warrant the addition of the feature "New to the Edition" located at the web site: http://nutrition.wadsworth.com. As always, the updated edition covers recently released DRIs, emerging topics (such as irritable bowel syndrome, insulin resistance, and metabolic syndrome in this edition); and an expanded glossary. It also includes:

- A stand-alone unit on diabetes (Diabetes Now, Unit 13)

- A new feature called "Reality Check" aimed at helping students learn to make healthful decisions about nutrition

- Updated content on types of fat and fat intake and health

- Expanded coverage of the DASH diet

- New recommendations for physical activity

- Expanded coverage on nutrient-gene interactions in health and disease

- A Nutrition Timeline

- An expanded table of measurement equivalents (facing the inside back cover)

- A more detailed explanation of digestion and absorption (Unit 7: Digestion and Absorption: How the Body Uses Food)

In addition, the fourth edition of *Nutrition Now* has gained an Instructor's Activity Book. Activities included in the Book are intended to get students involved in the topics being covered in class through hands-on and interactive experiences. The activities can be undertaken in classes ranging in size from tens to hundreds. Activities include taste testing to identify genetically determined sensitivity to bitterness, developing a dietary behavioral change plan, anthropometry lab, designing fraudulent nutrition products, a physical activity assessment, and an assessment of three days of dietary intake. The Instructor's Activity Book, as well as the forms used to conduct and submit activities, may be accessed by Instructors using *Nutrition Now* at the web address: http://nutrition.wadsworth.com.

Although a number of elements have changed in this new edition, many of the basic tenets of the text's approach have stayed the same. The text remains focused on meeting the needs of instructors offering introductory nutrition courses to non-majors. The 33 units in the text stand alone, and can be covered in the order of the instructor's choosing. Instructors may choose to customize their selection of units to be included in the text. The text remains heavily illustrated, and updated and revised figures, tables, and photographs have been added to the new edition.

Pedagogical Features

Nutrition Now continues to be oriented toward helping students build a foundation of knowledge about nutrition that will serve them well throughout life. Units are concise, focused on key facts and concepts, and provide ample real-life examples intended to enhance students' understanding of the material presented. Students are asked to apply their newly gained knowledge about nutrition in decision-making activities and exercises incorporated throughout the units.

Features included in this edition of *Nutrition Now* aimed at enhancing students' understanding of nutrition include:

- **Reality Check.** A new pedagogical feature, Reality Check presents brief, real-life scenarios and asks students to given a thumbs up or down to the optional solutions posed.

- **Nutrition Scoreboard.** Each unit begins with a 3-5 question pretest. Answers to the questions are given on the second page of the units.

- **Key Concepts and Facts.** Each unit begins with a listing of key concepts and facts related to the central topics covered.

- **On the Side.** Boxed inserts containing interesting facts related to nutrition topics covered in the units are sprinkled throughout the text.

- **Health Action.** Every unit has at least one Health Action box that relates to the personal application of information covered.

- **Margin definitions.** Unfamiliar terms are highlighted in bold in the text, and defined in nearby margins. The fourth edition includes more pronunciation guides for terms than previous editions.

- **Glossary.** Terms defined in the margins are listed in the Glossary near the end of the text. Approximately 20 new terms have been added to the glossary of the fourth edition.

- **Nutrition Up Close.** Each unit closes with an activity that gives students an opportunity to relate nutrition knowledge gained to their daily lives and experiences.

- **WWW links.** Internet sources of reliable nutrition information are listed at the end of each unit. These have been thoroughly updated in the fourth edition.

- **Appended materials.** Resources included in the appendices have been updated for the fourth edition of *Nutrition Now:* The Food Composition Table, Canadian nutrition labeling and food guide standards, the Reliable Sources of Nutrition Information list, and the Food Exchange System.

- **Nutrition Timeline.** New to the fourth edition, the Nutrition Timeline high-lights major developments in the science of nutrition.

- **Daily Values for Food Labels and Glossary of Nutrient Measures.** These appear on the last book page. The glossary has been expanded.

Resources for the Instructor

- **Instructor's Manual and Test Bank.** Features lecture outlines, suggested classroom activities, Web resources, discussion questions, transparency masters, and chapter-by-chapter test questions linked to the ExamView Computerized Testing program.

- **ExamView® Computerized Testing.** An easy-to-use assessment and tutorial system that facilitates creation, delivery, and customizing of tests and study guides, both print and online.

- **Multimedia Manager CD-ROM.** Includes book-specific Microsoft®, PowerPoint®, lecture slides with teaching points, graphics from the book, electronic versions of the Instructor's Manual and Test Bank, ABC video clips, and links to nutrition resources on the Web.

- **CNN Today: Nutrition Videos.** Three volumes of engaging video clips for launching lectures.

- **Instructor's Activity Book.** Features classroom activities that can be used by varying sizes of student groups.

- **Transparency Acetates.** Includes 80 full-color transparencies of key illustrations in the text.

- **Diet Analysis Plus 6.1.** Software that enables students to track and assess their food choices and create personal nutrition and activity profiles. Available on CD-ROM or online, and may be packaged with the text.

Acknowledgments

Authors are one component of an integrated matrix of activities that surround the development of a new edition of a textbook. Thomson Higher Education Publishing has the good fortune of being the home of the best in the nutrition textbook field. Peter Marshall began his outstanding career building better nutrition textbooks through collaboration with Dr. Ellie Whitney on her breakthrough textbooks on nutrition. I am indeed fortunate to have his wisdom and experience applied to *Nutrition Now.* I am also indebted to Dr. Whitney, who served on my doctoral committee and taught me to teach Introductory Nutrition when I was one of her teaching assistants at Florida State University.

Beth Howe, development editor, was a major force behind the successful completion of the fourth edition of *Nutrition Now.* She organized the work related to the new edition and made sure high-quality jobs were done all around. Beth is a joy to work with and deserves a really big raise.

The art and photo program underlying *Nutrition Now* came together under the creative direction of Ann Borman. Ann has been with *Nutrition Now* from the beginning and it shows. Anne Sheroff tirelessly searched for photos, and made sure everything was credited properly.

Elesha Feldman managed the ancillaries that accompany this edition. Contributors to the ancillaries include Stan Andrews, Judy Kaufman, Sara Long, and Fred

Wolfe. Chris Thillen copyedited this edition of *Nutrition Now* and brought a new eye for detail to the text.

My applause goes out to the sales representatives who are fans of *Nutrition Now* and who work hard to introduce faculty to its contents and features. You do a terrific job of communicating with instructors, and with me when you send instructors' thoughts my way.

It is said that instructors adopt a specific textbook but that students play a major role in instructors' decision to keep it. I am honored that you chose to adopt *Nutrition Now* and deeply pleased with the thought that students are helping you decide to keep it.

Reviewers' feedback is the lifeline of text writing, and the reviewers for the new edition conveyed very useful advice. The advice led me to some very interesting places on specific topics (GMO foods and fats, for example) that changed my thinking and writing. One reviewer asked, "Who is Max Brown—the person who is quoted here and there within the text?" Max is my son; he's the middle guy in Illustration 30.7. He has a way of summing up nutrition situations that I find irresistible.

Judith E. Brown

Reviewers of the Fourth Edition

Cassandra E. August
Baldwin-Wallace College

Laura Calderon
California State University, Los Angeles

Jasia (Jayne) Chitharanjan
University of Wisconsin, Stevens Point

Dorothy Coltrin
De Anza College

Gary Fosmire
The Pennsylvania State University

Susan E. Helm
Pepperdine University

Linda S. Kolb
Wabash Valley College

Marcia Magnus
Florida International University

Lisa Sasson
New York University

Tasleem A. Zafar
Purdue University

Key Nutrition Concepts and Terms

Nutrition Scoreboard

	TRUE	FALSE
1 Calories are a component of food.		
2 Nutrients are substances in food that are used by the body for growth and health.		
3 The Dietary Reference Intakes (DRIs) specify minimal levels of nutrients people should consume in their diet each day.		
4 Both high and low intakes of nutrients threaten health.		
5 Foods can basically be divided into two groups: those that are "good" for you and those that are "bad" for you.		

Answers on next page

[KEY CONCEPTS AND FACTS]

- At the core of the science of nutrition are concepts that represent basic "truths" and serve as the foundation of our understanding about normal nutrition. (They are listed in Table 1.4 on p. 1–13.)

- Most nutrition concepts relate to nutrients.

Answers to *Nutrition Scoreboard*

		TRUE	FALSE
1	Calories are a measure of the amount of energy supplied by food. They're a property of food, not a component of food.		✔
2	That's the definition of nutrients.	✔	
3	There are no "minimum dietary intake" standards. The DRIs represent nutrient intake levels appropriate for nearly all healthy people.		✔
4	Excessive as well as inadequate intake levels of vitamins, minerals, and other nutrients can be harmful to health.	✔	
5	There are *no* good or bad foods, but there are healthy and unhealthy diets.		✔

The Meaning of Nutrition

To be surprised, to wonder, is to begin to understand.
—José Ortega y Gasset

What is nutrition? It can be explained by situations captured in photographs as well as by words. This introduction presents a photographic tour of real-life situations that depict aspects of the study of nutrition.

Before the tour begins, take a moment to make yourself comfortable and clear your mind of clutter. Take a careful look at the photographs, pausing to mentally describe in two or three sentences what each shows.

David Young-Wolff/PhotoEdit

Photo Disc

David Young-Wolff/PhotoEdit

Photo Disc

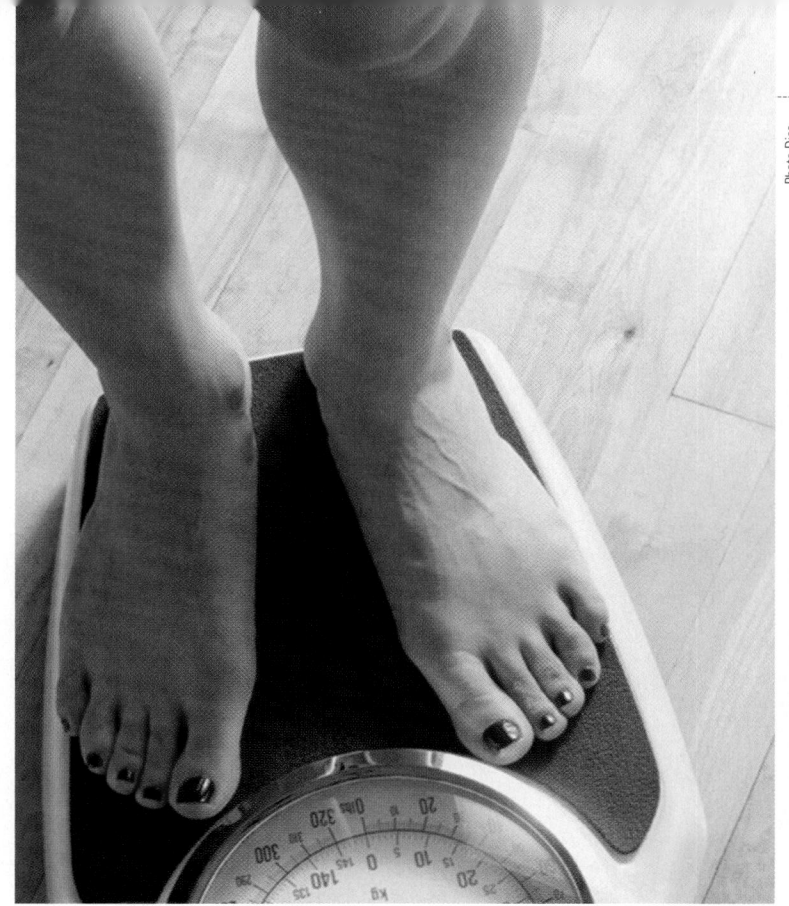

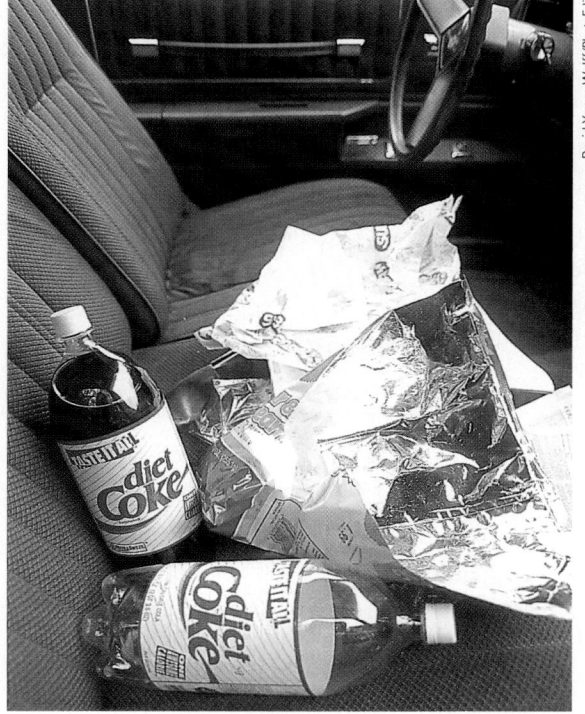

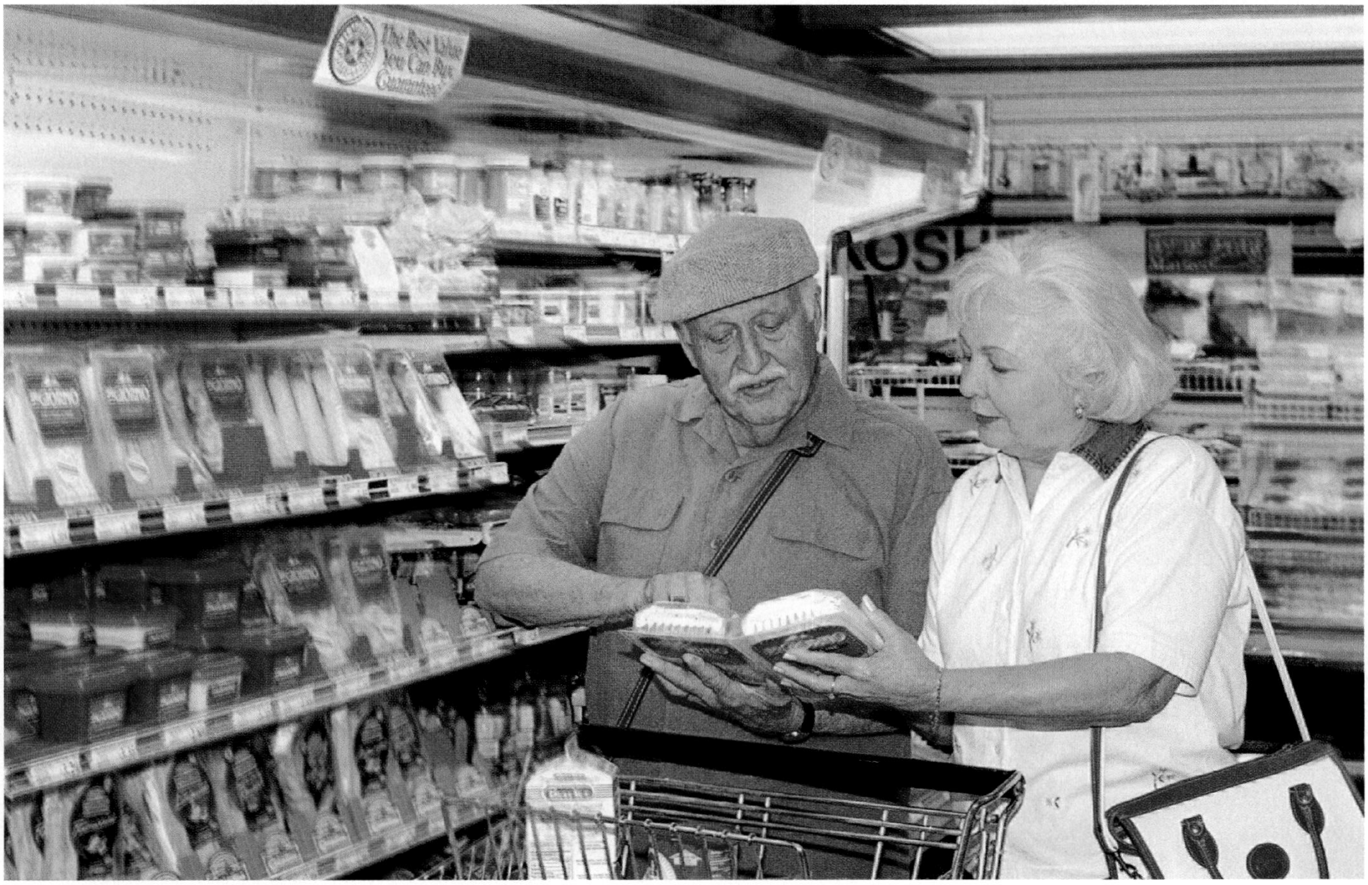

Royalty-Free/CORBIS

AP/Wide World Photos

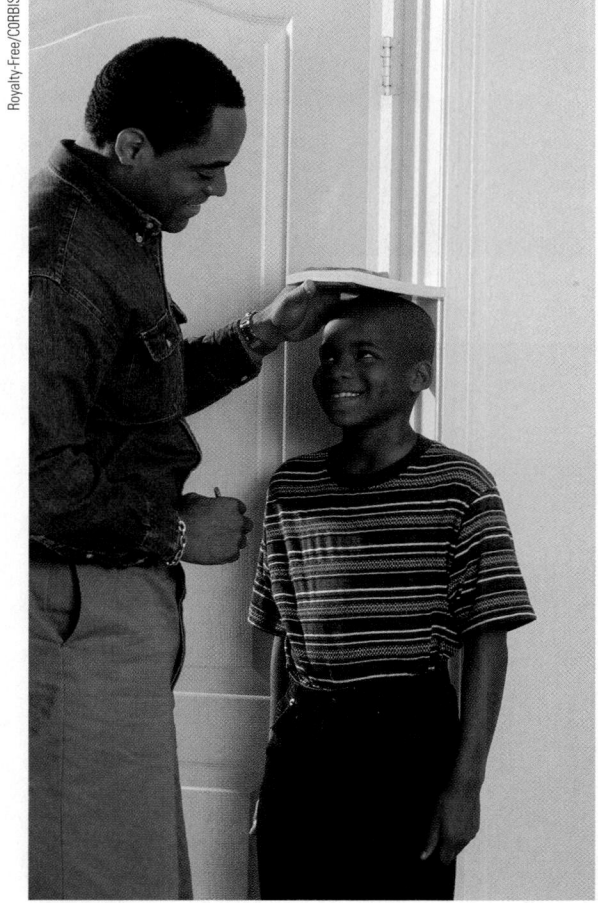

Photo Disc

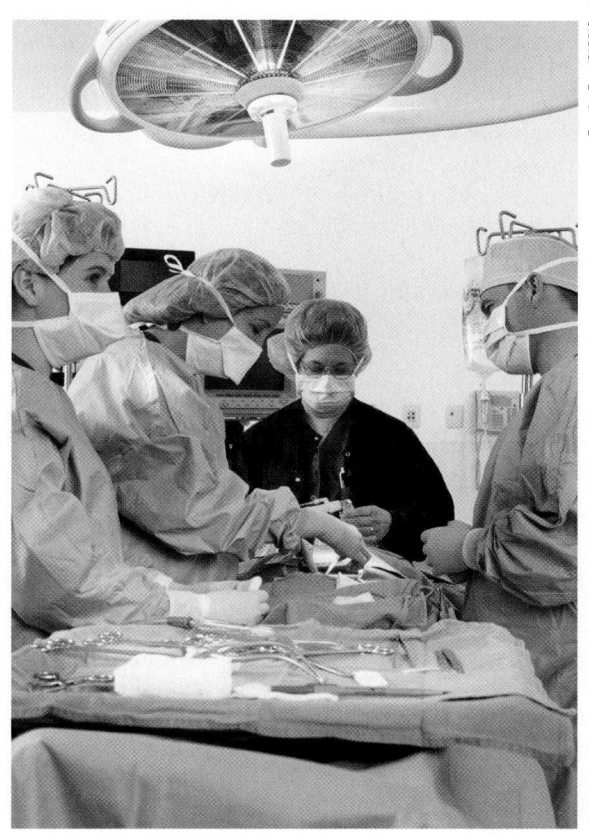

Royalty-Free/CORBIS

Photo Disc

APWide World Photos

Not everyone who looks at the photographs will describe them in the same way. Reactions will vary somewhat due to personal experiences, interests, attitudes, and beliefs. An individual trying to gain weight will probably react differently to the photograph of the person on the scale than someone who is trying to lose weight. If you grew up in a family that farmed for a living, the picture of pesticides being sprayed on a crop may mean increased food production to you. But another person's reaction may be that pesticide residues on foods are harmful to health. Although knowledge about nutrition is generated by impersonal and objective methods, it can be a very personal subject.

Nutrition Defined

nutrition
The study of foods, their nutrients and other chemical constituents, and the effects of food constituents on health.

In a nutshell, *nutrition* is the study of foods and health. It is a science that centers on foods, their nutrient and other chemical constituents, and the effects of food constituents on body processes and health. The scope of nutrition extends from food choices to the effects of specific components of foods on health.

Nutrition Is a "Melting Pot" Science

The broad scope of nutrition makes it an interdisciplinary science. Knowledge provided by the behavioral and social sciences, for example, is needed in studies that examine how food preferences develop and how they may be changed. Information generated by the biological, chemical, physical, and food sciences is required to propose and explain diet and disease relationships. Methods developed by quantitative scientists such as mathematicians and statisticians are needed to guide decision making about the significance of results produced by nutrition research. The study of nutrition will bring you into contact with information from a variety of disciplines (Illustration 1.1).

Nutrition Knowledge Is Applicable

As you study this science, you will discover answers to a number of questions about your own diet, health, and eating behavior. Is obesity primarily due to eating behaviors, physical inactivity, or your genes? How do you know whether new informa-

Illustration 1.1
Nutrition is an interdisciplinary science.

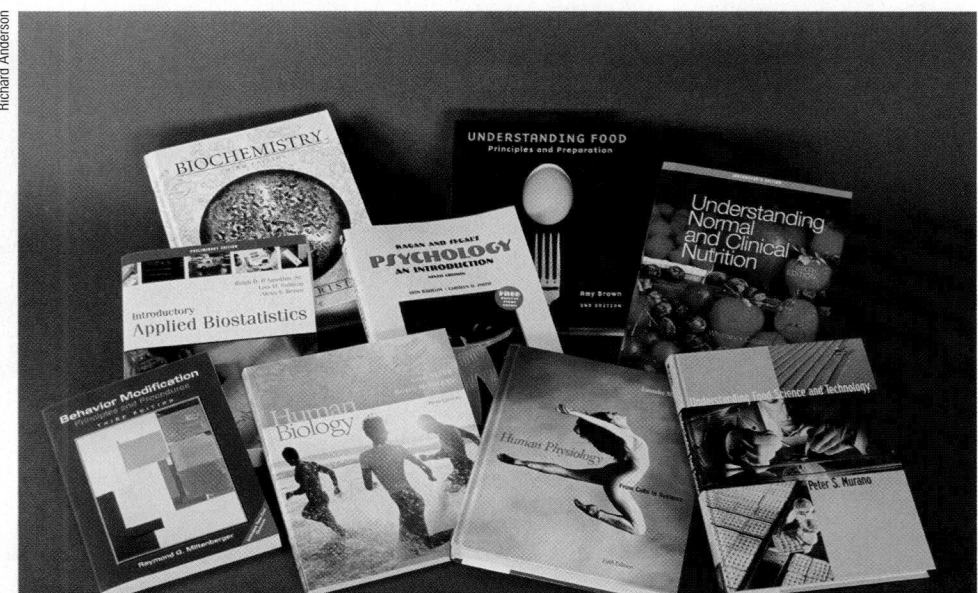

tion you hear about nutrition is true? Can sugar harm more than your teeth? Can the right diet or supplement give you a competitive edge? What is a healthful diet, how do you know if you have one, and, if improvements seem warranted, what's the best way to go about changing your diet for the better? These are just a few of the questions that nutrition and your course of study into it will address. If all goes well, you will take from this learning experience not only knowledge about nutrition and health, but skills that will keep the information and insights working to your advantage for a long time to come.

Foundation Knowledge for Thinking about Nutrition

Common sense requires a common knowledge.

You don't have to be a bona fide nutritionist to think like one. What you need is a grasp of the language and the basic concepts of the science. It's the purpose of this unit to give you this background. The essential topics covered here are explored in greater depth in units to come, and they build upon this foundation knowledge. With a working knowledge of nutrition terms and the concepts that provide the foundation of our understanding about nutrition, you will have an uncommonly good sense of nutrition.

NUTRITION CONCEPT #1

Food is a basic need of humans.

Humans need enough food to live and the right assortment of foods for optimal health. In the best of all worlds, the need for food is combined with the condition of *food security.* People who experience food security have access at all times to a sufficient supply of safe, nutritious foods needed for an active and healthy life. They are able to acquire acceptable foods in socially acceptable ways—without having to scavenge or steal food, for example, in order to eat or to feed their families. *Food insecurity* exists whenever the availability of safe, nutritious foods, or the ability to acquire them in socially acceptable ways, is limited or uncertain (Illustration 1.2).[1] Over time, food insecurity leads to poor health, increased risk for certain diseases, and in children, lowered academic achievement.[2]

food security
Access at all times to a sufficient supply of safe, nutritious foods.

food insecurity
Limited or uncertain availability of safe, nutritious foods.

Soup kitchen

Illustration 1.2
"It is possible to go an entire lifetime without knowing about people's experiences with hunger."
—Meghan LeCates, Capitol Area Food Bank

Photo Disc

Ricin Ricin, a deadly nerve toxin, comes from the seeds of the beautiful castor bean plant. The oil from the seeds (sometimes called beans) is used commercially in castor oil, motor oil, nylon, and paint. Castor plants are becoming increasingly abundant in southwestern parts of the United States, although their cultivation is discouraged. Ricin poisoning occurs when seeds with broken coatings or seed contents are ingested. 70 micrograms—the amount of a grain of salt—is enough to kill an adult. Exposure of the skin to the seeds causes irritation at the contact point.[7] The U.S. Senate offices were closed for several days in 2004 when powdered ricin was discovered in a letter sent to Senator Bill Frist. No one was injured.

©W.P. Armstrong 2000

calorie
A unit of measure of the amount of energy supplied by food. (Also known as a kilocalorie, or the "large Calorie" with a capital "C.")

nutrients
Chemical substances found in food that are used by the body for growth and health. The six categories of nutrients are carbohydrates, proteins, fats, vitamins, minerals, and water.

Illustration 1.3
Foods provide nutrients. "Please pass the complex carbohydrates, thiamin, and niacin . . . I mean, the bread!"

Brand X Pictures

Food insecurity exists in about 10% of U.S. and 4% of Canadian households.[2] It is most likely to occur in poor, female-headed households with young children living in inner-city areas. Compared to whites, African American and Hispanic households are twice as likely to be food insecure.[3]

Food Terrorism

The term *food security* now means more than it used to. Food is a potential weapon of bioterrorism. Although public concerns related to bioterrorism center on disease threats such as smallpox and anthrax, food and water could also be used to intentionally spread illness. Simply recalling the latest announcement about a foodborne illness outbreak makes it easy to imagine the panic and health consequences that could result from intentional contamination of food or water supplies.[4]

Toxic substances that could be introduced into food and water supplies include botulism toxin, ricin, radioactive particles, and microorganisms such as *Salmonella, E. coli 0517:H7,* and *Shigella.* Botulism toxin is the single most poisonous substance known. It takes at least 15 minutes of boiling to destroy this toxin. Most bacteria are killed when heated above 160°F.[5] Ricin (pronounced ryesin) is a widely available toxin found in the seeds of the castor plant, which is also widely available in tropical and warm areas of the world.[6] Castor plant seeds are sufficiently interesting to warrant being featured in the "On the Side" box nearby.

NUTRITION CONCEPT #2

Foods provide energy (calories), nutrients, and other substances needed for growth and health.
People eat foods for many different reasons. The most compelling reason is that we need the *calories, nutrients,* and other substances supplied by foods for growth and health.

Calories
A calorie is a unit of measure of the amount of energy in a food—and of how much energy will be transferred to the person who eats it. Although we often refer to the number of calories in this food or that one, calories are not a substance present in food. And, because calories are a unit of measure, they do not qualify as a nutrient.

Nutrients
Nutrients are chemical substances present in food that are used by the body (Illustration 1.3). Essentially everything that's in our body was once a nutrient in food we consumed.

There are six categories of nutrients (Illustration 1.4), and each category (except water) consists of a number of different substances used by the body for growth and health. The carbohydrate category includes simple sugars and complex carbohydrates (starches and dietary fiber). The protein category includes 20 amino acids, the chemical units that serve as the "building blocks" for protein. Several different types of fat are included in the fat category. Of primary concern are the saturated fats, unsaturated fats, essential fatty acids, and cholesterol (read more about them in Illustration 1.4). The vitamin category consists of 13 vitamins, and the mineral category includes 15 minerals. Water makes up a nutrient category by itself.

Carbohydrates, proteins, and fats supply calories and are called the "energy nutrients." Although each of these three types of nutrients performs

Illustration 1.4
Nutrients are grouped into six categories. Here the major types of nutrients in each category are listed, along with a description of each.

SIX CATEGORIES OF NUTRIENTs

1. CARBOHYDRATES are substances in food that consist of a single sugar molecule or of multiples of them in various forms. They provide the body with energy.

Simple sugars are the most basic type of carbohydrates. Examples include glucose (blood sugar), sucrose (table sugar), and lactose (milk sugar). **Starches** are complex carbohydrates consisting primarily of long, interlocking chains of glucose units. **Dietary fiber** consists of complex carbohydrates found principally in plant cell walls. Because dietary fiber cannot be broken down by human digestive enzymes, it cannot be used for energy.

2. PROTEINS are substances in food made up of amino acids. Amino acids are specific chemical substances from which proteins are made. Of the 20 amino acids, 9 are "essential," or a required part of our diet.

3. FATS are substances in food that are soluble in fat and not water.

Saturated fats are found primarily in animal products, such as meat, butter, and cheese, and in palm and coconut oils. Diets high in saturated fat may elevate blood cholesterol. **Unsaturated fats** are found primarily in plant products, such as vegetable oil, nuts, and seeds, and in fish. Unsaturated fats tend to lower blood cholesterol levels. **Essential fatty acids** are two specific types of unsaturated fats that are required in the diet. **Cholesterol** is a fat-soluble, colorless liquid found in animals but not in plants. It can be manufactured by the liver.

4. VITAMINS are chemical substances found in food that perform specific functions in the body. Humans require 13 different vitamins in their diet.

5. MINERALS are chemical substances that make up the "ash" that remains when food is completely burned. Humans require 15 different minerals in their diet.

6. WATER is essential for life. Most adults need about 11–15 cups of water each day from food and fluids.

ON THE SIDE

The distinctive odor garlic produces when chopped or crushed is actually this plant's defense against insect predators. Garlic cloves contain an odorless, sulfur-containing phytochemical called "alliin." When the clove is disrupted, alliin is released and reacts with an enzyme located in neighboring cells that converts the alliin to the odoriferous "allicin." Allicin is garlic's bug repellant—and the "people repellant" that makes many individuals shy about eating it.[9]

Photo Disc

phytochemicals (phyto = *plant*)
Chemical substances in plants, some of which perform important functions in the human body. Phytochemicals give plants color and flavor, participate in processes that enable plants to grow, and protect plants against insects and diseases.

antioxidants
Chemical substances that prevent or repair damage to cells caused by exposure to oxidizing agents such as environmental pollutants, smoke, ozone, and oxygen. Oxidation reactions are also a normal part of cellular processes.

a variety of functions, they share the property of being the body's only sources of fuel. Vitamins, minerals, and water are chemicals the body needs for converting carbohydrates, proteins, and fats into energy and for building and maintaining muscles, blood components, bones, and other parts of the body.

Other Substances in Food

Food also contains many other substances, some of which are biologically active in the body. The *phytochemicals* are a major type of these substances, and there are over two thousand of them in plants. Illustration 1.5 presents examples of plant foods that are particularly rich sources of phytochemicals. Phytochemicals provide plants with color, give them flavor, and foster their growth and health. A specific example of how one phytochemical works is described in the "On the Side" box on this page. In humans, consumption of phytochemicals in diets high in vegetables and fruits is strongly related to a reduced risk of developing certain types of cancer, heart disease, infections, and other disorders.[8]

Specific phytochemicals have names that are often hard to pronounce and difficult to remember. Nevertheless, here are a few examples. Plant pigments, such as lycopene (like-o-peen), which helps make tomatoes red, anthocyanins (an-tho-sigh-an-ins), which give blueberries their characteristic blue color, and beta-carotene (bay-tah-kar-o-teen), which imparts a dark yellow color to carrots, are phytochemicals that act as *antioxidants*. They protect plant cells—and in some cases, human cells, too—from damage that can make them susceptible to disease. Various types of sulfur-containing phytochemicals are present in cabbage, broccoli, cauliflower, brussels sprouts, and other vegetables of the same family. These substances may help prevent a number of different types of cancer.

Some Nutrients Must Be Provided by the Diet

Many nutrients are required for growth and health. The body can manufacture some of these from raw materials supplied by food, but others must come assembled. Nutrients that the body cannot make, or generally produce in sufficient quantity, are referred to as *essential nutrients*. Here "essential" means "required in the diet." Vitamin A, iron, and calcium are examples of essential nutrients. Table 1.1 lists all the known essential nutrients.

Nutrients used for growth and health that can be manufactured by the body from components of food in our diet are considered nonessential. Cholesterol, cre-

Illustration 1.5
Examples of good food sources of phytochemicals.

Photo Disc

TABLE 1.1

ESSENTIAL NUTRIENTS FOR HUMANS: A REFERENCE TABLE.

ENERGY NUTRIENTS	VITAMINS	MINERALS	WATER
Carbohydrates	Biotin	Calcium	Water
Fats[a]	Folate	Chloride	
Proteins[b]	Niacin (B$_3$)	Chromium	
	Pantothenic acid	Copper	
	Riboflavin (B$_2$)	Fluoride	
	Thiamin (B$_1$)	Iodine	
	Vitamin A	Iron	
	Vitamin B$_6$ (pyroxidine)	Magnesium	
	Vitamin B$_{12}$	Manganese	
	Vitamin C (ascorbic acid)	Molybdenum	
	Vitamin D	Phosphorus	
	Vitamin E	Potassium	
	Vitamin K	Selenium	
		Sodium	
		Zinc	

[a]Fats supply the essential nutrients linoleic and alpha-linolenic acid.
[b]Proteins are the source of 9 "essential amino acids": histidine, isoleucine, leucine, lysine, methionine, phenylalanine, threonine, tryptophan, and valine. The other 11 amino acids are not a required part of our diet; they are considered "nonessential."

atine, and glucose are examples of nonessential nutrients. **Nonessential nutrients** are present in food and used by the body, but they are not required parts of our diet because we can produce them.

Both essential and nonessential nutrients are required for growth and health. The difference between them is whether we need a dietary source. A dietary deficiency of an essential nutrient will cause a specific deficiency disease, but a dietary lack of a nonessential nutrient will not. People develop scurvy (the vitamin C–deficiency disease), for example, if they don't consume vitamin C. But you could have zero cholesterol in your diet and not become "cholesterol deficient," because your liver produces cholesterol.

Our Requirements for Essential Nutrients

The amount of essential nutrients humans need each day varies a great deal, from amounts measured in cups to micrograms. (See Table 1.2 to get a notion of the amount represented by a gram, milligram, and other measures.) Generally speaking, adults need 11 to 15 cups of water from fluids and foods, 9 tablespoons of protein, one-fourth teaspoon of calcium, and only one-thousandth teaspoon (a 30-microgram speck) of vitamin B$_{12}$ each day.

We all need the same nutrients, but not always in the same amounts. The amounts needed vary among people based on:

- Age
- Sex
- Growth status
- Body size
- Genetic traits

and the presence of conditions such as:

- Pregnancy

essential nutrients
Substances required for normal growth and health that the body *cannot* generally produce, or produce in sufficient amounts; they must be obtained in the diet.

nonessential nutrients
Nutrients required for normal growth and health that the body *can* manufacture in sufficient quantities from other components of the diet. We do not require a dietary source of nonessential nutrients.

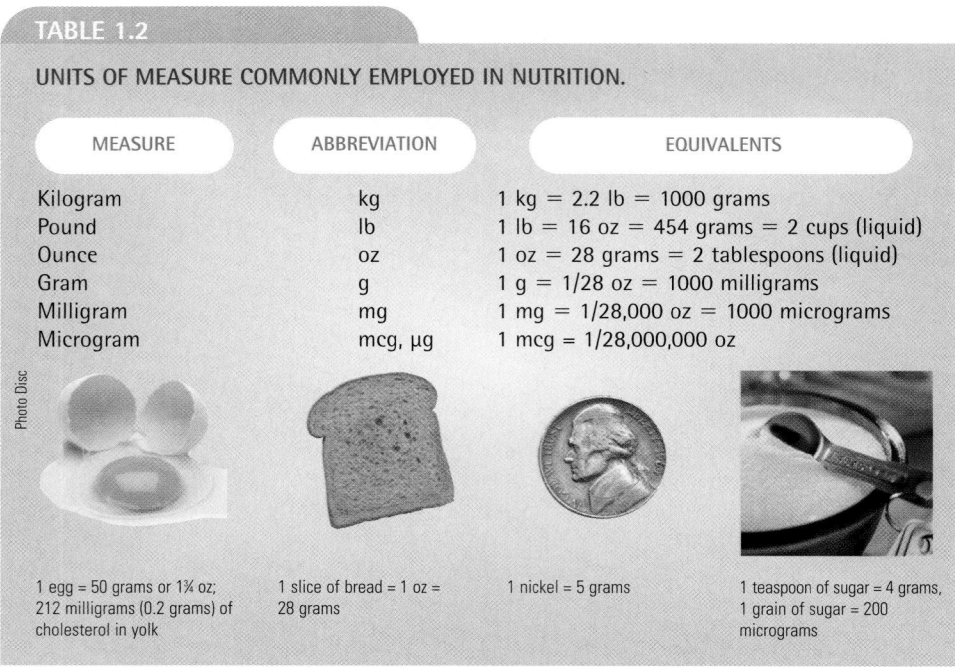

TABLE 1.2

UNITS OF MEASURE COMMONLY EMPLOYED IN NUTRITION.

MEASURE	ABBREVIATION	EQUIVALENTS
Kilogram	kg	1 kg = 2.2 lb = 1000 grams
Pound	lb	1 lb = 16 oz = 454 grams = 2 cups (liquid)
Ounce	oz	1 oz = 28 grams = 2 tablespoons (liquid)
Gram	g	1 g = 1/28 oz = 1000 milligrams
Milligram	mg	1 mg = 1/28,000 oz = 1000 micrograms
Microgram	mcg, μg	1 mcg = 1/28,000,000 oz

Photo Disc

1 egg = 50 grams or 1¾ oz; 212 milligrams (0.2 grams) of cholesterol in yolk

1 slice of bread = 1 oz = 28 grams

1 nickel = 5 grams

1 teaspoon of sugar = 4 grams, 1 grain of sugar = 200 micrograms

- Breastfeeding
- Illnesses
- Drug use
- Exposure to environmental contaminants

Each of these factors, and others, can influence nutrient requirements. General recommendations for diets that provide all the essential nutrients usually make allowances for major factors influencing the level of nutrient need, but they cannot allow for all of the factors.

The Recommended Dietary Allowances (RDAs) are the most widely used standard for identifying desired levels of essential nutrient intake in healthy people. Different RDAs are provided based on age and gender and according to whether a woman is pregnant or breastfeeding.

The first edition of the RDAs was published in 1943. It was prompted by the high rejection rate of World War II recruits due to underweight and nutrient deficiencies. The most recent update, completed in 2004 by U.S. and Canadian scientists, has given the RDAs a whole new look (Illustration 1.6)—and a new name: Dietary Reference Intakes (DRIs).

Illustration 1.6
One of the six reports on the new DRIs.

Food & Nutrition Board

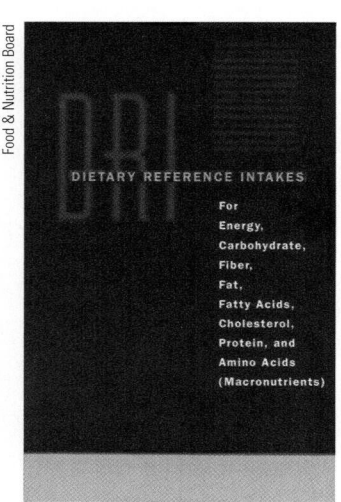

The Revolutionary DRIs

Previous editions of the RDAs were based on levels of essential nutrients that protected people from deficiency diseases such as scurvy and rickets. As the science of nutrition advances, it is becoming abundantly clear that other components of food besides essential nutrients affect health. It is also becoming apparent that levels of nutrient intake associated with deficiency disease prevention may be too low to help prevent cancer, heart disease, and osteoporosis. Excessively high intakes of nutrients from fortified foods and supplements are another problem that was not considered when previous editions of the RDAs were prepared.

Recommended intake levels of essential nutrients and other biologically active components of food have been updated, and safe, upper levels of intake developed.

The recommended daily levels of intake not only meet the nutrient needs of almost all healthy people (97 to 98%) but also promote health and help reduce the risk of chronic disease.[10] Table 1.3 provides examples of endpoints aimed at health promotion and disease prevention used by the DRI committees to estimate recommended levels of nutrient intake.

Recommended intakes of some nutrients, such as calcium, vitamin D, and fluoride, for which too little conclusive evidence on disease prevention exists, are being represented by an "Adequate Intake," or AI, level (Table 1.4). Regardless of which label is attached, the RDAs and AIs represent the best estimates of nutrient intake levels that promote optimal health. These values are represented in the main DRI tables presented on the inside front cover of this book. The term Estimated Average Requirement (EAR) has been introduced in the new dietary intake standards. It represents the intake level of a nutrient that is estimated to meet the requirement of 50% of the individuals within a group.

Estimates of safe upper limits of nutrient intake are called Tolerable Upper Intake Levels, abbreviated ULs. The ULs do not represent recommended or desired

TABLE 1.3

EXAMPLES OF PRIMARY ENDPOINTS USED TO ESTIMATE DRIS.[11]

Carbohydrate: Amount needed to supply optimal levels of energy to the brain.

Total Fiber: Amount shown to provide the greatest protection against heart disease.

Folate: Amount that reduces the risk of heart disease and newborn abnormalities.

Iodine: Amount that corresponds to optimal functioning of the thyroid gland.

Selenium: Amount that maximizes its function in protecting cells from damage.

TABLE 1.4

TERMS AND ABBREVIATIONS USED IN THE 1997–2003 DRIS AND A GRAPHIC REPRESENTATION OF THEIR MEANING.
Source: Adapted from Yates et al. J Am Diet Assoc 1998; 98:702

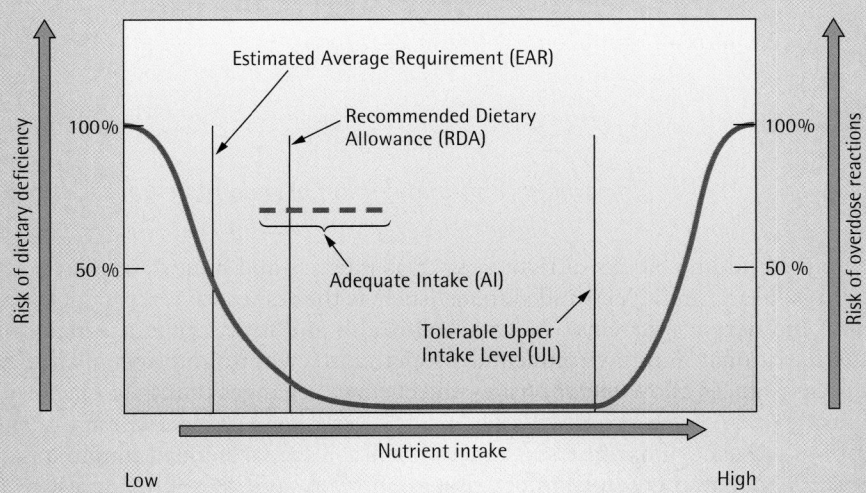

- **Dietary Reference Intakes (DRIs).** This is the general term used for the new nutrient intake standards for healthy people.
- **Recommended Dietary Allowances (RDAs).** These are levels of essential nutrient intake judged to be adequate to meet the known nutrient needs of practically all healthy persons while decreasing the risk of certain chronic diseases.
- **Adequate Intakes (AIs).** These are "tentative" RDAs. AIs are based on less conclusive scientific information than are the RDAs.
- **Estimated Average Requirements (EARs).** These are nutrient intake values that are estimated to meet the requirements of half the healthy individuals in a group. The EARs are used to assess adequacy of intakes of population groups.
- **Tolerable Upper Levels of Intake (ULs).** These are upper limits of nutrient intake compatible with health. The ULs do not reflect desired levels of intake. Rather, they represent total, daily levels of nutrient intake from food, fortified foods, and supplements that should not be exceeded.

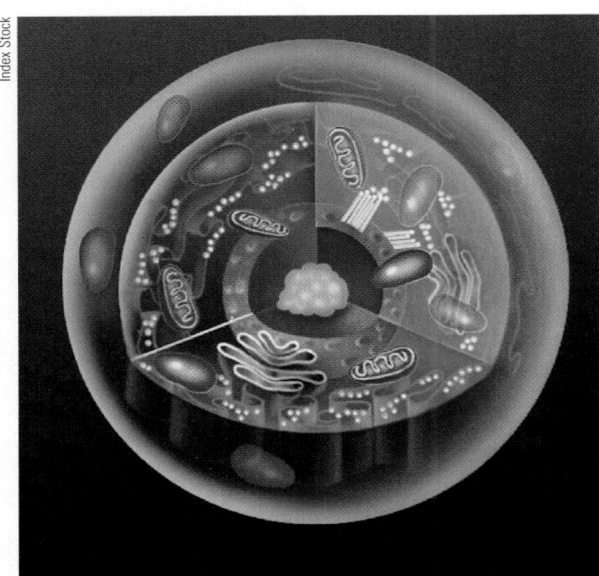

Illustration 1.7
Schematic representation of the structure of a human cell.

levels of intake. Instead, they represent total daily levels of nutrient intake from food, fortified food products, and supplements that should not be exceeded. To reflect the fact that recommended nutrient intakes now include the RDAs, AIs, EARs, and ULs, the name of the recommendations has been broadened to Dietary Reference Intakes.

The new DRIs are also revolutionary in that they are being developed by experts from the United States and Canada, and will apply to the people in both countries. (For more information on the DRI reports, visit the Web site listed in "www Links" at the end of the unit.)

NUTRITION CONCEPT #3

Health problems related to nutrition originate within cells. The main employers of nutrients are cells (Illustration 1.7). This is because all body processes required for growth and health take place within cells and the fluid that surrounds them. The human body contains more than one hundred trillion (100,000,000,000,000) cells. (Which type is the most common? See the "On the Side" box on this page.) The functions of each cell are maintained by the nutrients it receives. Problems arise when a cell's need for nutrients differs from the amount of nutrients supplied.

Nutrient Functions at the Cellular Level

Maximum health and lifespan require metabolic harmony.
—B. Ames[12]

Cells are the building blocks of tissues (such as muscles and bones), organs (such as the kidneys, heart, and liver), and systems (such as the respiratory, reproductive, circulatory, and nervous systems). Normal cell health and functioning are maintained when a nutritional and environmental utopia exists within and around the cells. Such circumstances allow ***metabolism***—the chemical changes that take place within and outside of cells—to proceed flawlessly. Disruptions in the availability of nutrients, or the presence of harmful substances in the cell's environment, initiate diseases and disorders that eventually affect tissues, organs, and systems. Here are two examples of how cell functions can be disrupted by the presence of low or high levels of nutrients:

- Folate, a B vitamin, participates in cellular reactions that keep the level of homocysteine (an amino-acid-like substance) normal. When too little folate is available, homocysteine accumulates in cells and spills into the blood. Once in the blood, high levels of homocysteine enhance plaque formation in arteries and increase the risk of heart disease.

- When too much iron is present in cells, the excess reacts with and damages cell components. If cellular levels of iron remain high, the damage spreads, impairing the functions of organs such as the liver, pancreas, and heart. Health problems in general begin with disruptions in the normal activity of cells. Humans are only as healthy as their cells.

metabolism
The chemical changes that take place in the body. The formation of energy from carbohydrates is an example of a metabolic process.

What's the most common cell in the body? It's the red blood cell. The human body contains more than 30 billion of them.

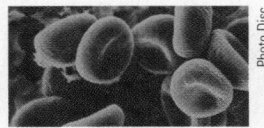

ON THE SIDE

NUTRITION CONCEPT #4

Poor nutrition can result from both inadequate and excessive levels of nutrient intake.

For each nutrient, every individual has a range of optimal intake that produces the best level for cell and body functions. On either side of the optimal range are levels of intake associated with impaired functions.[13] This concept is presented in Illustration 1.8. Inadequate essential nutrient intake, if prolonged, results in obvious deficiency diseases. Marginally deficient diets produce subtle changes in behavior or physical condition. If the optimal intake range is exceeded, mild to severe changes in mental and physical functions occur, depending on the amount of the excess and the nutrient. Overt vitamin C deficiency, for example, produces bleeding gums, pain upon being touched, and a failure of bone to grow. A marginal deficiency may cause delayed wound healing. On the excessive side, high intakes of vitamin C cause diarrhea.[14] Nearly all cases of vitamin and mineral overdose result from the excessive use of supplements—they are almost never caused by foods. For nutrients, "enough is as good as a feast."

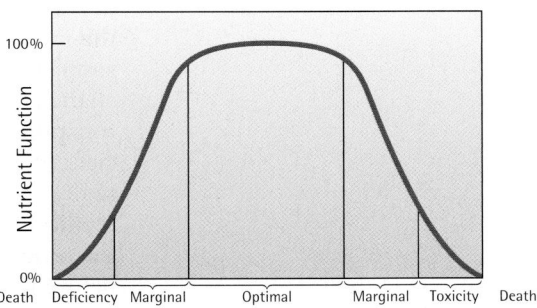

Illustration 1.8
For every nutrient, there is a range of optimal intake that corresponds to the optimal functioning of that nutrient in the body.

Steps in the Development of Nutrient Deficiencies and Toxicities

Poor nutrition due to inadequate diets generally develops in the stages outlined in Illustration 1.9. To help explain the stages, this illustration includes an example of how vitamin A deficiency develops. After a period of deficient intake of an essential

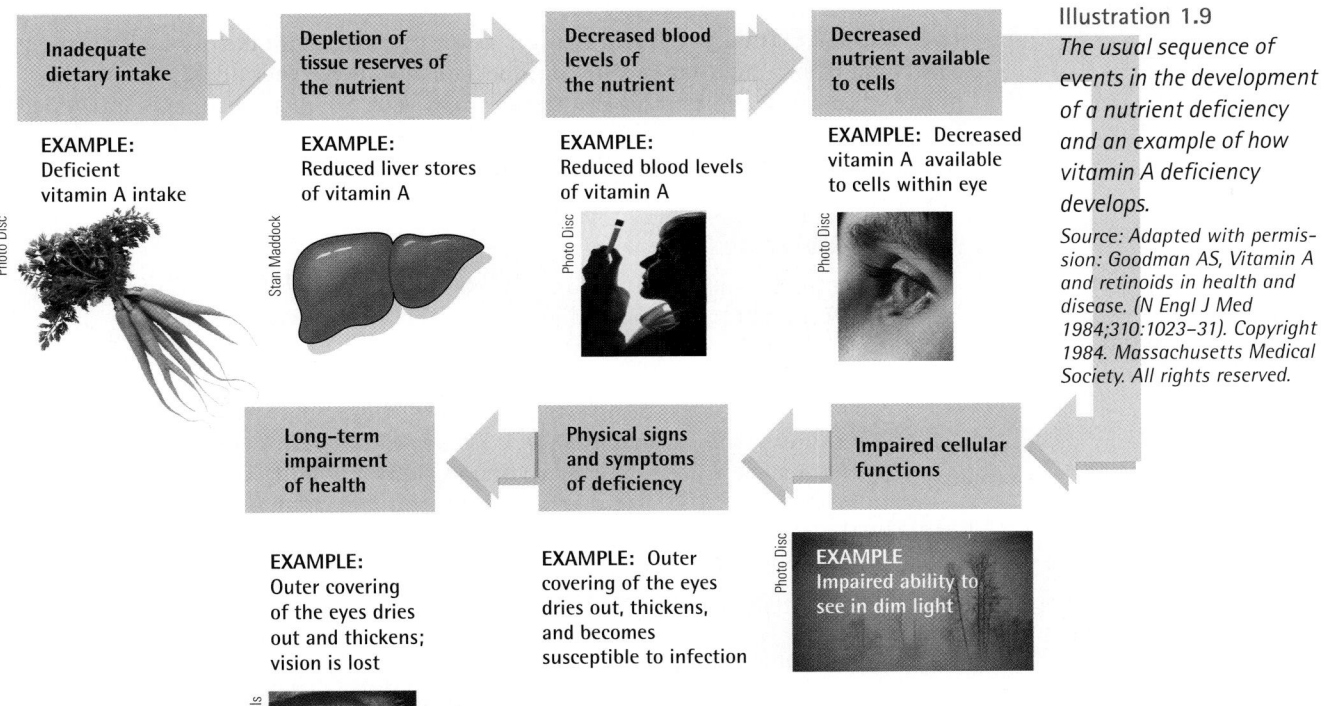

Illustration 1.9
The usual sequence of events in the development of a nutrient deficiency and an example of how vitamin A deficiency develops.

Source: Adapted with permission: Goodman AS, Vitamin A and retinoids in health and disease. (N Engl J Med 1984;310:1023–31). Copyright 1984. Massachusetts Medical Society. All rights reserved.

Inadequate dietary intake

EXAMPLE: Deficient vitamin A intake

Depletion of tissue reserves of the nutrient

EXAMPLE: Reduced liver stores of vitamin A

Decreased blood levels of the nutrient

EXAMPLE: Reduced blood levels of vitamin A

Decreased nutrient available to cells

EXAMPLE: Decreased vitamin A available to cells within eye

Impaired cellular functions

EXAMPLE Impaired ability to see in dim light

Physical signs and symptoms of deficiency

EXAMPLE: Outer covering of the eyes dries out, thickens, and becomes susceptible to infection

Long-term impairment of health

EXAMPLE: Outer covering of the eyes dries out and thickens; vision is lost nutrient, the body's tissue reserves of the nutrient become depleted. Blood levels of the nutrient then decrease because there are no reserves left to replenish the blood supply. Without an adequate supply of the nutrient in the blood, cells get short-changed. They no longer have the supply of nutrients needed to maintain normal function. If the dietary deficiency is prolonged, the malfunctioning cells cause sufficient impairment to produce physically obvious signs of a deficiency disease. Eventually, some of the problems produced by the deficiency may no longer be repairable, and permanent changes in health and function may occur. In most cases, the problems resulting from the deficiency can be reversed if the nutrient is supplied before this final stage occurs.

Excessively high intakes of many nutrients such as vitamin A and selenium produce toxicity diseases. The vitamin A toxicity disease is called "hypervitaminosis A," and the disease for selenium toxicity is called "selenosis." Signs of the toxicity disease stem from increased levels of the nutrient in the blood and the subsequent oversupply of the nutrient to the cells. The high nutrient load upsets the balance needed for normal cell function. The changes in cell functions lead to the signs and symptoms of the toxicity disease.

For both deficiency and toxicity diseases, the best time to correct the problem is usually at the level of dietary intake, before tissue stores are adversely affected. In that case, no harmful effects on health and cell function occur—they are prevented.

Nutrient Deficiencies Are Usually Multiple] Most foods contain many nutrients, so poor diets will affect the intake level of more than one nutrient. Inadequate diets generally produce a spectrum of signs and symptoms related to multiple nutrient deficiencies. For example, protein, vitamin B_{12}, iron, and zinc are packaged together in many high-protein foods. The protein-deficient children you may see in news reports on television are rarely deficient just in protein. They likely have iron, zinc, and other deficiencies in addition to protein.

The "Ripple Effect"] Dietary changes affect the level of intake of many nutrients. Switching from a high-fat to a low-fat diet, for instance, generally results in a lower intake of calories, cholesterol, and vitamin E as well. So, dietary changes introduced for the purpose of improving the intake level of a particular nutrient produce a ripple effect on the intake of other nutrients.

NUTRITION CONCEPT #5

Humans have adaptive mechanisms for managing fluctuations in nutrient intake. Healthy humans are equipped with a number of adaptive mechanisms that partially protect the body from poor health due to fluctuations in dietary intake. In the context of nutrition, adaptive mechanisms act to conserve nutrients when dietary supply is low, and to eliminate them when they are present in excessively high amounts. Dietary surpluses of energy and some nutrients—such as iron, calcium, vitamin A, and vitamin B_{12}—are stored within tissues for later use. In the case of iron and calcium, the body regulates the amounts absorbed in response to its need for them. The body has a low storage capacity for other nutrients, for instance vitamin C and water, and eliminates any excesses through urine or stools.

Here are some examples of how the body adapts to changes in dietary intake:

- When caloric intake is reduced by fasting, starvation, or dieting, the body adapts to the decreased supply by lowering energy expenditure. Declines in body temperature and the capacity to do physical work also act to decrease the body's need for calories. When caloric intake exceeds the body's need for energy, the excess is stored as fat for energy needs in the future.

- The ability of the gastrointestinal tract to absorb dietary iron increases when the body's stores of iron are low. To protect the body from iron overdose, the mechanisms that facilitate iron absorption in times of need shut down when enough iron has been stored.

- The body protects itself from excessively high levels of vitamin C from supplements by excreting the excess in the urine.

Although these built-in mechanisms do not protect humans from all the consequences of poor diets, they do provide an important buffer against the development of nutrient-related health problems.

NUTRITION CONCEPT #6

Malnutrition can result from poor diets and from disease states, genetic factors, or combinations of these causes.

Malnutrition means "poor" nutrition and results from both inadequate and excessive availability of calories and nutrients in the body. Vitamin A toxicity, obesity, vitamin C deficiency (scurvy), and underweight are examples of malnutrition.

Malnutrition can result from poor diets and also from diseases that interfere with the body's ability to use the nutrients consumed. Diarrhea, alcoholism, cancer, bleeding ulcers, and HIV/AIDS, for example, may be primarily responsible for the development of malnutrition in people with these disorders.

In addition, a percentage of the population is susceptible to malnutrition due to genetic factors. For example, people may be born with a tendency to produce excessive amounts of cholesterol, absorb high levels of iron, or utilize folate poorly. Some cases of obesity and underweight appear to be related to a combination of genetic predisposition and dietary factors.

malnutrition
Poor nutrition resulting from an excess or lack of calories or nutrients.

Illustration 1.10
Women who are pregnant, or breastfeeding, or recovering from illness are among the people who are at a higher risk of becoming inadequately nourished.

NUTRITION CONCEPT #7

Some groups of people are at higher risk of becoming inadequately nourished than others.

Women who are pregnant or breastfeeding, infants, growing children, the frail elderly, the ill, and those recovering from illness have a greater need for nutrients than people who fit into "none of the above." As a result, they are at higher risk of becoming inadequately nourished than other people (Illustration 1.10). In cases of widespread food shortages, such as those induced by natural disasters or war, the health of these nutritionally vulnerable groups is compromised the soonest and the most.

Within the nutritionally vulnerable groups, certain people and families are at particularly high risk of malnutrition. These are people and families who are poor and least able to secure food, shelter, and high-quality medical services. The risk of malnutrition is not shared equally among all persons within a population.

NUTRITION CONCEPT #8

Poor nutrition can influence the development of certain chronic diseases.

Poor nutrition does not result only in nutrient deficiency or toxicity diseases. Faulty diets play important roles in the development of heart disease,

Photo Disc

1–17

chronic diseases
Slow-developing, long-lasting diseases that are not contagious (e.g., heart disease, diabetes, and cancer). They can be treated but not always cured.

nutrient-dense foods
Foods that contain relatively high amounts of nutrients compared to their calorie value. Broccoli, collards, bread, cantaloupe, and lean meats are examples of nutrient-dense foods.

empty-calorie foods
Foods that provide an excess of calories in relation to nutrients. Soft drinks, candy, sugar, alcohol, and fats are considered empty-calorie foods.

hypertension, cancer, osteoporosis, and other **chronic diseases.** Diets high in animal fat, for example, are related to the development of heart disease; low vegetable and fruit diets to cancer; low-calcium diets to osteoporosis; and high-sugar diets to tooth decay. The harmful effects of negative dietary practices on the development of certain diseases and disorders may take years to become apparent.

NUTRITION CONCEPT #9

Adequacy, variety, and balance are key characteristics of a healthy diet. Healthy diets contain many different foods that together provide calories and nutrients in amounts that promote the optimal functioning of the body. A variety of food is required to obtain all the different essential nutrients we need in our diet. No one food (except breast milk for young infants) contains all the nutrients we need, and most single foods don't come close.

Many different combinations of foods can make up a healthy diet. It could include milk, broccoli, chicken, and apples; or ants, snails, sea cucumbers, and burdock root. What's important is that the contributions of a variety of individual foods add up to an adequate and balanced diet.

Nutrient Density

Adequate diets are most easily achieved by including foods that are good sources of a number of nutrients but not packed with calories. Foods that provide multiple nutrients in appreciable amounts relative to calories are considered **nutrient-dense.** Those that provide calories and low amounts of nutrients are called **empty-calorie foods.** Illustration 1.11 shows an example of a nutrient-dense food and an empty-calorie food. It is obviously easier to build an adequate diet around nutrient-dense foods such as fruits, vegetables, breads, cereal, lean meats, and milk (Illustration 1.12) than around foods such as soft drinks, candy, pastries, chips, and alcoholic beverages.

Illustration 1.11

An example of a food with a high nutrient density and a food with a low nutrient density. Percentages given represent percent contributions to adult female RDAs.

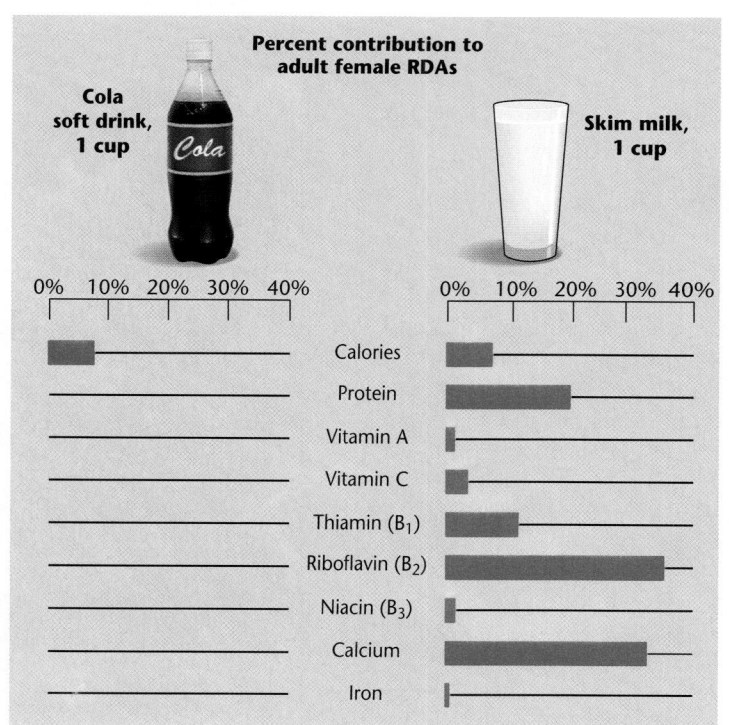

NUTRITION CONCEPT #10

There are no "good" or "bad" foods.

All things in nutriment are good or bad relatively.
—Hippocrates

Nearly 8 of 10 U.S. adults believe there are "good foods" and "bad foods."[15] Unless we're talking about spoiled stew, poison mushrooms, or something similar, however, no foods can be labeled as either good or bad. There are, however, combinations of foods that add up to a healthy or unhealthy diet. Consider the case of an adult who eats only foods thought of as "good"—for example, raw broccoli, apples, orange juice, boiled tofu, and carrots. Although all these foods are nutrient-dense, they do not add up to a healthy diet because

Illustration 1.12
"Basic foods" are nutrient-dense.

Photo Disc

Photo Disc

TABLE 1.5

NUTRITION CONCEPTS

1. Food is a basic need of humans.
2. Foods provide energy (calories), nutrients, and other substances needed for growth and health.
3. Health problems related to nutrition originate within cells.
4. Poor nutrition can result from both inadequate and excessive levels of nutrient intake.
5. Humans have adaptive mechanisms for managing fluctuations in nutrient intake.
6. Malnutrition can result from poor diets and from disease states, genetic factors, or combinations of these causes.
7. Some groups of people are at higher risk of becoming inadequately nourished than others.
8. Poor nutrition can influence the development of certain chronic diseases.
9. Adequacy, variety, and balance are key characteristics of a healthy diet.
10. There are no "good" or "bad" foods.

they don't supply a wide enough variety of the nutrients we need. Or, take the case of the teenager who occasionally eats fried chicken, but otherwise stays away from fried foods. The occasional fried chicken isn't going to knock his or her diet off track. But the person who eats fried foods every day, with few vegetables or fruits, and loads up on supersized soft drinks, candy, and chips for snacks has a bad diet. It's not any single food or group of foods that define a bad diet; it's the combination of foods generally consumed.

The basic nutrition concepts presented here are listed in Table 1.5. It may help you to remember the concepts and to start "thinking like a bona fide nutritionist," if you go back over each concept and give several examples related to it. If you understand these concepts, you will have gained a good deal of insight into nutrition.

Nutrition **UP CLOSE**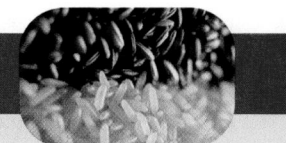

Nutrition Concepts Review

FOCAL POINT: Nutrition concepts apply to diet and health relationships.
Write the number of the nutrition concept from Table 1.5 that applies to the situation described. Use each concept and do not repeat concept numbers in your responses.

Nutrition Concept Number	Situation
1. _____	The Irish potato famine caused thousands of deaths.
2. _____ and _____	Otis mistakenly thought that as long as he consumed enough calories, protein, vitamins, and minerals, he would stay healthy no matter what he ate.
3. _____	I feel guilty every time I eat potato chips. I wish they weren't bad for me.
4. _____	Phyllis was relieved to learn that her chronic diarrhea was due to the high level of vitamin C supplements she had been taking.
5. _____	A low amount of iron in Tawana's red blood cells was the reason for her loss of appetite and low energy level.
6. _____	Far more young children than soldiers died as a result of the 10-year civil war in the Sudan.
7. _____	For the past 20 years, Don's idea of dinner was a big steak and potatoes. His recent heart attack changed his view of what's for dinner.
8. _____	During the two weeks they were backpacking in the Netherlands, Tomás and Ozzie ate very few vegetables and fruits. Their health remained robust, however.
9. _____	Zhang wasn't aware that he had the inherited condition hemochromatosis until he began taking iron supplements and developed iron overload symptoms.

FEEDBACK (answers to these questions) can be found at the end of Unit 1.

Key Terms

antioxidants, page 1-10

calorie, page 1-8

chronic diseases, page 1-18

empty-calorie foods, page 1-18

essential nutrients, page 1-11

food insecurity, page 1-7

food security, page 1-7

malnutrition, page 1-17

metabolism, page 1-14

nonessential nutrients, page 1-11

nutrients, page 1-8

nutrient-dense foods, page 1-18

nutrition, page 1-7

phytochemicals, page 1-10

www links

www.fns.usda.gov
The Food and Nutrition Service's main site provides information in English and Spanish on federal food and nutrition programs such as WIC and Food Stamps (including income eligibility, and how to apply), responses to FAQs, links to the Dietary Guidelines, and nutrition education materials.

www.nutrition.gov
Your guide to nutrition and health information from the U.S. government. Includes information on food facts, food safety, life-cycle nutrition, food assistance, and nutrient databases.

www.nal.usda.gov/fnic/foodcomp
This terrific site from the National Agricultural Library is stuffed with information on nutrition labeling, composition of foods, lists of food sources of nutrients, the DRI tables, dietary supplements, and answers to FAQs.

http://books.nap.edu/books/0309085373/html/related.html
Presents a list of DRI publications that you can read online.

Notes

1. Anderson SA. Core indicators of nutritional state for difficult-to-sample populations. J Nutr 1990;120:1598.

2. Vozoris NT, Tarasuk VS. Household food security is associated with poor health, J Nutr 2003;133:20–6; Picciano MF et al., The National Nutrition Summit: history and continued commitment to the nutritional health of the U.S. population, J Nutr 2003;133:1949–52.

3. Gaffield BE, et al. Position of the American Dietetic Association: domestic food and nutrition security. J Am Diet Assoc 2002;102:1840–47.

4. Khan AS, et al. Precautions against biological and chemical terrorism directed at food and water supplies. Pub Health Repts 2001;116:3–14.

5. Bruemmer B. Food biosecurity. J Am Diet Assoc 2003:688–92.

6. Ricin toxin from the castor bean plant. www.ansci.cornell.edu/plants/toxicagents, and waynesword.palomar.edu, accessed 6/03.

7. The castor bean. A plant named after a tick. http://waynesword.palomar.edu/plmar99.htm#elixir, accessed 6/03.

8. Craig WJ. Phytochemicals: guardians of our health. J Am Diet Assoc 1997;97 (suppl 2):S199–S204.

9. Herber D. The stinking rose: oregano-sulfur compounds and cancer. Am J Clin Nutr 1997;66:425–6.

10. Standing Committee on the Scientific Evaluation of Dietary Reference Intakes (Food and Nutrition Board, National Academy of Sciences). Washington, DC: National Academy Press; 1997.

11. Recommended Dietary Allowances, www.iom.edu/fnb, accessed 5/04.

12. Ames BN. The metabolic tune-up: metabolic harmony and disease prevention. J Nutr 2003;133:1544S–48S.

13. Based on Mertz W. The essential trace elements. Science 1981;213:1332–8.

14. Recommended Dietary Allowances, www.iom.edu/fnb, accessed 6/03.

15. Position of the American Dietetic Association: total diet approach to communicating food and nutrition information. J Am Diet Assoc 2002;102:100.

Nutrition UP CLOSE

Nutrition Concepts Review

Feedback for Unit 1

The nutrition concepts
apply to the situations as
follows:

1. 1
2. 2, 9
3. 10
4. 4
5. 3
6. 7

7. 8
8. 5
9. 6

The Inside Story about Nutrition and Health

Nutrition Scoreboard

	TRUE	FALSE

1 How long people live and how healthy they are depend on four factors: lifestyle behaviors, the environment to which people are exposed, genetic makeup, and access to quality health care.

2 Diet is related to the top two causes of death in the United States.

3 The body of modern humans was designed over 40,000 years ago.

Answers on next page

[KEY CONCEPTS AND FACTS]

- Health and longevity are affected by diet. Other lifestyle behaviors, genetic makeup, the environment to which we are exposed, and access to quality health care also affect health and longevity.

- Dramatic changes in the types of foods consumed by modern humans compared with early humans are related in some ways to the development of today's leading health problems.

- The health status of a population changes for the better or worse as diets change for the better or worse.

- The diets and health of Americans are periodically evaluated by national studies.

Answers to *Nutrition Scoreboard*

		TRUE	FALSE
1	There *are* no secrets to a long, healthy life.	✔	
2	Diet is associated with the development of heart disease and cancer (which, by the way, cause over half of all deaths in the United States).		✔
3	Hairstyles may be different, but our bodies are the same as they were 40,000 years ago.		✔

Nutrition in the Context of Overall Health

Think of your body as a machine. How well this machine performs depends on a number of related factors: the quality of its design and construction, the appropriateness of the materials used to produce it, and how well it is maintained.

A photocopying machine designed to produce 10,000 copies a day will break down sooner if it is used to make 20,000 copies a day. The repair call will, in all probability, come earlier if the machine is overused *and* poorly maintained, or if it has a part that doesn't work well. On the other hand, chances are good the copy machine will function at full capacity if it is free from design flaws, skillfully constructed from appropriate materials, properly used, and kept in good shape through regular maintenance.

Although much more complex and sophisticated, the human body is like a machine in some important ways. How well the body works and how long it lasts depend on a variety of interrelated factors. The health and fitness of the human machine depend on genetic traits (the design part of the machine), the quality of the materials used in its construction (your diet), and regular maintenance (your diet, other lifestyle factors, and health care).

Lifestyles exert the strongest overall influence on health and longevity (Illustration 2.1).[1] Behaviors that constitute our lifestyle—such as diet, smoking habits, illicit drug use or excessive drinking, level of physical activity or psychological stress, and the amount of sleep we get—largely determine whether we are promoting health or disease.[2] Of the lifestyle factors that affect health, our diet is one of the most important.[3] In a sense, it is fortunate that diet is related to disease development and prevention. Unlike age, gender, and genetic makeup, our diets are within our control.

People have an intimate relationship with food—each year we put over a thousand pounds of it into our bodies! Food supplies the raw materials the body needs for growth and health, and these, in turn, are affected by the types of food we usually eat. The diet we feed the human machine can hasten, delay, or prevent the onset of an impressive group of today's most common health problems.

The Neighborhood

Half way through his "Hearty Man" breakfast, Dwayne thought he heard some of his smaller arteries slamming shut.
Reprinted with special permission of King Features Syndicate.

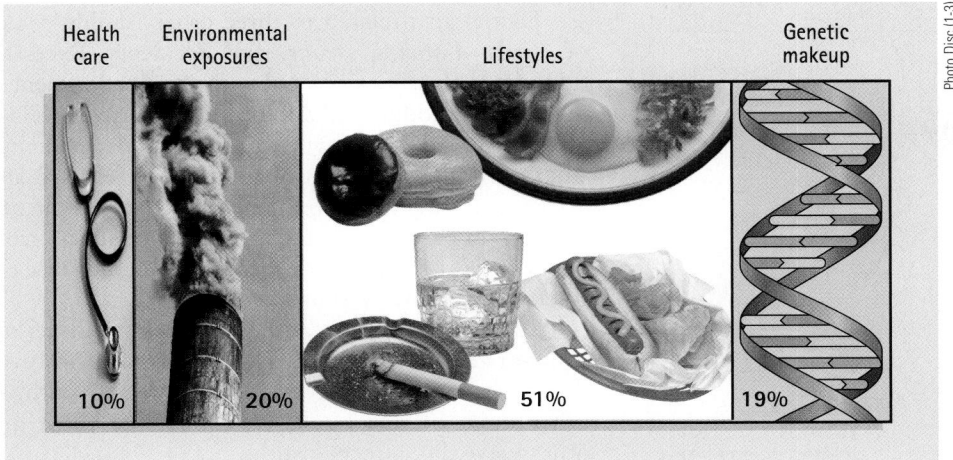

Photo Disc (1-3)

Illustration 2.1
Conditions that contribute to death among adults under the age of 75 years in the United States.
Health care refers to access to quality care; environmental exposures include the safety of one's surroundings and the presence of toxins and disease-causing organisms in the environment; lifestyle factors include diet, exercise, smoking, and alcohol and drug use; and genetic makeup consists of inherited health traits.

Source: Centers for Disease Control, Disease Prevention and Health Promotion Unit. Atlanta; 1989.

The Nutritional State of the Nation

Whosoever was the father of a disease, an ill diet was the mother.
—George Herbert, 1660

Since early in the twentieth century, researchers have known that what we eat is related to the development of vitamin and mineral deficiency diseases, to compromised growth and impaired mental development in children, and to the body's ability to fight off infectious diseases. Seventy years ago in the United States, widespread vitamin deficiency diseases filled children's hospital wards and contributed to serious illness and death in adults (Illustration 2.2). Now, however, dietary excesses are filling hospital beds and reducing the quality of life for millions of Americans.

Today, the major causes of death in Americans are slow-developing, lifestyle-related ***chronic diseases***. As Illustration 2.3 indicates, most Americans die from heart disease or cancer. Together these diseases account for 52% of all deaths. Diets high in saturated fat (the type found mainly in animal fat) and low in certain vitamins, vegetables, and fruits are linked to the development of heart disease. Certain types of cancer, such as breast and colon cancer, are related to low fruit and vegetable consumption and excessive levels of body fat.[4]

chronic diseases
Slow-developing, long-lasting diseases that are not contagious (e.g., heart disease, cancer, diabetes). They can be treated but not always cured.

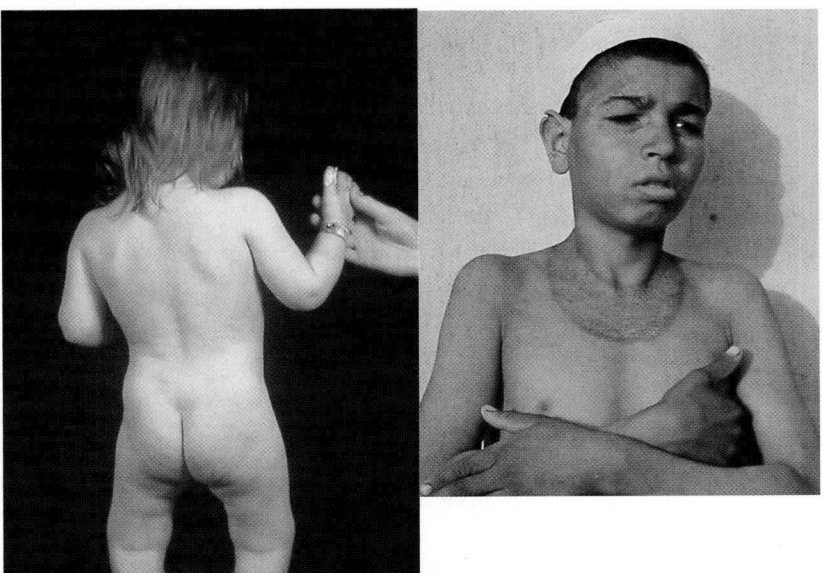

Photo Researchers, Inc.

Biophoto/Photoresearchers

Illustration 2.2
Vitamin D deficiency (rickets) and niacin deficiency (pellagra) were leading causes of hospitalization of children in the United States in the 1930s.

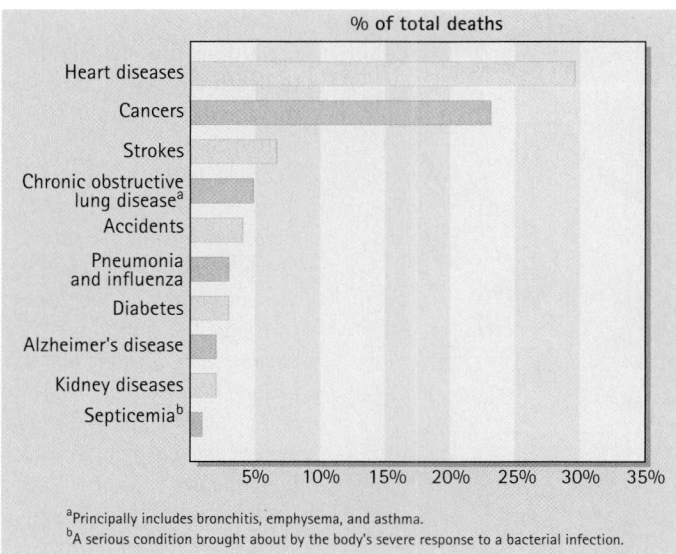

% of total deaths

Heart diseases
Cancers
Strokes
Chronic obstructive lung disease[a]
Accidents
Pneumonia and influenza
Diabetes
Alzheimer's disease
Kidney diseases
Septicemia[b]

5% 10% 15% 20% 25% 30% 35%

[a]Principally includes bronchitis, emphysema, and asthma.
[b]A serious condition brought about by the body's severe response to a bacterial infection.

Illustration 2.3

Percentage of total deaths for the top 10 leading causes of death in the United States, 2001
Source: Data from www.cdc.gov/nchs.

Diet is also related to three other leading causes of death: *diabetes, stroke,* and accidents. Excessive body fat is closely associated with the development of diabetes in middle-aged adults, and excessive levels of alcohol intake contribute to stroke and accidents.[5] That's not all. The types of food people eat are also associated with the development of overweight and obesity, hypertension (high blood pressure), iron-deficiency anemia, tooth decay and gum disease, *osteoporosis,* and *cirrhosis of the liver.*[6]

The following diet and health relationships are summarized in Table 2.1. This discussion, however, doesn't cover all of the ways diet affects health. In other units of this book, we'll examine these and other ways diet affects health.

- Obesity is the most common nutritional disorder in the United States. Approximately two-thirds of adults are overweight or obese, and one in nearly six children is overweight, too.[7] Being overweight—or, more correctly, being overly fat—increases the likelihood that diabetes, heart disease, hypertension, and certain types of cancer will occur.[8]

- High sodium and low vegetable and fruit intakes are associated with hypertension in many adults. Hypertension is one of the most common disorders in adults in the United States, and it is closely linked to heart disease and stroke.

- Low iron intake is a common problem among women and young children in the United States. Consequently, iron deficiency remains a public health problem in the United States today.

- The frequent consumption of sugars, especially in the form of "sticky sweets," is related to the development of tooth decay and gum disease.

TABLE 2.1

EXAMPLES OF DISEASES AND DISORDERS LINKED TO DIET.

DISEASE OR DISORDER	DIETARY CONNECTIONS
Heart disease	High animal (saturated) fat and cholesterol intakes; low intakes of certain vitamins, and fish, vegetables, and fruits; excessive body fat
Cancer	Low vegetable, fruit, and fiber intakes; excessive body fat and alcohol intake
Stroke	Low vegetable and fruit intake; excessive alcohol intake
Diabetes (primarily in adults)	Excessive body fat; low vegetable and fruit intake; high saturated fat intake
Cirrhosis of the liver	Excessive alcohol consumption; poor overall diets
Hypertension (high blood pressure)	High sodium (salt) and alcohol intake; low vegetable and fruit intake, excessive levels of body fat
Iron-deficiency anemia	Low iron intake
Tooth decay and gum disease	Excessive and frequent sugar consumption; inadequate fluoride intake
Osteoporosis	Inadequate intakes of calcium and vitamin D
Obesity	Excessive calorie intake

Sources: World Health Organization, Chapter 4: Quantifying Selected Major Risks to Health, at www.who.int.whr/2002/chapter 4 and ML McCullough et al., Diet quality and major chronic disease risk in men and women: moving forward toward improved dietary guidance (Am J Clin Nutr 2002;76:1261–71).

Adequate fluoride intakes help prevent tooth decay.

- Low intake of calcium, especially during the teen and young adult years, is associated with the development of osteoporosis and bone fractures. Nearly half of all women in the United States fail to get enough calcium in their diets.[9]

- Cirrhosis of the liver is a leading cause of death among adult men. Its development is closely related to long-term, excessive alcohol consumption.[10]

Photo Disc

Illustration 2.4
Lopsided, all-American food choices.

The Importance of Food Choices

People are not born with a compass that directs them to select a healthy diet—and it shows.[11] If given access to a food supply like that available in the United States, people show a marked tendency to choose a diet that is high in animal fat, and low in complex carbohydrates (Illustration 2.4). Such a diet also tends to include processed foods that are often high in salt or sugar and low in fiber, vegetables, and fruits. It is this type of diet that poses the greatest risks to the health of Americans.

Diet and Diseases of Western Civilization

Why is the U.S. diet—a "Western" style of eating—hazardous to our health in so many ways? What is it about a diet that is high in animal fat, salt, and sugar and low in vegetables, fruits, and fiber that promotes certain chronic diseases? A good deal of evidence indicates that the chronic diseases now prevalent in the United States and other Westernized countries have their roots in dietary and other changes that have taken place over centuries.

Our Bodies Haven't Changed

Biological processes that control what the body does with food today were developed more than 40,000 years ago. These hardwired, evolution-driven processes exist today because they are firmly linked to the genetic makeup of humans and genes change very little over great spans of time.[12]

Then . . .] For the first 200 centuries of existence, humans survived by hunting and gathering (Illustration 2.5).[13] They were constantly on the move, pursuing wild game or following the seasonal maturation of fruits and vegetables. Meat, berries, and many other plant products obtained from successful hunting and gathering journeys spoiled quickly, so they had to be consumed in a short time. Feasts would be followed by famines that lasted until the next successful hunt or harvest.[14]

. . . and Now] The bodies of modern humans, adapted to exist on a diet of wild game, fish, fruits, roots, vegetables, and grubs, to survive periods of famine, and to sustain a physically demanding lifestyle, are now exposed to a different set of circumstances. The foods we eat bear little resemblance to the foods available to our early ancestors (Illustration 2.6). Sugar, salt, alcohol, food additives, and fats, such as oils, margarine, and butter, were not a part of their diets. These ingredients came

diabetes
A disease characterized by abnormally high levels of blood glucose.

stroke
The event that occurs when a blood vessel in the brain suddenly ruptures or becomes blocked, cutting off blood supply to a portion of the brain. Stroke is often associated with "hardening of the arteries" in the brain. (Also called a *cerebral vascular accident.*)

osteoporosis
A condition in which bones become fragile and susceptible to fracture due to a loss of calcium and other minerals.

cirrhosis of the liver
Degeneration of the liver, usually caused by excessive alcohol intake over a number of years.

Illustration 2.5
Hunter-gatherers still exist in the world, but their numbers are diminishing. It is estimated that hunter-gatherers consume approximately 3000 calories daily due to their physically demanding way of life.

with Western civilization. Furthermore, we do not have to engage in strenuous physical activity to obtain food, and our feasts are no longer followed by famines.

The human body developed other survival mechanisms that are not the assets they used to be. Mechanisms that stimulate hunger in the presence of excess body fat stores, conserve the body's supply of sodium, and confer an innate preference for sweet-tasting foods, as well as a digestive system that functions best on a high-fiber diet, were advantages for early humans. They are not advantageous for modern humans, however, because diets and lifestyles are now vastly different.

Although the human body has a remarkable ability to adapt to changes in diet, health problems of modern civilization such as heart disease, cancer, hypertension,

Illustration 2.6
The disconnect between high animal fat, high salt, high sugar, and processed foods in Western-type diets (right) and wild plants and animal foods consumed by our early ancestors (left).
Foods consumed by hunter-gatherers shown in the photograph include bird's eggs, wild cucumbers, roots, nut, and berries. Not shown are grubs and other insects, which might be consumed as quickly as they are discovered.

and diabetes are thought to result partly from diets that are greatly different from those of our early ancestors.[15] It appears that the human body was built to function best on a diet that is low in sugar and sodium, contains lean sources of protein, and is high in fiber, complex carbohydrates, and vegetables and fruits (Illustration 2.6).[16] Strong evidence for this conclusion is provided by studies that track how disease rates change as people adopt a Western style of eating.

Changing Diets, Changing Disease Rates

Many countries are adopting the Western diet and the pattern of disease that accompanies it. People in Japan, for example, live longer than anyone else in the world—until they move to the United States (Table 2.2). In Japan, diets consist mainly of rice, vegetables, fish, shellfish, and meat (Illustration 2.7).[17] When Japanese people move to the United States, their diets change to include, on average, more meat and fat and fewer complex carbohydrates. Japanese living in the United States are much more likely to develop diabetes (Illustration 2.8), heart disease, breast cancer, and colon cancer than people who remain in Japan.[18]

Dietary habits in Japan are rapidly becoming similar to those in the United States. Hamburgers, fries, steak, ice cream, and other high-fat foods are gaining in popularity. Rates of diabetes, heart disease, and cancer of the breast and colon are on the rise in Japan.[19] Similarly, the "diseases of Western civilization" are occurring at increasing rates in Russia, Greece, Israel, and other countries adopting the Western diet.[20]

TABLE 2.2

LIFE EXPECTANCY AT BIRTH FOR COUNTRIES WITH HIGH LIFE EXPECTANCIES.

COUNTRY	LIFE EXPECTANCY (YEARS)
Japan	80.7
Australia	79.8
Sweden	79.6
Switzerland	79.6
Canada	79.4
Iceland	79.4
Italy	79.0
France	78.8
Spain	78.8
Norway	78.7
Greece	78.4
Netherlands	78.3
New Zealand	77.8
Belgium	77.8
United Kingdom	77.7
Austria	77.7
Finland	77.4
Germany	77.4
United States	77.2
Luxembourg	77.1
Denmark	76.5
Portugal	75.8

Source: Data from www.cdc.gov/nchs (accessed 7/03).

Photo Disc

Illustration 2.7
Typical Japanese foods. Compare these to the "all-American" food choices shown in Illustration 2.4.

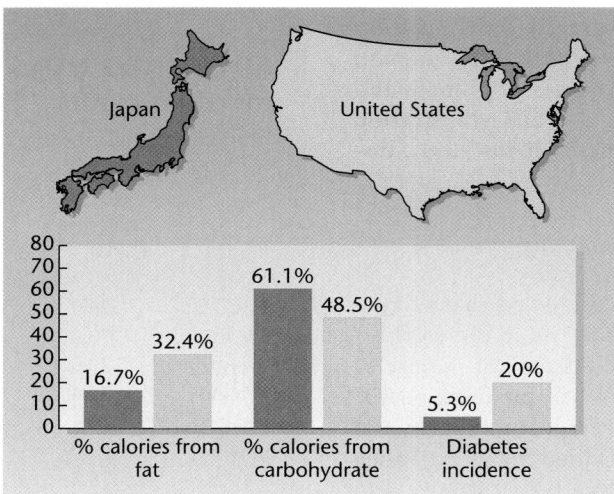

Illustration 2.8

An increased rate of diabetes in Japanese men immigrating to Seattle corresponds to dietary changes.

Source: CH Tsunehara, DL Leonetti, and WY Fujimoto, Diet of second-generation Japanese-American men with and without non-insulin-dependent diabetes (Am J Clin Nutr 1990;52:731–8).

Eating more like our early ancestors did given today's food supply is a bit challenging. What types of foods would you choose if you wanted to shape a diet that is closer to that of hunter-gathers? Join Sabrina and Shandra in making decisions about which food choices would get them closer to the basic diets of early humans in the "Reality Check."

The Power of Prevention

Heart disease, cancer, and other chronic diseases are not the inevitable consequence of Westernization. High-animal fat diets and lifestyle behaviors (such as smoking) that promote chronic disease can be avoided or changed. Although heart disease is still the leading cause of death in the United States, its rate has declined by 49% in the last 35 years.[21] Reductions in animal fat and cholesterol intake, along with reductions in the proportion of people who smoke, are largely credited with this striking improvement in deaths from heart disease.[22]

Improving the American Diet

Many efforts are under way to improve the diet and health status of Americans. Like some other countries with high rates of "Western" diseases, the United States has set national health goals and implemented programs aimed at improving health. Goals for changes in health status in the United States are presented in the report, *Healthy People 2010: Objectives for the Nation.*[23] Its objectives for improvements in nutrition are outlined in Table 2.3.

TABLE 2.3

A SUMMARY OF NUTRITION OBJECTIVES FOR THE NATION.

IMPROVEMENTS IN HEALTH STATUS	IMPROVEMENTS IN HEALTH PRACTICES	IMPROVEMENTS IN FOOD SERVICES AND PROGRAMS
Reduce rates of: • Overweight • Heart disease and stroke • Hypertension • Cancer • Osteoporosis • Diabetes • Congenital (in-born) abnormalities • Growth retardation • Iron deficiency • Baby bottle tooth decay • Food allergy deaths	*Increase rates of:* • Breastfeeding • Safe and effective weight-loss practices (diet and exercise) • Healthy weight gain in pregnancy *Reduce dietary intake of:* • Fat • Saturated fat • Sodium *Increase dietary intake of:* • Vegetables • Fruits • Grain and grain products (especially whole grains) • Calcium	• Improve food safety practices • Increase food security • Increase nutrition education in schools • Improve nutritional quality of school meals and snacks • Increase work site programs in nutrition and weight management • Increase nutrition counseling for patients with heart disease, diabetes, and other nutrition- and diet-related conditions

Source: Data from Healthy People 2010—Objectives for the Nation, www.nutrition.gov (accessed 2/00).

Sabrina:
wheat tortillas, white rice, organic hot dogs, soy protein bars, corn flakes, raw sugar, gummy bears, sweet potato chips

REALITY CHECK

Getting back to the basics—but how?

Sabrina and Shandra have been roommates for a year and usually shop for groceries together. For the next trip to the grocery store, they decide that each of them will make up a shopping list consisting of foods that resemble what their early ancestors might have eaten. Here are the results:

Which list do you think comes closest to matching the basic foods consumed by our early ancestors

?

Answers on page 2–11

Shandra:
barley, lentil beans, white fish, sunflower seeds, carrots, romaine lettuce, plums, strawberries

Photo Disc

What Should We Eat?

Attempts to improve the U.S. diet have led to the development of a set of guidelines for healthy eating. Known as the "Dietary Guidelines for Americans" (Illustration 2.9), this document identifies 10 key areas in need of improvement and provides practical suggestions for implementing dietary changes. The Dietary Guidelines, which are updated every five years, are a very useful tool for planning and evaluating dietary intake.

In addition to the Dietary Guidelines, recommendations for the selection of foods that contribute to a healthy diet are available. These recommendations are based on the food groups presented in the "Food Guide Pyramid." The Food Guide Pyramid is an updated version of the Basic Four Food Groups. It is shown in Illustration 2.10 (p. 2–10). Foods you won't see pictured in the Food Guide Pyramid groups are processed meats, salted and high-fat snack foods, soft drinks, and sugar-rich desserts. Overall, the Guide urges consumers to select basic foods that have been modified little by processing or have added fat, sugar, or salt.

Nutrition Surveys: Tracking the American Diet

The food choices people make and the quality of the American diet are regularly evaluated by national surveys. Table 2.4 on p. 2–10 summarizes the major surveys conducted by the federal government. The first survey was started in 1936. It was conducted in conjunction with the original national

Illustration 2.9
Dietary Guidelines for Americans.
An updated version will be released late in 2005 and will be available at www.nal.usda.gov/fnic

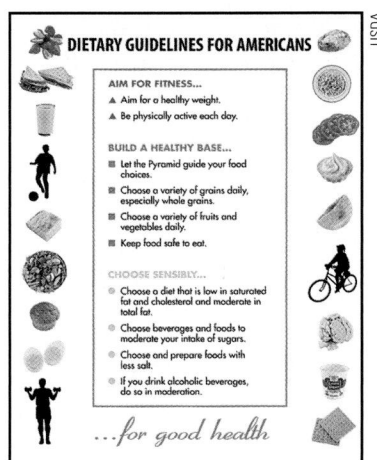

TABLE 2.4

PERIODIC, NATIONAL SURVEYS OF DIET AND HEALTH IN THE UNITED STATES.

SURVEY	PURPOSE
1. **National Health and Nutrition Examination Survey (NHANES)**	• Assesses dietary intake, health, and nutritional status in a sample of adults and children in the United States on a continual basis
2. **Nationwide Food Consumption Survey (NFCS)**	• Performs regular surveys of food and nutrient intake, and understanding of diet and health relationships among a national sample of individuals in the United States
3. **Total Diet Study** (sometimes called Market Basket Study)	• Ongoing studies that determine the levels of various pesticide residues, contaminants, and nutrients in foods and diets

program aimed at reducing hunger, poor growth in children, and vitamin and mineral deficiency diseases in the United States. Results of nutrition surveys are used to identify problem areas within the food supply, characteristics of diets consumed by people in the United States, and the nutritional health of the population. They provide information ranging from the amount of lead and pesticides in certain foods to the adequacy of diets of low-income families. Together with the results of studies conducted by university researchers and others, they provide the information needed to give direction to food and nutrition programs and efforts aimed at improving the availability and quality of the food supply.

Illustration 2.10
Making daily food choices with the USDA Food Guide Pyramid.
Originally released in 1992; an updated version of the Food Guide Pyramid may be released late in 2005. It would be available at www.nal.usda.gov/fnic

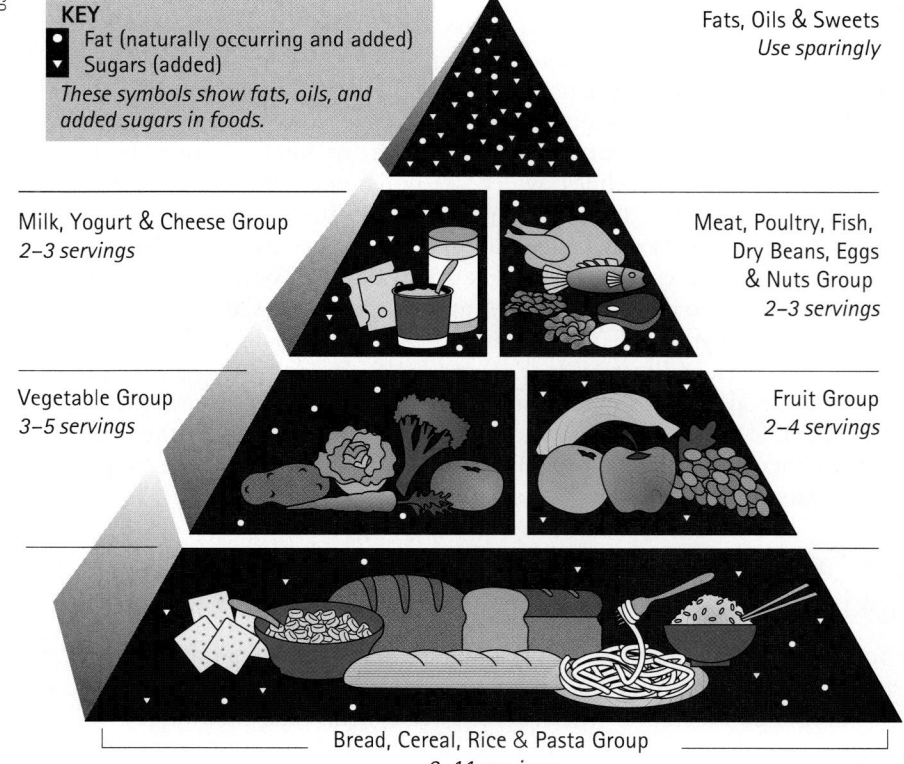

2-10

ANSWERS TO REALITY CHECK

Sabrina

Shandra's list contains more high-fiber and complex carbohydrate foods, more vegetables and fruits, and lower fat, sugar, and salt food choices than Sabrina's list.

Shandra

Photo Disc

Nutrition UP CLOSE

Rate Your Nutrition Knowledge

FOCAL POINT: How well-informed are you? Compare your knowledge of nutrition against that of the typical American consumer.

Researchers at the Centers for Disease Control (CDC) contacted a sample of U.S. adults and asked them a set of questions about diet and health relationships. If you answer all of these questions correctly, you know more than most.

1. Which of the following choices represent the *two* best ways to lose weight and keep it off?

   a. Do not eat at bedtime.
   b. Eat fewer calories.
   c. Take diet pills.
   d. Increase physical activity.
   e. Do not eat any fat.
   f. Eat a grapefruit at each meal.

2. Eating a diet high in animal fat _____ the risk of heart disease.

   a. increases
   b. decreases
   c. does not affect

3. A high blood cholesterol level _____ the risk of heart disease.

   a. increases
   b. decreases
   c. does not affect

4. Which of the following substances in food is most often associated with high blood pressure?

   a. cholesterol
   b. sugar
   c. sodium (or salt)

5. Which of the following aspects of diet is related to the development of osteoporosis?

   a. not enough dietary fiber
   b. too much fat
   c. not enough calcium

FEEDBACK (answers to these questions) can be found at the end of Unit 2.

Source: Adapted from material from the U.S. Government—Centers for Disease Control and Prevention.

Key Terms

www links

www.healthfinder.gov
Gateway for nutrition and health information.

www.nutrition.gov
One-stop shopping for information on government-sponsored food and nutrition programs, health statistics, and diet and health relationships.

www.nal.usda.gov/fnic
Get updates on the Food Guide Pyramid (plus related educational materials) and the Dietary Guidelines for Americans at this Web site.

www.cdc.gov/nchs
Your gateway to information on Nutrition Monitoring in the United States.

www.healthypeople.gov
Presents the Healthy People 2010 Objectives for the Nation.

www.hc-sc.gc.ca
Search the word *nutrition* and gain access to diet and health information, food and nutrition programs and resources in Canada.

www.nlm.nih.gov/medlineplus
Provides excellent information on diet-disease relationships.

www.cfsan.fda.gov
This site will lead you to additional information on the Total Diet Studies.

Notes

1. Centers for Disease Control, Disease Prevention and Health Promotion Unit. Atlanta; 1989.

2. Promoting Health: Intervention Strategies for Social and Behavioral Research, Institute of Medicine, National Academy of Sciences, July 2000.

3. German JB et al. Genomics and metabolomics as markers for the interaction of diet and health: lessons from lipids. J Nutr 2003;133:2078S–83S.

4. Weisburger JH, Approaches for chronic disease prevention based on current understanding of underlying mechanisms, Am J Clin Nutr 2000;71:1710S–4S; discussion 1715S–9S; and Blanchini F et al., Overweight, obesity, and cancer risk, The Lancet 2002;3:565–74.

5. Weisburger, Approaches for chronic disease prevention.

6. Blanchini, Overweight, obesity, and cancer risk.

7. www.cdc.gov/nchs, accessed 7/03.

8. Mokdad AH et al. Prevalence of obesity, diabetes, and obesity-related health risk factors, 2001. JAMA 2003;289:76-9.

9. Deckelbaum RJ, Fisher EA, Winston M, et al. Summary of a scientific conference on preventive nutrition: pediatrics to geriatrics. Circulation 1999;100:450–6.

10. www.cdc.gov/nchs, accessed 7/03.

11. Story M, Brown JE. Do young children instinctively know what to eat? N Engl J Med 1987;316:103–6.

12. Herber D and Bowerman S. Applying science to changing dietary patterns. J Nutr 2001;307:8S–81S.

13. Sebastian A et al. Estimation of the net acid load of the diet of ancestral preagricultural *Homo sapiens* and the hominid ancestors. Am J Clin Nutr 2002;76:1308–16.

14. Milton K. Hunter-gatherer diets—a different perspective [editorial]. Am J Clin Nutr 2000;71:665–7.

15. Milton, Hunter-gatherer diets.

16. Sebastian et al., Net acid load of *Homo sapiens*.

17. Sugano M, Hirahara F. Polyunsaturated fatty acids in the food chain in Japan, Am J Clin Nutr 2000;71:189S–96S; and Matsuzaki T. Longevity, diet, and nutrition in Japan: epidemiological studies, Nutr Rev 1992;50:355–9.

18. Kato H, Haybuchi H. Study of the epidemiology of health and dietary habits of Japanese and Japanese Americans. Japan Journal of Nutrition 1989;47:121–30.

19. Kato and Haybuchi, Study of the epidemiology of health.

20. Milton, Hunter-gatherer diets; Cooper R, Schatzkin A, Recent trends in coronary risk factors in the U.S.S.R., Am J Public Health 1982;72:431–40; and Trichopoulou AD, Efstathiadis PP, Changes of nutrition patterns and health indicators at the population level in Greece, Am J Clin Nutr 1989;49(suppl):1042–7.

21. www.cdc.gov/nchs, accessed 7/03.

22. Ernst ND, Sempos CT, et al. Consistency between US dietary fat intakes and fetal serum cholesterol concentrations. Am J Clin Nutr 1997;66(suppl):965S–72S.

23. Healthy People 2010—Objectives for the Nation. Food and Drug Administration and the National Institutes of Health, U.S. Department of Health and Human Services, 1999, www.nutrition.gov, accessed 2/00.

Nutrition **UP CLOSE**

Rate Your Nutrition Knowledge

Feedback for Unit 2		
The number in parentheses indicates the percentage of U.S. adults in the sample who responded correctly.	b and d (73%)	c (58%)
	a (80%)	c (43%)
	a (86%)	

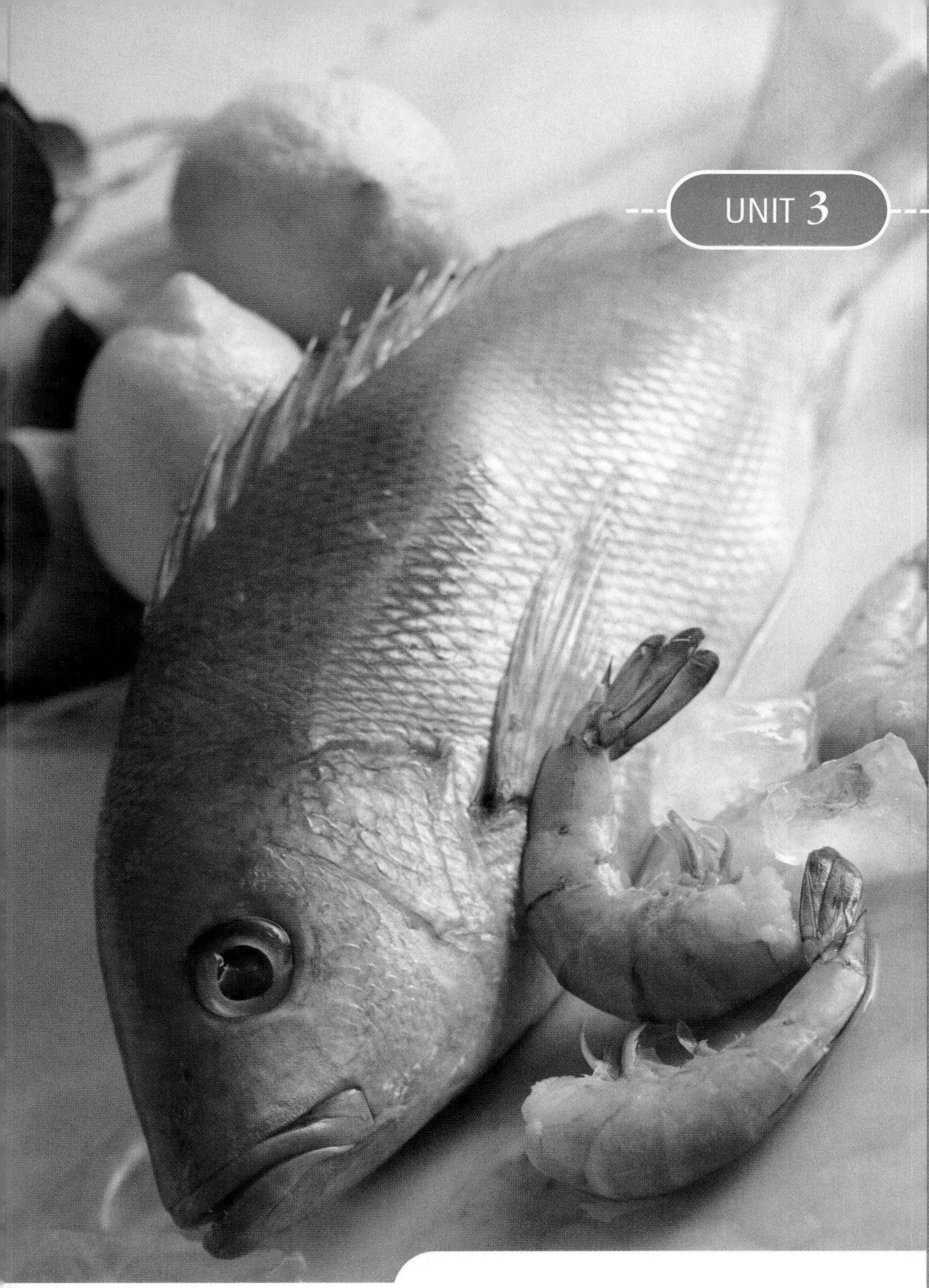

Ways of Knowing about Nutrition

Nutrition Scoreboard

	TRUE	FALSE
1 It is illegal to convey false or misleading information about nutrition in magazine and newspaper articles and on television.		
2 Knowledge about nutrition is gained by scientific studies.		
3 "Double-blind" studies are used to diminish the "placebo effect."		

Answers on next page

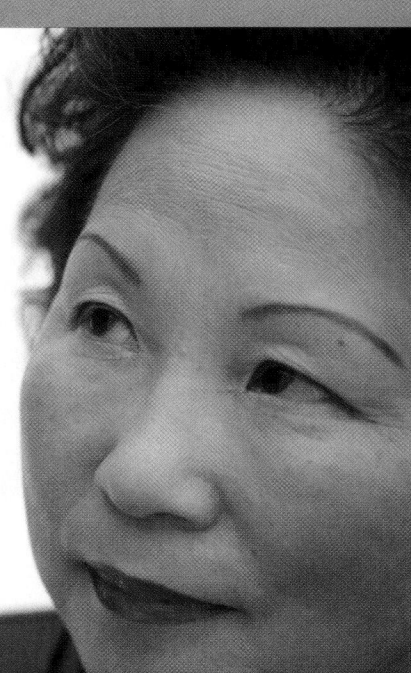

[KEY CONCEPTS AND FACTS]

- For the most part, nutrition information offered to the public does *not* have to be true or even likely true.

- Nutrition information offered to the public ranges in quality from sound and beneficial to outrageous and harmful.

- Science is knowledge gained by systematic study. Reliable information about nutrition and health is generated by scientific studies.

- Misleading and fraudulent nutrition information exists primarily because of financial interests and personal beliefs and convictions.

Answers to *Nutrition Scoreboard*

		TRUE	FALSE
1	Freedom of speech applies to information about nutrition in articles, speeches, pamphlets, and broadcasts. However, it is illegal to make false or misleading claims about nutrition in advertisements or on product labels and packaging.		✔
2	The "way of knowing" for the field of nutrition is science.	✔	
3	In a double-blind study, neither the research staff nor the research subjects know which subjects are getting the real treatment and which are receiving the fake treatment. This reduces the placebo ("sugar pill") effect, or changes in health that are due to the expectation that a specific treatment will have a particular impact on health.	✔	

How Do I Know if What I Read or Hear about Nutrition Is True?

Maria, a college student: "You really ought to try this new Herbal Melt Down Diet I found on the Internet. I used the herbs for a week and lost five pounds!"

Jessie, a premed student: "If I were you, I'd treat that urinary tract infection with cranberry juice."

Newspaper headline: "Eating Cauliflower Daily Prevents Breast Cancer!"

Infomercial: "Lose fat, gain muscle and energy! Eat Complete Cereal Pro bars every day!"

Photo Disc

How do you know if what you read or hear about nutrition is true? With only the information given in these examples, you couldn't know. In reality, however, this is how much of the information we receive about nutrition comes to us—in bits and pieces. The nutrition information offered to the public is a mix of truths, half-truths, and gossip. The information does not have to meet any standard of truth before it can be represented as true in books, magazines, newspapers, TV and radio reports and interviews, pamphlets, the Internet, and speeches.

Opinions expressed about nutrition are protected by the freedom of speech provisions of the U.S. Constitution. Although a tabloid article announcing that 19 foods have negative calories is misleading and fraudulent, it is not illegal. The promotion of nutritional remedies that are not known to work, such as amino acid supplements for hair growth and beef extract for the treatment of cancer, is likewise protected by freedom of speech. It is illegal, however, to put false or misleading information about a product or service on a product label, in a product insert, or in an advertisement. In addition, the U.S. and Canadian mail systems cannot be used to send or to receive payments for products that are fraudulent.

With so much misinformation available, it can be difficult to know what to do when we hear or read something about nutrition that may benefit us personally or perhaps help a friend or relative. Why does such a mix of nutrition sense and non-sense exist? How can you separate sound information from the highly questionable? Where does nutrition information you can trust come from? These questions are addressed in this unit.

Why Is There So Much Nutrition Misinformation?

In addition to protections afforded to marketers by freedom of speech, there are other reasons why consumers are bombarded with nutrition misinformation and ineffective or untested products and services. These reasons can be grouped into two categories: profit, and personal beliefs and convictions. The first and the more important reason, not surprisingly, is the profit motive.

Motivation for Nutrition Misinformation #1: Profit] As long as consumers seek quick and easy ways to lose weight, build muscle, slow aging, and reduce stress—goals that cannot be achieved quickly or easily—there will continue to be a huge market for nutritional products and services that offer assistance. The financial incentives are in place:

> . . . [join] a special company with 22nd century breakthrough nutritional products and unequaled compensation plan!
> —USA Today business advertisement

> Teaming up nutritional science with network marketing makes an awesome twosome!
> —Promotional article for "Nutritional System" in an advertising magazine

Not everyone who is in the business of nutrition has the goal of maintaining or improving people's health. Many are in the business to make money from people who are willing to believe their advertisements and buy products or services just in case they work.

People believe nutrition nonsense and buy fraudulent nutrition products for many reasons:

- Many people believe what they want to hear.
- People tend to believe what they see in print.
- Promotional materials are made to sound scientific and true.
- The products offer solutions to important problems that have few or no solutions in orthodox health care.
- Promotional materials often appeal to people who are disenchanted with traditional medical care and the side effects of many medications, or to those who want a "natural" remedy.
- Laws that govern truth in advertising are often not enforced.
- New bogus products and services find a ready market because existing bogus products and services don't work.

Instead of evidence and facts, profit-oriented companies use paid testimonials ("It worked for me, it will work for you!"), "medical experts," and Hollywood stars

Max, a fifty-year old engineer: "If I hadn't been following a low-fat diet for the last 10 years, I would have had a heart attack by now—just like my dad did when he was forty-eight."

Photo Disc

and sports heroes to promote their products. Their advertisements promote ideas like "a wonder to science" or "miraculous" that appeal to some people's inclinations toward the mysterious or the divine. Real remedies aren't advertised in such terms. Nor are they portrayed as being effective because they come from Europe, the Ecuadoran highlands, the ancient Orient, or organic algae ponds.

Illustration 3.1 offers a formula for developing and marketing a fraudulent nutrition product. Try following the steps and make up your own miraculous nutritional cure. Once you've devised your own product, you'll find it easy to detect when other people are doing it for real.

Controlling Profit-Motivated Nutrition Frauds] The Federal Trade Commission (FTC) has authority to remove from the airwaves and Internet advertisements that make false claims. In the past, the FTC has exerted its authority by removing blatantly false and misleading advertisements including those for the "European Weight Loss Patch," shark cartilage, juices with herbal extracts, coral calcium, and bee pollen. Nevertheless, misleading and inaccurate advertisements still appear because enforcement efforts are weak and are concentrated on very dangerous products. The FTC and other federal agencies do respond when several consumers register complaints about a nutritional product or advertisement.

Illustration 3.1
Create Your Own Fraudulent Nutrition Product

of Tennessee made $7,000 in commission. No Gimmicks – No competition – Big money every week! For more information call

(800) 555-1234

HAVE YOU EVER SEEN A FAT PLANT?

Scientists in our laboratory have discovered the secret. For a lifetime of being reed thin, send $29.95 to:

Obesity Research Center
P.O. Box 281752,
Miami, FL

HOT! HOT! HOT!

Banks & Financial Institutions made HUGE PROFITS trading foreign currencies & so can you! 10K invested

A Step-by-Step Guide

1. Identify a common problem people really want fixed that cannot easily or quickly be fixed another way.
   EXAMPLES: Obesity, low energy, weak muscles.

2. Make up a nutritional remedy and connect it to a biological process in the body. Try to think of a remedy that probably won't harm anyone. While you are at it, create a catchy name for your product.
   EXAMPLES: An herb that burns fat by speeding up metabolism, vitamins that boost the body's supply of energy, protein supplements that go directly to your muscles.

3. Develop a scientific-sounding explanation for the effect the product has on the body. Refer to the results of scientific studies.
   EXAMPLES: "Research has shown that the combination of herbs in this product stimulates the production of chemicals that trigger the energy-production cycle in the body." "As nutritional scientists have known for a century, the body requires B vitamins to form energy. The more you consume of this special formulation of B vitamins, the more energy you can produce and the more energy you will have!" "The unique combination of amino acids in this product is the same as that found in the jaw muscles of African lions. It is well known that an African lion can lift a 600-pound animal in its teeth."

4. Dream up testimonials from bogus previous users of the product, before-and-after photos, or "expert" opinions to quote. Use terms like "magical," "miraculous," "suppressed by traditional medicine," "secret," or "natural" as much as possible.
   EXAMPLES: Jane Fondu, Indianapolis: "I didn't think this product was going to work. The other products I tried didn't. My doctor couldn't help me lose weight. Thank God for [insert name of product]! I miraculously lost 20 pounds a week while eating everything I wanted." Dr. J. R. Whatsit, DM, NtD, Director of Nutritional Research: "We discovered this secret herb in our laboratories after years of looking for the substance in plants that keeps them from getting fat. Voilà! We found it. This discovery may win us a Nobel Prize." Or, show photos of a skinny person and a beefed-up person. (The photos don't have to be of the same person; the people only need to look similar.)

5. Offer customers a money-back guarantee.

Science for Sale] Here's a true story:

Friday, 2:00 p.m. A call came into the Nutrition Department from a man who wanted to talk to a nutrition expert. He had heard on last night's news that zinc lozenges were good for treating cold symptoms. Since he had a cold and didn't think the zinc would hurt, he purchased and consumed a whole roll of lozenges. Now he had a horrible, metallic taste in his mouth that he couldn't get rid of, no matter how often he brushed his teeth or gargled with mouthwash. He was worried that the taste was going to stay in his mouth forever. The nutrition expert assured him that it wouldn't. He had overdosed on zinc and the taste would go away slowly.

This man was not the only one who heard the newscast about zinc and consumed too many lozenges the next day. Sales of zinc lozenges skyrocketed after the newscast, and as it happened, the author of the research study profited handsomely from the increased sales. After completing the study, the author bought shares of stock in the company that made the lozenges.[1] Such a financial tie should have been reported in the article. It is possible that a financial incentive could have influenced how the author presented the study's results.

Many researchers have a vested interest in their research results (Illustration 3.2). A study of one thousand Massachusetts scientists who had published research articles found that one-third held a patent for the product tested, were paid industry consultants, or had another form of financial stake in the research.[2] Another study found that authors of research supportive of a new artificial fat were four times more likely to have financial affiliations with the company producing the product than were authors with no such company ties.[3]

Who Is Conducting the Research?] *Nutrition research is often conducted by people or companies that have a stake in the results.* Although this doesn't mean the

Illustration 3.2
Researchers may have a financial interest in their research.

A checklist for identifying nutrition misinformation

1. Is something being sold? _____ Yes _____ No

2. Does the product or service offer a new remedy for problems that are not easily or simply solved (e.g., obesity, cellulite, arthritis, poor immunity, weak muscles, low energy, hair loss, wrinkles, aging, stress)? _____ Yes _____ No

3. Are such terms as "miraculous," "magical," "secret," "detoxify," "energy restoring," "suppressed by organized medicine," "immune boosting," or "studies prove" used? _____ Yes _____ No

4. Are testimonials, before-and-after photos, or expert endorsements used? _____ Yes _____ No

5. Does the information sound too good to be true? _____ Yes _____ No

6. Is a money-back guarantee offered? _____ Yes _____ No

Illustration 3.3
If a red flag comes up as you read about a product or service, beware.

research results are invalid, it does mean that the study design should be carefully scrutinized before it is published in a scientific journal and broadcast on the news. People and companies with vested interests in the results of studies tend not to report findings that reflect negatively on a product, or their promotional materials may neglect to mention studies that produced different results. For these reasons, it is important to consider whether the study results or other presentations of nutrition facts may be tainted by a financial motive.

A Checklist for Identifying Nutrition Misinformation]
Illustration 3.3 lists some common features of fraudulent information for nutrition products and services. If you find any of these characteristics in articles, advertisements, Web sites, or pamphlets or hear them on infomercials or TV and radio interviews, beware of the product or service. People offering legitimate information are cautious and don't exaggerate nutritional benefits to health. They usually aren't selling anything but rather are trying to inform people about new findings so they can make better decisions about optimal nutrition.

The Business of Nutrition News]
Newspapers, TV stations, magazines, books, and other sources of information make money when the number of viewers or readers is high. To increase viewers or readers, the media attempt to present information that will pique people's interest. Nutrition breakthroughs tend to do that. The media have access to a large number of nutrition studies; more than eight thousand nutrition-related research articles are published each year (Illustration 3.4).

Studies reporting new results related to obesity, cancer, cholesterol, heart disease, vitamins, and food safety are particularly hot topics. To keep people interested, the media may sensationalize and oversimplify nutrition-related stories. They may

Illustration 3.4
On average, over eight hundred nutrition-related research articles are published monthly in scientific journals such as those shown here.

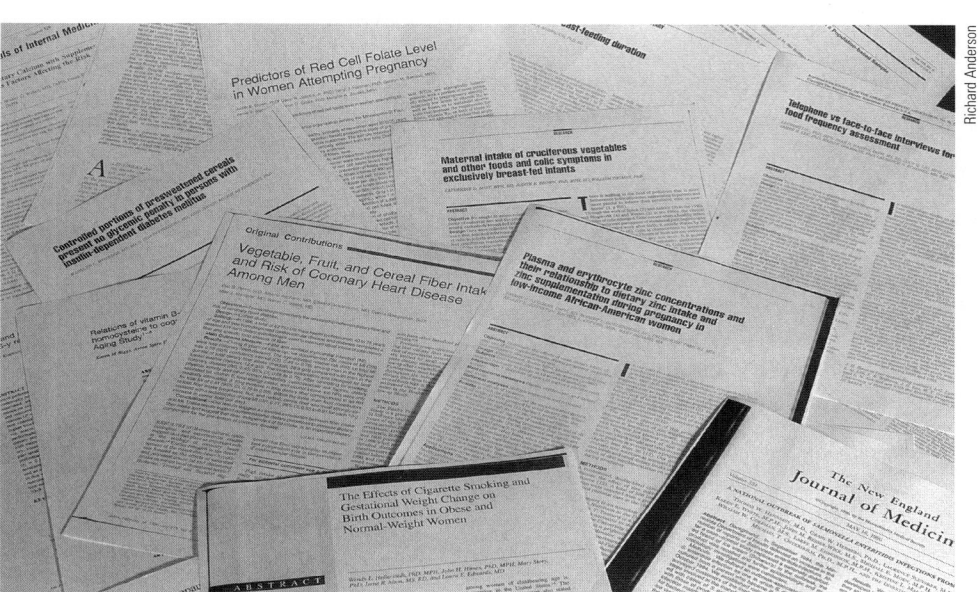

Just a fiber fable?
Researchers say oat bran doesn't really stack up as a miracle cholesterol fighter

Oat-Bran Added to Diet Lowers Cholesterol Levels

Oat Bran Muffins Lower Serum Cholesterol of Healthy Young People

Study of studies:
Oat bran helps modestly
Analysis of 19 previous efforts shows 3-4% drop in cholesterol

Oats: A New Look at the Original Heart Healthy Food
By Cathy Kapica, Ph.D, R.D.

Illustration 3.5
How to confuse the public: a lesson delivered by the head-lines about oats and blood cholesterol.
Despite some conflicting reports, a consensus of scientific opinion about the effects of oat bran on blood cholesterol levels has been reached: regular consumption of oat bran or oatmeal lowers LDL cholesterol (the so-called bad cholesterol) by 5% or more.[4]

report on one study one day and describe another study with opposite results the next. A classic example of headlines about nutrition that totally confused the public is shown in Illustration 3.5. Such study-by-study coverage of nutrition news leaves consumers not knowing what to believe or what to do. A Nutrition Trends Survey by the American Dietetic Association uncovered a strong consumer preference (81% of those sampled) for learning about the latest nutrition breakthroughs only *after* they had been generally accepted by nutrition and health professionals.

Nutrition studies are complex, and the results of one study are almost never enough to prove a point. Decisions about personal nutrition should be based on accumulated evidence that is broadly supported by nutritionists and other scientists.

Motivation for Nutrition Misinformation #2: Personal Beliefs and Convictions

Numerous alternative health practitioners (such as nutripaths, irridologists, electrotherapists, scientologists, and faith healers) use nutrition remedies. Although their approaches to and philosophies about health may vary, these practitioners often have strong beliefs in the benefits of what they do. These firm beliefs may be enough to "talk" some people out of their problems—problems that might have gone away anyway.

There's no way to say whether most of the remedies used by alternative nutrition practitioners work. (Some of the unproven remedies employed are shown in Illustration 3.6.) Strong belief in these remedies, rather than proof that they work or a desire for profits, may rule these practitioners' distribution of nutrition advice and products. The nutritional cures offered are not based on science, and they have not been scientifically evaluated. Until they are evaluated, the logical conclusion is that they are neither safe nor effective.

Illustration 3.6
Supplements that may be used to treat a variety of health problems.
The safety and effectiveness of many of these products are yet to be proved.

Professionals with Embedded Beliefs]
Professionals who work in health care and research are not immune to the pull of deeply rooted convictions about diet and health relationships. Although they are scientists, they sometimes remain wedded to theories about diet and health relationships even after they have been disproved.

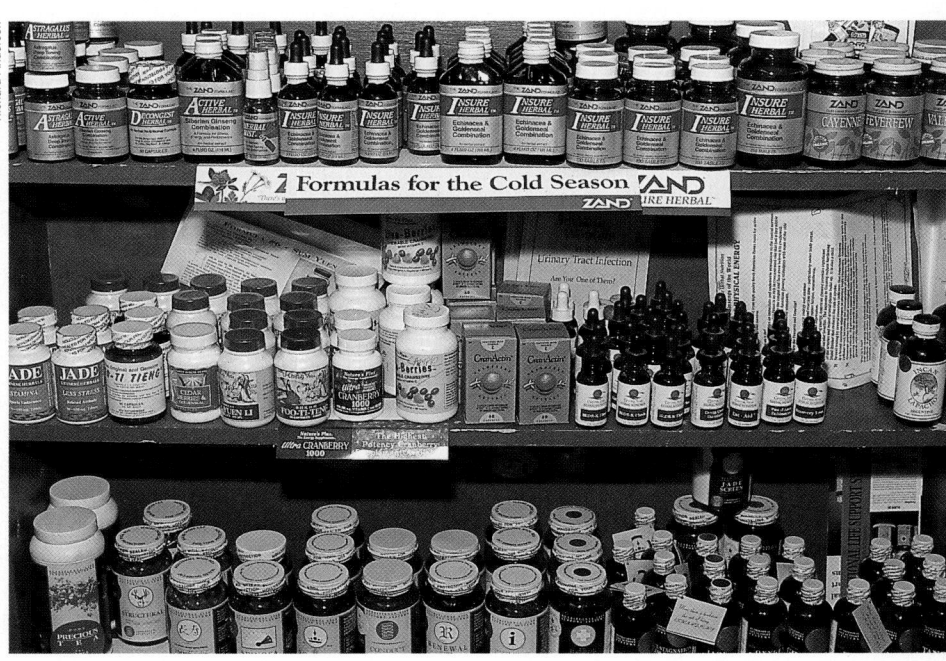

Richard Anderson

Illustration 3.7
Source: © Sidney Harris

The nurse or doctor who holds onto the belief that salt intake should be restricted in pregnancy and the university professor who is convinced that pesticides on foods pose no risk to health (Illustration 3.7) are two examples of professional sources of misinformation. The failure of such professionals to give up erroneous convictions about diet and health adds to the flow of nutrition misinformation as well as to consumer confusion and increased health risk.

How to Identify Nutrition Truths

The true method of knowledge is experiment.
—William Blake, 1788

Where does nutrition information you can count on come from? There is only one way to identify sound nutrition information—you have to put it to the test and see if it survives the dispassionate, systematic examination dictated by science.

Science is a unique and powerful explanatory system. It produces information based on facts and evidence. Our understanding of nutrition is based on scientifically determined facts and evidence obtained from laboratory, animal, and human studies. These types of studies provide information that qualifies for use when developing public policies about nutrition and health (Illustration 3.8). Science delivers information eligible for inclusion in textbooks about nutrition. (Many faculty members review the information in nutrition textbooks for scientific accuracy.)

Sources of Reliable Nutrition Information

Reliable sources of information about nutrition use the standards of proof required by science. They report decisions about nutrition and health relationships that are based on multiple studies and arrived at by "scientific consensus." These decisions represent the majority opinion of scientists knowledgeable about a particular nutrition topic.

Nutrition recommendations made to the public, such as the Dietary Guidelines for Americans and the Dietary Reference Intakes, are based on the consensus of scientific opinion. The information is made available to the public not to sell a product or to pass on an ideology, but to inform consumers honestly about nutrition and to help people use the information to maintain or improve their health. Organizations and individuals offering reliable nutrition information on the Internet and in publications are listed in Table 3.1 (p. 3–10).

Who Are Qualified Nutrition Professionals?

Many people refer to themselves as "nutritionists," but only some of them are qualified based on education and experience. These individuals are registered, licensed, or certified dietitians or nutritionists who meet qualifications established by national and state regulations. They have demonstrated a mastery of knowledge about the science of nutrition and appropriate clinical practices. Table 3.2 (p. 3–11) describes the qualifications of those who legitimately use the title dietitian or nutritionist. The specific titles vary somewhat depending on state regulations and laws governing the practice of nutrition and dietetics.

It is difficult for anyone to have all the information needed to make sound decisions about every nutrition question or issue that arises. When trying to make sense out of the information you read or hear, don't hesitate to get help. Visit the Web sites at the end of this unit, check out the index of this book for the relevant topic, or call

REALITY CHECK

Microwaves and Plastic

You've got mail: DO NOT microwave foods in plastic! Hot plastic releases highly toxic dioxins that can cause cancer . . . No source for the facts stated is given in the e-mail.

Cyndi's and Scott's responses:

Who gets the "thumbs up"

?

Cyndi:
I'll check it out at the FDA Web site.

Scott:
I was afraid of that. Now I know.

Photo Disc

Answers on next page

The *Warning Signs* of poor nutritional health are often overlooked. Use this checklist to find out if you or someone you know is at nutritional risk.

Read the statements below. Circle the number in the yes column for those that apply to you or someone you know. For each yes answer, score the number in the box. Total your nutritional score.

DETERMINE YOUR NUTRITIONAL HEALTH

	YES
I have an illness or condition that made me change the kind and/or amount of food I eat.	2
I eat fewer than two meals per day.	3
I eat few fruits or vegetables, or milk products.	2
I have three or more drinks of beer, liquor, or wine almost every day.	2
I have tooth or mouth problems that make it hard for me to eat.	2
I don't always have enough money to buy the food I need.	4
I eat alone most of the time.	1
I take three or more different prescribed or over-the-counter drugs a day.	1
Without wanting to, I have lost or gained 10 pounds in the last 6 months.	2
I am not always physically able to shop, cook and/or feed myself.	2
TOTAL	

Total Your Nutritional Score. If it's . . .

0–2   **Good!** Recheck your nutritional score in 6 months.

3–5   **You are at moderate nutritional risk.** See what can be done to improve your eating habits and lifestyle. Your office on aging, senior nutrition program, senior citizens center, or health department can help. Recheck your nutritional score in 3 months.

6 or more   **You are at high nutritional risk.** Bring this checklist the next time you see your doctor, dietitian, or other qualified health or social service professional. Talk with them about any problems you may have. Ask for help to improve your nutritional health.

These materials developed and distributed by the Nutrition Screening Initiative, a project of:

AMERICAN ACADEMY OF FAMILY PHYSICIANS   THE AMERICAN DIETETIC ASSOCIATION   NATIONAL COUNCIL OF THE AGING

Remember that warning signs suggest risk, but do not represent diagnosis of any condition.

Illustration 3.8
This evidence-based tool for assessing poor nutritional health was developed as part of the Nutrition Screening Initiative by the American Academy of Family Physicians, the American Dietetics Association, and the National Council on the Aging.

Cyndi

ANSWERS TO **REALITY CHECK**

Microwaves and Plastic

Good thinking, Cyndi. Health scares in your inbox happen. The FDA Web site at www.cfsan.fda.gov reports that plastic does leach into food cooked in microwave ovens, but not enough of it enters food to cause harm. It also notes that plastic wraps and containers do not contain dioxins.

Scott

your local health department or area Food and Drug Administration office. The more solid your information, the better your decisions will be.

The Methods of Science

Does the term *"scientific method"* conjure up an image of beady-eyed, white-coated scientists working diligently in windowless basement laboratories? Although we tend to assume that scientists are busy advancing our knowledge of nutrition and many other fields, for most people the methods of science are a mystery. They should not be (Illustration 3.9). Consumers are big users of nutrition information.

Illustration 3.9
The scientific method is not mysterious at all, but a carefully planned process for answering a specific question.

TABLE 3.1

RELIABLE SOURCES OF NUTRITION INFORMATION*

SOURCE OF NUTRITIONAL INFORMATION	EXAMPLES
Voluntary health organizations	American Heart Association American Cancer Society National Kidney Foundation American Red Cross
Scientific organizations	National Academy of Sciences American Society for Clinical Nutrition American Society of Nutrition Scientists
Government publications: nutrition, diet, and health reports	National Institutes of Health Surgeon General Food and Drug Administration Centers for Disease Control U.S. Department of Agriculture
Registered dietitians	Hospitals Public health departments Extension service Universities
Nutrition textbooks	College and university nutrition courses Nutrition faculty of accredited universities

*See Appendix B for more details about reliable sources of nutrition information.

TABLE 3.2

WHO'S WHO IN NUTRITION AND DIETETICS: LAWS AND REGULATIONS GOVERNING PRACTICE.[5]

TITLE	QUALIFICATIONS
Registered dietitian	Individual who has acquired knowledge and skills necessary to pass a national registration examination and participates in continuing professional education. Qualification is conferred by the Commission of Dietetic Registration.
Licensed dietitian/nutritionist	Individual (usually a registered dietitian) who is qualified to practice nutrition counseling based on education and experience. Nonlicensed individuals who practice nutrition counseling can be prosecuted for practicing without a license. Licenses are conferred by state law.
Certified dietitian/nutritionist	Individual who meets certain educational and experience qualifications. Unqualified individuals may practice nutrition counseling but have to call themselves something else (e.g., "nutritional counselor"). Certification is established by state regulation.
Dietitian/nutritionist	Individuals with any type of education or experience may register with the state of California as a dietitian or nutritionist.

To make sound judgments about nutrition and health, people need to know how to distinguish results produced by scientific studies from those generated by personal opinion, product promotions, and bogus studies.

There are many established methods of science. The specific methods employed vary from study to study depending on the type of research conducted. Nevertheless, all types of scientific studies have one feature in common: they are painstakingly planned. Planning is the *most* important, and often the most time-consuming, part of the entire research process.

Developing the Plan

The first part of the planning process entails clearly stating the question to be addressed—and, one hopes, answered—by the research. For the purposes of illustration, let's say our research will address the effects of vitamin X supplements on hair loss (Table 3.3 and Illustration 3.10). The idea for the research came from a study that hinted that users of vitamin X supplements had increased hair loss. In that study, adults were given 25,000 international units (IU) of vitamin X for three months to test its safety. Although the study found that vitamin X had no adverse effects on health and offered some benefits, several subjects who had received the supplements complained of hair loss. Accordingly, our study poses the question, "Do vitamin X supplements increase hair loss?"

The Hypothesis: Making the Question Testable

The question is then transformed into an explicit hypothesis that can be proved or disproved by the research. (*Hypothesis* and other research terms used in this unit are

Illustration 3.10
Do high doses of vitamin X increase hair loss?

TABLE 3.3

OVERVIEW OF A HUMAN NUTRITION RESEARCH STUDY: VITAMIN X SUPPLEMENTS AND HAIR LOSS.*

A. **Pose a clear question:** "Do vitamin X supplements increase hair loss?"

B. **State the hypothesis to be tested:** "Vitamin X supplements of 25,000 IU per day taken for three months increase hair loss in healthy adults."

C. **Design the research:**

1. *What type of research design should be used?* "In this study, a clinical trial will be used. The supplement and placebo will be allocated by a double-blind procedure."

2. *Who should the research subjects be?* "The study will exclude subjects who may be losing or gaining hair due to balding, hair treatments, medications, or the current use of vitamin X supplements."

3. *How many subjects are needed in the study?* "The required sample size is calculated to be 20 experimental and 20 control subjects."

4. *What information needs to be collected?* "Information on hair loss, conditions occurring that might affect hair loss, and the use of supplements and placebos will be collected."

5. *What are accurate ways to collect the needed information?* "The study will use the 'measure the hairs in a square inch of scalp' technique."

6. *What statistical tests should be used to analyze the results?* "Appropriate tests identified."

D. **Obtain approval for the study from the committee on the use of humans in research:** "Approval obtained."

E. **Implement the study design:** "Implemented."

F. **Evaluate the findings:** "Subjects receiving the 25,000 IU of vitamin X for three months lost significantly more hair than subjects receiving the placebo. Hypothesized relationship found to be true."

G. **Submit paper on the research for publication in a scientific journal or other document.**

*This is a fictitious study used for illustrative purposes only.

---

defined in Table 3.4.) The results of the research must provide a true or false response to the hypothesis, and not an explanatory sentence or paragraph. So, in this example, "vitamin X increases hair loss" wouldn't do as a hypothesis because it leaves too many questions about the relationship unanswered. The effect of vitamin X on hair loss may depend on the amount of vitamin X given; how long it is taken; or on the research subjects' health, age, and other characteristics. The hypothesis "vitamin X supplements of 25,000 IU per day taken for 3 months by healthy adults increase hair loss" is concrete enough to be addressed by research.

The Research Design: Gathering the Right Information

Poor research design, or the lack of a solid plan on how the research will be conducted, is often a weak link that renders studies useless (Illustration 3.11). It's the "oops, we forgot to get this critical piece of information!" or the "how did all the measurements come out wrong?" at the end of a study that can ruin months or even

TABLE 3.4

A SHORT GLOSSARY OF RESEARCH TERMS.

Association	The finding that one condition is correlated with, or related to another condition, such as a disease or disorder. For example, diets low in vegetables are associated with breast cancer. Associations do *not* prove that one condition (such as a diet low in vegetables) *causes* an event (such as breast cancer). They indicate that a statistically significant relationship between a condition and an event exists.
Cause and effect	A finding that demonstrates that a condition causes a particular event. For example, vitamin C deficiency causes the deficiency disease scurvy.
Clinical trial	A study design in which one group of randomly assigned subjects (or subjects selected by the "luck of the draw") receives an active treatment and another group receives an inactive treatment, or "sugar pill," called the placebo.
Control group	Subjects in a study who do not receive the active treatment or who do *not* have the condition under investigation. Control periods, or times when subjects are not receiving the treatment, are sometimes used instead of a control group.
Double blind	A study in which neither the subjects participating in the research nor the scientists performing the research know which subjects are receiving the treatment and which are getting the placebo. Both subjects and investigators are "blind" to the treatment administered.
Epidemiological studies	Research that seeks to identify conditions related to particular events within a population. This type of research does *not* identify cause-and-effect relationships. For example, much of the information known about diet and cancer is based on epidemiological studies that have found that diets low in vegetables and fruits are associated with the development of heart disease.
Experimental group	Subjects in a study who receive the treatment being tested or have the condition that is being investigated.
Hypothesis	A statement made prior to initiating a study of the relationship sought to be proved (found to be true) by the research.
Meta-analysis	An analysis of data from multiple studies. Results are based on larger samples than the individual studies and are therefore more reliable. Differences in methods and subjects among the studies may bias the results of meta-analyses.
Peer review	Evaluation of the scientific merit of research or scientific reports by experts in the area under review. Studies published in scientific journals have gone through peer review prior to being accepted for publication.
Placebo	A "sugar pill," an imitation treatment given to subjects in research.
Placebo effect	Changes in health or perceived health that result from expectations that a "treatment" will produce an effect on health.
Statistically significant	Research findings that likely represent a true or actual result and not one due to chance.

"We don't get involved with things like double-blind tests and peer review. We're just a little Mom-and-Pop laboratory."

Illustration 3.11
Source: © *Sidney Harris.*

"I think you should be more explicit here in step two."

Illustration 3.12
Source: © Sidney Harris.

years of work. Each step in the research process must be thoroughly planned. In research, there are no miracles (Illustration 3.12). If something can go wrong because of incomplete planning, it probably will.

Research designs are often based on the answers to the following questions:

- What type of research design should be used?
- Who should the research subjects be?
- How many subjects are needed in the study?
- What information needs to be collected?
- What are accurate ways to collect the needed information?
- What statistical tests should be used to analyze the findings?

What Type of Research Design Should Be Used?] Several different research designs can be used to test hypotheses. We could use an *epidemiological study* design to determine if hair loss is more common among people who take vitamin X supplements than among people who do not. Researchers commonly use this type of study to identify conditions that are related to specific health events in humans. To provide preliminary evidence, we could use animal studies that determine hair loss in supplemented and unsupplemented animals. Or we could use another design, such as a *clinical trial.* Since this design would work well for the proposed hypothesis about vitamin X supplements, we'll follow the rules for conducting a clinical trial in this example. (You can see a map of the research design used for this hypothetical study in Illustration 3.13.)

The purpose of clinical trials is to test the effects of a treatment or intervention on a specific biological event. In our example, we will test the effect of vitamin X supplements on hair loss.

The Experimental and Control Groups} Clinical trials, as well as other research studies that address questions about nutrition, require an *experimental group* (the group of subjects who receive vitamin X in this example) and a *control group* (the comparison group that receives a *placebo* and not vitamin X). Never trust the results of a study that didn't employ both and here's why. How do we know the effect of a certain treatment isn't due to something *other* than the treatment? What if we gave a group of adults vitamin X supplements and they lost hair? Does this mean the vitamin X caused the hair loss? Could it be that the subjects lost no more hair during the treatment period than they would have lost without the vitamin X? We can't know whether vitamin X supplements produce hair loss if we don't know how much hair is lost without vitamin X.

After measuring their usual hair loss, we will randomly assign (by the "luck of the draw") people in the study to serve in either the experimental or the control group. Individuals assigned to the control group will be given pills that look, taste, and feel like the vitamin X supplements, but have no effect on hair loss or gain. Neither the research staff who will be assessing the hair loss nor the people in the study will know who is getting vitamin X and who is getting the placebo (Illustration 3.14). Scientists use this *double-blind procedure* because knowing which group is which may affect people's expectations and change the results. This important topic deserves additional attention.

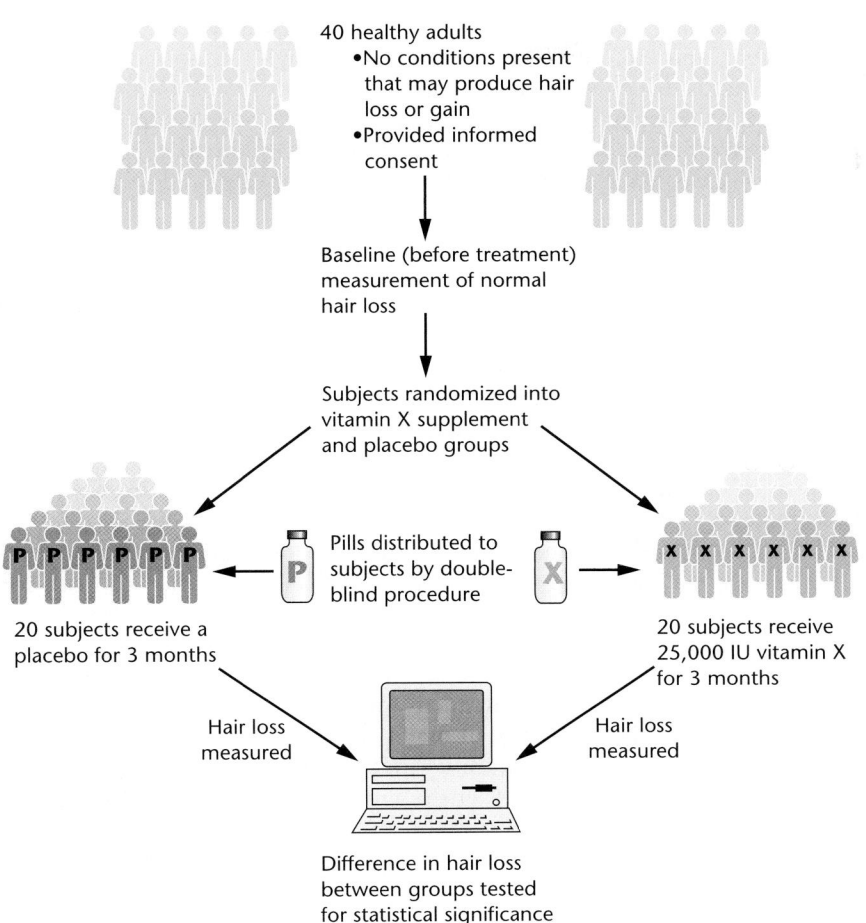

40 healthy adults
- No conditions present that may produce hair loss or gain
- Provided informed consent

Baseline (before treatment) measurement of normal hair loss

Subjects randomized into vitamin X supplement and placebo groups

P P P P P P

Pills distributed to subjects by double-blind procedure

X X X X X X

20 subjects receive a placebo for 3 months

20 subjects receive 25,000 IU vitamin X for 3 months

Hair loss measured

Hair loss measured

Difference in hair loss between groups tested for statistical significance

Illustration 3.13
Map of the design of the vitamin X supplement study.

The "Placebo Effect"} The *placebo effect* can cause a good deal of confusion in research. That's because people tend to have expectations about what a treatment will do, and those expectations can influence what happens.

A good example of the placebo effect occurred in a study that tested the effectiveness of a medication in reducing binge eating among people with bulimia.[6] After the usual number of binge-eating episodes was determined, 22 women with bulimia were given either the medication or a placebo in a double-blind fashion. The number of binge-eating episodes was then reassessed. At the end of the study period, a whopping 78% reduction in binge-eating episodes was found among women taking the medication. But binge-eating episodes dropped by 70% in the control group. Was the medication effective? No. Was the expectation that a medication would help effective? Yes, at least for the time covered by the study. Due to the placebo effect, a reduction of greater than 90% in episodes of binge eating would have been needed to conclude that the medication had a real effect on binge eating.

Who Should the Research Subjects Be?] Although expedient, it is not enough to say, "Well, let's use 10 faculty members in the study." The type and number of subjects employed by the research are important considerations.

The type of subjects involved in research is important, because we need to exclude people who have conditions that might produce the problem the research is examining. For example, hair loss may result from periodic balding, a bad perm, the

Illustration 3.14
Which is the vitamin X supplement? Neither the subject nor the investigator should know which is vitamin X and which is the placebo.

Stan Maddock

use of mousse or hair spray, or illnesses or medications. If we include people with these conditions and circumstances in the study, it will be difficult to determine whether hair loss is due to vitamin X or something else. In addition, it is important to exclude from the study people who already take vitamin X supplements or are bald. An inappropriate or biased selection of subjects could make the results useless or, in the worst case, make the study's results come out in a predetermined way.

How Many Subjects Are Needed in the Study?]   The number of subjects needed for a study is a mathematical question, and we won't go into the formulas here. The number of subjects is based upon the number needed to separate true differences between the experimental and control groups from differences that are due to chance (Illustration 3.15). Suppose that hair loss normally varies from 40 to 80 strands per week. We'll need to include enough subjects in the study to make sure differences in hair loss between the groups are not due to normal fluctuations in hair loss. This point is important because studies employing too few subjects lead to inconclusive results and a great deal of controversy about nutrition. One or a few subjects are never enough to prove a point about nutrition.

Let's assume that the mathematical formulas applied to the vitamin X study show that 20 people are needed in both the experimental group and the control group.

What Information Needs to Be Collected?]   Information recorded by researchers must represent an evenhanded approach to discovering the facts. We must identify not only the findings that may support the hypothesis, but *those that may refute it*. In order to know if vitamin X supplements increase hair loss, we need to know how much hair is lost, whether the subjects faithfully took the supplement, and whether something came up during the study (such as a bad perm) that might alter hair loss. The presence of such conditions should be determined by methods known to provide accurate results.

Illustration 3.15
How many subjects are needed in the study? It's important to get the number right.

Photo Disc

Nutrition UP CLOSE

Checking Out a Fat-Loss Product

FOCAL POINT: How to make an informed decision about the truthfulness of an advertisement for a fat-loss product.

Obesity Research Journal
P.O. Box 281752,
Miami, FL

**Lose Over 2 Inches
In 3 Weeks**

**A Scientifically Proven Fat
Reduction Cream That
Actually Works!**

A scientific, double-blind, placebo-controlled research study of both sexes demonstrated that LIPID MELT fat reduction cream reduces inches from the thigh area. Subjects lost an average of 2 inches from each thigh in only 3 weeks!

Subjects from a preliminary pilot study reported that LIPID MELT cream reduced inches from their abdomen. Both studies reported no skin rashes, discomfort or sensitivity. Formulated with patented liposome technology, LIPID MELT contains no drugs, only natural active ingredients. Initially sold only in salons and spas.

Six-week supply: $25 + $4.95 s/h. We guarantee that you will lose inches or your money will be refunded.

Obesity Research

Read the following advertisement for a fat-loss product. Then, check out the information using the checklist for identifying nutrition misinformation.

1. Is something being sold?

   Yes   No

2. Is a new remedy for problems that are not easily or simply solved being offered?

   Yes   No

3. Are terms such as *miraculous, magical, secret, detoxify, energy restoring, suppressed by organized medicine, immune boosting,* or *studies prove* used?

   Yes   No

4. Are testimonials, before-and-after photos, or expert endorsements used?

   Yes   No

5. Does the information sound too good to be true?

   Yes   No

6. Is a money-back guarantee offered?

   Yes   No

Optional: Repeat the activity using a nutrition-related advertisement from the Internet.

FEEDBACK (answers to these questions) can be found at the end of Unit 3.

What Are Accurate Ways to Collect the Needed Information?] "Garbage in—garbage out." If the information obtained on and from subjects is inaccurate, then the study is worthless. To avoid this problem, good research employs methods of collecting information known to produce accurate results.

Let's consider the problem of measuring hair loss. How can we do that accurately? First, we look for a method that has already been demonstrated by research to be accurate. Otherwise, we'll first have to prove that our method for measuring hair loss is accurate. For this example, assume an accurate method is used.

What Statistical Tests Should Be Used to Analyze the Findings?] Before collecting the first piece of information, we'll select appropriate tests for identifying ***statistically significant results*** of the research. Many different statistical tests can be used to identify significant differences between the findings from the experimental and control groups. The trick here is to make sure to select the appropriate tests for proving or disproving the hypothesis. Without such tests, it is often very hard to decide what the findings mean. What if the study shows the group that took vitamin X lost an average of 5% more hair than the control group? Is that 5% difference due to the vitamin X or to something else, such as chance fluctuations in normal hair loss? Well-chosen statistical tests will tell us whether the differences between groups are in all probability real or due to chance occurrences or coincidence.

Obtaining Approval to Study Human Subjects

An important step in the planning process is applying for approval to conduct the proposed research on human subjects. Universities and other institutions that conduct research have formal committees, called institutional review boards, that scrutinize plans to make sure proposed studies follow the rules governing research on human subjects. For this study, we will have to show to the committee's satisfaction that the level of vitamin X employed is safe. As a part of this process, the committee usually requires that human subjects consent, in writing, to participate in the study.

Implementing the Study

With the design in place and the appropriate approvals obtained, it is time to implement the study. Assume that, due to the study's solid design and importance, we have been awarded a grant to fund the research. We can now recruit subjects for the study and enroll them if they are eligible and consent to participate. Measurements of hair loss are performed according to schedule, use of the pills by subjects is monitored, results are checked for errors and entered into a computerized data file, and statistical tests are applied. This process usually ends with computer printouts that exhibit the findings.

Making Sense of the Results

Let's say subjects who received the supplement lost 30% more hair than the control subjects did. Having applied the appropriate statistical test, we find that 30% is a highly significant difference. Does that mean that vitamin X supplements *cause* hair loss? *Cause* is a strong word in research terms. It implies that a **cause-and-effect relationship** (that the vitamin X supplements caused the hair loss) exists. But many factors can contribute to hair loss, and since many causes may be unknown or not measured by the study, it is difficult to conclude with absolute certainty that vitamin X by itself caused the hair loss. Assume 4 of the 20 subjects in the experimental group who were known to have taken the vitamin X supplement as indicated lost less hair than usual. The vitamin X supplements *didn't* cause them to lose hair.

We could conclude from this research (if, as a reminder, there was such a thing as vitamin X) that vitamin X supplements, given to healthy adults at a dose of 25,000 IU per day for three months, are strongly *associated* with hair loss. The term *associated* as used here means that the vitamin X supplements were strongly *related* to hair loss but may not have *caused* the hair loss. The hypothesized relationship between supplemental vitamin X and hair loss would be found to be true. Although the research strongly indicates a cause-and-effect relationship, additional studies would be needed to prove the relationship exists in other groups of people and at different doses of vitamin X.

After determining the results, we can write a paper describing the research and its conclusions and submit it for publication in a scientific journal. If judged to be acceptable after review by other scientists, the paper is published.

Complete coverage of how nutrition questions are addressed through research would take a three-course sequence. Nevertheless, it is hoped that the information presented here makes it easier to judge the likely accuracy of reports about nutrition studies in the popular media. There is a lot more to a "Vitamin X Supplements Linked to Hair Loss" story than meets the eye.

Science and Personal Decisions about Nutrition] Science is based on facts and evidence. The grounding ethic of scientists is that facts and evidence are more sacred than any other consideration. These characteristics of science and scientists are strong assets for the job of identifying truths.

Although imperfect because they are undertaken by humans in an environment of multiple constraints, only scientific studies produce information about nutrition you can count on. Evidence is the single best ingredient for decision making about nutrition and your health.

Key Terms

association, page 3–13

cause-and-effect relationship,
    page 3–18

clinical trial, page 3–14

control group, page 3–14

double-blind procedure, page 3–14

epidemiological study, page 3–14

experimental group, page 3–14

hypothesis, page 3–11

meta-analysis, page 3–13

peer review, page 3–13

placebo, page 3–14

placebo effect, page 3–15

scientific method, page 3–10

statistically significant results,
    page 3–17

www links

www.altmedicine.com
A frequently updated site on alternative health, diet and health, and other nutrition topics based on recent scientific reports.

www.eatright.org
The American Dietetic Association's Web site is an excellent source of information about diet, health, and qualified dietetics personnel in your area.

www.nlm.nih.gov/medlineplus
Maintained by the National Library of Medicine, this site is a sure bet for authoritative information on food and nutrition topics, ongoing clinical trials, and other information. Also offers a medical dictionary and glossary.

www.navigator.tufts.edu
Summarizes and rates the scientific quality of nutrition information on the Web.

www.cfsan.fda.gov
A huge site from FDA's Food Safety and Applied Nutrition program. Contains search feature, up-to-date discussions of controversial and consumer nutrition topics, and instructions for reporting fraudulent nutrition products and labels.

www.quackwatch.com
Uncovers health frauds, including Nutrition scams, on "NutriWatch."

Notes

1. Who's minding the lab? Tufts University Health and Nutrition Letter 1997, May:4.

2. Who's minding the lab?

3. Levine J, Gussow JD et al. Authors' financial relationships with the food and beverage industry and their published positions on the fat substitute Olestra. Am J Pub Health 2003;93:664–9.

4. Schneeman BO. Building scientific consensus: the importance of dietary fiber. Am J Clin Nutr 1999;69:1.

5. Pathways to excellence. Chicago: The American Dietetic Association; 1997.

6. Alger SA, Schwalberg MD, Bigaouette JM, Michalek AV, Howard LJ. Effect of a tricyclic antidepressant and opiate antagonist on binge-eating behavior in normal weight, bulimic, and obese, binge-eating subjects. Am J Clin Nutr 1991;53: 865–71.

Nutrition UP CLOSE

Checking Out a Fat-Loss Product

Feedback for Unit 3

The product advertised earns five "yes" responses, making it highly unlikely that the product works. One or two "yes" responses provide a strong clue that advertised products may not work.

Understanding Food and Nutrition Labels

Nutrition Scoreboard

	TRUE	FALSE
1 Nutrition labels are required on all foods and dietary supplements sold in the United States.		
2 Nutrition labeling rules allow health claims to be made on the packages of certain food products.		
3 Nutrition labels contain all of the information people need to make healthy decisions about what to eat.		

Answers on next page

[KEY CONCEPTS AND FACTS]

- People have a right to know what is in the food they buy.

- The purpose of nutrition labeling is to give people information about the composition of food products so they can make informed food-purchasing decisions.

- Nutrition labeling regulations cover the type of foods that must be labeled and set the standards for the content and format of labels.

Answers to *Nutrition Scoreboard*

		TRUE	FALSE
1	Labeling is now required for almost all processed foods and for dietary supplements, but remains largely voluntary for fresh fruits and vegetables and raw meats.[1]		✔
2	Health claims for food products are allowed on many food packages. The claims must be truthful and adhere to FDA standards.	✔	
3	Nutrition labels are necessarily short and can't tell the whole story about food and health. They help people make several key decisions about a food's composition.		✔

Nutrition Labeling

Misleading messages, hazy health claims, and the slippery serving sizes that characterized food labels of the past have led to a revolution in nutrition labeling. Consumers—especially those responsible for buying food for the family; people with weight concerns, food allergies, or diabetes; and health-conscious consumers—made it clear they wanted to end the mystery about what's in many foods. Passage of the 1990 Nutrition Labeling and Education Act by Congress indicated that their concerns had been heard. In 1993 the Food and Drug Administration published rules for nutrition labeling, and implementation and revisions of the new standards have been ongoing since then.

Illustration 4.1

Many supermarkets make nutrition information available in posters, cards, or pamphlets in areas where produce and meats are sold.

What Foods Must Be Labeled?

According to nutrition labeling regulations, almost all multiple-ingredient foods must be labeled with nutrition information.[3] Labeling of fresh vegetables and fruits and raw meats is voluntary because these foods are often sold individually and without packaging. It is recommended, however, that supermarkets make nutrition labeling information available to consumers through appropriately placed charts or pamphlets (Illustration 4.1).[4] Nutrition labeling is not required on foods sold by local bakeries and foods with packaging that is too small to fit a label (bite-sized candy bars, for example).

Labeling is not required on restaurant menus unless a nutrition claim is made for an item. When an item is labeled, for example "low fat" or "low calorie," then restaurants are required to display nutrition information for the nutrient for which the claim is made.[5]

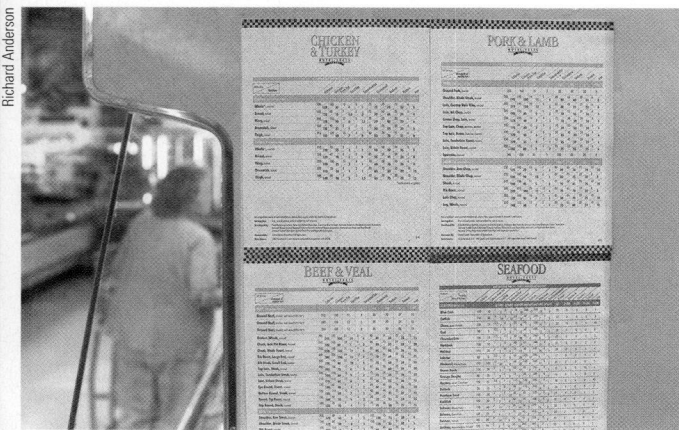

Richard Anderson

Jared:
They each have about
400 calories.

REALITY CHECK

Fast food nutrition labeling

Jared and Ronald think they have a pretty good idea of how many calories are in their favorite fast foods: Mandarin Chicken™ Salad with dressing, Crispy Chicken California Cobb Salad with dressing, and the 6" Italian BMT sub.

Who gets thumbs up

?

Photo Disc

Ronald:
I'm thinking a
Mandarin Chicken Salad
has about 300 calories, the
Crispy Chicken Cobb Salad
over 500, and the Italian
BMT sub around
500 calories.

Answers on next page

Due to the obesity epidemic, and because people in the United States spend on average half of their food dollars in restaurants, the pressure on fast food and other restaurants to include nutrition information on menus is increasing.[6] Would you find such labeling useful? The results of the "Reality Check" may influence your opinion.

What's on the Nutrition Label?

The nutrition information label is known as the "Nutrition Facts" panel. The panel, shown in Illustration 4.2, highlights a product's content of fat, saturated fat, cholesterol, sodium, dietary fiber, two vitamins (A and C), and two minerals (calcium and iron). Trans fat became a required component of Nutrition Facts panels in 2003. Food companies have until January 1, 2006, to implement this requirement. Trans fats are found primarily in shortenings, margarines, frying oils used in fast-food restaurants, and bakery goods. High intake of trans fat is related to the development of heart disease, and consumers are being urged to consume as little of it as possible.[7] The food's content of these nutrients must be based on a standard serving size as defined by the Food and Drug Administration (FDA). Standard serving sizes, which represent portion sizes generally consumed, are established for 131 types of food. For example, a serving of fruit juice is 6 ounces (¾ cup), and a serving of cooked vegetables is ½ cup

All labeled foods must provide the information shown on the Nutrition Facts panel in Illustration 4.2. Additional information on specific nutrients can be added to the panel on a voluntary basis (Table 4.1). However, if the package makes a claim about the food's content of a particular nutrient not on the "mandatory" list, then information about that nutrient *must* be added to the Nutrition Facts panel. Nutrition labels contain a column headed *% Daily Value (%DV)*. Figures given in this column are intended to help consumers answer such questions as "Does a serving of this macaroni and cheese contain more fat than the other brand?" and "How much fiber is in this cereal?"

Daily Values (DVs) are standard levels of dietary intake of nutrients developed specifically for use on nutrition labels. They are based on earlier editions of the Recommended Dietary Allowances and scientific consensus recommendations.[8] The %DV figures listed on labels represent the percentages of the standard nutrient

% Daily Value (%DV)
Scientifically agreed-upon standards of daily intake of nutrients from the diet developed for use on nutrition labels. The "% Daily Values" listed in nutrition labels represent the percentages of the standards obtained from one serving of the food product.

Daily Values (DVs)
Scientifically agreed-upon daily dietary intake standards for fat, saturated fat, cholesterol, carbohydrate, dietary fiber, and protein intake compatible with health. DVs are intended for use on nutrition labels only and are listed in the Nutrition Facts panel. "% Daily Value" on Nutrition Facts panels is calculated as the percentage of each DV supplied by a serving of the labeled food.

Jared

Did you know that a Mandarin Chicken™ Salad (150 calories) with one ounce of salad dressing (140 calories) has 290 calories, a Crispy Chicken California Cobb Salad (380 calories) with one ounce of salad dressing (140 calories) has 520, and a 6" BMT Italian sub 480 calories? They do.

Ronald

Illustration 4.2
What are you having for dinner? The Nutrition Facts panel lets you know.

Nutrition Facts

Serving Size 1 cup (253g)
Serving Per Container 4

Lists a standardized, reasonable portion size.

Amount Per Serving

Calories 260   Calories from Fat 72

Up-front listing of total calories and calories from fat.

	% Daily Value*
Total Fat 8 g	12%
Saturated Fat 3 g	15%
Trans Fat 0g	
Cholesterol 130 g	43%
Sodium 1010 mg	42%
Total Carbohydrate 22 g	7%
Dietary Fiber 9 g	36%
Sugars 4 g	
Protein 25 g	

Grams (g) are counted in "Total Fat."

Grams (g) are counted in "Total Carbohydrates."

The % Daily Value column shows how a food fits into the overall diet. It indicates the percentage of the recommended daily amounts contributed by a serving of the food.

Vitamin A 35%   •   Vitamin C 2%

Calcium 6%   •   Iron 30%

Lists % Daily Value for 2 vitamins and 2 minerals most likely to be lacking in the diet of today's consumers.

*Percent Daily Values are based on a 2000 calorie diet. Your daily values may be higher or lower depending on your calorie needs:
**Consume as little as possible

Important to note if you don't consume 2000 calories per day.

		Calories:	2,000	2,500
Total Fat	Less than		65 g	80 g
Sat Fat	Less than		20 g	25 g
Cholesterol	Less than		300 mg	300 mg
Sodium	Less than		2400 mg	2400 mg
Total Carbohydrate			300 g	375 g
Dietary Fiber			25 g	30 g

Calories per gram:
Fat 9 • Carbohydrate 4 • Protein 4

Translates the % Daily Values into amounts of key nutrients recommended daily. Daily Values are based on 30% of total calories from fat, 60% from carbohydrate, and 10% from protein.

Reference material. Useful for calculating percentage of total calories from fat, carbohydrate, and protein.

TABLE 4.1

MANDATORY AND VOLUNTARY COMPONENTS OF THE NUTRITION FACTS PANEL
AND ASSIGNED DAILY VALUES (DVS).

Components are listed in the order in which they must appear on the nutrition panel;
unapproved components may not be listed.*

MANDATORY	DV	VOLUNTARY	DV
Total calories	—	Calories from saturated fat	
Calories from fat	30%	Polyunsaturated fat	
Total fat	65 g	Monounsaturated fat	
Saturated fat	20 g	Stearic acid	
Trans fat	none		
Cholesterol	300 mg	Insoluble fiber	
Sodium	2400 mg		
Total carbohydrate	300 g	Other carbohydrates	
Dietary fiber	25 g	Soluble, Insoluble fiber	
Sugars	—	Sugar alcohols (xylitol,	
Protein	—	mannitol, sorbitol)	
Vitamin A	5000 IU	Vitamin D	400 IU
Vitamin C	60 mg	Vitamin E	30 IU
Calcium	1000 mg	Vitamin K	80 mcg
Iron	18 mg	Thiamin	1.5 mg
		Riboflavin	1.7 mg
		Niacin	20 mg
		Vitamin B6	2.0 mg
		Folate	400 mcg
		Vitamin B12	6 mcg
		Biotin	300 mcg
		Pantothenic acid	10 mg
		Phosphorus	1000 mg
		Iodine	150 mcg
		Magnesium	400 mg
		Zinc	15 mg
		Selenium	70 mcg
		Copper	2 mg
		Manganese	2 mg
		Chromium	120 mcg
		Molybdenum	75 mcg
		Chloride	3400 mg
		Potassium	3500 mg

*If the food package makes a claim about any of the voluntary components, or if the food is enriched or fortified with any of
them, they become a mandatory part of the nutrition panel.

amounts obtained from one serving of the food product. Standard values for total
fat, saturated fat, and carbohydrate are based on a daily intake of 2000 calories.
The %DV for total fat intake is based on 30% of total calories from fat (65 grams),
the saturated fat standard on 10% of total calories (20 grams), and the standard for
carbohydrate intake on 60% of total calories (300 grams). If, for example, a serving of a food product contained 10 grams of saturated fat, the %DV listed on the
nutrition label would be 50% because the standard for saturated fat intake is 20
grams.

Food labels that make a claim about a particular nutrient or ingredient naturally contained in the food or added to it must include that nutrient or ingredient
on the label. Other rules for including nutrition and health claims on food product
labels exist.

The "Jelly Bean Rule"

Foods high in fat, saturated fat, cholesterol, or sodium and low in certain vitamins, minerals, or fiber cannot be labeled with a nutrition claim. That means food manufacturers can't add, say, calcium to jelly beans, soft drinks, or similar foods to earn a claim that implies the product is "healthy."[11]

Must contain 3 grams or less fat per serving

Must contain fewer than 20 milligrams cholesterol per serving

Must contain fewer than 140 milligrams sodium per serving

Illustration 4.3
Eye-catching nutrition claims on food labels must conform to standard definitions.

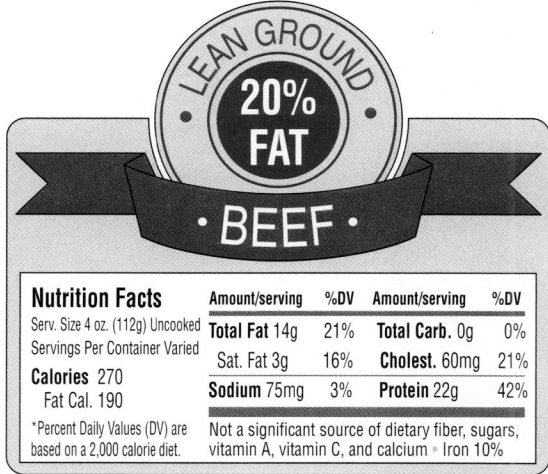

Nutrition Facts	Amount/serving	%DV	Amount/serving	%DV
Serv. Size 4 oz. (112g) Uncooked Servings Per Container Varied	**Total Fat** 14g	21%	**Total Carb.** 0g	0%
	Sat. Fat 3g	16%	**Cholest.** 60mg	21%
Calories 270 Fat Cal. 190	**Sodium** 75mg	3%	**Protein** 22g	42%
*Percent Daily Values (DV) are based on a 2,000 calorie diet.	Not a significant source of dietary fiber, sugars, vitamin A, vitamin C, and calcium • Iron 10%			

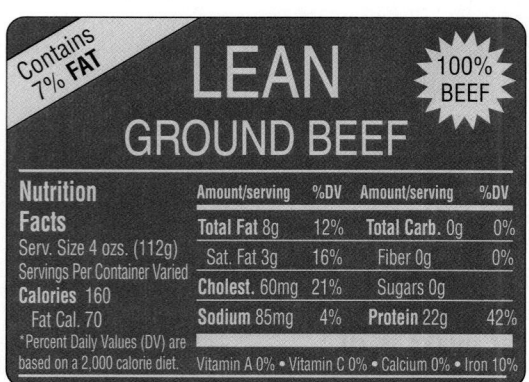

Nutrition Facts	Amount/serving	%DV	Amount/serving	%DV
Serv. Size 4 ozs. (112g) Servings Per Container Varied	**Total Fat** 8g	12%	**Total Carb.** 0g	0%
	Sat. Fat 3g	16%	Fiber 0g	0%
Calories 160 Fat Cal. 70	**Cholest.** 60mg	21%	Sugars 0g	
*Percent Daily Values (DV) are based on a 2,000 calorie diet.	**Sodium** 85mg	4%	**Protein** 22g	42%
	Vitamin A 0% • Vitamin C 0% • Calcium 0% • Iron 10%			

Illustration 4.4 *Two examples of labels claiming "lean" ground beef. Only the 7% fat beef is actually lean. "20% fat" ground beef is far from lean, actually providing 14 grams and 55% of total calories from fat per serving.*

Nutrition and Health Claims Must Be Truthful

Approximately 40% of food products sold in grocery stores make one or more nutrition claims on the packaging.[9] Can these claims be trusted? If they are among the nutrition claims approved by the FDA, the answer to this question is almost always yes.

Nutrition Claims] Statements used to label foods—such as "High fiber" and "Healthy"—must conform to standard definitions (Illustration 4.3). For example, foods labeled "low fat"—the most popular nutrition claim made on food packages[10]—must contain 3 grams of fat or less per serving. Low-fat foods can be labeled with a "percent fat free" label, such as "98% fat free" turkey. This label means that the product contains approximately 2% fat on a weight basis. Meat products labeled "lean" must contain less than 10 grams of fat, 4.5 grams of saturated fat and trans fat combined, and 95 milligrams of cholesterol per serving. Some labels promote meat as lean based on the percentage of the meat's weight that consists of fat. So, a meatloaf mix that is 16% fat on a weight basis might be labeled as "16% lean." It would not be lean according to nutrition labeling standards. Illustration 4.4 provides an example of an appropriately, and an inappropriately labeled meat to expand upon this point.

A variety of approved nutrition claims are in use, and this list of claims will grow longer. A key feature of many of the approved claims is that they cannot be used to make certain foods appear to be healthier for you than they really are. Foods often have to meet several standards for nutritional value before they qualify for a claim. You may see some food products labeled with claims like "pure," "10% fat," or "all natural" that are not listed in the "Health Action" Chart. Because these

HEALTH ACTION | *Some Examples of What "Front of the Package" Nutrition Terms Must Mean*

Term	Examples	Means That a Serving of the Product Contains:
Extra lean	Extra-lean pork, extra-lean hamburger	Fewer than 5 grams of fat, fewer than 2 grams of saturated fat and trans fat combined, *and* fewer than 95 milligrams of cholesterol (applies to meats only).
Lean	Lean beef, lean turkey	Fewer than 10 grams of fat, fewer than 4.5 grams of saturated fat and trans fat combined, *and* fewer than 95 milligrams of cholesterol (applies to meats only).
Extra, More	Bread with added fiber, fortified foods	At least 10% more of the Daily Value of a nutrient per serving than in a similar food.
Fat Free	Skim milk, no-fat salad dressing	Less than 0.5 grams of fat.
Free	Sugar-free, sodium-free	No—or negligible amounts—of sugars, sodium, or fat.
Good source	Good source of fiber, good source of calcium	From 10 to 19% of the Daily Value for a particular nutrient.
Healthy	Healthy burritos, canned vegetables	No more than 60 milligrams of cholesterol, 3 grams of fat, and 1 gram of saturated fat; and more than 10% of the Daily Value of vitamin A, vitamin C, iron, calcium, protein, or fiber. "Healthy" foods must also contain 360 milligrams or less of sodium.
High	High in iron, high in vitamin C	20% or more of the Daily Value for a particular nutrient.
Less	Less saturated fat, less cholesterol	25% less of a nutrient than a comparable food.
Light or lite	Light in sodium, lite in fat, light brown sugar, light and fluffy	33% fewer calories or half the fat as in the regular product, or 50% or less sodium than usual in a low-calorie, low-fat food. "Light" can also be used on labels to describe the texture or color of a food.
Low calorie	Low-calorie cookies, low-calorie fruit drink	40 calories or fewer.
Low cholesterol	Low-cholesterol egg product	20 milligrams or less cholesterol (applies to animal products only)
Low fat	Low-fat cheese, low-fat ice cream	3 grams or less fat.
Low saturated fat	Low-saturated-fat pancake mix, low-saturated-fat eggnog	1 gram or less saturated fat and 0.5 grams or less trans fat.
Low sodium	Low-sodium soup, low-sodium hot dogs	140 milligrams or less sodium.
Percent fat free	95% fat free, 98% fat free	The specified percentage of fat on a weight basis (only low-fat foods can use this label).
Reduced	Reduced calorie/cholesterol/ saturated fat bacon	25% less calories, cholesterol, or saturated fat and trans fat than the regular product.
Trans fat free	Trans fat–free margarine	Less than 0.5 grams trans fat and less than 0.5 grams saturated fat.

TABLE 4.2

APPROVED NUTRITION CLAIMS FOR DISEASE PREVENTION, AND CLAIMS NOT APPROVED OR PENDING.

TOPICS COVERED BY APPROVED CLAIMS

1. Calcium and osteoporosis
2. Fats and cancer
3. Saturated fat and cholesterol and heart disease
4. Fruits and vegetables and heart disease, cancer
5. Sodium and hypertension
6. Folate or folic acid and neural tube defects (newborn malformations such as spina bifida)
7. Whole grains and heart disease
8. Soluble fiber in oats and psyllium seed husks and heart disease
9. Sugar alcohols (e.g., xylitol, sorbitol) and tooth decay
10. Soy protein and heart disease
11. Whole grain foods and heart disease and certain cancers
12. Plant stanols and sterols (used in spreads such as Take Control and Benecol) and heart disease
13. Potassium and high blood pressure and stroke
14. Fruits, vegetables, high-fiber grain products and heart disease.

TOPICS OF CLAIMS NOT APPROVED/PENDING

1. Zinc and infection
2. Dietary fiber and colorectal cancer
3. 800 mcg folic acid more effective for reducing risk of neural tube defects than lower amounts
4. Calcium and hypertension
5. Alcohol and heart disease
6. Flaxseed and cancer
7. Omega-3 fatty acids and heart disease
8. Nuts and heart disease

TABLE 4.3

EXAMPLES OF MODEL HEALTH CLAIMS APPROVED BY THE FDA FOR LABELS OF FOODS THAT QUALIFY BASED ON NUTRIENT CONTENT. MODEL CLAIMS ARE OFTEN ABBREVIATED ON FOOD LABELS.

FOOD AND RELATED HEALTH ISSUES	MODEL HEALTH CLAIM
Whole grain foods and heart disease, certain cancers	Diets rich in whole grain foods and other plant foods and low in fat, saturated fat, and cholesterol may reduce the risk of heart disease and certain cancers.
Soy protein and heart disease	Diets low in saturated fat and cholesterol that include 25 grams of soy protein a day may reduce the risk of heart disease.
Sugar alcohols and tooth decay	Frequent between-meal consumption of food high in sugars and starches promotes tooth decay. The sugar alcohols in this food do not promote tooth decay.
Saturated fat, cholesterol, and heart disease	Development of heart disease depends on many factors. Eating a diet low in saturated fat and cholesterol and high in fruits, vegetables, and grain products that contain fiber may lower blood cholesterol levels and reduce your risk of heart disease.
Calcium and osteoporosis	Regular exercise and a healthy diet with enough calcium help maintain good bone health and may reduce the risk of osteoporosis later in life.
Fruits and vegetables and cancer	Low-fat diets rich in fruits and vegetables may reduce the risk of some types of cancer, a disease associated with many factors.
Folate and neural tube defects	Women who consume adequate amounts of folate daily throughout their child-bearing years may reduce their risk of having a child with a neural tube birth defect.

claims have not been approved, they may not represent benefits and have no standard definition. These are the types of claims that are hard to trust.

Health Claims] Upon approval by the FDA, foods or food components with scientifically agreed-upon benefits to disease prevention can be labeled with a health claim (Table 4.2). The health claim used on qualifying product labels, however, must be based on the FDA's "model claim" statements. For instance, scientific consensus holds that diets high in fruits and vegetables may lower the risk of cancer, and a health claim to this effect is allowed. The FDA's model claim for labeling fruits and vegetables is "Low fat diets rich in fruits and vegetables may reduce the risk of some types of cancer, a disease associated with many factors." The FDA approves health claims only for food products that are not high in fat, saturated fat, cholesterol, or sodium. See Table 4.3 for examples of model health claims approved by the FDA.

Labeling Foods as Enriched or Fortified] The vitamin and mineral content of foods can be increased by *enrichment* and *fortification*. Definitions for these terms were established more than 50 years ago. Enrichment pertains only to refined grain products, which lose vitamins and minerals when the germ and bran are removed during processing. Enrichment replaces the thiamin, riboflavin, niacin, and iron lost in the germ and bran. By law, producers of bread, cornmeal, pasta, crackers, white rice, and other products made from refined grains must use enriched flours. Beginning in 1998, federal regulations mandated that folate (a B vitamin) in the form of folic acid be added to refined grain products. This new regulation was put into effect to help reduce the risk of a particular type of inborn, structural problem in children (neural tube defects) related to low blood levels of folate early in pregnancy.

Any food product can be fortified with vitamins and minerals—and many are. One of the few regulations governing the fortification of foods is that the amount of vitamins and minerals added to the food must be listed in the Nutrition Facts panel. Illustration 4.5 shows some examples of enriched and fortified foods.

Food enrichment and fortification began in the 1930s to help prevent deficiency diseases such as rickets (vitamin D deficiency), goiter (from iodine deficiency), pellagra (niacin deficiency), and iron-deficiency anemia.[12] Today, foods are increasingly being fortified for the purpose of reducing the risk of chronic diseases such as osteoporosis, cancer, and heart disease. Other foods, such as energy bars, sports drinks and hydration fluids, and margarines are being fortified with an array of vitamins and minerals principally for sales appeal. Regular consumption of fortified foods increases the risk that people will exceed Tolerable Upper Intake Levels (ULs) of nutrients designated in the Recommended Dietary Allowances (RDAs). Regular use of multiple vitamin and mineral supplements along with liberal intake of fortified foods enhances the likelihood that excessive amounts of some nutrients will be consumed.[13]

enrichment
The replacement of thiamin, riboflavin, niacin, and iron lost when grains are refined.

fortification
The addition of one or more vitamins and/or minerals to a food product.

In 1984 the Kellogg Company launched an ad campaign for All Bran® cereal that announced, "eating the right foods may reduce your risk of some kinds of cancer." Sales of the high-fiber cereal increased 37% in one year, but then Kellogg's had to withdraw the ads. The FDA ruled the All Bran® statement was equivalent to a claim for a drug. The campaign, however, started the nutrition and health claims revolution.[2]

ON THE SIDE

The Ingredient Label

Still more useful information about the composition of food products is listed in ingredient labels. The label of any food that contains more than one ingredient must list the ingredients in order of their contribution to the weight of the food (Table 4.4). The ingredient that makes up the greatest portion of the product's weight must be listed first. Ingredients such as milk solids, peanuts, sulfites, or egg whites, which may cause allergic reactions in some people, should also be listed on the ingredient label.[14]

Richard Anderson

Illustration 4.5 *Examples of foods that are enriched (left) and fortified (right).*

food additives
Any substances added to food that become part of the food or affect the characteristics of the food. The term applies to substances added both intentionally and unintentionally to food.

Food Additives on the Label] Specific information about *food additives* must be listed on the ingredient label. Nearly three thousand chemical substances may be added to food to enhance its flavor, color, texture, cooking properties, shelf life, or nutrient content. Food additives on the FDA's GRAS (Generally Recognized As Safe) list can be used in food without preapproval. New additives must be approved by the FDA prior to use. Table 4.5 provides examples of functions of some additives used in food. The most common food additives are sugar and salt, but trace amounts of polysorbate, potassium benzoate, and many other additives that are not so familiar are also included in foods (Illustration 4.6). Although the new labeling regulations are more comprehensive than the old ones, they don't help consumers understand what some additives do. Appendix D lists many of the most common additives and indicates their function in foods.

the NOOZ   by Peter Kohlsaat

EXcuse me, but What is this third ingredient?

It's a double-bonded ester. It's for consistency.

K.hlsaat

Meet Phil, the supermarket's in-house chemist.

Illustration 4.6
Source: Tribune Media SErvices, Inc. All rights reserved. Reprinted with permission.

TABLE 4.4

ALL FOODS WITH MORE THAN ONE INGREDIENT MUST HAVE AN INGREDIENT LABEL. HERE ARE THE RULES.

1. Ingredients must be listed in order of their contribution to the weight of the food, from highest to lowest.

2. Beverages that contain juice must list the percentage of juice on the ingredient label.

3. The terms "colors" and "color added" cannot be used. The name of the specific color ingredients must be given (for example, caramel color, tumeric).

4. Cow's milk protein-based food additives or other ingredients to which some people are allergic must be listed on the label. (For example, if the protein casein is used in coffee lightener the ingredient label must state that the casein was derived from cow's milk.)

TABLE 4.5

SOLVING THE MYSTERY OF INGREDIENT LABEL TERMS.

CAKE MIX INGREDIENT LABEL	ADDITIVE	FUNCTION
Ingredients: Sugar, enriched flour bleached (wheat flour, niacin), iron thiamin Mononitrait, (vitamin B_1), riboflavin (vitamin B_2), vegetable shortening (contains partially hydrogenated soybean cotton-seed oil), sodium aluminum phosphate, dextrose, leavening, (baking soda, monocalcium phosphate, diacalcium phosphate, aluminum sulphate), wheat starch, propylene glycol monoesters, modified corn starch, salt, egg white, vanilla, dried corn syrup, polysorbate 60, nonfat milk mono and diglycerides, sodium caseetrate, xanthan gum, soy lecithin.	Sodium aluminum phosphate Propylene glycol monoesters Mono- and diglycerides Xanthan gum	Gives baked products a light texture Helps blend ingredients uniformly, enhances moisture content and texture Maintains product softness after baking Thickening agent, helps hold product together after baking

Trace amounts of substances such as pesticides; hormones and antibiotics given to livestock; fragments of packaging materials such as plastic, wax, aluminum, or tin; very small fragments of bone; and insects may end up in foods. These are considered "unintentional additives" and do not have to be included on the label.

Nutrition Labeling in Canada]   Food labeling standards in Canada are similar to those in the United States but include only 5 approved health claims.[15] Appendix F shows an example food label and points out required components on Canadian food labels.

Dietary Supplement Labeling

The wide assortment of **dietary supplements** available on the market must be labeled and are regulated by a common set of rules. Dietary supplements differ from drugs in that they do not have to undergo vigorous testing and obtain FDA approval before they are sold. In return, dietary supplement labels cannot claim that the products treat, cure, or prevent disease.[16]

According to FDA regulations, dietary supplements must be labeled as such and include a "Supplement Facts" panel that lists serving size, ingredients, and percent Daily Value (%DV) of essential nutrient ingredients (Illustration 4.7). Products can be labeled with a health claim, such as "high in calcium" or "low fat," if the product qualifies according to nutrition labeling regulations.

Structure or Function Claims]   Supplements can be labeled with "structure or function" health claims such as "improves circulation," "supports the immune system," and "helps maintain mental health." If a function claim is made on the label or package inserts, the label or insert must acknowledge that the FDA does not support the claim: *"This statement has not been evaluated by the FDA. This product is not intended to diagnose, treat, cure, or prevent any disease."* That's why you can see this acknowledgment printed on the

dietary supplements
Any products intended to supplement the diet, including vitamin and mineral supplements; proteins, enzymes, and amino acids; fish oils and fatty acids; hormones and hormone precursors; and herbs and other plant extracts. Such products must be labeled "Dietary Supplement."

Illustration 4.7
Dietary supplement label.

This statement has not been evaluated by the Food and Drug Administration
intended to diagnose, treat, cure
SUGGESTED USE: As a dietary softgel daily, with a meal. Keep in a cool, dry place, out of reach
Do not use if imprinted seal under
Our pure gelatin shell has no arti you may notice a natural color va
INGREDIENTS: Ascorbic Aci Acetate, Gelatin, Cottonseed Oil Calcium Phosphate, Partially H and Soybean Oils, Lecithin Oil, Z Beeswax, Manganese Sulfate, Selenate.
1-800-276-2878 M-F 7am-4pm P.T
Oil- and Water-soluble Vitamins with M
Mfd. By **Nature Made Nutritional**
Mission Hills, CA 91346-9606   Mad

Nutrition Facts Serving Size 1 softgel		
Amount Per Softgel		**% Daily Value**
Vitamin A 10,000 I.U.		200%
100% as Beta Carotene		
Vitamin C 250 mg		417%
Vitamin E 200 I.U.		667%
Zinc 7.5 mg		50%
Selenium 15 mcg		21%
Copper 1 mg		50%
Manganese 1.5 mg		75%

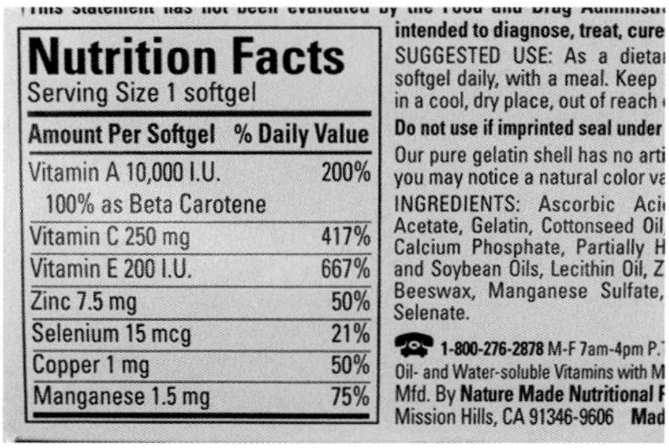

Irradiation is sometimes referred to as "cool pasteurization" because it destroys microorganisms in food without heat. Pasteurization was considered highly suspect by many consumers in the late nineteenth century, just as irradiation is now.[23]

Photo Disc

Illustration 4.8

The "radura," as the symbol is called, must be displayed on irradiated foods. The words "treated by irradiation, do not irradiate again" or "treated with radiation, do not irradiate again" must accompany the symbol.

labels of dietary supplements that make such claims. Structure and function claims don't have to be approved before they can appear on product labels. Dietary supplements labeled with misleading or untruthful information, and those that are not safe, can be taken off the market and the manufacturers fined.

Recently, manufacturers of dietary supplements are being allowed to label products with health claims supported to some extent by scientific studies. For example, EPA, DHA, and omega-3 fatty acid supplements (types of fat found primarily in fish and fish oils) can be labeled with a health claim that suggests these substances can reduce the risk of heart disease.[17]

Irradiated Foods

Food irradiation is an odd example of a food additive. It is actually a process that doesn't add anything to foods. Irradiation uses X rays, gamma rays, or electron beams to kill insects, bacteria, molds, and other microorganisms in food. Food irradiation enhances the shelf life of food products and decreases the risk of foodborne illness.[18] It must be performed according to specific federal rules. There is little evidence that irradiation adversely affects food quality, flavor, or safety.[19] Nevertheless, the notion of beaming gamma rays through foods may conjure up all sorts of concerns—in particular, the fear that radioactive particles stay in irradiated foods and end up in your body if you eat them.

Unlike foods exposed to radioactive fallout (such as that released after the nuclear power plant accident at Chernobyl in 1987), irradiated foods retain no radioactive particles. The process is like having your luggage X-rayed at the airport. Your luggage doesn't become radioactive, nor has anything in it changed. The process leaves no evidence of having occurred. Actually, this lack of change creates a challenge because inspection agencies can't determine whether a food has been irradiated or not. All irradiated foods except spices added to processed foods are required to display the international "radura" symbol and to indicate in writing that the food has been irradiated (Illustration 4.8).

Irradiated foods have a long history but have not yet won the trust of some American consumers.[20] Today, irradiated products face many of the same hurdles microwaved foods confronted in the 1960s. The desire to limit radioactive processes of all sorts, as well as radioactive waste products, is widespread; many people fear anything that involves radioactivity, especially when it comes to food. Acceptance of irradiated foods and even a preference for them tend to develop as consumers learn more about the safety and benefits of the process. Irradiation is approved for use on chicken, turkey, pork, beef, eggs, grains, fresh fruits and vegetables, and other foods in the United States.[21] Currently the process is used mainly to sterilize food for astronauts and for hospital patients who easily get infections; it is also used to sterilize paper and plastic medical supplies, tampons, and germ-free surgical products.

Food irradiation is increasingly viewed as an important part of efforts to control outbreaks of *E. coli*, salmonella, and other foodborne illnesses. It does not, however, destroy all toxins, bacterial spores, or viruses, and irradiated food can become contaminated after packages are opened.[22]

Organic Foods

For many years, consumers and producers of organic foods urged Congress to set criteria for the use of the term "organic" on food labels (Illustration 4.9). Consumers wanted to be assured that foods were really organic, and producers wanted to keep the business honest. Standards were needed because consumers cannot distinguish organically produced foods from others by looking at them, by tasting them, or by reading their nutrient values.[24] The U.S. Department of Agriculture (USDA) has developed and is implementing standards for organic foods.

Labeling Organic Foods] Rules that qualify foods as organic are shown in Illustration 4.10. If organic growers and processors qualify according to USDA approved certifying organizations, they can place the green-and-white USDA Organic seal on product labels. The USDA can impose financial penalties on companies that use the seal inappropriately. Organically grown and produced foods can be labeled in four other ways:

- "100% Organic" if they contain entirely organically produced ingredients

- "Organic" if they contain at least 95% organic ingredients

- "Made with organic ingredients" if they contain at least 70% organic ingredients

- "Some Organic Ingredients" if the product contains less than 70% organic ingredients

Many people choose organic foods not so much for what is in them, but for what is not in them: hormones and antibiotics, pesticide and herbicide residues (Figure 4.11). Most organic foods, however, are not totally free of some of these ingredients. Due to "pesticide drift" from sprayed crops, past use of pesticides on farmland, and the leaching of chemicals used on crops into groundwater, organically grown plants may contain traces of pesticides. However, conventionally grown crops are six times more likely to have traces of several pesticides than are organic crops.[25]

Illustration 4.9
Organic foods are going mainstream. More and more national brands are offering organic selections.

Illustration 4.10
USDA rules for qualifying foods as organic.

1. *Plants*
   - *Must be grown in soils not treated with synthetic fertilizers, pesticides, and herbicides for at least three years*
   - *Cannot be fertilized with sewer sludge*
   - *Cannot be treated by irradiation*
   - *Cannot be grown from genetically modified seeds, or contain genetically modified ingredients*

2. *Animals*
   - *Cannot be raised in "factory-like" confinement conditions*
   - *Cannot be given antibiotics or hormones to prevent disease or promote growth*

Scott Bauer/ARS/USDA

Illustration 4.11
Many people choose organic foods not so much for what is in them, but for what is not in them: hormones and antibiotics, pesticide and herbicide residues.

Beyond Nutrition Labels: We Still Need to Think

Even with the new food labels, consumers can't be stupid.
—Max Brown

Understanding and applying the information relayed in nutrition labels calls for more, not less, nutrition knowledge on the public's part. People need to know a good bit about nutrition before they can understand nutrition labels and incorporate labeled foods appropriately into an overall diet. Good diets include more than just foods with nutrition labels. The ice cream cone from the stand in the mall; the orange, potato, or fish we buy in the store; and the pizza delivered to the dorm are unlabeled parts of many diets. We need to know enough about the composition of unlabeled foods to fit them into a healthy diet. Nutrition labels often list only two vitamins and two minerals, but many more are required for health, so it's particularly important for people to know if their diet is varied enough to supply needed vitamins and minerals. Use of the Food Guide Pyramid should go hand in hand with label reading and food choices.

Healthy diets consist of a wide variety of foods that may include high-fat or high-sodium foods on occasion. Not every food we eat has to have the "right" label profile. Serving mostly low-fat or low-calorie foods to children, for example, might have unintended, unhealthy effects. Young children need calories and fat for growth and development. If diets are severely restricted, growth and development will be impaired. Nutrition labels are an important tool for helping people make informed food-purchasing decisions. However, labels do not now—nor will they ever—provide all the information needed to make wise decisions about food. Only people who are well informed about nutrition can do that.

Nutrition UP CLOSE

Comparison Shopping

FOCAL POINT: Comparing label information can help you choose the most nutritious product for your money.

Rate the nutrition value of these two cereals by completing the second paragraph below using information from the Nutrition Facts panels. The first paragraph has been done for you.

FEEDBACK (answers to these questions) can be found at the end of Unit 4.

Whole Grain Cereal

Nutrition Facts

Serving Size **1 cup (50g)**
Serving Per Container **10**

Amount Per Serving

Calories **170**   Calories from Fat **5**

	% Daily Value*
Total Fat 0.5 g	1%
Saturated Fat 0 g	0%
Trans Fat 0 g	
Cholesterol 0 g	0%
Sodium 0 mg	0%
Total Carbohydrate 41 g	14%
Dietary Fiber 5 g	21%
Sugars 0 g	
Protein 5 g	

Vitamin A 0%	Vitamin C 0%
Calcium 2%	Iron 8%

*Percent Daily Values are based on a 2000 calorie diet. Your daily values may be higher or lower depending on your calorie needs:
**Intake should be as low as possible.

		Calories:	2,000	2,500
Total Fat	Less than		65 g	80 g
Sat Fat	Less than		20 g	25 g
Cholesterol	Less than		300 mg	300 mg
Sodium	Less than		2400 mg	2400 mg
Total Carbohydrate			300 g	375 g
Dietary Fiber			25 g	30 g

Health Granola Cereal

Nutrition Facts

Serving Size **1 cup (50g)**
Serving Per Container **10**

Amount Per Serving

Calories **210**   Calories from Fat **30**

	% Daily Value*
Total Fat 3 g	5%
Saturated Fat 0 g	0%
Trans Fat 0 g	
Cholesterol 0 g	0%
Sodium 120 mg	5%
Total Carbohydrate 43 g	14%
Dietary Fiber 3 g	12%
Sugars 16 g	
Protein 5 g	

Vitamin A 2%	Vitamin C 0%
Calcium 2%	Iron 10%

*Percent Daily Values are based on a 2000 calorie diet. Your daily values may be higher or lower depending on your calorie needs:
**Intake should be as low as possible.

		Calories:	2,000	2,500
Total Fat	Less than		65 g	80 g
Sat Fat	Less than		20 g	25 g
Cholesterol	Less than		300 mg	300 mg
Sodium	Less than		2400 mg	2400 mg
Total Carbohydrate			300 g	375 g
Dietary Fiber			25 g	30 g

The **Whole Grain Cereal**, providing 170 **calories** per serving, contributes 5 **calories from fat** (or 1% of the % Daily Value). Each serving also provides 0 mg of **sodium** (or 0 % of the % Daily Value), 5 g of **dietary fiber** (or 21% of the % Daily Value), and 0 g of **sugars**.

The **Health Granola Cereal**, providing _____ **calories** per serving, contributes _____ **calories from fat** (or _____ % of the % Daily Value). Each serving also provides _____ mg of **sodium** (or 5% of the % Daily Value), _____ g of **dietary fiber** (or _____ % of the % Daily Value), and _____ g of **sugars**.

Which cereal provides the better nutritional value?

Key Terms

Daily Values (DVs), page 4-3

dietary supplements, page 4-11

enrichment, page 4-9

food additives, page 4-10

fortification, page 4-9

% Daily Value (%DV), page 4-3

www links

www.cfsan.fda.gov/~dms/flquiz1.html
Test your food label knowledge, and get answers to questions about identifying healthy food choices based on nutrition label information.

www.ams.usda.gov
USDA's National Organic Program website offers details about the Organic Foods Pro-

duction Act, organic food requirements, use of the USDA Organic Seal, and answers to FAQs.

www.navigator.tufts.edu
Search terms such as nutrition labeling, food additives, organic foods, food irradiation to find the best websites.

vm.csfan.fda.gov

Provides nutrition and dietary supplement labeling information and updates.

www.nal.usda.gov/fnic
Search "food irradiation" and "food labeling."

Notes

1. Brecher SJ et al. Status of nutrition labeling, health claims, and nutrient content claims for processed foods. J Am Diet Assoc 2000;100:1057–62.

2. Marquart L et al. Solid science and effective marketing for health claims. Nutr Today 2001;36:107–14.

3. Brecher et al., Status of nutrition labeling.

4. Pennington JA, Hubbard VS. Derivation of daily values used for nutrition labeling. J Am Diet Assoc 1999;97:1407–12.

5. Brecher et al., Status of nutrition labeling.

6. Eateries pressed for nutrition data. Wall Street Journal, 2/13/03, p. A3.

7. Questions and answers about trans fat nutrition labeling. www.cfsan.fda.gov/label.html, accessed 7/03

8. Pennington and Hubbard, Derivation of daily values.

9. Brecher et al., Status of nutrition labeling.

10. Brecher et al., Status of nutrition labeling.

11. Marquart et al., Solid science.

12. Park YE et al. History of cereal-grain product fortification in the United States. Nutr Today 2001;36:124–36.

13. Hunt J et al. Position of the American Dietetic Association: food fortification and dietary supplements. J Am Diet Assoc 2001;101:115–25.

14. Brecher et al., Status of nutrition labeling.

15. Canada's food labels. Nutrition Week 2003;23:1.

16. Ross S. Functional foods: the Food and Drug Administration perspective. Am J Clin Nutr 2000;71(suppl):1735S–8S.

17. FDA announces decision on another health claim for dietary supplements. Press release, 2000 Nov. 1.

18. Shea KM. Technical report: irradiation of food. Pediatrics 2000;106:1505–9.

19. Position of the American Dietetic Association: food irradiation. J Am Diet Assoc 2000;100:246–52.

20. Irradiated beef roundup: USDA could approve use in schools soon, confab causes flap. Nutrition Week 2003;May 19:3.

21. Shea, Technical report: irradiation of food; Position of the ADA: food irradiation.

22. Shea, Technical report: irradiation of food; www.ams.usda.gov, accessed 7/03.

23. Shea, Technical report: irradiation of food.

24. New organic foods standards announced by USDA. USDA press release, 2000 Dec. 21.

25. Consumer Union and the Organic Materials Review Institute study on pesticide residues in organic foods. Consumer Reports, August 2002.

Nutrition UP CLOSE

Comparison Shopping

Feedback for Unit 4

The **Health Granola Cereal**, providing 210 **calories** per serving, contributes 30 **calories from fat** (or 5% of the % **Daily Value**). Each serving also provides 120 mg of **sodium** (or 5% of the % **Daily Value**), 3 g of **dietary fiber** (or 12% of the % **Daily Value**), and 16 g of **sugars**. The **Whole Grain Cereal** provides the better nutritional value because it is lower in calories, total fat, sodium, and sugar and higher in fiber than the **Health Granola Cereal**. Don't be fooled by an attractive-sounding product. The proof of nutritional value is in the label, not the name.

Nutrition, Attitudes, and Behavior

Nutrition Scoreboard

	TRUE	FALSE
1 Food preferences are genetically determined.		
2 Three things in life never change: death, taxes, and food habits.		
3 Skipping breakfast does not affect school performance in well-nourished children.		

Answers on next page

[KEY CONCEPTS AND FACTS]

- Most food preferences are learned.

- The value a person assigns to eating right has more effect on dietary behaviors than does knowledge about how to eat right.

- Food habits can and do change.

- The smaller and more acceptable the dietary change, the longer it lasts.

- Behavior and mental performance can be affected by diet.

Answers to *Nutrition Scoreboard*

		TRUE	FALSE
1	Although genetics plays a role in food preferences, the predominant influences are environmentally determined.[1]		✔
2	True, true, false. Death and taxes are for certain, but the idea that food habits don't change is a myth.		✔
3	Well-nourished children who don't eat breakfast tend to score lower on problem-solving tests than children who eat breakfast.[2] (If you got this one wrong, answer one more question: Did you skip breakfast this morning?)		✔

Origins of Food Choices

Horse meat is a favorite food in a large area of north-central Asia. Pork, which is widely consumed in North and South America, Europe, and other areas, is rigidly avoided by many people in Islamic countries. Bone-marrow soup and sautéed snails are delicacies in France, while kidney pie is traditional in England. Dog is a popular food in Borneo, New Guinea, the Philippines and other countries, whereas snake is a delicacy in China. In some countries, people enjoy insects (Illustration 5.1) but consider corn and soybeans fit only for animal feed. And then there are steamed clams and raw oysters—food passions for some, but absolutely disgusting to others.[3]

When did you first think "yecck!"? The food choices just described would elicit that response among people from a variety of cultures. But they would not be responding to the same foods.

Why do people eat what they do? People learn from the society they live in which animals and plants are considered food and which are not.[4] Once items are identified as food, they develop a legacy of strong symbolic, emotional, and cultural meanings. Comfort foods, health foods, junk foods, fun foods, soul foods, fattening foods, mood foods, and pig-out foods, for example, have been identified in the

Illustration 5.1

Maggots (right) and grasshoppers (left) are Mexican delicacies, served at Girasoles Restaurant in Mexico City.

Jorge Uzon/AFP/Getty Images

Culture
- Acceptable foods
- Customs
- Food symbolism
- Religious beliefs

Nutrition Knowledge and Beliefs
- Health concerns
- Nutritional value of foods
- Attitudes and values
- Education
- Experience

Food Preferences
- Food taste, smell, color, texture, and temperature
- Heredity
- Familiarity

FOOD SELECTION
Dietary quality

Practical considerations
- Food cost
- Convenience
- Level of hunger
- Food availability
- Health status

Illustration 5.2
Factors influencing food selection and dietary quality. *Each of these sets of factors interacts with the others.*[8]

United States. And all cultures have their "super food." In Russia and Ireland, it's potatoes. In Central America, corn and yucca (a starchy root, also called manioc) are super foods, and in Somalia it's rice. The designation refers to the cultural significance of the food and not to its nutritional value.[5]

In countries where a wide variety of foods are available and people have the luxury of selecting which foods they will eat, food choices are influenced by a wide range of factors (Illustration 5.2). Of these factors, food preference has the largest impact on food choices for most people.[6] Rather than being inborn, food preferences are primarily learned.[7]

People Are Not Born Knowing What to Eat

The food choices people make are not driven by a need for nutrients or guided by food selection genes. People deficient in iron, for example, do not seek out iron-rich foods. If we're overweight, no inner voice tells us to reject high-calorie foods. Women who are pregnant don't instinctively know what to eat to nourish their growing fetuses. No evidence indicates that young children, if offered a wide variety of foods, would select and ingest a well-balanced diet.[9]

Humans are born with mechanisms that help them decide when and how much to eat, however.[10] An inborn attraction to sweet-tasting foods, a dislike for bitter foods, and the response of thirst when water is needed all influence food and fluid intake to an extent (Illustration 5.3).[11] There is evidence to suggest that people deficient in sodium experience an increased preference for salty foods.[13] Most people have a strong aversion to foods they think "smell bad," which are likely spoiled. Whether this response is inborn or learned is not clear.

Inborn traits that affect food choices are not very fine-tuned. People may overdo (consume excessive sugar and salt, for example) and underdo (for instance, drink too little water in response to thirst) their response to internal cues.

Why do we prefer the foods we do?

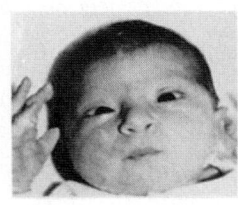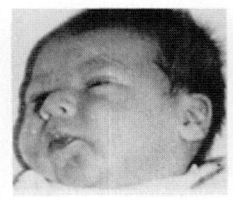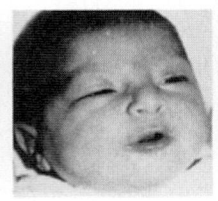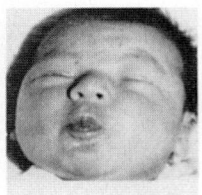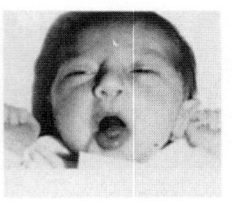

Illustration 5.3
Newborn infants respond to different tastes: (a) Baby at rest, (b) tasting distilled water, (c) tasting sugar, (d) tasting something sour, and (e) tasting something bitter.[12]
Source: Taste-induced facial expressions of neonate infants from the classic studies of J. E. Steiner, in Taste and Development: The Genesis of Sweet Preference, ed. J. M. Weiffenbach, HHS Publication no. NIH 77-1068 (Bethesda, Md.: U.S. Department of Health and Human Services, 1977), pp. 173–189, with permission of the author.

Food Choices and Preferences

"You're going to eat that! Do you have any idea what's in that hot dog?"

"Yeah, I do. There's barbecue in the backyard, ball games with my mom and dad, and parties at my friend's house. The memories taste great!"

The strong symbolic, emotional, and cultural meanings of food come to life in the form of food preferences. We choose foods that, based on our cultural background and other learning experiences, give us pleasure.[14] Foods give us pleasure when they relieve our hunger pains, delight our taste buds, or provide comfort and a sense of security. We find foods pleasurable when they outwardly demonstrate our superior intelligence, our commitment to total fitness, or our pride in our ethnic heritage. We reject foods that bring us discomfort, guilt, and unpleasant memories and those that run contrary to our values and beliefs.

The Symbolic Meaning of Food

Food symbolism, cultural influences, and emotional reasons for food choices are broad concepts that may become clearer with concrete examples. Here are a few examples to consider.

Status Foods] Vance Packard, in his book *The Status Seekers*, provided a memorable example of the symbolic value of food:

> As a lad, this man had grown up in a poor family of Italian origin. He was raised on blood sausages, pizza, spaghetti, and red wine. After completing high school, he went to Minnesota and began working in logging camps, where—anxious to be accepted—he soon learned to prefer beef, beer, and beans, and he shunned "Italian" food. Later, he went to a Detroit industrial plant, and eventually became a promising young executive. . . . In his executive role he found himself cultivating the favorite foods and beverages of other executives: steak, whiskey, and seafood. Ultimately, he gained acceptance in the city's upper class. Now he began winning admiration from people in his elite social set by going back to his knowledge of Italian cooking, and serving them, with the aid of his manservant, authentic Italian treats such as blood sausage, spaghetti, and red wine![15]

Comfort Foods]

The first thing I remember liking that liked me back was food.
—Rita Rudner, comedian

Ice cream, apple pie, chicken noodle soup, boxed chocolates, meat loaf and mashed potatoes: these are the most popular comfort foods in the United States.[16] The feel-

Men tend to like olives and anchovies more than women do.[19]

ings of security and love that came along with the tea and honey or chicken soup that your mother or father gave you when you had a cold, or with the ice cream and popsicles lovingly given to soothe a sore throat, are renewed with comfort foods. Once the symbolic value of a food is established, its nutritional value will remain secondary.[17] Food status is a strong determinant of food choices; and after all, as a noted nutritionist once said, "life needs a little bit of cheesecake" (Illustration 5.4).[18]

"Discomfort Foods"] *"When I was a kid, my parents made me eat the brussels sprouts they put on my plate. I finally ate them, but then I threw up. Nobody has ever made me eat brussels sprouts again."*

Memories of bad experiences with food, and expectations that certain foods will harm us in some way, each contribute to our learning about food and affect our food preferences. Eating a piece of blueberry pie right before an attack of the flu hits or overdosing on sweet pickles or olives, for example, may take these foods off your preferred list for a long time.

Cultural Values Surrounding Food

A team of scientists observed that the diet of certain groups in the Chin States of Upper Burma was seriously deficient in animal protein. After considerable study a way was found to improve the situation by cross-breeding the small, local black pigs raised by the farmers with an improved strain to obtain progeny giving a greater yield of meat. The entire operation, however, completely failed to benefit the nutrition of the population because of one fact which had been viewed as irrelevant. The cross-bred pigs were spotted. And it was firmly believed—as firmly as we believe that to eat, say, mice, would be disgusting—that spotted pigs were unfit to eat.[20]

Dietary change introduced into a culture for the purpose of improving health can be successful only if it is accepted by the culture. Cultural norms are not easily modified.

Other Factors Influencing Food Choices and Preferences

Food preferences and selections are also affected by the desire to consume foods considered healthy. Reducing fat intake, eating more fruits and vegetables, and cutting down on sweets bring rewards and pleasures such as weight loss and maintenance of

Illustration 5.4
Source: Reprinted by permission: Tribune Media Services.

Americans are introduced to approximately 5,000 new food items in grocery stores each year.[32]

Photo Disc

the lost weight, an end to constipation, lower blood cholesterol level, and a newly discovered preference for basic foods.

Food Cost and Availability] Food choices are also affected by the cost and availability of food. Researchers found that college students eating in dining halls have better diets when they prepay for their meals for the entire term rather than paying at each meal.[21] Grocery shoppers tend to select more low-fat and high-fiber foods when presented with a wide selection of those foods rather than just a few.[22]

Genetic Influences] The primary role genetics plays in food preference development relates to taste. Some people are born with a strong sensitivity to bitterness and may reject foods such as brussels sprouts and broccoli (especially if they are overcooked!), bitter-tasting teas and wines, and tonic water. Some people with this genetic trait will continue to eat strong-flavored vegetables—particularly after they cover up the bitterness with seasonings or sauces.

Food Choices Do Change

Who says old dogs can't learn new tricks? Most Americans aren't eating the way they used to. Per person consumption of pasta has more than doubled since 1970, while fresh egg intake has fallen from an average of 5⅓ eggs per week to 3½. Low-fat milk sales have risen 165%, beef consumption dropped 9%, and broccoli consumption is skyrocketing. Per person consumption of broccoli has risen 386% since 1970—a bigger gain than for any other food. Americans are eating 76% more cheese and 25% more vegetables and fruits than in 1970.[23] For examples of changes in food choices since 1990, see Illustration 5.5. Food choices are largely learned and do change as we learn more about foods. Perhaps your food choices have changed over time. How do the food choices you make now compare with the choices you made five years ago?

How Do Food Choices Change?

What are the ingredients for change in food choices? Why do some people succeed in improving their food choices while other people find that very hard to do? Nutrition knowledge, attitudes, and values have a lot to do with changing food choices for the better.

Nutrition Knowledge and Food Choices] Sound knowledge about good nutrition necessarily precedes the selection of a healthful diet. But is knowledge enough to ensure that healthy changes in diet will be made? The answer is "yes" for some people and "no" for others. Students at James Madison University in Virginia

Illustration 5.5
Changes in Americans' food choices since 1990.[24]

Photo Disc

Sugars +10%   Skim milk +11%   Chicken -9%   Margarine -31%   Oranges - 31%

improved their diets after taking a course in nutrition. At the end of the course, students reported eating, on average, less fat and sodium than at the beginning of the course (Illustration 5.6).[25] Changes toward healthier diets among students taking a nutrition course at the University of Texas at Austin have also been reported.[27] For some Americans, news that animal fats raise blood cholesterol levels was enough to convince them to cut down on beef and whole milk. Knowledge that diets containing an adequate amount of fiber reduce the risk of colon cancer led many Americans to switch to high-fiber breakfast cereals. Information about good nutrition leads some people to modify their eating behavior some of the time and is more likely to change the food choices of women than men.[28] Knowledge alone is often not enough, however.

Daily Tribune

Nutrition Course Improves Diets of College Students

The diets of college students improved after taking a basic nutrition course

Results indicated that students decreased the amount of total fat from 82 to 68 grams and made other changes putting their diets more in line with the Dietary Guidelines.

Illustration 5.6
A true story.[26]

When Knowledge Isn't Enough] Many people know far more about the components of a good diet than they put into practice. The problem is that between knowledge and practice lie multiple beliefs and experiences that act as barriers to change (Illustration 5.7). Change of any type is most likely to succeed when the benefits of making the change outweigh the disadvantages. This makes changes in food choices a very individual decision. Individuals decide whether a change is in their best interests. But what sorts of circumstances, in addition to increased knowledge, make changes in food choices worthwhile for individuals—and even highly desired?

Nutrition Attitudes, Beliefs, and Values] The value individuals place on diet and health is reflected in the food choices they make. A survey of restaurant patrons found that food choices varied according to the consumer's perceptions of the importance of diet to health:[29]

- "Unconcerned" consumers—people who are unconcerned about the connection between diet and health and who tend to describe themselves as "meat and potato eaters"—select foods for reasons other than health.

- "Committed" consumers believe that a good diet plays a role in the prevention of illness. They tend to consume a diet consistent with their commitment to good nutrition.

- "Vacillating" consumers—people who describe themselves as concerned about diet and health but who do not consistently base food choices on this concern—tend to vary their food choices depending on the occasion. These consumers are likely to abandon diet and health concerns when eating out or on special occasions, but they generally adhere to a healthy diet.

Avoiding illness and curing or diminishing current health problems are likewise strong incentives for changing food choices (Table 5.1).[30] One study found that

Illustration 5.7
Why knowledge about a good diet may not be enough to improve food choices.

Photo Disc

"I feel guilty about eating the foods I like."

"Eating right is too expensive."

"I tried eating better, but I didn't stick with it."

"I don't have the time to eat right."

"The vegetables I like aren't available."

"I'm healthy now... Why should I worry about my diet?"

TABLE 5.1

FACTORS THAT ENHANCE FOOD CHANGES.

- Attitude that nutrition is important
- Belief that diet affects health
- Perceived susceptibility to diet-related health problems
- Perception that benefits of change outweigh barriers to change

nurses and dietitians based dietary changes primarily on the benefits of good diets to future health.[31] In almost all instances, the key to lasting improvements in diet is to make changes you can live with—changes that bring more pleasure than inconvenience or discomfort. Painful changes, such as those that require willpower to maintain, rarely last.

Successful Changes in Food Choices

Perhaps the key in making dietary changes is to determine which changes are the easiest.

The primary reason most efforts to improve food choices fail is that they are too drastic. Improvements that last tend to be the smallest acceptable changes needed to do the job.[33]

The Process of Changing Food Choices] Assume you need to lose weight and want to keep it off by modifying your food choices. A promising plan to accomplish this goal would begin by identifying food choices you would like to change and lower-calorie food options you would be willing and able to eat (Table 5.2). Then you could make the plan more specific by identifying the changes that would be easiest to implement. For example, assume the low-calorie foods you like and would eat include frozen nonfat yogurt and oranges. You might decide to eat yogurt or an orange in place of your usual bedtime snack of ice cream. A specific dietary change such as this is much easier to implement than a broad notion, such as "eat less." Although weight loss will take a while, such a small acceptable change has a much better chance of working than a drastic change in diet.

TABLE 5.2

CHANGING FOOD CHOICES FOR THE BETTER.[34]

THE PROCESS	AN EXAMPLE
1. Identify a healthful change in your diet you'd like to make.	1. I'd like to lower my fat intake.
2. Identify food choices you'd like to change.	2. I eat at fast-food restaurants five times a week. I always have a large order of fries, and several times a week I eat fried chicken.
3. Identify specific, acceptable options for more healthful food choices.	3. Options identified: • Order tossed salad instead of fries. • Eat grilled chicken sandwiches once a week instead of fried chicken. • Eat Mexican fast-food more often.
4. Decide which option is easiest to accomplish and requires the smallest change to get the job done.	4. I love Mexican food. It would be easy to eat it more often.
5. Plan how to incorporate the change into your diet.	5. Mondays and Fridays, I'll eat Mexican.
6. Implement the change. Be prepared for midcourse corrections.	6. Midcourse correction: On Fridays, when I'm with my friends, it's easier to eat at the restaurant they like. I'll order the grilled chicken sandwich and a salad.

Planning for Relapses] When making a change in your diet, be prepared for relapses. Relapses happen for a number of reasons, and they don't mean the attempt has failed. People often return to old habits because the change they attempted was too drastic or because they tried to make too many changes at once. If the change undertaken doesn't work out, rethink your options and make a midcourse correction.

Does Diet Affect Behavior?

Food affects behavior in some rather striking ways. Irritable, crying infants rapidly change into cooing, sleepy angels after they are fed. Low-on-sleep employees perk up after their morning coffee. A high-calorie lunch makes many people feel calm and sleepy.[35] Not only do our behaviors affect our diet, but our diets also affect our behaviors in some ways. Behaviors shown to be affected by various components of the diet, as well as behaviors for which the effects of diet are unsubstantiated but nevertheless widely believed to exist, are listed in Table 5.3. One common belief about food and behavior is the subject of this unit's "Reality Check."

Malnutrition and Mental Performance

The World Health Organization estimates that 40 to 60% of the world's children suffer mild to moderate malnutrition.[38]

Like growth and health, mental development and intellectual capacity can be affected by diet. The effects range from mild and short term to serious and lasting, depending on when the malnutrition occurs, how long it lasts, and how severe it is. The effects are most severe when malnutrition occurs while the brain is growing and developing.

Severe deficiency of protein, calories, or both early in life leads to growth retardation, low intelligence, poor memory, short attention span, and social passivity. When the nutritional insult is early and severe, some or all of these effects may be lasting (Illustration 5.8). In Barbados, for example, children who

---

TABLE 5.3

EXAMPLES OF RELATIONSHIPS BETWEEN DIET AND BEHAVIOR.[36]

Established relationships between diet and behavior:
- Severe protein-calorie malnutrition early in life and impaired mental development
- Breakfast skipping and reduced problem-solving skills in children
- Fetal exposure to high levels of alcohol and mental retardation and behavioral problems
- Iron-deficiency anemia and reduced motor skills, attention span, and problem-solving skills
- Overexposure to lead and poor academic performance
- Folate deficiency and depression
- High-carbohydrate and high-calorie meals and calmness, sleepiness

Unsubstantiated claims about diet and behavior relationships:
- Sugar intake, hyperactivity, and criminal behavior
- Vitamin supplements and increased intelligence
- Meat and aggressive behavior
- High-carbohydrate intake and increased brain serotonin level, decreased appetite, and weight loss

---

Glenda:
Chocolate and vanilla are natural love potions, there's no doubt about it. If I eat a chocolate truffle and spritz on vanilla flavoring like perfume before a date, the guy always goes nuts for me!

REALITY CHECK

Can food be a love potion?

Toward the end of a friend's birthday party, Glenda and Cassell got into a heated debate about the existence of food aphrodisiacs. Part of the conversation went like this:

Is it all in Glenda's head

?

Photo Disc

Cassell:
Are you kidding? I've heard about oysters and this tree bark that are supposed to work miracles, too. It's all in your head.

Answers on next page

Glenda

ANSWERS TO REALITY CHECK

The idea that food can act as an aphrodisiac has been around since ancient times. Although many have looked and others have tried, no one has found a food that acts like a love potion.[37]

Cassell

Illustration 5.8
Malnutrition in early childhood has long lasting effects. Some children never fully recover.

experienced protein-calorie malnutrition in the first year of life did not fully recover even with nutritional rehabilitation. Growth improved but academic performance did not. Compared to well-nourished children, those experiencing protein and calorie deficits during infancy were more likely to drop out of school and were four times more likely to have symptoms of ADHD (attention deficit hyperactivity disorder).[39]

Protein-calorie malnutrition that occurs later in childhood, after the brain has developed, produces behavioral effects that can be corrected with nutritional rehabilitation. Correction of other deficits that often accompany malnutrition, such as the lack of educational and emotional stimulation and harsh living conditions, hastens and enhances recovery.[40]

Protein-calorie malnutrition severe enough to cause permanent delays in mental development rarely occurs in the United States. When it does, malnutrition is usually due to neglect or inadequate caregiving. More common dietary events that impair learning in U.S. children are breakfast skipping, fetal exposure to alcohol, iron deficiency, and lead toxicity.

Does Breakfast Help You Think Better?] Short-term fasting, such as skipping breakfast, reduces the late-morning problem-solving performance of children.[41] About one in seven young children in the United States doesn't eat breakfast regularly.[42] No one has yet studied the effects of breakfast skipping on college students' midterm exam scores . . .

Early Exposure to Alcohol Affects Mental Performance] Mental development can be permanently delayed by exposure to alcohol during fetal growth. Although growth is also retarded, the most serious effects of fetal exposure to alcohol are permanent delays in mental development and behavioral problems associated with them. Women are advised not to drink if they are pregnant or may become pregnant.[43]

Iron Deficiency Impairs Learning] *On a global basis, an estimated 20% of men, 35% of women, and 40% of all children are anemic, primarily due to iron deficiency.[44]*

Most cases of iron-deficiency anemia in children result from inadequate intake of dietary iron. Iron-deficiency anemia in children is a widespread problem in developed and developing countries and likely is the most common single nutrient deficiency.[45] The potential impact of iron-deficiency anemia on the functional capacity of humans represents staggering possibilities.

Until recently, it was thought that the effects of iron-deficiency anemia on intellectual performance were short term and could be corrected by treating the anemia. It now appears that some of the effects may be lasting. Five-year-old children in Costa Rica treated for iron-deficiency anemia during infancy scored lower on hand-eye coordination and other motor skill tests than similar children without a history of anemia. Studies in the United States have detected shortened attention span and reduced problem-solving ability in iron-deficient children.[46]

Overexposure to Lead]   Our concern about exposure to lead has recently increased, due in large measure to information indicating that low exposure to lead has long-term behavioral effects.[47] There are many opportunities for overexposure to lead. Approximately 84% of U.S. houses built before 1980 contain some lead-based paint.[48] Children living in or near these houses may eat the paint flakes (they taste sweet), or the old paint may contaminate the soil near the houses (Illustration 5.9). Lead also ends up in soil from industrial and agricultural chemicals, in water from lead-based pipes and solder, and in the air from the days when leaded gas was used.

Although the use of lead in cans, pipes, and gasoline has decreased dramatically (Illustration 5.10), lead remains in the environment for long periods. It also stays in the body, stored principally in the bones, for a long time—20 years or more. The effects of excessive exposure to lead include increased absenteeism from school, longer reaction times, impaired reading skills, and higher dropout rates. Blood lead levels in children have dropped substantially in recent decades. Despite this drop, however, half a million young children in the United States still have elevated blood lead levels.[49]

Carbohydrates and Behavior

Carbohydrates are credited with improving mood, decreasing food intake, and inducing sleepiness. Simple sugars are blamed for causing hyperactivity in children and criminal behavior in adults. Where do the truths related to carbohydrate intake and behavior lie?

Carbohydrate Intake, Mood, and Appetite]   *Serotonin* is a chemical messenger produced from *tryptophan,* an essential amino acid. High blood levels of tryptophan increase the production of serotonin in the brain. Serotonin produced in the

Illustration 5.9
There are many opportunities for overexposure to lead. Young children are especially vulnerable.

PhotoEdit, Felicia Martinez

serotonin (pronounced sare-uh-tone-in)
A neurotransmitter, or chemical messenger, for nerve cell activities that excite or inhibit various behaviors and body functions. It plays a role in mood, appetite regulation, food intake, respiration, pain transmission, blood vessel constriction, and other body processes.

tryptophan (pronounced trip-tuh-fan)
An essential amino acid that is used to form the chemical messenger serotonin (among other functions). Tryptophan is generally present in lower amounts in food protein than most other essential amino acids. It can be produced in the body from niacin, a B vitamin.

Illustration 5.10
Trends in children's lead exposure. A phasedown of leaded gasoline from 1976 to 1985 and near elimination of leaded food cans have contributed to a reduction in children's lead levels.
Source: www.cdc.gov.2003.

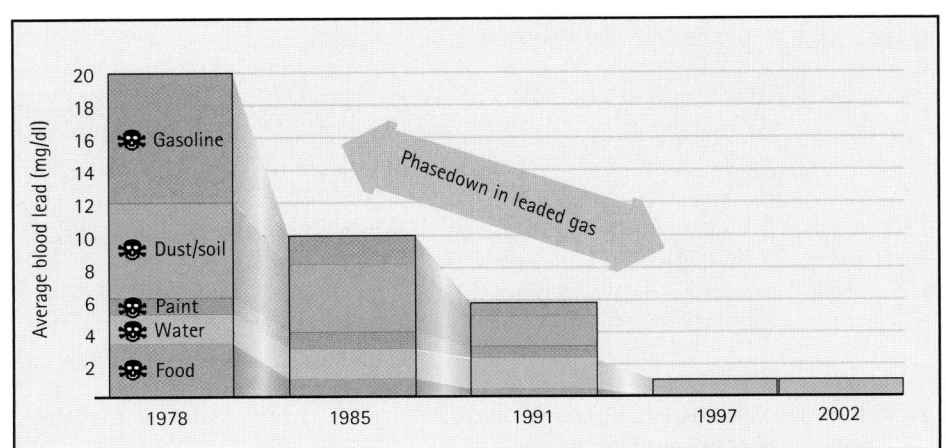

Phasedown in leaded gas

brain plays important roles in mood and the regulation of appetite. Due to its influence on appetite regulation, serotonin is a target of action for a number of weight-loss drugs and popular weight-loss diets. Consumed by itself, carbohydrate increases blood levels of tryptophan and serotonin production in the brain.[50]

For years it was thought that high-carbohydrate meals may decrease appetite and food intake by increasing levels of serotonin in the brain. Recent research indicates, however, that increases in serotonin related to carbohydrate intake are small, and that any increase is quickly blocked by the simultaneous intake of small amounts of protein. Although eating food can erase a bad mood in hungry people, it does not appear that high-carbohydrate foods decrease appetite by increasing brain levels of serotonin.[51] High-carbohydrate diets are also purported to decrease fat intake, but this notion is controversial.[52] High-carbohydrate meals may induce drowsiness or calmness.[53]

Sugar, Hyperactivity, and Criminal Behavior] Sugar has been wrongly blamed for all sorts of behavioral problems. A number of studies have investigated the relationship of sugar intake to hyperactivity[54] and criminal behavior,[55] and none has shown an effect. In fact, studies generally indicate that sugars and other carbohydrates *decrease* activity level in children and adults.[56]

The myth that sugar causes hyperactivity in children lingers on in the minds of many parents, however. The excitement that often accompanies high-sugar eating occasions such as Halloween and birthday parties, or the expectation that children will be more active after they eat sugary foods, may sustain the myth. It has also been claimed that food additives, especially artificial food colors, cause hyperactivity in children. Well-controlled studies do not support this claim, either.[57]

The Future of Diet and Behavior Research

Identifying the effects of nutrition on behavior is a tricky business. Many factors in addition to diet influence behavior, making it difficult to separate dietary from social, economic, educational, and genetic influences. We still have much to learn, and many assumptions about diet and behavior must await confirmation through research.

Key Terms

serotonin, page 5-11

tryptophan, page 5-11

www links

www.nal.usda.gov/fnic
Provides ethnic and cultural resources related to food and nutrition and to nutrition and learning/behavior information.

www.nimh.nih.gov
The National Institute of Mental Health provides information on attention deficit hyperactivity disorder.

healthfinder.gov
Use this site to find reliable information about iron deficiency, fetal alcohol syndrome, and other disorders presented in this unit.

www.ncemch.org
Connects to the National Center for Education in Maternal and Child Health;

provides information on lead poisoning and searchable databases.

www.ipm.iastate.edu/misc/insectsasfood.html
Go to this site for insect recipes from Iowa State University.

Nutrition UP CLOSE

Improving Food Choices

FOCAL POINT: Developing a plan for healthier eating.

Identify a change in your diet that you would like to make. Then develop a plan for making the change by thinking through and responding to each element of the dietary change process listed. (Refer to Table 5.2 for examples of responses.)

DIETARY CHANGE PROCESS	YOUR RESPONSE
1. Identify a healthful change in your diet you'd like to make.	1. _____
2. Identify food choices you'd like to change to make your diet more healthful.	2. _____
3. Identify specific, acceptable options for the healthful food choices.	3. _____
4. Decide which option is easiest to accomplish and requires the smallest change to get the job done.	4. _____
5. Plan how to incorporate the change into your diet.	5. _____

FEEDBACK (answers to these questions) can be found at the end of Unit 5.

Notes

1. van den Bree MBM et al. Genetic and environmental influences on eating patterns of twins aged ≥50 y. Am J Clin Nutr 1999;70:456–65.

2. Murphy JM et al. The relationship of school breakfast to psychosocial and academic functioning. Arch Pediatr Adol Med 1998;152:899–907.

3. Pyke M. Man and food. New York: McGraw-Hill; 1972.

4. Beauchamp GK, Mennella JA. Sensitive periods in the development of human flavor perception and preference. Ann Nestle 1998;56:19–31.

5. Pyke, Man and food.

6. Rolls B. Obesity and Weight Control Symposium, Experimental Biology Annual Meeting, San Diego, CA, April 14, 2003.

7. van den Bree, Eating patterns of twins.

8. Birch LL, Johnson SI, Jones MB, Peters JC, Effects of a nonenergy fat substitute on children's energy and macronutrient intake (Am J Clin Nutr 1993;58: 326–33); and V Packard, The status seekers (New York: Random House, 1959), 146.

9. Story M, Brown JE. Do young children instinctively know what to eat? The studies of Clara Davis revisited. N Engl J Med 1987;316:103–6.

10. Birch LL, Johnson SI, Jones MB, Peters JC. Effects of a nonenergy fat substitute on children's energy and macronutrient intake. Am J Clin Nutr 1993;58: 326–33.

11. Steiner JE. The gustofacial response: observation on normal and anencephalic newborn infants. In: Bosma JF, ed. Fourth symposium on oral sensation and perception: development in the fetus and infant. Bethesda (MD): National Institute of Dental Research, DHEW Publication No. 73-546, US Dept HEW, NIH; 1973:254.

12. Steiner, The gustofacial response.

13. Factors influencing food choices in humans. Nutr Rev 1990;48:442.

14. Factors influencing food choices in humans.

15. Packard, The status seekers.

16. Yankelovich Survey results reported in USA Today, 2000 Feb. 7:D1.

17. Parraga IM. Determinants of food consumption. J Am Diet Assoc 1990; 90:661–3.

18. Zallen DT, Brown ML. Food fights: deciding about diet and disease. Am J Clin Nutr 1990;52:944–5.

19. Ullrich N, Tepper BJ. Food preferences are influenced by gender and genetic taste sensitivity to 6-n-propylthiouracil (PROP).

20. Pyke, Man and food.

21. Beerman KA. Variation in nutrient intake of college students: a comparison by students' residence. J Am Diet Assoc 1991;91:343–4.

22. Cheadle A, Psaty BM, Curry S, et al. Community-level comparisons between the grocery store environment and individual dietary practices. Prev Med 1991; 20:250–6.

23. Economic Research Service. Food consumption (per capita) system. www.ers. usda.gov/data/foodconsumption, accessed 8/03.

24. Economic Research Service (www.ers. usda.gov/data/foodconsumption).

25. Brevard PB. Nutrition education improves the diet of college students. J Am Diet Assoc 1991, abs. A-47.

26. Brevard, Nutrition education.

27. Davis JN et al. Is nutrition knowledge reflected in healthier dietary practices of college students? Experimental Biology 2003: Meeting Abstract 709.8, San Diego, April 2003.

28. Ankeny K, Oakland MJ, Terry RD. Dietary fat: sources of information used by men. J Am Diet Assoc 1991;91:1116–7.

29. Shepherd R, Stockley L. Nutrition knowledge, attitudes, and fat consumption. J Am Diet Assoc 1987;87:615–9.

30. Becker MH, Maiman LA, Kirscht JP, Haefner DP, Drachman RH. The health belief model and prediction of dietary compliance: a field experiment. J Health Soc Behav 1977;18:348–66.

31. Holdt CS, Gates GE, Lassa S. Knowledge and behaviors of allied health professionals regarding meat (abs.). J Am Diet Assoc 1991;91S, abs. A-13.

32. Tillotson JE. Food brands: friend or foe? Nutr Today 2—2;37:78–80.

33. Holdt et al., Knowledge and behaviors of allied health professionals; Cullen KW et al., Using goal setting as a strategy for dietary behavioral change, J Am Diet Assoc 2110;101:562–66; Mitchell SJ, Changes after taking a college basic nutrition course, J Am Diet Assoc 1990;90:955–61; and van Beurden E, James R, Christian J, Church D, Dietary self-efficacy in a community-based intervention: implications for effective dietary counselling, Austr J Nutr Diet 1991;48:64–67.

34. Green GW, Rossi SR. Stages of change for reducing fat intake over 18 months. J Am Diet Assoc 1998;98:529–34.

35. Young SN. Nutrition 3. The fuzzy boundary between nutrition and psy-

chopharmacology. Can Med Assoc J, Jan. 22, 2002, vol. 155.

36. Young SN, Nutrition 3: The fuzzy boundary between nutrition and psychopharmacology (Can Med Assoc J, vol. 155, Jan. 22, 2002); Wolraich ML et al., The effect of sugar on behavior or cognition in children: a meta-analysis (JAMA 1995;274:1617); I deAndraca et al., Psychomotor development and behavior in iron-deficient anemic infants (Nutr Rev 1997;55:125); Child Health USA, 1997 (Washington, DC: US Maternal and Child Health Bureau, PHS), 29; and Fernstrom JD, Can nutrient supplements modify brain function? (Am J Clin Nutr 2000;71 (suppl.):1669S–73S).

37. Ferguson L. Eating for energy and ecstasy. Toronto, Ontario: Malibu Consulting; 2000.

38. Bellamy C. The state of the world's children. Oxford (UK): Oxford University Press; 1996.

39. Allen LH. Functional indicators of nutritional status of the whole individual or the community. Clin Nutr 1984; 3:169–75.

40. Allen, Functional indicators of nutritional status.

41. Murphy et al., School breakfast; Cueto S, Breakfast and performance, Pub Health Nutr 2001;4:1429–31.

42. Nicklas TA, Boa W, Webber LS, Berenson GS. Breakfast consumption affects adequacy of total daily intake in children. J Am Diet Assoc 1993;93:886–91.

43. Tittman HG. What's the harm in just a drink? Alcohol and Alcoholism 1990;25:287–91; and Dietary Guidelines for Americans, US Dept of HHS and USDA, 2000.

44. Lazoff B. Nutrition and behavior. Am Psychologist 1989;44:231–6.

45. Lazoff, Nutrition and behavior.

46. Lazoff B, Jimenez E, Wolf AW, Long-term developmental outcome of infants with iron deficiency, N Engl J Med 1991;325:687–94; and Hurade EK et al., Early childhood anemia and mild or moderate mental retardation, Am J Clin Nutr 1999;69:115–9.

47. Canfield L et al, Intellectual impairment in children with blood lead concentrations below 10 μg per deciliter. N Engl J Med 2003;34:1517–26.

48. Child Health USA, 1997. US Maternal and Child Health Bureau, PHS, Washington, DC, p. 29.

49. Rogan WJ, Ware JH. Exposure to lead—how low is low enough? N Engl J Med 2003;348:1515–16.

50. Gans DA, Harper AE, Bachorowski J-A, Newman JP, Shrago ES, Taylor SL. Sucrose and delinquency: oral sucrose tolerance test and nutritional assessment. Pediatrics 1990;86:254–62.

51. Young, Fuzzy boundary between nutrition and psychopharmacology.

52. Young, Fuzzy boundary; Fernstrom JD, Can nutrient supplements modify brain function? Am J. Clin Nutr 2000;71 (suppl):1669S–73S.

53. Young, Fuzzy boundary.

54. Young, Fuzzy boundary.

55. Gans et al., Sucrose and delinquency.

56. Young, Fuzzy boundary.

57. Ferguson HB, Stoddart C, Simeon JG. Double-blind challenge studies of behavioral and cognitive effects of sucrose-aspartame ingestion in normal children. Nutr Rev 1986;44:144–50; Regalado M. Busting the sugar-hyperactivity myth. WebMD, www.webMD.com, 1999, accessed 1/01.

Nutrition **UP CLOSE**

Improving Food Choices

Feedback for Unit 5 There are no right or wrong answers to this unit's "Nutrition Up Close." Plans for dietary change that are specific, easy to accomplish, and highly acceptable are most likely to work in the long run.

What's a Healthful Diet?

This Is Just to Say
I have eaten
the plums
that were in
the icebox
and which
you were probably
saving
for breakfast
Forgive me
they were delicious
so sweet
and so cold

—William Carlos Williams

Nutrition Scoreboard

	TRUE	FALSE
1 Three out of four U.S. adults consume the recommended five or more servings of vegetables and fruits a day.		
2 The basic food groups include a "healthy snack" food group.		
3 "Supersized" fast-food meals often provide two to three times more calories than regular-sized fast food meals.		

Answers on next page

[KEY CONCEPTS AND FACTS]

- Healthy diets are characterized by adequacy and balance.

- There are many types of healthy diets.

- There are no good or bad foods, only diets that are healthy or unhealthy.

- The Dietary Guidelines for Americans and the Food Guide Pyramid provide foundation information for healthy diets.

Answers to *Nutrition Scoreboard*

		TRUE	FALSE
1	Only one in two U.S. adults consumes "five a day."[1]		✔
2	The basic food groups do not include a "healthy snack" food group.		✔
3	Supersizing fast-food meals piles on calories.	✔	

Healthy Eating: Achieving the Balance between Good Taste and Good for You

May I have your attention please? For a moment, think about the foods in Illustration 6.1. If your mouth is watering and you're ready to go out and buy some ripe peaches, you have found the balance between good taste and good for you. Who said foods that taste good aren't good for you?

Illustration 6.1
Can you smell it? Can you taste it?
A plump, golden peach. It's so ripe that juice spurts from it and drips down your chin when you take a bite.... A golden brown turkey just taken out of the oven. The wonderful smell fills the kitchen. A steaming loaf of homemade bread just set out to cool. A perfect ripe tomato just picked from the garden. It melts in your mouth.

California Tree Fruit Agreement

Photo Disc

Photo Disc

Photo Disc

Characteristics of Healthy Diets

Healthy diets come in a variety of forms. They may be based on bread, olives, nuts, fruits, beans, vegetables, lamb, and chicken (as in Greece); rice, vegetables, and small amounts of fish and other meats (as in China); or black beans, rice, meat, and tropical fruits (as in Cuba and Costa Rica). Illustration 6.2 gives examples of the diverse foods that can be part of a healthy diet.

Although the types of foods that go into them can vary substantially, healthy diets all share two characteristics: they are adequate and balanced. *Adequate diets* include a wide variety of foods that together provide sufficient levels of calories and *essential nutrients.* What's sufficient? For calories, it's the number that maintains a healthy body weight. For essential nutrients, sufficiency corresponds to intakes that are in line with recommended intake levels represented by the *Recommended Dietary Allowances (RDAs)* and *Adequate Intakes (AIs).* (These are shown in the

Illustration 6.2
Foods that contribute to healthy diets in different countries.
a. Dinner in Italy might be linguine primavera (pasta with vegetables). b. Pad Thai, rice noodles and vegetables, is a favorite dish in Thailand. c. Tamales are a celebration food in many Latin cultures. d. Dal, curry dishes, vegetables, and chicken are popular parts of the cuisine of India.

a.

b.

c.

d.

Terms, from page 6-3
adequate diet
A diet consisting of foods that together supply sufficient protein, vitamins, and minerals and enough calories to meet a person's need for energy.

essential nutrients
Substances the body requires for normal growth and health but cannot manufacture in sufficient amounts; they must be obtained in the diet.

Recommended Dietary Allowances (RDAs)
Intake levels of essential nutrients that meet the nutritional needs of practically all healthy people while decreasing the risk of certain chronic diseases.

Adequate Intakes (AIs)
Provisional RDAs developed when there is insufficient evidence to support a specific level of intake.

Illustration 6.3
Acceptable Macronutrient Distribution Ranges (AMDRs) for percent of total calories in the diet that should consist of carbohydrate, protein, and fat.[3]

Dietary Reference Intake tables inside the front cover of this book.) Recommended amounts of essential nutrients should be obtained from foods to reap the benefits offered by the variety of naturally occurring substances in foods that promote health. Following the pattern of food intake recommended in the Food Guide Pyramid helps ensure that the need for essential nutrients as well as the need for other beneficial components of food are met.

A *balanced diet* provides calories, nutrients, and other components of food in the right proportion—neither too much nor too little. Diets that contain too much sodium or too little fiber, for example, are out of balance. Diets that provide more calories than needed to maintain a healthy body weight are also out of balance.

Current dietary intake standards include guidelines to help individuals balance their intake of carbohydrate, protein, and fat—the *macronutrients.* Called the "Acceptable Macronutrient Distribution Ranges," or AMDRs, the guidelines indicate percentages of total caloric intake that should consist of carbohydrate, protein, and fat. These recommendations apply to individuals over the age of 4 years. Ranges are recommended because the prevention of heart disease, diabetes, and obesity as well as the assurance of dietary adequacy are related to various levels of intakes rather than to one specific intake level of these macronutrients. In the previous edition of the Dietary Reference Intakes (DRIs), recommendations were given for the percentage of total calories that should consist of *unsaturated* and *saturated fats.* These recommendations have been removed from the new standards. It is now recommended that diets be kept low in saturated fat and provide sufficient amounts of linoleic acid and alpha-linoleic acid. Both of these fatty acids are unsaturated and considered essential nutrients.[2]

How Balanced Is the American Diet?

Results of assessments of what Americans actually consume compared to recommendations for healthy diets given in the DRIs and the Food Guide Pyramid are shown in Illustrations 6.3 and 6.4. Compared to earlier studies, results reported in these illustrations show that the American diet is improving in some respects. Over the past few years, intake of vegetables and fruits has increased, and we are eating more dairy and grain products.

Over-consumption of Added Sugar and Saturated Fat] On average, the U.S. diet remains overloaded with added sugars and low in essential fatty acids, dairy products, and vegetables and fruits. Added sugar intake from soft drinks, fruit

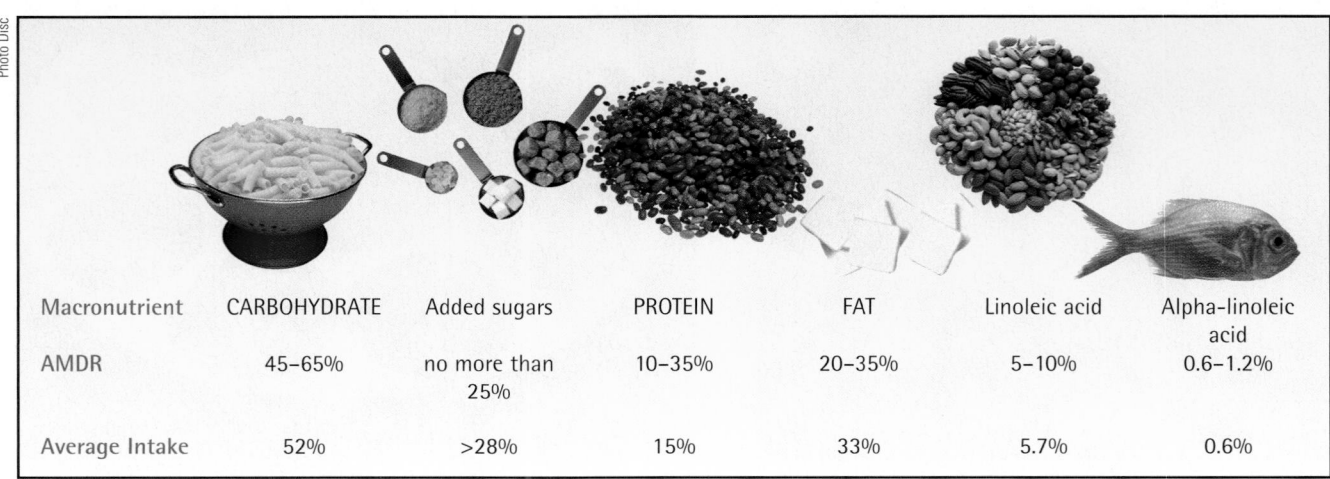

Macronutrient	CARBOHYDRATE	Added sugars	PROTEIN	FAT	Linoleic acid	Alpha-linoleic acid
AMDR	45–65%	no more than 25%	10–35%	20–35%	5–10%	0.6–1.2%
Average Intake	52%	>28%	15%	33%	5.7%	0.6%

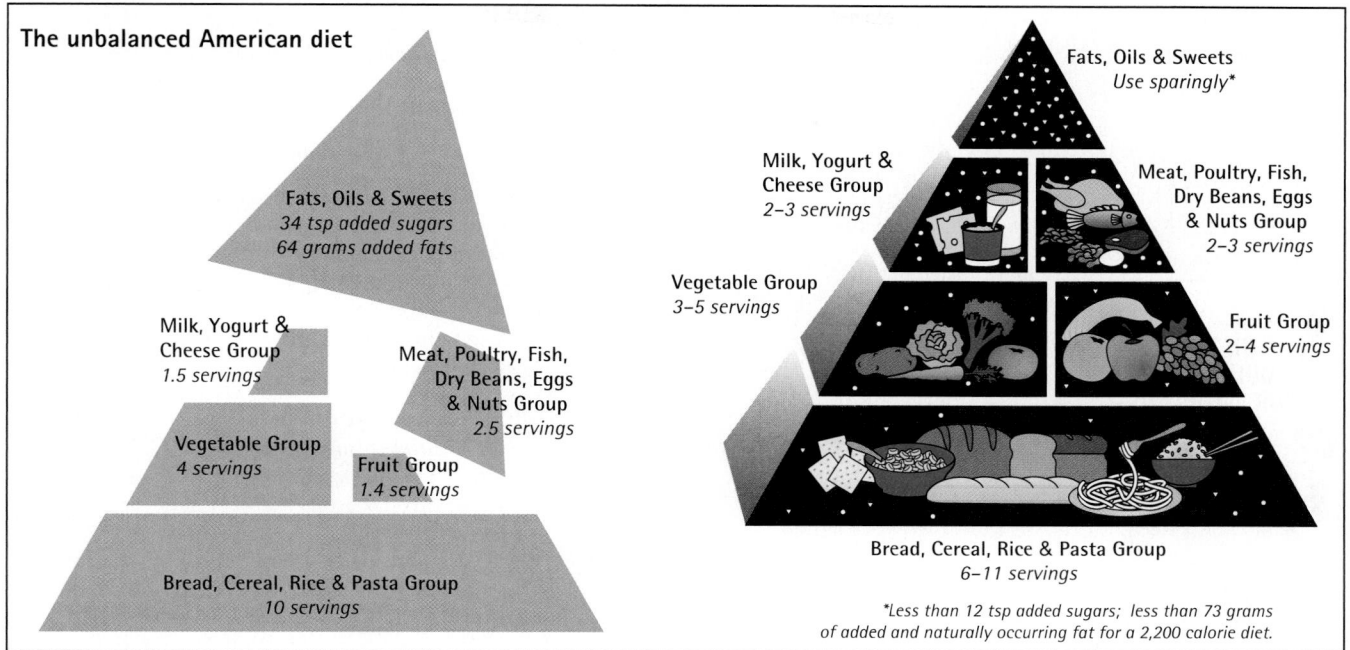

The unbalanced American diet

Fats, Oils & Sweets
34 tsp added sugars
64 grams added fats

Milk, Yogurt &
Cheese Group
1.5 servings

Meat, Poultry, Fish,
Dry Beans, Eggs
& Nuts Group
2.5 servings

Vegetable Group
4 servings

Fruit Group
1.4 servings

Bread, Cereal, Rice & Pasta Group
10 servings

Fats, Oils & Sweets
*Use sparingly**

Milk, Yogurt &
Cheese Group
2–3 servings

Meat, Poultry, Fish,
Dry Beans, Eggs
& Nuts Group
2–3 servings

Vegetable Group
3–5 servings

Fruit Group
2–4 servings

Bread, Cereal, Rice & Pasta Group
6–11 servings

*Less than 12 tsp added sugars; less than 73 grams
of added and naturally occurring fat for a 2,200 calorie diet.*

drinks, candy, pastries, and other sweetened foods is increasing, bringing with it an increased risk of tooth decay and reduced overall diet quality. Continuing high intakes of saturated fat place many Americans at risk for heart disease, and low intake of dairy products increases the risk of osteoporosis.[5]

Under-consumption of Vegetables and Fruits] Nearly one in two adults fail to consume at least three servings of vegetables daily; and three out of four adults consume less than the minimal, recommended number of servings of fruits.[6] Only a small proportion of Americans consumes vegetables and fruits at the upper end of the ranges for daily servings in the Food Guide Pyramid. Risk of heart disease, stroke, and cancer decrease as vegetable and fruit intakes increase to 5–9 servings daily.[7] Approximately 6% of adults regularly consume the types of vegetables most closely associated with reduced risk of cancer: broccoli; cauliflower; brussels sprouts; and dark, leafy, green vegetables. The most commonly consumed vegetable in the United States is french fries.[8]

More Whole Grains Are Needed] Within the Food Guide Pyramid, it is recommended that at least 3 of the 6–11 servings of grain products consist of whole grain products. Ample consumption of whole grains is related to reducing the risk of certain types of cancer and heart disease. Yet, Americans consume an average of one serving of whole grain products daily.[9] How do we strengthen the weak points of the American diet while maintaining its healthiest features? The answer is at hand.

Guides to Healthy Diets

Healthy diets for people in the United States are described in the "Dietary Guidelines for Americans" and the Food Guide Pyramid. The Dietary Guidelines primarily address the issue of dietary balance, and the Food Guide Pyramid focuses upon food choices for adequate diets. Both are updated periodically as knowledge about diet and health expands.[11]

Illustration 6.4
Food Guide Pyramid recommended intake of standard servings of foods from the various groups compared with average actual intakes in the United States.[4]
Source: USDA, at www.ers.usda.gov/ data/foodcomposition (accessed 8/03).
USDA

Terms, from page 6-4:
balanced diet
A diet that provides neither too much nor too little of nutrients and other components of food such as fat and fiber.

saturated fats
The type of fat that tends to raise blood cholesterol levels and the risk for heart disease. They are solid at room temperature and are found primarily in animal products such as meat, butter, and cheese.

continued next page

HEALTH ACTION | *The 2000 Dietary Guidelines for Americans Explained*

AIM, BUILD, CHOOSE—FOR GOOD HEALTH

Dietary Guideline	Explanation
AIM for fitness • Aim for a healthy weight. • Be physically active each day.	Following these two guidelines will help keep you and your family healthy and fit. Healthy eating and regular physical activity enable people of all ages to work productively, enjoy life, and feel their best. They also help children grow, develop, and do well in school.
BUILD a healthy base • Let the Pyramid guide your food choices. • Eat a variety of grains daily, especially whole grains. • Eat a variety of fruits and vegetables daily. • Keep food safe to eat.	Following these four guidelines builds a base for healthy eating. Let the Food Guide Pyramid guide you so that you get the nutrients your body needs each day. Make grains, fruits, and vegetables the foundation of your meals. This forms a base for good nutrition and good health and may reduce your risk of certain chronic diseases. Be flexible and adventurous—try new choices from these three groups in place of some less nutritious or higher-calorie foods you usually eat. Whatever you eat, always take steps to keep your food safe to eat.
CHOOSE sensibly • Choose a diet that is low in saturated fat and cholesterol and moderate in total fat. • Choose beverages and foods that limit your intake of sugars. • Choose and prepare foods with less salt. • If you drink alcoholic beverages, do so in moderation.	These four guidelines help you make sensible choices that promote health and reduce risk of certain chronic diseases. You can enjoy all foods as part of a healthy diet as long as you don't overdo on fat (especially saturated fat), sugars, salt, and alcohol. Read labels to identify foods that are high in saturated fats, sugars, and salt (sodium).

macronutrients
The group name for the energy-yielding nutrients of carbohydrate, protein, and fat. They are called macronutrients because we need relatively large amounts of them in our daily diet.

unsaturated fats
The type of fat that tends to lower blood cholesterol level and the risk of heart disease. They are liquid at room temperature and found in foods such as nuts, seeds, fish, shellfish, and vegetable oils.

The Dietary Guidelines for Americans

The Dietary Guidelines were introduced in 1980 and are revised every 5 years. They address broad diet and health concerns and present advice on selecting balanced and, to a lesser degree, adequate diets. The 2000 edition of the Dietary Guidelines departs from tradition by categorizing the 10 guidelines into three groups: fitness and body weight, basic diet, and food choices. It's the first time the Dietary Guidelines have given a major focus to physical fitness and weight and to food safety. The rationale for including fitness and weight and food safety is based on evidence that high rates of overweight and obesity in the United States and foodborne illnesses are major public health problems.[12] Because the recommendations made in the Dietary Guidelines are broad, they are easier to put into practice if explained further. The "Health Action" briefly explains each guideline.

Compared to the current American diet, that recommended in the Dietary Guidelines is lower in saturated fat, cholesterol, salt, sugars, and alcohol (if warranted) and higher in vegetables, fruits, and whole grain products. Foods recommended for inclusion in the diet reflect a "back to basics" theme: they include foods without added fat, salt, or sugars; fresh vegetables and fruits and whole grain products; low-fat dairy products; and fish, poultry, unprocessed meats, and dried beans.[13] Such diets tend to be low in saturated fat and cholesterol. Excessive intakes of these two ingredients are linked to heart disease.

The new DRIs set the bar for intakes of fat. Maximum fat intake levels recommended by the DRI are somewhat higher than the level recommended in the 2000 Dietary Guidelines. The DRI recommendations focus on the types of fat consumed and prevention of excessive caloric intake related to high fat diets. The DRIs recommend minimal intake of foods high in saturated fats, **trans fats,** and **cholesterol.** Saturated fat intake can be reduced by selecting lean meats, fish, low-fat milk, and vegetable oils over high-fat animal products such as fatty meats, whole milk, and butter. Trans fat intake can be lowered by limiting intake of foods made or cooked with hydrogenated oils (such as margarine, pastry, crackers, and french fries). New nutrition labeling regulations requiring food companies to list the trans fat content of foods in Nutrition Facts Panels give consumers an additional way to identify foods high in trans fat. The DRI report also urges Americans to limit intake of organ meats, eggs, and other foods high in cholesterol.[14]

Other Dietary Guidelines around the World

Many countries have developed guides for healthy eating. Table 6.1 presents examples of guidelines from Canada and Japan. The specific recommendations in each set of dietary guidelines address priority health problems that can be improved by dietary changes. Although the guidelines differ a bit, certain themes are evident in all of them: "eat a wide variety of foods," "maintain a healthy weight," and "consume foods low in total fat, saturated fat, and salt."

Mexican is the most popular ethnic cuisine in the United States, followed by Italian and Chinese.[36]

Photo Disc

trans fats
A type of unsaturated fat present in hydrogenated oils, margarine, shortening, pastries, and some cooking oils that increase the risk of heart disease.

cholesterol
A fat-soluble, colorless liquid found in many animal products but not in plants. High cholesterol intake is related to the development of heart disease.

TABLE 6.1

GOOD ADVICE FROM THE DIETARY GUIDELINES OF CANADA AND JAPAN.

CANADIAN DIETARY GUIDELINES

A. Summary of Nutrition Recommendations

The Canadian diet should:

- provide energy consistent with the maintenance of body weight within the recommended range.
- include essential nutrients in amounts specified in the Recommended Nutrient Intakes.
- include no more than 30% of energy as fat (33 g/1000 kcal or 39 g/5000 kJ) and no more than 10% as saturated fat (11 g/1000 kcal or 13 g/5000 kJ).
- provide 55% of energy as carbohydrate (138 g/1000 kcal or 165 g/5000 kJ) from a variety of sources.
- be reduced in sodium content.
- include no more than 5% of total energy as alcohol, or 2 drinks daily, whichever is less.
- contain no more caffeine than the equivalent of 4 cups of regular coffee per day.
- Community water supplies containing less than 1 mg/litre should be fluoridated to that level.

B. Guidelines for Healthy Eating

- Enjoy a VARIETY of foods.
- Emphasize cereals, breads, other grain products, vegetables, and fruits.
- Choose low-fat dairy products, lean meats, and foods prepared with little or no fat.
- Achieve and maintain a healthy body weight by enjoying regular physical activity and healthy eating.
- Limit salt, alcohol, and caffeine.

Table 6.1 continues on next page

TABLE 6.1

GOOD ADVICE FROM THE DIETARY GUIDELINES OF CANADA AND JAPAN, continued.

JAPAN'S DIETARY GUIDELINES FOR HEALTH PROMOTION

1. Enjoy your meals
   - Have delicious and healthy meals that are good for your mind and body.
   - Enjoy communication at the table with your family or other people and participate in the preparation of meals.
2. Establish a healthy rhythm by keeping regular hours for meals.
3. Eat well-balanced meals with staple food, as well as main and side dishes.
4. Eat enough grains such as rice and other cereals.
5. Combine vegetables, fruits, milk products, beans and fish in your diet.
6. Avoid too much salt and fat.

7. Learn your healthy body weight and balance the calories you eat with physical activity.
8. Chew your food well and do not eat too quickly.
9. Enjoy nature's bounty and the changing seasons by using local food products and ingredients in seasons, and by enjoying holiday and special-occasion dishes.
10. Reduce leftovers and waste through proper cooking and storage methods.
11. Assess your daily eating habits.
    - Learn and practice healthy eating habits at school and at home.
    - Promote appreciation of good eating habits from an early stage of life.

The Food Guide Pyramid: The Food Group Approach to an Adequate Diet

Food group guides for adequate diets have existed in the United States since 1916. In 1992 a substantially modified basic food guide, the Food Guide Pyramid, was released (Illustration 6.5). The Food Guide Pyramid consists of five basic food groups and recommended ranges of daily servings for each group. It represents an outline of types of food to eat each day rather than a rigid prescription of what to eat. This food guide

Illustration 6.5
The Food Guide Pyramid: A guide to daily food choices.

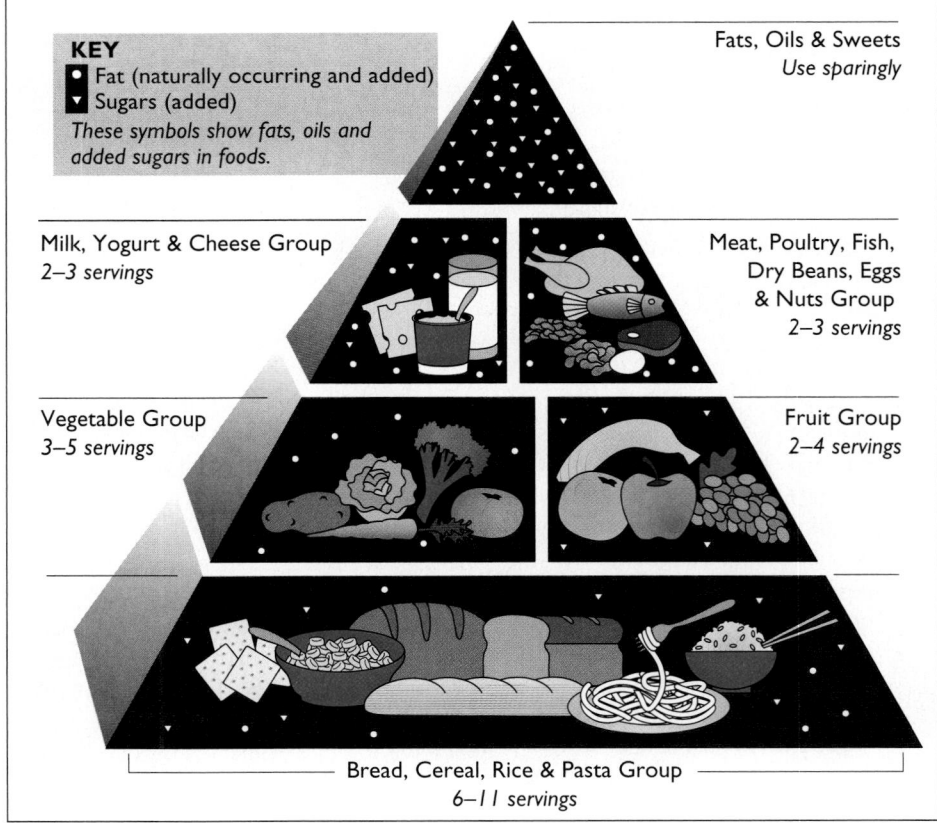

KEY
- ● Fat (naturally occurring and added)
- ▼ Sugars (added)

These symbols show fats, oils and added sugars in foods.

Fats, Oils & Sweets
Use sparingly

Milk, Yogurt & Cheese Group
2–3 servings

Meat, Poultry, Fish, Dry Beans, Eggs & Nuts Group
2–3 servings

Vegetable Group
3–5 servings

Fruit Group
2–4 servings

Bread, Cereal, Rice & Pasta Group
6–11 servings

Mohammad:
I feel like I just ate about 6 servings from the grain group.

REALITY CHECK

Portion Distortion

Mohammad and Isaac decided to eat at an Italian restaurant after soccer practice. Just for the fun of it, they agreed to guess how many servings from the grain products group they would consume if they ate the portion of spaghetti that was served to them.

Isaac:
I bet they gave us 1 serving worth of grain products.

Answers on page 6-11

encourages decision making about which foods within each group fit best into an individual's diet.

Food groups are located in the pyramid based on their recommended number of servings and not on their relative importance to diet or health. Since each food group consists of foods that make particular contributions to nutrient intake, foods from one group cannot be substituted for those from another group. Healthy diets consist of foods from all of the basic groups. At the top of the pyramid is not a food group, but a message about fats and sweets. These foods are assigned the smallest area in the pyramid, sending the message that added fat, oils, and sweets should be a small part of our diet.

In general, the more types of food an individual eats in a day, the wider the array of nutrients and other beneficial components of food, such as fiber and antioxidants, are provided. People in the United States tend to consume 18 to 20 different foods a day. Dietary guidelines developed for Japan recommend that people eat 30 different foods daily.

Additional information provided with the Food Guide Pyramid (Table 6.2) defines serving sizes and differences in servings per day needed by women, men, children, and active people. This information is important to know because serving sizes and numbers of servings recommended are major sources of confusion about the Food Guide Pyramid.

As portion sizes have grown, people don't really know what a cup of spaghetti looks like on a plate. We have no conception of what portions are anymore.
—Ellen Schuster, Ph.D.[15]

Portion Distortion] Standard serving sizes developed by the USDA are based on portions of food consumed by people studied in national surveys.[16] The portions tend to be smaller than people think, and much smaller than marketplace portion sizes. Table 6.3 compares marketplace portion sizes to standard serving sizes used in the Food Guide Pyramid. If you find yourself thinking that, for example, 3 to 5 servings of vegetables—or 6 to 11 servings of grain products—is a lot, stop to think about serving size. A half-cup of cooked vegetables or rice is just a bit bigger than a regular cell phone. Read through the "Reality Check" to see if one idea you may have about serving size comes close to the truth.

Is Supersizing Leading to Supersized Americans?] Supersizing fast foods can double or triple the caloric content of the foods compared to their regular-sized

90% of all french fries consumed in the United States are from fast-food restaurants.[10]

TABLE 6.2

SERVING SIZES AND NUMBERS OF SERVINGS RECOMMENDED IN THE FOOD GUIDE PYRAMID.

What counts as one serving?

BREAD, CEREAL, RICE, AND PASTA GROUP

1 slice of bread
½ cup of cooked rice or pasta
½ cup of cooked cereal
1 ounce of ready-to-eat cereal
½ cup grits
¼ cup granola
1 tortilla
1 chapati
½ bagel
½ English muffin
1 small muffin (1½ oz.)
1 small pancake (4")
1 small roll (2½" × 2")

VEGETABLE GROUP

½ cup of chopped raw or cooked vegetables
1 cup of leafy raw vegetables
¾ cup vegetable juice

FRUIT GROUP

1 piece of fruit
1 slice cantaloupe (⅛ whole)
¾ cup of juice
½ cup of canned fruit
¼ cup of dried fruit
1 cup fresh berries
½ cup applesauce

MILK, YOGURT, AND CHEESE GROUP

1 cup of milk, soymilk, or yogurt
1½ ounces of natural cheese
2 ounces of process cheese
½ cup cottage cheese

MEAT, POULTRY, FISH, DRY BEANS, EGGS, AND NUTS GROUP

2½ to 3 ounces of cooked lean meat, poultry, or fish. Count ½ cup of cooked beans, or 1 egg, or 2 tablespoons of peanut butter, 2 oz. vegeburger, ½ cup tofu, ¼ cup nuts or seeds, as 1 ounce of lean meat

FATS AND SWEETS

LIMIT CALORIES FROM THESE especially if you need to lose weight.

FATS AND ADDED SUGARS

The small tip of the pyramid shows fats and sweets. These are foods such as salad dressings, cream, butter, margarine, sugars, soft drinks, candies, and sweet desserts. Alcoholic beverages are also part of this group. These foods provide calories but few vitamins and minerals. Most people should go easy on foods from this group.

Some fat or sugar symbols are shown in the other food groups. That's to remind you that some foods in these groups can also be high in fat and added sugars. When choosing foods for a healthful diet, consider the fat and added sugars in your choices from all the food groups, not just fats and sweets from the pyramid tip.

HOW MANY SERVINGS DO YOU NEED EACH DAY?

Calorie Level[a]	Bread Group	Vegetable Group	Servings Fruit Group	Milk Group	Meat Group
Women and some older adults (about 1600 calories)	6	3	2	2–3[b]	2 (for a total of 5 ounces)
Children, teenage girls, active women, and most men (about 2200 calories)	9	4	3	2–3[b]	2 (for a total of 6 ounces)
Teenage boys and active men (about 2800 calories)	11	5	4	2–3[b]	3 (for a total of 7 ounces)

[a]These are the calorie levels if you choose low-fat, lean foods from the five major food groups and use foods from the fats and sweets group sparingly.
[b]Women who are pregnant or breastfeeding, teenagers, and young adults to age 24 need three servings.
Source: U.S. Department of Agriculture and the U.S. Department of Health and Human Services.

Mohammad 👍

ANSWERS TO REALITY CHECK

Portion Distortion

A portion is different than a standard serving. The average portion size of pasta served by restaurants is nearly 3 cups—or 6 standard servings of pasta according to the Food Guide Pyramid.[18] A random survey conducted in 2002 by the American Dietetic Association showed that 55% of adults overestimate what constitutes a standard serving of pasta.

Isaac 👎

counterparts. Larger portions don't cost restaurants much more than smaller portions, they increase sales volume, and they also encourage people to eat more. Many Americans are already eating plenty, and it is suspected that rising rates of obesity may be partly related to increased portion sizes.[19] Some health activists and legislators think it's time for fast-food and other restaurants to let consumers know the caloric cost, as well as the price, of menu items (Illustration 6.6).[20]

TABLE 6.3

FOOD GUIDE PYRAMID STANDARD SERVING SIZES, TYPICAL PORTION SIZES IN THE MARKETPLACE, AND THE NUMBER OF FOOD GUIDE PYRAMID SERVINGS REPRESENTED BY MARKETPLACE PORTION SIZES.[17]

	MARKETPLACE PORTION SIZE	STANDARD SERVING SIZE	FOOD GUIDE PYRAMID SERVINGS
Bagel	5.8 oz	1.0 oz	5.8
Muffin	6.5 oz	1.5 oz	4.3
Hamburger bun	2.2 oz	2.0 oz	1.1
Hamburger	3.9 oz	2.5 oz	1.6
Steak	8.1 oz	2.5 oz	3.2
French fries	5.3 oz	2.0 oz	2.7
Pasta	2.9 c	0.5 c	5.8

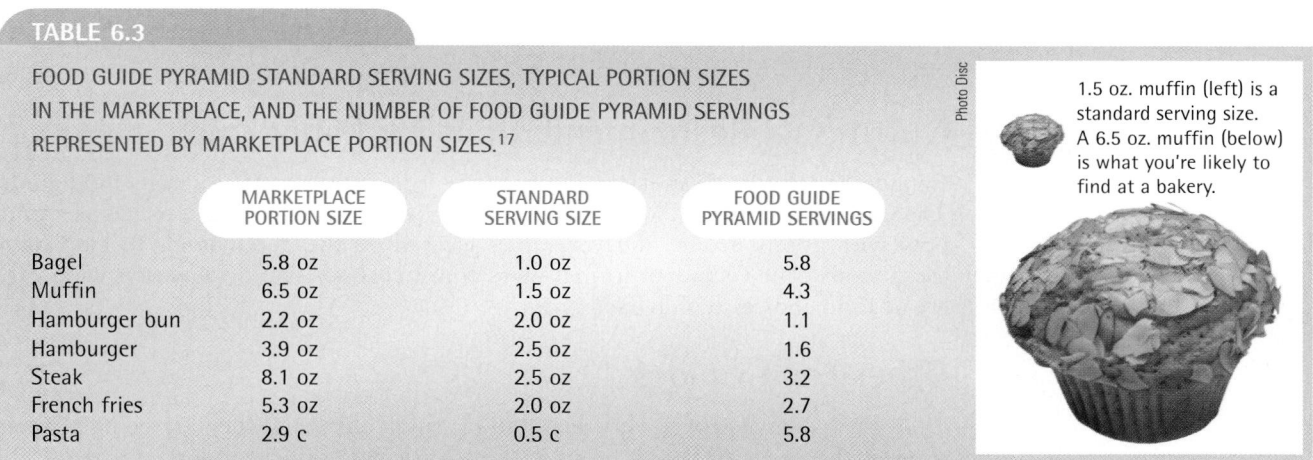

1.5 oz. muffin (left) is a standard serving size. A 6.5 oz. muffin (below) is what you're likely to find at a bakery.

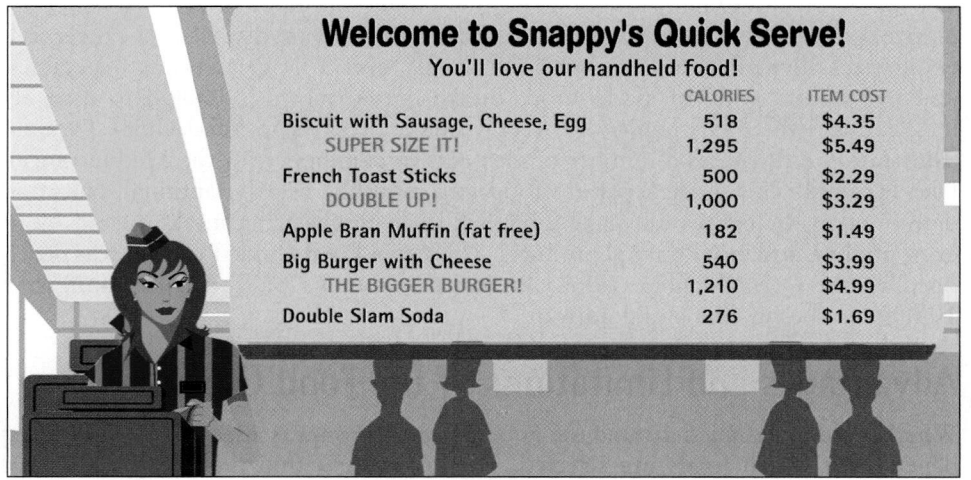

Welcome to Snappy's Quick Serve!
You'll love our handheld food!

	CALORIES	ITEM COST
Biscuit with Sausage, Cheese, Egg	518	$4.35
SUPER SIZE IT!	1,295	$5.49
French Toast Sticks	500	$2.29
DOUBLE UP!	1,000	$3.29
Apple Bran Muffin (fat free)	182	$1.49
Big Burger with Cheese	540	$3.99
THE BIGGER BURGER!	1,210	$4.99
Double Slam Soda	276	$1.69

Illustration 6.6
Should fast-food and chain restaurants be required to disclose the caloric content of menu items?
It could happen.[21]

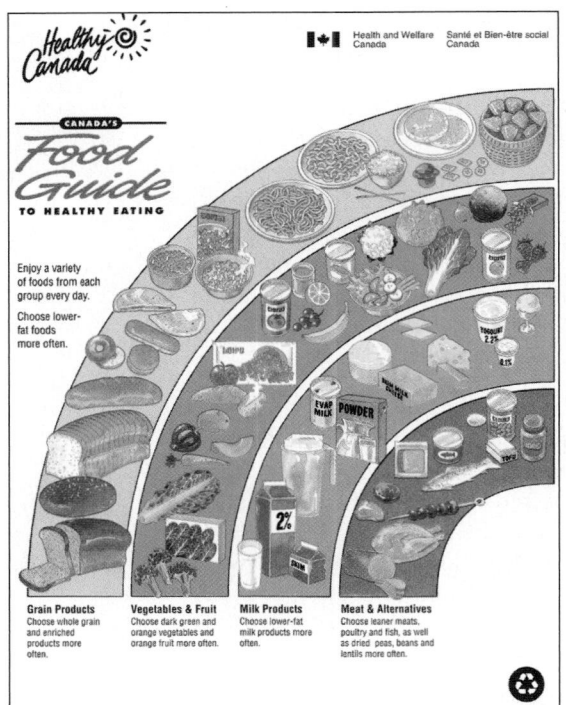

Illustration 6.7
The Canadian Food Guide to Healthy Eating

Health and Welfare Canada

Nutritional Recommendations for Canadians

Canada has chosen to stress grain products, fruits, and vegetables in its food guide. The Canadian Food Guide to Healthy Eating (Illustration 6.7) is depicted as a rainbow with grains, fruits, and vegetables located on the outer bands to emphasize these foods. The Canadian format gives consumers a visual display of a wide variety of food choices within each group.

Other Food Guides

Food groups have proven very useful in planning diets, and they are being adapted to meet the needs of people in various countries and ethnic groups. Of these, the most well known is the "Mediterranean Diet Pyramid" shown in Illustration 6.8. The World Health Organization (WHO) developed this guide in 1994 to help popularize a type of diet associated with reduced risk of heart disease and cancer.[22] It emphasizes olive oil, breads, whole grain cereals, nuts, fish, dried beans, vegetables, and fruits—and wine in moderation. Intake of red meats is limited to monthly intake, and sweets and poultry to weekly intake. The Asian Food Guide Pyramid (Illustration 6.9) shares a number of elements in common with the Mediterranean Diet Pyramid. This guide separates food into monthly, weekly, optional daily, and daily intakes. It focuses on plant foods while de-emphasizing intake of red meat, eggs, poultry, and other animal products. The Asian Food Guide Pyramid was developed by the Cornell-China-Oxford Project from results of studies on more than 10,000 families in China and Taiwan.[24]

Advantages and Limitations of the Food Groups

Whether a diet designed around the Food Pyramid groups is adequate and balanced depends on which foods are selected. For example, within the bread and cereal

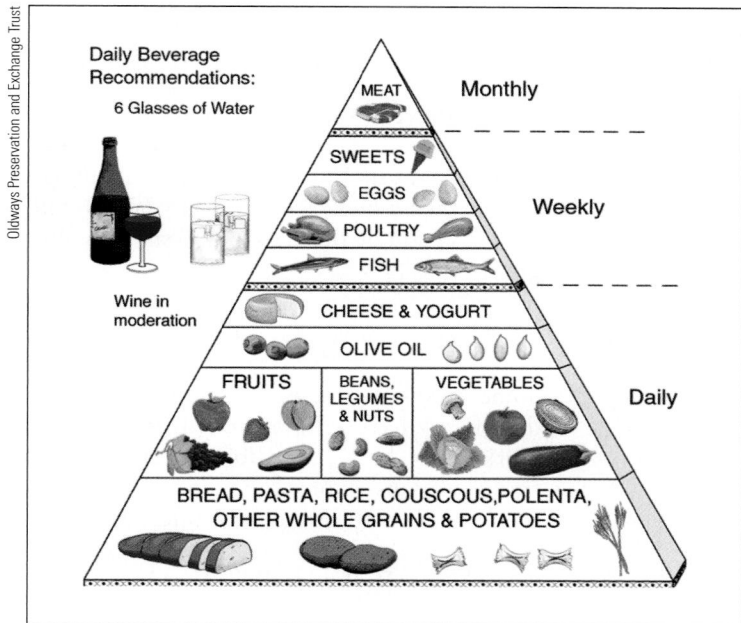

Illustration 6.8   The Mediterranean Diet Pyramid.

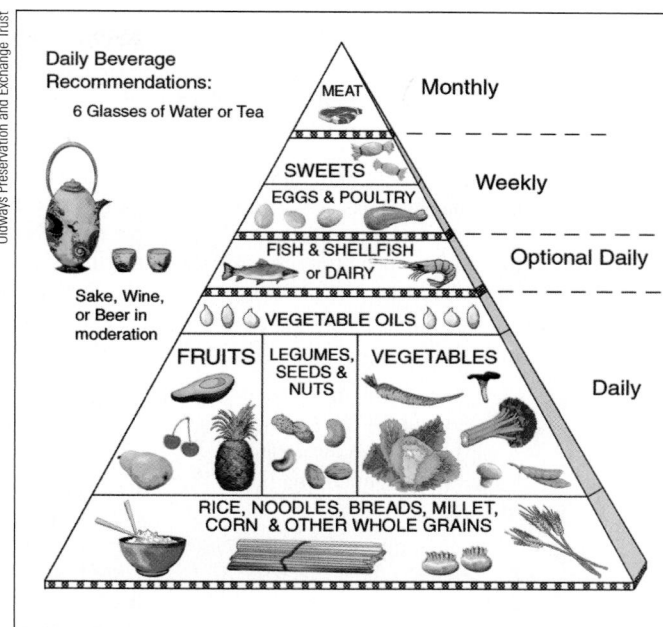

Illustration 6.9   Asian Food Guide Pyramid.[23]

group, you could choose white bread or whole grain bread. Both are breads, but their fiber content differs considerably. Or you might regularly choose to eat ice cream, fried chicken, and processed meats high in fat and sodium. Such a diet may lack balance (Illustration 6.10). See Table 6.4 for other limitations, as well as advantages, of the food group approach to healthy diets.

Illustration 6.10
Pat Byrnes for USA Weekend

"I'm trying this thing called the 'balanced' diet."

6-13

TABLE 6.4

ADVANTAGES AND LIMITATIONS OF THE BASIC FOOD GROUP APPROACH TO HEALTHY DIETS.[25]

ADVANTAGES	LIMITATIONS
• People eat foods and not nutrients, making it easier to relate to foods in groups rather than to recommended nutrient intake levels (i.e., the RDAs).	• Applicability of the food groups to vegetarians and people from a variety of ethnic groups may be unclear.
• The food groups are fairly easy to remember and teach.	• The food groups fail to directly address important nutritional concerns such as overweight or alcohol intake.
• Diets that include the recommended number of servings from the food groups are more likely to be adequate than diets that don't.	• The food groups do not include combination dishes such as pizza, soups, and stews that many people often eat. They do not include water (an essential nutrient).
• The food groups provide leeway for personal food choices within each group.	• People may end up with an inadequate diet if they select the same foods from each of the groups regularly.

Dietary Guidelines, Food Groups, and Your Diet

When you read the information on adequate and balanced diets, did you wonder how your diet rates? Illustration 6.11 encourages you to take a close look at your diet by running it through the Diet Analysis Plus Program.

Good Diets by Design

Some people find it easier to plan a healthy diet if they have a framework from which to work (see Table 6.5, for example). This framework distributes foods from the various food groups into three meals and a snack.

Are You Better Off Eating Breakfast?

People who eat breakfast *are* better off in a number of ways than people who don't.[26] Yet breakfast is the most frequently missed meal (Illustration 6.12). Approximately 15% of U.S. adults and 29% of people over the age of 15 in Canada skip it.[27] According to a U.S. Department of Agriculture study, males aged 20 to 29 are the least likely to eat anything before lunch (72%).[28]

Illustration 6.11

Analyzing your diet.

A convenient method is available for finding out how well your diet hits the targets recommended and where it misses. Keep track of your usual food and beverage intake for three days (include one weekend day) and enter the information into the Diet Analysis Plus Program.

The program works best when the dietary records prepared for input are complete and detailed. Record foods and beverages right after you eat them and write down the amounts. The quantities listed on food labels can often help you determine how much of a particular food you consumed.

TABLE 6.5

A FRAMEWORK FOR PLANNING ADEQUATE DIETS BASED ON THE FOOD GUIDE PYRAMID.

Include one or more servings of foods from each group listed.

ADEQUATE DIET PLAN	EXAMPLE
Meal one	
Fruits	Strawberries, ½ cup, with
Milk, yogurt, and cheese	low-fat yogurt, 1 cup
Bread, cereal, rice, and pasta	Whole-wheat toast, 2 slices
Meal two	
Vegetables	Two bean tostadas with
Bread, cereal, rice, and pasta	tomatoes, lettuce, and cheese
Meat, poultry, fish, dry beans, eggs, and nuts	
Fruits	Pear
Milk, yogurt, and cheese	Frozen yogurt, 1 cup
	Iced tea
Meal three	
Vegetables	Tossed salad, 2 cups
	Green beans, ½ cup
Bread, cereal, rice, and pasta	Rice, ½ cup
	Rolls, 2
Meat, poultry, fish, dry beans, eggs, and nuts	Salmon, 3 ounces
Milk, yogurt, and cheese	Skim milk, 1 cup
Snack	
Bread, cereal, rice, and pasta	Popcorn, 2 cups
Vegetables or fruits	Apple

Photo Disc

Eating breakfast pays off in these ways:

- No midmorning hunger pangs—you don't regret *not* eating breakfast at 10:30 a.m.
- Lower blood cholesterol levels—breakfast skippers tend to have higher cholesterol levels and higher fat intake than people who eat breakfast.[29]
- Lower body weight—breakfast skippers tend to weigh more than nonskippers.[30]
- Alertness—students who eat breakfast (especially children) tend to do better on math tests.[31]
- Higher-quality diets—skipping breakfast may lower the nutritional adequacy of your diet.[32]

Photo Disc

SALLY FORTH   **BY GREG HOWARD**

Illustration 6.12
Source: Reprinted with special permission of King Features Syndicate.

Photo Disc

Eating a full meal—like toast, cereal, juice, and milk—at breakfast is ideal. But when you don't have time for a complete meal, eating something nutritious is better than nothing. A carton of yogurt with fruit, a peanut butter sandwich, a raisin bagel, or an orange and crackers eaten on the way to class or work will give you the edge breakfast can provide.

Can You Still Eat Right When Eating Out?

The question about what to eat often boils down to choosing a restaurant. According to recent USDA data, 50% of Americans eat out every day.[33] In general, foods eaten away from home have lower nutrient content and are higher in fat than foods eaten at home.[34] In addition, children and teenagers who eat dinner with their families most days tend to have more healthful diets—including more vegetables, fruits, vitamins, minerals, and fiber, and less fat—than others who never or occasionally eat dinner with the family.[35]

Staying on Track while Eating Out] You'll find it easier to stick to a healthy diet if you decide what to eat before you enter a restaurant and look over the menu. You could make the decision to order soup and a salad, broiled meat, a half-portion of the entrée, or no dessert *before* entering the restaurant. "Impulse ordering" is a hazard that can throw diets out of balance. If you're going to a party or business event where food will be served, decide before you go what types of food you will eat and what you will drink. If only high-calorie foods are offered, plan on taking a small portion and stopping there.

Decide what to eat before you enter a restaurant. Impulse ordering is a hazard that can throw your diet out of balance.

Can Fast Foods Be Part of a Healthy Diet?] Many fast foods deserve their reputation for being high in calories, fat, saturated fat, and salt. (Table 6.6 lists the calorie and fat contributions of a variety of foods sold in fast-food restaurants). Tostadas, bean burritos, grilled chicken sandwiches, and baked potatoes are among the lowest-fat options available at fast-food restaurants. In response to negative publicity about the caloric and nutritional value of foods traditionally offered, many fast-food chains are adding specialty salads to their menus. The salads offer nutrient-dense vegetables; but additions like crispy chicken, bacon, and salad dressing can increase their caloric value to around 500 and increase fat calories to 40 to 50% of total calories.

You can eat an occasional cheeseburger, fries, or specialty sandwich at your favorite fast-food restaurant and still maintain a healthy diet. Usual dietary intake is more important to long-term health than are occasional dinners at a fast-food restaurant.

ON THE SIDE

Do you recognize these vegetables?

Royalty-free CORBIS

Royalty-free CORBIS

Royalty-free CORBIS

Photo Disc

Photo Disc

a. Parsnips, b. Turnips, c. Kale, d. Leeks, e. Swiss chard

TABLE 6.6

THE FAT CONTENT OF SOME FAST FOODS.*

FOOD	CALORIES	% OF CALORIES FROM FAT	FOOD	CALORIES	% OF CALORIES FROM FAT
Sausage Breakfast Croissanwich®	538	69	Burrito Supreme®	457	43
Fried chicken thigh	278	62	Fillet-o-fish	373	43
Fried chicken wing	181	60	Egg McMuffin®	340	42
Bacon cheeseburger	610	58	Hamburger	350	41
Chicken McNuggets® (6)	323	56	Beef tostada	239	41
Fried chicken breast	276	55	Beef n' Cheddar®	490	39
Fried chicken drumstick	147	55	Taco	186	39
Quarter Pounder with cheese®	525	55	Thin-n-Crispy Cheese Pizza® (2 slices)	398	38
Big Mac®	570	55	Beef burrito	357	38
Sausage McMuffin®	427	55	Roast beef sandwich	353	38
Whopper with cheese®	711	54	Pan Pizza with pepperoni® (2 slices)	540	37
Fried onion rings	274	53	Pan Pizza with cheese® (2 slices)	492	33
Whopper®	628	52	Chili, regular	240	30
Coleslaw	103	52	Tostada	179	30
French fries, regular	227	52	Grilled chicken breast on bun	320	28
Hot dog	280	51	Pizza with pepperoni (2 slices)	380	28
Whaler Fish Sandwich®	488	50	Bean burrito	357	25
Quarter Pounder®	427	48	Pizza with cheese (2 slices)	376	24
Pan Supreme Pizza® (2 slices)	589	46	Vanilla shake	280	19
Cheeseburger	318	45	Chocolate shake	447	16
Ham and Cheese Specialty Sandwich®	471	44	Mashed potatoes, without gravy	62	15
Thin-n-Crispy Pizza with pepperoni® (2 slices)	413	44	Baked potato, plain	250	7
Thin-n-Crispy Pizza Supreme® (2 slices)	459	43			

*Calories and fat content may vary somewhat depending on the fast-food restaurant or chain.

Slow Food

An interesting trend in food preparation and consumption is making its way across the globe. The trend is away from fast food and junk food, and toward "homespun" foods eaten in a casual atmosphere.[37] The "Slow Food USA" (Illustration 6.13) movement represents some aspects of this trend. Part of an international group, Slow Food USA is an educational organization that supports ecologically sound food production; the revival of the kitchen and the table as centers of pleasure, culture, and community; and living a slower and more harmonious rhythm of life.[38] The trend is placing the topic of healthy eating in a new light for many individuals, and it may help bring people closer to family, friends, and the environment.

What If You Don't Know How to Cook?

With so many convenience foods available and time at a premium, there is growing concern that we're becoming a nation of cooking illiterates. Cooking at home gives you control over what you eat and how it's prepared. Some people immensely enjoy cooking and get a real thrill out of making their specialties for friends

Illustration 6.13
Blacklash to fast foods and fast lives? *New food trends are under way, and here's an example.*

Illustration 6.14
Some good starter cookbooks and recipe sources.
Most bookstores offer a wide variety of basic cookbooks.

and family. It's becoming a popular leisure-time activity: 43% of U.S. adults have taken it up for enjoyment.[39]

If you don't know how to cook, there are several ways to learn. You could start on your own by using the recipes on food packages like pasta, tomato sauce, or dried beans. Or you could buy a basic cookbook and make simple dishes like salads, tacos, shish kebab, and lentil soup. You could even take a community education course. Illustration 6.14 shows some examples of good starter cookbooks. Read the sections on basic cooking skills, and learn what types of equipment and utensils you need to prepare basic dishes. Get the foods and other supplies you need. Select the recipes that look doable, and be sure they pass your taste and nutritional standards tests. Voilà! You're cooking.

Bon Appétit!

Japan's dietary guidelines contain one other rule of healthy eating that would serve Americans well: "Enjoy your meals." Eating a healthy diet has to be enjoyable. If it's too much of a struggle, the healthy diet won't last. The best diets are those that keep us healthy and enhance our sense of enjoyment and well-being. The trick is to remember the broad array of nutritious foods we like that give us good taste and enjoyment when we eat them. And remember not to feel guilty when you occasionally eat hamburgers or ice cream. Enjoy them to the utmost—as much as ripe oranges, antipasto salads, homemade soups, roast turkey, hummus, and countless other nutritious delicacies.

Key Terms

adequate diet, page 6-3

Adequate Intakes (AIs), page 6-3

balanced diet, page 6-4

cholesterol, page 6-7

essential nutrients, page 6-3

macronutrients, page 6-4

Recommended Dietary Allowances (RDAs), page 6-3

saturated fats, page 6-4

trans fats, page 6-7

unsaturated fats, page 6-4

www links

www.nal.usda.gov/fnic
Provides links to new releases of the Food Guide Pyramid and the Dietary Guidelines, international food guides and dietary guidelines, food guides for different cultural groups, and culturally specific food and nutrition guidance. Look up the nutrient content of foods from this site, too.

www.nasa.gov
Go to image search to view photos of space cuisine.

www.ers.usda.gov
A great source for information on the U.S. food supply, diet, consumption, health, and agricultural policy.

www.dietitians.ca
Dietitians of Canada home page. Go from there to find consumer information on advice for healthy eating and food preparation, visit a virtual grocery store, assess your nutrition profile, and get menu-planning help.

www.afic.com
The Asian Food Information Center. Provides nutrition news, tips for healthy eating, and other advice for Asians.

www.nal.usda.gov/fnic/etext/000023.html
Sends you to Cornell's information on the Asian Food Guide Pyramid.

http://www.usda.gov/cnpp
This is the USDA's Center for Nutrition Policy and Promotion site. It gives you access to the Healthy Eating Index for dietary assessment, and to multiple U.S. food, nutrition, and health guides. The Healthy Eating Index calculates number of Food Guide Pyramid servings from reported food portion sizes consumed.

http://hin.nhlbi.nih.gov/portion
Presents interactive quiz for evaluation of food portion sizes versus those of 20 years ago, calories in today's portions, and how much physical activity it takes to burn up the calories from food in today's portion sizes.

www.usda.gov/cnpp/Pubs/Brochures/HowMuchAreYouEating.pdf
Translate food portion sizes into Food Guide Pyramid servings using this Web site.

www.nal.usda.gov/fnic/etext/fnic.html
Visit the fruit and vegetable of the month page for ideas and recipes of how to eat a colorful variety every day.

www.navigator.tufts.edu
This online rating and review guide is designed to help Web users quickly find accurate nutrition information.

Nutrition UP CLOSE

The Food Guide Pyramid: A Balancing Act

FOCAL POINT: Discover how well-balanced your current diet is.

Think about what you typically eat each day for breakfast, lunch, dinner, and snacks. Next, think about the amount of food you normally consume at each meal. Now you are ready to compare your eating habits with the Food Guide Pyramid recommendations.

Check the appropriate boxes below, indicating the average number of servings you usually consume daily in each food group. For help in calculating serving sizes, refer to Table 6.2. If you consumed a mixed dish such as spaghetti with tomato sauce and meatballs, estimate how many Food Guide Pyramid standard servings you consumed as spaghetti noodles, tomato sauce, and meatballs. Then mark the appropriate food groups for each ingredient. Repeat the same process for other mixed-food items such as sandwiches, casseroles, and soups. Compare your responses with the Food Guide Pyramid recommendations in the Feedback section.

				Daily Servings			
Food Guide Pyramid Group	0	1	2	3	4	5	6 or more
How many daily servings of **bread, cereal, rice, and pasta?**	○	○	○	○	○	○	○
How many daily servings of **fruit?**	○	○	○	○	○	○	○
How many daily servings of **vegetables?**	○	○	○	○	○	○	○
How many daily servings of **meat, poultry, fish, dry beans, eggs, and nuts?**	○	○	○	○	○	○	○
How many daily servings of **milk, yogurt, and cheese?**	○	○	○	○	○	○	○
How many daily servings of **fats, oils, and sweets?**	○	○	○	○	○	○	○

Special note: You can use the Diet Analysis Plus software to determine how your diet matches the Food Guide Pyramid recommendations. Enter your food intake for one day; then, in the analysis/reports section, select "Pyramid." This report will show you whether you are getting enough, or too much, of each of the food groups.

FEEDBACK (answers to these questions) can be found at the end of Unit 6.

Notes

1. Food consumption and spending. Food Review, vol. 23, no. 3, www.ers.usda.gov/data/foodcomposition, accessed 8/03.

2. Dietary Reference Intakes. Energy, carbohydrate, fiber, fat, fatty acids, cholesterol, protein, and amino acids. Institute of Medicine, National Academy of Sciences. Washington, DC: National Academies Press, 2002.

3. Food consumption and spending (www.ers.usda.gov/data/foodcomposition); and Dietary Reference Intakes, 2002.

4. Food consumption and spending (www.ers.usda.gov/data/foodcomposition).

5. Dietary Reference Intakes, 2002.

6. Food consumption and spending (www.ers.usda.gov/data/foodcomposition).

7. Weisburger JH. Approaches for chronic disease prevention based on current understanding of underlying mechanisms. Am J Clin Nutr 2000;71(suppl):1710S–4S.

8. Food consumption and spending (www. ers.usda.gov/data/foodcomposition); and Johnston CS et al. More Americans are eating "5-A-Day" but intake of dark green cruciferous vegetables remains low. J Nutr 2000;130:3063–67.

9. Dietary Reference Intakes, 2002; and Food consumption and spending (www. ers.usda.gov/data/foodcomposition).

10. Tillotson JE. Fast-casual dining: our next eating passion? Nutr Today 2003;38:91–4.

11. For information on new releases of the Food Guide Pyramid and the Dietary Guidelines expected late in 2005, go to www.nal.usda.gov/fnic

12. Johnson RK, Kennedy E. The 2000 Dietary Guidelines for Americans: what are the changes and why were they made? J Am Diet Assoc 2000;100: 769–74.

13. Weisburger, Approaches for chronic disease prevention.

14. Dietary Reference Intakes, 2002.

15. Ellen Schuster is a professor of nutrition at Oregon State University. The quote was taken from a *Wall Street Journal* article published 5/20/03, p. D1.

16. Food portions and servings: how do they differ? Nutrition Insights, USDA, 1999 March 11.

17. Young LR, Nestle M. Expanding portion sizes in the US marketplace: implications for nutrition counseling. J Am Diet Assoc 2003;103:231–4.

18. Young and Nestle, Expanding portion sizes.

19. Young and Nestle, Expanding portion sizes; and Fisher JO et al. Children's

bite size and intake of an entrée are greater with large portions than with age-appropriate or self-selected portions. Am J Clin Nutr 2003;77: 1164–70.

20. Food consumption and spending (www. ers.usda.gov/data/foodcomposition); and Fast-food chains face pressure for healthier menus. CNI Nutrition Week 2003;33:1.

21. Food consumption and spending (www. ers.usda.gov/data/foodcomposition).

22. de Lorgeril M, et al. Mediterranean diet, traditional risk factors, and the rate of cardiovascular disease complications after myocardial infarction. Circulation 1999;99:779–85.

23. Asian Food Guide Pyramid; available at www.nal.usda.gov/fnic/etext/000023 .html

24. Asian Food Guide Pyramid (www.nal. usda.gov/fnic/etext/000023.html).

25. Johnson et al., More Americans are eating "5-A-Day."

26. Heaton KW. Breakfast—do we need it? J Royal Society of Med 1989;82:770–1.

27. Eating breakfast greatly improves school children's diet quality. Nutrition Insights, USDA 1999 Dec.; and Chao ESM, Vanderkooy PS. An overview of breakfast nutrition. J Can Diet Assoc 1989;50:225–8.

28. Breakfast Consumption. CNI Nutrition Newsletter 1999; April 12:7.

29. Stanton JL Jr., Keast DR. Serum cholesterol, fat intake, and breakfast consumption in the United States adult population. J Am Coll Nutr 1989;8: 567–72.

30. Heaton, Breakfast—do we need it?

31. Murphy JM, et al. The relationship of school breakfast to psychosocial and academic functioning. Arch Pediatr Adol Med 1998;152:899–907.

32. Eating breakfast greatly improves school children's diet quality. Nutrition Insights, USDA 1999 Dec.

33. Food consumption and spending (www. ers.usda.gov/data/foodcomposition).

34. Guthrie JH et al. Role of food prepared away from home in the American diet. J Nutr Behav 2002;34:140–50.

35. Guthrie, Role of food prepared away from home; and Gillman MW et al. Family dinner and diet quality among older children and adolescents. Arch Fam Med 2000;9:235–40.

36. Economic trends in ethnic food restaurants. Results presented on *Wall Street Week,* 7/7/00.

37. Tillotson, Fast-casual dining; Fast-food chains face pressure for healthier menus; Maine anti-obesity legislation defines "junk food." CNI Nutrition Week, March 3, 2003, p. 3. Sloan AE. Top 10 trends to watch and work on: 3rd biannual report. Food Tech 2001;55:38–58; and Junk food calories compromise good diets. Nutr Today 2000;35:205.

38. Slow Food International Organizations, www.slowfood.com.

39. Cooking. CNI Nutrition Newsletter 1999;April 12:8.

Nutrition UP CLOSE

The Food Guide Pyramid: A Balancing Act

Feedback for Unit 6

You met the Food Guide Pyramid recommendations if your answers indicate that daily you eat 6–11 servings in the **bread, cereal, rice, and pasta** group; 2–4 servings in the **fruit** group; 3–5 servings in the **vegetable** group; 2–3 servings in the **meat, poultry, fish, dry beans, eggs, and nuts** group; and 2–3 servings in the **milk, yogurt, and cheese** group, while practicing moderation in the **fats, oils, and sweets** group.

If you did not meet the recommendations, balance your diet by choosing more foods from the group(s) in which you scored low, while cutting back on foods eaten in excess.

How the Body Uses Food: Digestion and Absorption

Nutrition Scoreboard

	TRUE	FALSE
1 Almost all of the carbohydrate and fat you consume in foods is absorbed by the body, but only about half of the protein is.		
2 Disorders of the digestive system are a leading cause of hospitalizations and medical visits in the United States and Canada.		
3 Lactose maldigestion is a common digestive disorder.		

Answers on next page

[KEY CONCEPTS AND FACTS]

- Our bodies are in a continuous state of renewal. Materials used to renew body tissues come from the food we eat in the form of nutrients.

- Digestion and absorption are processes that make nutrients in foods available for use by the body.

- Digestive disorders are common and often related to dietary intake.

Answers to *Nutrition Scoreboard*

		TRUE	FALSE
1	Over 90% of all the carbohydrates, fats, *and* proteins consumed in food are absorbed and become part of the body.		✔
2	Digestive disorders are the leading cause of hospitalizations among 20- to 44-year-olds in the United States and Canada. They account for over 70 million medical visits in the United States each year.[1]	✔	
3	Over half of the world's population digests lactose from milk and milk products incompletely or not at all.[2]	✔	

Photo Disc

My Body, My Food

La vie est une fonction chimique. ("Life is a chemical process.")
—Antoine Lavoisier, late eighteenth century

You are not the same person you were a month ago. Although your body looks the same and you don't notice the change, the substances that make up the organs and tissues of your body are constantly changing. Tissues we generally think of as solid and permanent, such as bones, the heart and blood vessels, and nerves, are continually renewing themselves. The raw materials used in the body's renewal processes are the nutrients you consume in foods.

Each day, about 5% of our body weight is replaced by new tissue. Existing components of cells are renewed, the substances in our blood are replaced, and body fluids are recycled. Taste cells, for example, are replaced about every 7 days, and cells lining the intestinal tract every 1 to 3 days. All of the cells of the skin are replaced every month. Red blood cells turn over every 120 days. If you thought it was hard to maintain a car, an apartment, or a house, just imagine what the maintenance is on a body! Maintenance is just one of the body's ongoing functions that require nutrients as raw material.

How Do Nutrients in Food Become Available for the Body's Use?

The human body is an amazing machine. Mine is, anyway. For example, I regularly feed my body truly absurd foods, such as cheez doodles, and somehow it turns them into useful body parts, such as glands. At least I assume it turns them into useful body parts.[3]
—Dave Barry, 1987

The components of food that make "useful body parts" are nutrients. Through the processes of *digestion* and *absorption,* they are made available for use by every cell in the body.

The Internal Travels of Food: An Overview] The "food processor" of the body is the digestive system shown in Illustration 7.1. It consists of a 25- to 30-foot-long

digestion
The mechanical and chemical processes whereby ingested food is converted into substances that can be absorbed by the intestinal tract and utilized by the body.

absorption
The process by which nutrients and other substances are transferred from the digestive system into body fluids for transport throughout the body.

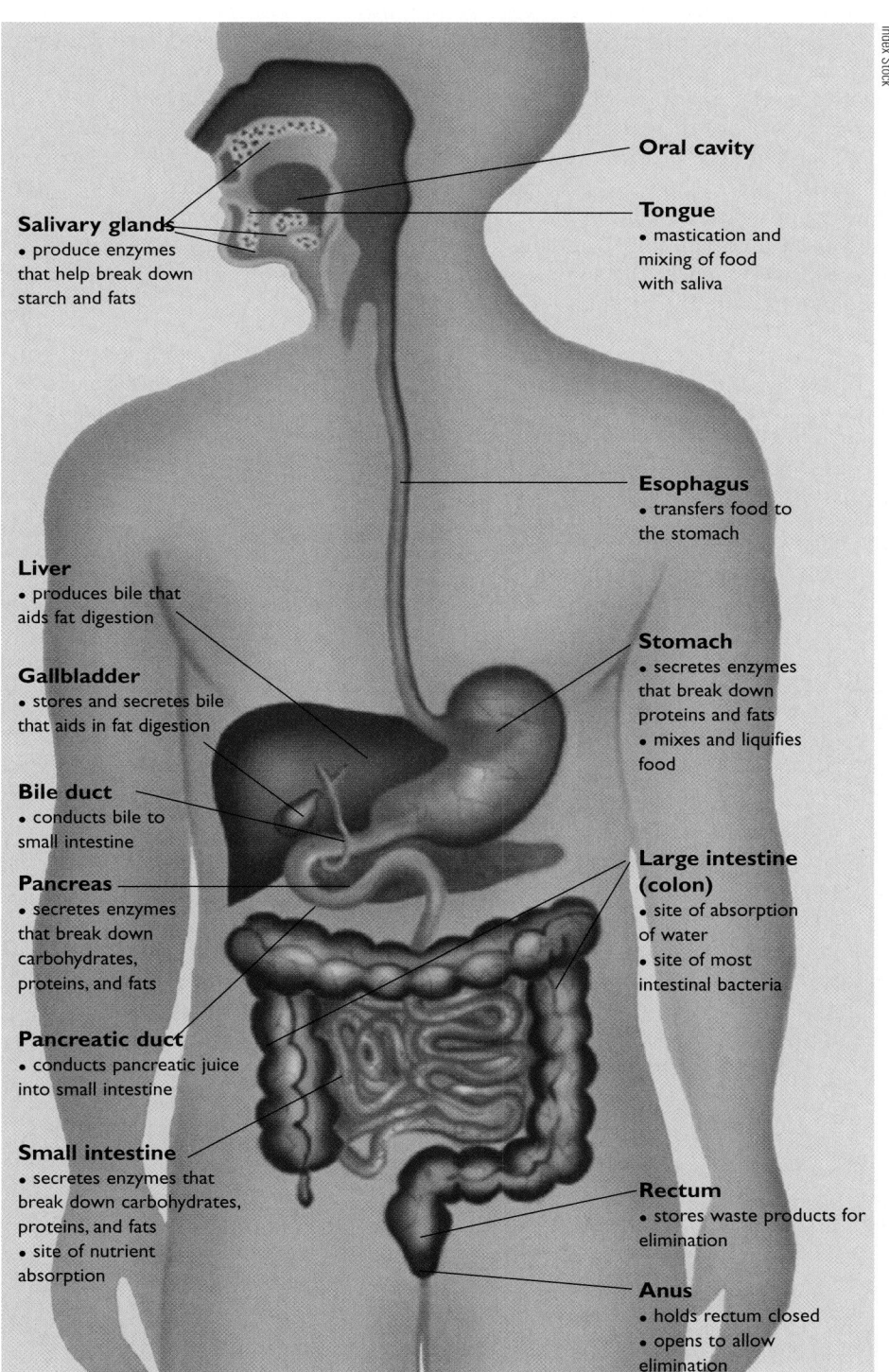

Index Stock

Oral cavity

Tongue
• mastication and mixing of food with saliva

Salivary glands
• produce enzymes that help break down starch and fats

Esophagus
• transfers food to the stomach

Liver
• produces bile that aids fat digestion

Stomach
• secretes enzymes that break down proteins and fats
• mixes and liquifies food

Gallbladder
• stores and secretes bile that aids in fat digestion

Bile duct
• conducts bile to small intestine

Pancreas
• secretes enzymes that break down carbohydrates, proteins, and fats

Large intestine (colon)
• site of absorption of water
• site of most intestinal bacteria

Pancreatic duct
• conducts pancreatic juice into small intestine

Small intestine
• secretes enzymes that break down carbohydrates, proteins, and fats
• site of nutrient absorption

Rectum
• stores waste products for elimination

Anus
• holds rectum closed
• opens to allow elimination

Illustration 7.1
The digestive system.

7-3

monosaccharides
(*mono* = one, *saccharide* = sugar) Simple sugars consisting of one sugar molecule. Glucose, fructose, and galactose are monosaccharides.

enzymes
Protein substances that speed up chemical reactions. Enzymes are found throughout the body but are present in particularly large amounts in the digestive system.

starch
Complex carbohydrates made up of complex chains of glucose molecules. Starch is the primary storage form of carbohydrate in plants. The vast majority of carbohydrate in our diet consists of starch, monosaccharides, and disaccharides.

disaccharide
Simple sugars consisting of two sugar molecules. Sucrose (table sugar) consists of a glucose and a fructose molecule, lactose (milk sugar) consists of glucose and galactose, and maltose (malt sugar) consists of two glucose molecules.

bile
A yellowish-brown or green fluid produced by the liver, stored in the gallbladder, and secreted into the small intestine. It acts like a detergent, breaking down globs of fat entering the small intestine to droplets, making the fats more accessible to the action of lipase.

muscular tube and organs such as the liver and pancreas that secrete digestive juices. The digestive juices break foods down into very small particles that can be absorbed and used by the body. The absorbable forms of carbohydrates are ***monosaccharides,*** such as glucose and fructose. Proteins are absorbed as amino acids, and fats as fatty acids and glycerol. Vitamins and minerals are not broken down before they are absorbed; they are simply released from foods during digestion.

Much of the work of digestion is accomplished by ***enzymes*** manufactured by components of the digestive system such as the salivary glands, stomach, and pancreas. Enzymes are complex protein substances that speed up reactions that break down food. A remarkable feature of enzymes is that they are not changed by the chemical reactions they affect. This makes them reusable.

Carbohydrates, proteins, and fat each have their own set of digestive enzymes. All together, over a hundred different enzymes participate in the digestion of carbohydrates, proteins, and fat. Table 7.1 presents information on some of the enzymes involved in digestion and highlights their specific roles. In Table 7.2 you will see these enzymes cited in the summary of the processes involved in the digestion of carbohydrate, fat, and protein.

A Closer Look]   As you chew food, glands under the tongue release saliva that lubricates food so that it can be swallowed and pass easily along the intestinal tract. Saliva also gets food digestion started. It contains salivary amylase and lipase that begin to break down carbohydrates and fats.

TABLE 7.1

PRIMARY FUNCTION OF SOME DIGESTIVE ENZYMES.

ENZYME	ENZYME FUNCTION	ENZYME SOURCE
A. Carbohydrate Digestion		
Amylase	Breaks down ***starch*** into smaller chains of glucose molecules	Produced in the salivary glands (salivary amylase) and the pancreas (pancreatic amylase)
Sucrase	Separates the ***disaccharide*** sucrose into glucose and fructose	Produced in the small intestine
Lactase	Splits the disaccharide lactose into glucose and galactose	Produced in the small intestine
Maltase	Separates maltose into two molecules of glucose	Produced in the small intestine
B. Fat Digestion		
Lipase	Breaks down fats into fragments of fatty acids and glycerol	Produced in salivary glands (lingual lipase), and the pancreas (pancreatic lipase). The action of lipase is enhanced by **bile**
C. Protein Digestion		
Pepsin	Separates protein into shorter chains of amino acids	Produced by the stomach
Trypsin	Splits short chains of amino acids into molecules containing, one, two or three amino acids	Produced by the pancreas

TABLE 7.2

SUMMARY OF THE DIGESTION OF CARBOHYDRATES, FATS, AND PROTEINS.

	MOUTH	STOMACH	SMALL INTESTINE, PANCREAS, LIVER, AND GALLBLADDER	LARGE INTESTINE (COLON)
Carbohydrates (excluding fiber)	The salivary glands secrete saliva to moisten and lubricate food; chewing crushes and mixes it with salivary amylase that initiates starch digestion.	Digestion of starch continues while food remains in the stomach. Some alcohol (a carbohydrate-like substance) is absorbed. Acid produced in the stomach aids digestion and destroys bacteria in food.	Pancreatic amylase continues starch digestion. Sucrase, lactase, and maltase break down disaccharides into monosaccharides. Some alcohol is absorbed here.	Undigested carbohydrates reach the colon and are partly broken down by intestinal bacteria.
Fiber	The teeth crush fiber and mix it with saliva to moisten it for swallowing.	No action.	Fiber binds cholesterol and some minerals.	Most fiber is excreted with feces; some fiber is digested by bacteria in the colon.
Fat	Fat-rich foods are mixed with saliva. Small amounts of lingual lipase accomplish some fat breakdown.	Fat tends to separate from the watery stomach fluid and foods and float on top of the mixture. Only a small amount of fat is digested. Fat is last to leave the stomach.	Bile readies fat for the action of lipase from the pancreas. Lipase splits fats into fatty acids and glycerol fragments.	A small amount of fatty materials escapes absorption and is carried out of the body with other wastes.
Protein	In the mouth, chewing crushes and softens protein-rich foods and mixes them with saliva.	Stomach acid works to uncoil protein strands and to activate the stomach's protein-digesting enzyme. Pepsin breaks the protein strands into smaller fragments.	Trypsin splits protein into molecules containing one, two, or three amino acids.	The large intestine carries undigested protein residue out of the body. Normally, almost all food protein is digested and absorbed.

After food is chewed, it is swallowed and passes down the esophagus to the stomach. Muscles that act as valves at the entrance and exit of the stomach ensure that the food stays there until it's liquified, mixed with digestive juices, and ready for the digestive processes of the small intestine. Solid foods tend to stay in the stomach for over an hour, whereas most liquids pass through it in about 20 minutes. When the stomach has finished its work, it ejects 1 to 2 teaspoons of its liquified contents into the small intestine through the muscular valve at its end. Stomach contents continue to be ejected in this fashion until they are totally released into the small intestine. These small pulses of liquified food stimulate muscles in the intestinal walls to contract and relax; these movements churn and mix the food as it is digested by enzymes. When the diet contains a lot of fiber and sufficient fluids, the bulge of digesting food in the intestine tends to be larger. Larger food bulges stimulate a higher level of intestinal muscle activity than do smaller food bulges. Thus, high-fiber meals pass through the digestive system somewhat faster than low-fiber meals.

Get Your Juices Flowing
You don't have to actually eat food to start your digestive juices flowing. You just have to think about food or see it.[4] Put this information to the test. Clear your mind, turn the page, and take a close look at Illustration 7.2.

ON THE SIDE

Illustration 7.2
Testing, testing. This is a test of your salivary secretions. Did the lemon speak directly to your salivary glands?

If you want to turn the digestive processes off, quit thinking about food. Finish your reading assignment!

Michael Newman/PhotoEdit

Digestion, as well as the absorption of nutrients, is greatly enhanced by the structure of the intestines (Illustration 7.3). Fingerlike projections called "villi" line the inside of the intestinal wall and increase its surface area tremendously. If laid flat, the surface area of the small intestine would be about the size of a baseball infield, or approximately 675 square feet. This large mass of tissue requires a high level of nutrients for maintenance. Much of this need (50% in the small intestine and 80% in the large intestine) is met by foods that are being digested.[5]

Digestion is completed when carbohydrates, proteins, and fats are reduced to substances that can be absorbed, and when vitamins and minerals are released from food. Most nutrients are absorbed in the small intestine. Water, sodium, and some of the end products of bacterial digestion, however, are absorbed from the large intestine. The large intestine is home to many strains of bacteria that consume undigested fiber and other types of complex carbohydrates that are not broken down by human digestive enzymes. These bacteria excrete gas as well as fatty acids that are partly absorbed in the large intestine. Substances in food that cannot be absorbed collect in the large intestine and are excreted in the stools.

lymphatic system
A network of vessels that absorb some of the products of digestion and transport them to the heart, where they are mixed with the substances contained in blood.

circulatory system
The heart, arteries, capillaries, and veins responsible for circulating blood throughout the body.

Absorption]   Absorption is the process by which the end products of digestion are taken up by the *lymphatic system* (Illustration 7.4) and the *circulatory system* (Illustration 7.5) for eventual distribution to cells of the body. Lymph vessels and

Illustration 7.3
Scanning electron micrographs of cross sections of the small intestine (left) and the large intestine (right).
Note the high density of villi in the small intestine and the relative flatness of the lining of the large intestine.

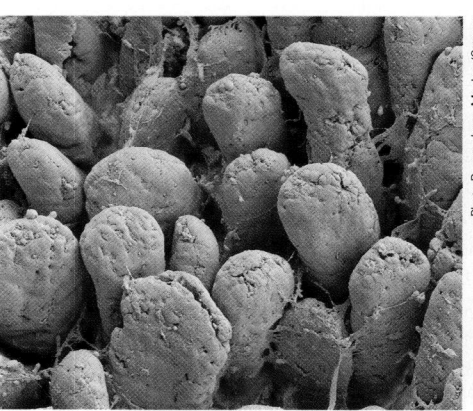

Photo Researchers, Meckes/Ottawa

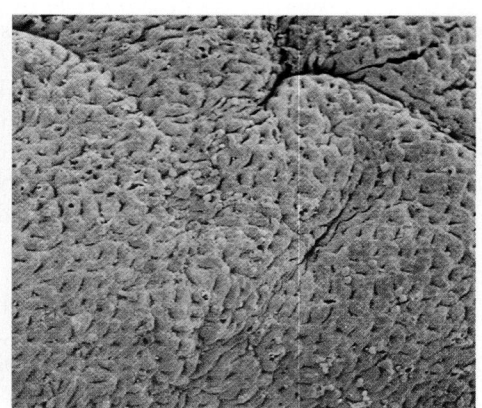

Photo Researchers, Prof. P. Motta

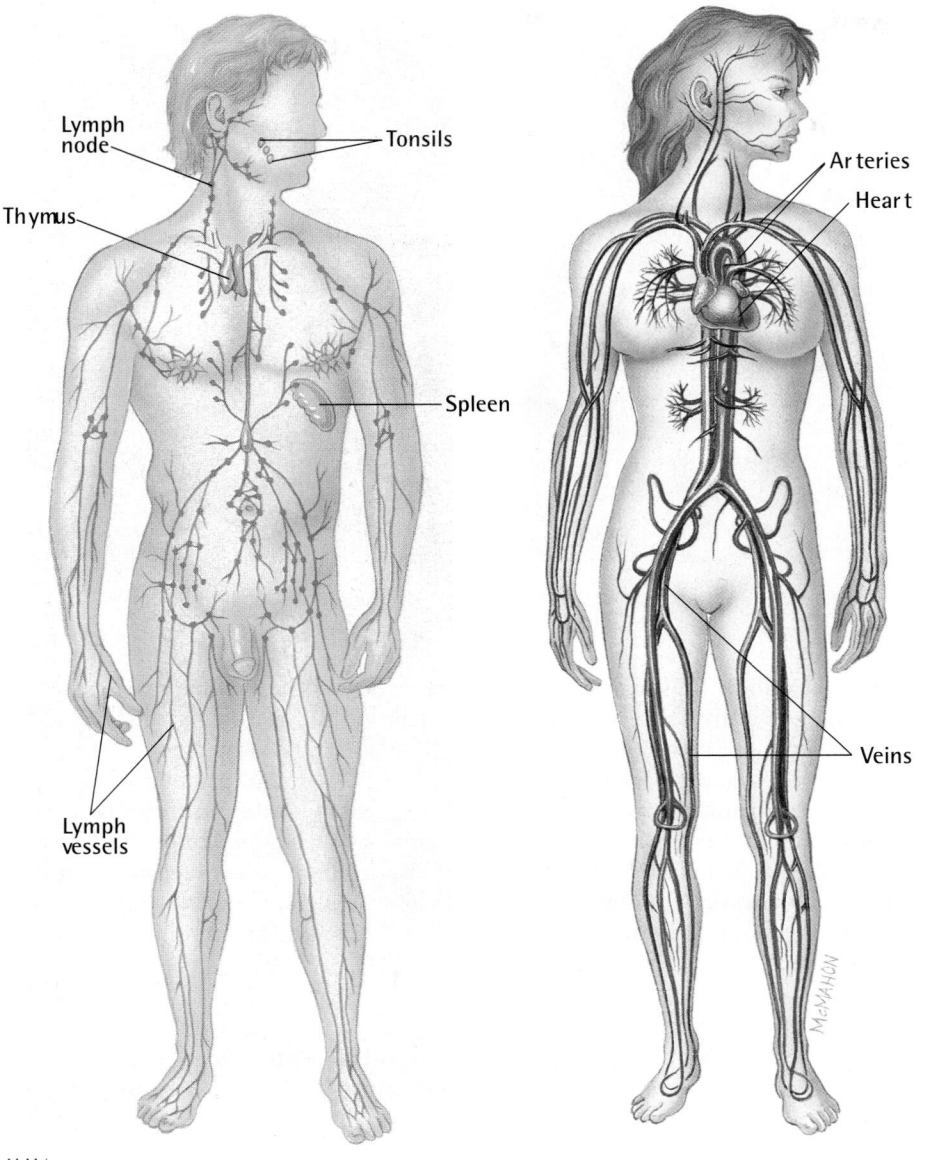

Illustration 7.4
(left) The lymphatic system.

Illustration 7.5
(right) The circulatory system includes the heart and blood vessels.
This system serves as the nutrient transportation system of the body.

McMahon

blood vessels infiltrate the villi that line the inside of the intestines (Illustration 7.6) and transport absorbed nutrients toward the major branches of the lymphatic and circulatory systems. The breakdown products of fat digestion are largely absorbed into lymph vessels, whereas carbohydrate and protein breakdown products enter the blood vessels.

The nutrient-rich contents of the lymphatic system are transferred to the bloodstream at a site near the heart where vessels from both systems merge into one vessel. From there the lymph and blood mixture is sent to the heart and subsequently throughout the body by way of the circulatory system. The circulatory system reaches every organ and tissue in the body, thereby supplying cells with nutrients obtained from food.

Illustration 7.6
Structure of villi, showing blood and lymph vessels.

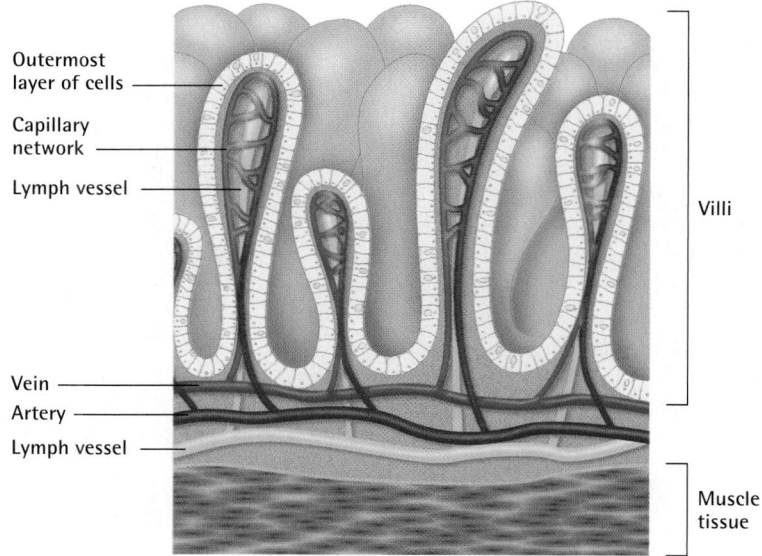

Outermost layer of cells

Capillary network

Lymph vessel

Villi

Vein

Artery

Lymph vessel

Muscle tissue

Beyond Absorption} Cells can use nutrients directly for energy, body structures, or the regulation of body processes, or convert them into other usable substances. For example, glucose delivered to cells can be used "as is" for energy formation or converted to glycogen and stored for later use. Fatty acids, an end product of fat digestion, can be incorporated into cell membranes or used in the synthesis of certain hormones. Vitamins and minerals freed from food by digestion can be used by cells to control enzyme activity or can be stored for later use. The body has a limited storage capacity for some vitamins and minerals. Consequently, excessive amounts of certain vitamins and minerals such as vitamin C, thiamin, and sodium are largely excreted in urine.

Digestion and Absorption Are Efficient} Of our intake of energy nutrients, approximately 99% of the carbohydrate, 92% of the protein, and 95% of the fat we consume in food are digested and absorbed. Dietary fiber, however, leaves the digestive system in much the same form as it entered. Humans don't have enzymes that break down fiber. It should be noted, however, that some fiber is digested in the large intestine by bacteria.

Digestive Disorders Are Common

Excluding childbirth, digestive disorders such as *heartburn, hemorrhoids, irritable bowel syndrome,* and *duodenal and stomach ulcers* are the leading cause of hospitalization among U.S. and Canadian adults aged 20–44 years. They account for over 70 million medical visits yearly in the United States alone.[6] Table 7.3 shows the percentages of U.S. adults who have common digestive disorders. Digestive disorders are common in children as well as adults. At least one-third of U.S. adults experience heartburn, and up to 28% of school children experience constipation.[8]

Constipation and Hemorrhoids]
Both constipation and hemorrhoids are often due to diets that provide too little fiber. Fiber intakes by adults of 25 to 30 grams per day along with plenty of fluids (8 to 12 cups a day) can help prevent constipa-

heartburn
A condition that results when acidic stomach contents are released into the esophagus, usually causing a burning sensation.

hemorrhoids (hem-or-oids)
Swelling of veins in the anus or rectum.

irritable bowel syndrome (IBS)
A disorder of bowel function characterized by chronic or episodic gas, abdominal pain, diarrhea or constipation, or both.

duodenal (do-odd-en-all) and stomach ulcers
Open sores in the lining of the duodenum (the uppermost part of the small intestine) or the stomach.

tion and hemorrhoids in healthy people. Some good sources of dietary fiber are shown in Illustration 7.7.

Ulcers and Heartburn] Ulcers develop when the protective barrier formed by cells lining the stomach and duodenum (the uppermost part of the small intestine) is damaged. This allows stomach acid and digestive enzymes to erode the lining of the stomach and duodenum and cause an "ulcer." Duodenal ulcers are ten times more common than stomach ulcers and are closely associated with the presence of *Helicobacter pylori (H. pylori)* bacteria.[9] *H. pylori* infects and irritates the lining of the stomach. The infection is acquired by the ingestion of foods and other substances contaminated with saliva, vomit, or feces from people harboring the bacteria in their stomach. Rates of *H. pylori* infection are highest in countries with poor sanitary conditions.[10] Excessive production of stomach acid can lead to ulcers, too, and may also cause heartburn. Heartburn is unrelated to heart conditions. It is called that because acid that escapes from the stomach causes a burning sensation in an area of the esophagus located near the heart.

Stress, anxiety, and frequent use of aspirin, ibuprofen, naproxen, and other medications appear to be related to the development of ulcers and heartburn in some people.[11] Fatty foods, coffee, alcohol, citrus fruits, soft drinks, and a variety of other foods have been implicated in the development of ulcers and heartburn. Rather than cause these disorders, certain foods appear to aggravate the symptoms of ulcers and heartburn. High-fiber diets, on the other hand, appear to be protective against the development of ulcers and heartburn.[12]

A number of medications are available for the treatment of *H. pylori* and the excessive production of stomach acid. Reduction of stress, judicious elimination of offending foods and beverages from the diet, and small, frequent meals also help relieve ulcers and heartburn in some people.[13]

Irritable Bowel Syndrome] Irritable bowel syndrome, abbreviated IBS, is not considered a disease but rather a persistent disorder in the way the colon functions. For some reason, the colon (or bowel) in people with IBS spasms, and that leads to

TABLE 7.3
COMMON DIGESTIVE DISORDERS.[7]

U.S. ADULTS AFFECTED*	
Heartburn	33.0%
Hemorrhoids	12.8
Irritable bowel syndrome	6–12%
Ulcers	3.5
Chronic constipation	3.0
Chronic diarrhea	1.2

*Noninstitutionalized adults experiencing the disorder in the past 12 months.

All Bran Cereal, 1/3 c. (10 g)
1 mango (4 g)
1/2 c. lima beans (5 g)
1 pear with skin (4 g)
1/2 c. corn (3 g)
1/2 c. carrots (2.8 g)

Richard Anderson

Illustration 7.7
Food sources of dietary fiber.
Together, the foods shown provide 29 grams of dietary fiber, an amount that helps prevent constipation.

probiotics
Non-harmful bacteria and some yeasts that help colonize the intestinal tract with beneficial microorganisms and that sometimes replace colonies of harmful microorganisms. Most common probiotic strains are *Lactobacilli* and *Bifidobacteria*.

diarrhea
The presence of three or more liquid stools in a 24-hour period.

painful cramps. Gas production, a feeling of being bloated, diarrhea, constipation, or both, are common features of IBS. Most commonly, mild IBS is treated with a low-fat diet (25% of calories), counseling, and over-the-counter fiber powders, antidiarrhea pills, and pain medications. More serious cases may respond to stress reduction, eating in relaxed surroundings, antidepressants, and drugs that reduce colon muscle spasms. *Probiotics,* or "friendly" bacteria that help colonize the colon with microorganisms that help protect and heal the colon lining, are being increasingly used for this disorder.[14]

Diarrhea] *Diarrhea* is a common problem in the United States and a leading public health problem in developing countries. Most cases of diarrhea are due to bacterial- or viral-contaminated food or water, lack of immunizations against infectious diseases, and vitamin A, zinc, and other nutrient deficiencies that make children particularly susceptible to diarrhea. Diarrhea can deplete the body of fluid and nutrients and produce malnutrition as well. If it lasts more than two weeks or is severe, diarrhea can lead to dehydration, heart and kidney malfunction, and death. An estimated 3.5 million deaths from diarrhea diseases occur each year to the world's population of children 5 years of age or under.[15]

The vast majority of cases of diarrhea can be prevented through food and water sanitation programs, immunizations, and adequate diets. The early use of oral rehydration fluids (for example, the formula provided by the World Health Organization and commercial formulas such as Pedialyte and Rehydralyte) shortens the duration of diarrhea. Rehydration generally takes 4 to 6 hours after the fluids are begun.[16]

Rather than "resting the gut" during diarrhea as used to be recommended, children and adults, once rehydrated, should eat solid foods. Foods such as yogurt, lactose-free or regular milk, chicken, potatoes and other vegetables, dried beans, and rice and other cereals are generally well tolerated and provide nutrients needed for the repair of the intestinal tract. It is best to avoid sugary fluids such as soft drinks. High-sugar beverages tend to draw fluid into the intestinal tract rather than increase the absorption of fluid.[17]

flatulence (flat-u-lens)
Presence of excess gas in the stomach and intestines.

Flatulence] Everyone experiences *flatulence*—it's normal. Gas can occur in the esophagus, stomach, small intestine, and large intestine due to swallowed air or bacterial breakdown of food in the large intestine. Air may be swallowed along with food and beverages or while chewing gum. Eating and drinking while in a rush generally increases air ingestion. Bacterial production of gas in the large intestine may be related to the ingestion of dried beans, broccoli, cauliflower, brussels sprouts, onions, corn, and other vegetables containing a type of complex carbohydrate that bacteria, but not humans, can break down. Fructose, which is used to sweeten a variety of food products and beverages, and sorbitol (used in some types of candy and gum) may lead to gas formation by bacteria that produce gas as a waste product of carbohydrate digestion. Heartburn and other gastrointestinal tract disorders and medications such as antibiotics are also associated with gas production.[18]

People often think they produce too much gas, even when they don't. The amount of gas swallowed and produced by gut bacteria varies a good deal among individuals, and within the same individual. Gas production changes depending on what foods are eaten, the types of bacteria populating the large intestine, the medications used, and the presence of gastrointestinal tract disorders. Severe and painful symptoms related to gas production may signal the presence of a digestive disorder.[19]

Stomach Growling] Gas in the stomach can make your stomach growl. When your stomach growls, you know that gas and food or fluids are mixing in your stomach. The growling tends to be louder when your stomach is empty, when there's no food to muffle the noise. It can occur anytime, whether your stomach is empty or full, but is more likely to happen when your stomach is empty.[20]

Lactose Maldigestion and Intolerance

I can't drink milk. It tears me up inside.

A very common digestive disorder is ***lactose maldigestion.*** The lactose found in milk and milk products presents a problem for most of the world's adults, who cannot digest it, either partially or completely (Table 7.4).[21] The condition occurs more commonly in population groups that have no historical links to dairy farming and milk drinking.[23] Early humans in central and northwestern Europe and the regions of Africa and China highlighted in Illustration 7.8 tended to raise dairy animals and drink milk.

Lactose maldigestion is caused by a genetically determined low production of lactase, the enzyme that digests lactose. People who lack this enzyme end up with free lactose in their large intestine after they consume milk or milk products that contain lactose. The presence of lactose in the large intestine produces the symptoms of ***lactose intolerance.*** These symptoms include a bloated feeling and diarrhea due to fluid accumulation, and gas and abdominal cramping caused by the excretion of gas by bacteria that digest lactose.

TABLE 7.4

ESTIMATED INCIDENCE OF LACTOSE MALDIGESTION AMONG OLDER CHILDREN AND ADULTS IN DIFFERENT POPULATION GROUPS.[22]

	INCIDENCE OF LACTOSE MALDIGESTION
Asian Americans	90%
Africans	70
African Americans	70
Asians	65 or more
American Indians	62 or more
Mexican Americans	53 or more
U.S. adults (overall)	25
Northern Europeans	20
American Caucasians	15

lactose maldigestion
A disorder characterized by reduced digestion of lactose due to the low availability of the enzyme lactase.

lactose intolerance
The term for gastrointestinal symptoms (flatulence, bloating, abdominal pain, diarrhea, and "rumbling in the bowel") resulting from the consumption of more lactose than can be digested with available lactase.

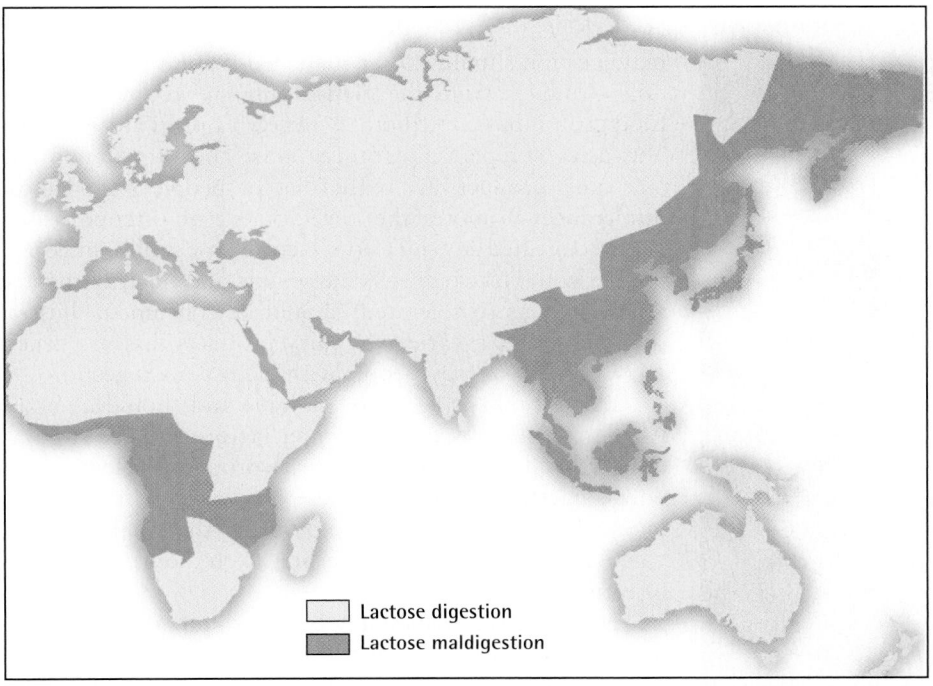

☐ Lactose digestion
■ Lactose maldigestion

Illustration 7.8
Lactose maldigestion is less common among descendants of people who consumed milk from domesticated animals during prehistoric times (light areas) than among people whose early ancestors did not drink milk (dark areas).[24]

Lactose maldigestion is rare in young children and affects adults to various degrees. Some adults produce little or no lactase and develop symptoms of lactose intolerance when they consume only small amounts of milk or milk products. Others produce some lactase and can tolerate limited amounts of lactose-containing milk and milk products, such as a cup of milk at a time or two cups of milk consumed with meals during the day. Regular consumption of milk may improve lactose digestion due to enhanced bacterial breakdown of lactose in the gut.[25]

Many people who are lactose maldigesters have no trouble eating yogurt and other fermented milk products such as cultured buttermilk, kefir, and aged cheese. The bacteria used to culture yogurt can digest half or more of the lactose. This reduction in lactose content is sufficient to prevent adverse effects in many people with lactose maldigestion.[26]

Milk solids, milk, and other lactose-containing components of milk may be added to foods you wouldn't expect. Consequently, it's best to examine food ingredient labels when in doubt. Milk, for instance, is a primary ingredient in some types of sherbet, and milk solids are added to many types of candy.

Do You Have Lactose Maldigestion?] The single most reliable indicator of lactose maldigestion is the occurrence of lactose intolerance within hours after consuming lactose.[27] If you consistently experience these symptoms (described earlier), visit your health care provider for a diagnosis. The symptoms could be due to lactose, other substances in milk, or another problem.

How Is Lactose Maldigestion Managed?] Lactose maldigestion should *not* be managed by omitting milk and milk products from the diet! Doing so would exclude a food group that contributes a variety of nutrients that cannot easily be replaced by other foods. The omission of milk and milk products from the diet of people with lactose intolerance promotes the development of osteoporosis.[28] Rather, fortified soy milk, low-lactose cow's milk, milk pretreated with lactase drops, and yogurt and other fermented milk products (if tolerated) should be consumed. Illustration 7.9 shows a variety of dairy products that are generally well tolerated by people with lactose maldigestion.

Lactase tablets are also available and should be taken within 30 minutes of consuming lactose.[29] Lactase is an enzyme, and enzymes are made of protein. Protein is partially digested in the stomach, so some of the lactase ingested in the tablets may not reach the small intestine where it is needed if the tablets are taken too far in advance of eating.

Digestion is a remarkably complex and efficient process that could be covered in much more detail than has been presented here. Readers are encouraged to consult the Web sites listed at the end of the unit for additional information.

"Brain freeze," or that splitting headache you can get from eating ice cream too fast, is caused by the quick drop in temperature in the back of your mouth. That causes vessels to constrict, and the result is an "ice cream headache."

Photo Disc

Richard Anderson

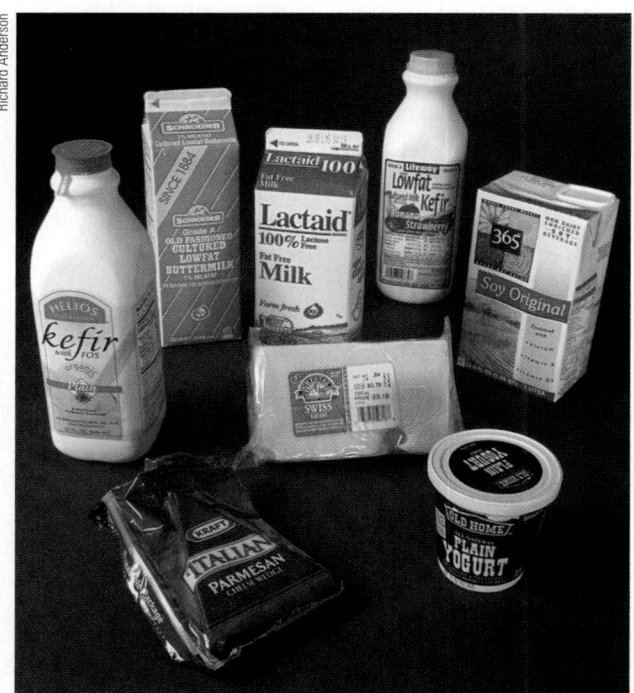

Illustration 7.9
Dairy products generally well tolerated by people with lactose maldigestion.

Key Terms

absorption, page 7–3

bile, page 7–4

circulatory system, page 7–6

diarrhea, page 7–10

digestion, page 7–3

disaccharide, page 7–4

duodenal and stomach ulcers,
    page 7–8

enzymes, page 7–4

flatulence, page 7–10

heartburn, page 7–8

hemorrhoids, page 7–8

irritable bowel syndrome (IBS),
    page 7–8

lactose intolerance, page 7–11

lactose maldigestion, page 7–11

lymphatic system, page 7–6

monosaccharides, page 7–4

probiotics, page 7–10

starch, page 7–4

www links

www.healthfinder.gov
Search digestive diseases by name, or search "digestion."

www.navigator.tufts.edu
Nutrition Navigator will help you find the most reliable sites on topics related to digestion.

digestive.niddk.nih.gov
NIH's National Digestive Disease Clearing-house homepage. This site provides links to information about specific digestive diseases, clinical trials, statistics, and answers to questions such as Why do I have gas?

www.nlm.nih.gov/medlineplus
Find out more about digestion, absorption, organs of the gastrointestinal tract, and digestive diseases at this site.

Notes

1. Digestive disease statistics, National Digestive Disease Clearinghouse, http://digestive.niddk.nih.gov/statistics/statistics.htm, accessed 8/03; and Leading causes of hospitalization in Canada. Health Canada, Population and Public Health Branch, www.hc-sc.gc.ca, accessed 8/03.

2. Hertzler SR, Clancy SM. Kefir improves lactose digestion and tolerance in adults with lactose maldigestion. J Am Diet Assoc 2003; 103:582–74.

3. Leading causes of hospitalization in Canada (www.hc-sc.gc.ca).

4. Schneeman B. Nutrition and gastrointestinal function. Nutr Today 1993;Jan/Feb:20–24.

5. Bengmark S. Econutrition and health maintenance: a new concept to prevent GI inflammation, ulceration, and sepsis. Clin Nutr 1996;15:1–10.

6. Digestive disease statistics (http://digestive.niddk.nih.gov/statistics/statistics.htm); Leading causes of hospitalization in Canada (www.hc-sc.gc.ca).

7. Digestive disease statistics (http://digestive.niddk.nih.gov/statistics/statistics.htm); Leading causes of hospitalization in Canada (www.hc-sc.gc.ca).

8. Borowitz SM et. al. Precipitants of constipation during early childhood. J Am Fam Pract 2003;16:213–8.

9. Current medical diagnosis and treatment, 35th ed. Stamford (CT): Appleton & Lange; 1997.

10. Suerbaum S, Michetti P. Helicobacter pylori infection. N Engl J Med 2002; 347:1175–86.

11. Kurata AN, et al. Dyspepsia in primary care: perceived causes, reasons for improvement, and satisfaction with care. J Fam Practice 1997;44:281–8.

12. The Merck manual of medical information. Whitehouse (NJ): Merck Research Laboratory; 2000; and Suerbaum and Michetti, Helicobacter pylori infection.

13. Kurata et al., Dyspepsia in primary care.

14. Irritable bowel syndrome. American Gastroenterologial Association Medical Position Statement. Gastoenterol 1997; 112:2118–9; and Barclay L. Highlights from digestive diseases week. An expert interview with Lawrence R. Schiller. Medscape Medical News, 2003, www.medscape.com.

15. Kilgore PE, et al. Trends in diarrheal disease associated mortality in US children, 1968 through 1991. JAMA 1995; 274:1143–8.

16. Kilgore et al., Trends in diarrheal disease; Goepp JG, Katz SA, Oral rehydration therapy, Am Fam Phys 1993;47: 843–8; and Meyers A. Oral rehydration

therapy: what are we waiting for? Am Fam Phys 1993;47:740–2.

17. Lima AAM, Guerrant RL. Persistent diarrhea in children: epidemiology, risk factors, pathophysiology, nutritional impact, and management. Epidemiol Rev 1992;34:222–42; and Goepp and Katz, Oral rehydration therapy.

18. Digestive disease statistics (http://digestive.niddk.nih.gov/statistics/statistics.htm).

19. Digestive disease statistics (http://digestive.niddk.nih.gov/statistics/statistics.htm.

20. Stomach growling. Scientific American. Com ask the expert. www.scientificamerican.com, accessed 8/03.

21. Hertzler and Clancy, Kefir improves lactose digestion.

22. Scrimshaw NS, Murray, EB. Prevalence of lactose maldigestion. Am J Clin Nutr 1988;48(suppl):1086–98; and Inman-Felton AE. Overview of lactose maldigestion (lactose nonpersistence). J Am Diet Assoc 1999;99:481–9.

23. Simmons FJ. Primary adult lactose intolerance and the milking habit: a problem in biological and cultural interrelationships. I. Review of the medical research. Am J Digestion 1981;14:819.

24. Simoons FJ. Primary lactose intolerance

and the milking habit: a problem in biological and cultural interrelationships. Am J Digestive Diseases 1970;15:696–710.

25. Tolstoi LG. Adult-type lactase deficiency. Nutr Today 2000;35:134–42; and Lactose intolerance: a self-fulfilling prophecy leading to osteoporosis. Nutr Rev 2003; 61:221–3.

26. Hertzler and Clancy, Kefir improves lactose digestion; and Martini MC, Smith DE, Savaiano DA. Lactose digestion from flavored and frozen yogurts, ice milk, and ice cream by lactase-deficient persons. Am J Clin Nutr 1987;46:636–40.

27. Simoons, Primary lactose intolerance.

28. Lactose intolerance: a self-fulfilling prophecy.

29. Lactose intolerance: a self-fulfilling prophecy.

Nutrition UP CLOSE

Personal History of Digestive Upsets

FOCAL POINT: Digestive disorders are common.

To bring you closer to your digestive system, review the following list and check the upsets you have experienced in the past month:

Digestive Upset	Experienced in the Past Month?
Heartburn or indigestion	_____
Diarrhea	_____
Constipation	_____
Stomach cramps	_____
Vomiting	_____
Abdominal bloating	_____
Hemorrhoids	_____

FEEDBACK: Most bouts of digestive upsets are brief and only bothersome. Some will be related to diet (inadequate or excessive dietary fiber, or bacterially contaminated food, for example) whereas others will be initiated by illness, stress, or another condition. Painful or prolonged episodes of digestive disorders should, of course, be brought to the attention of your health care provider.

UNIT 8

Calories! Food, Energy, and Energy Balance

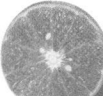

Nutrition Scoreboard

		TRUE	FALSE
1	Gram for gram, carbohydrates provide the body with more energy than protein does.		
2	A teaspoon of butter has a higher caloric value than a teaspoon of margarine.		
3	Energy can be neither created nor destroyed. It can, however, change from one form to another.		

Answers on next page

[KEY CONCEPTS AND FACTS]

- A calorie is a unit of measure of energy.

- The body's sources of energy are carbohydrates, proteins, and fats (the "energy nutrients").

- Fats provide over twice as many calories per unit weight as carbohydrates and proteins do.

- Most foods contain a mixture of the energy nutrients as well as other substances.

- Weight is gained when caloric intake exceeds the body's need for energy. Weight is lost when caloric intake is less than the body's need for energy.

Answers to *Nutrition Scoreboard*

		TRUE	FALSE
1	Carbohydrates and protein both provide 4 calories per gram. Fat, on the other hand, provides 9 calories per gram, and alcohol 7 calories per gram.		✔
2	Butter and margarine contain the same amount of fat, so they provide the same number of calories (35 per teaspoon).		✔
3	Stated is the conservation of energy principle, the first law of thermodynamics.	✔	

Energy!

It's high in energy, you say? Sure, I'll eat one, thanks. I thought those snack bars were loaded with calories!

When you think of calories, do you think of energy? That is the "scientifically correct" way to think about them. Energy is what calories are all about.

Illustration 8.1

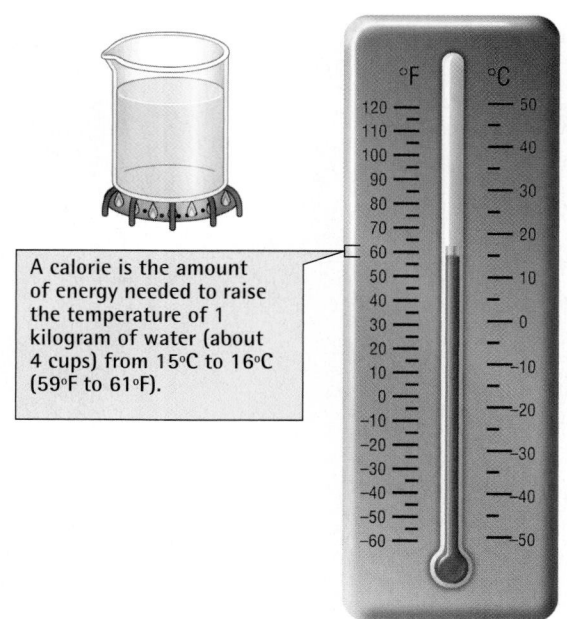

A calorie is the amount of energy needed to raise the temperature of 1 kilogram of water (about 4 cups) from 15°C to 16°C (59°F to 61°F).

calorie (*calor* = heat)
A unit of measure used to express the amount of energy produced by foods in the form of heat. The calorie used in nutrition is the large "Calorie," or the "kilocalorie" (kcal). It equals the amount of energy needed to raise the temperature of 1 kilogram of water (about 4 cups) from 15 to 16°C (59 to 61°F). The term *kilocalorie*, or "calorie" as used in this text, is gradually being replaced by the "kilojoule" (kJ) in the United States; 1 kcal = 4.2 kJ.

Calories Are a Unit of Measure

The *calorie* is like a centimeter or pound in that it is a unit of measure. Rather than serving as a measure of length or weight, the calorie is used as a measure of energy. Specifically, a calorie is the amount of energy needed to raise the temperature of 1 kilogram of water (about 4 cups) from 15°C to 16°C (59°F to 61°F) (Illustration 8.1). This amount of energy is used as a standard for assigning caloric values to foods. Because calories are a unit of measure, they are not a component of food like vitamins or minerals. When we talk about the caloric content of a food, we're really talking about the caloric value of the food's energy content.

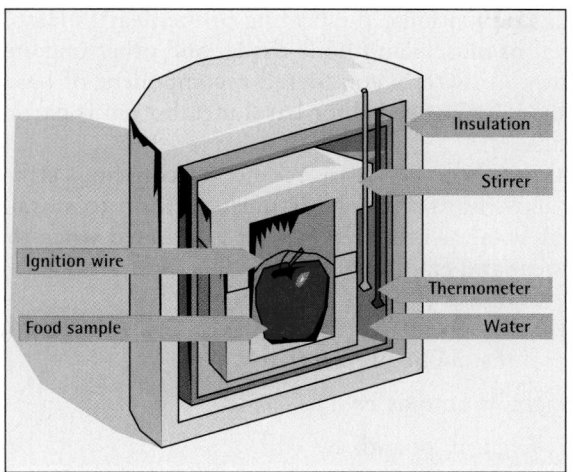

Illustration 8.2
How the energy content of food is measured.
The caloric value of a food is determined by the amount of heat it releases and transfers to water when completely burned.

The caloric value of food is determined by burning it completely in a container surrounded by a specific amount of water (Illustration 8.2). The energy released by the food in the form of heat raises the temperature of the surrounding water. The rise in temperature indicates how many calories were released from the portion of food. Although the body doesn't literally burn food, the amount of heat released by food while burning is approximately the same as the amount of energy it supplies the body.

The Body's Need for Energy

The body uses energy from foods to fuel muscular activity, growth, and tissue repair and maintenance; to chemically process nutrients; and to maintain body temperature (to name a few examples). These needs for energy are subdivided into three categories: basal metabolism, physical activity, and dietary thermogenesis (Table 8.1 and Illustration 8.3).

The largest single contributor to energy need is basal metabolism. It accounts for 60–80% of the total need for calories in the vast majority of people. Energy-requiring

---

TABLE 8.1

THE THREE ENERGY-REQUIRING PROCESSES OF THE BODY.

Basal metabolism
- Energy required to maintain normal body functions while at rest

Physical activity
- Energy needed for muscular work

Dietary thermogenesis
- Energy use related to food ingestion. (The process gives off heat.)

basal metabolism

physical activity

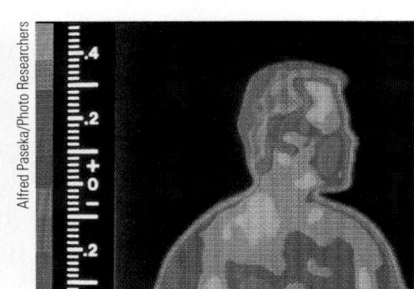

dietary thermogenesis

Illustration 8.3
Examples of the three types of energy-requiring processes in the body.

basal metabolism
Energy used to support body processes such as growth, health, tissue repair and maintenance, and other functions. Assessed while at rest, basal metabolism includes energy the body expends for breathing, the pumping of the heart, the maintenance of body temperature, and other life-sustaining, ongoing functions.

processes of *basal metabolism* include breathing, the beating of the heart, maintenance of body temperature, renewal of muscle and bone tissue, and other ongoing activities that sustain life and health. Growth is considered a component of basal metabolism. The proportion of total calories needed for basal metabolism is particularly high during the growing years.

Energy-using activities of basal metabolic processes require no conscious effort on our part; they are continuous activities that the body must perform to sustain life. The energy needed to carry out basal metabolic functions is assessed when the body is in a state of complete physical and emotional rest.

How Much Energy Do I Expend for Basal Metabolism?]

You can quickly estimate the calories needed for basal metabolic processes:

- For men: Multiply body weight in pounds by 11.
- For women: Multiply body weight in pounds by 10.[1]

Thus, a man who weighs 170 pounds needs approximately 170×11, or 1870 calories per day for basal metabolic processes. A 135-pound woman needs 135×10, or 1350 calories.

This formula gives an estimate of calories used for basal metabolism based on sex and weight. Other factors, particularly physical activity level, muscle mass, height, health status, and genetic traits, also affect how efficiently the body uses calories for basal metabolism. Consequently, results obtained using this quick formula may be 10 to 20% lower or higher than the true number of calories required for basal metabolism.[2]

How Much Energy Do I Expend in Physical Activity?]

The caloric level needed for physical activity can vary a lot, depending on how active a person is. It usually accounts for the second highest amount of calories we expend. The energy cost of supporting a physically inactive lifestyle (Table 8.2) is about 30% of the number of calories needed for basal metabolism. An "average" activity level requires roughly 50%, and an "active" level requires approximately 75% of the calories needed for basal metabolism.[4] A physically inactive person needing 1500 calories a day for basal metabolism, for example, would require about 450 calories (1500 calories $\times$ 0.30—for 30%) for physical activity.

A Common Source of Error} People have a tendency to overestimate time spent in physical activity. This in turn tends to overestimate calories needed for physical activity, and depending on the amount of overestimation, can lead to highly inac-

TABLE 8.2

ENERGY EXPENDITURE BY USUAL LEVEL OF ACTIVITY.[3]

ACTIVITY LEVEL	PERCENTAGE OF BASAL METABOLISM CALORIES
Inactive. Sitting most of the day; less than 2 hours of moving about slowly or standing	30%
Average. Sitting most of the day; walking or standing 2 to 4 hours, but no strenuous activity	50
Active. Physically active 4 or more hours each day; little sitting or standing; some physically strenuous activities	75

curate estimates of total calorie need. Physical activity level should be based on time spent actually engaged in physical activity and not include time spent getting ready for the activity, on breaks, or between activities.

How Many Calories Does Dietary Thermogenesis Take?] A portion of the body's energy expenditure is used for chewing and swallowing foods, digesting foods, absorbing and utilizing nutrients, and transporting nutrients into cells. Some of the energy involved in such activities escapes as heat. These processes are referred to as *dietary thermogenesis.* Calories expended for dietary thermogenesis are estimated as 10% of the sum of basal metabolic and usual physical activity calories. For instance, say a person's basal metabolic need is 1500 calories and 450 calories are required for usual activity: 1500 calories + 450 calories = 1950 calories. Calories expended for dietary thermogenesis would equal approximately 10% of the 1950 calories, or 195 calories.

Adding It All Up] Your estimated total daily need for calories is the sum of calories used for basal metabolism, physical activity, and dietary thermogenesis. In the preceding example, total calorie need would be 2145 calories (1500 + 450 + 195). A complete example of calculations involved in estimating a person's total calorie need is provided in Table 8.3. Although the caloric level calculated won't be exactly right, it should provide a good estimate of your total caloric need.

Where's the Energy in Foods?

Any food that contains carbohydrates, proteins, or fats (the "energy nutrients") supplies the body with energy. (That *includes* the foods listed in Table 8.4!) Carbohydrates and proteins supply the body with 4 calories per gram, and fat provides 9 calories per gram. Alcohol also serves as a source of energy. There are 7 calories in each gram of alcohol in the diet (Table 8.5).

dietary thermogenesis
Thermogenesis means "the production of heat." Dietary thermogenesis is the energy expended during food ingestion, the digestion of food, and the absorption and utilization of nutrients. Some of the energy escapes as heat. It accounts for approximately 10% of the body's total energy need. Also called *diet-induced thermogenesis* and *thermic effect of foods or feeding.*

TABLE 8.4

FOODS THAT DO HAVE CALORIES.

1. A candy bar eaten with a diet soda
2. Celery and grapefruit
3. Hot chocolate, cheesecake, or soft drinks consumed to make you feel better
4. Cookie pieces
5. Foods "taste-tested" during cooking
6. Foods you eat while on the run
7. Foods you eat straight from their original containers (like ice cream from the carton, milk from a jug, or chips from a large bag)

TABLE 8.3

SUMMARY OF CALCULATIONS FOR ESTIMATING TOTAL CALORIE NEED OF A 130–POUND, INACTIVE WOMAN.

	CALORIES
1. **Basal metabolism** Multiply body weight in pounds by 10. (For men the figure is 11.)	130 × 10 = 1300
2. **Physical activity** Multiply basal metabolism calories by 0.30 (for 30%) based on the usual energy expenditure level of "inactive" (see Table 8.2).	1300 × 0.30 = 390
3. **Dietary thermogenesis** Add calories needed for basal metabolism and physical activity together: 1300 + 390 = 1690 Multiply the result by 0.10 (for 10%).	1690 × 0.10 = 169
4. **Total calorie need** Add calories needed for basal metabolism, physical activity, and dietary thermogenesis together. 1300 + 390 + 169 =	**Total calorie need = 1859**

TABLE 8.5

CALORIC VALUES OF THE ENERGY NUTRIENTS AND ALCOHOL.

	cals/gm
Carbohydrate	4
Protein	4
Fat	9
Alcohol	7

If you enjoy grilling foods on an outdoor barbecue, you have probably observed firsthand the high level of stored energy in fats (Illustration 8.4). Unlike the drippings of low-fat foods such as shrimp or vegetables, drips from high-fat foods cause bursts of flames to shoot up from the grill. The high-energy content of alcohol can be seen in the alcohol-fueled flames that adorn cherries jubilee and baked Alaska.

If you know the carbohydrate, protein, fat, and (if present) alcohol content of a food or beverage, you can calculate how many calories it contains. For example, say a cup of soup contains 15 grams of carbohydrate, 10 grams of protein, and 5 grams of fat. To calculate the caloric value of the soup, multiply the number of grams of carbohydrate and protein by 4 and the number of grams of fat by 9. Then add the results together:

$$\begin{aligned}
15 \text{ grams carbohydrate} \times 4 \text{ calories/gram} &= 60 \text{ calories} \\
10 \text{ grams protein} \times 4 \text{ calories/gram} &= 40 \text{ calories} \\
5 \text{ grams fat} \times 9 \text{ calories/gram} &= \underline{45 \text{ calories}} \\
& 145 \text{ calories}
\end{aligned}$$

You can calculate the percentage of total calories from carbohydrate, protein, and fat in the soup by dividing the number of calories supplied by each nutrient by the total number of calories and then multiplying by 100:

- Carbohydrate: $\dfrac{60 \text{ calories}}{145 \text{ calories}} = 0.41 \times 100 = 41\%$

- Protein: $\dfrac{40 \text{ calories}}{145 \text{ calories}} = 0.28 \times 100 = 28\%$

- Fat: $\dfrac{45 \text{ calories}}{145 \text{ calories}} = 0.31 \times 100 = \underline{31\%}$
  $ 100\%$

Given this information about the caloric value of carbohydrates, proteins, and fats, which of the items in Illustration 8.5 would you expect to be highest in calories?

College students have been found to miss this question 47% of the time.[5] It's the margarine! Margarine contains the most fat; it is made primarily from oil. *High-fat foods provide more calories ounce for ounce than foods that contain primarily carbohydrate or protein.* That means bread and potatoes (rich in carbohydrates), catsup (rich in water and a low-calorie vegetable), sugar, and other low-fat or no-fat foods provide fewer calories than equal amounts of foods that contain primarily fat.

Illustration 8.4
If you have observed the flames produced by fat dripping from a steak or hamburger on a grill, you have seen the powerhouse of energy stored in fat.
The carbohydrate and protein contents of grilled foods don't burn with nearly the same intensity. They have less energy to give.

Felicia Martinez/PhotoEdit

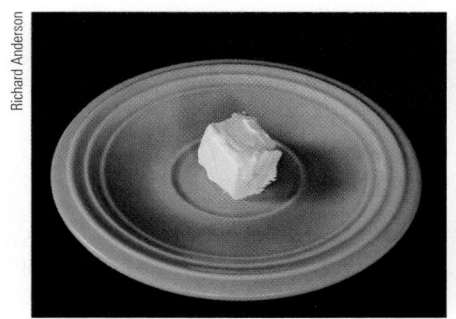
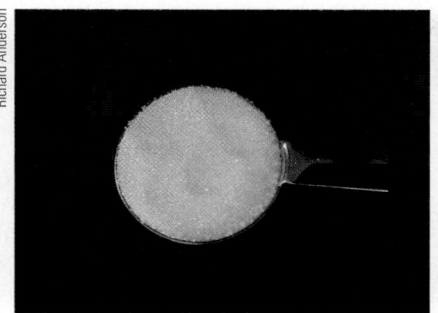

Illustration 8.5
Which contains the most calories—a tablespoon of margarine, sugar, or pork?
(Turn the page to find the answer.)

Most Foods Are a Mixture

Some foods (such as oil and table sugar) consist almost exclusively of one energy nutrient, but most foods contain many substances in varying amounts.

Bread is high in complex carbohydrates ("starch"), but it also contains many other substances such as protein, water, vitamins, minerals, and fiber. Likewise, steak is not all protein. Although protein constitutes about 32% of the total weight of lean sirloin steak, fats make up about 8%, and the largest single ingredient is water—60% of the weight. Yet we don't think of steak as a "high-water food"; instead, it's often thought of as pure protein.

So, even though some foods provide relatively more carbohydrate, protein, or fat than other foods, most foods contain a mixture of energy nutrients.

The Caloric Value of Foods: How Do You Know?

Some people say you can estimate the caloric value of food by how it tastes or by how appetizing it looks. Actually, it's not that simple. How many calories do you think are contained in a half-cup of peanuts, a boiled, medium-sized potato, and a cup of rice (Illustration 8.6)?

Taste, appearance, and reputation do not make good criteria for determining the caloric value of foods. Here's an example:

I used to order the fish sandwich at fast-food restaurants because I was trying to avoid all the calories in hamburgers. Then I found out the fish sandwich had about the same number of calories as the quarter-pound hamburger! Where did I get the idea that fried fish loaded with tartar sauce has fewer calories than a hamburger?

Illustration 8.6
What's the caloric value of these foods?
One contains 420 calories, another 205, and the third 118 calories. Try matching the caloric values with the foods, and then check your answers on p. 8–9.

Answer to Illustration 8.5: Margarine. There are about 101 calories in one tablespoon of margarine, 46 in a tablespoon of sugar, and 40 in a tablespoon of relatively lean pork.

hunger
Unpleasant physical and psychological sensations (weakness, stomach pains, irritability) that lead people to acquire and ingest food.

satiety
A feeling of fullness or of having had enough to eat.

appetite
The desire to eat; a pleasant sensation that is aroused by thoughts of the taste and enjoyment of food.

People expend an average of 11 calories an hour chewing gum. If you chewed gum for an hour every day and didn't change any other component of energy balance, you'd lose about a pound a year.[9]

Photo Disc

This doesn't have to happen to you! By referring to Appendix A, you could determine that a typical fast-food fish sandwich and a quarter-pound hamburger weigh in at about 400 calories each. If you're interested in the caloric value of foods you frequently eat, check them out.

How Is Caloric Intake Regulated by the Body?

Because energy is critical to survival, the body has a number of mechanisms that encourage regular caloric intake. It has less effective means of discouraging an excessive intake of calories.[6]

Mechanisms that encourage food intake don't depend on weight status.[7] Rather, they are keyed to encouraging eating on a regular basis so that food, if available, will be consumed and carry the body through times when food isn't around. Humans developed on a schedule of "feast and famine." Those who could store enough fat to see them through the times when food was scarce had an advantage. So, no matter how thin or fat a person is, she or he experiences *hunger* if food is not consumed several times throughout the day. The "hungry" signal is thought to be sent by a series of complex mechanisms when cells run low on energy nutrients supplied by the last meal or snack.[8]

When we eat, we reach a point when we feel full and are no longer interested in eating. The signal is due to hormones and internal sensors in the brain, stomach, liver, and fat cells that indicate *satiety*—the feeling that we've had enough to eat.[10]

For most people, hunger and satiety mechanisms adjust energy intake to match the body's need for energy. However, internal signals that urge us to eat or to stop eating can be overridden. People can resist eating, no matter how strong the hunger pains. Even after the "I'm full" siren has sounded, people can go on eating.[11] And sometimes people eat because they have an *appetite* for specific foods and desire the pleasure foods can bring.

Appetite may or may not be related to being hungry. It can be triggered when we smell or see a tasty food right after a meal or when we're really hungry. Ever see those TV commercials for a juicy burger and fries that air about 11:00 p.m.? How many people who jump into their cars and head to the carry-out window are actually hungry? Or have you noticed how appealing the idea of eating is and how good food tastes when you're very hungry? That's appetite at work.

The Question of Energy Balance] Unless you are currently losing or gaining weight, the number of calories you need is the number you usually consume in your diet. Adults who maintain their weight are in a state of energy balance (Illustration 8.7). Because they are not losing weight (using fat and other energy stores) or gaining it (storing energy), their body's expenditure of energy and its intake of energy are balanced. When energy intake is less than the amount of energy expended, people are in negative energy balance. In this case, energy stores are used and people lose weight. When a positive energy balance exists, weight and fat stores are gained because more energy is available from foods than is needed by the body.

Sometimes, a positive energy balance is a healthy and normal circumstance. For example, a positive energy balance is normal when growth is occurring, as in childhood or pregnancy, or when a person is regaining weight lost during an illness.

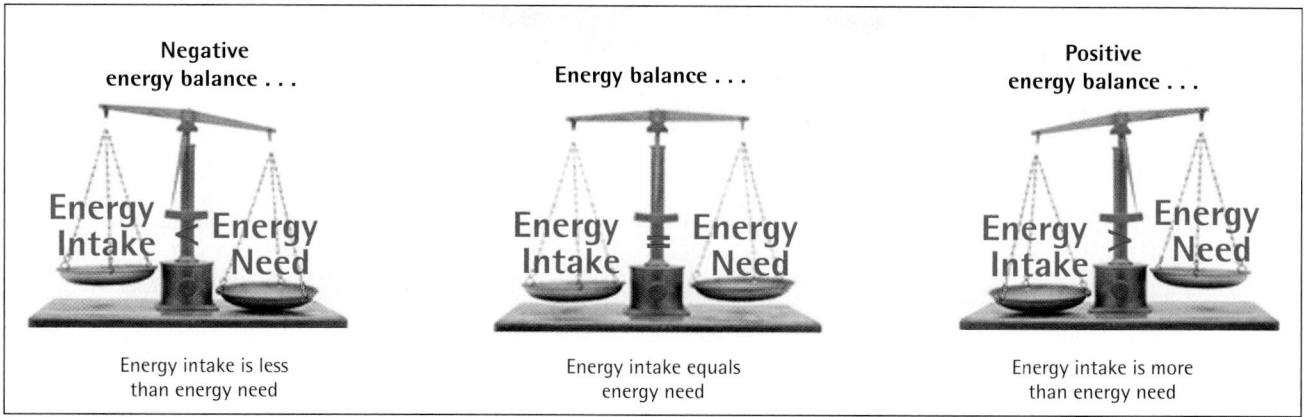

Illustration 8.7
The body's energy status.

A Few Final Notes about Energy Balance] A person's total caloric need, and therefore the number of calories needed to maintain energy balance and weight, is affected somewhat by:

- Smoking

- Lean muscle mass

- Genetic makeup

People who smoke tend to burn slightly more calories than people who don't, and muscular individuals need more energy to maintain their body weight than do individuals of the same weight with less muscle.[12] A person's genetic makeup may increase or decrease caloric need, but such effects appear to be quite small.[13] It used to be thought that "yo-yo" dieting, or repeated bouts of weight loss and gain, lowers a person's need for calories. Well-controlled studies indicate this is not the case, however.[14] Slight changes in caloric need may not amount to much on a day-to-day basis. Over the course of a year, however, a *positive* or *negative* daily energy balance of 50 calories would result in a weight gain or loss of approximately 5 pounds.

Keep Calories in Perspective

Calories is not a word that means "fattening" or "bad for you." Calories are a life- and health-sustaining property of food. Diets compatible with health contain a mixture of foods providing various amounts of calories. It's not the caloric content of individual foods that makes them good or not. It's the sum of calories and nutrients in foods that make up our total diets.

Caloric values of the foods in Illustration 8.6:

	Calories
½ cup of peanuts	420
1 medium boiled potato	118
1 cup of white rice	205

Nutrition UP CLOSE

Food as a Source of Calories

FOCAL POINT: Examine the distribution of calories in the foods you eat.

What are the sources of calories in food? Determine the caloric contribution of the fat, carbohydrate, and protein content of the following snack foods using the composition data and calorie conversion factors listed. Then answer the two questions that follow.

Which snack is lower in total calories? _____
Which snack is lower in fat calories? _____

FEEDBACK (answers to these questions) can be found at the end of Unit 8.

Photo Disc

Calorie conversion factors

1g fat = 9 calories

1g carbohydrate = 4 calories

1 g protein = 4 calories

Potato Chips

Serving size: 1 oz (about 20 chips)

Fat: 10g X ____ = ____ calories from fat

Carbohydrate: 15g X ____ = ____ calories from carbohydrate

Protein: 2g X ____ = ____ calories from protein

____ Total calories

% of calories from fat: ____ = _____ = ____

Mini Pretzels

Serving size: 1 oz (about 17 pieces)

Fat: 0g X ____ = ____ calories from fat

Carbohydrate: 24g X ____ = ____ calories from carbohydrate

Protein: 3g X ____ = ____ calories from protein

____ Total calories

% of calories from carbohydrate: ____ = _____ = ____

Key Terms

appetite, page 8-8

basal metabolism, page 8-4

calorie, page 8-2

dietary thermogenesis, page 8-5

hunger, page 8-8

satiety, page 8-8

www links

www.nal.usda.gov/fnic/foodcomp
Look up the calorie, fat, and other nutrient values of food at this Web site. It's a bit slow. You can also use Appendix A in this text.

www-users.med.cornell.edu/~spon/picu/calc/beecalc.htm
Use this site from Cornell University to estimate your basal energy expenditure.

www.brianmac.demon.co.uk/energyexp.htm
This site presents tables of calories expended in exercise. It also includes a "Calories Counter" that can be used to estimate calories expended when performing different physical activities.

www.cyberdiet.com
Select "Fast Food Quest" to look up and compare caloric and nutrient values of foods from fast-food restaurants.

Notes

1. Harris JA, Benedict FG. A biometric study of basal metabolism in men. Publication no. 279 of the Carnegie Institute of Washington, 1919.

2. Sjodin AM, et al., The influence of physical activity on BMR, Med Sci Sports Exercise 1996;28:85–91; Klausen B, et al., Age and sex effects on energy expenditure, Am J Clin Nutr 1997;65: 895–907; and Frankenfield DC, et al., The Harris-Benedict studies of human basal metabolism: history and limitations, J Am Diet Assoc 1998;98:439–45.

3. Boothby WM, Berksen J, Dunn HL. Studies of the energy of metabolism of normal individuals: a standard for basal metabolism, with a nomogram for clinical application. Am J Physiol 1936;116: 468–84.

4. Boothby et al., Studies of the energy of metabolism of normal individuals.

5. Melby CL, Femea PL, Sciacca JP. Reported dietary and exercise behaviors, beliefs and knowledge among university undergraduates. Nutr Res 1986;6: 799–808.

6. Mattes RD. Hunger and satiety. Presentation at the American Dietetic Association annual meeting, Boston: 1997 Oct. 27.

7. Raben A et al. Meals with similar energy densities but rich in protein, fat, carbohydrate, or alcohol have different effects on energy expenditure and substance metabolism but not on appetite and energy intake. Am J Clin Nutr 2003;77:91–100.

8. Raben et al., Meals with similar energy densities.

9. Levine J, et al. The energy expended in chewing gum. N Eng J Med 1999;341: 2100.

10. McCrory MA, et al. Dietary determinants of energy intake and weight regulation in healthy adults. J Nutr 2000; 130:276S–9S.

11. Marmonier C et al. Snacks consumed in a nonhungry state have poor satiating efficiency. Am J Clin Nutr 2002;76: 518–28.

12. Perkins KA, Effects of tobacco smoking on caloric intake, Br J Addiction 1992;87:193–205; and Poehlman ET, Horton ES, The impact of food intake and exercise on energy expenditure, Nutr Rev 1989;47:129–37.

13. Ravussin E, Bogardus C. Relationship of genetics, age, and physical fitness to daily energy expenditure and fuel utilization. Am J Clin Nutr 1989;49 (suppl):968–75.

14. Bray GA. Physiology and consequences of obesity. Medscape Diabetes and Endocrinology Clinical Management Modules, http://womenshealth. medscape.com/medscape/endocrinology/ ClinicalMgmt/CM.v03/pnt-CM.v03. html, 2001, accessed 1/01.

Nutrition UP CLOSE

Food as a Source of Calories

Feedback for Unit 8

Potato Chips

Serving size: 1 oz (about 20 chips)

Fat: 10 g 3 __9__ = __90__ calories from fat

Carbohydrate: 15 g 3 __4__ = __60__ calories from carbohydrate

Protein: 2 g 3 __4__ = __8__ calories from protein

__158__ Total calories

% of calories from fat: $\dfrac{90}{158}$ = **0.57** × **100** = **57%**

Mini Pretzels

Serving size: 1 oz (about 17 pieces)

Fat: 0 g 3 __9__ = __0__ calories from fat

Carbohydrate: 24 g 3 __4__ = __96__ calories from carbohydrate

Protein: 3 g 3 __4__ = __12__ calories from protein

__108__ Total calories

% of calories from carbohydrate: $\dfrac{96}{108}$ = **0.89** × **100** = **89%**

The pretzels are a better snack choice because they are lower in
total calories and fat calories than the potato chips.

The Highs and Lows of Body Weight

Nutrition Scoreboard

	TRUE	FALSE
1 There is usually little difference between the ideal body weights defined by cultural norms and those defined by science.		
2 Fat stores concentrated around the stomach present a greater hazard to health than do fat stores located around the hips and thighs.		
3 High-calorie diets are the cause of obesity.		
4 Healthy underweight people often find it nearly impossible to gain weight.		

Answers on next page

[KEY CONCEPTS AND FACTS]

- Ideal body weight and shape are defined by culture and by health measures. The operating definition should be based on health.

- Rates of overweight and obesity are increasing worldwide. Disease and disorders related to excess body fat are also increasing.

- The location of body fat stores, as well as the amount of body fat, are important to health.

- The causes of obesity are complex and not completely understood. Diet, physical activity, environmental exposures, and genetic factors influence the development of obesity.

- Underweight in the United States usually results from a genetic tendency to be thin, or from poverty, illness, or the voluntary restriction of food intake.

Answers to *Nutrition Scoreboard*

		TRUE	FALSE
1	Cultural norms of ideal body weights are often at odds with the "healthy" weights defined by science.		✔
2	Health is affected by the location of body fat stores as well as by obesity.[1] Adults who store body fat in the stomach, or central area of the body, are at higher risk for a number of health problems than are adults who store fat primarily in their hips and thighs.	✔	
3	Obesity has many causes, and high-calorie diets are just one of them.		✔
4	Gaining weight is as difficult for many underweight people as losing weight is for many overweight people.	✔	

Variations in Body Weight

Wouldn't it be terrific if your body adjusted your food intake based on a healthy level of body fat stores? Then you wouldn't have to worry about being too thin or getting too fat. Why doesn't that happen?

Photo Disc

Obesity appears to represent a weak link in the biological evolution of humans. Body processes that regulate food intake developed over 40,000 years ago when "feast and famine" cycles were common. Being underweight was a distinct disadvantage. The more body fat people could store after a feast, the better their chances of surviving the subsequent famines. Consequently, multiple mechanisms that favored food intake and body fat storage developed.[2] These body mechanisms continue to encourage food intake even when food is constantly available and obesity poses a greater threat to survival than does famine. In the words of Theodore Van Itallie, a noted obesity researcher, in environments with an abundant supply of food and no requirement for vigorous physical activity, "perhaps thin people are the ones who are abnormal."[3]

Circumstances today are very different from those existing when the human genetic endowment and its food intake regulatory mechanisms were established. These changes may go a long way to explaining why obesity is—or is becoming—a major problem in many industrialized countries.

How Is Weight Status Defined?

Culture and science define the appropriateness of body size. In the fifteenth century, moon-faced and pear-shaped adults were considered beautiful.[4] In the 1930s, full-figured women such as Jean Harlow were box office stars, and Marilyn Monroe was an icon of the 1950s. Then came Twiggy, Calista Flock-hart, and Lara Flynn Boyle. Thinness beyond health boundaries became the standard of beauty. With "too thin" as the cultural ideal, many people regard weights within the normal range as being too high. Men feel the pressure, too. For them, visible body fat is culturally taboo.[5]

Science defines standards for body weight for adults based on the risk of death from all causes. Death rates are highest among adults who have very high body weights for height, and lowest among adults who are normal weight for height (Illustration 9.1).[6] You can get an idea of what a normal weight for height is by following the steps detailed in the "Health Action."

In the past, standards used to identify whether a person was underweight, normal weight, overweight, or obese were based on weight for height tables developed by the insurance industry to estimate life expectancy and the risk of death of adults. These tables have been replaced by standards that employ *body mass index*.[9]

Body Mass Index

Most commonly referred to as BMI, body mass index is a measure of weight for height that provides a fairly good estimate of body fat content.[10] Ranges of BMI are used to define weights for height that correspond to underweight, normal weight, and obesity in adults (Table 9.1). BMI has the advantage of being calculated the same way for adult males and females. Calculating BMI involves dividing weight in pounds by height in inches, then dividing this result by height in inches again, and then multiplying that result by 703. An example of BMI calculation is given in Table 9.2. You can use the chart shown in Illustration 9.2 to identify weight status based on BMI from information on weight and height.

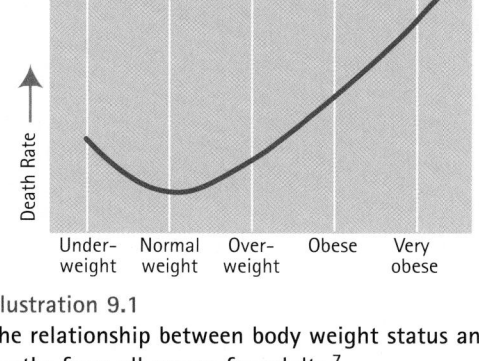

Illustration 9.1

The relationship between body weight status and deaths from all causes for adults.[7]

body mass index (BMI)
An indicator of the appropriateness of a person's weight for their height. It is calculated by dividing weight in kilograms by height in meters squared. It can also be calculated using the method shown in Table 9.2.

TABLE 9.1

CLASSIFYING WEIGHT STATUS BY
BODY MASS INDEX.[11]

	BODY MASS INDEX
Underweight	under 18.5 kg/m²
Normal weight	18.5–25 kg/m²
Overweight	25–30 kg/m²
Obese	30 kg/m² or higher*

*Obesity is subdivided for some purposes into the BMI groups of moderate obesity (30–35 kg/m²), severe obesity (35–40 kg/m²), and very severe obesity (40+ kg/m²).

HEALTH ACTION I *Estimating Normal Weight for Height*

There is a quick way to estimate what a normal, or healthy weight for height in adults is. It's the Hamwi[8] method, and it is done as follows.

WOMEN
Begin with 5 feet equals 100 pounds, and then add 5 pounds for each additional inch of height. Here's an example of how you would estimate a healthy weight for a woman who is 5 feet, 7 inches tall:

5 feet	= 100 pounds
7 inches × 5 pounds	= 35 pounds
100 pounds + 35 pounds	= 135 pounds

MEN
For men, five feet equals 106 pounds, and each additional inch of height adds on 6 pounds. So, for example, we would estimate a healthy weight for a man who is 5 feet 10 inches tall the following way:

5 feet	= 106 pounds
10 inches × 6 pounds	= 60 pounds
100 pounds + 66 pounds	= 166 pounds

TABLE 9.2

CALCULATING BMI: AN EXAMPLE.

Say Chris weighs 140 pounds and is 5 feet, 3 inches tall. To calculate BMI:

Figure out how many inches tall Chris is:

5 feet × 12 inches per foot = 60 inches
60 inches + 3 inches = 63 inches.

Divide Chris's weight by his height in inches:

$$\frac{140 \text{ pounds}}{63 \text{ inches}} = 2.22$$

Divide the result (2.22) by Chris's height in inches again:

$$\frac{2.22}{63 \text{ inches}} = 0.035$$

Multiply this result by 703:

0.035 × 703 = 24.8 This is Chris's BMI.

Assessing Weight Status in Children and Adolescents] Standards used to assess weight status in children and adolescents employ BMI percentile ranges for girls and boys (Illustration 9.3). Percentiles of BMI are based on the proportion of children and adolescents who have different levels of BMI at given ages. For example, if a child's BMI is at the 50th percentile for his or her age, then half of children will have BMIs that are below, and half will have values above this child's BMI level. Table 9.3 shows BMI percentile ranges that correspond to underweight, at risk of underweight, normal weight, at risk of overweight, and overweight in children and adolescents. Obesity is not included in this table. Body fat content, rather than BMI percentile ranges, should be used to diagnose obesity in children and adolescents.[14]

BMI percentile ranges for children and adolescents do not provide information on growth progress in terms of height. Consequently, growth in height cannot be assessed using BMI. Other standards for assessment of height in children and adolescents are available.

Illustration 9.2
BMI Chart

Body Mass Index (BMI)

Height	18	19	20	21	22	23	24	25	26	27	28	29	30	31	32	33	34	35	36	37	38	39	40
											Body Weight (pounds)												
4'10"	86	91	96	100	105	110	115	119	124	129	134	138	143	148	153	158	162	167	172	177	181	186	191
4'11"	89	94	99	104	109	114	119	124	128	133	138	143	148	153	158	163	168	173	178	183	188	193	198
5'0"	92	97	102	107	112	118	123	128	133	138	143	148	153	158	163	168	174	179	184	189	194	199	204
5'1"	95	100	106	111	116	122	127	132	137	143	148	153	158	164	169	174	180	185	190	195	201	206	211
5'2"	98	104	109	115	120	126	131	136	142	147	153	158	164	169	175	180	186	191	196	202	207	213	218
5'3"	102	107	113	118	124	130	135	141	146	152	158	163	169	175	180	186	191	197	203	208	214	220	225
5'4"	105	110	116	122	128	134	140	145	151	157	163	169	174	180	186	192	197	204	209	215	221	227	232
5'5"	108	114	120	126	132	138	144	150	156	162	168	174	180	186	192	198	204	210	216	222	228	234	240
5'6"	112	118	124	130	136	142	148	155	161	167	173	179	186	192	198	204	210	216	223	229	235	241	247
5'7"	115	121	127	134	140	146	153	159	166	172	178	185	191	198	204	211	217	223	230	236	242	249	255
5'8"	118	125	131	138	144	151	158	164	171	177	184	190	197	203	210	216	223	230	236	243	249	256	262
5'9"	122	128	135	142	149	155	162	169	176	182	189	196	203	209	216	223	230	236	243	250	257	263	270
5'10"	126	132	139	146	153	160	167	174	181	188	195	202	209	216	222	229	236	243	250	257	264	271	278
5'11"	129	136	143	150	157	165	172	179	186	193	200	208	215	222	229	236	243	250	257	265	272	279	286
6'0"	132	140	147	154	162	169	177	184	191	199	206	213	221	228	235	242	250	258	265	272	279	287	294
6'1"	136	144	151	159	166	174	182	189	197	204	212	219	227	235	242	250	257	265	272	280	288	295	302
6'2"	141	148	155	163	171	179	186	194	202	210	218	225	233	241	249	256	264	272	280	287	295	303	311
6'3"	144	152	160	168	176	184	192	200	208	216	224	232	240	248	256	264	272	279	287	295	303	311	319
6'4"	148	156	164	172	180	189	197	205	213	221	230	238	246	254	263	271	279	287	295	304	312	320	328
6'5"	151	160	168	176	185	193	202	210	218	227	235	244	252	261	269	277	286	294	303	311	319	328	336
6'6"	155	164	172	181	190	198	207	216	224	233	241	250	259	267	276	284	293	302	310	319	328	336	345
Underweight (<18.5)		Healthy Weight (18.5–24.9)						Overweight (25–29.9)									Obese (≥30)					

Find your height along the left-hand column and look across the row until you find the number that is closest to your weight. The number at the top of that column identifies your BMI. The area shaded in green represents healthy weight ranges.

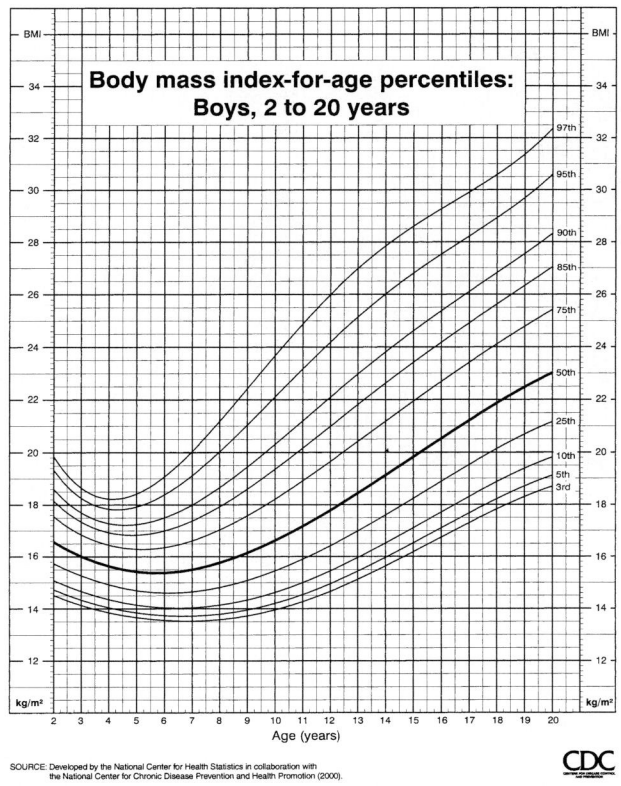

Illustration 9.3

An example of the CDC's 2000 Growth Charts: BMI-for-age. *Copies of the growth charts can be obtained from www.cdc.gov/growthcharts*

TABLE 9.3

WEIGHT STATUS STANDARDS FOR 2- TO 20-YEAR-OLDS BASED ON BMI-FOR-AGE.[13]

	PERCENTILE RANGE(S)
Underweight	5th or less
At risk of underweight	5th–15th
Normal weight	15th–85th
At risk of overweight	85th–95th
Overweight	95th or higher

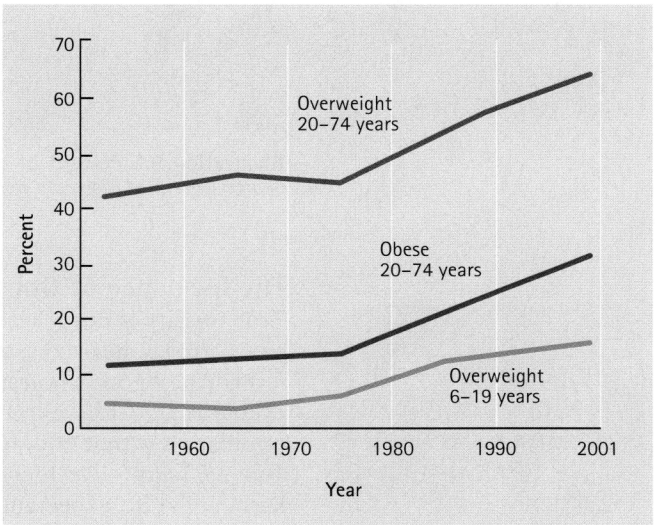

Illustration 9.4   Overweight and obesity by age in the United States, 1960–2001.
Source: Centers for Disease Control and Prevention, National Center for Health Statistics, CDC Fast Stats, 2003.

Most Adults in the United States Weigh Too Much

Being *overweight* or obese *is* the norm in the United States. The combined incidence of overweight and obesity among adults is nearly 65% and is on its way up (Illustration 9.4).

Over 1 in 7 children and adolescents in the United States are overweight, and this percentage is rising as you read this.[15] Overweight and obesity is becoming America's number one health problem. Obesity, which represents the largest risks to health, varies considerably among U.S. states (Illustration 9.5). In 2001, for example, rates of obesity were highest (24–26%) in Mississippi, West Virginia, and Michigan; and lowest (14–17%) in Colorado, Massachusetts, and Vermont.[16]

The trend toward higher rates of overweight and *obesity* in the United States is now shared by almost all other countries of the world. In Europe, 10–25% of adults are obese, and over a billion adults worldwide are overweight or obese.[17] A sampling of obesity rates by country is shown in Table 9.4. Obesity is a major, worldwide public health problem.

The high and growing incidence of overweight and obesity in the United States and around the world represents a time bomb for a future explosion in obesity-related disease rates.

overweight
A high weight-for-height.

obesity
A condition characterized by excess body fat.

TABLE 9.4

OBESITY RATES IN A SAMPLING OF COUNTRIES.[18]

PERCENT OBESE	
West Germany	37
Czech Republic	36
England	32
Canada	30
Australia	21
Netherlands	21
France	10
Japan	4
China	3

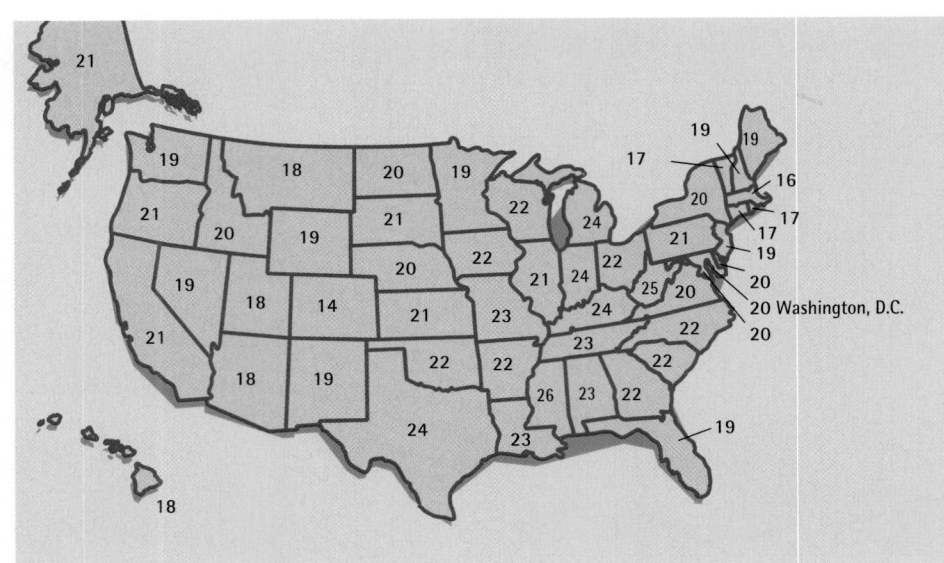

Illustration 9.5 **Percent of obese adults (BMI = 25 kg/m^2) by state.**
Source: Generated from data from CDC, 2003.

TABLE 9.5

THE MISMATCH BETWEEN OBESITY AND HEALTH. ALL THESE PROBLEMS ARE ASSOCIATED WITH OBESITY.[21]

- Diabetes
- Hypertension
- Stroke
- Elevated cholesterol level
- Low HDL-cholesterol level
- Heart disease
- Certain types of cancer
- Gallbladder disease
- Shortened life expectancy
- Discrimination
- Depression
- Infertility
- Accidents
- Skin disorders
- Sleep disorders

The Influence of Obesity on Health

True obesity is not a healthy state. It carries an increased risk of diabetes, hypertension, stroke, heart disease, elevated total cholesterol levels, low HDL-cholesterol levels (the "good cholesterol"), certain types of cancer, and other health problems (Table 9.5).[20] Life expectancy is also influenced by body weight. Contrary to the popular view that underweight people live the longest, adults with BMIs between 22 and 25 kg/m^2 live longer on average than do underweight adults (BMI < 18.5 kg/m^2).[22] Life expectancy in overweight adults is approximately 3 years, and in obese persons 6 years, shorter than average.[23] The more obese a person is, the more common health problems related to obesity become.[24] In the category of good news is the fact that loss of 5–10% of body weight in obese adults decreases risks to health substantially. Weight loss lowers blood pressure, lowers LDL cholesterol, increases HDL cholesterol (the one you want to be high), and decreases the risk of diabetes or high glucose levels in people with diabetes.[25]

Another consequence of obesity is the ingrained cultural prejudice to which obese people are subjected. Children who are obese are more likely to suffer unfair or indifferent treatment from teachers than are nonobese children. They experience more isolation, rejection, and feelings of inferiority than other children. Obese adults are likely to be discriminated against in hiring and promotion decisions and to be thought of as lazy or lacking in self-control, even by health professionals who care for them.[26] Society's prejudice against people who don't conform to the cultural ideal of body size may be the most injurious consequence of obesity.[27]

Importance of Body Fat Content

Although closely related,[28] weight-for-height and percent body fat do not always correspond. Take a 165-pound, 30-year-old woman who is 5 feet 6 inches tall and weight trains heavily. Her body weight for height would indicate obesity, but she may have a low body fat content and be healthy. Sometimes people who are classified as normal weight or underweight by BMI standards have too much body fat because they are physically inactive. Certain medications make people retain fluid.

Their weight-for-height may qualify them as overweight, but their body fat content may actually be very low. Obviously, measures of body fat content are better estimators of health status than are measures of weight-for-height.

Methods for Assessing Body Fat Content] Death and disease rates are more closely related to body fat content than to weight-for-height.[29] Yet, the most commonly used standards for assessing health risks are based on weight-for-height and not body fat content. Tools for assessing body fat content that are easy to use, accurate, and inexpensive are now available and their use is spreading. This development is starting a trend for the determination of healthy body weights based on body fat content.[30] Standards for classifying percent body fat, or the percent of weight that consists of fat, have been developed (Table 9.6) and will be refined as additional studies on the relationships between body fat and health risk are done.

Here are the most common methods for determining body fat content:

- skinfold thickness measures
- bioelectrical impedance analysis (BIA)
- underwater weighing
- magnetic resonance imaging (MRI)
- computerized axial tomography (CT or CAT scans)
- dual-energy X-ray absorptiometry (DEXA)
- whole body air displacement (BOD POD)

The theories underlying each of these methods and its advantages and limitations are presented in Table 9.7. Each body fat test has its advantages and drawbacks. The tests are most likely to provide accurate results when they are performed by skilled, experienced technicians using proper, well-maintained equipment, and when the measurements are converted into percent body fat by the appropriate formulas.

Everybody Needs Some Body Fat] A certain amount of body fat—3 to 5% for men and 10 to 12% for women—is needed for survival. Body fat serves essential roles in the manufacture of hormones; it's a required component of every cell in the body; and it provides a cushion for internal organs. Fat that serves these purposes is not available for energy formation no matter how low energy reserves become. Low body fat levels are associated with delayed physical maturation during adolescence, infertility, accelerated bone loss, and problems that accompany starvation.[32]

Body Fat Location Is Important to Health] The location of body fat stores, as well as total body fat content, is important to health.[33] You may be better off health-wise if you're a "pear" rather than an "apple" (Illustrations 9.6 and 9.7). Pear-shaped individuals store fat primarily in their hips and thighs, whereas apples store fat centrally—around the waist. The apple shape is more common among obese men than the pear shape is. Obese women exhibit both shapes.[34]

Fat cells that make up central fat deposits are larger than those around the hips and "resistant" to insulin. This decreases the ability of insulin to lower blood glucose levels. As a result of this state of **insulin resistance**, blood levels of insulin increase, as do levels of glucose over time. Increased blood levels of insulin are related to increased blood triglyceride levels and blood pressure, and reduced levels

TABLE 9.6

RANGES OF HEALTHY PERCENTAGES OF BODY FAT.[31]

AGE IN YEARS	% BODY FAT
Females	
20–39	21–33
40–59	23–34
60–79	24–36
Males	
20–39	8–20
40–59	11–22
60–79	13–25

ON THE SIDE

On average, men in the United States are 5 feet, 9 inches tall and weigh 180 pounds. Women are 5 feet, 3½ inches tall and weigh 152 pounds.[12]

insulin resistance
A condition in which cells "resist" the action of insulin in facilitating the passage of glucose into cells.

TABLE 9.7

COMMONLY USED METHODS FOR ASSESSING BODY FAT CONTENT.

SKINFOLD MEASUREMENT

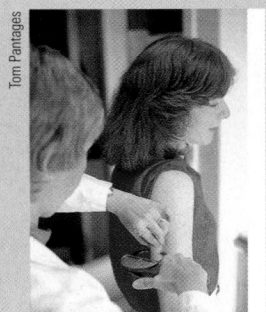

Body fat content can be estimated by measuring the thickness of fat folds that lie underneath the skin. Calipers are used to measure the thickness of fat folds, preferably over several sites on the body. Body fat content is estimated by "plugging" the thicknesses into the appropriate formula.

Advantages: Calipers are relatively inexpensive (they cost about $250–$350). The procedure is painless if done correctly and can yield a good estimate of percent body fat.

Limitations: This technique is often performed by untrained people. Skinfolds may be difficult to isolate in some individuals.

BIOELECTRICAL IMPEDANCE ANALYSIS (BIA)

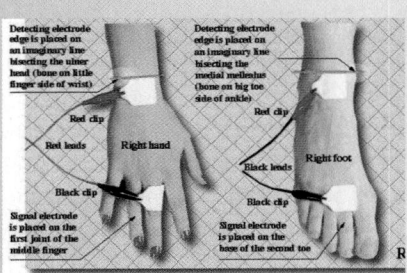

Because fat is a poor conductor of electricity, and water and muscles are good conductors, body fat content can be estimated by determining how quickly electrical current passes from the ankle to the wrist.

Advantages: The equipment required is portable, and the test is easy to do and painless. The results are fairly accurate for people who are not at the extremes of weight-for-height.

Limitations: Equipment may be expensive; inferior equipment produces poor results. Hydration status and meal ingestion may affect electrical conductivity and produce inaccurate results, as may inaccurate formulas used to calculate body fat content from test results.

UNDERWATER WEIGHING

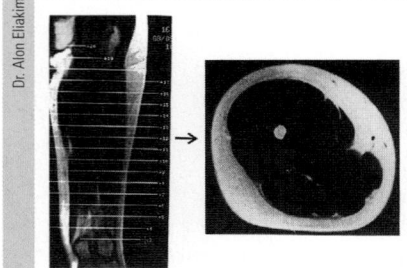

The subject is first weighed on dry land; next he or she is submerged in water and exhales completely; then his or her weight is measured. The less the person weighs under water compared to the weight on dry land, the higher the percent body fat. (Fat, but not muscle or bone, floats in water.)

Advantages: If undertaken correctly and if appropriate formulas are used in calculations, this technique gives an accurate value of percent body fat.

Limitations: The equipment required is expensive and not easily moved. The test doesn't work well for people who don't swim or who are ill or disabled in some way.

MAGNETIC RESONANCE IMAGING (MRI)

MRI and CT or CAT scans provide similar results. Using this technology, a person's body fat and muscle mass can be photographed from cross-sectional images obtained when the body, or parts of it, is exposed to a magnetic field (or, in the case of CAT scans, to radiation). Based on the volume of fat and muscle observed, total fat and muscle content can be determined.

Advantages: Provides highly accurate assessment of fat and muscle mass.

Limitations: Expensive ($500–$1000 per assessment). Largely used for research purposes.

TABLE 9.7

COMMONLY USED METHODS FOR ASSESSING BODY FAT CONTENT. (CONTINUED)

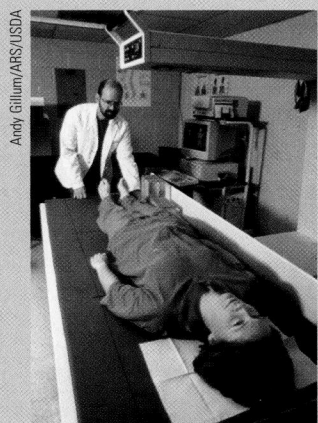

Andy Gillum/ARS/USDA

DUAL-ENERGY X-RAY ABSORPIOTMETRY (DEXA)

DEXA (or DXA) is based on the principle that various body tissues can be differentiated by the level of X-ray absorption. The measure is made by scanning the body with a small dose of X rays (similar to the level of exposure from a transcontinental flight) and then calculating body fat content based on the level of X-ray absorption.

Advantages: Provides highly accurate results when measurements are undertaken correctly. DEXA is safe and "user friendly" for people being measured, and can also be used to assess bone mineral content and lean tissue mass.

Limitations: DEXA machine is expensive and so is the cost of individual assessments ($150 per person). Machine must be operated by a trained and certified radiation technologist in many states.

WHOLE BODY AIR DISPLACEMENT

This is an established method that has become practical for broader use due to development of the BOD POD. The method is similar to that of underwater weighing but uses air displacement for determining percent body fat. Individuals sit in an enclosed "cabin" for about 5 minutes while wearing a tight-fitting swim suit and cap. Computerized sensors determine body weight and the amount of air that is displaced by the body.

Advantages: This method provides a quick, comfortable, automated, and reasonably accurate way to assess body fat content. It is suitable for disabled individuals, the elderly, and children.

Disadvantages: Results can be modified by drinking, eating, or exercising before testing; a full bladder, and failure to adhere to test procedures, may lead to error in results. Costly, but less expensive than DEXA.

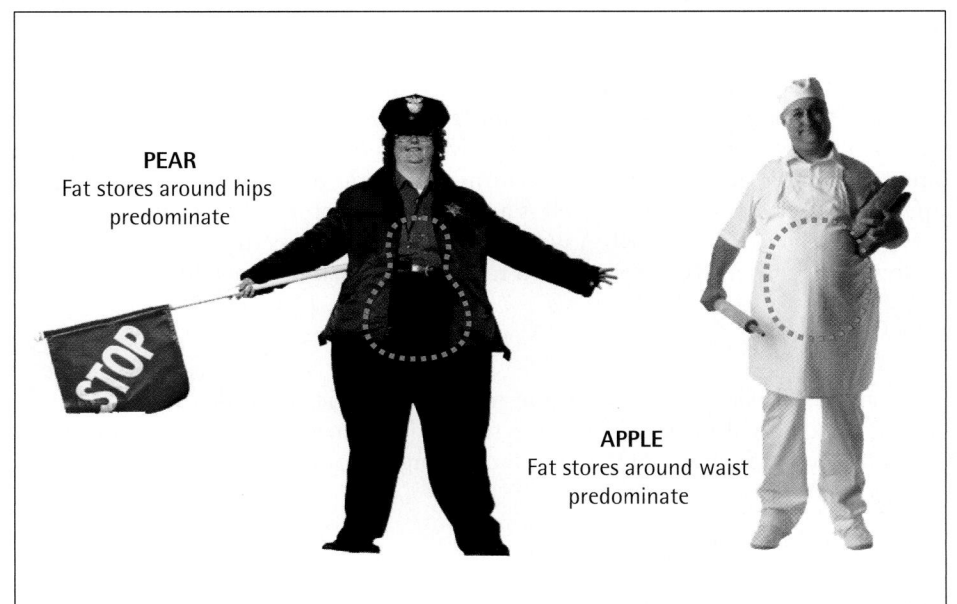

PEAR
Fat stores around hips predominate

APPLE
Fat stores around waist predominate

Illustration 9.6
Basic body shapes.
The pear normally has narrow shoulders, a small chest, and an average-size waist. Fat is concentrated in the hips and thighs. Weight loss in these areas is usually difficult.

The apple looks round in the middle. Fat is concentrated in the waist and can be lost with diet and exercise. Apple-shaped people are at increased risk for diabetes, hypertension, high blood cholesterol, and heart disease.

Illustration 9.7
The man on the left is 6 feet tall and weighs 240 pounds; the man on the right is 6 feet 1 inch tall and weighs 230 pounds.

Although similar in weight and height, the man on the left is at higher risk for diabetes, hypertension, and heart disease due to his "apple" shape and body fat content.

of beneficial HDL cholesterol. These changes increase the risk of hypertension and heart disease. Increasing levels of blood glucose promote the development of diabetes.[35]

Weight loss in people with high levels of body fat reduces disease risk by lowering both central and lower body fat stores. Central body fat stores decline first with weight loss, making it easier for most people to lower central rather than lower body fat stores.[36]

Assessing Body Fat Distribution] Body fat distribution can be determined by assessing a person's waist-to-hip ratio (Illustration 9.8). People with the pear shape have lower waist-to-hip ratios than apples. Waist-to-hip ratios of more than 0.80 in women and 0.90 in men indicate central body fat distribution—or the apple shape.[38] Waist circumference by itself is a stronger predictor of heart disease, stroke, and type 2 diabetes than is BMI. A waist circumference of less than 40 inches in men or below 35 inches in women is related to decreased risk of these diseases.[39]

What Causes Obesity?

Simply stated, obesity results when the intake of calories exceeds caloric expenditure. But the cause of obesity is not that simple. Whether people accumulate excess body fat or not is due to complex and interacting factors that include

- diet
- physical activity
- environmental exposures
- genetic background[40]

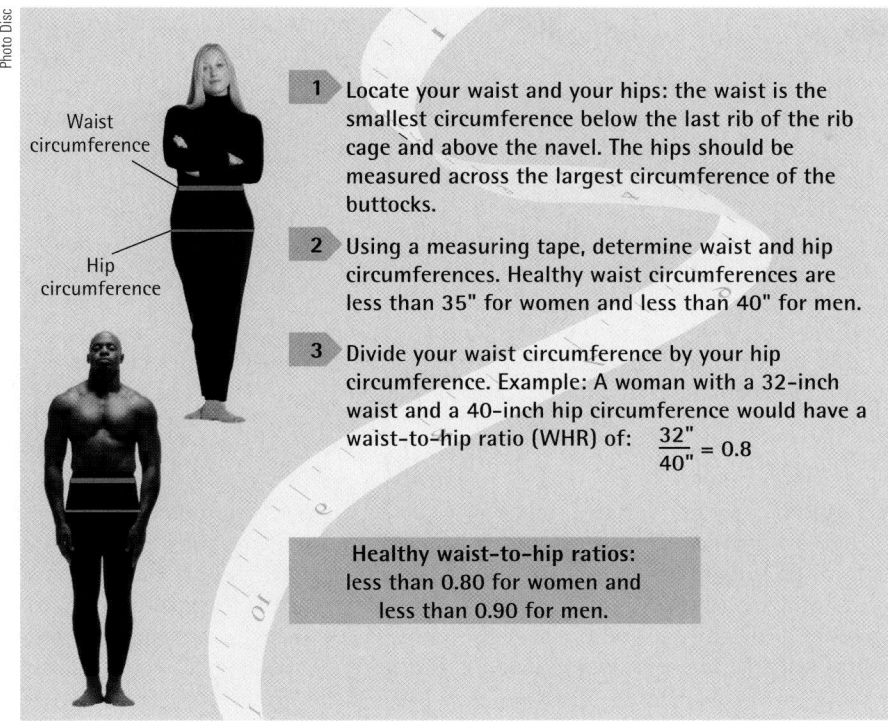

Illustration 9.8
Determining your waist-to-hip ratio.[37]

Waist circumference

Hip circumference

1 Locate your waist and your hips: the waist is the smallest circumference below the last rib of the rib cage and above the navel. The hips should be measured across the largest circumference of the buttocks.

2 Using a measuring tape, determine waist and hip circumferences. Healthy waist circumferences are less than 35" for women and less than 40" for men.

3 Divide your waist circumference by your hip circumference. Example: A woman with a 32-inch waist and a 40-inch hip circumference would have a waist-to-hip ratio (WHR) of: $\dfrac{32"}{40"} = 0.8$

Healthy waist-to-hip ratios: less than 0.80 for women and less than 0.90 for men.

Between 25 and 40% of domestic cats seen in veterinary practices are overweight or obese.[19]

ON THE SIDE

Several diseases and medications are also associated with the development of obesity, but their contribution to the overall incidence of obesity is small (although meaningful to the affected people).

Are Some People Born to Be Obese?

Accumulating evidence indicates that genetics plays important roles in the development and maintenance of obesity. Heredity may account for approximately 25 to 40% of the incidence of obesity.[41] Genetic influences on obesity take several paths. Some people are born with errors in *metabolism* that produce obesity (this appears to be rare). Others are born with genetic traits that predispose them to becoming obese (probably common). A predisposing genetic trait is expressed when the right *environmental trigger* exists. For example, a person with a genetic predisposition to becoming obese may maintain normal weight as long as she or he is physically active or consumes a low-fat diet. If activity level becomes low or the diet changes to one high in fat (two types of environmental triggers), the genetically susceptible person then gains weight.[42] Genetically based differences in biological processes affecting food intake and energy utilization appear to influence a person's susceptibility to obesity. Individuals lacking leptin, a hormone involved in appetite regulation, or who secrete high levels of insulin may be at high risk for obesity.[43]

The propensity to gain weight when excessive levels of calories are consumed may be affected by genetic traits. Claude Bouchard and his colleagues at Lavel University in Canada fed adult identical twins 840 more calories a day than each twin needed to maintain his or her weight.[44] At the end of the 100-day feeding period, weight gain among the twins ranged from 9 to 29 pounds! (The average weight gain was 24 pounds.) Related twins tended to gain similar amounts of weight, but the weight gains varied widely even though all subjects consumed the same number of excess calories. Another study showed that identical twins reared apart are less likely to have similar body weights than identical twins who are reared together.[45]

metabolism
The chemical changes that take place in the body. The conversion of glucose to energy or to body fat is an example of a metabolic process.

environmental trigger
An environmental factor, such as inactivity, a high-fat diet, or a high sodium intake, that causes a genetic tendency toward a disorder to be expressed.

Other Potential Genetic Influences]   Other hereditary factors have also been theorized to be related to obesity. The "set-point theory," for example, suggests that individuals are programmed to weigh a certain amount and that body weight will return to that level after weight is lost or gained. It has also been proposed that the number and size of fat cells predispose some people to obesity. Neither of these two popular theories has been verified by research.[46]

Recent increases in the incidence of obesity in the United States and other developed countries indicate that genetic influences in combination with environmental factors, such as inactivity, high-calorie diets, and a continuous supply of highly palatable foods, play key roles in the development of obesity.[47] We are just beginning to understand the nature of the interactions between environmental factors and genes. The day will come, however, when people with inborn tendencies toward obesity can be identified by an examination of their genes.

Do Obese Children Become Obese Adults?

The link between early and later obesity is very weak for obese infants and children under the age of three years.[48] It becomes stronger among older children, especially if one or both parents are obese. Only 8% of obese children who are heavy at one to two years of age and who do not have an obese parent are obese as adults. However, nearly 80% of children who are obese between the ages of ten and fourteen and have at least one obese parent are obese as adults.[49]

The Role of Diet in the Development of Obesity

Regardless of the cause of obesity, weight gain results when more energy is consumed than expended. Americans are consuming more energy than in previous years: caloric intake is up by an average of 340 per day while physical activity level has remained about the same. Although Americans are consuming a lower proportion of total calories from fat, total fat intake is up due to increased caloric intake. Fruits, vegetables, and fiber are often missing from diets that provide too many calories.[50]

The generous availability of inexpensive, calorie-dense foods; eating out, and large portions of favorite foods tend to increase caloric intake. Adults served large portions of food consume an average of 30 to 50% more food than when presented with small food portions. When offered a lot of tasty food, people tend to eat beyond the feeling of satiety and past the point where food continues to taste good.[51]

Do High-Fat Diets Promote Obesity?]   It does not appear that high-fat diets are more fattening than other types of diets. It's caloric intake and energy expenditure that count. Fats do provide over twice the calories as do equal amounts of carbohydrate or protein, so it may take less food to reach high-calorie intakes from diets high in fat.[52]

Low Levels of Physical Activity Promote Obesity]   Low levels of physical activity are thought to be related to the high and increasing incidence of obesity among Americans.[53] To a large extent, physical activity has become voluntary. Many farmers now plow, sow, and reap in air-conditioned tractors with power steering. Lawn mowers propel themselves and sometimes their operators, too. Instead of walking or biking two and a half miles to the store and back, we drive. An average-size adult driving 2.5 miles burns about 17 calories. Biking that distance would use seven times more calories (122). But walking 2.5 miles would burn around 210 calories, more than 12 times the calories it takes to drive! We have traded physical activity for convenience, time, and, perhaps, personal fat stores. Too much television watching, in particular, has been blamed for obesity in children and adults; but it is not clear which comes first—obesity, sedentary lifestyles, or too much TV.[54]

Obesity: The Future Lies in Its Prevention

Whether obesity is related to genetic predisposition, environmental factors, or a combination of both, certain steps can be taken to help prevent it.

Preventing Obesity in Children

For children, the prevention of obesity includes the early development of healthy eating and activity habits. Parents should offer a nutritious selection of food, but children themselves should be allowed to decide how much they eat.[55] Physical activities that are fun for every child—not just those who show athletic promise—should be routine in schools and summer programs.

Interactions between parents and children around eating and body weight can set the stage for the prevention or the promotion of over-weight in children. Parents who overreact to a child's weight by focusing on it, restricting food access, and making negative comments to the child may increase the likelihood that eating and weight problems will develop or endure. Lifestyle changes for the whole family—such as incorporating fun physical activities into daily schedules; making a wide assortment of nutritious foods available in the home; and decreasing a focus on eating, foods, and weight—are some of the positive changes families can make to promote healthy eating and exercise habits, and normal weight in children.[56]

THE LOCKHORNS

"DON'T BE DOWNHEARTED. THINK HOW MUCH LESS YOU WEIGH NOW THAN YOU WILL NEXT YEAR!"

Illustration 9.9
Source: Reprinted with special permission of King Feature Syndicate.

Preventing Obesity in Adults

Action needs to be taken to prevent weight gain during the adult years. Many adults gain weight at a slow pace (about three-fourths of a pound per year) as they age (Illustration 9.9), whereas others gain substantial amounts of weight over short periods of time.[57] Data from a national nutrition and health survey indicate that major gains in weight are most likely to occur in adults between the ages of 25 and 34 years.[58] Regular, vigorous exercise may prevent or lessen the amount of weight gain that occurs with age, as may decreased portion sizes at home and in restaurants.[59] Paying attention to the "I'm full" feeling can help moderate food intake. For more on this topic, visit the "Reality Check on page 9-15."

Regular intake of 2–4 servings per day of milk and milk products may also help prevent weight gain. A number of studies have shown that people who routinely consume milk, yogurt, and other dairy products weigh less and have less body fat than people who don't.[61] In situations where obesity cannot be prevented, society itself should take action. It would help if people acknowledged and accepted the fact that people come in all sizes. In reality, why *should* one size fit all (Illustration 9.10)?

Some People Are Underweight

Worldwide, underweight due to inadequate diets is much more common than obesity and a good deal more life-threatening.[62] There are many preventable reasons why people fail to get enough to eat. The lows of body weight discussed here relate to underweight as it occurs in many economically developed nations.

Illustration 9.10
People come in many different sizes and shapes.

In contrast to the desperate situations faced by many people in economically emerging countries, underweight in developed nations largely results from illnesses such as HIV/AIDS, pneumonia, and cancer; an eating disorder (anorexia nervosa); and the voluntary restriction of food. An important and preventable cause of underweight in the United States and some economically developed nations is poverty.[63]

Underweight Defined

underweight
Usually defined as a low weight-for-height. May also represent a deficit of body fat.

People who are **underweight** have too little body fat, or less than 20% body fat in adult females and 8% body fat in males. They have BMIs below 18.5 kg/m^2 as indicated in Table 9.1. A portion of people classified as underweight by BMI will not actually be underweight, just as a subset of people categorized as obese by BMI are not really obese.

Some people assessed as underweight for height are healthy and have a healthy body composition. Like the person in Illustration 9.11, they are probably genetically thin.[64] People who are naturally thin often have as much difficulty gaining weight as obese people have losing it.[65] Unless health is compromised, as indicated by fatigue, frequent illness, impaired concentration, apathy, or extreme intolerance to cold temperatures, there is little reason to be concerned about being underweight.

Illustration 9.11
Some underweight people are genetically thin and are healthy.

An Avoidable Cause of Underweight: Middle-Class Malnutrition

Underweight related to low-calorie, low-fat diets has been observed in children of middle- to upper-income families in the United States.[66] Some of the parents say they implement this sort of diet to help their children form good lifelong eating habits and avoid obesity and heart disease. Parents' fear that their children will become obese appears to be the major motivation for restricting their children's food intake.

About 2% of adolescents from middle- and upper-income families limit their diets in order to prevent obesity.[67] While not having anorexia or bulimia, these adolescents moderately and continually restrict their weight gain. Unlike people with anorexia nervosa or bulimia, the goal is not to lose weight, but to avoid *gaining* weight. The failure to gain weight during periods of growth has an undesired consequence: it may halt growth in height. Counseling has been effective in getting adolescents to eat enough to allow growth in height to proceed.[68]

REALITY CHECK

The "I'm full" feeling

How do you know when you've had enough to eat

?

Moonesong:
I'm on automatic pilot for eating until my plate is clean.

Shankuan:
I stop eating when, like boom!—the "I'm full" signal hits.

Answers on next page

Underweight and Longevity in Adults

Longevity can be extended in adult mice and monkeys by feeding them a nutritious, calorie-restricted diet that produces underweight.[69] Although it is not known whether caloric restriction and underweight would serve as a fountain of youth for humans, a growing number of adults believe it will. Devotees of calorie restriction for longer life tightly control their food intake and activity level to maintain a lean body.

Adults following calorie-restricted diets are at risk of iron deficiency, osteoporosis, infertility, infections, and of becoming irritable. The theory that life expectancy would be increased by calorie restriction is challenged by data on humans that show increased disease and death rates among underweight compared to normal weight people.[70]

Toward a Realistic View of Body Weight

A widespread belief among Americans is that individuals can achieve any body weight or shape they desire if they just diet and exercise enough. It's a myth. People naturally come in different weights and shapes, and these can only be modified so much.[71] Half of all women in the United States wear size 14 to 26, yet many clothing models are very underweight. Many men, no matter how hard they work out, will never have a washboard stomach or fit into "slim jeans."

Size Acceptance

The U.S. obsession with body weight and shape is spreading to other industrialized countries as part of popular culture. Ironically, strong societal bias against certain body sizes may contribute broadly to weight and health problems. Intolerance of overweight and obese children and adults tends to increase discrimination against them. This type of societal bias lowers the individual's feeling of self-worth, and may promote eating disorders, including the consumption of too much food. Females are hardest hit by negative attitudes about body size. Although the incidence of overweight and obesity tends to be higher in males, obesity in females carries with it many more negative stereotypes.[72] Overreactive parents make things worse.

ANSWERS TO REALITY CHECK

The "I'm full" feeling

Even favorite foods lose some or all of their appeal after we're full. The "I'm full" feeling will let you know when that happens—if you pay attention.[60]

Moonesong

Shankuan

Of all the things societies can do to prevent obesity, acceptance of people of different sizes and a more realistic view of obtainable body weights and shape may be two of the most important.

Key Terms

body mass index (BMI), page 9-3

environmental trigger, page 9-11

insulin resistance, page 9-7

metabolism, page 9-11

obesity, page 9-5

overweight, page 9-5

underweight, page 9-14

www links

www.cdc.gov/nchs/fastats/overwt.htm
Rates of overweight and obesity are increasing fast enough to make it difficult to keep textbooks updated between revisions. For the latest statistics, go to this site.

www.cdc.gov
Select weight loss/dieting under health topics and you'll be met by a wealth of information and resources.

www.nhlbi.nih.gov/subsites/index.htm
Provides information that targets the Dietary Guideline "Aim for Healthy Weight." Includes interactive BMI calculator and other applications.

www.nhlbi.nih.gov/guidelines/obesity/ob_home.htm
Leads you to a BMI calculator, the report "Clinical Guidelines for Identifying, Evalu-

ating, and Treating Overweight and Obesity in Adults."

www.naafa.org
The National Association to Advance Fat Acceptance is dedicated to improving the quality of life for fat people. The organization takes on issues of size discrimination and policies against the interest of fat people, and it promotes fat acceptance through media and legislative routes.

www.sizewise.com
Provides information, updates, and resources related to size acceptance.

www.healthfinder.gov
Search "obesity" to find clinical guidelines for diagnosis and treatment, the role of genes in the development of obesity, national efforts aimed to reduce obesity, and more.

www.shapeup.org
This site offers information, advice, and products for weight management. Provides quick access to an interactive calculator for estimation of resting metabolic rate, appropriate portion sizes, and an "All About Me" button that allows you to get a rough estimate of percent body fat and to calculate BMI.

www.cdc.gov/growthcharts
The 2000 growth charts for children and adolescents 2 to 20 years old, including BMI-for-age graphs, are available at this site.

Notes

1. Lean MJ, Han TS. Waist worries. Am J Clin Nutr 2002;76:699–70.

2. Brown PJ, Konner M. An anthropological perspective on obesity. Annals of NY Acad Sci 1987;499:29–46.

3. Van Itallie TB. Obesity: the American disease. Food Technol 1979;(Dec):43–47.

4. Rossner S. Ideal body weight—for whom? (editorial). Acta Medica Scand 1984;216:241–2.

5. Fallon A, Rozin P. Sex differences in perception of desirable body shape. J Abnormal Psychology 1985;94:102–5.

6. Stevens J et al. Evaluation of WHO and

NHANES II standards for overweight using mortality rates. J Am Diet Assoc 2000;100:825–7; and Calle EE et al. Overweight, obesity, and mortality from cancer in a prospectively studied cohort of U.S. adults. N Engl J Med 2003;348: 1625–38.

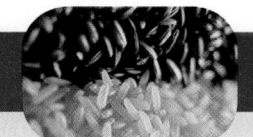

Nutrition **UP CLOSE**

Are You an Apple or a Pear?

FOCAL POINT: Determining your waist-to-hip ratio.

Follow the directions given in Illustration 9.8 for determining your waist-to-hip circumference ratio. If you don't have a measuring tape, use a string. Mark the string where the two ends intersect when placed around your waist and your hips. Then use a ruler to measure the distance between the end and each mark on the string.

FEEDBACK can be found at the end of Unit 9.

7. Calle et al., Overweight, obesity, and mortality from cancer.

8. Roth JH. What is optimum body weight? (letter) J Am Diet Assoc 1995;95:856–7.

9. Defining overweight and obesity, www.cdc.gov/nccdphp/dnpa/obesity/defining.htm, accessed 7/03.

10. Fernández JR et al. Is percentage body fat differentially related to body mass index? Am J Clin Nutr 2003;77:71–5.

11. Defining overweight and obesity (www.cdc.gov/nccdphp/dnpa/obesity/defining.htm).

12. Centers for Disease Control and the National Center for Health Statistics, www.cdc.gov/nchs/fastats/overweight.htm, accessed 8/25/03.

13. CDC growth charts: United States. Advance Data, No. 314, 12/4/00 (Revised), www.cdc.gov/nchs.

14. CDC growth charts: United States (www.cdc.gov/nchs).

15. CDC growth charts: United States (www.cdc.gov/nchs).

16. CDC growth charts: United States (www.cdc.gov/nchs).

17. European nations failing obese patients—survey. Survey results from health professionals at the 12th European Congress in Obesity in Helsinki, www.medscape.com, accessed 6/13/03; and Aitman TJ, Genetic medicine and obesity. N Engl J Med 2003;348:2138–39 (P 9-2, para 2).

18. European nations failing obese patients-survey (www.medscape.com).

19. Kanchuk ML et al. Weight gain in gonadectomized normal and lipoprotein-lipase-deficient male domestic cats. J Nutr 2003;133:1866–74.

20 Bray GA. Physiology and consequences of obesity. Medical Education Collaborative, Diabetes and endocrinology clinical management, www.medscape.com/Medscape/endocrinology/Clinical Mgmt/Cm.v03/public/index.CM.v03.html, accessed 1/8/01.

21. Bray, Physiology and consequences of obesity; and Peeters A et al., Obesity in adulthood and its consequences for life expectancy. Ann Int Med 2003;138:24–32.

22. Calle EE et al., Body mass index and mortality in a prospective cohort of U.S. adults. N Engl J Med 1999;341:1097–1105.

23. Peeters et al., Obesity in adulthood.

24. Bray, Physiology and consequences of obesity.

25. Harrard WR. Weight control and exercise. JAMA 1995;274:1964–5.

26. Mayer J, Overweight: causes, cost, and control, Englewood Cliffs (NJ): Prentice-Hall; 1968; Patterson RE, et al., Factors related to obesity in preschool children, J Am Diet Assoc 1986;86:1376–81; Kanarek RB, Orthen-Gambill N, Marks KR, Mayer J, Obesity: possible psychological and metabolic determinants, in: Galler JR, ed. Human nutrition: a comprehensive treatise, Vol. 5, New York: Plenum; 1984; and Price J, Desmond S, Krol R, Snyder F, O'Connell J, Family practice physicians' beliefs, attitudes and practices regarding obesity, Am J Prev Med 1987;3:339–45.

27. Taylor RW et al. Body fat percentages measured by dual-energy X-ray absorptiometry corresponding to recently recommended body mass index cutoffs for overweight and obesity in children and adolescents aged 3–18 y. Am J Clin Nutr 2002;76:1416–21.

28. Fernández, Is percentage body fat differentially related to body mass index?

29. Prentice AM, Jebb SA. Beyond body mass index. Obes Rev 2001;2:141–7.

30. Prentice et al., Beyond body mass index; and Taylor et al., Body fat percentages.

31. Gallagher GD et al. Healthy percentage body fat ranges: an approach for developing guidelines based on body mass index. Am J Clin Nutr 2000;72:694–701.

32. Oppliger RA, Nielsen DH, Vance CG, Wrestlers' minimal weight: anthropometry, bioimpedance, and hydrostatic weighing compared, Med Sci Sports 1991;23:247–53; and Gibson RS, Principles of nutritional assessment, New York: Oxford University Press; 1990.

33. Lean and Han, Waist worries.

34. Bouchard C, Bray GA, Hubbard VS. Basic and clinical aspects of regional fat distribution. Am J Clin Nutr 1990;52:946–50.

35. Nicklas BJ et al. Level of visceral obesity defined for elevated coronary heart disease risk in women. Diabetes Care 2003;26:1413–20.

36. Bray GA. Physiology and consequences of obesity, Medscape Diabetes and Endocrinology Clinical Management Modules, http://womenshealth.medscape.com/medscape/endocrinology/ClinicalMgmt/CM.v03/pnt-CM.v03.html, 2001; and Pi-Sunyer FX.

Health implications of obesity. Am J Clin Nutr 1991;53: 1595S–603S.

37. Lean and Han, Waist worries; Lakka HM et al., Abdominal obesity is associated with increased risk of acute coronary events in men, Eur Heart J 2002;23:706–13; and Purness JQ, Obesity, www.medscape.com/viewarticle/45726, accessed 6/03.

38. Lakka et al., Abdominal obesity.

39. Lean and Han, Waist worries; and Zhu SK et al. Waist circumference and obesity-associated risk factors among whites in the third National Health and Nutrition Examination Survey:clinical action thresholds. Am J Clin Nutr 2002;76:743–49.

40. Purness, Obesity (www.medscape.com/viewarticle/45726).

41. Bouchard C. Heredity and the path to overweight and obesity. Med Sci Sports 1991;23:285–91.

42. Perusse L, Bouchard C. Role of genetic factors in childhood obesity and in susceptibility to dietary variations. Ann Med 1999;31(suppl):19–25.

43. Purness, Obesity (www.medscape.com/viewarticle/45726).

44. Bouchard C et al. The response to long-term overfeeding in identical twins. N Engl J Med 1990;322:1477–82.

45. Macdonald A, Stunkard AJ. Body-mass indexes of British separated twins (letter). N Engl J Med 1990;322:1530.

46. Purness, Obesity (www.medscape.com/viewarticle/45726).

47. Rolls BJ. The super-sizing of America. Nutr Today 2003;38:42–53.

48. Bray, Physiology and consequences of obesity.

49. Whitaker JA et al. Predicting obesity in young adulthood from childhood parental obesity. N Engl J Med 1997;337:869–73.

50. McCrory MA et al., Biobehavioral influences on energy intake and adult weight gain, J Nutr 2002;132:383OS–34S; and McCrory M, Adult weight gain: causes and implications for obesity prevention. Experimental Biology Annual Conference, New Orleans, 4/22/02.

51. Rolls, The super-sizing of America.

52. Gillis J et al, Relationship between juvenile obesity, dietary energy and fat intake, and physical activity, Int J Obes Rel Meta Disord 2002;26:458–63; and Dietary reference intakes: energy, carbohydrate, fiber, fat, protein, and amino acids, Institute of Medicine, National Academies of Sciences, Ch. 11, 2002.

53. McCrory et al., Biobehavioral influences on energy intake.

54. Gortmaker SL, Dietz WH Jr., Cheung LWY, Inactivity, diet, and the fattening of America, J Am Diet Assoc 1990;90:1247–55; and Andersen RE et al, Relationship of physical activity and television watching with body weight and level of fatness among children, JAMA 1998;279:938–42.

55. Rolls, The super-sizing of America.

56. Rolls, The super-sizing of America.

57. Brown JE, Kaye SA, Folsom AR, Parity-related weight change in women, Int J Obesity 1992;16:627–31; and Jeffery RW et al, Prevalence and correlates of large weight gains and losses in adults, Int J Obes 2002;26:969–72.

58. Costanzo PR, Schiffman SS. Thinness—not obesity—has a genetic component. Neuroscience and Biobehavioral Reviews 1989;13:55–58.

59. Rolls, The super-sizing of America; and McCrory et al., Biobehavioral influences on energy intake.

60. Rolls, The super-sizing of America.

61. Heaney RP et al. Calcium and weight: clinical studies. J Am Coll Nutr 2002; 21:152S–155S.

62. Onis M et al. Prevalence and trends of overweight among preschool children in developing countries. Am J Clin Nutr 2000;72:1032–9.

63. Committee on Diet and Health (Food and Nutrition Board; National Research Council). Diet and health. Implications for reducing chronic disease risk. Washington, DC: National Academy Press; 1989.

64. Costanzo and Schiffman, Thinness.

65. Bouchard, Heredity and the path to overweight and obesity.

66. Lifshitz F, Nutrition and growth, Clin Nutr 1985;4:40–47; and Pugliese MT, Weyman-Dunn M, Moses N, Lifshitz F, Parental health beliefs as a cause of nonorganic failure to thrive, Pediatrics 1987;80:175–82.

67. Lifshitz, Nutrition and growth.

68. Lifshitz, Nutrition and growth; and Pugliese et al., Parental health beliefs.

69. Masoro EJ. Caloric restriction and aging. Exp Gerontol 2000;35:299–305.

70. Casper RC. Nutrition and its relationship to aging. Exp Gerontol 1995;30:299–314.

71. Satter EM. Internal regulation and the evaluation of normal growth as the basis for prevention of obesity in children. J Am Diet Assoc 1996;96:860–4.

72. Schuster K, The dark side of nutrition, Food Mang 1999;34:34–9; and UK panel calls for media to end pressure on girls to be thin, Report of the UK's Cabinet's Body Image Summit, Reuter Health 2000; June 22.

Nutrition UP CLOSE

Are You an Apple or a Pear?

Feedback for Unit 9

If your waist-to-hip ratio is 0.8 or greater if you're a female, and 0.9 if male, you're an apple. If it is less than 0.8 or 0.9, you're a pear. (You are all a peach for doing this exercise.) Most men will turn out to be apples, and women will tend to be pears.

Weight Control: The Myths and Realities

Nutrition Scoreboard

	TRUE	FALSE
1 Anybody who really wants to can lose weight and keep it off.		
2 Weight loss is the cure for obesity.		
3 In almost all cases, the quicker weight is lost, the quicker it's put back on.		
4 Weight-loss products and services must be shown to be safe and effective before they can be marketed.		

Answers on next page

[KEY CONCEPTS AND FACTS]

- The effectiveness of weight-control methods should be gauged by their ability to prevent weight regain.

- That a weight-loss product or service is widely publicized and utilized doesn't mean it works.

- Some people cannot lose weight and keep it off, no matter how hard they try.

- Successful weight control is characterized by gradual weight loss from small, acceptable, and individualized changes in eating and activity.

Answers to *Nutrition Scoreboard*

		TRUE	FALSE
1	If this claim were true, hardly anybody would be obese.		✔
2	Maintenance of weight loss is the cure for obesity.		✔
3	Quick weight-loss diets almost never work because weight lost is regained after the diet ends.	✔	
4	Unfortunately, many weight-loss products and services on the market have not been shown to be safe or effective. Laws and regulations do not fully protect the consumer from the introduction of bogus products and services.		✔

Photo Disc

Baseball, Hot Dogs, Apple Pie, and Weight Control

Americans are preoccupied with their weight. On any given day, 40% of adults, and a majority of individuals who are overweight or obese, are trying to lose weight. To help get the weight off, Americans spend over *$33 billion* annually (an average of approximately $120 per person each year) on weight-loss products and services. Yet Americans are gaining weight faster than they are losing it, and the incidence of obesity in the United States is on the rise.[1] Roughly 5 to 10% of people who lose weight keep it off.[2] Consumers are paying handsomely for weight loss without experiencing the desired cosmetic changes or health benefits that come when weight loss is maintained (Illustration 10.1).

Why do so many Americans fail at weight control? For some people, achieving permanent weight reduction on their own may be truly impossible.[3] For others, the problem is the ineffective methods employed, not the people who use them.

Illustration 10.1
Source: Reprinted by permission of Universal Press Syndicate.

Weight Loss versus Weight Control

"Jess—did I hear you say you wanted to lose some weight? It happens I know about a terrific diet! A couple of months ago I went on Dr. Quick's Amino Acid Diet and lost 15 pounds! You eat nothing but fish, papaya, and broccoli and the pounds just melt away!"

Jess was shocked. Selma looked heavier than she did last year when they were in class together.

"But Selma," Jess inquired, "what happened to the weight you lost?"

"Oh, I gained it back. I ran out of willpower and started eating everything in sight. I'm going to get back on track, though. Tomorrow I start the Slim Chance Diet."

Dietary torture and weight loss are not the cure for overweight. If losing weight was all it took to achieve the cultural ideal of thinness, nobody would be overweight. Any popularized approach to weight loss (and some get pretty spectacular) that calls for a reduction in caloric intake can produce weight loss in the short run. These methods fail in the long run, however, because they become too unpleasant. Feelings of hunger, deprivation, and depression that often occur while on a weight-loss diet eventually lead to a breakdown of control, eating binges, a return to previous habits, and weight regain.[4] Humans are creatures of pleasure and not pain. Any painful approach to weight control is bound to fail. Improved and enjoyable eating and exercise habits are needed to keep excess weight off, and quick weight-loss approaches don't change habits.

Unfortunately, many dieters think weight-control methods are successful if they lead to a rapid loss of weight.[5] When the lost weight is regained, dieters tend to blame themselves and not the faulty method. Blaming themselves for the failure, many people are ready to try other quick weight-loss methods. But they usually fail, too.[6]

The Business of Weight Loss] More than 29,000 weight-loss products and services are available. Some of these are shown in Illustration 10.2. Most of them either don't work at all or don't prevent weight regain.[7] "Quick fix" weight-loss approaches that don't lead to long-term changes in behavior are the primary reason approximately 90 to 95% of people who lose weight gain it back. The demand for such products and services is so great, however, that many are successful—financially.

One reason so many weight-loss products and services are available is that almost none of them work. If any widely advertised approach helped people lose weight and keep it off, manufacturers of bogus methods would go out of business. The weight-loss industry also thrives because of the social pressure to be thin. Many people try new weight-loss methods even though they sound strange or too good to be true. Often people believe that a product or service must be effective, or it wouldn't be allowed on the market.[8] Although reasonable, this belief is incorrect.

The Lack of Consumer Protection] The truth is that general societal standards for consumer protection do not apply to the weight-loss industry.[9] No laws require a product to be effective: in most cases, companies do not have to show that weight-loss products or services actually work before they can be sold. That's why products like herbal remedies, forks with stop and go lights, weight-loss skin patches, colored "weight-loss" glasses, electric cellulite dissolvers, mud and plastic wraps, and

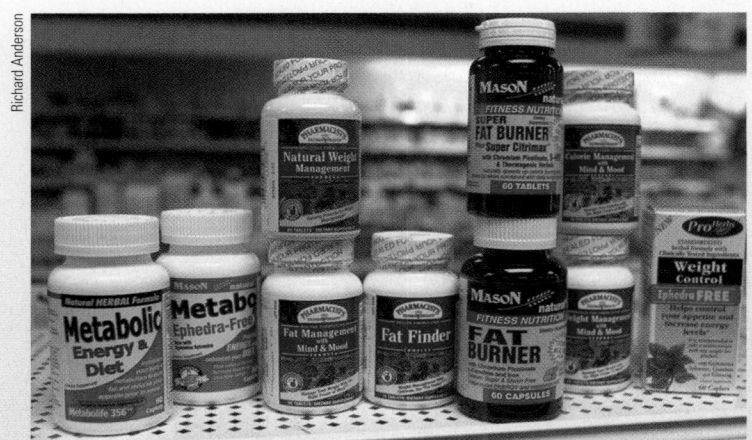

Illustration 10.2
There is no lack of weight-loss books and products.
There is a lack of popular approaches that help people keep weight off.

inflatable pressure pants are available for sale. These, like many other weight-loss products, do not work, and there is no reason why they should. Promotions for the products just have to make it *seem* as though they might work wonders on appetite or body fat.

Furthermore, weight-loss products and services are usually not tested for safety before they reach the market. The history of the weight-loss industry is littered with abject failures: fiber pills that can cause obstructions in the digestive tract, very low-calorie diets and intestinal bypass surgeries that lead to nutrient-deficiency disease, amphetamines that can produce physical addiction, diet pills that cause heart valve problems, and liquid protein diets that have led to heart problems and death from heart failure.[11] A brief history of weight-loss method failures is chronicled in Table 10.1.

REALITY CHECK

Do some foods have negative calories?

You're looking for a way to shed 10 pounds you gained during freshman year and are seriously thinking about adding grapefruit and vinegar to your diet. You have heard these foods have "negative calories" because they make the body burn fat.

Juanita:
I know vinegar cleans the grease off windows. Maybe it will melt away my fat, too.

Chuck:
Vinegar and grapefruit are so sour—I bet they rev up the metabolism so you burn fat.

Answers on next page

Juanita

ANSWERS TO REALITY CHECK

Do some foods have negative calories?

The vinegar, grapefruit, negative calorie myth lives on because it strikes many people as reasonable. Neither vinegar nor grapefruit, nor any other food has negative calories or causes the body to gear up metabolism and burn fat.[10]

Chuck

Photo Disc

Pulling the Rug Out] Fraudulent weight-loss products and services may be investigated and taken off the market. Currently, the Federal Trade Commission (FTC) monitors deceptive practices and weight-loss claims on a case-by-case basis. It has filed suits against companies making exaggerated claims, and has identified the most common dubious claims made by the industry (Table 10.2). The FTC performs most investigations in response to consumer complaints. Illustration 10.3 shows three examples of products that were taken off the market and explains why.

Requirements for truth in labeling keep many bogus products from including false or misleading information on their labels. These laws, however, do not keep outrageous claims from being made on television or printed in pamphlets, books, magazines, and advertisements. Few companies get caught if they include false or misleading information in weight-loss product ads.

TABLE 10.1

A BRIEF HISTORY OF DISCONTINUED WEIGHT-LOSS METHODS.
It is possible that ephedra will be added to this list.

YEAR	METHOD	REASON FOR DISCONTINUATION
1940s–1960s	Amphetamines	Highly addictive, heart and blood pressure problems
	Vibrating machines	Did not work
1970s	Jejunoileal; bypass surgery	Often caused chronic diarrhea, vitamin and mineral deficiencies, kidney stones, liver failure, arthritis
	Liquid protein diet	Poor-quality protein caused heart failure, deaths
1990	Oprah Winfrey liquid diet success (she lost 67 pounds)	
1991	Oprah Winfrey gains 67 pounds and declares "No more diets!"	
1997	Phen-Fen (Redux)	Heart valve defects, hypertension in lung vessels

TABLE 10.2

LOADS OF FALSE AND MISLEADING WEIGHT-LOSS ADVERTISEMENTS APPEAR IN THE MEDIA. HERE IS A LIST OF THE TOP SIX FEATURES OF WEIGHT-LOSS ADS THAT MAKE FALSE OR MISLEADING CLAIMS.[12]

1. Use testimonials, before-and-after photos.
2. Promise rapid weight loss.
3. Require no special diet or exercise.
4. Guarantee long-term weight loss.
5. Include a "clinically proven" or "doctor approved" statement.
6. Make a "safe," "natural," or "easy" claim.

What if the Truth Had to Be Told?] Suppose the weight-loss industry had to inform consumers about the results of scientific tests of the effectiveness and safety of weight-loss products and services. What if this information had to be routinely included on weight-loss product labels (Illustration 10.4)? What do you think would happen to consumer choices and the weight-loss industry? Increased federal enforcement of truth-in-advertising laws may help put an element of honesty into promotions for many weight-loss products. In addition, the FTC has proposed that claims about long-term weight loss "must be based on the experience of patients followed for at least two years after they complete the [weight-loss] program."[13] If this proposal ever becomes law, it will change the weight-loss industry in the United States.

Illustration 10.3
A few examples of bogus weight-loss products that were removed from the market.

Diet Pills] Two types of prescription diet pills are currently approved for long-term use in obese people in the United States. One is Meridia (sibutramine), and the other is Xenical (orlistat).[14] Meridia works by enhancing satiety (which reduces food intake), and Xenical works by partially blocking fat absorption in the intestines.

Stan Maddock

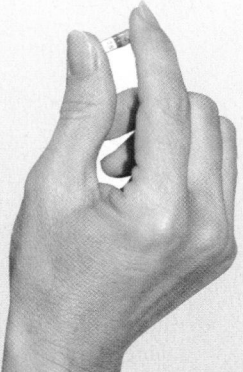

Fat magnet pills were purported to break into thousands of magnetic particles once swallowed. When loaded with fat, the particles simply flushed themselves out of the body. The FTC found the product too hard to swallow. The company took the product off the market and made $750,000 available for customer refunds.

"Blast" away 49 pounds in less than a month with Slim Again, Absorbit-All, and Absorbit-AllPlus pills claimed ads in magazines, newspapers, and on the Internet. The company got blasted by the FTC to the tune of $8 million for making fraudulent claims.

Take a few drops of herbal liquid, put them on a bandage patch, and voilà a new weight-loss product! The herbal liquid was supposed to reach the appetite center of the brain and turn the appetite off. Federal marshals weren't impressed. They seized $22 million worth of patch kits and banned the sale of others.

Both have side effects. Meridia may increase blood pressure and heart rate, cause headaches, and lead to dry mouth. Xenical's primary side effect is oily stools, which are particularly bothersome if high-fat meals are consumed. Malabsorption of fat caused by Xenical reduces absorption of a number of fat-soluble nutrients such as vitamin D, E, and beta-carotene.[15]

It is recommended that Meridia and Xenical be used in conjunction with reduced-calorie diets. Both effectively lower weight by an average of about 10% over a period of six months to a year. Weight regain after pill use stops is common but not universal.[16]

Many other diet drugs are under development or testing. Much activity currently centers around leptin, a hormone produced in fat cells, that may modify caloric intake or energy expenditure, growth hormone, and other hormones that modify the sensations of hunger or satiety. Even though people using prescribed diet drugs are carefully screened and monitored by medical professionals, serious side effects can develop. No diet drug is absolutely safe, and none are known to cure obesity forever.[17]

Ephedra] Ephedra, or ma huang, was an over-the-counter diet pill and energy-booster. It was sold as a herbal product, often in combination with caffeine. With caffeine, ephedra appears to promote weight loss for at least 6 months, but serious side effects (hypertension, irregular heartbeat, stroke, heart attack, seizure, and sudden death) are related to its use. The National Institutes of Health does not recommend ephedra for weight control, and it can no longer be sold for weight loss.

Illustration 10.4
Just imagine what would happen if weight-loss approaches were required to divulge their effectiveness and risks.

The Amazing BEE POLLEN-GRAPEFRUIT DIET
Dr. Quickfix

MELTS AWAY POUNDS OF CELLULITE IN JUST 10 DAYS!

Consumer Protection Label

Effectiveness: Ineffective for 95% of users

Risks: Bee pollen may cause adverse reactions in allergic people. Weight regain may exceed weight loss. If the diet fails, people may blame themselves thereby precipitating anxiety and self-doubt.

High-Protein Diets

"The high protein diet probably gets its allure from the naughtiness of its approach."
—G. Taubes, journalist[19]

Do high-protein diets help people lose weight? Yes. Do they pose risks to health? Some. Do they help people maintain weight loss in the long run? No.

High-protein, low-carbohydrate diets can lead to weight loss, and may improve health status as a result (Illustration 10.5).[20] A series of studies have concluded that such diets have no adverse effect, or a somewhat positive effect on blood lipids, glucose, insulin, and blood pressure.[21] High-protein diets promote weight loss by decreasing appetite. When the body is deprived of carbohydrates, fats must be used as the body's primary source of energy. Using fat to meet much of the body's energy needs leads to increased blood levels of by-products of fat utilization, called "ketone bodies," which decreases appetite. People often experience rapid weight loss when they first start high-protein diets, and that can be very encouraging. The weight loss is primarily due to water excretion. High-protein diets increase water need because the nitrogen component of much of the protein consumed is excreted from the body, diluted in urine.[22]

All is not well with high-protein diets, however. They are not recommended for people with liver or kidney problems, may promote heart disease and prostate cancer, may leave glycogen stores nearly empty, and

The high-protein, low-carbohydrate diet for weight loss has been around for over a century. The first "Dr. Atkins" was William Banting, a London undertaker. He published a book on his terrific new diet in 1864. It was so popular that dieting to this day is referred to as "Banting" in England.[23]

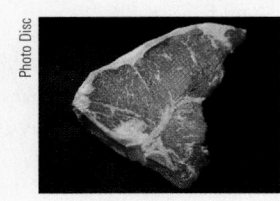

Photo Disc

ON THE SIDE

Illustration 10.5
High-protein diets don't work in the long run, and their long-term consequences on health are not clear.

Richard Anderson

may cause nausea, headache, decreased stamina, and a reduced ability to concentrate. Long-term consequences of high-protein diets on health are far from clear.[24]

What's the primary limitation of high-protein diets as far as we know now? It's a big one: they don't work in the long run. As with other weight-loss diets requiring substantial changes in food intake, people revert to more pleasurable, usual diets. As soon as carbohydrates are returned to the diet in normal amounts, appetite returns and weight gain follows.

Organized Weight-Loss Programs

The mark of successful weight-loss programs is maintenance of weight loss in the long term. Whether organized weight-loss programs such as Weight Watchers, Jenny Craig, Overeaters Anonymous, or services provided by many medically supervised weight-loss clinics hit this mark is not known. Results of studies of the effectiveness of privately run weight-loss programs are rarely published. Programs requiring the largest changes in diet and lifestyle would be expected to have the lowest success rates in terms of weight-loss maintenance. No weight-loss product, diet, program, or clinic that substantially alters usual eating and activity patterns has been shown to result in a long-term loss of excess body fat in more than 5 to 10% of people who use it.[25]

Drastic Measures

Fasting and surgery represent extreme approaches to body fat reduction. These approaches are riskier than weight-loss methods that produce a gradual loss in weight.

Fasting] Sometimes people fast to lose weight, but this method is not recommended unless the fast is closely supervised by medical professionals. Abruptly stopping food intake not only does not foster long-term weight-loss maintenance, but it produces a high level of wear and

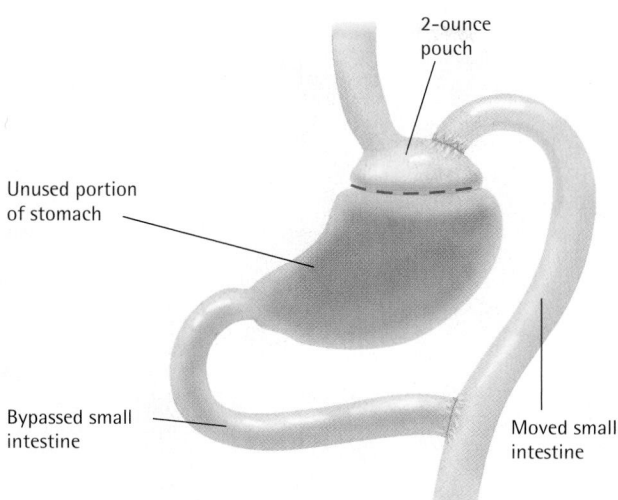

2-ounce
pouch

Unused portion
of stomach

Bypassed small
intestine

Moved small
intestine

Illustration 10.6
Gastric bypass surgery.

tear on the gastrointestinal tract. Fasting causes "intestinal starvation" because it removes a primary source of nutrients from the intestines—foods that are being digested. Without sufficient nutrients to maintain the gastrointestinal tract, mucus and cells lining the tract are lost and infection is more likely. Far from "cleansing the system," as the saying goes, fasting can clog up the lower intestines. Fasting can also lead to a preoccupation with food and eating and a rapid rebound in weight once the fast ends.[26]

Surgical Techniques]   Weight control surgery is a method of last resort. It is reserved for very obese persons (BMI over 40 kg/m^2) or for people with marked obesity (BMI over 35 kg/m^2) whose weight jeopardizes their immediate health. Surgery for obesity is experiencing an increase in demand. Some people who don't qualify are gaining weight to get past BMI thresholds.[27] Surgical techniques are expensive and carry the risks associated with surgical interventions, such as infection or adverse reaction to anesthesia.[28]

Stomach Surgery} The idea for stomach surgery to control obesity came as a eureka moment to surgeons who noticed that people having large segments of their stomachs removed for cancer or severe ulcers lost a good deal of weight.[29] Several surgical techniques reduce the size of the stomach and thereby limit the amount of food people can eat. The technique that has the lowest failure rate (gastric bypass surgery) is shown in Illustration 10.6.[30] In gastric bypass surgery, most of the stomach is stapled shut, leaving a pouch at the top that can hold about 2 to 4 tablespoons of food. A section of the small intestine that connects to the bottom of the stomach is cut off and attached as a drain for the stomach pouch. The loose end of the small intestine is then reattached to the section of the small intestine that leads from the pouch. This way, digestive juices from the stomach and the upper part of the small intestine are made available for digesting food. People who have their stomachs stapled can eat only a small amount of food. If they eat too much, they feel nauseous and vomit.

Gastric bypass surgery typically costs $20,000–$30,000 and leads to an average weight loss of 100 pounds or more in very obese people. Most people (70%) maintain a 50% loss in body weight 5 years after the surgery.[31] The surgery has other benefits: it can cure type 2 diabetes and reduce the risk for heart disease and hypertension.[32] The procedure is not foolproof. People can eat enough to separate

Illustration 10.7
The effects of liposuction.
The photograph on the left shows the thighs of a woman before liposuction, and the photo on the right shows the same woman's thighs after liposuction.

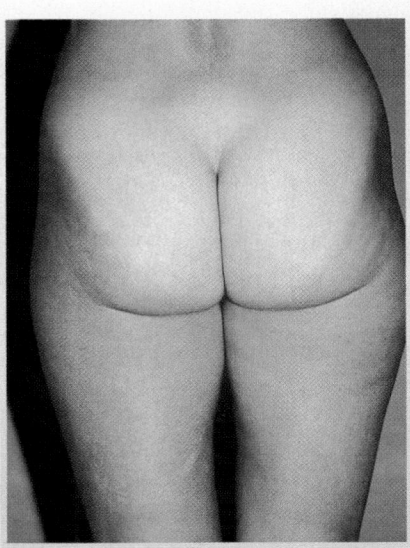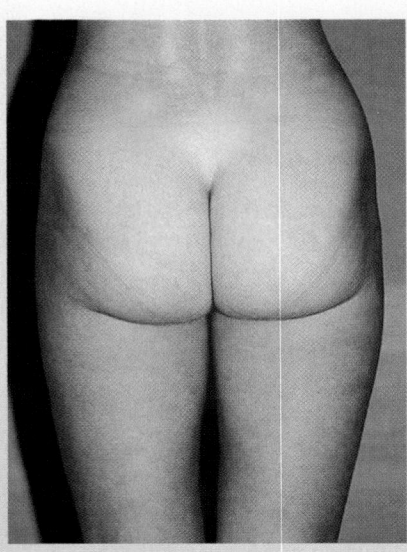

the staples and end up back in surgery. In addition, blocking off most of the stomach reduces vitamin B_{12} absorption and increases the risk of osteoporosis later in life.[33]

Liposuction} At a cost of over $3000 per surgery (prices vary by fat deposit site, inflation, and surgeon), fat deposits in the thighs, hips, arms, back, or chin can be partially removed by liposuction. The procedure is the most common type of cosmetic surgery performed in the United States. Considered cosmetic, it is not intended for weight loss or to diminish health risks associated with obesity (Illustration 10.7). Surgical standards require that no more than 8 pounds of fat be removed by liposuction.[34] If a person gains a good deal of weight after liposuction, fat will be deposited to some extent in the breasts and other areas not operated on. These deposits can lead to a return of an undesired body shape.[35] In addition, surgery always carries a risk of infection and other complications, so it cannot be taken lightly.

Weight Loss: Making It Last

People come in all sizes, and not everyone can change his or her body shape or weight. Some can, however, and the most effective ways to lose weight and keep it off are known. Many weight-loss methods lead to a temporary reduction in body weight, but this is not the test of a successful method. An approach to weight loss can be considered successful only if it is safe, healthful, and prevents weight regain. Table 10.3 summarizes characteristics of weight-loss programs that promote maintenance of weight loss. An approach that fosters dietary and exercise patterns that prevent weight regain is best.[37]

Such methods take time, however, so they are not among the more popular approaches. Because they cannot be sold as a product, the weight-loss industry isn't very interested in them. Nevertheless, these methods work in the long run for many overweight people. The key ingredients of these methods are small, acceptable changes in diet and activity. These approaches focus on healthy eating and exercise for a lifetime, rather than "dieting." The goal is to achieve improvements in lifestyle, fitness, and health—not losing a certain number of pounds in a week.[38] Simple though they sound, such methods are the difference between those who lose weight and keep it off and those who don't (see the "Health Action").

The Importance of Small, Acceptable Changes

For most people, excess body fat accumulates slowly over time. Consuming just an extra 50 calories a day, for example, will lead to a gain of approximately 5 pounds in a year. (Approximately 3500 excess calories will produce a weight gain of 1 pound, and a 3500-calorie deficit will produce a loss of 1 pound in body weight in many people.) Excess weight is rarely all gained over the course of a few weeks or months. Fat is put on slowly, and that's the best way to take it off.

Gradual losses in body fat do not require dramatic changes in diet or activity level. Only small changes in diet and activity are needed. By cutting back food intake by 100 calories per day, a person could lose 10 pounds in a year. Jogging 20 minutes three times a week in addition to usual activity could take off another 10 pounds in a year. Routinely engaging in more vigorous forms of exercise such as handball or rowing can contribute to body fat loss, weight-loss maintenance, and fitness.[40] The secret to success is including foods and activities that are enjoyed in the lifestyle change package. To be lasting, the changes have to be acceptable, even preferable, to existing patterns (Illustration 10.8). Often this means that the changes have to be small.[41]

TABLE 10.3

CHARACTERISTICS OF WEIGHT-LOSS PROGRAMS THAT PROMOTE MAINTENANCE OF WEIGHT LOSS.[36]

1. Offer long-term approach to beneficial changes in diet and physical activity.

2. Employ reduced-calorie, nutritionally adequate diets.

3. Recommend foods from all food groups.

4. Do not exclude certain foods.

5. Employ relatively small and acceptable changes in usual lifestyle, dietary, and other behaviors.

6. Do not rely solely on drugs or herbal remedies.

HEALTH ACTION | *Weight-Loss Maintainers versus Weight Regainers*[39]

Weight-Loss Maintainers
- Lose weight slowly by making small, acceptable changes in eating and activity.
- Consume regular meals.
- Exercise regularly.
- Make conscious efforts to avoid regaining weight.
- Avoid feeling deprived while changing habits (i.e., eat foods they like and undertake activities they enjoy).
- Use available sources of social support.

Weight Regainers
- Are not committed to a gradual weight loss through behavioral changes.
- Change diet radically to lose weight.
- Exercise little. Have low muscle strength.
- Eat unconsciously in response to stress.
- Take diet pills.
- Do not seek out or have social support.
- Cope with problems by escape and avoidance.

Photo Disc

Illustration 10.8
Small, acceptable changes are the key to a successful weight-loss/weight-maintenance program.
Try to find activities that you enjoy; get out there and play!

It should be mentioned that a minority of people are able to make drastic changes in dietary intake (adopt a very low-fat diet, for example) or increase their exercise level substantially and maintain these changes and lower body weight over time.[42] Such people tend to be the exceptions, however.

Identifying Small, Acceptable Changes] To identify changes in eating and activity that have staying power, first list the weak points in your diet and activity. Dietary weak points might include the consumption of high-fat foods due to eating out often, or relying on high-fat convenience foods that can be heated up in seconds. Another weak point might be skipping breakfast and overeating later in the day because of extreme hunger. Weak points in physical activity might include driving instead of walking, not engaging in sports, or spending too little time playing outside.

For each weak point, identify options that seem acceptable and enjoyable. A person who enjoys broiled chicken with barbecue sauce might not mind eating that at restaurants instead of fried chicken. That's a change people can make if they plan ahead. A person who gets too full from a large serving of fries might be happier ordering a small serving and not eating so many. A breakfast skipper might find grabbing a piece of fruit and a slice of cheese for breakfast acceptable and doable. People who enjoy walking may not mind leaving the car or bus behind and letting their feet carry them to class, the grocery store, or a friend's house.

Many acceptable options for making small improvements in diet and activity may be available (see the "Health Action—Small Changes That Make a Big Difference"). The easiest changes to accomplish are the ones that should be incorporated into the overall lifestyle improvement plan. Some people, for example, lose weight and keep it off simply by consciously cutting down on portion sizes. Others avoid eating too much at any meal and walk more. Simply adding breakfast helps some

HEALTH ACTION I *Small Changes That Make a Big Difference*[44]

- Use 50% less margarine or butter and mayonnaise on foods.
- Choose low-fat varieties of foods (cheese, milk, yogurt, fish, poultry) that you like.
- Trim the fat off meats or buy lean meats.
- Eat small portions of high-calorie foods.
- Eat fried foods infrequently.
- Eat breakfast.
- Eat when you are hungry.
- Stop eating when you feel full.
- Take home chunks of the large portions of foods served in restaurants. Safely store the food and make another meal (or two) out of it.

- Regular-size your fast-food meals.
- Eat all the vegetables and fruits you want.
- Seek out and eat your favorite high-fiber foods.
- Cook at home.
- Walk to school.
- Bike to work.
- Take the stairs.
- Go outside and play.

people lose weight and maintain the loss.[43] The easier the changes are to follow, the more likely they are to succeed.

Individualized plans that don't work out often include unacceptable or unenjoyable changes. The changes may be too large or too different from the usual. In that case, go back to the drawing board and modify the plan to include small changes that are acceptable in the long run. Perhaps the original plan included jogging, but it turns out that jogging isn't any fun. In that case, take jogging out of the plan! Replace it by any other physical activity that would be enjoyed. Midcourse corrections should be expected. Some experimentation may be required to identify the small changes that will last.

What to Expect for Weight Loss] If all goes well, weight loss will be gradual but lasting. The pattern of loss should be somewhat like that graphed in Illustration 10.9, where the person lost 18 pounds over 30 weeks. The pattern will include peaks, valleys, and plateaus—not a straight downward curve. Sometimes a bit of weight will be gained, and other times more weight than expected will be lost. It's more important to enjoy and continue improved eating and activity patterns than to concentrate on the number of pounds lost.

If diet and exercise behaviors are improved in acceptable ways, there is little need to become preoccupied with the number of calories consumed, the number of calories burned off in a bout of exercise, or the number of pounds lost last week. The goal is reached when the small changes become an enjoyable part of life on a day-to-day basis. Improved eating and activity patterns offer many benefits. Weight loss is only one of them.

Illustration 10.9
People who lose weight gradually are more likely to keep it off than those who lose weight rapidly.
The weight loss graphed here averages half a pound per week.

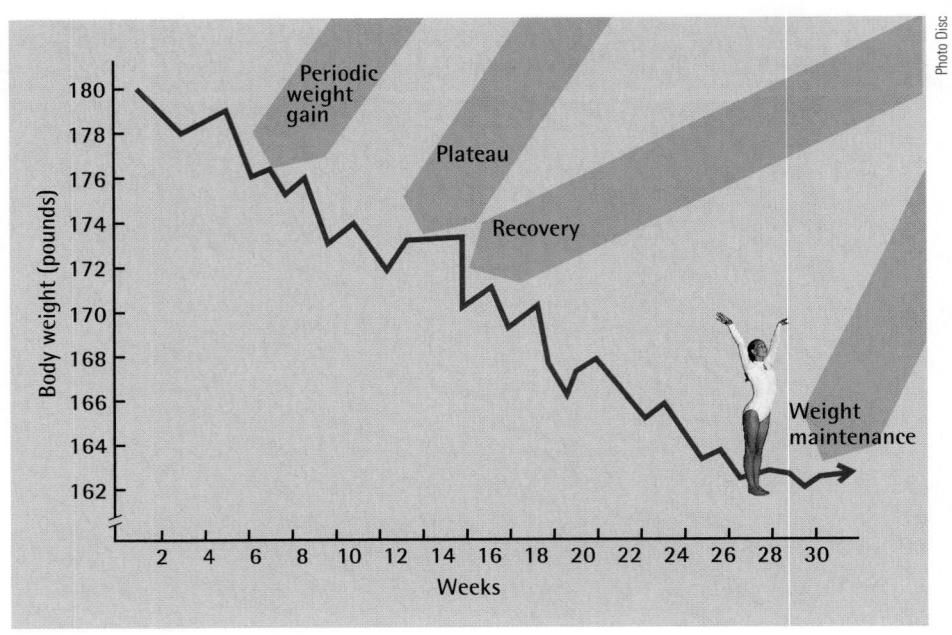

www links

www.4woman.gov
The home page of the National Women's Health Initiative can start your search for information on safe and effective commercial weight-loss programs.

www.mayoclinic.org
Provides reliable information on nutrition, health, and weight control. Check out "My Health Interests" if you want to lose weight.

www.ediets.com
A commercial site offering personalized diet plans and other resources developed largely by dietitians.

www.niddk.nih.gov/health/nutrit/win.htm
The Weight Control Information Network provides science-based information on weight control and nutrition. Call toll-free: 1-877-946-4627.

www.shapedown.com
This commercial site provides background information and methods for the tested Shapedown Program for weight management in children and adolescents.

www.cyberdiet.com
Construct a personal menu plan and get an estimate of calories burned in physical activity at this site. Get free diet profile, join for access to weight-loss services including menu planning, online support group.

www.quackwatch.com
Check out "quacky" weight-loss products and services.

Nutrition UP CLOSE

Will You Maintain or Regain?

FOCAL POINT: People who lose weight and keep the pounds off share similar successful behaviors.

Have you recently lost weight? Answer the following questions to find out if your current eating and activity behaviors promote making the weight loss permanent.

	YES	NO
1. Do you usually eat breakfast?		
2. Do you avoid skipping meals?		
3. Do you plan ahead for snack and mealtime foods?		
4. Do you prepare most of your meals at home?		
5. Do you limit consumption of high-fat foods like butter, margarine, mayonnaise, and fried foods?		
6. Do you consume high-calorie foods like desserts infrequently and in small portions?		
7. Do you walk instead of drive whenever possible?		
8. Do you engage in regular physical activities that you enjoy?		
9. Do you seek social support to help maintain your weight loss?		
10. Do you eat when you're hungry and stop eating when you are full?		

FEEDBACK (answers to these questions) can be found at the end of Unit 10.

Notes

1. Obesity Trends, www.cdc.gov/nccdphp/dnpa/obesity/trend/prev_bmi.htm, accessed 9/03.

2. Sarlio-Lahteenkorva SS, et al. A descriptive study of weight loss maintenance: 6 and 15 year follow-up of initially overweight adults. Int J Obesity 2000;24:116–25.

3. Marcus MD, Wing RR, Lamparski DM. Binge eating and dietary restraint in obese patients. Addict Behav 1985;10:163–8.

4. Goodrick GK, Foreyth JP. Why treatments for obesity don't last. J Am Diet Assoc 1991;91:234–47.

5. Womble L et al. Unrealistic weight loss goals and feelings of failure. Presentation at the North American Association for the Study of Obesity, Long Beach, CA, 2000 Nov.

6. Wooley SC, Garner DM. Obesity treatment: the high cost of false hope. J Am Diet Assoc 1991;91:1248–51.

7. Wooley and Garner, Obesity treatment.

8. Lustig A. Weight loss programs: failing to meet ethical standards? J Am Diet Assoc 1991;91:1252–4.

9. Wooley and Garner, Obesity treatment.

10. Cunningham A, Marcason W. Is it possible to burn calories by eating grapefruit or vinegar?

11. Wadden TA, Stunkard AJ, Brownell KD. Very low calorie diets: their efficacy,

safety, and future. Ann Intern Med 1983;99:675–84.

12. Current trends in weight-loss advertising. J Am Diet Assoc 2003;103:150.

13. Taubes G. Dietary approaches to weight control. What works? Experimental Biology Annual Meeting, San Diego, April 14, 2003.

14. Yanovski SZ, Yanovski JA. Obesity. N Engl J Med 2002;346:591–602.

15. Yanovski and Yanovski, Obesity.

16. Yanovski and Yanovski, Obesity.

17. Yanovski and Yanovski, Obesity.

18. Yanovski and Yanovski, Obesity.

19. Taubes, Dietary approaches to weight control.

20. Farnsworth E et al., Effect of high protein, energy-restricted diet on body composition, glycemic control, and lipid concentrations in overweight and obese hyperinsulinemic men and women, Am J Clin Nutr 2003;78:31–9; and Layman DK et al., A reduced ratio of dietary carbohydrate to protein improves body composition and blood lipid profiles during weight loss in adult women, J Nutr 2003;133:411–7.

21. Bravata DM et al. Efficacy and safety of low-carbohydrate diets. JAMA 2003; 289:1837–50.

22. Bravata et al., Efficacy and safety of low-carbohydrate diets.

23. Low-carb diet plans, Harvey-Banting Diet. www.lowcarb.ca/atkins-diet-and-low-carb-plans/harvey-banting.html, accessed 9/03.

24. Bravata et al., Efficacy and safety; Bonow RO, Eickel RH, Diet, obesity, and cardiovascular risk, N Engl J Med 2003;348:2057–8; and Reddy ST et al. Effect of low-carbohydrate high protein diet and acid-base balance, stone-forming propensity, and calcium metabolism. Am J Kidney Dis 2002;40:265–74.

25. Goodrick and Foreyth, Why treatments for obesity don't last.

26. Polivy J. Psychological consequences of food restriction. J Am Diet Assoc 1996;96:589–92.

27. Parker-Pope T. A desperate diet: putting on pounds to qualify for weight loss surgery. Wall Street Journal, 7/8/03, p. D1.

28. Gastrointestinal surgery for severe obesity. The Weight-Control Information Network, www.niddk.nih.gov/health/nutrit/nutrit.htm, accessed 8/03.

29. Gastrointestinal surgery for severe obesity (www.niddk.nih.gov/health/nutrit/nutrit.htm).

30. Gastrointestinal surgery (www.niddk.nih.gov/health/nutrit/nutrit.htm).

31. Gastrointestinal surgery (www.niddk.nih.gov/health/nutrit/nutrit.htm); and Parker-Pope, A desperate diet.

32. Perugini RA et al. Weight loss surgery effective but can involve complications: case series. Arch Surg 2003;138:541–6.

33. Gastrointestinal surgery (www.niddk.nih.gov/health/nutrit/nutrit.htm); and Perugini et al., Weight loss surgery.

34. Liposuction comes of age, and New liposuction technique faster, less invasive. www.healthfinder.gov, accessed 9/03.

35. Liposuction comes of age (www.healthfinder.gov); and Cohen S. Lipolysis: pitfalls and problems in a series of 1246 procedures. Aesthetic Plast Surg 1985;9:207.

36. Weinsier RL et al., Free-living energy expenditure in women successful and unsuccessful at maintaining a normal body weight, Am J Clin Nutr 2002; 75:499–504; St Jeor ST et al., A classification system to evaluate weight maintainers, gainers, and losers, J Am Diet Assoc 1997;97:481–8; and Position of the American Dietetic Association: weight management, J Am Diet Assoc 2002;102:1145–55.

37. Position of the ADA: weight management.

38. Position of the ADA: weight management.

39. Weinsier et al., Free-living energy expenditure; St Jeor et al., A classification system; and Position of the ADA: weight management.

40. Schoeller DA et al., How much physical activity is needed to minimize weight gain in previously obese women? Am J Clin Nutr 1997;66:551–6.

41. National Institutes of Health. Methods for voluntary weight loss and control. Nutr Today 1992;27:27–33.

42. Schoeller et al., How much physical activity is needed; and Shick SM et al., Persons successful at long-term weight loss and maintenance continue to consume a low-energy, low-fat diet, J Am Diet Assoc 1998;98:408–14; and Cullinen K, Caldwell M, Weight training increases fat-free mass and strength in untrained young women, J Am Diet Assoc 1998;98:414–18.

43. Schlundt DG, Hill JO, Sbrocco T, Pope-Cordle J, Sharp T. The role of breakfast in the treatment of obesity: a randomized trial. Am J Clin Nutr 1992;55:645–51.

44. Rolls BJ., The super-sizing of America, Nutr Today 2003;38:42–53; and McCrory MA et al., Biobehavioral influences on energy intake and adult weight gain, J Nutr 2002;132:3830S–24S.

Nutrition **UP CLOSE**

Will You Maintain or Regain?

Feedback for Unit 10

The more checks you made in the "Yes" column, the greater the chances that you will maintain your weight loss. Reflect for a moment upon any checks you made in the "No" column. Enacting small, acceptable changes in your eating and activity plan now may make the difference between maintaining your weight loss or regaining the pounds.

Disordered Eating: Anorexia Nervosa, Bulimia, and Pica

Nutrition Scoreboard

	TRUE	FALSE
1 The United States has one of the world's highest rates of anorexia nervosa.		
2 Eating disorders result from psychological, and not biological, causes.		
3 People in many different cultures may consume clay, dirt, and other nonfood substances.		

[KEY CONCEPTS AND FACTS]

- Anorexia nervosa, bulimia nervosa (bulimia), binge-eating disorder, and pica are four specific eating disorders. They may seriously threaten health.

- Eating disorders are much more common in females than males.

- The incidence of eating disorders in a society is related

to the value placed on thinness by that society.

- An important route to the prevention of anorexia nervosa and bulimia is to change a society's cultural ideal of thinness and to eliminate biases against people (especially women) who are not thin.

Answers to *Nutrition Scoreboard*

	TRUE	FALSE
1 Anorexia nervosa is most common in the United States and other Westernized countries. It is rarely observed in developing, non-Westernized nations.[1]	✔	
2 The cause (or causes) of eating disorders is not yet known. Both psychological and biological factors may play a role.		✔
3 Although not recommended for health reasons, people in many different cultures practice pica—the regular ingestion of nonfood items such as clay and dirt.	✔	

The Eating Disorders

Three square meals a day, an occasional snack or missed meal, and caloric intakes that average out to match the body's need for calories—this set of practices is considered "orderly" eating. Self-imposed semi-starvation, feast and famine cycles, binge eating, *purging,* and the regular consumption of nonfood substances such as paint chips and clay—these behaviors are symptoms of disordered eating.

Four specific types of disordered eating patterns are officially recognized as eating disorders and have been assigned diagnostic criteria. They are (1) anorexia nervosa, (2) bulimia nervosa, (3) binge-eating disorder, and (4) pica.[2] Other forms of disordered eating such as compulsive overeating, restrained eating, and food preoccupation have been observed, but too little research exists to establish criteria for diagnosis.

Anorexia Nervosa

It's about 9:30 on a Tuesday night. You're at the grocery store picking up sandwich fixings and some milk. Although your grocery list contains only four items, you arrive at the checkout line with a half-filled cart. The woman in front of you has only five items: a bag with about 10 green beans, an apple, a bagel, a green pepper, and a 4-ounce carton of nonfat yogurt. As she carefully places each item into her shopping bag, you notice that she is dreadfully thin.

The woman is Alison. She has just spent half an hour selecting the food she will eat tomorrow. Alison knows a lot about the caloric value of foods and makes only low-calorie choices. Otherwise, she will never get rid of her excess fat. To Alison, weight is everything—she cannot see the skeleton-like appearance others see when they look at her.

As a child, Alison enjoyed little independence. Decisions about her life were made for her, a situation that contributed to her low self-esteem. During her teen years, Alison was overweight and clearly remembers the painful

purging
The use of self-induced vomiting, laxatives, or diuretics (water pills) to prevent weight gain.

Illustration 11.1
A day's diet? *For a person with anorexia nervosa, it was. The foods shown provide approximately 562 calories.*

anorexia nervosa
An eating disorder characterized by extreme weight loss, poor body image, and irrational fears of weight gain and obesity.

teasing and ridicule she had to endure. Now Alison is on her own, away from home and in control in an out-of-control way.

You didn't know this about Alison when you saw her. There is much more to anorexia nervosa than meets the eye.

Individuals with *anorexia nervosa* starve themselves (Illustration 11.1). They can never be too thin—no matter how emaciated they may be. As shown in Illustration 11.2, people with anorexia nervosa look extraordinarily thin from the neck down. The face and the rest of the head usually look normal because the head is the last part of the body to be affected by starvation.

Instead of the normal amount of body fat (20–25% of body weight), people with anorexia nervosa have little fat (7–13% of body weight).[3] They become cold easily and have unusually low heart rates and sometimes an irregular heartbeat, dry skin, low blood pressure, absent or irregular menstrual cycles, infertility, and poor pregnancy outcomes (Table 11.1).[4] Approximately 9 in 10 women with anorexia nervosa have significant bone loss, and 38% have osteoporosis. The extent of bone loss correlates strongly with undernutrition: the lower the body weight, the lower the bone density. Improving calcium and vitamin D intakes and use of bone-density drugs have limited effectiveness in rebuilding bones in females with current or past anorexia nervosa.[5]

The Female Athlete Triad] Pediatricians, nutritionists, and coaches are beginning to be on the lookout for eating disorders, menstrual cycle dysfunction, and decreased bone mineral density in young, female athletes. Low caloric intakes and underweight related to eating disorders can lower estrogen levels and disrupt menstrual cycles. The lack of estrogen decreases calcium deposition in bones and reduces bone density at a time when peak bone mass is accumulating.

Illustration 11.2
Eating disorders occur in males as well as females, but females make up approximately 95% of all cases.

The average size of female gymnasts on the U.S. Olympic team shrank from 5 feet 3 inches tall and 105 pounds in 1976 to 4 feet 9 inches tall and 88 pounds in 1992.

Photo Disc

TABLE 11.1

FEATURES OF ANOREXIA NERVOSA.
Reprinted with permission from the Diagnostic and Statistical Manual of Mental Disorders, Fourth Edition, Text Revision. Copyright 2000 American Psychiatric Association.

A. Essential Features
1. Refusal to maintain body weight at or above 85% of normal weight for age and height.
2. Intense fear of gaining weight or becoming fat, despite being underweight.
3. Disturbance in the way in which body weight or shape is experienced, undue influence of body weight or shape on self-evaluation, or denial of the seriousness of current low body weight.
4. Lack of menstrual periods in teenage females and women (missing at least three consecutive periods).

Restricting type: Person does not regularly engage in binge-eating or purging behavior.

Binge-eating type: Person regularly engages in binge-eating or purging behavior (self-induced vomiting; laxative, diuretic, or enema use).

B. Common Features in Females
1. Low-calorie diet, extensive exercise, low body fat
2. Soft, thick facial hair, thinning scalp hair
3. Loss of heart muscle, irregular, slow heartbeat
4. Low blood pressure
5. Increased susceptibility to infection
6. Anemia
7. Constipation
8. Low body temperature (hypothermia)
9. Dry skin
10. Depression
11. History of physical or sexual abuse
12. Low estrogen levels
13. Low bone density
14. Infertility, poor pregnancy outcome

C. Common Features in Males
1. Most of the common features in females
2. Substance abuse
3. Mood and other mental disorders

Irregular or absent menstrual cycles used to be thought of as "no big deal." That attitude has changed, however, due to research results indicating that abnormal cycles in young females are related to delayed healing of bone and connective tissue injuries, and to bone fractures and osteoporosis later in life.[6]

Motivations Underlying Anorexia Nervosa] The overwhelming desire to become and remain thin drives people with anorexia nervosa to refuse to eat, even when ravenously hungry, and to exercise intensely. Half of the people with anorexia turn to ***binge eating*** and purging—features of bulimia nervosa—in their efforts to lose weight.[7] Preoccupied with food, people with anorexia may prepare wonderful meals for others, but eat very little of the food themselves. Family members and friends, distressed by their failure to persuade the person with anorexia to eat,

binge eating
The consumption of a large amount of food in a small amount of time.

Richard Anderson

Illustration 11.3

There is a need for a more realistic body shape on television and in fashion magazines.
—Vivienne Nathanson, British physician, 2000

report high levels of anxiety. Although adults often describe people with anorexia as "model students" or "ideal children," their personal lives are usually marred by low self-esteem, social isolation, and unhappiness.[8]

What Causes Anorexia Nervosa?] The cause of anorexia nervosa isn't yet clear. It is likely that many different conditions, both psychological and biological, predispose an individual to become totally dedicated to extreme thinness.[9] The value that Western societies place on female thinness, the need to conform to society's expectations of acceptable body weight and shape, low self-esteem, and a need to control some aspect of one's life completely are commonly offered as potential causes for this disorder (Illustration 11.3).[10]

How Common Is Anorexia Nervosa?] It is estimated that 1% of adolescent and young women in the Western world and less than 0.1% of young males have anorexia nervosa. The disorder has been reported in girls as young as 5 and in women through their forties;[11] however, it usually begins during adolescence. It is estimated that one in ten females between the ages of 16 and 25 has "subclinical" anorexia nervosa, or exhibits some of the symptoms of the disorder.[12]

Certain groups of people are at higher risk of developing anorexia nervosa than others (Table 11.2). People at risk come from all segments of society, but tend to be overly concerned about their weight and food, and have attempted weight loss from an early age.[14]

Treatment] There is no "magic bullet" treatment that cures anorexia nervosa quickly and completely. In all but the least severe cases, the disorder generally takes a good deal of time and professional help to correct. Treating the disorder is often difficult because few people with anorexia believe their weight needs to be increased.[15]

Treatment programs for anorexia nervosa generally focus on restoring nutritional health and body weight, psychological counseling to improve self-esteem and

TABLE 11.2

RISK GROUPS FOR ANOREXIA NERVOSA.[13]

- Dieters
- Ballet dancers
- Competitive athletes (gymnasts, figure skaters)
- Fitness instructors
- Dietetics majors
- People with type 1 (insulin dependent) diabetes

attitudes about body weight and shape, antidepressant or other medications, family therapy, and normalizing eating and exercise behaviors. These programs are successful in 50% of people, and partially successful in most other cases.[16] One-third of people with full recovery from anorexia nervosa will relapse within 7 years or less. By 8 years after diagnosis, 3% of people with anorexia nervosa will have died from the disorder, and it claims the lives of 18% 33 years later. Results of treatment are often excellent when the disorder is treated early.[17] Unfortunately, many people with the condition deny that problems exist and postpone treatment for years. Initiation of treatment is often prompted by a relative, coach, or friend.[18]

Bulimia Nervosa

Finally home alone, Lisa heads to the pantry and then to the freezer. She has carefully controlled her eating for the last day and a half and is ready to eat everything in sight.

It's a bittersweet time for her. Lisa knows the eating binge she is preparing will be pleasurable, but that she'll hate herself afterward. Her stomach will ache from the volume of food she'll consume, she'll feel enormous guilt from losing control, and she'll be horrified that she may gain weight and will have to starve herself all over again. Lisa is so preoccupied with her weight and body shape that she doesn't see the connection between her severe dieting and her bouts of uncontrolled eating. To get rid of all the food she is about to eat, she will do what she has done several times a week for the last year. Lisa avoids the horrible feelings that come after a binge by "tossing" everything she ate as soon as she can.

In just 10 minutes, Lisa devours 10 peanut butter cups (the regular size), a 12-ounce bag of chocolate chip cookies, and a quart of mint chocolate chip ice cream. Before 5 more minutes have passed Lisa will have emptied her stomach, taken a few deep breaths, thrown on her shorts, and started the 5-mile route she jogs most days. As she jogs, she obsesses about getting her 138-pound, 5-foot 5-inch frame down to 115 pounds. She will fast tomorrow and see what news the bathroom scale brings.

bulimia nervosa
An eating disorder characterized by recurrent episodes of rapid, uncontrolled eating of large amounts of food in a short period of time. Episodes of binge eating are often followed by purging.

Lisa is not alone. ***Bulimia nervosa*** occurs in 1 to 3% of young women and in about 0.5% of young males in the United States.[19] The disorder is characterized by regular episodes of dieting, binge eating, and attempts to prevent weight gain by purging; use of laxatives, diuretics, or enemas; dieting; and sometimes exercise. In most cases, bulimia nervosa starts with voluntary dieting to lose weight. At some point, voluntary control over dieting is lost, and people feel compelled to engage in binge eating and vomiting.[20] The behaviors become cyclic: food binges are followed by guilt, purging, and dieting. Dieting leads to a feeling of deprivation and intense hunger, which leads to binge eating, and so on. Once a food binge starts, it is hard to stop.

Table 11.3 lists the features of bulimia nervosa. Approximately 86% of people with this condition vomit to prevent weight gain and avoid postbinge anguish. A smaller proportion of people use laxatives, diuretics (water pills), or enemas alone or in combination with vomiting.[22] Laxatives, enemas, and diuretics do not prevent weight gain, however, and their regular use can be harmful. The habitual use of laxatives and enemas causes "laxative dependency"—these products become necessary for bowel movements. Diuretics can cause illnesses by depleting the body of water and certain minerals and disturbing its fluid balance.[23]

The lives of people with bulimia nervosa are usually dominated by conflicts about eating and weight. Some affected individuals are so preoccupied with food

TABLE 11.3

FEATURES OF BULIMIA NERVOSA.[21]
Reprinted with permission from the Diagnostic and Statistical Manual of Mental Disorders, Fourth Edition, Text Revision. Copyright 2000 American Psychiatric Association.

A. Essential Features

 1. Recurrent episodes of binge eating. An episode of binge eating is characterized by both of the following:

   a. Eating an amount of food within a two-hour period of time that is definitely larger than most people would eat in a similar amount of time and under similar circumstances.

   b. A sense of lack of control over eating during the episode; a feeling that one cannot stop eating or control what or how much one is eating.

 2. Recurrent inappropriate compensatory behavior in order to prevent weight gain, such as self-induced vomiting; misuse of laxatives, diuretics, enemas, or other medications; fasting; or excessive exercise.

 3. The binge eating and inappropriate compensatory behaviors both occur, on average, at least twice a week for 3 months.

 4. Self-evaluation is unduly influenced by body weight and shape.

 5. The disturbance does not occur exclusively during episodes of anorexia nervosa.

Purging type: The person regularly engages in self-induced vomiting or the misuse of laxatives, diuretics, or enemas.

Nonpurging type: The person regularly engages in fasting or excessive exercise but does not regularly engage in self-induced vomiting or the misuse of laxatives, diuretics, or enemas.

B. Common Features

 1. Weakness, irritability
 2. Abdominal pain, constipation, bloating
 3. Dental decay, tooth erosion
 4. Swollen cheeks and neck
 5. Binge on high-calorie foods
 6. Eat in secret
 7. Normal weight or overweight
 8. Guilt and depression
 9. Substance abuse
 10. Dehydration
 11. Impaired fertility
 12. History of sexual abuse

that they spend days securing food, bingeing, and purging. Others experience only occasional episodes of binge eating, purging, and fasting.[24]

Unlike those with anorexia nervosa, people with bulimia usually are not underweight or emaciated. They tend to be normal weight or overweight.[25] Like anorexia nervosa, bulimia nervosa is more common among athletes (including gymnasts, weight lifters, wrestlers, jockeys, figure skaters, physical trainers, and distance runners) and ballet dancers than in other groups.[26]

Bulimia nervosa leads to major changes in metabolism. The body must constantly adjust to feast and famine cycles and mineral and fluid losses. Salivary glands become enlarged, and teeth may erode due to frequent vomiting of highly acidic foods from the stomach.[27]

Is the Cause of Bulimia Nervosa Known?]

"Do you follow a special diet?" asks the dietitian at the eating disorder clinic.

"Yes," answers the client with bulimia. "Feast or famine."

The cause of bulimia nervosa is not known with certainty, but the scientific finger is pointing at depression, abnormal mechanisms for regulating food intake, and feast-and-famine cycles as possible causes.[28] Fasts and **restrained eating** may prompt feelings of deprivation and hunger that may trigger binge eating.[29] The ideal thinness may become more and more difficult to achieve as the feast-and-famine cycles continue.

Treatment]   The cornerstone of bulimia treatment is nutrition and psychological counseling to break the feast-and-famine cycles. Replacing the disordered pattern of eating with regular meals and snacks often reduces the urge to binge and the need to purge. Psychological counseling aimed at improving self-esteem and attitudes toward body weight and shape goes hand in hand with nutrition counseling. In many cases, antidepressants are a useful component of treatment.[30] The full recovery of women with bulimia nervosa is higher than that for anorexia nervosa. Nearly all women with bulimia achieve partial recovery, but one-third will relapse into bingeing and purging within 7 years.[31] Bulimia nervosa usually improves substantially during pregnancy; about 70% of women with the condition will improve their eating habits for the sake of their unborn baby.[32]

Binge-Eating Disorder

Psychiatrists now recognize an eating disorder called **binge-eating disorder** (Table 11.4). People with this condition tend to be overweight or obese, and one-third are male.[34] Like individuals with bulimia nervosa, people with binge-eating disorder eat several thousand calories' worth of food within a short period of time during a solitary binge, feel a lack of control over the binges, and experience distress or depression after the binges occur. People must experience eating binges twice a week on average over a period of 6 months to qualify for the diagnosis. Unlike individuals with bulimia nervosa, however, people with binge-eating disorder don't vomit, use laxatives, fast, or exercise excessively in an attempt to control weight gain.[35]

It is estimated that 9 to 30% of people in weight-control programs and 30 to 90% of obese people have binge-eating disorder.[36] The condition is far less common (2 to 5%) in the general population.[37] Stress, depression, anger, anxiety, and other negative emotions appear to prompt binge-eating episodes. Preliminary evidence indicates that binge-eating disorder may be related to a genetic mutation that impairs control of normal eating behavior.[38]

The Treatment Approach to Binge-Eating Disorder]   The treatment of binge-eating disorder focuses on both the disordered eating and the underlying psychological issues.[39] Persons with this condition will often be asked to record their food intake, indicate bingeing episodes, and note feelings, circumstances, and thoughts related to each eating event (Illustration 11.4). This information is used to identify circumstances that prompt binge eating and practical alternative behaviors that may

restrained eating
The purposeful restriction of food intake below desired amounts in order to control body weight.

binge-eating disorder
An eating disorder characterized by periodic binge eating, which normally is not followed by vomiting or the use of laxatives. People must experience eating binges twice a week on average over a period of 6 months to qualify for the diagnosis.

TABLE 11.4

FEATURES OF BINGE-EATING DISORDER.[33]

1. Rapid consumption of extremely large amounts of food (several thousand calories) in a short period of time

2. Two or more such episodes of binge eating per week over a period of 6 months

3. Binge eating by oneself

4. Lack of control over eating or an inability to stop eating during a binge

5. Postbinge-eating feelings of self-hatred, guilt, and depression or disgust

6. Purging, fasting, excessive exercise, or other compensation for high-calorie intakes not present

REALITY CHECK

Close to home

Although she hides it, you are sure your sister has bulimia nervosa and that she is not getting help. You are deeply concerned for her health and well-being, but don't know what to do about it. Here's what Heather and Crystal say they would do:

Who do you think has the better idea

?

Heather:
I'd talk with her about getting help.

Crystal:
I'd spend more time with her to let her know I love her.

Photo Disc

Answers on next page

prevent it. Individuals being treated for binge-eating disorder are usually given information about it, attend individual and group therapy sessions, and receive nutrition counseling on normal eating, hunger cues, and meal planning. Antidepressants may be part of the treatment. Treatment is successful in 85% of women treated for binge-eating disorder.[40]

Resources for Eating Disorders

Information and services related to eating disorders are available from a variety of sources. Services are best delivered by health care teams specializing and experienced in the treatment of eating disorders. Contact with a primary care physician, dietitian, or nurse practitioner is often a good start to the process of identifying qualified health care teams. Reliable sources of information about eating disorders, support groups, nearby treatment centers, and hot lines can be found on the Internet. (See "WWW Links" at the end of the unit.)

One of the most important resources for people with an eating disorder may be a trusted friend or relative. This unit's "Reality Check" explores this resource in a very personal way—by putting you in the shoes of a person whose sister has bulimia.

Illustration 11.4
Example of a food diary of a person with binge-eating disorder.

Daily Food Record

Date_____

Time	Type and amount of food and beverage	Meal, Snack, Binge?	Eating triggers (feelings, situation)
7:30 am	coffee, 2 cups sugar 2 tsp cornflakes, 2 cups skim milk, 1 cup	M	Hunger!
11:30 am	tuna sandwich ice tea, 2 cups	M	Bored, hungry
7:30 pm	3 hamburgers 2 large fries 24 oreos 1/2 gallon ice cream	B	Stressed out, angry at my coach

Heather

ANSWERS TO **REALITY CHECK**

Close to home

Both ideas are admirable and deserve a thumbs-up. Heather's idea is aimed directly at helping her sister consider treatment and may be the appropriate action to take sometimes. There is a way to talk to a relative or friend about your concerns for them that may help both of you. Learn more about it from the information presented in Table 11.5.

Crystal

Undieting: The Clash between Culture and Biology

There is a saying that women underreport their weight and men overreport their height. Clearly there are cultural norms at work here.
—L. Cohen, 2001[42]

The pressure to conform to society's standard of beauty and acceptability is thought to be a primary force underlying the development of eating disorders.[43] Children acquire prevailing cultural values of beauty before adolescence. As early as age 5, American children learn to associate negative characteristics with people who are overweight and positive characteristics with those who are thin.[44] Standards of beauty defined by models and movie and television stars often include thinness, but

TABLE 11.5

HELPING A FAMILY MEMBER OR FRIEND WITH AN EATING DISORDER.

Whether at work, home, or play, many of us experience anxiety and a sense of helplessness when someone we love is living with an eating disorder. We may feel compelled to take action to help, but aren't sure what to do or how to do it. Here are some tips on how to express your concerns to a friend or relative with an eating disorder.

1. Gather information about services for people with eating disorders to share with your friend or relative.

2. Talk with your friend or relative privately when there is enough time to fully discuss the issue. Tell them you are worried and that they may need to seek help.

3. Encourage your friend or relative to express his or her feelings, and then listen intently. Be accepting about the feelings that are expressed. Be ready to talk to them more about it in the future.

4. Do not argue with your friend or relative about whether she or he has an eating disorder. Let your friend or relative know you heard what was said, but that you are concerned that he or she may not get better without treatment.

5. Seek emergency medical help in life-threatening situations.

Only individuals with an eating disorder can make the decision to get help. Knowledge that people who love them will be around to support them and their decision to seek treatment may help encourage the person with an eating disorder to take action.[41]

Photo Disc

Photo Disc

Illustration 11.5
The trend toward size acceptance. *Acceptance of a realistic standard of body weight and shape—one that corresponds to health and physical fitness—and respect for people of all body sizes may be the most effective measures that can be taken to prevent anorexia nervosa, bulimia nervosa, and binge-eating disorder.*

the body shape portrayed as best is often unhealthfully thin and unattainable by many. The disparity between this ideal and what people normally weigh has led to widespread discontent. Approximately 50% of normal-weight adult women are dissatisfied with their weight; and many diet, binge, and purge occasionally or fast in an attempt to reach the standard of beauty set for them.[45]

A movement toward acceptance of body size, fashionable attire for larger people, full-size models, and a more realistic view of individual differences in body shapes is emerging slowly in America (Illustration 11.5). Acceptance of a realistic standard of body weight and shape—one that corresponds to health and physical fitness—and respect for people of all body sizes may be the most effective measures that can be taken to prevent anorexia nervosa, bulimia nervosa, and binge-eating disorder.

Pica

When did I start eating clay? I know it might sound strange to you, but I started craving clay in the summer of '58. It was a beautiful spring morning—it had just rained. I smelled something really sweet in the breeze coming in my bedroom window. I went outside and knew instantly where the sweet smell was coming from. It was the wet clay that lies all around my house. I scooped some up and tasted it. That's when and how I started my craving for that sweet-smelling clay. I keep some in the fridge now because it tastes even better cold.

A most intriguing type of eating disorder, **pica** has been observed in chimpanzees and in humans in many different cultures since ancient times.[46] The history and persistence of pica might suggest that the practice has its rewards. Nevertheless, important health risks are associated with eating many types of nonfood substances.

pica (pike-eh)
The regular consumption of nonfood substances such as clay or laundry starch.

TABLE 11.6

CHARACTERISTICS OF PICA.

A. **Essential features:** Regular ingestion of nonfood substances such as clay, paint chips, laundry starch, paste, plaster, dirt, or hair.

B. **Other common features:** Occurs primarily in young children and pregnant women in the southern United States.

geophagia (ge-oh-phag-ah)
Clay or dirt eating.

pagophagia (pa-go-phag-ah)
Ice eating.

amylophagia (am-e-low-phag-ah)
Laundry starch or cornstarch eating.

plumbism
Lead (primarily from old paint flakes) eating.

The characteristics of pica are summarized in Table 11.6. Young children and pregnant women are most likely to engage in the practice; for unknown reasons, it rarely occurs in men.[47] It most commonly takes the form of *geophagia* (clay or dirt eating), *pagophagia* (ice eating), *amylophagia* (laundry starch and cornstarch eating), or *plumbism* (lead eating). A potpourri of nonfood substances, listed in Table 11.7, may be consumed.

Pica has a forceful calling card:

> *Pica permits the mind no rest until it is satisfied.*[48]

It is not clear why pica exists, although several theories have been proposed.

Geophagia

Some people very much like to eat certain types of clay or dirt. Those who do often report that the clay or dirt tastes or smells good, quells a craving, or helps relieve nausea or an upset stomach. The belief that certain types of clay provide relief from

TABLE 11.7

A PARTIAL LIST OF NONFOOD SUBSTANCES REPORTED TO BE CONSUMED BY INDIVIDUALS WITH PICA.

Animal droppings	Coffee grounds	Leaves	Plaster
Baking soda	Cornstarch	Mothballs	Sand
Burnt matches	Crayons	Nylon stockings	String
Cigarette butts	Dirt	Paint chips	Wool
Clay	Foam rubber	Paper	
Cloth	Hair	Paste	
Coal	Laundry starch	Pebbles	

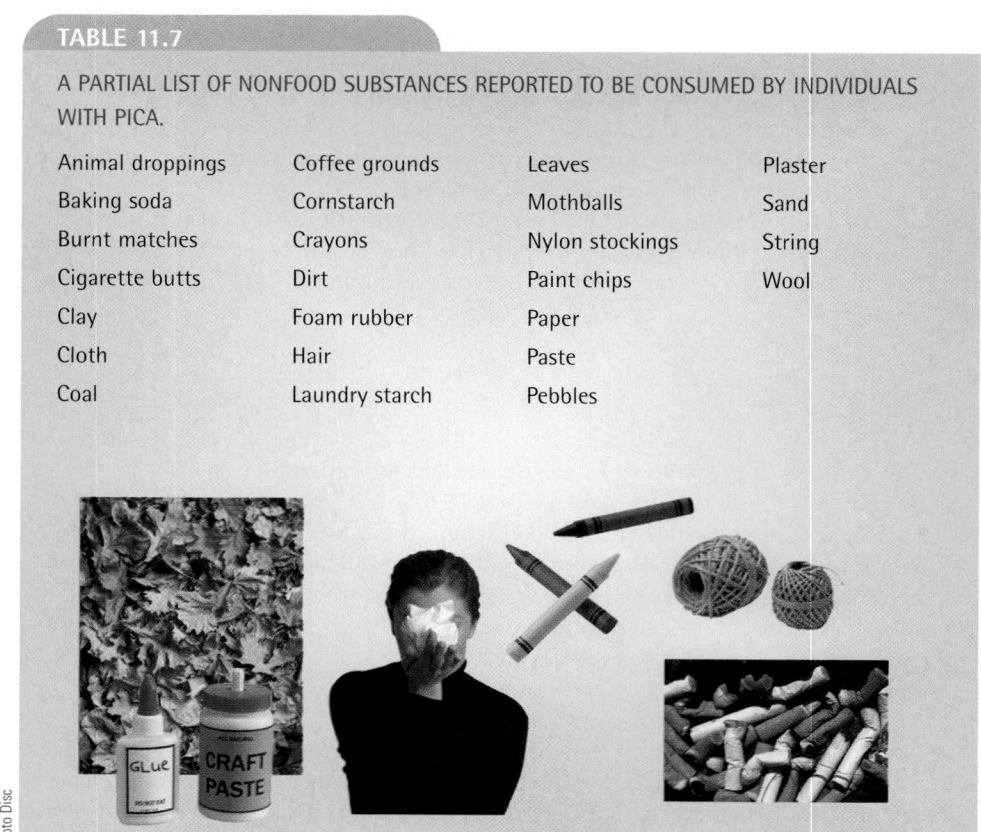

Photo Disc

stomach upsets may have some validity: a component of some clays is used in nausea and diarrhea medicines. There is no evidence that geophagia is motivated by a need for minerals found in clay or dirt, however.[49]

Although the reasons given for clay and dirt ingestion make the practice understandable, the consequences to health outweigh the benefits. Clay and dirt consumption can block the intestinal tract and cause parasitic and bacterial infections.[50] The practice is also associated with iron-deficiency and sickle-cell anemia in some individuals.[51]

Pagophagia

Have you ever known somebody who constantly crunches on ice? That person may have a 9-in-10 chance of being iron deficient. Regular ice eating, to the extent of one or more trays of ice cubes a day, is closely associated with an iron-deficient state. Ice eating usually stops completely when the iron deficiency is treated.[52]

Ice eating may be common during pregnancy. In one study of women from low-income households in Texas, 54% of pregnant women reported eating large amounts of ice regularly. Ice eaters had poorer iron status than other pregnant women who did not eat ice.[53]

Amylophagia

The sweet taste and crunchy texture of flaked laundry starch are attractive to a small number of women, especially during pregnancy. If the laundry starch preferred is not available, cornstarch may be used in its place. Laundry starch is made from unrefined cornstarch. The taste for starch almost always disappears after pregnancy.[54]

Laundry starch and cornstarch have the same number of calories per gram as do other carbohydrates (4 calories per gram). Consequently, starch eating provides calories and may reduce the intake of nutrient-dense foods. In addition, starch may contain contaminants because it is not intended for consumption. Starch eaters' diets are generally inferior to the diets of pregnant women who don't consume starch, and their infants are more likely to be born in poor health.[55]

Plumbism

The consumption of lead-containing paint chips poses a major threat to the health of children in the United States and many other countries (Illustration 11.6). Many older homes and buildings, especially those found in substandard housing areas, are covered with lead-based paint and its dried-up flakes. Children may develop lead poisoning if they eat the sweet-tasting paint flakes or inhale lead from contaminated dust and soil near the buildings. An estimated 1 million young children in the United States have elevated blood lead levels.[56]

High levels of exposure to lead can cause profound mental retardation and death in young children. Low levels of exposure can lead to hearing problems, growth retardation, reduced intelligence, and poor classroom performance. Children with lead poisoning are more likely to fail or drop out of school than children not exposed to lead in their environment.[57]

Eating disorders affect the health and well-being of over a million people in the United States. Although there are treatment strategies, such as counseling and the removal of lead-based paints from old houses and apartments, the solution to eating disorders lies in their prevention. With the exception of certain types of pica discussed here, the most effective way to prevent eating disorders may be to adjust our expectations and cultural norms to reflect reality.

Illustration 11.6
The regular consumption of lead-based paint chips from old houses is a major cause of lead poisoning in young children.

Nutrition UP CLOSE

Eating Attitudes Test

FOCAL POINT: Discover if your eating attitudes and behaviors are within a normal range.

Date _____ Age _____ Gender _____
Height Present weight _____ How long at present weight? _____
Highest past weight _____ How long ago? _____
Lowest past weight _____ How long ago? _____

Answer the following questions using these responses:

      A = always   S = sometimes   U = usually   R = rarely   O = often   N = never

_____ 1. I am terrified of being overweight.

_____ 2. I avoid eating when I am hungry.

_____ 3. I find myself preoccupied with food.

_____ 4. I have gone on eating binges where I feel that I may not be able to stop.

_____ 5. I cut my food into very small pieces.

_____ 6. I am aware of the calorie content of the foods I eat.

_____ 7. I particularly avoid foods with a high-carbohydrate content.

_____ 8. I feel that others would prefer that I ate more.

_____ 9. I vomit after I have eaten.

_____ 10. I feel extremely guilty after eating.

_____ 11. I am preoccupied with a desire to be thinner.

_____ 12. I think about burning up calories when I exercise.

_____ 13. Other people think I am too thin.

_____ 14. I am preoccupied with the thought of having fat on my body.

_____ 15. I take longer than other people to eat my meals.

_____ 16. I avoid foods with sugar in them.

_____ 17. I eat diet foods.

_____ 18. I feel that food controls my life.

_____ 19. I display self-control around food.

_____ 20. I feel that others pressure me to eat.

_____ 21. I give too much time and thought to food.

_____ 22. I feel uncomfortable after eating sweets.

_____ 23. I engage in dieting behavior.

_____ 24. I like my stomach to be empty.

_____ 25. I enjoy trying new rich foods.

_____ 26. I have the impulse to vomit after meals.

Feedback (including scoring) can be found at the end of Unit 11.

Source: McSherry JA. Progress in the diagnosis of anorexia nervosa. Journal of the Royal Society of Health 1986;106:8–9. (Eating Attitudes Test developed by Dr. Paul Garfinkel.)

Key Terms

amylophagia, page 11-12

anorexia nervosa, page 11-3

binge eating, page 11-4

binge-eating disorder, page 11-8

bulimia nervosa, page 11-6

geophagia, page 11-12

pagophagia, page 11-12

pica, page 11-11

plumbism, page 11-12

purging, page 11-2

restrained eating, page 11-8

www links

www.anad.org
This site, from the National Association of Anorexia Nervosa and Associated Disorders, provides free hot line counseling, a national network of support groups, and health care referrals.

www.nedic.ca
The National Eating Disorder Info Center in Toronto provides information and resources on eating disorders and weight preoccupation, and a telephone support line, information on support groups, and listings of Canada-wide treatment resources.

www.nationaleatingdisorders.org
Declare independence from a weight-obsessed world on this site. Click on "seeking treatment" to learn about components of treatment of eating disorders and location of nearby care providers.

www.edreferral.com
This site includes basic information on eating disorders along with specific information on treatment and recovery for men, pregnant women, and others with eating disorders.

www.naafa.org
From the National Association to Advance Fat Acceptance, Inc., this site is dedicated to improving the quality of life for fat people. It takes on policies and practices related to size discrimination and promotes size acceptance by individuals and society.

www.mirror-mirror.org/eatdis.htm
Subject categories allow you to select topics such as myths and realities of eating disorders, where to get help, recovery, and links to other Web sites.

www.hedc.org
Harvard's Eating Disorders Center's site provides facts about eating disorders, answers to FAQs (frequently asked questions), advice on how to help a friend, child, or self with an eating disorder, and where to find help.

Notes

1. Lake AJ et al. Effect of Western culture on women's attitudes to eating and perceptions of body shape. Int J Eat Disord 2000;27:83–9.

2. American Psychiatric Association. Diagnostic and statistical manual of mental disorders: DMS-IV, 4th ed., Text Revision. Washington, DC: 2000.

3. Mazess RB, Barden HS, Ohlrich ES. Skeletal and body-composition effects of anorexia nervosa. Am J Clin Nutr 1990; 52:438–41.

4. Eating disorders. Facts about eating disorders and the search for solutions. National Institute of Mental Health, Bethesda, MD, 2001. www.nimh.nih.gov.

5. Grinspoon S et al. Prevalence and predictive factors for regional osteopenia in women with anorexia nervosa. Ann Intern Med 2000;133:790–4.

6. Beals KA, Manore MM. Disorders of the female athlete triad among collegiate athletes. Int J Sports Nutr Exer Metab 2002;12:281–93.

7. APA, DMS-IV.

8. Omizo SA, Oda EA. Anorexia nervosa: psychological considerations for nutrition counseling. J Am Diet Assoc 1988; 88:49–51.

9. Tamburrino MB, McGinnis RA. Anorexia nervosa: a review. Panminerva Med 2002;44:301–11.

10. Eating disorders (www. nimh.nih.gov); and Patel DR et al., Eating disorders, Indian J Pediatr 1998;65:487–94.

11. Eating disorders (www. nimh.nih.gov); and Feldman W, Feldman E, Goodman JT, Culture versus biology: children's attitudes toward thinness and fatness, Pediatrics 1988;81:190–4.

12. Grinspoon, Prevalence and predictive factors for regional osteopenia.

13. Grinspoon, Prevalence and predictive factors for regional osteopenia; and Worobey J, Schoenfeld D, Eating disordered behavior in dietetics students and students in other majors, J Am Diet Assoc 1999;99:1100–2.

14. Grinspoon, Prevalence and predictive factors for regional osteopenia.

15. APA, DMS-IV.

16. Robinson PH, Recognition and treatment of eating disorders in primary and secondary care, Alimentary Pharmacol & Thera 2000;14:367–77; and Patel et al., eating disorders.

17. Herzog DB et al., Recovery and relapse in anorexia and bulimia nervosa: a 7.5 year follow-up study, J Am Acad Child & Adol Psychiatr 1999;38:829–37; Herzog DB et al, Mortality in eating disorders: a descriptive study, Int J Eat Disord 2000;28:20–6; and Tamburrino and McGinnis, Anorexia nervosa: a review.

18. Treating eating disorders, Harvard Women's Health Watch 1996 May:4–5; and When eating goes awry: an update on eating disorders, Food Insight 1997 Jan/Feb:35.

19. Eating disorders (www. nimh.nih.gov).

20. Faris PL et al. Effect of decreasing afferent vagal activity with ondansetron on symptoms of bulimia nervosa: a randomized, double-blind trial. Lancet 2000;355:792–70.

21. APA, DMS-IV.

22. APA, DMS-IV.

23. Robinson, Recognition and treatment of eating disorders.

24. APA, DMS-IV.

25. APA, DMS-IV.

26. Edell D, Beware of personal trainers with their looks, IDEA Health and Fitness Source. www.healthcentral.com/drdean/ DeanFullTextTopics.cfm?ID=41044&src=n2, accessed 9/2000; and When eating goes awry: an update on eating disorders.

27. APA, DMS-IV.

28 Faris et al., Effect of decreasing afferent vagal activity; and Tolstoi LG. The role of pharmacotherapy in anorexia nervosa and bulimia. J Am Diet Assoc 1989;89:1640–6.

29. Herzog DB, Copeland PM. Bulimia nervosa—psyche and satiety (editorial). N Engl J Med 1988;319:716–8.

30. Robinson, Recognition and treatment of eating disorders; and Eating disorders III, Disease definition, epidemiology, and natural history, www.mentalhealth.com, accessed 12/2000.

31. Herzog et al., Recovery and relapse in anorexia and bulimia nervosa.

32. Hohlstein LA. Eating disorders program. American Dietetic Association annual meeting, Boston, 1997 Oct 27.

33. APA, DMS-IV.

34. Eating disorders (www.nimh.nih.gov).

35. Fairburn CG et al. Distinctions between binge eating disorder and bulimia nervosa. Arch Gen Psychiatry 2000;57:659–65.

36. Branson R et al. Binge eating as a major phenotype of melanocortin 4 receptor gene mutations. N Engl J Med 2003;348:1096–103.

37. Hohlstein, Eating disorders program; and Basdevant A et al. Prevalence of binge eating disorder in different populations of French women. Int J Eat Disord 1999;18:309–15.

38. Eating disorders III, disease definition; and Branson et al., Binge eating.

39. Hay PJ, Bacaltchuk J. Psychotherapy for bulimia nervosa and bingeing (Cochrane Review). In: The Cochrane Library, Issue 2, 2003.

40. Fairburn et al., Distinctions between binge eating disorder and bulimia nervosa.

41. Eating Disorder Referral and Information Center, www.edreferral.com and Harvard's Eating Disorders Center, www.hedc.org, accessed 6/03.

42. Cohen, LA, Nutrition and cancer prevention. Nutr Today 2001;36:78–9.

43. Lake et al., Effect of Western culture on women's attitudes.

44. Feldman et al., Culture vs. biology.

45. Branson et al., Binge eating as a major phenotype; and Zuckerman DM, Colby A, Ware NC, Layerson JS. The prevalence of bulimia among college students. Am J Public Health 1986;76:1135–7.

46. Cooper M. Pica: a survey of the historical literature as well as reports from the fields of veterinary medicine and anthropology. Springfield (IL): Charles C. Thomas; 1957.

47. Edwards CH, McDonald S, Mitchell JR, et al. Clay- and cornstarch-eating women. J Am Diet Assoc 1959;35:810–15.

48. Craign FW. Observations on cachexia Africana or dirt-eating. Am J Med Sci 1935;17:365.

49. Johns T, Duquette M. Detoxification and mineral supplementation as functions of geophagy. Am J Clin Nutr 1991;53: 448–56.

50. APA, DMS-IV.

51. Korman SH, Pica as a presenting symptom in childhood celiac disease, Am J Clin Nutr 1990;51:139; and Hackworth SR, Williams LL. Pica for foam rubber in patients with sickle-cell disease. South Med J 2003; 96:81–3.

52. Reynolds RD, Binder HJ, Miller MB, Chang WWY, Horan S, Pagophagia and iron deficiency anemia, Ann Intern Med 1968;69:435–40; and Coltman CA Jr., Pagophagia and iron lack, JAMA 1969;207:513–16.

53. Rainville AJ. Pica practices of pregnant women are associated with lower maternal hemoglobin level at delivery. J Am Diet Assoc 1998;98:293–6.

54. Edwards et al., Clay- and cornstarch-eating women.

55. Edwards et al., Clay- and cornstarch-eating women.

56. Child health USA. Washington, DC: Maternal and Child Health Bureau, U.S. PHS;2000.

57. Child health USA.

Nutrition **UP CLOSE**

Eating Attitudes Test

Feedback for Unit 11

Never = 3
Rarely = 2
Sometimes = 1
Always, usually, and often = 0

A total score under 20 points may indicate abnormal eating behavior. If you think you have an eating disorder, it is best to find out for sure. Careful evaluation by a qualified health professional is necessary to exclude any possible underlying medical reasons for your symptoms. Contacting a physician, nurse practitioner, dietitian, or the student health center is an important first step. You may wish to show your Eating Attitudes Test to the health professional.

Useful Facts about Sugars, Starches, and Fiber

Nutrition Scoreboard

		TRUE	FALSE
1	Pasta, bread, and potatoes are good sources of complex carbohydrates.		
2	Ounce for ounce, presweetened breakfast cereals and unsweetened cereals provide about the same number of calories.		
3	A 12-ounce can of soft drink contains about 3 tablespoons (9 teaspoons) of sugar.		
4	Lettuce, onions, and celery are high in dietary fiber.		
5	Cooking vegetables destroys their fiber content.		
6	Sugar consumption is the leading cause of tooth decay.		

Answers on next page

[KEY CONCEPTS AND FACTS]

- Simple sugars, the "starchy" complex carbohydrates, and dietary fiber are members of the carbohydrate family.

- Ounce for ounce, sugars and the starchy complex carbohydrates supply fewer than half the calories of fat.

- Tooth decay and poor-quality diets are related to high sugar intake.

- Fiber benefits health in a number of ways.

Answers to *Nutrition Scoreboard*

		TRUE	FALSE
1	If you get this right, you may be in the minority. In one study of college students,[1] only 38% could identify good sources of complex carbohydrates. Rice, crackers, grits, dried beans, corn, peas, tortillas, biscuits, and oatmeal are also good sources of complex carbohydrates.	✔	
2	That's true! Both sweetened and unsweetened cereals consist primarily of carbohydrate. A gram of carbohydrate provides 4 calories whether the source is sugar or flakes of corn.	✔	
3	That's a lot of sugar!	✔	
4	Although these are healthy food choices, they do not contain very much dietary fiber. (Not all that goes "crunch" is a good source of fiber.)		✔
5	Cooking doesn't destroy dietary fiber.		✔
6	Rates of tooth decay increase in populations as sugar intake increases.[2]	✔	

The Carbohydrates

carbohydrates
Chemical substances in foods that consist of a simple sugar molecule or multiples of them in various forms.

Carbohydrates are the major source of energy for people throughout the world. They are the primary ingredient of staple foods such as pasta, rice, cassava, beans, and bread. On average, Americans consume less carbohydrate than people in much of the world: approximately 51% of total calories. This level of intake is on the low end of the recommended range of carbohydrate intake of 45 to 65% of total calories.[3]

The carbohydrate family consists of three types of chemical substances:

1. Simple sugars

2. Complex carbohydrates ("starch")

3. Total fiber

Some food sources of these different types of carbohydrate are shown in Illustration 12.1. With the exception of fiber, members of the carbohydrate family perform one role in common in the body: they provide energy. Simple sugars and starchy complex carbohydrates supply 4 calories per gram. Dietary fiber cannot be digested by enzymes produced by humans and is not considered a source of energy. Nevertheless, dietary fiber plays a number of important roles in maintaining health.

Carbohydrates have two sets of "chemical relatives" of significance to human nutrition. Both are derived from carbohydrates and have an alcohol group as part of their chemical structure. The first of these, the *alcohol sugars*, are presented in this unit. *Alcohol* (or, more precisely, ethanol), the other chemical relative of carbohydrates, supplies 7 calories per gram. It is covered in a separate unit.

Illustration 12.1
The carbohydrate family.
Some food sources of simple sugars (left), and some food sources of starch and dietary fiber (right) are shown.

Simple Sugar Facts

Simple sugars are considered "simple" because they are small molecules that require little or no digestion before they can be used by the body. They come in two types: ***monosaccharides*** and ***disaccharides.*** The monosaccharides consist of one molecule and include glucose ("blood sugar" or "dextrose"), fructose ("fruit sugar"), and galactose. Disaccharides consist of two molecules. The combination of a glucose molecule and a fructose molecule makes sucrose (or "table sugar"); maltose ("malt sugar") is made from two glucose molecules; and lactose ("milk sugar") consists of a glucose molecule plus a galactose molecule. Honey, by the way, is a disaccharide. It is composed of glucose and fructose just as sucrose is, but it's a liquid rather than a solid because of the way the two molecules of sugar are chemically linked together. Disaccharides are broken down into their monosaccharide components during digestion; only glucose, fructose, and galactose are absorbed into the bloodstream.

High-fructose corn syrup—a liquid sweetener used in some soft drinks, fruit drinks, and other products—and alcohol sugars are also considered simple sugars.[4] Most of the simple sugars have a distinctively sweet taste.

The only simple sugar the body can use to form energy is glucose. That's not a problem because both fructose and galactose are readily converted by the body to glucose. When the body has more glucose than it needs for energy formation, it converts the excess to fat or ***glycogen,*** the body's storage form of glucose. Glycogen is a type of complex carbohydrate. It consists of chains of glucose units linked together in long strands. Glycogen is produced only by animals and is stored in the liver (Illustration 12.2) and muscles. When the body needs additional glucose, glycogen is broken down, making glucose available for energy formation. Glucose can also be derived from certain amino acids and the glycerol component of fats. Illustration 12.3 shows the various ways glucose becomes available to the body. A constant supply of glucose is needed because the brain, red blood cells, white blood cells, and specific cells in the kidneys require glucose as an energy source.[5]

Where's the Sugar?] Sugars enter our diets primarily from foods that have added sugar.[6] Although used primarily because of their sweet taste, simple sugars improve

simple sugars
Carbohydrates that consist of a glucose, fructose, or galactose molecule; or a combination of glucose and either fructose or galactose. High-fructose corn syrup and alcohol sugars are also considered simple sugars. Simple sugars are often referred to as "sugars."

monosaccharides
(*mono* = one, *saccharide* = sugar): Simple sugars consisting of one sugar molecule. Glucose, fructose, and galactose are monosaccharides.

disaccharides
(*di* = two, *saccharide* = sugar): Simple sugars consisting of two molecules of monosaccharides linked together. Sucrose, maltose, and lactose are disaccharides.

glycogen
The body's storage form of glucose. Glycogen is stored in the liver and muscles.

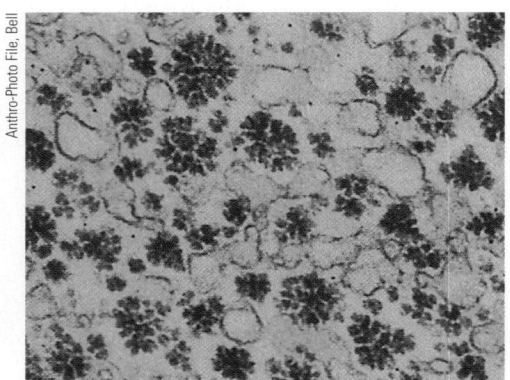

Illustration 12.2
Glycogen in a liver cell. *The black "rosettes" are aggregates of glycogen molecules. This cell was photographed under an electron microscope at a magnification of 65,000X.*

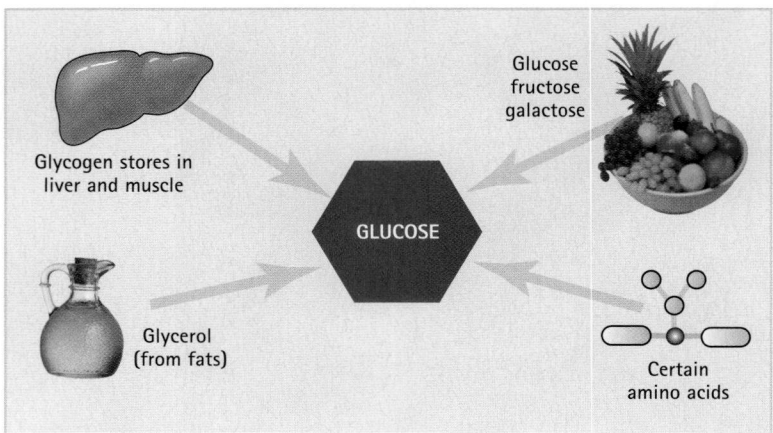

Illustration 12.3 **The body's sources of glucose.**

the appearance, consistency, and cooking properties of many food products. Consequently, they are the most commonly used food additive. On average, Americans consume 84 grams of added sugar per day, and that equals about 17% of the average caloric intake.[7] Sugar consumption has increased 23% in the past 30 years. The biggest source of sugar in many diets is soft drinks,[8] but sucrose is added to a lot of other foods, too. Sugar is a very common ingredient of breakfast cereals, for example (Table 12.1). Simple sugars are also present in fruits, and small amounts occur in some vegetables (Table 12.2). Unlike many foods with added sugar, both fruits and vegetables provide an array of nutrients. With the exception of milk, which contains the simple sugar lactose, animal products are nearly devoid of simple sugars.

Added Sugars] The major source of simple sugars in most diets is sugar added to food during food processing or preparation (Table 12.3). Nutrition labels must list the total amount of sugar per serving under the heading "sugars" (Illustration 12.4). In addition, in the ingredient list, all simple sugars contained in the product must be lumped together under "sweeteners" followed by a parenthetical list itemizing each type in order of weight.[9] Labels contain information on total sugars per serving and do not distinguish between sugars naturally present in foods and added sugars.

What's So Bad about Sugar?] Foods to which simple sugars have been added are often not among the top sources of nutrients. Simple sugars are among the few foods that provide only calories. Many foods high in simple sugars, such as cake, sweet rolls, cookies, pie, candy bars, and ice cream, are also high in fat. The like-

TABLE 12.1

THE SIMPLE SUGAR CONTENT OF A 1-OUNCE SERVING OF SOME BREAKFAST CEREALS.

CEREAL	SIMPLE SUGARS (grams)	% TOTAL CALORIES FROM SIMPLE SUGARS
Apple Jacks	13	52%
Corn Pops	14	47
Frosted Cheerios	13	43
Raisin Bran	19	40
Frosted Flakes	11	40
All Bran	5	29
Frosted MiniWheats	6	24
Bran Flakes	5	22
Life	6	20
Rice Krispies	3	13
Grape-Nuts	3	11
Product 19	3	11
Special K	3	11
Wheaties	3	11
Wheat Chex	2	8
Cornflakes	2	7
Cheerios	1	4
Shredded Wheat	0	0

TABLE 12.2

THE SIMPLE SUGAR CONTENT OF SOME COMMON FOODS.

	AMOUNT	SIMPLE SUGARS (grams)*	% TOTAL CALORIES FROM SIMPLE SUGARS
Sweeteners			
Corn syrup	1 tsp	5	100%
Honey	1 tsp	6	100
Maple syrup	1 tsp	4	100
Table sugar	1 tsp	4	100
Fruits:			
Apple	1 medium	16	91
Peach	1 medium	8	91
Watermelon	1 wedge (4" × 8")	25	87
Orange	1 medium	14	86
Banana	1 medium	21	85
Vegetables:			
Broccoli	½ cup	2	40
Corn	½ cup	3	30
Potato	1 medium	1	4
Beverages:			
Fruit drinks	1 cup	29	100
Soft drinks	12 oz	38	100
Skim milk	1 cup	12	53
Whole milk	1 cup	11	28
Candy:			
Gumdrops	1 oz	25	100
Hard candy	1 oz	28	100
Caramels	1 oz	21	73
Fudge	1 oz	21	73
Milk chocolate	1 oz	16	44

*4 grams sucrose = 1 teaspoon.

TABLE 12.3

ADDED SUGARS.

Brown sugar

Confectioner's sugar

Corn syrup

Corn syrup solids

Crystal corn syrup

Dextrose

Fructose

High-fructose corn syrup

Honey

Liquid fructose

Malt syrup

Maple syrup

Molasses

Pancake syrup

Powdered sugar

Raw sugar

White sugar

lihood that diets will provide insufficient amounts of vitamins and minerals increases along with sugar intake. Furthermore, it is perfectly clear that the frequent consumption of sticky sweets causes tooth decay.[11]

Simple sugars have been blamed for far more problems than tooth decay. Sugar has been accused of depleting the body of nutrients (it doesn't), causing violent and

Illustration 12.4
Labeling the sugar content of breakfast cereals: two examples.

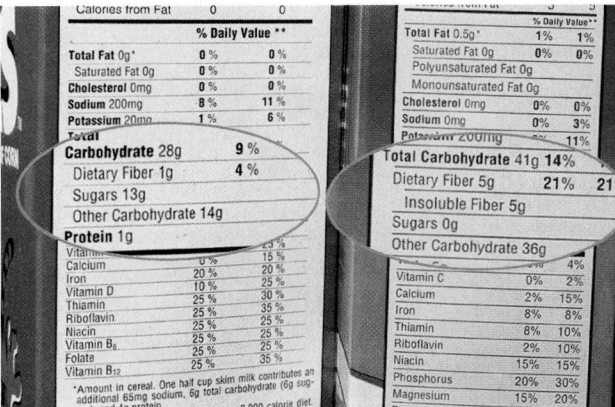

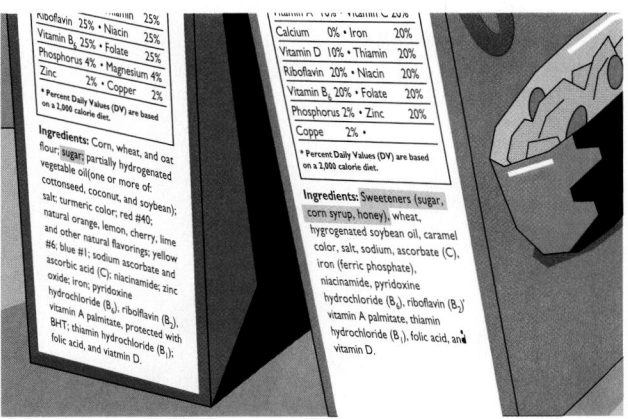

criminal behavior (this has been disproved by a number of studies), and producing diabetes (also untrue).[12] Nor is there any evidence that sugar intake causes hyperactivity in children.[13] Actually, high sugar intakes may not only decrease children's activity level but also tend to make adults sleepy.[14] Sugary foods have a dampening effect on appetite in the short term.[15] In other words, your parents were right when they told you not to eat that cookie before dinner.

Advice on Sugar Intake] What's the bottom line on eating sugary foods? Enjoy them in limited amounts. The Dietary Reference Intakes call for limiting intake of added sugars to less than 25% of total caloric intake. Other health authorities, including the World Health Organization and those of 26 countries, recommend limiting added sugars to 10% of calories or less.[16] Brush your teeth after eating sticky sweets, and think about replacing regular soft drinks with water. And don't let sweet, high-fat foods like candy bars, pastries, and ice cream replace nutrient-dense foods such as vegetables, fruits, and whole grain products in your diet!

The Alcohol Sugars—What Are They?

alcohol sugars
Simple sugars containing an alcohol group in their molecular structure. The most common are xylitol, mannitol, and sorbitol.

Nonalcoholic in the beverage sense, the ***alcohol sugars*** are like simple sugars except that they include a chemical component of alcohol. Like simple sugars, the alcohol sugars have a sweet taste. Xylitol is by far the sweetest alcohol sugar—it's much sweeter than the other two common alcohol sugars, mannitol and sorbitol.

Alcohol sugars are found naturally in very small amounts in some fruits. Their presence in foods is most often due to their use as sweetening agents in gums and candy (Illustration 12.5). Unlike the simple sugars, xylitol, mannitol, and sorbitol do not promote tooth decay. The extensive use of alcohol sugars in food is limited by their tendency to cause diarrhea if consumed in large amounts.[17]

Artificial Sweetener Facts

Unwanted calories in simple sugars, the connection of sucrose with tooth decay, the need for a sugar substitute for people with diabetes, and sugar shortages such as occurred during the two world wars have all provided incentives for developing sugar substitutes. Four artificial sweeteners are currently on the market in the United States, and a fifth has been approved by the FDA. Many more artificial sweeteners are being developed.

Illustration 12.5
Examples of products sweetened with xylitol, mannitol, and sorbitol.
Alcohol sugars provide 4 calories per gram.

Richard Anderson

Illustration 12.6
Some of the thousands of foods that contain artificial sweeteners.

Artificial sweeteners, also known as "intense sweeteners," are contained in a wide assortment of prepared foods (Illustration 12.6). None of these sweeteners are significant sources of calories, and they do not promote tooth decay or raise blood glucose levels. But, none developed so far provide the optimal taste qualities of sucrose.[18] Whether the use of artificial sweeteners leads to weight loss or the prevention of weight gain is much debated. Research results can be cited that support or refute a role of artificial sugars in weight loss.[19]

Saccharin] Saccharin was the first artificial sweetener developed. Did you know that it was discovered in a laboratory in the late 1800s? That's right—saccharin is over 100 years old.[20] The availability of this artificial sweetener, which is 300 times as sweet as sucrose, helped relieve the sugar shortages that occurred during World Wars I and II.

In 1977 saccharin was taken off the market after very high doses were found to cause cancer in laboratory animals. At that time, however, saccharin was the only no-calorie, artificial sweetener available, and its removal sparked a public outcry. After many people complained to Congress, saccharin was returned to the market by congressional mandate. Saccharin was deemed safe in 2000 after scientists concluded there was no clear evidence that it causes cancer in humans.[22]

Aspartame] Early in the 1980s, the artificial sweetener aspartame was approved for use in the United States and more than 90 other countries. Known as Nutrasweet, this artificial sweetener is about 200 times sweeter than sucrose. Aspartame is used in more than 4000 products worldwide, including soft drinks, whipped toppings, jellies, cereals, puddings, and some medicines.[23] Products containing aspartame must carry a label warning people with **phenylketonuria** (an inherited disease) and others with certain liver conditions about the presence of phenylalanine. People with these disorders are unable to utilize the amino acid phenylalanine, causing it to build up in the blood. Because high temperatures tend to break down aspartame, it is not used in baked or heated products.

Aspartame Was Developed by Chemists, Not Nature} Aspartame is made from two amino acids. Both are found in nature, but it took chemists to arrange their chemical partnership. Because aspartame is made from amino acids (the building blocks

ON THE SIDE

A few drops of sucrose solution on an infant's tongue has a pain-relieving effect that persists for several minutes. Sugar solution is being increasingly used in hospitals to control pain in infants undergoing heel pricks for blood collection and other minor procedures.[10]

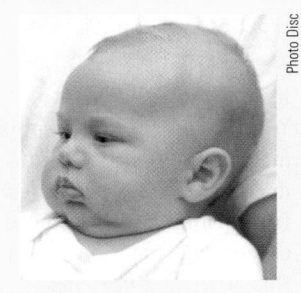

phenylketonuria (feen-ol-key-tone-u-re-ah), PKU
A rare genetic disorder related to the lack of the enzyme phenylalanine hydroxylase. Lack of this enzyme causes the essential amino acid phenylalanine to build up in blood.

| NEOTAME |
| SUCRALOSE |
| SACCHARIN |
| ACESULFAME K |
| ASPARTAME |
| FRUCTOSE |
| SUCROSE |
| XYLITOL |
| GLUCOSE |
| SORBITOL |
| MANNITOL |
| GALACTOSE |
| MALTOSE |
| LACTOSE |

Illustration 12.7
A ranking of various types of artificial sweeteners and naturally occurring sugars in order of sweetness.

Illustration 12.8
You would have to consume more than 20 cans of soft drinks sweetened with aspartame (Nutrasweet) a day to exceed the safe limit set for this artificial sweetener.

of protein), it supplies 4 calories per gram. Aspartame is so sweet, however, that very little is needed to sweeten products. Illustration 12.7 shows the relative sweetening power of various artificial sweeteners and naturally occurring sugars.

Is Aspartame Safe?} A safe level of aspartame intake is defined as 50 milligrams per kilogram of body weight per day in the United States and as 40 milligrams per kilogram of body weight in Canada.[24] In food terms, the limit in the United States is equivalent to approximately 20 aspartame-sweetened soft drinks or 55 desserts per day (Illustration 12.8). The average intake of aspartame in the United States, Canada, Germany, and Finland, for example, ranges from 2 to 10 milligrams per day, well below the level of intake considered safe.[25] A small proportion of individuals, however, report that they are sensitive to aspartame and develop headaches, dizziness, or anxiety when they consume small amounts. Studies have failed to confirm these effects.[26]

Sucralose] This noncaloric, intense sweetener is made from sucrose, is safe, is very sweet (600 times sweeter than sucrose), and does not leave a bitter aftertaste. Known as "Splenda" on product labels, it is used in both hot and cold food products, including soft drinks, baked goods, frosting, pudding, and chewing gum.

Acesulfame Potassium] The fourth artificial sweetener currently available in the United States is acesulfame potassium, also known as acesulfame K, "Sunette" and "Sweet One." Approved by the FDA in 1988, acesulfame K is added to at least 4000 foods and is used in food production in about 90 countries. It is 200 times as sweet as sucrose, provides zero calories, and does not break down when heated. The FDA has approved "neotame," an artificial sweetener deemed safe. It delivers over 7000 times the sweetness of sucrose, so very small amounts are needed in foods.[27]

Cyclamate: Still banned in the USA] Cyclamate was widely used in food products in the United States from the 1940s through most of the 1960s. It was banned in 1969 because large doses of a cyclamate-saccharin mixture caused cancer in laboratory mice. Cyclamates are available in over 50 countries, some of which (including Canada) have banned the use of saccharin! Several national groups of experts have recently concluded that cyclamates are safe.[28]

Richard Anderson

REALITY CHECK

Carbohydrate craving

Have you ever pined or whined for something sweet? Is a meal not complete unless there's a "starch"? Do you think you can get addicted to carbohydrates?

Who gets thumbs up

?

Tyrell:
Maybe you could really love to eat them, but addicted? I don't think so.

Wolfgang:
The more I don't eat candy, the more I want to eat some. I swear, I'm addicted.

Photo Disc

Answers on next page

Complex Carbohydrate Facts

Starches, glycogen, and dietary fiber constitute the **complex carbohydrates** known as **polysaccharides.** Only plant foods such as grains, potatoes, dried beans, and corn that contain starch and dietary fiber are considered dietary sources of complex carbohydrates (Table 12.4). Very little glycogen is available from animal products.

Which Foods Have Carbohydrates?] Food sources of complex carbohydrates include whole grain breads, cereals, pastas, and crackers, as well as these same foods produced from refined grains. Whole grain products provide more fiber and beneficial substances naturally present in grains than do refined grain products. Whole grain foods reduce the risk of heart disease and some types of cancer.[29]

complex carbohydrates
The form of carbohydrate found in starchy vegetables, grains, and dried beans and in many types of dietary fiber. The most common form of starch is made of long chains of interconnected glucose units.

polysaccharides
(*poly* = many, *saccharide* = sugar): Carbohydrates containing many molecules of sugar linked together. Starch, glycogen, and dietary fiber are the three major types of polysaccharides.

TABLE 12.4

THE COMPLEX CARBOHYDRATE CONTENT OF SOME COMMON FOODS.

	AMOUNT	COMPLEX CARBOHYDRATE (grams)	% TOTAL CALORIES FROM COMPLEX CARBOHYDRATES
Grain and grain products:			
Rice (white), cooked	½ cup	21	83%
Pasta, cooked	½ cup	15	81
Cornflakes	1 cup	11	76
Oatmeal, cooked	½ cup	12	74
Cheerios	1 cup	11	68
Whole wheat bread	1 slice	7	60
Dried beans (cooked):			
Lima beans	½ cup	11	64
White beans	½ cup	13	63
Kidney beans	½ cup	12	59
Vegetables:			
Potato	1 medium	30	85
Corn	½ cup	10	67
Broccoli	½ cup	2	40

Photo Disc

Tyrell

Carbohydrate craving

The popular theory about carbohydrate craving and addiction goes like this: Carbohydrates increase blood levels of glucose, then insulin level surges to drop blood glucose levels. Resulting low blood glucose levels send a message to your brain telling you that carbohydrates are needed to bring your blood glucose levels back up. The truth is there is no voice inside us, or a genetic predisposition, that leads us to consume candy, pie, rice, or bread when insulin increases and blood glucose levels are lowered. Any source of calories will relieve hunger.[21]

Wolfgang

Photo Disc

ON THE SIDE

Have you ever had buckwheat pancakes (3 grams of fiber per serving)? Did you know that buckwheat is a member of the rhubarb family—not a grain?

Photo Disc

functional fiber
Specific types of nondigestible carbohydrates and connective tissues that have beneficial effects on health. Two examples of functional fibers are psyllium and pectin.

dietary fiber
Naturally occurring, intact forms of nondigestible carbohydrates in plants and "woody" plant cell walls. Oat and wheat bran, and raffinose in dried beans, are examples of this type of fiber.

total fiber
The sum of functional and dietary fiber.

Are They Fattening?] Starchy foods are caloric bargains (Illustration 12.9, next page). A medium baked potato weighs in at only 122 calories, a half-cup of corn at 85 calories, and a slice of bread at 70 calories. You can expand the caloric value of complex carbohydrates quite easily by adding fat, sauces, and cheese. One cup of macaroni (about 200 calories) gains around 180 calories when it comes as macaroni and cheese. Adding a quarter-cup of gravy to potatoes elevates calories by 150.

Dietary Fiber Facts

What has no calories; prevents constipation; may lower the risk of heart disease, obesity, and diabetes; and is generally underconsumed by people selecting a Western-type diet? The answer is dietary fiber.[30]

Total fiber intake by U.S. children and adults is well below the amount recommended (Table 12.5).[31] People who consume the recommended amount of fiber tend to select whole grain breads, high-fiber cereal, and dried beans most days and eat at least five servings of vegetables and fruits daily.[33] Food sources of dietary fiber are listed in Table 12.6. It doesn't matter whether the fiber foods are mashed, chopped, cooked, or raw. They retain their fiber value through it all. Note, though, that fast foods tend to be poor sources of fiber.

Types of Fiber] A new classification system for defining edible fibers based on source of fiber and effects on body processes has been developed.[34] Fibers are classified as *functional fiber, dietary fiber,* and *total fiber*. All fiber shares the property of not being digested by human digestive enzymes.

Functional fibers perform specific, beneficial functions in the body, including decreasing food intake by providing a feeling of fullness, reducing post-meal rises in blood glucose levels, preventing constipation, and decreasing fat and cholesterol absorption.[35] This type of fiber can be extracted from foods, or produced commercially for use in fortifying foods with fiber, and in fiber supplements. Connective tissues from animals included in some meat products, psyllium, pectin, gels, and seed and plant gums are classified as functional fibers.

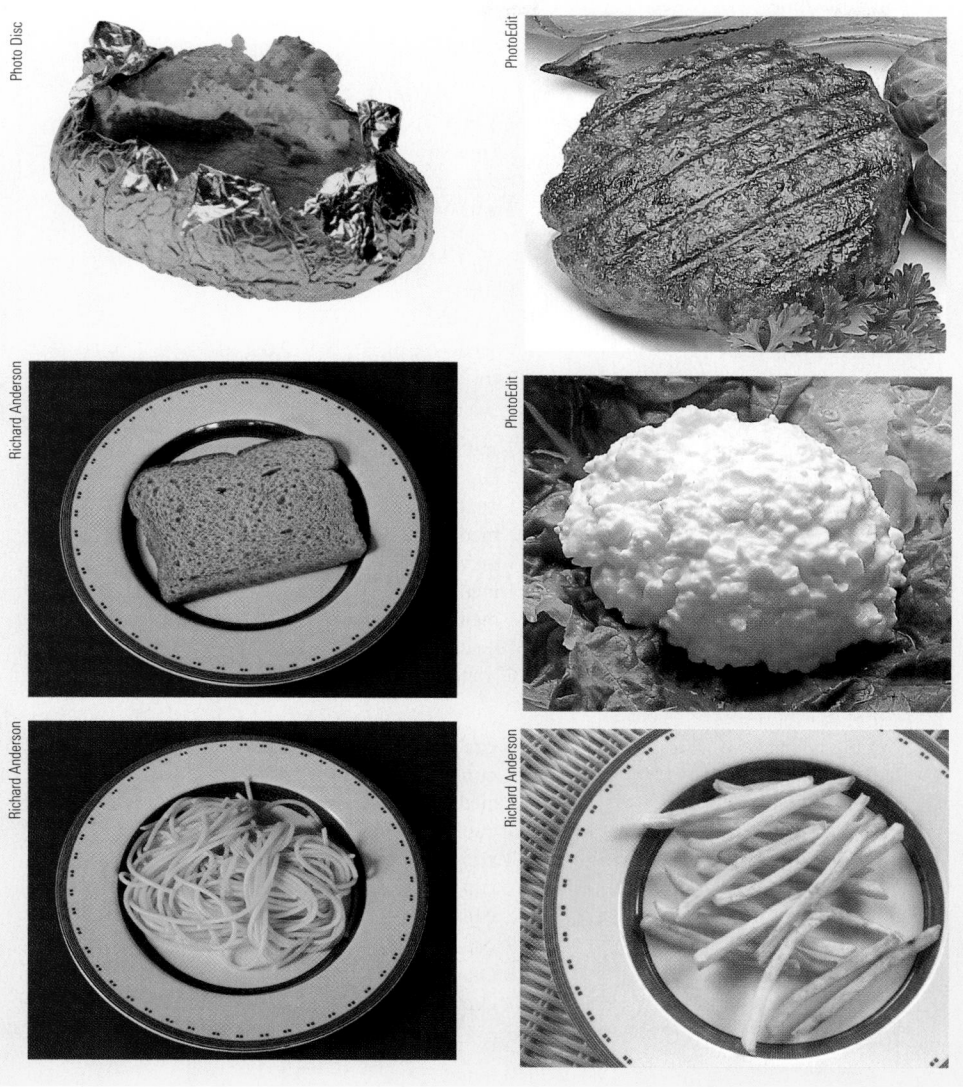

Illustration 12.9

Which has more calories?
Check your answers below.*

One medium baked potato (4 ounces) OR 3 ounces of lean hamburger?

One slice of bread OR a half-cup of low-fat cottage cheese?

One cup of spaghetti noodles OR 17 french fries (3 ounces)?

*Answers:
Potato = 122 calories;
Lean hamburger = 239 calories.
Bread = 70 calories;
Cottage cheese = 102 calories.
Spaghetti = 197 calories;
French fries = 265 calories.

Functional fibers can be found in foods, but when they are consumed in foods they are considered dietary fiber. Dietary fiber consists of nondigestible carbohydrates found in plant foods. Because functional fibers are components of plant foods, this type of fiber comes with all of the other beneficial nutrients and other substances found in plants. Dietary fibers are found in the bran component of oats and wheat

TABLE 12.5

AVERAGE AND RECOMMENDED DAILY INTAKES OF TOTAL FIBER.[32]

AGE-GROUP	AVERAGE TOTAL FIBER INTAKE (GM)	RECOMMENDED INTAKE OF TOTAL FIBER (gm)
4–8 years	12.0	19
men, 19–50 years	13.7	38
women, 19–50 years	13.2	25

TABLE 12.6

EXAMPLES OF GOOD SOURCES OF DIETARY FIBER.
The recommended intake of total fiber for men is 38 grams per day and 25 grams per day for women.

	AMOUNT	DIETARY FIBER (grams)
Grain and grain products:		
Bran Buds	½ cup	12.0
All Bran	⅓ cup	10.0
Raisin Bran	1 cup	7.0
Granola (homemade)	½ cup	6.0
Bran Flakes	¾ cup	5.0
Oatmeal	1 cup	4.0
Spaghetti noodles	1 cup	4.0
Shredded Wheat	1 biscuit	2.7
Whole wheat bread	1 slice	2.0
Bran (dry; wheat, oat)	2 tbs	2.0
Fruits:		
Avocado	½ medium	7.0
Raspberries	1 cup	5.0
Mango	1 medium	4.0
Pear (with skin)	1 medium	4.0
Orange (no peel)	1 medium	3.0
Apple (with skin)	1 medium	2.4
Peach (with skin)	1 medium	2.3
Strawberries	10 medium	2.1
Banana	6" long	1.8
Vegetables:		
Lima beans	½ cup	5.0
Green peas	½ cup	4.0
Potato (with skin)	1 medium	3.5
Broccoli	½ cup	3.4
Brussels sprouts	½ cup	3.0
Collard greens	½ cup	3.0
Corn	½ cup	3.0
Carrots	½ cup	2.8
Green beans	½ cup	2.7
Cauliflower	½ cup	2.0
Nuts:		
Almonds	¼ cup	4.5
Peanuts	¼ cup	3.3
Peanut butter	2 tbs	2.3
Dried beans (cooked):		
Pinto beans	½ cup	10.0
Black beans (turtle beans)	½ cup	8.0
Black-eyed peas	½ cup	8.0
Kidney or navy beans	½ cup	6.2
Peas	½ cup	5.7
Lentils	½ cup	5.0
Fast foods:		
Big Mac	1	3
French fries	1 regular serving	3
Whopper	1	3
Cheeseburger	1	2
Taco	1	2
Chicken sandwich	2	1
Egg McMuffin	1	1
Fried chicken, drumstick	1	1

(Illustration 12.10), in cellulose (a rigid component of plant cell walls), in vegetables and fruits, and in the nondigestible starch components of dried beans. The recommended daily intake of fiber is based on total fiber, which is the sum of functional plus dietary fiber intake.[36]

Be Cautious When Adding More Fiber to Your Diet] Newcomers to adequate fiber diets often experience diarrhea, bloating, and gas for the first week or so of increased fiber intake. These side effects can be avoided. They occur when too much fiber is added to the diet too quickly. Although humans do not produce enzymes that can break down fiber, bacteria that live in the lower intestines do. The bacteria consume dietary fiber as food and excrete gas and other products as wastes. The gas produced by the bacteria causes the bad feelings that can accompany unusually high fiber intakes.[37] Adding sources of dietary fiber to the diet gradually can prevent these side effects. In addition, dietary fiber can be constipating if consumed with too little fluid. Your fluid intake should increase along with your intake of dietary fiber. You know you've got the right amount of fiber in your diet when stools float and are soft and well formed.

The "Health Action" on the next page suggests some food choices that will put fiber into your meals and snacks. Fiber should be added carefully to children's diets, because the large volume of some high-fiber diets can fill the children up before they consume enough calories.[38]

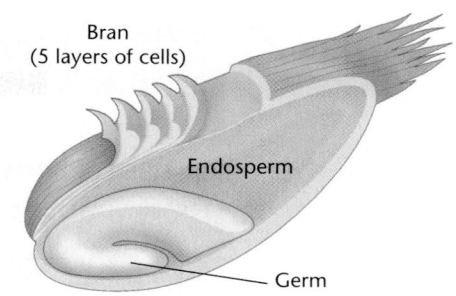

Bran
(5 layers of cells)

Endosperm

Germ

Illustration 12.10
Diagram of a grain of wheat showing the bran that is a rich source of dietary fiber. *The germ contains protein, unsaturated fats, thiamin, niacin, riboflavin, iron, and other nutrients. (The bran and germ are removed in the refining process.) The endosperm primarily contains starch, the storage form of glucose in plants.*

tooth decay
The disintegration of teeth due to acids produced by bacteria in the mouth that feed on sugar. Also called *dental caries* or *cavities*.

Carbohydrate-Related Disorders

Carbohydrates are related to a number of health problems. Carbohydrate-related disorders include diabetes and hypoglycemia (Unit 13), lactose intolerance (Unit 7), and tooth decay and other oral health problems.

Carbohydrates and Your Teeth

We tried the white man's food, and we liked it.
Now we have tooth aches.
—Chief Senebura, Xavante people in Pimentel Barbosa, Brazil[39]

The relationship between sugar and ***tooth decay*** is very close, and the history of tooth decay closely parallels the availability of sugar. The incidence of tooth decay is estimated to have been very low (less than 5%) among hunter-gatherers who had minimal access to sugars.[40] Tooth decay did not become a widespread problem until the late seventeenth century, when great quantities of sucrose were exported from the New World to Europe and other parts of the world. When sugar shortages occurred in the United States and Europe during World War I and World War II, rates of tooth decay declined; they rebounded when sugar became available again.[41] Rates of tooth decay in children vary substantially among countries, but the highest rates are in countries where sugar is widely available in processed foods and beverages (Illustration 12.11). Tooth decay is spreading rapidly in developing countries where sugar, candy, soft drinks, and fruit drinks are becoming widely available.[42]

Sweets are not the only culprit. Simple sugars that promote tooth decay can also come from starchy foods, especially pretzels, crackers, and breads that stick to your gums and teeth. Some of the starch is broken down to simple sugars by enzymes in the mouth.

Bacteria on the tongue have recently been identified as the primary cause of bad breath. Mouthwash gets rid of bad breath for about an hour; brushing teeth with toothpaste takes care of it in 25% of people, but brushing the tongue eliminates it in 70 to 80%.[46]

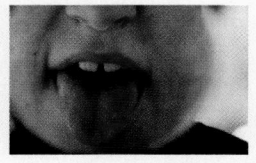

HEALTH ACTION | *Putting the Fiber into Meals and Snacks*

HIGH–FIBER OPTIONS FOR BREAKFAST

Whole grain toast		2 g per slice
Bran cereal:		
Bran flakes	1 cup	7 g
All Bran	$^{1}/_{3}$ cup	10 g
Raisin bran	$^{3}/_{4}$ cup	5 g
Oat bran	$^{1}/_{3}$ cup	5 g
Bran muffin, with fruit:	1 small	3 g
Strawberries	10	2 g
Raspberries	$^{1}/_{2}$ cup	3 g
Bananas	1 medium	2 g

LUNCHES THAT INCLUDE FIBER

Whole grain bread		2 g per slice
Baked beans	$^{1}/_{2}$ cup	10 g
Carrot	1 medium	2 g
Raisins	$^{1}/_{4}$ cup	2 g
Peas	$^{1}/_{2}$ cup	4 g
Peanut butter	2 tablespoons	2 g

FIBER ON THE MENU FOR SUPPER

Brown rice	$^{1}/_{2}$ cup	2 g
Potato	1 medium	3 g
Dried cooked beans	$^{1}/_{2}$ cup	8 g
Broccoli	$^{1}/_{2}$ cup	3 g
Corn	$^{1}/_{2}$ cup	3 g
Tomato	1 medium	2 g
Green beans	$^{1}/_{2}$ cup	3 g

FIBER–FILLED SNACKS

Peanuts	$^{1}/_{4}$ cup	3 g
Apple	1 medium	2 g
Pear	1 medium	4 g
Orange	1 medium	3 g
Prunes	3	2 g
Sunflower seeds	$^{1}/_{4}$ cup	2 g
Popcorn	2 cups	2 g

*Prunes contain fiber, but their laxative effect is primarily due to a naturally occurring chemical substance that causes an uptake of fluid into the intestines and the contraction of muscles that line the intestines.

To reduce the incidence of tooth decay, a number of countries have developed campaigns to help inform consumers about cavity-promoting foods. Switzerland, for example, labels foods safe for the teeth with a "happy tooth" symbol (Illustration 12.12) and encourages the use of alcohol sugars (which don't promote tooth decay) in sweets. Other countries recommend that sweets be consumed with meals or that teeth be brushed after sweets are eaten.

There's More to Tooth Decay Than Sugar Per Se] How frequently sugary and starchy foods are consumed and how long they stick to gums and teeth make a difference in their tooth-decay-promoting effects. Marshmallows, caramels, and taffy, for example, are much more likely to promote tooth decay than are apples and milk chocolate. Nevertheless, all of the foods listed in Table 12.7 can promote tooth decay if allowed to remain in contact with the gums and teeth. Drinking coffee or tea with sugar throughout the day or consuming three or more regular soft drinks between meals hastens tooth decay (more than if these beverages are consumed with meals) Candy, cookies, and crackers eaten between meals are much more likely to promote tooth decay than are the same foods consumed as part of a meal. Chewing as few as two sticks of sugar-containing gum a day also significantly increases tooth decay.[43]

Why Does Sugar Promote Tooth Decay?] Sugar promotes tooth decay because it is the sole food for certain bacteria that live in the mouth and excrete acid that dissolves teeth. In the presence of sugar, bacteria in the mouth multiply rapidly and form a sticky, white material called *plaque*. Tooth areas covered by plaque are prime locations for tooth decay because they are dense in acid-producing bacteria. Acid production by bacteria increases within 5 minutes of exposure to sugar. It continues for 20 to 30 minutes after the bacteria ingest the sugar.[44] If teeth are frequently exposed to sugar, the acid produced by bacteria may erode the enamel, producing a cavity. If the erosion continues, the cavity can extend into the tooth and allow bacteria to enter the inside of the tooth. That can cause an infection and the loss of the tooth. It can be prevented if the plaque is removed before the acid erodes much of the enamel. Teeth are capable of replacing small amounts of minerals lost from enamel.[45]

The Remarkable Success of Water Fluoridation] In the early 1930s, lower rates of tooth decay were observed among children living in areas where water naturally contained fluoride. This provided the initial evidence that led to the fluoridation of many community water supplies. Fluoridated water reduces the incidences

"Just pull my sweet tooth."

Illustration 12.11
Source: © 2001 H.L. Schwadron.
Reprinted with permission.

plaque
A soft, sticky, white material
on teeth; formed by bacteria.

TABLE 12.7

THE "STICKINESS" VALUE OF SOME FOODS.
The stickier the food, the worse it is for your teeth.

VERY STICKY	STICKY	SOMEWHAT STICKY	BARELY STICKY
Caramels	Doughnuts	Bagels	Apples
Chewy cookies	Figs	Cake	Bananas
Crackers	Frosting	Cereal	Fruit drinks
Cream-filled cookies	Fudge	Dry cookies	Fruit juices
Granola bars	Hard candy	Milk chocolate	Ice cream
Marshmallows	Honey	Rolls	Oranges
Pretzels	Jelly beans	White bread	Peaches
Taffy	Pastries		Pears
	Raisins		
	Syrup		

Illustration 12.12
Switzerland's "happy tooth"
symbol.

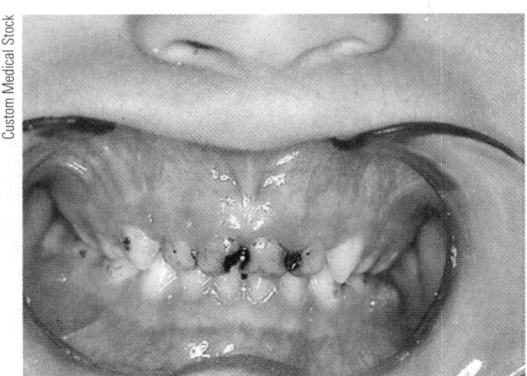

Custom Medical Stock

Illustration 12.13
"Baby bottle caries" (also called "nursing bottle syndrome") occurs in infants who habitually receive sweet fluids or milk in bottles when they go to sleep.
Cavities occur first in the upper front teeth because that's where fluid pools when babies sleep.

of tooth decay by 50% or more and is primarily responsible for declining rates of tooth decay and loss.[47]

Credit for declines in tooth decay and tooth loss in the United States is also shared by fluoride supplements, toothpastes, rinses and gels, protective sealants, and improved dental hygiene and care.[48] Further improvements in rates of dental caries will occur with reduced intake of sugars and sticky carbohydrates. Fluoridation is a safe, effective, and cheap method of controlling dental disease; providing fluoridated community water supplies costs about 50 cents per person per year.[49] Despite the advantages of fluoride, more than a million Americans consume water from a less than optimally fluoridated water supply.[50]

Baby Bottle Caries] A startling example of the effect that frequent and prolonged exposure to sugary foods can have is "baby bottle caries" (Illustration 12.13). Infants and young children who routinely fall asleep while sucking a bottle of sugar water, fruit drink, milk, or formula—or while breastfeeding—may develop severe decay. After the child falls asleep, the fluid may continue to drip into the mouth. A pool of the fluid collects between the tongue and the front teeth, bathing the teeth in the sweet fluid for as long as the child sleeps. The upper front teeth become decayed first because the tongue protects the lower teeth. Baby bottle caries occur in 5 to 10% of infants and young children and can lead to the destruction of all baby teeth.[51]

Foods with little carbohydrate such as cheese and peanut butter, protein (eggs, meats, yogurt), and dietary fiber (vegetables and whole grain products, for example) inhibit tooth decay (Table 12.8).[52]

TABLE 12.8

FOODS THAT DON'T PROMOTE TOOTH DECAY.[53]

Artificial sweeteners	Gum and candy sweetened with alcohol sugars	Peanut butter
Cheese		Tea
Coffee (no sugar)	Meats	Water
Eggs	Milk	Yogurt (plain)
Fats and oils	Nuts	

Nutrition UP CLOSE

Does Your Fiber Intake Measure Up?

FOCAL POINT: Approximate the amount of fiber your diet contains.

Are you meeting your fiber quota, or do you consume the typical low-fiber American diet? To determine if your fiber intake is adequate, award yourself the allotted number of points for **each serving of the following foods that you eat in a typical day.** For example, if you normally eat 1 slice of whole grain bread each day, give yourself 2 points. If you eat 2 slices daily, give yourself 4 points. After tallying your score, refer to the Feedback section for the results.

High-fiber food choices

Fruits **2 points for each serving**

   1 whole fruit (e.g., apple, banana) _____

   ½ cup cooked fruit _____

   ¼ cup dried fruit _____

Grains:

   ½ cup cooked brown rice _____

   1 whole grain slice of bread, roll, muffin, or tortilla _____

   ½ cup hot whole grain cereal (e.g., oatmeal) _____

   ¾ cup cold whole grain cereal (e.g., Cheerios) _____

   2 cups popcorn _____

Nuts and seeds:

   ¼ cup seeds (e.g., sunflower) _____

   2 tablespoons peanut butter _____

Vegetables:**3 points for each serving**

   1 whole vegetable (e.g., potato) _____

   ½ cup cooked vegetable (e.g., green beans) _____

Bran cereals:**7 points for each serving**

   ½ cup cooked oat bran cereal _____

   1 cup cold bran cereal _____

Legumes:**8 points for each serving**

   ½ cup cooked beans (e.g., baked beans, pinto beans) _____

                   Total score _____

Special note: You can also calculate your fiber intake using the Diet Analysis Plus software. Input your food intake for one day. Then go to the Analyses/Reports section to view the total number of grams of fiber in your diet on that day.

FEEDBACK (including scoring) can be found at the end of Unit 12.

Key Terms

alcohol sugars, page 12-6

carbohydrates, page 12-2

complex carbohydrates, page 12-9

dietary fiber, page 12-10

disaccharides, page 12-3

functional fiber, page 12-10

glycogen, page 12-3

monosaccharides, page 12-3

phenylketonuria (PKU), page 12-7

plaque, page 12-15

polysaccharides, page 12-9

simple sugars, page 12-3

tooth decay, page 12-13

total fiber, page 12-10

www links

http://cis.nci.nih.gov/fact/3_19.htm
The National Cancer Institute provides information on artificial sweeteners and cancer from this site.

www.nlm.nih.gov/medlineplus/dietaryfiber.html
This site offers specific suggestions on how to fit more fiber into your diet, fiber and health information, fiber content of foods, and the latest research on fiber.

http://search.nih.gov
This short address leads you to a universe of reliable information on an unimaginable variety of health topics, including those covered in this unit.

www.ada.gov
It's not the American Dietetic Association, it's the American Dental Association, they have a site called "The Public" that provides information on oral health, finding a dentist, and tips for teachers.

www.nlm.nih.gov/medlineplus/dentalhealth.html
Long menu of topics includes answers to FAQs about caring for your teeth and gums; information on diet and dental health, gum chewing and caries prevention, causes of periodontal disease, facts about fluoride, and an atlas of teeth.

www.adha.org
The American Dental Hygienists Association Web site. Get dental health education tools and information on dental health and care here.

www.ific.org
An industry-sponsored site offering information on artificial sweeteners, glycemic index, and oral health.

www.mchoralhealth.org
The Bureau of Maternal and Child Health offers information on practices and programs aimed at improving oral health in children.

Notes

1. Shaw JH. Diet and dental health. Am J Clin Nutr 1985;41:1117–31.

2. Shaw, Diet and dental health.

3. Dietary reference intakes: energy, carbohydrate, fiber, fat, fatty acids, cholesterol, protein, and amino acids. Institute of Medicine, National Academies of Sciences. Washington, DC: National Academies Press, chapter 11, 2002.

4. Oberrieder HK, Fryer EB. College students' knowledge and consumption of sorbitol. J Am Diet Assoc 1991;91:715–7.

5. Dietary reference intakes, NAS, chap. 11.

6. Dietary reference intakes, NAS, chap. 11.

7. Committee on Diet and Health (Food and Nutrition Board; National Research Council). Diet and Health. Implications for reducing chronic disease risk. Washington, DC: National Academy Press; 1989; and Dietary reference intakes, NAS, chap. 11.

8. Frazao E, Allshouse J. Strategies for intervention: commentary and debate. J Nutr 2003;133:844S–47S.

9. Oberrieder and Fryer, College students' knowledge; and Food and Drug Administration (Department of Health and Human Services), Nutrition labeling, Federal Register 1991 Nov 27.

10. Young SN. Nutrition 3. The fuzzy boundary between nutrition and psychopharmacology. Can Med Assoc J, Jan. 22, 2002; 1:66.

11. Dietary reference intakes, NAS, chap. 11; and National Institutes of Health Consensus Development Conference Statement: Diagnosis and management of dental caries throughout life. March 28, 2001; 18:1–24.

12. Bachorowski J-A, Newman JP, Nichols SL, Gans DA, Harper AE, Taylor SL, Sucrose and delinquency: behavioral assessment, Pediatrics 1990;86:244–53; and Janket S-J et al., A prospective study of sugar intake and the risk of type 2 diabetes in women, Diab Care 2003;26:1008–15.

13. Wolraich M et al., The effect of sugar on behavior or cognition in children: a meta-analysis, JAMA 1995;274:

1617–21; and Wolraich ML, Lindgren SD, Stumbo PJ et al., Effects of diets high in sucrose or aspartame on the behavior and cognitive performance of children, N Engl J Med 1994;330:301–7; and Kinsbourne M, Sugar and the hyperactive child, N Engl J Med 1994;330:355–6.

14. Wolraich et al., The effect of sugar; and Liberman HR, Sugars and behavior, Clin Nutr 1985;4:195–9; and Pivonka EEA, Grunewald KK, Aspartame- or sugar-sweetened beverages: effects on mood in young women, J Am Diet Assoc 1990;90:250–4.

15. Jones JM, Elam K. Sugars and health: Is there an issue? J Am Diet Assoc 2003; 103:1058–60.

16. Dietary reference intakes, NAS, chap. 11; and Joint WHO/FAO Expert Consultation on Diet, Nutrition, and the Prevention of Chronic Diseases, Geneva, Switzerland; 2002.

17. Oberrieder and Fryer, College students' knowledge; and Badiga MS, Jain NK, Casanova C, Pitchumoni CS, Diarrhea

in diabetics: the role of sorbitol, J Am Coll Nutr 1990;9:578–82.

18. Coulston AM et al. Sugar and sugars: myths and realities. J Am Diet Assoc 2002;102:351–53.

19. Usefulness of artificial sweeteners for body weight control. Nutr Rev 2003;61:219–220.

20. Oberrieder and Fryer, College students' knowledge.

21. Freedman MR et al. Popular diets: a scientific review. Obes Res 2001;9:1S–40S.

22. Saccharin deemed safe. Community Nutrition Institute, 2000; June 2:8.

23. Verni C. Artificial sweeteners: review and update for practitioner. SCAN's Pulse, Summer 1999.

24. Aspartame safe. Nutr Today 2003;38:40.

25. Butchko HH, Kotsonis FN. Acceptable daily intake vs. actual intake: the aspartame example. J Am Coll Nutr 1991;10:258–66.

26. Spiers PA et al. Aspartame: neuropsychologic and neurophysiologic evaluation and chronic effects. Am J Clin Nutr 1998;68:531–7.

27. Neotame: a new sweetener. Nutr Today 2002;37:84.

28. Verni, Artificial sweeteners.

29. Johnson RK, Kennedy E. The 2000 Dietary Guidelines for Americans: what are the changes and why were they made? J Am Diet Assoc 2000;100:769–74.

30. Marlett JA, Slavin JL. Position of the American Dietetic Association: health implications of dietary fiber. J Am Diet Assoc 1997;97:1157–9.

31. Dietary reference intakes, NAS, chap. 11.

32. Dietary reference intakes, NAS, chap. 11.

33. Marlett and Slavin, Health implications of dietary fiber.

34. Dietary reference intakes, NAS, chap. 11.

35. Dietary reference intakes, NAS, chap. 11.

36. Dietary reference intakes, NAS, chap. 11.

37. Topping DL. Soluble fiber polysaccharides: effects on plasma cholesterol and colonic fermentation. Nutr Rev 1991;49:195–206.

38. Williams CL. A summary of conference recommendations on dietary fiber in childhood. Pediatrics 1995;96:1023–8.

39. Jordan M. Colgate brings dental care to Brazilian Indian tribes; ravages of tobacco and rice. Wall Street Journal, 7/23/02, page B1.

40. Cornero S, Puche RC. Diet and nutrition of prehistoric populations at the alluvial banks of the Parana River. Medicina 2000;60:109–14.

41. Shaw, Diet and dental health.

42. Parajas IL. Sugar content of commonly eaten snack foods of school children in relation to their dental health status. J Phillipine Dent Assoc 1999;51:4–21.

43. Ismail A, Burt BA, Eklund SA, The cariogenicity of soft drinks in the United States, J Am Dental Assoc 1984;109:241–5; Ismail A, Food cariogenicity in Americans aged from 9 to 29 years accessed in a national cross-sectional study, 1971–1974, J Dental Res 1986;65:1435–40; and Edgar WM, Sugar substitutes, chewing gum, and dental caries—a review, British Dental J 1998;184:29–32; and Rugg-Gunn AJ, The benefits of using sugar-free chewing gum: a proven anti-caries effect, British Dental J 1998;184:26.

44. Schachtele CF, Harlander SK. Will the diets of the future be less cariogenic? J Canadian Dental Assoc 1984;3:213–9.

45. Palmer CA, Watson LM. Position of the American Dietetic Association: the impact of fluoride on dental health. J Am Diet Assoc 1994;94:1428–31.

46. Bad breath. Nutr Today 2000;35:6.

47. Position of the American Dietetic Association: oral health and nutrition. J Am Diet Assoc 2003;103:615–25.

48. NIH Statement, Diagnosis and management of dental caries.

49. Trends in children's oral health. National Maternal and Child Oral Health Resource Center, www.ncemch.org, 1999.

50. Populations receiving optimally fluoridated public drinking water—United States, 2000. MMWR 2002;51:144–7.

51. Trends in children's oral health (www.ncemch.org).

52. Position of the ADA: oral health and nutrition.

53. Dietary reference intakes, NAS, chap. 11.

Nutrition **UP CLOSE**

Does Your Fiber Intake Measure Up?

Feedback for Unit 12

The total number of points you scored approximates the **grams** of total fiber you typically consume daily.* Use this scale to find out if your fiber intake meets the recommended goal:

- **0–10 grams:** You consume less than the average American. Increase your fiber intake by including more fruits, vegetables, whole grains, and legumes in your diet overall.

- **11–15 grams:** Like other Americans, you consume too little fiber. Increase the number of servings of high-fiber foods you already enjoy, while substituting more high-fiber foods for refined food products. A quick way to add fiber to your diet is to consume more of the two fiber powerhouses: legumes and bran cereal.

- **15–20 grams:** You currently consume more fiber than the average American. Make sure you're including 5 or more servings of fruits and vegetables. Eat 6–11 servings of bread, cereal, rice, and pasta daily; choose whole grain versions of these foods often.

- **20–40 grams:** Congratulations! Your dietary fiber intake is in the vicinity of that recommended. Keep up the good work.

*Because different foods within a food group contribute varying amounts of dietary fiber, the point values have been averaged. Make sure to check the Nutrition Facts panel on bran cereals because these cereals vary in the amount of dietary fiber they contain.

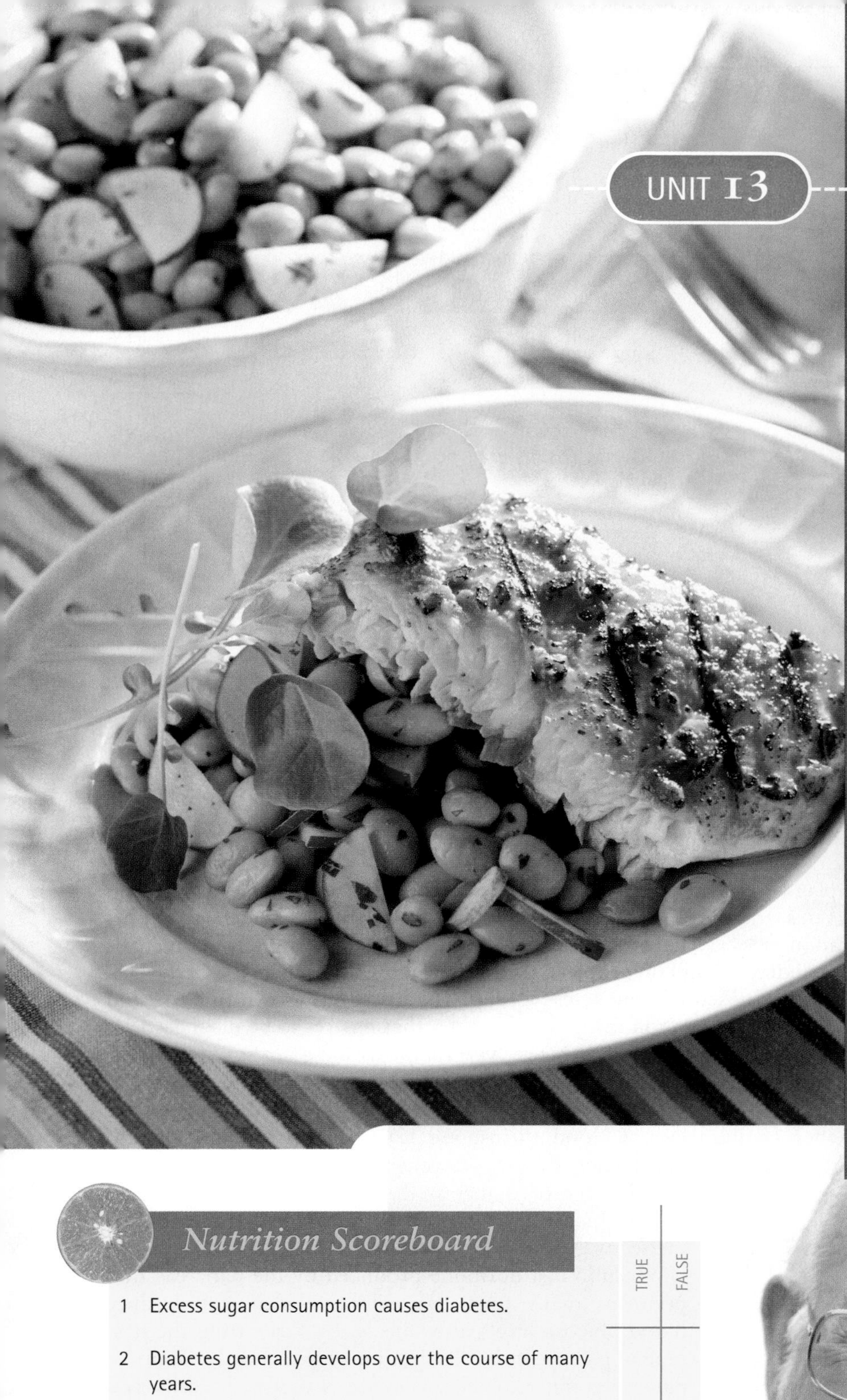

Diabetes Now

Nutrition Scoreboard

	TRUE	FALSE
1 Excess sugar consumption causes diabetes.		
2 Diabetes generally develops over the course of many years.		
3 Glycemic index is a measure of the simple sugar content of foods.		

Answers on next page

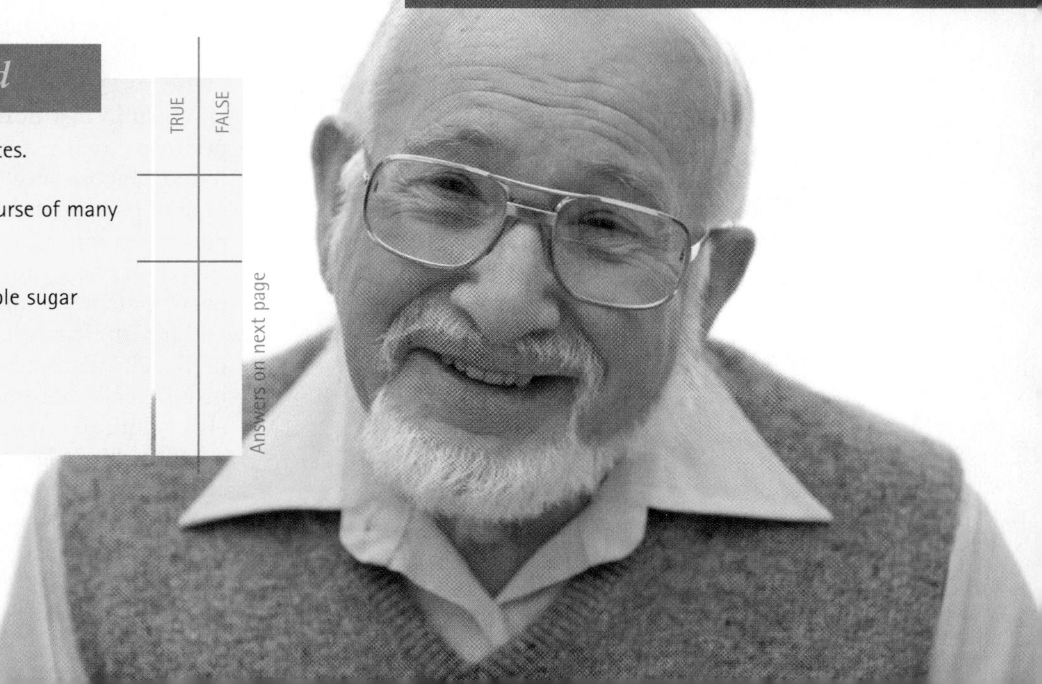

[KEY CONCEPTS AND FACTS]

- Diabetes is related to abnormal utilization of glucose by the body.

- The three main forms of diabetes are type 1, type 2, and gestational diabetes.

- Rates of type 2 diabetes increase as obesity does.

- Weight loss and physical activity can prevent or delay the onset of type 2 diabetes in many people.

Answers to *Nutrition Scoreboard*

		TRUE	FALSE
1	Simple sugar intake is not related to the risk of developing diabetes.[1]		✔
2	Disease processes underlying the development of diabetes exist for years before the onset of diabetes in most cases.[2]	✔	
3	Glycemic index is a measure of the potential of carbohydrate-containing foods to raise blood glucose level. Only a few simple sugars have a high glycemic index.[3]		✔

diabetes

A disease characterized by abnormal utilization of carbohydrates by the body and elevated blood glucose levels. There are three main types of diabetes: type 1, type 2, and gestational diabetes. The word *diabetes* in this unit refers to type 2, which is by far the most common form of diabetes.

The Diabetes Epidemic

What AIDS was in the last 20 years of the 20th century, diabetes is to be in the first 20 years of this century.
—Paul Zimmet, International Diabetes Institute[4]

It's not the plague, yellow fever, or heart disease. The latest worldwide disease epidemic is *diabetes*, and the rising rates are directly related to the global increase in obesity (Illustration 13.1). Diabetes affects approximately 200 million individuals worldwide, including 17 million people in the United States.[4] Less than 1% of the U.S. population was diagnosed with diabetes in 1960. The figure has grown to 8.6% and includes 2.4% of the population living with diabetes that has yet to be diagnosed.[5]

There are three major forms of diabetes: *type 1, type 2,* and *gestational diabetes.* Type 2 diabetes is the most common by far and is fueling the diabetes epidemic. Table 13.1 summarizes key features of type 1 and type 2 diabetes. Both types are diagnosed when fasting levels of blood glucose are 126 milliliters/deciliter (mg/dl) and higher; these types generally take years to develop.[7] In all cases of diabetes, the central defect is elevated blood glucose level caused by an inadequate supply or ineffective utilization of insulin.[8]

Insulin is a hormone produced by the pancreas that performs many functions, one of which is to reduce blood glucose levels after meals. By facilitating the passage of glucose into cells, insulin keeps a steady supply of glucose going into cells. Glucose is needed by cells as a source of energy for thousands of chemical reactions that participate in the maintenance of ongoing body functions and health. If insulin is produced in insufficient amounts, or if cell membranes are not sensitive to the action of insulin, cells become starved for glucose. Functional levels of multiple tissues and organs in the body degrade as a result. High levels of blood glucose have adverse side effects on the body, too, such as elevated blood levels of triglycerides, increased blood pressure, and hardening of the arteries.[9]

Illustration 13.1
Diabetes headlines say it all.

HEALTH
The Continuing Epidemics of Obesity and Diabetes in the United States
Associated Press
WASHINGTON

WORLD & NATION
Diabetes Threat on the Rise among U.S. Children, Specialists Say
Associated Press
WASHINGTON, D.C. —

CDC's Forecast for Diabetes is Grim

By Maria Elena Baca
Star Tribune Staff Writer

In a recent government study, the Centers for Disease Control and Prevention (CDC) estimated that obesity is fast approaching tobacco as the underlying

Thirty-four percent of U.S. adults are considered overweight, and an additional 31 percent are obese. Anyone with a body mass index (a ratio between your height and weight) of 25 or above -- that's someone, for example, who is 5-foot-4 and 145 pounds --

considered overweight, according to the National Institutes of Health. Anyone with a body mass index of 30 or above -- such as someone who is 5-foot-6 and 186 pounds -- is considered obese. Check your body mass ind

TABLE 13.1

KEY CHARACTERISTICS OF TYPE 1 AND TYPE 2 DIABETES.[6]

CHARACTERISTIC	TYPE 1 DIABETES	TYPE 2 DIABETES
Insulin deficiency?	Yes	Possible in advanced stages of the disease
Proportion of cases	10%	90%
Risk factors	Viral infection early in life (or other triggers in genetically sensitive individuals) that destroys part of the pancreas. Young birth, age, certain medications.	Obesity (especially abdominal fatness), sedentary lifestyle, insulin resistance, low weight at certain ethnicities, family history, older age
Treatment	Insulin, diet, exercise	Weight loss (in most cases), increased physical activity, sometimes oral medications, and/or insulin

Health Consequences of Diabetes

Health effects of diabetes vary depending on how well blood glucose levels are controlled and on the presence of other health problems such as hypertension or heart disease. In the short run, poorly controlled and untreated diabetes produces blurred vision, frequent urination, weight loss, increased susceptibility to infection, delayed wound healing, and extreme hunger and thirst. In the long run, diabetes may contribute to heart disease, hypertension, blindness, kidney failure, stroke, and the loss of limbs due to poor circulation. The number one cause of death among people with diabetes is heart disease. Many of the side effects of diabetes can be prevented or delayed if blood glucose levels are maintained within the normal range.[10]

Type 2 Diabetes

The development of this common type of diabetes is most likely to occur in overweight and obese, inactive people (Illustration 13.2). Although most often diagnosed in people over the age of 40, type 2 diabetes is becoming increasingly common in children and adolescents. There is a genetic component to this disease, as evidenced by the fact that it tracks in families and is more likely to occur in certain groups (Hispanic American, African American, Asian and Pacific Islanders, and Native Americans) than others.[12] People who develop type 2 diabetes usually have a condition known as *prediabetes* years before being diagnosed with type 2 diabetes.

Prediabetes and Insulin Resistance]   Elevated fasting blood glucose levels that are somewhat below the cut point used to diagnose type 2 diabetes characterize *prediabetes*.[13] Approximately 6% of U.S. adults, and 314 million people worldwide, are at risk of type 2 diabetes due to this condition. The presence of prediabetes increases a person's odds of developing type 2 diabetes by 10% per year.[14] Most people diagnosed with prediabetes have a condition known as *insulin resistance.* Obesity, low levels of physical activity, and a genetic predisposition are common risk factors for insulin resistance. Abdominal obesity, or high levels of central-body fat, is a particularly potent risk factor for insulin resistance.[15]

Insulin resistance is due to abnormalities in the way the body uses insulin.[16] Normally, insulin is able to lower blood glucose levels after meals by binding to receptors on cell membranes. These receptors are activated by insulin and allow glucose

type 1 diabetes
A disease characterized by high blood glucose levels resulting from destruction of the insulin-producing cells of the pancreas. This type of diabetes was called juvenile-onset diabetes and insulin-dependent diabetes in the past, and its official medical name is type 1 diabetes mellitus.

type 2 diabetes
A disease characterized by high blood glucose levels due to the body's inability to use insulin normally, or to produce enough insulin. This type of diabetes was called adult-onset diabetes and non-insulin-dependent diabetes in the past, and its official medical name is type 2 diabetes mellitus.

gestational diabetes
Diabetes first discovered during pregnancy.

prediabetes
A condition in which blood glucose levels are higher than normal but not high enough for the diagnosis of diabetes. It is characterized by impaired glucose tolerance, or fasting blood glucose levels between 110 and 126 mg/dl.

insulin resistance
A condition in which cell membrane have reduced sensitivity to insulin so that more insulin than normal is required to transport a given amount of glucose into cells. It is characterized by elevated levels of serum insulin, glucose, and triglycerides, and increased blood pressure.

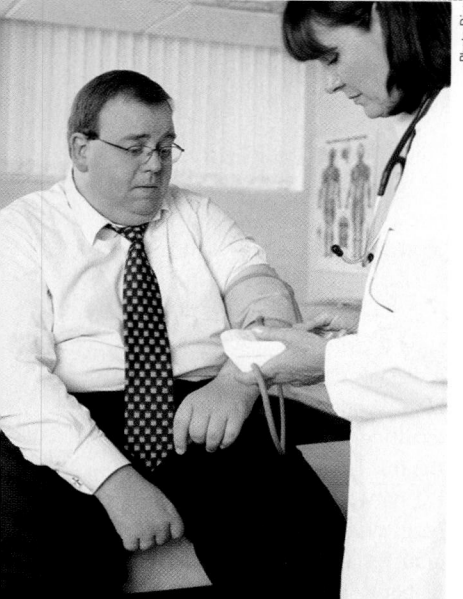

Illustration 13.2
Obesity characterized by central-body fat stores and physical inactivity are strong risk factors for type 2 diabetes.[11]

metabolic syndrome
A constellation of metabolic abnormalities that increase the risk of heart disease and type 2 diabetes. Metabolic syndrome is characterized by insulin resistance, abdominal obesity, high blood pressure and triglycerides levels, low levels of HDL cholesterol, and impaired glucose tolerance. It is also called *Syndrome X* and *insulin resistance syndrome*.

to pass into cells. With insulin resistance, cell membranes "resist" the effects of insulin, and that lowers the amount of glucose transported into cells. To prevent blood glucose levels from becoming too high, the pancreas produces and secretes additional insulin. Higher-than-normal levels of insulin are generally sufficient to keep blood glucose levels under control for a number of years. Cells in the pancreas may become exhausted, however, from the years of overwork. In such cases, production of insulin slows down or stops, and glucose accumulates in the blood as a result. When fasting blood glucose levels consistently reach 126 mg/dl or higher, type 2 diabetes is under way.[17]

Insulin resistance is also related to the development of a spectrum of metabolic abnormalities that have far-reaching effects. Collectively, the adverse effects of insulin resistance are included in a disorder called *metabolic syndrome.*

Metabolic Syndrome] Physicians have known for decades that obese people with hypertension and type 2 diabetes are at high risk of heart disease. What they didn't know is why. Over time, research studies discovered that a large part of the answer to the "why" question was insulin resistance.[18] This condition is related to a cluster of metabolic abnormalities that increase the risk of heart disease and include

- High levels of central-body fat
- High blood insulin levels
- High blood pressure (130/85 mm/Hg or higher)
- Elevated blood triglycerides levels (150 mg/dl or higher)
- Low levels of protective HDL cholesterol (less than 50 mg/dl in women and 40 mg/dl in men)
- High blood glucose levels (110 mg/dl or higher)

The diagnosis of metabolic syndrome is made when three or more abnormalities are identified. Individuals with four or five metabolic abnormalities have a 3.7 times greater risk of heart disease, and a 25 times higher risk for diabetes than do people with no abnormalities.[19] It is estimated that 25% of men and women in the United States have metabolic syndrome, with the bulk of cases being made up of overweight and obese inactive adults.[20] Weight loss, exercise, and the other factors listed in Illustration 13.3 are the key components of preventing and managing metabolic syndrome.

Type 1 Diabetes

Type 1 diabetes results from a deficiency of insulin and accounts for about 10% of all cases. The diagnosis of type 1 peaks around the ages of 11 to 12 years, and it usually occurs before the age of 40.[21]

Illustration 13.3
Key components of the prevention and management of metabolic syndrome.

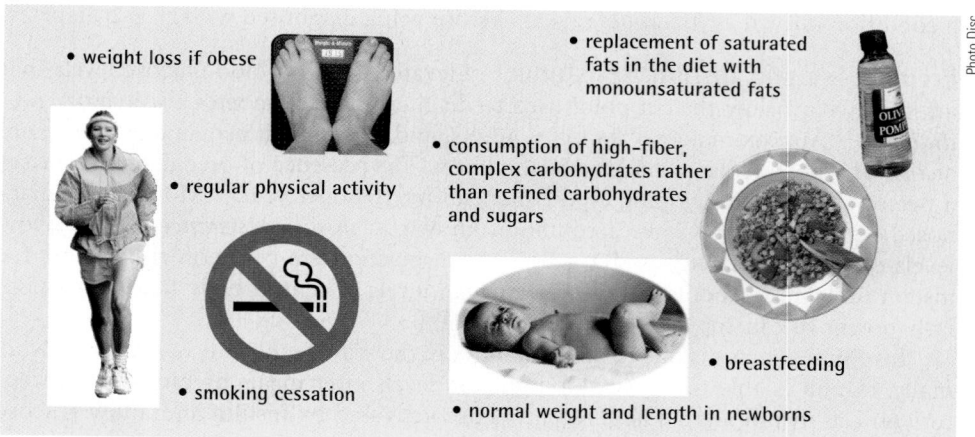

- weight loss if obese
- regular physical activity
- smoking cessation
- replacement of saturated fats in the diet with monounsaturated fats
- consumption of high-fiber, complex carbohydrates rather than refined carbohydrates and sugars
- breastfeeding
- normal weight and length in newborns

Illustration 13.4
The first insulin pumps in the 1960s were heavy and cumbersone—so large they had to be worn as a backpack. Today's pumps allow much greater freedom of movement.

People with type 1 diabetes are instructed to measure blood glucose levels several times daily and to adjust insulin dose according to the results. Both the skin pricks required to test blood and delivery of insulin by injection may be painful, however, and are a feared part of the disease.[22] Technological advances such as the insulin pump (Illustration 13.4) are taking out some of the sting. Insulin pumps have a glucose sensor and an insulin pump that releases insulin in response to blood glucose level. It is expected that over 40% of people with type 1 diabetes will eventually use the pump. Such devices have been found to improve blood glucose control. Weight loss, even as little as 10 to 15 pounds, benefits blood glucose control in overweight and obese individuals with type 1 diabetes.[23]

Insulin deficiency appears to stem from a viral infection (such as mumps, rubella, or the flu), or from allergic reactions in genetically susceptible people that eventually destroy the portion of the pancreas that produces insulin. Medications used to treat high blood pressure, arthritis, and other disorders may also contribute to the development of type 1 diabetes. Breastfeeding for the first 4 months of life or so may confer an element of protection against type 1 diabetes.[24]

Environmental factors appear to be more important than genetic background in the development of this type of diabetes. The incidence of type 1 diabetes varies 36-fold among countries (Illustration 13.5), suggesting that environmental factors play a key role in its development.[26]

ON THE SIDE

Ouchless blood glucose sampling systems are on their way. Watch-like gadgets containing sensors that indirectly measure blood glucose levels are available. The device not only tells time but also displays blood glucose level at the touch of a button.[27]

Ou ch!

Gestational Diabetes

Approximately 3 to 6% of women develop gestational diabetes during pregnancy, but the incidence varies a good deal based on age, body weight, and ethnicity. Native Americans, African Americans, women over the age of 35 years, obese women, and those with habitually low levels of physical activity are at higher risk than other women.[28] Infants born to women with poorly controlled diabetes may be very fat at birth and have blood glucose

Illustration 13.5
Incidence of type 1 diabetes in various countries.[25]

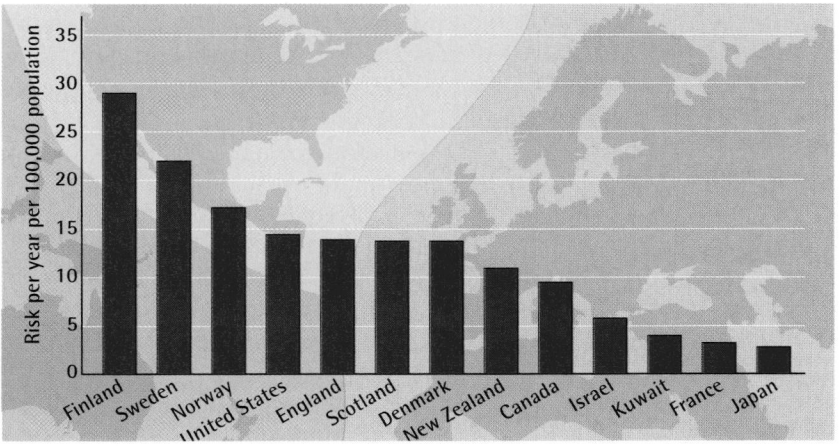

control problems after delivery. They are at greater risk for developing diabetes later in life, and 6 to 20% will have a physical abnormality that may threaten survival or a high quality of life.[29]

As is the case for type 2 diabetes, women with gestational diabetes are insulin resistant and can control their blood glucose levels with an individualized diet and exercise plan. Some women require daily insulin injections for blood glucose control.[30]

Gestational diabetes disappears after delivery, but type 2 diabetes may appear later in life. Exercise, maintenance of normal weight, and consumption of a healthy diet reduce the risk that diabetes will return.[31]

Managing Type 2 Diabetes

Diet and exercise are the cornerstones of the treatment of type 2 diabetes.[32] Modest weight loss alone (5–10% of body weight) has been repeatedly shown to significantly improve blood glucose control in overweight and obese people with type 2 diabetes.[33]

In general, diets developed for diabetes emphasize

- Complex carbohydrates including whole grain breads and cereals, and other high-fiber foods, vegetables, fruits, low-fat milk and meats, and fish

- Unsaturated fats

- Regular meals and snacks[34]

Chromium supplements (500–1000 µg per day) are sometimes recommended. It appears the essential mineral chromium improves blood glucose and lipid levels in many people with type 2 diabetes.[35]

Dietary management of diabetes should focus on heart disease risk reduction as well as blood glucose control. Food sources of monounsaturated fats, such as vegetable oils, nuts, seeds, and lean meats and seafoods are recommended over foods high in saturated or trans fats. Monounsaturated fats raise blood levels of HDL cholesterol while reducing LDL cholesterol level, improve insulin resistance, and lower blood glucose levels somewhat.[36]

Diet and weight loss interventions may be supplemented by oral medications that decrease insulin resistance and blood lipids, and by insulin if needed.[37]

Knowledge of the role of diet, exercise, insulin, and other factors in the management of diabetes is expanding so rapidly that the American Diabetes Association updates management recommendations yearly. Dietary recommendations are currently not consistent across developed countries, indicating that scientific consensus is yet to be reached on a number of important issues related to diet and diabetes.[38]

Glycemic Index and Glycemic Load

Carbohydrate-containing foods have a range of effects on blood glucose levels—some cause a rapid rise and others do not. Foods that increase blood glucose to relatively high levels require more insulin to move glucose into cells than do foods that produce lower levels of glucose. Over the past 25 years, many carbohydrate-containing foods have been tested for their effect on blood glucose level and assigned a *glycemic index* value. Compared to high glycemic index (GI) carbohydrate sources, foods with low GI values increase blood glucose levels to a lower extent, and decrease insulin need.[39]

The glycemic index of carbohydrate-containing foods is determined by assessing the elevation in blood glucose level caused by ingestion of 50 grams of a carbohydrate-containing food compared to the rise in blood glucose level that results from con-

glycemic index (GI)
A measure of the extent to which blood glucose level is raised by a 50-gram portion of a carbohydrate-containing food compared to 50 grams of glucose or white bread.

REALITY CHECK

Will the real whole grain please stand up?

Which bread is made from whole grains?

Who gets thumbs up

?

Wheat Bread

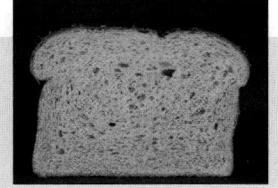

Whole Wheat Bread

Richard Anderson

Answers on next page

suming 50 grams of glucose. (Sometimes the standard for comparison is white bread). Table 13.2 shows the glycemic index of a number of foods, using 50 grams of glucose as the standard for comparison. As you read over the list of foods and their glycemic index values, you may find some surprises. Sucrose, honey, fructose, and other simple sugars are not high glycemic index foods, and many fruits we think of as sweet do not cause a relatively high rise in blood glucose level. Coarse-ground whole wheat breads

TABLE 13.2

GLYCEMIC INDEX (GI) OF SELECTED FOODS.[40]

HIGH GI		MEDIUM GI		LOW GI	
glucose	100	Cheerios	74	muesli	48
French bread	95	popcorn	72	green peas	48
scone	92	watermelon	72	pasta	48
potato, baked	85	Grape Nuts	71	carrots, raw	47
potato, instant mashed	85	wheat bread	70	cassava	46
Corn Chex	83	white bread	70	lactose	46
pretzel	83	orange soda	68	milk chocolate	43
Rice Krispies	82	sucrose	68	All Bran	42
cornflakes	81	croissant	67	orange	42
Corn Pops	80	Cream of Wheat	66	peach	42
Gatorade	78	couscous	65	apple juice	40
jelly beans	78	chapati	62	plum	39
doughnut, cake	76	sweet potato	61	apple	38
waffle, frozen	76	muffin, blueberry	59	pear	38
French fries	75	Coca-Cola	58	tomato juice	38
Shredded Wheat	75	rice, white or brown	60	yam	37
		honey	55	dried beans	25
		oatmeal	54	grape fruit	25
		corn	53	cherries	22
		cracked wheat bread	53	fructose	19
		orange juice	52	xylitol	8
		banana	52	hummus	6
		mango	51		
		potato, boiled	50		

Wheat Bread

Whole Wheat
Bread

ANSWERS TO **REALITY CHECK**

Will the real whole grain please stand up?

You can't tell by looking, but the slice of bread on the right contains more fiber (4 vs. 2 grams). Want to find whole grain breads? Look for the term "whole grain" or "whole wheat" on the label. Bread products carrying that label contain 51% or more whole grains. The ingredient label on whole grain products will list one or more whole grains before listing enriched flour.

glycemic load (GL)
A measure of the extent to which blood glucose level is raised by a given amount of a carbohydrate-containing food. GL is calculated by multiplying a food's GI by its carbohydrate content.

and dried beans have medium-to-low glycemic index values. It probably comes as no surprise that glucose has the highest glycemic index, but it is infrequently found by itself in foods.

Some high-GI foods such as baked potatoes, French bread, and cornflakes are good sources of a number of nutrients. Just because a food has a high glycemic index doesn't mean it should not be consumed as part of a balanced diet.[41] Adjusting food choices toward selection of mainly low-GI foods is most helpful for people attempting to prevent or control type 2 diabetes, or to diminish the effects of insulin resistance.[42]

Because of the way glycemic index is calculated, it is difficult to know the extent to which blood glucose will be raised by consumption of a particular amount of a food. A new index of the blood-glucose-raising potential of carbohydrate-containing foods has been developed to help straighten out this confusion. It's called *glycemic load,* and it represents the blood-glucose-raising potential of the specific amount of food consumed. Glycemic load is calculated by multiplying the grams of carbohydrate in a specific amount of food times the food's glycemic index. This result is then divided by 100 to calculate glycemic load. A carrot provides approximately 7 grams of carbohydrate and has a glycemic index of 47. Its glycemic load would be calculated as:

$$7 \times 47 = 329,$$
$$329/100 = 3.29$$
$$\text{Glycemic load} = 3.29$$

The blood-glucose-raising effect of one carrot doesn't amount to much. If you consumed 4 slices (4 ounces) of French bread, the result would tell a different story. The glycemic load supplied by this amount of French bread is 49.4, a level that raises blood glucose and insulin levels far more than a carrot or 1 slice of French bread.

Consumption of low-GI foods is a recommended component of the dietary management of type 2 and gestational diabetes in a number of countries, but is not a primary recommendation in the United States.[43] Consumption of low-GI foods is viewed as a useful part of the management of insulin resistance and metabolic syndrome, and as a secondary aid to blood glucose control among people with diabetes.[44]

Sugar Intake and Diabetes

Does sugar intake cause diabetes? The answer to this reasonable and often asked question is no. Results of multiple studies indicate that high intakes of simple sugars do not cause diabetes.[45] Should people with diabetes exclude sugars from their diet? Sugar does not have to be eliminated from the diet of people with diabetes. Intake of total carbohydrates, rather than sugar intake specifically, is most strongly related to blood glucose levels. However, high-sugar diets increase blood triglycerides levels in people with metabolic syndrome, and that may increase the risk of heart disease.[46]

Prevention of Type 2 Diabetes

The effects of weight loss and exercise in preventing type 2 diabetes can be quite dramatic.[47] In one large study that took place over a 3-year period, people with pre-diabetes reduced their risk of developing type 2 diabetes by over 50% with losses in body weight of around 7% and 150 minutes a week of exercise.[48] Diets rich in whole grain and high-fiber foods are protective against the development of type 2 diabetes and appear to aid weight loss.[49] Components of high-fiber, whole grain foods raise blood glucose levels marginally and appear to provide nutrients and other biologically active substances that lessen the risk of this disease.[50]

It is anticipated that 800,000 new cases of type 2 diabetes may develop each year in the United States due to rising rates of obesity.[51] The additional burdens such an increase would place on individuals and health care costs clearly convey the message that prevention is urgently needed. Public health campaigns are now under way to encourage people to lose weight if overweight, to exercise regularly, and to select whole grain products and other high-fiber foods along with ample intake of vegetables and fruits.

Photo Disc

To prevent type 2 diabetes, select whole grain products and other high-fiber foods along with ample servings of vegetables and fruits.

Hypoglycemia: The Low Blood Sugar Blues?

Hypoglycemia is due to abnormally low blood glucose levels. It is thought to be a rare disorder because it is not often diagnosed. The diagnosis is tricky—blood tests for glucose should be conducted when the symptoms of hypoglycemia are present rather than during a scheduled appointment.[52] The true incidence of hypoglycemia is not known.

Hypoglycemia is most often caused by an excessive availability of insulin in the blood. The oversupply of insulin may be caused by certain tumors that secrete insulin, by other health problems, by high alcohol intake on an empty stomach, or, in people with diabetes, by an insulin dose that is too high.[53] Hypoglycemia also occurs during prolonged starvation, but blood glucose levels become very low only when starvation threatens life.

Symptoms of hypoglycemia include weakness, sweating, nervousness, confusion, and irritability. They may appear before blood glucose levels are low enough to be clinically considered abnormal. Apparently, people vary considerably in their response to low-normal glucose levels.[54] Symptoms of hypoglycemia disappear within 5 to 15 minutes after glucose, candy, orange juice, or similar food is consumed. Increased blood glucose levels need to be maintained by food after the drop in blood glucose level has been corrected.[55]

The notion that high sugar intakes by themselves cause hypoglycemia has not been verified by research.[56] People who suspect that they have this disorder should consult a physician and have their blood glucose level checked at a time when the symptoms normally occur.[57]

hypoglycemia
A disorder resulting from abnormally low blood glucose levels. Symptoms of hypoglycemia include irritability, nervousness, weakness, sweating, and hunger. These symptoms are relieved by consuming glucose or foods that provide carbohydrate.

Standard diet therapy for hypoglycemia calls for people to consume five to six small meals that include a variety of complex carbohydrates and high-protein foods and to avoid alcohol and snacks high in simple sugars.[58]

Diabetes in the Future

The anticipated surge in the worldwide incidence of type 2 diabetes is not inevitable. It could be lowered substantially by environmental and lifestyle changes that reduce the risk for, and incidence of, overweight and obesity. Increased awareness of the connection between diabetes and body weight may help. Only a small proportion of people are aware of the connection now.[59] The hoped-for future of diabetes would be the one that negates the dire forecasts of the experts.

Key Terms

diabetes, page 13-2

gestational diabetes, page 13-3

glycemic index, page 13-6

glycemic load, page 13-8

hypoglycemia, page 13-9

insulin resistance, page 13-3

metabolic syndrome, page 13-4

prediabetes, page 13-3

type 1 diabetes, page 13-3

type 2 diabetes, page 13-3

www links

www.webmd.com
Go to the "Conditions Center," select **diabetes,** and hit "go." Get the latest information on the treatment, a risk assessment for diabetes, and test your diabetes IQ.

www.diabetes.com
Part of WebMD. Provides diabetes facts, tips for eating well, information on insulin resistance, and more.

www.niddk.nih.gov

Search diabetes and receive information for people newly diagnosed with diabetes, the low-down on diabetes treatment, diet planning, and a lot more.

http://diabetes.niddk.nih.gov/dm/pubs/hypoglycemia/index.htm
From the National Institutes of Health, this site describes hypoglycemia and its causes and care.

www.healthfinder.gov
Search insulin resistance, diabetes, metabolic syndrome, and hypoglycemia to find

reliable reports.

www.intelihealth.com/IH/ihtIH/EMIHC0000/21054/21054.html
Search **diabetes,** then select from a list of interactive tools such as symptom scout, diabetes quiz, diabetes dictionary, and ask an expert.

www.nhlbi.nih.gov/health/public/heart/other/ktb_recipebk/index.htm
A source for heart health recipes.

Notes

1. Liu S et al. Sugar intake and diabetes risk. Diabetes Care 2003;26:1008–15.

2. Nathan DM. Initial management of glycemia in type 2 diabetes mellitus. N Engl J Med 2002;347:1242–50.

3. Foster-Powell K et al. International table of glycemic index and glycemic load values: 2002. Am J Clin Nutr 2002;76: 5–56.

4. World seen facing diabetes catastrophe, impact may outpace AIDS. International Diabetes Federation Conference, Paris 2003. Reported in www.medscape.com, 9/6/03.

5. Cowie CC, MMWR, CDC Surveillance System, 2003:52:833–7; and Diabetes rose slightly in the 1990s, Centers for Disease Control Press Release, 9/12/03.

6. Diabetes overview, www.intelihealth .com/diabetes, accessed 6/03; and Gautier J-F et al., Gestational diabetes and the risk of type 2 diabetes, Lancet 2003: 361:1839,1861–5.

7. Nathan, Initial management of glycemia; and Gale EAM et al. Can we change the course of beta-cell destruction in type 1 diabetes? N Eng J Med 2002;346:1740–1.

8. Diabetes overview (www.intelihealth .com/diabetes).

9. Alexander CM et al. NCEP-defined metabolic syndrome, diabetes, and prevalence of coronary heart disease among NHANES III participants age 50 years and older. Diabetes 2003;52: 1210–14.

10. Diabetes overview (www.intelihealth .com/diabetes); and Solomon CG, Reducing cardiovascular risk in type 2 diabetes, N Engl J Med 248:2003;457–60.

11. Diabetes overview (www.intelihealth .com/diabetes); and Diabetes Prevention Program Research Group, Reduction in the incidence of type 2 diabetes with lifestyle intervention or metformin, N Engl J Med 2002;346;393-403; and

Nutrition UP CLOSE

Calculating Glycemic Load

FOCAL POINT: To gain an appreciation of the effect of source and amount of carbohydrate consumed on blood glucose levels.

Glycemic index provides an estimate of the rise in blood glucose level expected from consuming 50 grams of a carbohydrate containing food. Glycemic load, on the other hand, is a measure of the expected rise in blood glucose related to the ingestion of other amounts of this food. Consequently, glycemic load estimates the blood-glucose-raising potential of the amount of carbohydrate containing food actually consumed.

1. Use Appendix A to look up the carbohydrate content of each food for the serving size listed.
2. Use Table 13.2 to identify the glycemic index for each food.
3. Calculate glycemic load based on this formula:

   Glycemic load = grams carbohydrate × glycemic index of the food, divided by 100

Here's an example for 2 tsp sucrose:

$$\text{grams carbohydrate in 2 tsp. sucrose} = 8$$
$$\text{Glycemic index of sucrose} = 68$$
$$8 \times 68 = 544$$
$$\text{Glycemic load} = 544 \div 100 = 5.44$$

Food	Serving Size	grams carbohydrate	×	Glycemic Index	÷	Glycemic Load 100
cola beverage	36 oz (3 cans)	_____		_____		_____
potato, baked, no skin	1	_____		_____		_____
apple juice, bottled	1 cup	_____		_____		_____
milk chocolate, plain	1 oz	_____		_____		_____
hummus	½ cup	_____		_____		_____

FEEDBACK Glycemic load answers can be found at the end of Unit 13.

Dietary Reference Intakes. Energy, carbohydrate, fiber, fat, fatty acids, cholesterol, protein, and amino acids. Institute of Medicine, National Academy of Sciences, Washington, DC: National Academies Press, 2002.

12. Cowie, MMWR, CDC Surveillance System.

13. Diabetes Prevention Program, Reduction in the incidence of type 2 diabetes.

14. Cowie, MMWR, CDC Surveillance System; and Diabetes overview (www.intelihealth.com/diabetes).

15. Men with 3 of 5 metabolic abnormalities risk diabetes, heart disease. www.nlm.nih.gov/medlineplus/heart disease.html, accessed 7/03.

16. Diabetes overview (www.intelihealth.com/diabetes).

17. Diabetes overview (www.intelihealth.com/diabetes).

18. Alexander et al., NCEP-defined metabolic syndrome.

19. Men with 3 of 5 metabolic abnormalities (www.nlm.nih.gov/medlineplus/heartdisease.html).

20. Park Y-W et al. The metabolic syndrome: prevalence and associated risk factors. Arch Int Med 2003;163:427–36.

21. Orban T. Food for thought in diabetes. Nutr Today 2001;26:238–48.

22. Diabetes overview (www.intelihealth.com/diabetes).

23. Weissberg-Benchell J et al. Insulin pump therapy. Diabetes Care 2003;26:1079–87; Intensive diabetes self-management: 2000 update, American Association of Clinical Endocrinologists, Endocrine Practice 2000;6:42–67.

24. Virtanen SM, Rasanen L, Aro A et al. Infant feeding in Finnish children over the age of 4 years with newly diagnosed IDDM. Diabetes Care 1991;14:415–7.

25. LaPorte et al., Geographic differences; and Muntoni S et al., Nutritional factors and worldwide incidence of childhood type 1 diabetes, Am J Clin Nutr 2000; 71:1525–9; Intensive diabetes self-management: 2000 update, American Association of Clinical Endocrinologists, Endocrine Practice 2000;6:42–67.

26. LaPorte RE, Taijima N, Akerblom HR, et al. Geographic differences in the risk of insulin-dependent diabetes mellitus. Updated in 2000. Diabetes Care 1985; 8(suppl):101–7.

27. Chase HP et al. Use of the GlucoWatch Biographer in children with type 1 diabetes. Pediatrics 2003;111:790–4.

28. Solomon CG et al. A prospective study of pregravid determinants of gestational diabetes mellitus. JAMA 1997;278:1078–83.

29. Carr SR. Effect of maternal hyperglycemia on fetal development. Annual meeting of the American Dietetic Association, Boston, 1997 Oct. 28.

30. Intensive diabetes self-management: 2000 update.

31. Gautier et al., Gestational diabetes.

32. Solomon, Reducing cardiovascular risk. Brand-Miller J et al, Low glycemic index foods and glycemic control in diabetics, Diabetes Care 2003;26:2261–7.

33. American Diabetes Association. The prevention or delay of type 2 diabetes. Diabetes Care 2002;25:vol.4.

34. Solomon, Reducing cardiovascular risk; and Franz MJ et al. Evidence-based nutrition principles and recommendations for the treatment and prevention of diabetes and related complications. Diabetes Care 2002;25:148–66.

35. Wong Z et al. Chromium supplements and glucose sensitivity in type 2 diabetes. 18th IDF Congress, abstract 154, 756, 762, Aug. 28, 2003.

36. Solomon, Reducing cardiovascular risk.

37. Nathan, Initial management of glycemia.

38. Franz et al., Evidence-based nutrition principles; and Foster-Powell et al., International table of glycemic index and glycemic load values.

39. Dietary Reference Intakes, Energy, . . . and amino acids; and Jones JJ et al. Sugars and health: is there an issue? J Am Diet Assoc 2003;103:1058–60.

40. Foster-Powell et al., International table of glycemic index and glycemic load values.

41. Schefrin R. Good carbs, bad carbs. Today's Dietitian 2003;April:36–9.

42. Brand-Miller et al., Low glycemic index foods; and Jenkins DJA et al. Glycemic index: overview of implications in health and disease. Am J Clin Nutr 2003;78:99–103.

43. Foster-Powell et al., International table of glycemic index and glycemic load values.

44. Foster-Powell et al., International table of glycemic index and glycemic load values; Dietary Reference Intakes, Energy, . . . and amino acids; and Schafer G et al., Comparison of the effects of dried peas with those of potatoes in mixed meals on type 2 diabetes, Am J Clin Nutr 2003;78:99–103.

45. Liu et al., Sugar intake and diabetes risk; and McKeown N et al., Whole-grain intake is favorably associated with metabolic risk factors for type 2 diabetes and cardiovascular disease, Am J Clin Nutr 2002;76:390–8.

46. Franz et al., Evidence-based nutrition principles; and Jones et al., Sugars and health.

47. Sheard NF. Moderate changes in weight and physical activity can prevent or delay the development of type 2 diabetes mellitus in susceptible individuals. Nutr Rev 2003;61:76–9.

48. Diabetes Prevention Program Research Group, Reduction in the incidence of type 2 diabetes.

49. McKeown et al., Whole-grain intake.

50. Dietary Reference Intakes, Energy, . . . and amino acids.

51. Diabetes overview (www.intelihealth.com/diabetes).

52. Hypoglycemia frequently "diagnosed" but rarely true. Envir Nutr 1998;Feb.7.

53. Amiel S. Reversal of unawareness of hypoglycemia. N Engl J Med 1993;329:876–7.

54. Diabetes overview (www.intelihealth.com/diabetes). Palard J, Havrankova J, LePage R, Matte R, Belander R, D'Amour P, Ste.-Marie L-G, Blood glucose measurements during symptomatic episodes in patients with suspected postprandial hypoglycemia, N Engl J Med 1989;321:1421.

55. Franz et al., Evidence-based nutrition principles.

56. Ryan CM, Atchison J, Puczynski S, Puczynski M, Arslanian S, Becker D. Mild hypoglycemia associated with deterioration of mental efficiency in children with insulin-dependent diabetes mellitus. J Pediatr 1990;117:32–38; Nelson RL. Hypoglycemia: fact or fiction? Mayo Clin Proc 1985;60:844–50.

57. Service FJ. Hypoglycemia and the postprandial syndrome. N Engl J Med 1989;321:1472-4.

58. Service, Hypoglycemia and the postprandial syndrome.

59. Few overweight and obese people aware of diabetes risk. www.medscape.com, accessed 9/03.

Nutrition **UP CLOSE**

Calculating Glycemic Load

Feedback for Unit 13

Food	Glycemic Load
cola beverage	67.9
baked potato	28.9
apple juice	11.6
milk chocolate	7.3
hummus	1.5

Effects of carbohydrate-containing foods on blood glucose levels vary depending on the glycemic index of these foods and the amount of them we consume.

Alcohol: The Positives and Negatives

Nutrition Scoreboard

	TRUE	FALSE
1 Alcohol is produced from carbohydrates.		
2 You can protect your body from the harmful effects of consuming excessive amounts of alcohol by eating a nutritious diet.		
3 Alcohol abuse plays a major role in injuries and deaths in the United States.		

Answers on next page

[KEY CONCEPTS AND FACTS]

- Alcohol is both a food and a drug and can have positive or negative effects on health.

- Alcohol is produced from carbohydrates.

- Alcohol abuse is harmful to the body and is associated with a high proportion of acts of violence and accidents.

- Both genetic and environmental factors are associated with the development of alcoholism.

Answers to *Nutrition Scoreboard*

		TRUE	FALSE
1	Alcohol (actually ethanol) is produced by the fermentation of carbohydrates in grains, fruits, and other foods.	✔	
2	High intakes of alcohol are harmful to the body, regardless of the quality of the diet.		✔
3	The statistics on alcohol abuse, injury, and death are startling. Alcohol abuse is a major personal, social, and public health problem in the United States.	✔	

Alcohol Facts

Alcohol is both a food and a drug. It's a food because alcohol is made from carbohydrates and the body uses it as an energy source. Alcohol is a drug because it modifies various body functions.

fermentation
The process by which carbohydrates are converted to ethanol by the action of the enzymes in yeast.

The type of alcohol people consume in beverages is ethanol. (We refer to ethanol by the broader term *alcohol* in this unit.) Alcohol is produced from carbohydrates in grains, fruits, and other foods by the process of *fermentation.* Wines, brews made from grains, and other alcohol-containing beverages are a traditional part of the food supply of many cultural groups. In high doses, however, alcohol is harmful to the body and can cause a wide variety of nutritional, social, and physical health problems.

The Positive

Whether alcohol has harmful effects on health depends on how much is consumed. The consumption of moderate amounts of alcohol by healthy adults who are not pregnant appears to cause no harm.[1] In fact, moderate alcohol consumption is associated with a significant level of protection against heart disease.[2] A moderate level of alcohol consumption is considered to be one standard-sized drink per day for women and two drinks for men (Illustration 14.1). Alcohol increases the body's production of high-density lipoprotein cholesterol (HDL).[3] HDL is known as the "good cholesterol" because it helps eliminate cholesterol from the body, and that reduces plaque build-up in arteries.

All types of alcohol-containing beverages reduce the risk of heart disease. But is there something special about red wine that reduces the risk of heart disease more than other alcohol-containing beverages do? The answer may be yes for women.[4] Red wine includes pigments that make grapes deep red, blue, or purple. A number of these pigments act as antioxidants and may decrease the tendency of blood to clot (which may happen during a heart attack) and may decrease the ability of LDL cholesterol to stick to plaque in the arteries. Purple grape juice contains these antioxidants, too, and may also help prevent heart disease.[5]

People don't have to consume alcohol to reduce their risk of heart disease. Diets low in saturated fat, liberal intakes of vegetables and fruits, ample physical activity, and not smoking also reduce the risk of heart disease.[6]

The Negative

Heavy drinking, often defined as the consumption of five or more drinks per day, poses a number of threats to the health of individual drinkers and often to other people as well. Although health can be damaged by the regular consumption of high amounts of alcohol, the ill effects of alcohol are most obvious in people with *alcoholism.*

Habitually high alcohol intakes and alcoholism increase the risk of developing high blood pressure, stroke, and cirrhosis of the liver; throat, stomach, and bladder cancer; central nervous system disorders; and vitamin and mineral deficiency diseases. Alcohol abuse is associated with a high proportion of deaths from homicide, drowning, fires, traffic accidents, and suicide (Illustration 14.2). It is also involved in a large proportion of rapes and assaults. Alcohol poisoning from the consumption of a large amount of alcohol in a short period of time can cause death—and does to a number of college students each year.[7]

Drinking during pregnancy may harm the fetus. Women who binge drink or drink regularly during pregnancy are at risk of delivering an infant with signs of fetal alcohol syndrome (Illustration 14.3).[8] Children with fetal alcohol syndrome experience long-term growth and mental retardation. The severity of the condition depends on how much alcohol was consumed during pregnancy and whether excessive intake occurred early or late in pregnancy.[9] Although small amounts of alcohol do not appear to be hazardous, it is recommended that women who are, or may become, pregnant not drink.[10]

Illustration 14.1
Standard serving sizes of alcohol-containing beverages.
Serving sizes shown contain 13 to 16 grams of alcohol.

alcoholism
An illness characterized by a dependence on alcohol and by a level of alcohol intake that interferes with health, family and social relations, and job performance.

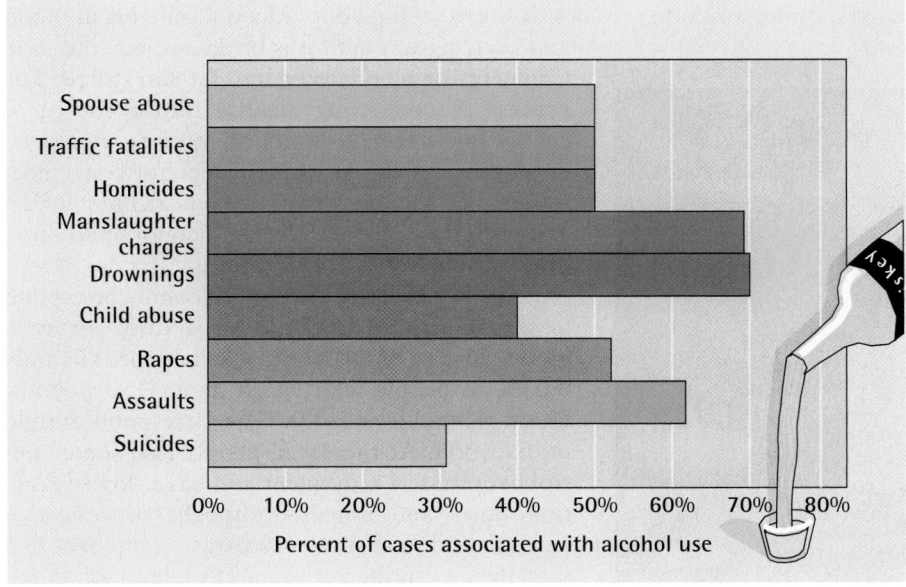

Illustration 14.2
Violence and injuries associated with alcohol.
Source: National Institute on Alcohol Abuse and Alcoholism, 2001.

Illustration 14.3

Children with fetal alcohol syndrome experience growth and mental retardation, in addition to specific facial characteristics.

Alcohol-containing beverages must show a warning state-ment on labels.

1.5L
CONTAINS SULFITES

GOVERNMENT WARNING:
(1) ACCORDING TO THE SURGEON GENERAL, WOMEN SHOULD NOT DRINK ALCOHOLIC BEVERAGES DURING PREGNANCY BECAUSE OF THE RISK OF BIRTH DEFECTS. (2) CONSUMPTION OF ALCOHOLIC BEVERAGES IMPAIRS YOUR ABILITY TO DRIVE A CAR OR OPERATE MACHINERY, AND MAY CAUSE HEALTH PROBLEMS. **0**

Alcohol Intake and Diet Quality

Alcohol provides 7 calories per gram, making alcohol-containing beverages rather high in caloric content (Illustration 14.4). Because many alcohol-containing beverages provide calories and few or no nutrients, they are considered "empty-calorie" foods. On average, alcohol accounts for 3 to 9% of the caloric intake of U.S. adults who drink. The average goes up to around 50% among heavy drinkers.[11] Although beer, wine, and mixed drinks are known to contain alcohol and to provide calories, there exists some confusion about whether calories from alcohol contribute to weight gain. This issue is addressed in the "Reality Check."

As caloric consumption from alcohol-containing beverages increases, the quality of the diet generally decreases. Diets of heavy drinkers frequently provide too little thiamin, vitamins A and C, calcium, and iron.[13] Deficiencies of nutrients, as well as direct, toxic effects of high levels of alcohol ingestion, produce most of the physical health problems associated with alcoholism. The lack of thiamin, for example, impairs the brain's utilization of glucose. When people with alcoholism initially withdraw from alcohol, the thiamin deficiency may result in "delirium tremens," a condition called the "DTs" by people who staff detoxification centers. People with delirium tremens experience convulsions and hallucinations and are severely confused. Thiamin injections are a key component of treatment for delirium tremens.[14] Because alcohol in excess is directly toxic to body tissues, consuming an adequate diet won't protect heavy drinkers from all of the harmful effects of alcohol.[15]

How the Body Handles Alcohol

Alcohol is easily and rapidly absorbed in the stomach and small intestine. Within minutes after it is consumed, alcohol enters the circulatory system and is on its way to the liver, brain, and other tissues throughout the body. Alcohol remains in blood and body tissues until it is broken down and used for energy or is converted into fat and stored. The process of converting alcohol into a source of energy takes several hours or more to complete, depending on the amount of alcohol consumed. Because of the lag time between alcohol intake and utilization, blood levels of alcohol build up as drinking continues (Table 14.1).

The intoxicating effects of alcohol correspond to blood alcohol levels. After a drink or two, blood levels of alcohol reach approximately 0.03% in people who weigh about 150 pounds. Blood alcohol levels of 0.03% correspond to mild intoxication. At this level, people lose some control over muscle movement and have slowed reaction times and impaired thought processes. A person's ability to drive or operate equipment in a safe manner is decreased at this level of blood

Illustration 14.4

Caloric value of common alcohol-containing beverages.

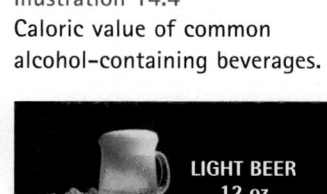

LIGHT BEER
12 oz.
95 calories

BEER
12 oz.
150 calories

MALT BEVERAGE
12 oz.
180 calories

WINE
5 oz.
110 calories

WINE COOLER
12 oz.
170 calories

80-PROOF LIQUOR
11/2 oz.
105 calories
(without mix)

Pedro:
I started drinking a beer at night over the summer, and my weight never changed.

REALITY CHECK

Do alcohol calories count?

Perhaps you've heard the popular scientific opinion that alcohol intake does not increase the risk of obesity. Is that the same as the not-so-popular opinion of scientists?

Do your thoughts side with Pedro or Erik

?

Erik:
The six-pack around my abdomen is really a six-pack.

Answers on next page

alcohol content (Illustration 14.5). Blood alcohol levels of around 0.06% are associated with an increased involvement in traffic accidents. The legal limit for intoxication according to most states' highway safety ordinances is 0.08 to 0.1%—beyond the point where driving is impaired. When blood alcohol content increases to 0.13%, speech becomes slurred, "double vision" occurs, reflexes are dulled, and body movements become unsteady.[16]

If blood alcohol level continues to increase, drowsiness occurs and people may lose consciousness. Levels of blood alcohol above 0.6% can cause death, especially in individuals who have not developed a tolerance for alcohol.

Over 150 medications, including sleeping pills, antidepressants, and painkillers, interact harmfully with alcohol.[17] Combining three or more drinks per day with aspirin or nonaspirin pain relievers (acetaminophen, ibuprofen) may cause stomach ulcers or liver damage.[18]

A given amount of alcohol intake among women produces higher blood levels of alcohol than for men of the same body weight. Pound for pound, women's bodies contain less water than men's bodies, so blood alcohol levels in women increase

Illustration 14.5
The legal limit for intoxication in most states' highway safety ordinances is 0.08 to 0.1%—beyond the point where driving is impaired.

TABLE 14.1

ALCOHOL DOSES AND BLOOD LEVELS.

NUMBER OF DRINKS*	PERCENT BLOOD ALCOHOL BY BODY WEIGHT				
	100 LB	120 LB	150 LB	180 LB	200 LB
2	0.08	0.06	0.05	0.04	0.04
4	0.15	0.13	0.10	0.08	0.08
6	0.23	0.19	0.15	0.13	0.11
8	0.30	0.25	0.20	0.17	0.15
12	0.45	0.36	0.30	0.25	0.23
14	0.52	0.42	0.35	0.34	0.27

*Taken within an hour or so; each drink equal to ½ ounce pure ethanol.
Source: Sizer F, Whitney EN. Nutrition: current concepts and controversies, eighth edition. Belmont, CA: Wadsworth; 2000, p. 172.

Pedro

Erik

ANSWERS TO **REALITY CHECK**

Do alcohol calories count?

Maybe you've never seen an obese person with alcoholism, so you're tempted to think alcohol calories don't count. Chronic alcohol abuse is associated with weight loss and muscle wasting, even though the calorie intake of heavy drinkers is high. The effect appears to be due to an inhibition of fat tissue accumulation. The calories do count for light and moderate drinkers, however.[12]

faster than in men.[19] Consequently, women may experience the intoxicating effects of alcohol on lower amounts of alcohol than men.

Acute Alcohol Poisoning

Very high blood levels of alcohol can be extremely dangerous. Such levels can be reached by drinking large amounts of alcohol (usually in the form of liquor such as rum or gin) in a short time. People who have overdosed on alcohol become unconscious; their pulse is rapid, their blood pressure is low, and their pupils remain dilated.[20] They require emergency medical care.

How to Drink Safely if You Drink] Many of the problems related to alcohol intake can be prevented by not drinking, or drinking responsibly. That means:

- Not drinking if you are or could become pregnant
- Not drinking on an empty stomach (which can make you intoxicated surprisingly fast)
- Slowly sipping rather than gulping drinks
- Limiting alcohol to an amount that doesn't make you lose control over your mind and body
  - Never driving a car or boat, hunting, or operating heavy equipment while under the influence of alcohol.

What Causes Alcoholism?

One in 13 adults in the United States abuse alcohol or has alcoholism.[21] Alcoholism tends to run in families, so there is a genetic component to the disease. Its development is also influenced by environmental factors. In general, the younger individuals are when they begin to drink, the greater likelihood that they will develop a drinking problem at some point in life. Individuals who begin drinking before the age of 15, for example, are four times more likely to become alcohol dependent than are people who do not drink before age 21 (Illustration 14.6). Close association with friends or peers who drink, high levels of stress, and availability of alcohol may also increase the risk of alcoholism.[22] Television ads depicting youth-oriented parties, fun, and beer may increase underage drinking.[23]

Illustration 14.6
The younger a person is when drinking begins, the higher the probability that a drinking problem will develop.

Nutrition UP CLOSE

Effects of Alcohol Intake

FOCAL POINT: Estimating blood alcohol levels and side effects.

Scenario: Ligia and Mark attend a wedding reception. Prior to the meal, they both drink a glass of champagne to toast the bride. Fiften minutes later, they drink another glass to toast the groom. Ligia weighs 150 pounds and Mark, 180.

Questions: Using the information in Table 14.1 and the information on "How the Body Handles Alcohol" (p. 14–4), answer the following questions:

A. After two glasses of champagne, what would be Ligia's estimated blood alcohol level?
_____ % blood alcohol. What would be Mark's? _____ % blood alcohol

B. List three side effects of these blood alcohol levels: 1)   2)   3)

FEEDBACK (answers to these questions) can be found at the end of Unit 14.

Alcohol Use Among Adolescents

Alcohol use among adolescents is increasing, and the age when teens begin drinking is going down. Underage drinking accounts for 20% of all the alcohol consumed in the United States. The average age when teens begin drinking is now 14 years. These trends are particularly disturbing because they may lead to higher rates of alcoholism and alcohol-related problems in the near future.[24]

Reduction in alcohol intake by adolescents is a major public health initiative of the Year 2010 Health Objectives for the Nation.[25]

Much remains to be learned about the prevention of alcohol abuse and alcoholism. Until then treatment programs, such as Alcoholics Anonymous and educational programs that stress safe drinking, will be relied upon to lessen the impact of alcohol abuse on personal and public health.

Key Terms

alcoholism, page 14-3

fermentation, page 14-2

www links

www.niaaa.nih.gov
The National Institute on Alcohol Abuse and Alcoholism offers this site for exploration of alcohol and health issues, quick facts, college drinking prevention programs, current research projects, and answers to common questions on alcohol abuse.

www.MayoClinic.org
Search "alcohol" and get the latest information on alcoholism, treatment programs,

taking control, pros and cons of alcohol use, and an alcohol quiz.

www.alcoholicsanonymous.net
Information about services for alcoholism available from Alcoholics Anonymous World Service is available at this address.

www.al-anon.alateen.org
Al-Anon/Alateen alcohol treatment services can be found at this address.

www.intelihealth.com
This site offers peer-reviewed information on topics including alcoholism, cirrhosis, alcohol and health, alcohol use in pregnancy, alcohol and heart disease, and alcohol use and aggression.

www.nlm.nih.gov/medlineplus
Search the word "alcohol" and be greeted by a large selection of topics related to alcohol: women and alcohol, fetal alcohol syndrome, Alcoholics Anonymous, and alcohol and youth.

Notes

1. Alcoholism: getting the facts. National Institute on Alcohol Abuse and Alcoholism. www.niaa.nih.gov, accessed 6/03.

2. Di Castelnuovo A et al. Meta-analysis of wine and beer consumption in relation to vascular risk. Circulation 2002;105: 2836–44.

3. Silva ERDE et al. Alcohol consumption raises HDL cholesterol levels by increasing the transport rate of apolipoproteins A-I and A-II. Circulation 2000;102: 2347–52.

4. Di Castelnuovo A et al. Alcohol and coronary heart disease (letter). N Engl J Med 2003;348:1720–1.

5. Freedman J et al. Select flavonoids and whole juice from purple grapes inhibit platelet function and enhance nitric oxide release. Clin Invest Reports, June 2001.

6. Goldberg IJ. To drink or not to drink? N Engl J Med 2003;348:163–64.

7. Alcohol Alert. National Institute on Alcohol Abuse and Alcoholism. www.niaaa.nih.gov, accessed 6/03.

8. Spohr H-L, Willms J, Steinhausen H-C, Prenatal alcohol exposure and long-term developmental consequences, Lancet 1993;241:907–10; and Mattson SN, et al., Heavy prenatal alcohol exposure with or without physical features of fetal alcohol syndrome leads to IQ deficits. J Pediatrics 1997;131:718–21.

9. Committee on Substance Abuse and Committee on Children with Disabilities. Fetal alcohol syndrome and alcohol-related neurodevelopmental disorders. Pediatrics 2000;106:358–61.

10. Knupfer G, Abstaining for fetal health: the fiction that even light drinking is dangerous, British J Addiction 1991;86:1063–73; and Alcohol Alert, NIAAA (www.niaaa.nih.gov).

11. Wannamethee SG and Shaper AG, Alcohol, body weight, and weight gain in middle-aged men, Am J Clin Nutr 2003;77:1312–7; and Kesse E et al., Do eating habits differ according to alcohol consumption? Am J Clin Nutr 2001;74: 322–7.

12. Levine JA, et al. Energy expenditure in chronic alcohol abuse. Euro J Clin Invest 2000;30:779–86.

13. Kesse et al., Do eating habits differ? and Levine JA, et al. Energy expenditure in chronic alcohol abuse. Euro J Clin Invest 2000;30:779–86.

14. Lieber CS. The influence of alcohol on nutritional status. Nutr Rev 1988;46: 241–54.

15. Klatsky AL. Diet, alcohol, and health: a story of connections, confounders, and cofactors. Am J Clin Nutr 2001;74: 279–80.

16. Charness ME, Simon RP, Greenberg DA. Ethanol and the nervous system. N Engl J Med 1989;321:442–53.

17. Alcohol and your health: the pros and cons. www.MayoClinic.org/alcohol, accessed 6/03.

18. FAQs on alcohol abuse and alcoholism. National Institution on Alcohol Abuse and Alcoholism. www.niaaa.nih.gov, accessed 6/03.

19. Alcohol Alert, NIAAA (www.niaaa.nih.gov).

20. Charness et al., Ethanol and the nervous system.

21. FAQs on alcohol abuse, NIAAA (www.niaaa.nih.gov); Grant BR, Dawson DA. Age of onset of alcohol use and its association with DSM-IV alcohol abuse and dependence: results from the National Longitudinal Alcohol Epidemiologic Survey. J Subst Abuse 1998;9: 103–10.

22. Alcohol consumption and expenditures for underage drinking and adult excessive drinking. Alcoholism & Drug Abuse Weekly 2003;15(9):3–4.

23. Steinberg B and Vranica S. Brewers are urged to tone down party. Wall Street Journal, 6/23/03, p. B4.

24. Alcohol consumption and expenditures.

25. Healthy People 2010 Objectives for the Nation. web/health.gov/healthy/people, 2001 Jan.

Nutrition UP CLOSE

Effects of Alcohol Intake

Feedback for Unit 14

A. Ligia's estimated % blood alcohol = 0.05%.

Mark's estimated % blood alcohol = 0.04%.

B. Three side effects of these blood alcohol levels:

1. Loss of some control over muscle movements

2. Slowed reaction time

3. Impaired thought processes

Proteins and Amino Acids

Nutrition Scoreboard

	TRUE	FALSE
1 The primary function of protein is to provide energy.		
2 "Nonessential amino acids" are not required for normal body processes. Only "essential amino acids" are.		
3 High-protein diets enhance muscle development in individuals who work out a lot.		

Answers on next page

- Proteins are made of amino acids. Some amino acids are "essential" (required in the diet), and some are "nonessential" (not a required part of diets).

- Although protein can be used for energy, its major functions in the body involve the construction, maintenance, and repair of protein tissues.

- Protein tissue construction in the body proceeds only when all nine essential amino acids are available.

- Appropriate combinations of plant foods can supply sufficient quantities of all the essential amino acids.

Answers to *Nutrition Scoreboard*

		TRUE	FALSE
1	Energy is a function of protein, but it's not the primary one.		✔
2	"Nonessential amino acids" are required by the body, but they are not required components of our diet. (Yes, it is confusing.)	✔	
3	Muscles contain protein, but you can't increase muscle mass by consuming a high-protein diet—whether you're a couch potato or a nationally ranked athlete.		✔

Protein's Image versus Reality

The term ***protein*** is derived from the Greek word *protos,* meaning "first." The derivation indicates the importance ascribed to this substance when it was first recognized. An essential structural component of all living matter, protein is involved in almost every biological process in the human body. Protein has a very positive image (Illustration 15.1). It's so positive that you don't have to talk about the importance of protein—people are already convinced of it.

Rich or poor, nearly all people in the United States get enough protein in their diets. Actually, most people consume more protein than they need. Average intakes of protein exceed the Recommended Dietary Allowance (RDA) level for all age and sex groups. Approximately 15% of total calories in the average U.S. adult diet are supplied by protein.

High-protein intakes are generally accompanied by high-fat and low-fiber intakes. That's because foods high in protein such as hamburger, cheese, nuts, and eggs are high in fat and contain little or no fiber. Even lean meats provide a considerable proportion of their total calories as fat (Illustration 15.2).

Functions of Protein

Proteins perform four major functions in the body (Table 15.1). They are an integral structural component of skeletal muscle, bone, connective tissues (skin, collagen, and cartilage), organs (such as the heart, liver, and kidneys), red blood cells and hemoglobin, hair, and fingernails. Proteins are the basic substance that make up digestive enzymes and the thousands of other enzymes in the human body, and they are a major component of hormones such as insulin. All protein-containing structures and tissues in the body require ongoing maintenance in the form of renewal of their content of protein. Tissue maintenance and the repair of organs and tissues damaged due to illness or injury are functions of different types of protein. Finally, protein serves as an energy source.[1]

The body of a 154-pound man contains approximately 24 pounds of protein. Nearly half of the protein is found in muscle, while the rest is present in the skin,

protein

Chemical substance in foods made up of chains of amino acids.

Illustration 15.1
The protein perception.

Protein

Other nutrients

(a) Hamburger (90% lean): 45%   (b) Tenderloin: 43%   (c) Top loin: 41%   (d) Round tip: 34%

(e) Sirloin: 33%   (f) Eye of round: 26%   (g) Top round: 25%   (h) Pork chop, lean: 48%

(i) Pork loin roast: 36%   (j) Pork tenderloin: 28%   (k) Chicken thigh, no skin: 47%   (l) Baked chicken breast, no skin: 19%

Illustration 15.2

The fat content of 3-ounce portions of "lean" meats. *The percentage of calories from fat is indicated for each portion. (A 3-ounce portion of meat is about the size of a deck of cards.) Each portion of meat provides approximately 21 grams of protein.*

collagen, blood, enzymes, and *antibodies;* organs such as the heart, liver, and intestines; and other body parts. All protein in the body is continually being turned over, or broken down and rebuilt. This process helps maintain protein tissues in optimal condition so they continue to function normally. The process of protein turnover utilizes roughly 9 ounces of protein each day. Yet, we consume only 2–3 ounces of protein daily. Most of the protein used for maintenance is recycled from protein tissues and substances being turned over. Proteins play key roles in the repair of body tissues by serving as substances such as fibrin that helps blood clot (Illustration 15.3) and by replacing tissue proteins damaged by illness or injury.[2]

antibodies
Blood proteins that help the body fight particular diseases. They help the body develop an immunity, or resistance, to many diseases.

TABLE 15.1

FUNCTIONS OF PROTEIN.

1. Serves as a structural material in muscles, connective tissue, organs, and hemoglobin

2. Serves as the basic component of enzymes and hormones

3. Maintains and repairs protein-containing tissues

4. Serves as an energy source

Illustration 15.3
Red blood cells enmeshed in fibrin in a color-enhanced microphotograph.
Red blood cells and fibrin (which helps stop bleeding by causing blood to clot) are made primarily from protein.

"FORGET ENLIGHTENMENT. I WANT YOU TO CONCENTRATE ON THE STRUCTURE OF THE PROTEIN MOLECULE."

Illustration 15.4
Source: © 2001 by Sidney Harris.

DNA (deoxyribonucleic acid)
Genetic material contained in cells that initiates and directs the production of proteins in the body.

essential amino acids
Amino acids that cannot be synthesized in adequate amounts by humans and therefore must be obtained from the diet. They are some-times referred to as "indispens-able amino acids."

nonessential amino acids
Amino acids that can be read-ily produced by humans from components of the diet. Also referred to as "dispensable amino acids."

Protein serves as a source of energy in healthy people, but not nearly to the extent that carbohydrates and fats do. Protein is unlike carbohydrate and fat in that it contains nitrogen and does not have a storage form in the body. In order to use protein for energy, amino acids that make up proteins must first be stripped of their nitrogen. The free nitrogen can be used as a component of protein formation within the body; or, if present in excess, it is excreted in urine. Excre-tion of nitrogen requires water, so high intake of protein increases water need. Amino acids missing their nitrogen component are con-verted to glucose or fat that then can be used to form energy. A small amount of protein (1%) can be obtained from the liver and blood and used to cover occasional deficits in protein intake.[3]

Amino Acids

The "building blocks" of protein are amino acids (Illustration 15.4). Protein consumed in food is broken down by digestive enzymes and absorbed into the bloodstream as amino acids. There are 20 common amino acids (Table 15.2) that form proteins when linked together. Every protein in the body is composed of a unique combination of amino acids linked together in chains (Illustration 15.5). The organization of amino acids into the chains is orchestrated by *DNA,* the genetic mate-rial within each cell that directs protein synthesis. Once formed, the chains of amino acids may fold up into a complex shape. Some proteins are made of only a few amino acids, while other proteins contain hundreds. Whatever the number of amino acids, the specific amino acids involved and their arrangement determine whether the pro-tein is an enzyme, a component of red blood cells, a muscle fiber, or another tissue made from protein. Nine of the 20 common amino acids are considered *essential,* and 11 are *nonessential.* Despite the labels, all 20 amino acids are required to build and maintain protein tissues. The essential amino acids are called "essential" because the body cannot produce them, or produce enough of them, so they must be provided by the diet. Proteins in foods contain both essential and nonessential amino acids.

Proteins Differ in Quality

The ability of proteins to support protein tissue construction in the body varies depending on their content of essential amino acids. How well dietary proteins sup-port protein tissue construction is captured by tests of the protein's "quality."

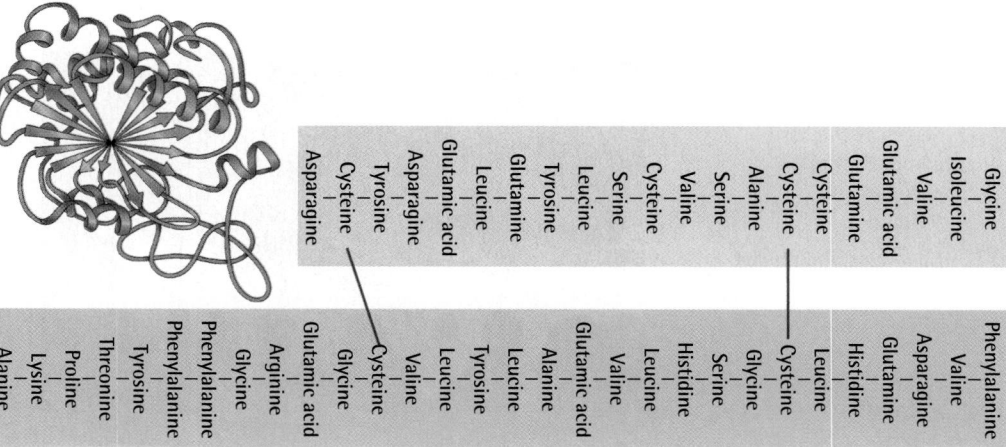

Illustration 15.5
Amino acid chains in the protein insulin (shown at right) and the structure of insulin (shown above).

TABLE 15.2

ESSENTIAL AND NONESSENTIAL AMINO ACIDS.

ESSENTIAL		NONESSENTIAL	
Histidine	Tryptophan	Alanine	Glutamine
Isoleucine	Valine	Arginine	Glycine
Leucine		Asparagine	Proline
Lysine		Aspartic acid	Serine
Methionine		Cysteine	Tyrosine
Phenylalanine		Glutamic acid	
Threonine			

Illustration 15.6
Animal sources of protein supply "complete proteins." *Each food shown is a source of complete protein.*

Proteins of high quality contain all the essential amino acids in the amounts needed to support protein tissue formation by the body. If any of the essential amino acids are missing in the diet, proteins are not formed—even those proteins that could be produced from available amino acids. Shutting off all protein formation for want of an amino acid or two may appear inefficient; but if the body did not cease all protein formation, cells would end up with an imbalanced assortment of proteins. This would seriously affect cell functions. When the required level of an essential amino acid is lacking, the remaining amino acids are primarily used for energy.

Amino acids cannot be stored very long in the body, so we need a fresh supply of essential amino acids daily. This means we need to consume foods that provide a sufficient amount of all essential amino acids every day.

complete proteins
Proteins that contain all of the essential amino acids in amounts needed to support growth and tissue maintenance.

incomplete proteins
Proteins that are deficient in one or more essential amino acids.

Complete Proteins] Food sources of high-quality protein (meaning they contain all the essential amino acids in the amount needed to support protein formation) are called *complete proteins*. Proteins in this category include those found in animal products such as meat, milk, and eggs (Illustration 15.6). *Incomplete proteins* are deficient in one or more essential amino acids. Proteins in plants are "incomplete," although soybeans are considered a complete source of protein for adults.[4] (Soybeans may not meet the essential amino acid requirements of young infants.) You can "complement" the essential amino acid composition of plant sources of protein by combining them to form a "complete" source of protein. Illustration 15.7 shows a few plant combinations that produce complete proteins.

Vegetarian Diets] Diets consisting only of plant foods can provide an adequate amount of complete proteins. The key to success is eating a variety of complementary sources of protein each day. Vegetarian diets have been practiced for centuries by some religious and cultural groups, bearing testimony to their general adequacy and safety. (Unit 16 on vegetarian diets expands on this topic.)

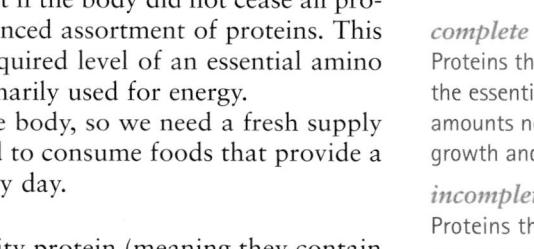

Illustration 15.7
Each of these combinations of plant foods is also a source of complete protein. *Shown are mixed vegetables, tofu, and brown rice; succotash (lima beans and corn), and rice and black beans.*

Illustration 15.8
A wide array of amino acid supplements are available over the counter, but the safety of these supplements is unknown.

Amino Acid Supplements

Because amino acids occur naturally in foods, people often assume they are harmless, no matter how much is taken (Illustration 15.8). Researchers have known for decades, however, that high intakes of individual amino acids can harm health. High amounts may disrupt normal protein production by overwhelming cells with a surplus of some amino acids and a relative deficit of others. Amino acid supplements have been known to cause loss of appetite, diarrhea, and other gastrointestinal upsets.[5] They increase the workload of the liver and kidneys and the likelihood of dehydration.[6] The safety of amino acid supplements began to gain the public's attention in 1989 when something went wrong.

The Tryptophan Supplement Scare]   Nature's original sleeping pill? Safe as a cup of warm milk? Supplements of the amino acid tryptophan (pronounced "trip-toe-fan") turned into heartbreaking stories of a serious illness that terrified many people.

Late in the summer of 1989, people from coast to coast developed a mysterious illness characterized by painful muscles and joints, weakness, fever, cough, a rash, and swelling. Some people who had the disease described the muscle pain as "like having a charley horse" in every muscle. Diagnosing the cause of the symptoms turned out to be difficult. Within a few months, however, the disease was diagnosed as eosinophilia-myalgia syndrome (EMS). It has no cure. The outbreak was quickly related to tryptophan supplement use when researchers confirmed that 98% of the patients with EMS used tryptophan.[7] In March 1990, tryptophan was taken off the market, and people were urged to stop taking the supplements. By late 1990, over 1500 cases of EMS had been reported and at least 37 people died from the disease. Over 15 years later, many people who developed EMS from the supplement still have the disease. Contaminants in certain batches of the supplement and tryptophan itself have been related to the development of EMS.[8] Although history now, the tryptophan scare taught us an important lesson about the potential hazards of amino acid supplements.

Illustration 15.9
Tryptophan supplements were banned in the United States on April 23, 1990, a month after their recall.
A close chemical relative, melatonin, is available. Other derivatives of tryptophan are also available, and safety is a concern.

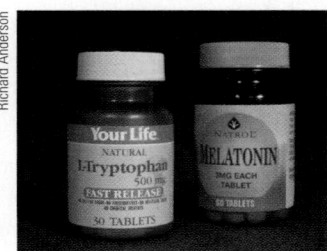

Melatonin: The Tryptophan Replacement?]   A derivative of tryptophan, the hormone melatonin (Illustration 15.9), entered the marketplace shortly after tryptophan supplements were banned. In addition to promoting drowsiness and sleep, melatonin has been touted as a cure for jet lag and a way to help your body switch to a night shift. Studies have produced conflicting results on the ability of melatonin to facilitate sleep in young adults, but doses of about 1 milligram per day may improve sleep in older adults lacking normal levels of brain melatonin. It does not appear to be effective against jet lag or helpful in the switch to night shifts.[9]

Low doses of melatonin (1 milligram per day) do not appear to be harmful, but the safety of melatonin supplements is still unclear. As was the case for tryptophan, there is little oversight of the purity or dose levels of melatonin in supplements.

Can Amino Acid Supplements and Protein Powders Build Muscle?]   You can't just consume amino acids or protein powders and watch your muscles grow (no matter how convincing the ads that sell such products are).[10] If that happened, everyone who wanted a rippled stomach and bulging triceps could have them. No amino acids or protein powders are delivered directly to our muscles and cause muscles to enlarge. Only exercise and a good diet build muscles (Illustration 15.10). Most athletes get more than enough protein from their usual diet to support their need for it—which is not extraordinarily high.[11]

Food as a Source of Protein

The average intake of protein in the United States is 75 grams per day, exceeding the RDA for men of 56 grams and that for women of 46 grams.[12] Approximately 70% of the protein consumed by Americans comes from meats, milk, and other animal products.[8] Dried beans and grains are not as well known for their protein, but are nevertheless good sources (Table 15.3). Plant sources of protein are generally low in fat, making them a wise choice for consumers who are trying to limit their intake of

Illustration 15.10
No amount of protein powders or amino acid supplements will build muscles like these. *Exercise and a good diet are needed to build muscles.*

TABLE 15.3

FOOD SOURCES OF PROTEIN.
The adult RDA is 46 grams for women aged 19 to 24 years and 58 grams for men aged 19 to 24 years.

FOOD	AMOUNT	PROTEIN CONTENT GRAMS	PERCENTAGE OF TOTAL CALORIES
Animal products			
Tuna (water packed)	3 oz	24	89%
Shrimp	3 oz	11	84
Cottage cheese (low-fat)	½ cup	14	69
Beef steak (lean)	3 oz	26	60
Chicken (no skin)	3 oz	24	60
Pork chop (lean)	3 oz	20	59
Beef roast (lean)	3 oz	23	45
Skim milk	1 cup	9	40
Fish (haddock)	3 oz	19	38
Leg of lamb	3 oz	22	37
Yogurt (low-fat)	1 cup	13	34
Hamburger (lean)	3 oz	24	34
Egg	1 medium	6	32
Swiss cheese	1 oz	8	30
Sausage (pork links)	3 oz	17	28
2% milk	1 cup	8	26
Cheddar cheese	1 oz	7	25
Whole milk	1 cup	8	23
Dried beans and nuts			
Tofu	½ cup	14	38
Soybeans (cooked)	½ cup	10	33
Split peas (cooked)	½ cup	5	31
Lima beans (cooked)	½ cup	6	27
Dried beans (cooked)	½ cup	8	26
Peanuts	¼ cup	9	17
Peanut butter	1 tbs	4	17
Grains			
Corn	½ cup	3	29
Egg noodles (cooked)	½ cup	4	25
Oatmeal (cooked)	½ cup	3	15
Whole-wheat bread	1 slice	2	15
Macaroni (cooked)	½ cup	3	13
White bread	1 slice	2	13
White rice (cooked)	½ cup	2	11
Brown rice (cooked)	½ cup	2	10

TABLE 15.4

IRON CONTENT IN A 3-OUNCE SERVING OF VARIOUS MEATS.
The RDA for women aged 19 to 24 years is 15 milligrams. The RDA for men aged 19 to 24 years is 10 milligrams.

MEAT	IRON CONTENT (MG)
Pork chop (lean)	3.4
Round steak (lean)	3.1
Hamburger (lean)	3.0
Shrimp	2.6
Tuna	1.6
Baked chicken (no skin)	1.4
Lamb (lean)	1.3

kwashiorkor (kwa-she-or-kor)
A deficiency disease primarily caused by a lack of complete protein in the diet. It usually occurs after children are taken off breast milk and given solid foods containing low-quality protein.

edema
Swelling due to an accumulation of fluid in body tissues.

marasmus
A condition of severe body wasting due to deficiencies of both protein and calories. Also called *protein-energy malnutrition* and *protein-calorie malnutrition*.

fat. Nearly all food sources of protein provide an assortment of vitamins and minerals as well. Beef and pork are particularly good sources of iron, a mineral often lacking in the diets of women (Table 15.4).

What Happens When a Diet Contains Too Little Protein?

Protein deficiency can occur by itself or in combination with a deficiency of calories and nutrients. Because food sources of protein generally contain essential nutrients such as iron, zinc, vitamin B_{12}, and niacin, diets that produce protein deficiency usually cause a variety of other deficiencies, too. Protein does not generally serve as an important source of energy, but body protein will be used as a major energy source during starvation. To meet the need for energy, the body will extract protein from the liver, intestines, heart, muscles, and other organs and tissues. Loss of more than about 30% of body protein results in reduced body strength for breathing, susceptibility to infection, abnormal organ functions, and death.[13]

Kwashiorkor is a severe form of protein deficiency in children. It usually develops after a child has been weaned from breast milk and given high-carbohydrate, low-protein foods such as cassava or a watery gruel made from oats, rice, or corn. A protein-deficient body cannot grow, maintain a normal level of blood volume, produce digestive enzymes, fight infections adequately, or maintain existing tissues in proper working order. As seen in Illustration 15.11, children with kwashiorkor may look fat due to *edema* (swelling), but they are actually very skinny. Children with protein deficiency are apathetic, irritable, small, and highly vulnerable to infection.

Protein deficiency may be accompanied by a lack of calories. In this case, a condition called *marasmus,* or protein-energy malnutrition, exists. Unfortunately, this condition is all too frequent in areas of the

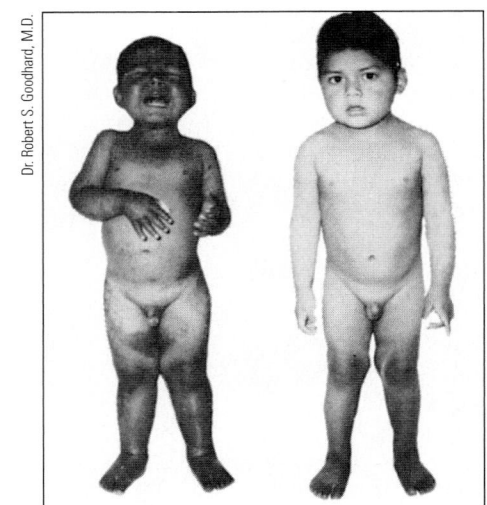

Illustration 15.11
Children suffering from a severe form of protein deficiency called kwashiorkor experience swelling in the arms, legs, and stomach area. *The swelling hides the devastating wasting that is taking place within their bodies. The child at left has the characteristic "moon face" (edema), swollen belly, and patchy dermatitis (from zinc deficiency) often seen with kwashiorkor. At right, the same child after nutritional therapy.*

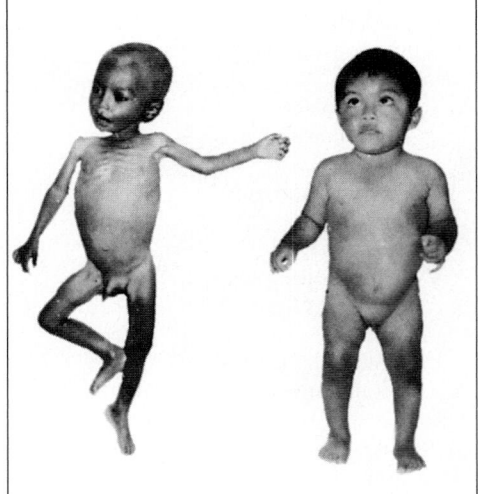

Illustration 15.12
Individuals with marasmus look as starved as they are. *People with this condition lack both protein and calories. The child at left is suffering from the extreme emaciation of marasmus. At right is the same child after nutritional therapy.*

Photo Disc

REALITY CHECK

Pure protein

You've got a nutrition exam coming up, so you and your classmate Scenario have gotten together to study. You get into a discussion about food sources of protein that goes like this:

Who gets the "thumbs up"

?

Scenario:
Lean meats are the best source. They're pure protein!

Pickles:
Pure protein? Even the driest, toughest meats contain more than protein.

Answers on next page

world where hunger and famine are common. It is also the type of malnutrition that occurs among people with severe anorexia nervosa and in certain diseases such as cancer and AIDS. Individuals with marasmus look starved and wasted (see Illustration 15.12). They have precious little body fat and must utilize protein from muscles, the liver, and other tissues as an energy source.

In addition to having a skin-and-bones appearance, children with marasmus are apathetic, highly susceptible to infection, and have dry skin and brittle hair. Use of the fat pads in the cheeks for energy (which are among the last fat deposits to be used for energy during starvation) gives children with advanced cases of marasmus the look of a very old person. Both kwashiorkor and marasmus are generally accompanied by multiple vitamin and mineral deficiencies due to limited food intake. The conditions are frequently complicated by infection, diarrhea, and dehydration.

How Much Protein Is Too Much?

Adults can consume a substantial amount of protein—approximately 35% of total calories—for months at a time without ill effects. This observation is based on studies of the diets of Eskimos, explorers, trappers, and hunters in northern America. The very high-protein diets would generally contain a good deal of fat in the form of whale blubber, lard, or fat added to dried meat. Consumption of 45% of total calories from protein is considered too high. Consumption of this level of protein is related to nausea, weakness, and diarrhea. Diets very high in protein result in death after several weeks. The disease resulting from excess protein intake was termed "rabbit fever" after it occurred in trappers attempting to exist on wild rabbit only.[14]

High-protein diets have been implicated in the development of weak bones, kidney stones, cancer, heart disease, and obesity. The National Academy of Sciences has concluded that the risk of such disorders does not appear to be increased among individuals consuming 10 to 35% of total calories from protein, and on average adults consume 15%.[15]

A Tolerable Upper Intake Level (UL) for protein has not been established. Because information on the effects of high-protein intakes is limited, people are cautioned not to consume high levels of protein from foods or supplements.

Photo Disc

Scenario

ANSWERS TO REALITY CHECK

Pure protein

Some people think muscle and lean meat consist only of protein. They don't. By weight, lean cooked sirloin steak is 29% protein, 8% fat, and 62% water. Lean pork is 29% protein, 9% fat, and 61% water, for example.

Pickles

Trends

In the early 1900s about half of the protein consumed in the United States came from plant sources and half from animal sources. Now, approximately two-thirds of the protein intake comes from animal products.[16] As countries develop economically, the proportion of dietary protein obtained from meats tends to increase. The increased intake of protein from meats is accompanied by an increased consumption of fat and by elevated rates of some of the "diseases of Western civilization" such as heart disease and certain cancers. Adequate intakes of protein without excess levels of fat can easily be obtained from diets that include dried beans, cereals, and other grains as sources of protein as well as lean meats and low-fat dairy products.

Key Terms

antibodies, page 15-3

complete proteins, page 15-5

DNA (deoxyribonucleic acid),
      page 15-4

edema, page 15-8

essential amino acids, page 15-4

incomplete proteins, page 15-5

kwashiorkor (kwa-she-or-kor),
      page 15-8

marasmus, page 15-8

nonessential amino acids, page 15-4

protein, page 15-2

www links

www.healthfinder.gov
Search topics related to protein.

www.nemsn.org
The National EMS Network provides information on symptoms, current issues, and answers to questions about EMS, tryptophan, and melatonin.

www.iom.edu
Select "Food & Nutrition" to gain access to Dietary Reference Intake report on macronutrients and the chapter on Protein.

www.nal.usda.gov/fnic/etext/macronut.html
You will find the topic "protein" listed and linked to a wealth of information on functions, structures, food sources, and more.

Nutrition UP CLOSE

My Protein Intake

FOCAL POINT: Determine the amount of protein in your diet yesterday.

For *each serving* of a food item you ate yesterday, write the grams of protein the food contains in the corresponding blank. For example, a standard serving of meat is 3 ounces (about the size of the palm of your hand or a deck of cards).

If you had *one* 3-ounce pork chop yesterday, write *20 grams* in the corresponding blank. If you had *two* 3-ounce pork chops, write *40 grams*. If a protein food you ate yesterday is not included, choose the item

on the list closest to it. Then, total the grams of protein you ate yesterday from both plant and animal sources. Finally, compare your protein intake with the RDA of 46 grams for women or 56 grams for men.

Food	One Serving	Protein in One Serving (grams)	Protein You Ate (grams)
Animal products			
Milk (whole)	1 c (8 oz)	8	_____
Yogurt	1 c (8 oz)	13	_____
Cottage cheese	½ c (4 oz)	14	_____
Hard cheese	1 oz	7	_____
Hamburger (lean)	3 oz	24	_____
Beef steak (lean)	3 oz	26	_____
Chicken (no skin)	3 oz	24	_____
Pork chop (lean)	3 oz	20	_____
Fish	3 oz	19	_____
Hot dog	1	6	_____
Sausage	3 oz	17	_____
		Subtotal from animal foods:	_____
Plant products			
Bread	1 slice	2	_____
Rice	½ c (4 oz)	2	_____
Pasta	½ c (4 oz)	3	_____
Cereals	½ c (4 oz)	3	_____
Vegetables	½ c (4 oz)	2	_____
Peanut butter	1 tbs	4	_____
Nuts	¼ c (2 oz)	7	_____
Cooked beans (legumes)	½ c (4 oz)	8	_____
		Subtotal from plant foods:	_____
		Total grams of protein from plant and animal foods:	_____
		Amount above/below RDA:	_____

Special note: You can also calculate your protein intake using Wadsworth Diet Analysis Plus software. Input your food intake for one day. Then go to the Analyses/Reports section to view the total number of grams of protein in your diet.

FEEDBACK can be found at the end of Unit 15.

Notes

1. Dietary Reference Intakes. Energy, carbohydrate, fiber, fat, fatty acids, cholesterol, protein, and amino acids. Institute of Medicine, National Academy of Sciences, Washington, DC: National Academies Press; 2002.

2. Dietary Reference Intakes, Energy, . . . amino acids.

3. Matthews DE. Proteins and amino acids. In: Modern nutrition in health and disease, 9th edition, Shils ME et al., eds. Philadelphia: Lippincott, Williams & Wilkins, 1998, pp. 11–48.

4. Dietary Reference Intakes, Energy, . . . amino acids.

5. Dietary Reference Intakes, Energy, . . . amino acids.

6. High protein intake harms the kidneys. www.healthfinder.gov/high protein diet, accessed 9/03.

7. Centers for Disease Control. Update: eosinophilia-myalgia syndrome associated with ingestion of L-tryptophan—United States, as of January 9, 1990. JAMA 1990;263–633.

8. Information paper on 5-hydroxytryptophan and 5-hydroxy-L-tryptophan, www.nal.usda.gov/fnic/etext/macronut.html, accessed 7/03.

9. Brzezinski A, Melatonin in humans, New Engl J Med 1997;336:186–95; and Arendt J, Jet-lag and shift work, Therapeutic uses of melatonin. J Royal Soc Med 1999;92:402–5.

10. Lund BC, Perry PJ, Nonsteroid performance-enhancing agents in athletic competition: an overview for clinicians, Medscape Pharmacotherapy: www.medscape.com/Medscape/pharmacology/journal/2000/v02,n05; accessed Sept. 25, 2000; and Ergogenic aids: reported facts and claims, SCAN's Pulse 1999;winter:12–21.

11. Dietary Reference Intakes, Energy, . . . amino acids.

12. Dietary Reference Intakes, Energy, . . . amino acids.

13. Matthews, Proteins and amino acids.

14. Dietary Reference Intakes, Energy, . . . amino acids.

15. Dietary Reference Intakes, Energy, . . . amino acids.

16. McDowell MA et al. Energy and macronutrient consumption of persons 2 months and over in the United States: third national Health and Nutrition Examination Survey, 1988–91. Hyattsville, MD: National Center for Health Statistics, 1994.

Nutrition **UP CLOSE**

My Protein Intake

Feedback for Unit 15

Compare your subtotals to find out which protein source you prefer, plant or animal. Protein from animal products is often accompanied by fat. If you are concerned about calories and fat in your diet, choose plant protein sources more often. And, if you are similar to many Americans, your intake of protein will exceed the RDA by quite a bit.

UNIT 16

Vegetarian Diets

Nutrition Scoreboard

	TRUE	FALSE
1 The human body developed to function best on a vegetarian diet.		
2 People who don't eat meat have more health problems than people who do.		
3 Macrobiotic diets cure some types of cancer.		
4 In order to consume enough high-quality protein, vegetarians need to consume combinations of plant foods that provide a complete source of protein at every meal.		

Answers on next page

[KEY CONCEPTS AND FACTS]

- Vegetarianism is more than a diet. It is often part of a value system that influences a variety of attitudes and behaviors.

- Appropriately planned vegetarian diets are health promoting.

- Vegetarian diets that lead to caloric and nutrient deficiencies generally include too narrow a range of foods.

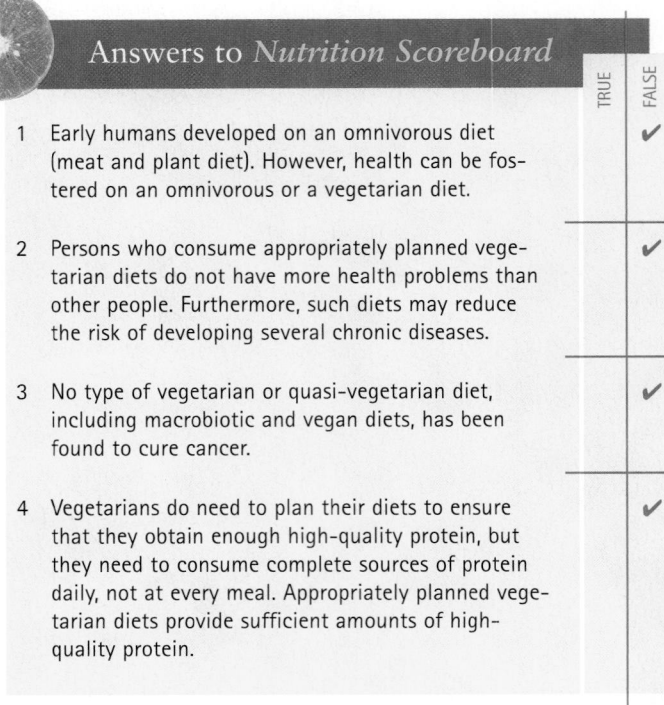

Answers to *Nutrition Scoreboard*

		TRUE	FALSE
1	Early humans developed on an omnivorous diet (meat and plant diet). However, health can be fostered on an omnivorous or a vegetarian diet.		✔
2	Persons who consume appropriately planned vegetarian diets do not have more health problems than other people. Furthermore, such diets may reduce the risk of developing several chronic diseases.		✔
3	No type of vegetarian or quasi-vegetarian diet, including macrobiotic and vegan diets, has been found to cure cancer.		✔
4	Vegetarians do need to plan their diets to ensure that they obtain enough high-quality protein, but they need to consume complete sources of protein daily, not at every meal. Appropriately planned vegetarian diets provide sufficient amounts of high-quality protein.		✔

Perspectives on Vegetarianism

. . . unless vegetarianism comes from the soul, it will just be a passing fad.
—Janis Barkas, 1978

It wouldn't be any fun to live without meat. . . . I do not believe in this meatless, dairyless, butterless society.
—Julia Child, 1991

Vegetarianism in the United States, Canada, and other economically developed countries is moving from the realm of counterculture to the mainstream.[1] Yet even with the increased acceptance of vegetarian diets, people tend to be for vegetarianism or against it, often without knowing much about it. Few of those opposed to vegetarianism have tried to learn about the vegetarian way of life or understand that vegetarian diets can be healthful. A small percentage of health professionals are vegetarians, and those who are not are often skeptical about how healthy a vegetarian diet can be. The possibility that something so different from the customary diet can be nourishing to the body may be rejected out of hand.

An objective look at vegetarianism reveals that both appropriately planned vegetarian diets and the lifestyle often followed by vegetarians can be very good for health.[2] Vegetarians tend to be health conscious, often avoid using alcohol, tobacco, and illicit drugs, and engage in regular physical activity.[3] Diet is usually one of several characteristics shared by people practicing particular types of vegetarianism.

Photo Disc

Reasons for Vegetarianism

Worldwide, vegetarians number in the hundreds of millions.[4] For many people, vegetarianism is not a choice (Table 16.1). Most of the world's population subsists on vegetarian diets because meat and other animal products are scarce or too expensive.[6] In other societies, people have the luxury of choosing a healthy assortment of food from an abundant and affordable food supply that includes a wide variety of items acceptable to vegetarians (Illustration 16.1). When food availability is not an issue, people tend to adopt vegetarian diets because of a desire to cause no harm to animals, a desire to preserve the environment and the world's food supply by "eating low on the food chain," or a belief that animal products are unhealthful or unsafe. They may avoid animal products as part of a value or religious belief system. Others follow vegetarian diets to keep their weight down or to lower the risk of developing specific diseases such as cancer or heart disease.[7]

Vegetarian Diets Come in Many Types]

> *Yeah, I guess I'm kind of a vegetarian, too. I try not to eat meat—well, I love chicken but I don't eat red meat. I mean, I eat hamburgers. I can't remember the last time I had a steak.*
>
> —A 17-year-old "vegetarian"

There is no one vegetarian diet. People who consider themselves to be vegetarians range from those who eat all foods except red meat (mainly beef) to those who exclude all foods from animal sources, including honey. According to the American Vegetarian Society, a person is a vegetarian only if she or he eats *no* meat.

Vegetarian Diet Options]

The types of foods included in common vegetarian diets are summarized in Table 16.2. The least restrictive form of vegetarian diet has been unofficially labeled "far vegetarian" because it excludes only red meats. This diet is very much like that consumed by omnivores, or meat and plant eaters. Quasi-vegetarian diets (also called semivegetarian diets) vary somewhat, but generally exclude beef, pork, and poultry while including fish, eggs, dairy products, and plant foods. The lacto-ovovegetarian diet includes only dairy foods, eggs, and plant foods.

TABLE 16.1
REASONS FOR VEGETARIANISM.[5]
• Lack of availability or affordability of animal products
• Desire not to cause harm to animals
• Religious beliefs
• Desire to "eat low on the food chain"
• Desire to preserve the world's food supply
• Health
• Desire to omit hormones, antibiotics, and possible contaminants in meats

Illustration 16.1
The growing selection of vegetarian foods in supermarkets.

Richard Anderson

TABLE 16.2

VEGETARIAN DIETS COME IN MANY TYPES.

	FOODS INCLUDED					
TYPE OF DIET	BEEF, LAMB, PORK ("RED MEAT")	POULTRY	FISH	EGGS	MILK AND MILK PRODUCTS	PLANT FOODS
"Far" vegetarian		x	x	x	x	x
Quasi-vegetarian		*	x	x	x	x
Lacto-ovovegetarian				x	x	x
Lactovegetarian					x	x
Macrobiotic			**			x
Vegan						x

*Quasi-vegetarian diets (also called semivegetarian) may include poultry; they tend to vary in the type of animal products consumed.
**Macrobiotic diets may include fish and other animal products.

What's in the bowl? It depends on what type of vegetarian diet you are following.

fruitarian
A form of vegetarian diet in which fruits are the major ingredient. Such diets provide inadequate amounts of a variety of nutrients.

Individuals practicing this type of diet exclude all meats. The lactovegetarian diet, as the name implies, includes only milk and milk products and plant foods. Vegan diets are a more restrictive type of vegetarian diet. Vegans eat only plant foods; in addition, they may avoid honey and clothes made from wool, leather, or silk.

Macrobiotic diets fall somewhere between quasi-vegetarian and vegan diets. The formulation of macrobiotic diets has changed dramatically over the past decade. In addition to brown rice, other grains, and vegetables, these diets may now include fish, dried beans, spices, fruits, and many other types of foods. No specific foods are prohibited, and locally grown and whole foods are emphasized.[8]

Each of these vegetarian diets has philosophical underpinnings. Persons adhering to the macrobiotic philosophy, for example, place value on consuming organic foods grown locally and balancing the intake of foods nonjudgmentally assigned as "yin" and "yang." Foods are classed as yin or yang based on beliefs about the food's relationship to the emotions and the physical condition of the body. Yin foods such as corn, seeds, nuts, fruits, and leafy vegetables are considered negative, dark, cold, and feminine. Yang foods represent opposing positive forces of light, warmth, and masculinity. Poultry, fish, eggs, and cereal grains such as buckwheat are yang foods.[9]

For a number of vegetarian regimes, the spiritual or emotional importance assigned to certain foods supersedes consideration of their contribution to an adequate diet.[10] Vegetarians adhering to the "living foods diet" consume uncooked and fermented plant foods only. This diet is inadequate in a number of nutrients, including vitamin B_{12}.[11] *Fruitarians* consume only fruit and olive oil. People adopting this type of diet rarely stick with it for long—it does not sustain health.

Vegetarian Diets and Health

Appropriately planned vegetarian diets have been shown to be healthful, nutritionally adequate, and beneficial in the prevention and treatment of certain diseases.
—Position of the American Dietetic Association and Dietitians of Canada, 2003[13]

Long-standing vegetarian dietary regimes that promote health in India and China have been tested by time (if not by science) and found to be adequate in essential nutrients.[14] All of the vegetarian diets listed in Table 16.2 are health promoting if implemented correctly. In fact, vegetarian diets are generally healthier than those of

REALITY CHECK

Reintroducing meat

Larry has been a vegetarian for the last 7 years and wants to eat a hamburger again to see if it tastes as good as he remembers. He's a bit nervous about doing it because he thinks meat might make him feel sick.

Who gets thumbs up

?

Susan:
Larry should eat the hamburger if he wants to. It's a food, not an indigestion time bomb.

Doug:
Larry's stomach isn't used to the heaviness of meat. It will make him nauseous.

Answers on next page

nonvegetarians.[15] On the other hand, highly restrictive vegetarian diets—such as fruitarian regimes, the raw food diet, and other popular vegetarian diets that stray from established regimes—can be harmful to health.[16] In general, the more restrictive the diet, the more likely it is to be inadequate and lead to health problems. This is especially true for pregnant women, children, and persons who are ill, all of whom have relatively high needs for nutrients. Reports of caloric and nutrient deficiencies in young children of vegan parents are more common than is the case with less restrictive vegetarian diets.[17] Vegan diets may supply too little vitamin B_{12}, vitamin D, calcium, and zinc—a mineral that is poorly absorbed from plant sources.

Availability of a large assortment of vegan foods in the United States and other developed countries is making it less likely that vegetarian diets of children and adults will be inadequate.[18] Vegetarians in economically developed countries are not at greater risk for iron deficiency than nonvegetarians.[19] They generally obtain sufficient iron from plants and have ample intake of vitamin C, which enhances absorption of iron from plants. In contrast, the diets of vegetarians in developing countries often include too few iron-rich plants, making iron deficiency common. Both the quality and the quantity of protein in vegetarian diets can also be a source of concern. However, vegetarians in developed countries generally have adequate protein intakes.[20] Vitamin B_{12} has consistently been shown to be the most likely nutrient to be lacking in the diets of individuals who do not consume animal products.[21] Increasingly, vegetarian diets are being recognized for their beneficial effects on health and disease prevention.[22]

What Are the Health Benefits of Vegetarian Diets?

I am a great eater of beef, and I believe that does great harm to my wit.
—William Shakespeare, 1589

Compared to the usual American diet and lifestyle, vegetarianism is associated with a lower risk of developing heart disease, stroke, hypertension, type 2 diabetes, chronic bronchitis, gallstones and kidney stones, and colon cancer. People who follow a vegetarian diet rarely become obese or develop high blood cholesterol levels.[23] For children and adults at risk of early heart disease, the low-fat and high-fiber

ANSWERS TO REALITY CHECK

Reintroducing meat

The thought that meat could make you sick may be enough to trigger indigestion. Many self-defined vegetarians consume meat on rare occasion without reported ill effects.[12]

Susan

Doug

vegetable and fruit content of vegetarian diets may help reduce blood cholesterol levels and the risk of heart disease.[24] Benefits of vegetarian diets, macrobiotic diets, and other plant-based diets in the treatment of cancer have not been demonstrated.[25]

Dietary Recommendations for Vegetarians

Because vegetarian diets exclude one or more types of foods, it's important that the foods included provide sufficient calories and the assortment and quantity of nutrients needed for health. No matter what the motivation underlying the assortment of foods included, vegetarian diets that fail to provide all the nutrients humans need in the required amounts will not sustain health.

How to Eat Vegetarian

The watchword for vegetarians striving for a nutritionally complete diet is *variety*. At the minimum, vegetarian diets should include an assortment of fruits, vegetables, grains and grain products (such as cereals, pasta, and breads), nuts, seeds, dried beans, and dairy products or fortified soy substitutes.

Illustration 16.2 shows the Food Guide Pyramid adapted by the USDA for use by vegetarians. Although a number of detailed vegetarian food guides have been developed, this one is considered the "official" guide. The specific foods selected from each group will vary depending on the type of vegetarian diet consumed. For example, lacto-ovovegetarians could include eggs from the "dry beans, eggs, and nuts group." Lactovegetarians would exclude eggs but could include food from the "milk, yogurt, and cheese group."

The American Dietetic Association has developed a set of guidelines for vegetarian meal planning. These guidelines are summarized in Table 16.3.

Special Considerations for Vegetarian Diets

Diets that include few or no animal products may be low in sources of **complete protein,** vitamin B_{12}, vitamin D, calcium, and zinc. With appropriate food selection, these potential nutrient inadequacies can be prevented.

Complementing Plant Protein Sources] Animal products such as meat, eggs, and milk provide all of the nine **essential amino acids** in sufficient quantity to qualify as a complete source of protein. In addition, tests have shown soy proteins to be complete protein sources for children and adults.[27] The diet must include complete

complete proteins
Proteins that contain all of the nine essential amino acids in amounts sufficient to support protein tissue construction by the body.

essential amino acids
Amino acids that cannot be synthesized in adequate amounts by the human body and must therefore be obtained from the diet.

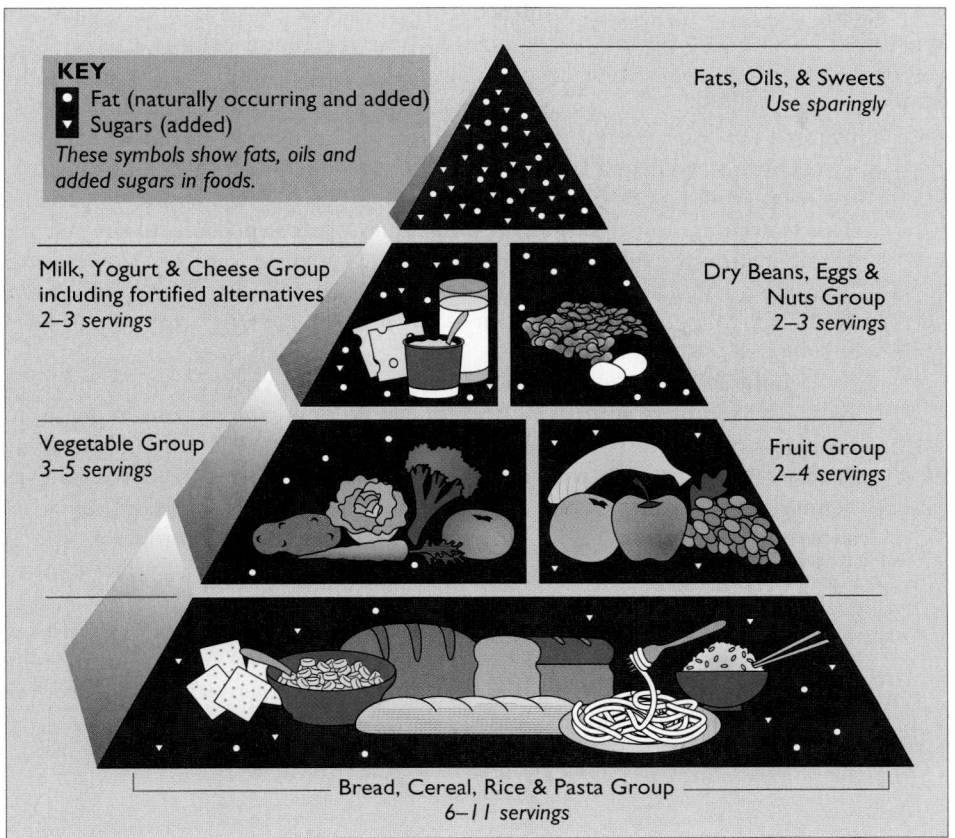

Illustration 16.2
The "official" vegetarian Food Guide Pyramid.

KEY
- ● Fat (naturally occurring and added)
- ▼ Sugars (added)

These symbols show fats, oils and added sugars in foods.

Fats, Oils, & Sweets
Use sparingly

Milk, Yogurt & Cheese Group including fortified alternatives
2–3 servings

Dry Beans, Eggs & Nuts Group
2–3 servings

Vegetable Group
3–5 servings

Fruit Group
2–4 servings

Bread, Cereal, Rice & Pasta Group
6–11 servings

TABLE 16.3

THE AMERICAN DIETETIC ASSOCIATION'S GUIDELINES FOR VEGETARIAN MEAL PLANNING.[26]

1. Choose a variety of foods, including whole grains, vegetables, fruits, legumes, nuts, seeds, and, if desired, dairy products and eggs.

2. Choose whole, unrefined foods often and minimize intake of highly sweetened, fatty, and highly refined foods.

3. Choose a variety of fruits and vegetables.

4. If animal foods such as dairy products and eggs are used, choose lower-fat versions of these foods. Cheeses and other high-fat dairy foods and eggs should be limited in the diet because of their saturated fat content and because their frequent use displaces plant foods in some vegetarian diets.

5. Vegans should include a regular source of vitamin B_{12} in their diets along with a source of vitamin D if sun exposure is limited.

6. Do not restrict dietary fat in children younger than two years. For older children, include some foods higher in unsaturated fats (e.g., nuts, seeds, nut and seed butters, avocado, and vegetable oils) to help meet nutrient and energy needs.

complementary protein sources
Plant sources of protein that together provide sufficient quantities of the nine essential amino acids.

There are more than 70 varieties of dried beans. They are an important part of the cuisine of almost all cultures. Bean cuisine includes chili, baked beans, beans and rice, hummus, and refried beans.

Photo Disc

proteins because the body needs a sufficient supply of each essential amino acid if it is to build and replace protein substances such as red blood cells and enzymes. If any of the essential amino acids are missing in the diet, protein tissue construction stops, and the available amino acids will be used for energy instead. Essential amino acids consumed in foods are not stored, so the body needs a fresh supply each day.

Vegetarians who don't consume animal products can meet their need for essential amino acids by combining plant foods to yield complete proteins. This is done by consuming plant foods that *together* provide all the essential amino acids although each individual food is missing some of these essential nutrients. The goal is to "complement" plant sources of essential amino acids, or to consume ***complementary protein sources*** from plant foods regularly.[28]

Many different combinations of plant foods yield complete proteins. Basically, complete sources of protein can be obtained by combining grains such as rice, bulgur (whole wheat), millet, or barley with dried beans, tofu, or green peas, or corn with lima beans or dried beans, or seeds with dried beans. Some examples of complementary sources of plant proteins are shown in Illustration 16.3. Milk, meat, and eggs contain complete proteins and will complement the essential amino acids profile of any plant source of protein.

It used to be thought that you had to consume complementary sources of protein in the same meal to meet the body's need for essential amino acids. They now conclude, however, that complementary protein sources should be consumed approximately daily.[29]

Illustration 16.3

Some combinations of plant foods that provide complete protein:

- *Rice and dried beans*
- *Rice and green peas*
- *Bulgur (wheat) and dried beans*
- *Barley and dried beans*

- *Corn and dried beans*
- *Corn and green peas*
- *Corn and lima beans (succotash)*

- *Soybeans and seeds*
- *Peanuts, rice, and dried beans*
- *Seeds and green peas*

Rice and black beans.

Hummus and bread.

Corn and black-eyed peas.

Bulgur (whole wheat) and lentils.

Tofu and rice.

Corn and lima beans (succotash).

Tortilla with refried beans (e.g., a bean burrito).

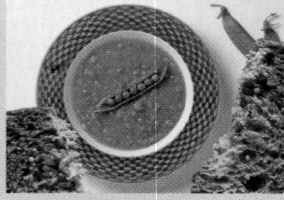

Pea soup and bread.

Getting Enough Vitamin B$_{12}$] Vitamin B$_{12}$ is of particular concern to vegetarians because it is present only in animal products. It was once believed that sea vegetables, algae, and fermented soybean products such as tamari, tempeh, and miso provided vitamin B$_{12}$. We now know that the vitamin B$_{12}$ in these foods is in a form that humans utilize poorly.[30] Vegetarians who don't consume animal products can obtain vitamin B$_{12}$ from fortified products such as soy milk and breakfast cereals and from vitamin B$_{12}$ supplements.

Where's the Vitamin D?] People get vitamin D from two primary sources—the sun and milk. Exposure of the skin to sunlight causes vitamin D to be produced from a substance that circulates in the blood. People can meet their need for vitamin D this way, but not everyone lives in a climate where the skin can be exposed to the sun regularly enough to produce all the vitamin D needed. Vitamin D is found in only a few foods, and most of our intake comes from the consumption of vitamin D–fortified milk. (Milk is fortified with vitamin D, but other dairy products such as yogurt and cheese generally are *not*.) If milk is not part of the vegetarian's diet plan, and exposure of the skin to sunlight is limited, vitamin D–fortified soy milk and breakfast cereals or a vitamin D supplement should be consumed. Because vitamin D is toxic at high levels, supplements should provide around 200 to 400 IU (or 5 to 10 micrograms) per day.

Calcium Sources in the Vegetarian Diet] Vegetarians who exclude milk and milk products need to rely on good plant sources of calcium (kale, broccoli, bok choy, and dried beans, for example) and calcium-fortified products such as soy milk, breakfast cereals, and orange juice. High intakes of soy products such as fermented soybeans, miso, and tofu appear to promote bone mass and protect against the development of osteoporosis.[31]

Solving the Zinc Problem] Many plant foods are good sources of zinc. The problem is that zinc from plants is poorly absorbed, so only a small portion of the zinc consumed becomes available for use by the body. To obtain enough zinc, vegans should regularly consume whole grains, dried beans, nuts, zinc-fortified breakfast cereals, and other sources of zinc.[32]

Where to Go for More Information on Vegetarian Diets

You can find information about vegetarian diets in a variety of sources. World Wide Web resources appear at the end of this unit, and several cookbooks that offer information about vegetarian nutrition and delicious recipes are pictured in Illustration 16.4. Unfortunately, some of the information available on vegetarianism is wrong or misconstrued. Some vegetarian organizations are more committed to selling particular beliefs, memberships, or books and magazines than to promoting healthy vegetarian diets. Beware of vegetarian diets that claim to cure cancer, AIDS, or other serious illnesses or promise you'll experience inner peace or spiritual renewal. Appropriately planned vegetarian diets are health promoting, but they are not magic bullets that will cure all the ills of body and soul.

Illustration 16.4

Many excellent vegetarian cookbooks are available; most provide information on diet as well as recipes.

The books pictured here are Moosewood Restaurant Low-Fat Favorites: Flavorful Recipes for Healthful Meals, edited by Pam Krauss; Vegetarian Times Complete Cookbook, by the editors of Vegetarian Times; Vegetarian Cooking for Everyone, by Deborah Madison; The Whole Soy Cookbook, by Patricia Greenberg; and Vegetable Heaven, by Mollie Katzen.

Richard Anderson

Richard Anderson

Nutrition UP CLOSE

Vegetarian Main Dish Options

FOCAL POINT: Serving up vegetarian alternatives to meat dishes.

Assume you are a member of the "vegetarian option" planning committee for your college. You are asked to identify three meatless main dishes that could be served in the dining halls for lunch and another three that could be served for dinner. The one stipulation is that they should be main dishes you would enjoy eating.

What meatless dishes would you identify as options for lunch and for dinner?

Lunch Dishes

1.

2.

3.

Dinner Dishes

1.

2.

3.

Key Terms

complementary protein sources, page 16-8

complete proteins, page 16-6

essential amino acids, page 16-6

fruitarian, page 16-4

www links

www.nal.usda.gov/fnic/etext/000058.html
Check this site for general information on vegetarian diets, vegetarian nutrition for kids, diets, and recipes.

www.vrg.org
Contact the Vegetarian Pages for dietary advice, healthy cooking and shopping tips, and other information for the vegetarian community.

www.kushiinstitute.org
Learn more about the macrobiotic philosophy, conferences, and events.

www.vegweb.com
Provides guide to vegetarianism, recipes and books, and much more (including ads).

www.healthfinder.gov
Search "vegetarian diets" and get tips for budding vegetarians, information on benefits, and guides for vegetarian diets for pregnant women and children.

Notes

1. Sabate J. The contribution of vegetarian diets to health and disease: a paradigm shift? Am J Clin Nutr 2003;78(suppl):501S–7S.

2. Sabate, Contribution of vegetarian diets.

3. Buffonge I. The vegetarian/vegan lifestyle. West Indian Med J 2000;49:17–19.

4. Antony AC. Vegetarianism and vitamin-12 (cobalamin) deficiency. Am J Clin Nutr 2003;78:3–6.

5. Antony, Vegetarianism and vitamin-12 deficiency; and Barr SL, Chapman GE, Perceptions and practices of self-defined current vegetarian, former vegetarian, and nonvegetarian women, J Am Diet Assoc 2002;102:354–6.

6. Havala S, Dwyer J. Position of the American Dietetic Association: vegetarian diets—technical support paper. J Am Diet Assoc 1988;88:352–5.

7. Barr and Chapman, Perceptions and practices.

8. Kushi LH et al. The macrobiotic diet and cancer. J Nutr 2001. 131:3056S–64S.

9. Kushi et al, The macrobiotic diet and cancer.

10. Mutch PB. Food guides for the vegetarian. Am J Clin Nutr 1988;48:913–19.

11. Rauma A-L et al. Vitamin B-12 status of long-term adherents of a strict, uncooked vegan diet ("living foods diet") is compromised. J Nutr 1995;125:2511–15.

12. Barr and Chapman, Perceptions and practices.

13. Position of the American Dietetic Association and Dietitians of Canada: vegetarian diets. J Am Diet Assoc 2003;103:748–65.

14. Sabate, Contribution of vegetarian diets.

15. Haddad EH, Tanzman JS. What do vegetarians in the United States eat? Am J Clin Nutr 2003;78(suppl):626S–32S.

16. Sabate, Contribution of vegetarian diets.

17. Sabate, Contribution of vegetarian diets; Kushi et al, The macrobiotic diet and cancer; and Position of the ADA and Dietitians of Canada, Vegetarian diets.

18. Position of the ADA and Dietitians of Canada, Vegetarian diets; and Messina V et al. A new food guide for North American vegetarians. J Am Diet Assoc 2003;103:771–5.

19. Position of the ADA and Dietitians of Canada, Vegetarian diets; and Haddad EH et al., Dietary intake and biochemical, hematologic, and immune status of vegans compared with nonvegetarians, Am J Clin Nutr 1999;70(suppl):586S–93S.

20. Position of the ADA and Dietitians of Canada, Vegetarian diets.

21. Antony, Vegetarianism and vitamin-12 deficiency.

22. Sabate, Contribution of vegetarian diets; and Messina et al., a new food guide.

23. Sabate, Contribution of vegetarian diets; and Position of the ADA and Dietitians of Canada, Vegetarian diets.

24. Resnicow K, et al. Diet and serum lipids in vegan vegetarians: a model for risk reduction. J Am Diet Assoc 1991;91:447–53.

25. Kushi et al, The macrobiotic diet and cancer.

26. Position of the ADA and Dietitians of Canada, Vegetarian diets.

27. Dietary reference intakes for carbohydrates, fiber, fat, fatty acids, cholesterol, protein, and amino acids. Washington, DC: National Academies Press; 1999.

28. Havala and Dwyer, Position of the ADA: vegetarian diets.

29. Dietary reference intakes for carbohydrates.

30. Rauma et al., Vitamin B-12 status.

31. Somekawa Y et al. Soy intake and bone mass in postmenopausal women, Obstet Gynecol 2000;97:109–15.

32. Smith CF et al. Vegetarian and weight-loss diets among young adults. Obesity Res 2000;8:123–9.

Nutrition UP CLOSE

Vegetarian Main Dish Options

Feedback for Unit 16

Vegetarians definitely had the advantage in completing this exercise! For the meat eaters, were the main dishes you identified meat-free?

Here are six vegetarian options for lunch and dinner main dishes:

- Eggplant parmesan

- Falafel

- Red beans and rice

- Pasta with mixed vegetables

- Melted cheese, tomato, and sprout sandwiches

- Bean burritos

- Veggie burgers

Food Allergies and Intolerances

Nutrition Scoreboard

	TRUE	FALSE
1 About one in every three Americans are allergic to at least one food.		
2 Food labels must list all allergenic ingredients.		
3 Skin prick tests are an accurate way to diagnose food allergies.		
4 Food intolerances cause less severe reactions than food allergies do.		

Answers on next page

[KEY CONCEPTS AND FACTS]

- True food allergies involve a response by the body's immune system to a particular substance in food.

- Food allergies may be caused by hundreds of different foods. However, 90% of food allergies are due to 8 foods: nuts, eggs, wheat, milk, peanuts, soy, seafood, and fish.[3]

- Food intolerances cover an array of adverse reactions that do *not* involve the body's immune system.

- The most accurate method of identifying food allergies and intolerances is the double-blind, placebo-controlled food challenge.

Answers to *Nutrition Scoreboard*

		TRUE	FALSE
1	About one in three Americans *believe* they are allergic to at least one food. Probably 1 to 2 in 100 adults and 8 in 100 young children actually are.[1]		✔
2	Listing all potential, relatively common allergens on food ingredient labels is voluntary. Congratulations are due to those companies that do list them all.		✔
3	Skin prick tests for food allergies are often inaccurate.[2]		✔
4	(Yikes! Four false answers in a row! There are a lot of myths about food allergies.) Food allergies and food intolerances have different causes, but both can produce life-threatening reactions.		✔

Food Allergy Mania

Talk about your controversial subjects. This one is very hot. People use the term *food allergy* to refer to virtually any type of problem they have with food. At one extreme are people who believe allergies to milk, wheat, and sugar are to blame for hyperactivity and a host of other behavioral problems in children. At the other extreme are some health professionals who think people who complain of food allergies need to have their heads examined rather than their bodies' reactions to foods. Despite the controversies, food allergies have become a major health concern in the United States and many other Westernized countries.[4]

Food allergies are real and can be very serious. At the minimum, true food allergies can cause a rash or an upset stomach. At the maximum, they can result in death. Unreal food allergies can cause problems, too. They can lead people to eliminate nutritious foods from their diet unnecessarily, resulting in inadequate diets and eventually in health problems. One of the most intriguing aspects of food allergies is the frequency and ease with which foods are falsely blamed for a variety of mental and physical health problems.

Adverse Reactions to Foods

People experience adverse reactions to food for three primary reasons. One is food poisoning. The other two involve the body's reaction to substances in food that are normally harmless. With *food allergies,* the body's *immune system* reacts to a substance in food (almost always a protein) that it identifies as harmful (Illustration 17.1). The immune system is not involved in *food intolerances,* which encompass other adverse reactions to normally harmless substances in food.

Why Do Food Allergies Involve the Immune System?] It may seem odd that a system that helps the body conquer bacteria and viruses is involved in "protecting" us from normal constituents of food. In people with allergies, however, the cells recog-

food allergy
Adverse reaction to a normally harmless substance in food that involves the body's immune system. (Also called *food hypersensitivity.*)

immune system
Body tissues that provide protection against bacteria, viruses, and other substances identified by cells as harmful.

food intolerance
Adverse reaction to a normally harmless substance in food that does *not* involve the body's immune system.

Illustration 17.1
What do these foods have in common?
Each food shown, plus many more not shown, may cause adverse reactions in some individuals.

Illustration 17.2
The development of allergic reactions to foods.

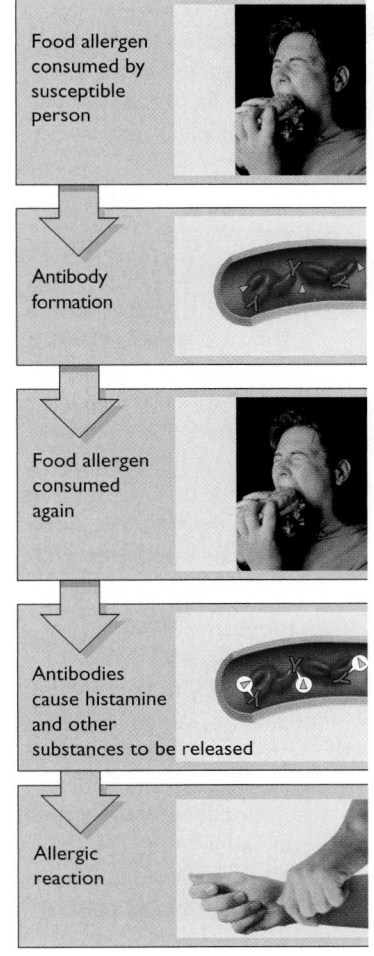

nize some food components as harmful—just as they recognize bacteria and viruses. Components of food that trigger the immune system are called *food allergens.*

In a small proportion of susceptible people, exposure to trace amounts of an allergen triggers an allergic reaction. Highly sensitive people can develop an allergic reaction simply by being in the room when the food to which they are allergic is prepared, or by touching the food. Allergic reactions to peanuts have been triggered by kissing someone who has recently consumed them and by breathing peanut vapors on airplanes.[5]

In response to the allergen, the immune system forms *antibodies* (Illustration 17.2). The antibodies attach to cells located in the nose, throat, lungs, skin, eyes, and other areas of the body. When the allergen appears again, the body is ready. The previously formed antibodies recognize the allergen, attach to it, and signal the body to secrete *histamine* and other substances that cause the physical signs of allergic reactions. Most commonly, allergic reactions cause a rash, diarrhea, congestion, or wheezing (Table 17.1), but the symptoms may be much more serious.

TABLE 17.1

THE THREE MOST COMMON TYPES OF SYMPTOMS CAUSED BY FOOD ALLERGIES.
Some people experience more than one type of reaction.[6]

SYMPTOM	PERCENTAGE OF PEOPLE WITH FOOD ALLERGIES WHO DEVELOP THE SYMPTOM
• Skin eruptions: rash, hives	84%
• Gastrointestinal upsets: diarrhea, vomiting, cramps, nausea	52
• Respiratory problems: congestion, runny nose, cough, wheezing, asthma	32

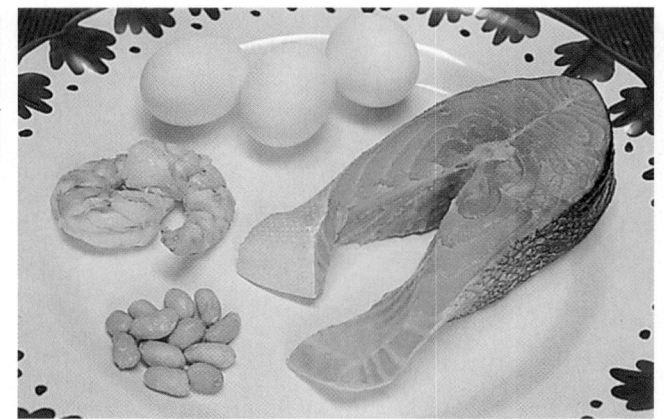

Illustration 17.3

These foods can cause anaphylactic shock, a severe allergic reaction, in highly sensitive people.
(Most food allergies cause milder symptoms.)

Exposure to trace amounts of an allergen in peanuts, nuts (peanuts are actually a legume, not a nut), fish, and shellfish can cause *anaphylactic shock* (Illustration 17.3). This massive reaction of the immune system can result in death, caused by the cutoff of the blood supply to tissues throughout the body. People experiencing anaphylactic shock can be revived with an injection of epinephrine. The incidence of peanut allergy in the United States is increasing—about 1 in 150 people are affected.[7] Although it's not clear why peanut allergy is increasing, some scientists suspect the reasons are related to expanded use of soy infant formulas, skin oils and creams that contain peanut oil, or consumption of peanuts very early in life.[8]

Symptoms of food allergies may occur within seconds after the body is exposed to the food allergen, or they may take up to 2 hours to occur. The symptoms often disappear within 2 hours after they begin.[9]

What Foods Are Most Likely to Cause Allergic Reactions?] Many foods may cause allergic reactions in susceptible individuals (Illustration 17.4 on p. 17–5). Nevertheless, approximately 90% of all food allergies are caused by eight foods: nuts, eggs, wheat, milk, peanuts, soy, seafood, and fish. The nearby "On the Side" feature gives you an idea for how to remember these foods. Together, these are considered "the big eight" food allergies.[11]

A Closer Look at Wheat Allergy}

Minnie is 46 years old and has three sisters and two brothers, all of whom are taller than she is. (She considers herself the "runt of the litter.") Since childhood, Minnie has had problems with diarrhea and cramps. Over the past 2 years, these problems have become worse, and she has been chronically tired. Assuming that she was lactose intolerant, Minnie stopped eating all dairy products. Her problems persisted, however. When she almost lost her job as a receptionist due to her frequent bathroom breaks and her obvious fatigue, she decided to see a doctor again.

Diagnosing anemia, the doctor sent Minnie to a registered dietitian. Hearing Minnie's story, the dietitian suspected celiac disease caused by an allergy to a component of wheat and other grains. After 2 weeks on an allergen-free diet, Minnie started to regain her strength and spent much less time in the bathroom. Diagnostic tests subsequently confirmed that Minnie had celiac disease, a condition that likely had existed since childhood.

An allergy to wheat, or more specifically to a component of gluten found in wheat and other grains such as barley, rye, and triticale has been underdiagnosed in the past. Called *celiac disease* (and also celiac sprue and gluten-sensitive enteropathy), it affects one in 120–300 people in many countries, including the United States.[12] It differs somewhat from traditional immunoglobulin E-mediated allergies in that the immune system reaction to gluten is localized in the lining of the small intestine. Celiac disease can be difficult to diagnose because the symptoms (diarrhea, weight loss, cramps, anemia) are similar to those of other diseases. With new awareness about the disorder, doctors are ordering diagnostic tests more frequently, and the rate of confirmed cases of celiac disease is increasing.[13]

food allergen
A substance in food (almost always a protein) that is identified as harmful by the body and elicits an allergic reaction from the immune system.

antibodies
Proteins the body makes to combat allergens.

histamine
(hiss-tah-mean) A substance released in allergic reactions. It causes dilation of blood vessels, itching, hives, and a drop in blood pressure and stimulates the release of stomach acids and other fluids. Antihistamines neutralize the effects of histamine and are used in the treatment of some cases of allergies.

anaphylactic shock
(an-ah-fa-lac-tic) Reduced oxygen supply to the heart and other tissues due to the body's reaction to an allergen in food. Symptoms of anaphylactic shock include paleness, a flush discoloration of the lips and fingertips, a "faceless" expression, weak and rapid pulse, and difficulty breathing.

NUTS

EGGS

PEANUTS

SOY

WHEAT

MILK

MILK

SEAFOOD

FISH

Photo Disc

Celiac disease is treated with a lifelong gluten-free diet. Because gluten is included in many food additives as well as grain products, it generally takes the help of an experienced, registered dietitian to design and monitor gluten-free diets.[14]

Illustration 17.4
The top 8 foods that cause allergies.[10]

Do Genetically Modified Foods Contain Allergens?] It's possible but unlikely. Genetically modified foods contain proteins from different types of plants or animals. Consequently, a protein that causes allergic reactions in some people could possibly end up in a genetically modified food. This possibility is recognized by manufacturers, and genetically modified foods that may contain allergens are tested for safety before they are sold. Pretests of soybeans with added Brazil-nut proteins, for example, found that people allergic to Brazil nuts were allergic to the genetically modified soybeans. Plans to make the soybeans commercially available were dropped, and the manufacturer halted further production.[15] Due to manufacturers' testing procedures and FDA regulations related to these foods, the possibility that genetically modified foods contain hidden allergens is limited to some extent. However, the true risks of allergic reactions from genetically modified food are unknown.[16]

How Common Are Food Allergies?

Infants and young children experience food allergies more often than adults do. Approximately 8% of infants and young children develop a food allergy, which most often is due to a protein in cow's milk.[17] To prevent allergic reactions, it is recommended that cow's milk and other potentially allergenic foods such as peanuts, nuts, fish, seafood, eggs, and wheat be excluded from the infant's diet during the first year of life. Infants from families with a food allergy to peanuts, soy, or nuts should not be given these foods for the first 3 years.[18] Most infants outgrow their allergy to cow's milk or to other foods such as milk, egg white, wheat, and soy as the digestive system matures. They often do not outgrow allergies to peanuts, nuts, fish, and shellfish.[19]

The incidence of food allergies in adults is estimated to be 1 to 2% and appears to be increasing. Nevertheless, around one-third of Americans believe they are allergic to one or more foods. About 60% of complaints of food allergy fail to be confirmed by testing.[20]

NEW Milk. Please Serve Safe Food. This image and saying may help you remember the top foods related to allergies: You're standing in front of a refrigerator and have just taken a big gulp of milk straight from the carton. In that second, it hits you that the milk was sour. To avoid that ever happening again, you think of the saying, "NEW Milk. Please Serve Safe Food." Translation:

N = NUTS
E = EGGS
W = WHEAT
MILK
PLEASE = PEANUTS
SERVE = SOY
SAFE = SEAFOOD
FOODS = FISH

ON THE SIDE

17–5

celiac disease
A disease characterized by inflammation of the small intestine lining resulting from a genetically based intolerance to gluten. The inflammation produces diarrhea, fatty stools, weight loss, and vitamin and mineral deficiencies. (Also called *celiac sprue* and *gluten-sensitive enteropathy.*)

Why the Difference between Reality and Perception?

One reason food allergies seem more widespread than they actually are is that popular wisdom endows various foods with certain properties. The following are some of the more widely held beliefs:

- Allergies to sugar or food colors lead to hyperactivity in children.
- Yeast causes chronic fatigue syndrome.
- Dairy products, especially milk, cause phlegm.

None of these associations has been documented by scientific studies.

Bad timing and mental attitudes about certain foods can also lead to erroneous diagnoses of food allergies. Here are two examples:

1. Isaiah, a lover of blueberries and blueberry pie, hasn't touched a blueberry since 1995. That year, Isaiah had a piece of blueberry pie in a restaurant and later became violently sick to his stomach. Bingo! Isaiah thought he must be allergic to blueberries.

   Actually, Isaiah gave up blueberries for no good reason. He was coming down with the flu when he ate the pie. Nevertheless, to this day when he thinks of blueberries, he gets a bit queasy.

2. For 11 years, Emilia rigidly avoided even small amounts of cow's milk. She was convinced that just a few drops of cow's milk would cause pressure in her head, blurred vision, dizziness, cramps, and nausea.

   Finally, Emilia's presumed allergy to cow's milk was put to the test. She was given liquid through a dark tube inserted into her stomach from her mouth. When she was told the fluid was cow's milk—even though it was actually water—the familiar symptoms appeared within 10 minutes. When she was given milk but told it was water, no symptoms appeared.[21]

   Emilia was shocked. For 11 years she had scrutinized nearly everything she ate and had wasted hundreds of dollars on special food products and supplements. Finishing a frosted brownie and a glass of milk, Emilia contemplated the practical realities of the power of suggestion.

Diagnosis: Is It a Food Allergy?

It is essential to realize that the [allergic] reactions occur in human beings and are therefore enmeshed in the interplay between body, mind and soul. . . . Only through experience with blind food challenges can the amazing power of self-deception be appreciated.
—Charles May, 1980[22]

double-blind, placebo-controlled food challenge
A test used to determine the presence of a food allergy or other adverse reaction to a food. In this test, neither the patient nor the care provider knows whether a suspected offending food or a placebo is being tested.

A variety of tests are used to diagnose food allergies, but there is only one "gold standard." That test is the **double-blind, placebo-controlled food challenge.**[25]

Suppose you suspect you are allergic to grapes because you got an annoying rash twice after eating them. Your doctor arranges an appointment for you at an allergy clinic. At the clinic, you will be given grapes, either in a concentrated pill form or blended into a liquid, and placebo pills (or a liquid) without being told which is the grapes and which the placebo. After an appropriate amount of time, your reaction to the grapes and to the placebo will be noted. If the rash appears after the grapes but not after the placebo, you have an adverse reaction to grapes that may be an allergy.

REALITY CHECK

I think I'm allergic to . . .

You and your coworkers decide to go to a Chinese restaurant for lunch. Trevor wants to go, but hesitates because, as he said, "I must be allergic to Chinese food. Every time I eat it I feel dizzy, sweat, and get this ringing in my ears."

Who's got the better idea

?

Haley:
Trevor, that's not an allergy. It's all in your head.

Darrel:
Have you ever asked them to leave out the MSG?

Answers on next page

Food challenges are best undertaken under medical supervision. True food allergies can cause serious reactions, and immediate help may be needed. Foods suspected of causing anaphylactic shock in the past should not be given in food challenge tests. They should be totally omitted from the diet.[26]

The Other Tests for Food Allergy

Other tests for food allergy are reliable to varying degrees. None of them is always accurate, and the results of bogus tests are not to be trusted at all.

Skin Prick Tests] Skin prick tests (also called prick skin tests) for food allergy are commonly employed and are useful for identifying *the absence* of a food allergy. A positive test result isn't proof positive that an allergy exists, however, because positive test results are inaccurate about 50% of the time.[27]

For this test, a few drops of food extract are placed on the skin, and the skin is then pricked with a needle. At approximately the same time, another area of skin is pricked using only water. If the area around the food skin prick becomes redder and more swollen than the area pricked with water (Illustration 17.5), it is concluded that the person *might* be allergic to the food tested. Skin prick tests are useful for ruling out allergies and for reducing the number of foods that may have to be tested by double-blind, placebo-controlled food challenges.

The RAST Test] The acronym RAST stands for "radioallergosorbent test" (pronounced radio-all-er-go-sor-bent). In this test, a sample of blood is examined for immune substances formed in response to an allergen. The test is not specific for individual food allergens, nor does it reliably indicate the presence of a food allergy.[28]

Bogus Tests] A number of companies offer food allergy tests through the mail. In one study, researchers sent five of these com-

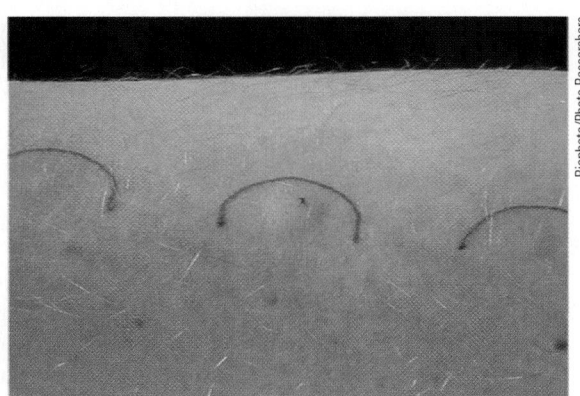

Illustration 17.5
A positive reaction to a skin prick test for a food allergy usually looks like this.
The result indicates that an allergy to the food tested may be present, but doesn't guarantee it.

Haley

ANSWERS TO **REALITY CHECK**

I think I'm allergic to . . .

Some people are sensitive to MSG but it doesn't elicit an immune system reaction, so it's not a food allergy.[23] Illness due to the flu, food poisoning, lactose, sulfites, and foods that may cause gas and bloating (like dried beans and cabbage) are often mistakenly attributed to food allergies.[24]

Darrel

panies the required samples, nine from adults who were allergic to fish and nine from adults who had no allergies. All five companies missed the fish allergy. When duplicate samples from the same adults were sent separately to each company, reports came back with different results for each of the pairs of samples. In addition, the companies reported their laboratories had identified various food allergies that none of the adults actually had.[29] Mail-order allergy tests are a waste of money; they do not use reliable assessment techniques. Other unreliable methods for identifying food allergies include hair analysis, cytotoxic blood tests, iridology, and sublingual (under the tongue) and food injection provocation tests.[30]

What's the Best Way to Treat Food Allergies?

After a food allergy is confirmed, the food is eliminated from the diet. This is the only treatment available for food allergies. "Allergy" shots for food allergens are not yet available but may be in the future.[31] If the eliminated food is an important source of nutrients or is found in many food products (such as milk or wheat), consultation with a registered dietitian is recommended. As noted earlier, for infants and young children, an allergy-producing food such as cow's milk, eggs, soy, or wheat can usually be reintroduced into the diet in a year or two. Most (but not all) infants outgrow food allergies by the age of 2 or 3 years. Children and adults with severe allergies to peanuts, nuts, fish, or seafood may have to eliminate the food from their diets for a lifetime.[32]

Food Intolerances

Food intolerances produce some of the same reactions as food allergies, but the reactions develop by different mechanisms. Food intolerance reactions do not involve the immune system; they are due to a missing enzyme or other cause.[33]

It is clear that some people are intolerant of lactose, sulfite, histamine (a component of red wine and aged cheese), and other foods (Table 17.2). It is also clear that not all food intolerances are real. Just as with food allergies, the best way to separate the real from the unreal is the double-blind, placebo-controlled food challenge. True food intolerances produce predictable reactions. Problems such as headache, diarrhea, swelling, or stomach pain will occur every time a person consumes a sufficient amount of the suspected food.

TABLE 17.2

FOODS AND SUBSTANCES IN THEM LINKED TO FOOD INTOLERANCE REACTIONS.

- Aged cheese
- Anchovies
- Aspartame
- Beer
- Catsup
- Chocolate
- Dried beans
- Lactose
- Pickled cabbage
- Red wine
- Sardines
- Sausage (hard, cured)
- Soy sauce
- Spinach
- Tomatoes
- Yeast extracts

Lactose Maldigestion and Intolerance

Lactose maldigestion, or the inability to break down lactose in dairy products due to the lack of the enzyme lactase, results in the condition known as *lactose intolerance.* Symptoms of lactose intolerance, such as flatulence, bloating, abdominal pain, diarrhea, and "rumbling in the bowel," occur in lactose maldigesters within several hours of consuming more lactose than can be broken down by the available lactase. These symptoms are due to the breakdown of undigested lactose by bacteria in the lower intestines (which produce gas as a by-product of lactose ingestion) and by fluid accumulation.[34]

Reduced intake of lactose-containing dairy products (milk, ice cream, and cottage cheese, for example) reduces lactose intolerance. However, foods such as hard cheese, low- or no-lactose milk, buttermilk, and yogurt without added milk solids contain low amounts of lactose and can generally be consumed. Small amounts of milk or other lactose-containing dairy products are generally well tolerated by people with lactose maldigestion. Care should be taken to ensure adequate intake of calcium and vitamin D if milk and dairy product intake is restricted.[35]

Lactose maldigestion is a very common disorder. It occurs in about 25% of U.S. adults, but the incidence varies dramatically by population group. By one set of estimates, 62 to 100% of Native American adults, 90% of Asian Americans, 80% of African Americans, 53% of Mexican Americans, and 15% of Caucasian Americans are lactose maldigesters.[36]

Sulfite Sensitivity

Sulfite is a food additive used by food manufacturers in the past to keep vegetables and fruits looking fresh and to prevent mold growth. It is also added to some beers, wines, processed foods, and medications as a preservative (see Table 17.3). Very small amounts of sulfite can cause anaphylactic shock and bring on an asthma attack in sensitive people.[37] Because many foods have added sulfite, people sensitive to sulfite should read food ingredient labels carefully. The FDA requires that processed foods containing sulfites list sulfites on the ingredient label. It also prohibits the use of sulfite on fresh vegetables and fruits.

Red Wine, Aged Cheese, and Migraines

Some people develop migraine headaches when they consume red wine. The presence of histamine in wine has been blamed for the headaches. People with this intolerance are unable to break down histamine during digestion, so it accumulates in the blood and causes headaches.[38] Histamine is also found in beer, sardines, anchovies, hard cured sausage, pickled cabbage, spinach, and catsup. Tyramine, a compound closely related to histamine and found in aged cheese, soy sauce, and other fermented products, causes migraine headaches in sensitive people. Chocolate, too, may bring on migraine headaches in sensitive individuals, although this finding is controversial.[39]

MSG

Monosodium glutamate (MSG) is related to dizziness, sweating, flushing, headache, and a ringing sound in the ears following the ingestion of foods, soups, and other food products that contain this flavor enhancer.[40] The reaction has been dubbed the "Chinese restaurant syndrome" because Chinese foods often contain MSG. Restaurants are often willing to prepare foods without MSG if requested.

lactose maldigestion
A disorder characterized by reduced digestion of lactose due to the low availability of the enzyme lactase.

lactose intolerance
The term for gastrointestinal symptoms (flatulence, bloating, abdominal pain, diarrhea, and "rumbling in the bowel") resulting from the consumption of more lactose than can be digested with available lactase.

TABLE 17.3

SOME FOODS THAT MAY CONTAIN SULFITE.

- Wine
- Beer
- Hard cider
- Tea
- Fruit juices
- Vegetable juices
- Guacamole
- Dried fruit
- Potato products
- Canned vegetables
- Baked goods
- Spices
- Gravy
- Soup mixes
- Jam
- Trail mix
- Fish and seafood

Photo Disc

Foods may be cross-contaminated with potential allergens during production or preparation. Yogurt-covered peanuts have been accidentally mixed with yogurt-covered raisins, and milk chocolate has been found to contain peanut fragments.[41]

Photo Disc

Precautions

People with food allergies or intolerances have to be very careful about what they eat and should have a plan of action ready in case they develop a serious reaction. Becoming highly knowledgeable about which foods and food products contain ingredients that cause an adverse reaction is essential. When eating out, people with allergies and intolerances should ask a lot of questions to make sure they know what is being served. Increasingly, food product ingredients such as cow's milk protein, egg white, sulfites, peanuts, and peanut oil are listed on food labels, and these should be checked carefully. Products that contain sulfites must be labeled. Genetically engineered foods (which contain proteins from another plant or animal) must thoroughly examine these foods and list known potential allergens such as soy or nut protein on the label.

Importantly, people who develop anaphylactic shock should always carry a preloaded syringe of epinephrine, or "epipen" and know how to use it. They should also know where to call for emergency help or be ready to dial 911.[42] People whose reactions to foods lessen with the use of asthma inhalers and antihistamines should keep them readily available.

One final thought about food allergies and intolerances—they are never caused by studying about them.

Key Terms

anaphylactic shock, page 17-4

antibodies, page 17-3

celiac disease, page 17-4

double-blind, placebo-controlled food challenge, page 17-6

food allergen, page 17-3

food allergy, page 17-2

food intolerance, page 17-2

histamine, page 17-3

immune system, page 17-2

lactose intolerance, page 17-9

lactose maldigestion, page 17-9

www links

www.nal.usda.gov/fnic/
Search food allergy and intolerance terms of interest and receive reliable information on specific types of food allergies and sensitivities, fact sheets, and statistics.

www.aaaai.org
Home page for the American Academy of Allergy, Asthma, and Immunology. Go from there to tips to remember, information on diagnosis, treatment, and prevention.

www.foodallergy.org
The Food Allergy and Anaphylaxis Network provides lots of information on food allergies and intolerances, how to avoid food allergens, advice on how to deal with school food service, support services, and recipes.

www.nlm.nih.gov/search.html
A great address for quickly getting to information from the National Institutes of Health on a variety of nutrition topics, including food allergies and intolerances.

www.mayoclinic.org
A great resource for scientific information and practical tips on food allergies.

www.healthfinder.gov
Search food allergy and food intolerance to find facts, specific information on peanut allergy, avoiding allergies, and treatment.

Nutrition UP CLOSE

Prevention of Food Allergies

FOCAL POINT: Foods that may cause allergies in the first year of life.

Scenario: Darla and Mitch have a 5-month-old son who has been breastfed exclusively. The time has come to add semisolid food to their son's diet. Mitch had food allergies as a child, so he and Darla are concerned that their son may develop allergies, too.

Question: If you were Darla or Mitch, what eight foods would you omit from your son's diet during the first year of life?

1. _____

2. _____

3. _____

4. _____

5. _____

6. _____

7. _____

8. _____

FEEDBACK (answers to these questions) can be found at the end of Unit 17.

Notes

1. Allergy statistics, www.nal.usda.gov/fnic/, accessed 9/03; and Some of the newest food allergies have nonfood triggers; how best to test, Envir Nutr Newsletter May 2000, vol. 23, no. 5, pp. 1, 6.

2. Skin testing for food allergy. WebMD Scientific American Online, www.WebMD.com, vol. 26, no. 9, Sept. 2003.

3. Food allergies at a glance. Today's Dietitian, Dec. 2002, p. 47.

4. Sampson HA. Peanut allergy. N Engl J Med 2002;346:1294–9.

5. Hallett R et al., Peanut allergies and kissing, N Engl J Med 2002;346:1833; and Koerner CB, Food allergies, Annual meeting of the American Dietetic Association, Boston, 1997 Oct. 28.

6. Sampson HA. State of the art theories on food allergies. Annual meeting of the American Dietetic Association, Boston, 1997 Oct. 28.

7. Leung DYM et al., Effect of anti-IgE therapy in patients with peanut allergy, N Engl J Med 2003;348:986–93; and Sicherer SH et al., Prevalence of peanut and tree nut allergy in the United States determined by a random digit dial telephone survey, J Allergy Clin Immunol 1999;103:559–62.

8. Leung et al., Effect of anti-IgE therapy; and Kuwayama SP. Peanut allergy. N Engl J Med 2002;347:1534.

9. Burks AW, Food hypersensitivity, Annual meeting of the American Dietetic Association, Boston, 1977 Oct. 28; and Leinhas JL et al., Food allergy challenges: guidelines and implications, J Am Diet Assoc 1987;87:604–8.

10. Allergy statistics (www.nal.usda.gov/fnic); and Food allergies at a glance, p. 47.

11. Allergy statistics (www.nal.usda.gov/fnic).

12. Farrell RJ, Kelley CP. Celiac sprue. N Engl J Med 2002;346:180–8.

13. Diagnosing gluten-sensitive enteropathy. WebMD Scientific American Online, www.WebMD.com, vol. 26, no. 9, Sept. 2003.

14. Murray JA. The widening spectrum of celiac disease. Am J Clin Nutr 1999;69: 354–65.

15. Taylor SL et al. Food allergies and avoidance diets. Nutr Today 1999;34: 15–22.

16. Falk MC et al. Food biotechnology: benefits and concerns. J Nutr 2002;132: 1384–90.

17. Allergy statistics (www.nal.usda.gov/fnic).

18. Sampson, Peanut allergy.

19. Sampson, Peanut allergy.

20. Allergy statistics (www.nal.usda.gov/fnic); Some of the newest food allergies; and Sampson, State of the art theories.

21. May CD. Food sensitivity: facts and fancies. Nutr Rev 1984;42:72–8.

22. May, Food sensitivity.

23. Dietary Reference Intakes. Energy, carbohydrate, fiber, fat, fatty acids, cholesterol, protein, and amino acids. Institute of Medicine, National Academy of Sciences; Washington, DC: National Academies Press, 2002.

24. Some of the newest food allergies; and What people mistake as food allergies, www.intelihealth.com, accessed 8/03.

25. Dutau G. Food allergies and diagnostic alternatives: labial food challenge, oral food challenge. Revue Francaise d'Allergologie et d'Immunologie Clinique. 2000;40:728–41.

26. Datau, Food allergies and diagnostic alternatives.

27. Skin testing for food allergy (www.WebMMD.com).

28. Sampson, State of the art theories.

29. May, Food sensitivity.

30. Ferguson A. Food sensitivity or self-deception? N Engl J Med 1990;323: 476–8.

31. Leung et al., Effect of anti-IIgE therapy.

32. Food allergies at a glance, p. 47.

33. Chandra RK, Food hypersensitivity and allergic disease: a selective review, Am J Clin Nutr 1997;66:526S–9S; and Fouchard T, Development of food allergies with special reference to cow's milk allergy, Pediatrics 1985;75(suppl): 177–81.

34. Tolstoi LG. Adult-type lactase deficiency. Nutr Today 2000;35:134–42.

35. Tolstoi, Adult-type lactase deficiency.

36. Inman-Felton AE. Overview of lactose maldigestion (lactose nonpersistence). J Am Diet Assoc 1999;99:481–89.

37. Yang WH, Purchase EC. Adverse reactions to sulfites. Can Med Assoc J 1985;133:865–7.

38. Jarisch R, Wantke F. Wine and headache. Int Arch Allergy Immunol 1996;110:7–12.

39. Hannington E, Preliminary report on tyramine headache, Br Med J 1967;2:550–1; and Chandra, Food hypersensitivity and allergic disease.

40. Dietary Reference Intakes.

41. What people mistake as food allergies (www.intelihealth.com)

42. Allergy statistics (www.nal.usda.gov/fnic).

Nutrition UP CLOSE

Prevention of Food Allergies

Feedback for Unit 17

Although it is not known whether their son would develop food allergies, Darla and Mitch would be wise to omit foods most likely to cause allergic reactions during the first year of life. These foods include:

1. nuts

2. eggs

3. wheat

4. milk

5. peanuts

6. soy

7. seafood

8. fish

Fats and Cholesterol in Health

Nutrition Scoreboard

		TRUE	FALSE
1	It is currently recommended that adults consume diets providing less than 30% of calories from fat.		
2	The types of fat consumed are more important to health than is total fat intake.		
3	Cholesterol is present in every food.		
4	Saturated fat intake has a much stronger influence on blood cholesterol levels than does cholesterol intake.		

Answers on next page

[KEY CONCEPTS AND FACTS]

- Fats are our most concentrated source of food energy. They supply 9 calories per gram.

- Dietary fats "carry" the essential fatty acids, fat-soluble vitamins, and healthful phytochemicals along with them in foods.

- Fats are not created equal. Some types of fat have positive effects, and some have negative effects on health.

- Saturated fats and trans fats raise blood cholesterol levels more than does dietary cholesterol or any other type of fat.

Answers to *Nutrition Scoreboard*

		TRUE	FALSE
1	Recommendations for fat intake have changed! The acceptable range of fat intake is now 20-35% of total calories.		✔
2	The types of fat consumed are more important to health than the total amount of fat.[1]	✔	
3	Cholesterol is present only in animal products.		✔
4	Saturated fat, which is found primarily in animal products, raises blood cholesterol levels a good deal more than does dietary cholesterol.	✔	

Changing Views about Fat Intake and Health

Scientific evidence and opinions related to the effects of fat on health have changed substantially in recent years—and so have recommendations about fat intake. In the past, it was recommended that Americans aim for diets providing less than 30% of total calories from fat. Evidence indicating that the *type* of fat consumed is more important to health than is total fat intake has changed this advice. The watchwords for thinking about fat have become "not all fats are created equal: some are better for you than others." American adults are being urged to select food sources of "healthy" or "good" fats while keeping fat intake within the range of 20–35% of total caloric intake. Concerns that high-fat diets encourage the development of obesity have been eased by studies demonstrating that excessive caloric intakes—and not just diets high in fat—are related to weight gain.[2]

New recommendations regarding fat intake do not encourage increased fat consumption. Rather, they emphasize that healthy diets include certain types of fat, and that total caloric intake and physical activity are the most important components of weight management. Diets providing as low as 20% of calories from fat, and those providing 30–35%, can be healthy—depending on the types of fat consumed and the quality of the rest of the diet.[3] This unit provides facts about fats, explains the reasons behind recent changes in recommendations for fat intake, and addresses the practical meaning of it all.

Facts about Fats

lipids
Compounds that are insoluble in water and soluble in fat. Triglycerides, saturated and unsaturated fats, and essential fatty acids are examples of lipids, or "fats."

Fats are a group of substances found in food. They have one major property in common: they are not soluble (or, in other words, will not dissolve) in water. If you have ever tried to mix vinegar and oil when making salad dressing, you have observed the principle of water and fat solubility firsthand.

Fats are actually a subcategory of the fat-soluble substances known as *lipids*. Lipids include fats, oils, and cholesterol. Dietary fats such as butter, margarine, and shortening are often distinguished from oils by their property of being solid at room temperature. This physical difference between fats and oils is due to their chemical structures.

Contributions of Dietary Fats

Fats in Foods Supply Energy and Fat-Soluble Nutrients]   Fats include the *essential fatty acids* (linoleic acid and alpha-linolenic acid) and provide the fat-soluble vitamins D, E, K, and A (the "deka" vitamins). So, part of the reason we need fats in our diet is to get a supply of the essential nutrients they contain (Table 18.1). Diets containing little fat (less than 20% of total calories) often fall short on delivering adequate amounts of essential fatty acids and fat-soluble vitamins.[4]

Fats Increase the Flavor and Palatability of Foods]   Although "pure" fats by themselves tend to be tasteless, they absorb and retain the flavor of substances that surround them. Thus, fats in meats and other foods pick up flavors from their environment and give those flavors to the food. This characteristic of fat is why butter, if placed next to the garlic in the refrigerator, tastes like garlic.

Fats Contribute to the Sensation of Feeling Full]   As they should, at 9 calories per gram! Fats tend to stay in the stomach longer than carbohydrates or proteins and are absorbed over a longer period of time. That's why foods with fat "stick to your ribs."[5]

A Crucial Role of Fat Is to Serve as a Component of Cell Membranes]   Some types of fats give cell membranes flexibility and help regulate the transfer of nutrients into and out of cells.[6]

Excess Dietary Fat, Carbohydrates, and Protein Are Stored as Fat

Fats in the body include those consumed in the diet and those produced from carbohydrates and proteins. Humans eat only a few times a day, but we need energy throughout the day. To ensure a constant supply of energy, the body converts carbohydrates and proteins, which have been supplied from foods and are not used to meet immediate needs, to storage forms of energy. Some of the excess carbohydrate and protein is converted to glycogen, the storage form of glucose under normal circumstances, but most of the excess is changed to fat and stored in fat cells (Illustration 18.1).

Body fat is not just skin deep. Fat is also located around organs such as the kidneys and heart. It's there to cushion and protect the organs and keep them insulated. Cold-water swimmers can attest to the effectiveness of fat as an insulation material. They purposefully build up body fat stores because they need the extra layer of insulation (Illustration 18.2).

Fats Come in Many Varieties

There are many types of fat in food and our bodies (Table 18.2). Of primary importance are *triglycerides* (or "triacylglycerols"), *saturated* and *unsaturated fats*, and *cholesterol* (for definitions, see Table 18.3). The different types of fats have different effects on health.

TABLE 18.1

ROLES OF DIETARY FAT.

- Provides a concentrated source of energy
- Carries the essential fatty acids, the fat-soluble vitamins, and certain phytochemicals
- Increases the flavor and palatability of foods
- Provides sustained relief from hunger
- Serves as a component of cell membranes

essential fatty acids
Components of fats (linoleic acid—pronounced lynn-oh-lay-ick and alpha-linolenic acid—lynn-oh-len-ick) required in the diet.

Illustration 18.1
A close look at fat cells (color-enhanced microphotograph).

Illustration 18.2
Although their body fat stores don't fit the image of the superathlete, cold-water swimmers need the fat to help stay warm. *Pictured here is the English swimmer Mike Read, who had swum the English Channel 20 times by age 39.*

Triglycerides, which consist of one *glycerol* unit (a glucose-like substance) and three fatty acids (Illustration 18.3), make up 98% of our dietary fat intake and the vast majority of our body's fat stores. Triglycerides are transported in blood attached to protein carriers and are used by cells for energy formation and tissue maintenance. A minority of fats take the form of *diglycerides* (glycerol plus two fatty acids) and *monoglycerides* (glycerol and one fatty acid). Diglycerides are present in some oils and small amounts are used in food products as emulsifiers—or to increase the blending of fat- and water-soluble substances. Monoglycerides are present in small amounts in some oils; we don't consume very much of them in foods.

As far as health is concerned, the glycerol component of fat is relatively unimportant. It's the fatty acids that influence what the body does with the fat we eat; and they are responsible, in part, for how fat affects health. Many different types of fatty acids are found in triglycerides. You've heard of the major ones: those that make fat "saturated" or "unsaturated."

Saturated and Unsaturated Fats] Fatty acids found in fats consist primarily of hydrogen atoms attached to carbon atoms (Illustration 18.4). When the carbons are attached to as many hydrogens as possible, the fatty acid is "saturated"—that is, saturated with hydrogen. Saturated fats tend to be solid at room temperature. Except for palm and coconut oil, only animal products are rich in saturated fats (Illustration 18.5 on p. 18–6).[7] Fatty acids that contain fewer hydrogens than the maximum are "unsaturated." They tend to be liquid at room temperature. By and large, plant foods are the best sources of unsaturated fats.

Unsaturated fats are classified by their degree of unsaturation. If only one carbon-carbon bond in the fatty acid is unsaturated, the fat is called "monounsaturated." If two or more carbon-carbon bonds are unsaturated with hydrogen, the fat qualifies as "polyunsaturated."

The Omega-3 and Omega-6 Fatty Acids

The essential fatty acids linoleic acid and alpha-linolenic acid are members of the fatty acid families of omega-6 (also called n-6 fatty acids) and omega-3 fatty acids (also known as n-3 fatty acids), respectively. Both are polyunsaturated, can be used as a source of energy, and are stored in fat tissue. Because they are essential, both linoleic and alpha-linolenic acid are required in the diet.

Linoleic acid is required for growth, maintenance of healthy skin, and normal functioning of the reproductive system. It is a component of all cell membranes and is found in particularly high amounts in nerves and the brain. A number of biologically active compounds, produced in the body, that participate in regulation of blood pressure and blood clotting are derived from linoleic acid. The major food sources of linoleic acid are sunflower, safflower, corn, and soybean oils.

TABLE 18.2

BASIC FACTS ABOUT THE TYPES OF FAT.

Fats can be:
- Monoglycerides
- Diglycerides
- Triglycerides

Fats can be:
- Saturated
- Monounsaturated
- Polyunsaturated

Unsaturated fats come in:
- "Cis" forms
- "Trans" forms

TABLE 18.3

A GLOSSARY OF FATS.

Triglycerides: Fats in which the glycerol molecule has three fatty acids attached to it; also called *triacylglycerol.* Triglycerides are the most common type of fat in foods and in body fat stores.

Saturated fats: Molecules of fat in which adjacent carbons within fatty acids are linked only by single bonds. The carbons are "saturated" with hydrogens; that is, they are attached to the maximum possible number of hydrogens. Saturated fats tend to be solid at room temperature. Animal products and palm and coconut oil are sources of saturated fats.

Unsaturated fats: Molecules of fat in which adjacent carbons are linked by one or more double bonds. The carbons are not saturated with hydrogens; that is, they are attached to fewer than the maximum possible number of hydrogens. Unsaturated fats tend to be liquid at room temperature and are found in plants, vegetable oils, meats, and dairy products.

Glycerol: A syrupy, colorless liquid component of fats that is soluble in water. It is similar to glucose in chemical structure.

Cholesterol: A fat-soluble, colorless liquid found in animals but not in plants. Cholesterol is used by the body to form hormones such as testosterone and estrogen and is a component of animal cell membranes.

Diglyceride: A fat in which the glycerol molecule has two fatty acids attached to it; also called diacylglycerol.

Monoglyceride: A fat in which the glycerol molecule has one fatty acid attached to it; also called monoacylglycerol.

Monounsaturated fats: Fats that contain a fatty acid in which one carbon-carbon bond is not saturated with hydrogen.

Polyunsaturated fats: Fats that contain a fatty acid in which two or more carbon-carbon bonds are not saturated with hydrogen.

Illustration 18.3
A triglyceride.

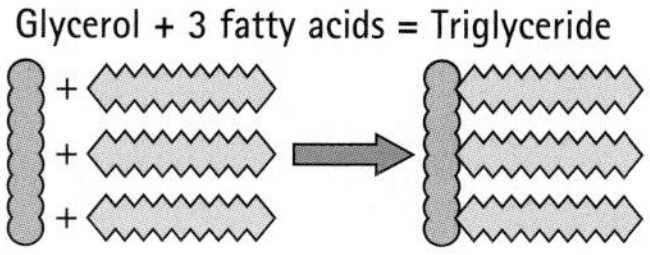

Glycerol + 3 fatty acids = Triglyceride

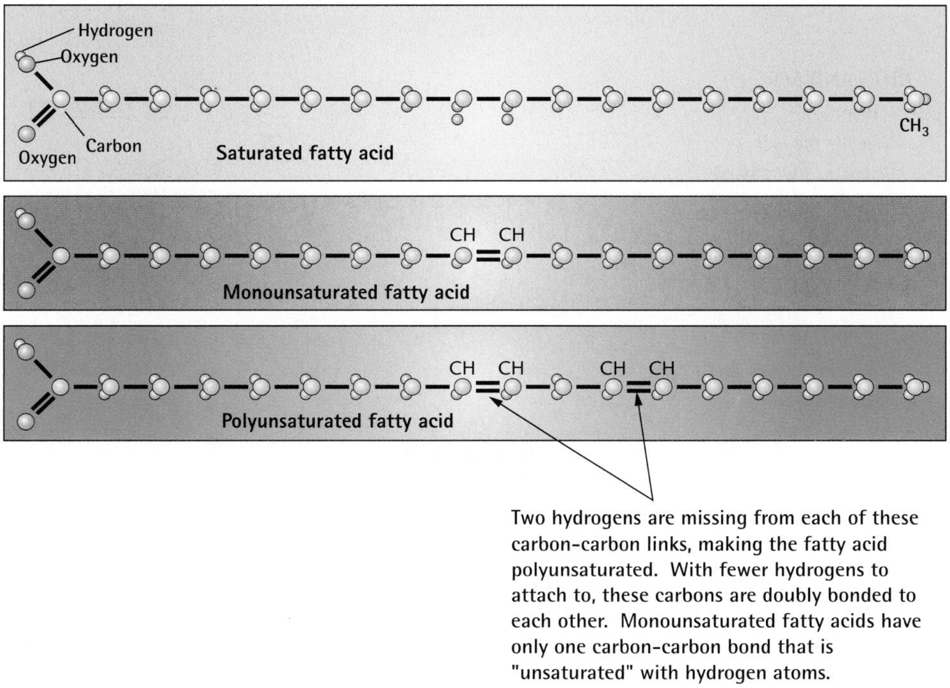

Illustration 18.4
A look at the difference between a saturated and an unsaturated fatty acid.

Two hydrogens are missing from each of these carbon-carbon links, making the fatty acid polyunsaturated. With fewer hydrogens to attach to, these carbons are doubly bonded to each other. Monounsaturated fatty acids have only one carbon-carbon bond that is "unsaturated" with hydrogen atoms.

Illustration 18.5
Fat profiles of selected foods.

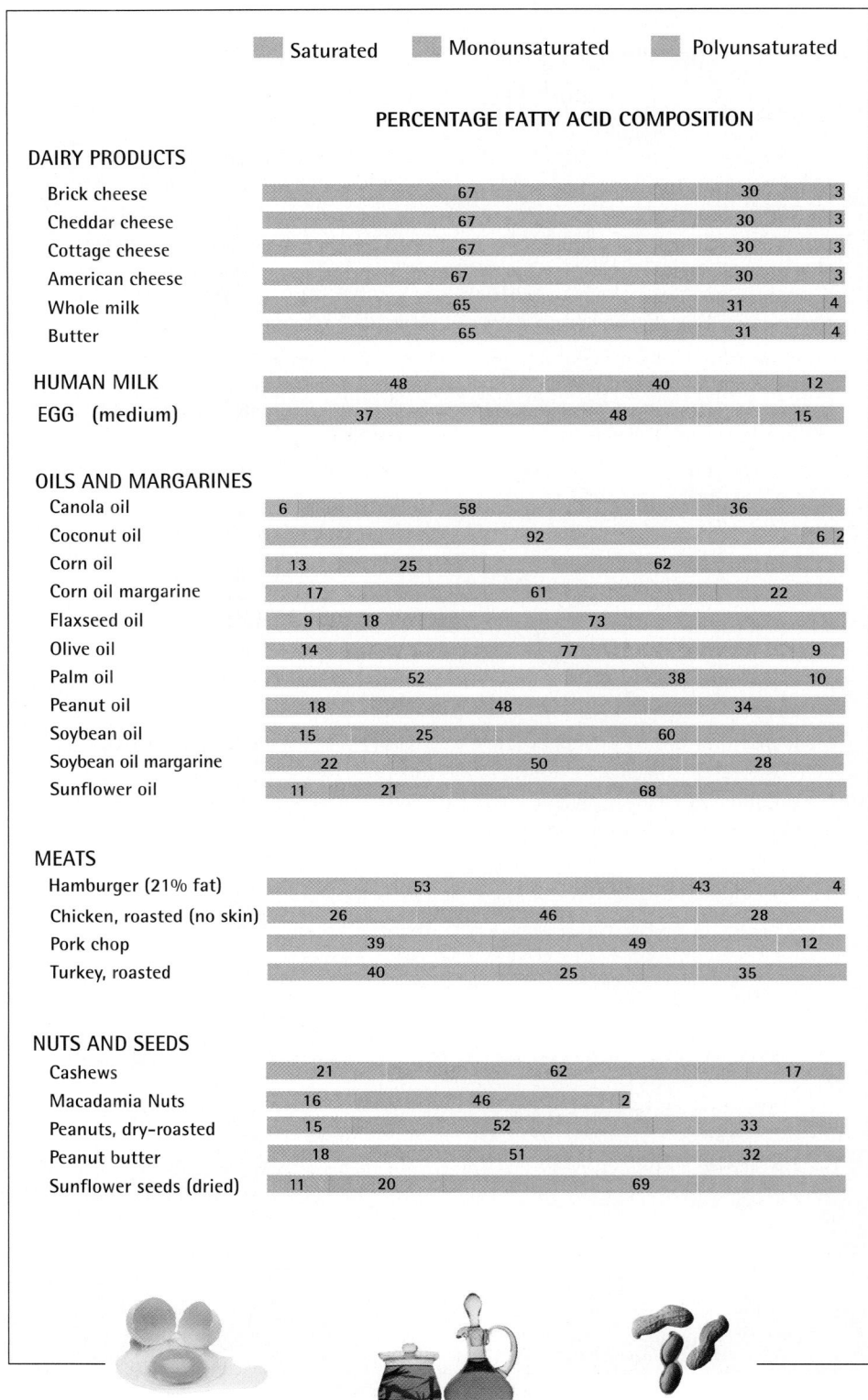

Alpha-linolenic acid, EPA (eicosapentaenoic acid—pronounced e-co-sah-pent-tah-no-ick) and DHA (docosahexaenoic acid: dough-cos-ah-hex-ah-no-ick) are primary members of the omega-3 fatty acid family. Alpha-linolenic acid is an essential nutrient, but EPA and DHA are not. These two members of the omega-3 fatty acid family can be produced in the body from alpha-linolenic acid. However, the conversion of this essential fatty acid to EPA and DHA is slow and results in the availability of relatively little EPA and DHA.[9] Alpha-linolenic acid is a structural component of all cell membranes and is found in high amounts in cells of the brain and eyes. Alpha-linolenic acid is also involved in the formation of biologically active compounds used in regulating blood pressure and blood clotting; but these compounds have the opposite effect on the blood pressure and blood clotting, as do linoleic acid derivatives.[10] Omega 3-fatty acids are found in a few foods such as walnuts, flaxseed, canola oil, and soybeans in the form of alpha-linolenic acid. The most beneficial sources of omega-3 fatty acids, however, are marine oils—due to their content of DHA and EPA.[11] These two omega-3 fatty acids play important roles in disease prevention and health promotion.

> *Omega-3 fatty acids are not just good fats,*
> *they affect heart health in positive ways.*
> —Penny Kris-Etherton, Distinguished Professor of Nutrition[12]

The Omega-3 Fatty Acids and Fish Oils] Many unsaturated fats contain fatty acids with one or two unsaturated carbon-to-carbon bonds. DHA and EPA found in fish oils, however, stand out in that they contain from 4 to 6 double bonds. This high level of unsaturation gives these omega-3 fatty acids the unique property of being highly fluid even at cold temperatures. This property allows the fatty acids to keep cell membranes and fat-containing tissues in fish flexible in cold water. Fatty fish and fish that live in cold water contain the highest amounts of DHA and EPA (Table 18.4).

In humans, DHA and EPA form biologically active compounds that reduce blood pressure and the tendency of blood to clot, and these properties confer health advantages. Regular consumption of these omega-3 fatty acids (two or more fish-containing meals per week) decreases the risk of heart attack, protects against irregular heartbeat and sudden death, decreases plaque formation in arteries, lowers high blood pressure, and decreases the risk of stroke. Fish oil capsules are being increasingly viewed as potential sources of EPA and DHA. They are sometimes recommended for people at risk of heart disease and should be taken under medical supervision. High levels of fish oils (over 3 grams per day) can prevent blood from clotting.[14]

Mercury in Fish and Fish Intake} Some fish contain high levels of mercury that can end up in fish oil supplements.[15] Mercury was found to be a potent fungicide over 100 years ago and was used extensively to prevent mold growth on crop seeds. Some of it ended up in streams, lakes, and oceans, and it lingers there for many decades. Several catastrophic outbreaks of mercury poisoning due to consumption of contaminated fish occurred during the mid-1900s and brought a halt to the use of mercury fungicides. Mercury is still found in some ocean and lake fish, but recent studies indicate that the benefits of fish intake among people over the age of 40 far outweigh risks related to mercury.[16]

Pregnant and breastfeeding women, and infants, are particularly susceptible to the harmful effects of mercury and are advised to omit from their diets the top four mercury-rich fish: shark, swordfish, king mackerel, and tilefish (also called golden snapper and golden bass). Intake of all other fish should be limited to 12 ounces or less per week among pregnant and breastfeeding women, and to 2 ounces or less for children under the age of 6 years.[17]

Local advisories related to mercury levels of fish in lakes and streams are available online (www.epa.gov/ost/fish), and advice on intake levels of fish caught in

ON THE SIDE

Fishy eggs? An odd source of fish oil has become part of North Americans' diets. Hens given feed with added cod liver oil produce eggs with yolks containing 100 to 350 milligrams of highly unsaturated omega-3 fatty acids.[8]

Eric Risberg/AP Photo

TABLE 18.4

OMEGA-3 FATTY ACID (EPA & DHA) CONTENT OF SELECTED SEAFOODS.[13]

SEAFOOD (3½-OUNCE SERVING)	OMEGA-3 FATTY ACIDS (GRAMS)	SEAFOOD (3½-OUNCE SERVING)	OMEGA-3 FATTY ACIDS (GRAMS)
Sardines in sardine oil	3.3	Oysters	0.6
Mackerel	2.6	Catfish	0.5
Salmon, Atlantic, farmed	2.2	Flounder	0.2
Lake trout	2.0	Shrimp	0.5
Herring	1.8	Halibut	0.5
Salmon, Atlantic, wild	1.8	Pollock	0.5
Tuna, white, canned	1.7	Scallops	0.5
Salmon, sockeye	1.5	Whiting	0.5
Whitefish, lake	1.5	Carp	0.3
Anchovies	1.4	Crab	0.3
Salmon, chinook	1.4	Pike, walleye	0.3
Bluefish	1.2	Tuna, fresh	0.3
Halibut	1.2	Catfish, wild	0.2
Trout, rainbow, farmed	1.2	Clams	0.2
Oysters	1.1	Fish sticks	0.2
Salmon, pink	1.0	Haddock	0.2
Trout, rainbow, wild	1.0	Lobster	0.2
Bass, striped	0.8	Salmon, red	0.2
Swordfish	0.8	Snapper, red	0.2

them should be heeded. Large fish (longer than 20 inches) in general contain higher concentrations of mercury than do smaller fish, so you may want to let the big ones get away . . .

Fish such as flounder, cod, trout, sole, salmon, tilapia, haddock, pollack, and "light" tuna—as well as shrimp and shellfish—generally contain low levels of mercury.[18] Other sources of DHA and EPA include eggs enriched with omega-3 fatty acids. The American Heart Association recommends that adults consume two or more servings of fish per week, or a total of 0.65 gram of DHA and EPA daily.[19]

Balancing Intake of Omega-6 and Omega-3 Fatty Acids] The ratio of omega-6 to omega-3 fatty acid intake is important because the functions of one are adversely modified by the presence of disproportionately high amounts of the other. Although an exact ratio has not been agreed upon, it is thought that people should consume omega-6 fatty acids and omega-3 fatty acids in a proportion of roughly 4 (or less) to 1. Many Americans regularly consume vegetable oils but eat fish infrequently. Consequently, the ratio between the intake of omega-6 and omega-3 fatty acids is over 9 to 1, indicating a need to increase intake of omega-3 fatty acids.[20]

Modifying Fats

Unsaturated fats aren't as stable as saturated fats. They are more likely to turn rancid with time and exposure to air and heat than are saturated fats. Additionally, solid fats are preferable to oils for some cooking applications. These problems with unsaturated fats have a solution. It's called hydrogenation.

hydrogenation
The addition of hydrogen to unsaturated fatty acids.

What's Hydrogenation?] *Hydrogenation* is a process that adds hydrogen to liquid unsaturated fats, thereby making them more saturated and solid. The shelf life, cooking properties, and taste of vegetable oils are improved in the process. Hydrogenation has two drawbacks, however. Hydrogenated vegetable oils contain more saturated fat than the original oil. Corn oil, for example, contains only 6% saturated fats; but corn oil margarine has 17%. The other negative is that hydrogenation causes a change in the structure of the unsaturated fatty acids. Specifically, hydrogenation converts some unsaturated fats into **trans fats.**

Trans Fatty Acids] Up to 30% of the fatty acids in unsaturated fat molecules may be converted from their naturally occurring "cis" to the "trans" form as a result of hydrogenation.[21] Fatty acids in this structural form raise blood cholesterol levels more than dietary cholesterol, saturated fats or any other type of fat do.[22] The bulk of trans fat in our diet comes from hydrogenated vegetable oils. Trans fats are more stable and have a longer shelf life than other fats and are preferred for use in margarine, snack foods, bakery products, and fried foods (Table 18.5). Ruminant animals like cows form some trans fat in the stomach; so milk, other dairy products, beef, and lamb contain small amounts of this type of fat. People in the United States consume an average of 2.6% of total calories from trans fat, or 5.8 grams per day.[23]

It is recommended that Americans consume as little trans fats as possible,[24] and new nutrition information labeling requirements are making that easier to accomplish. Nutrition Facts panels must include the trans fat content of food products by January 1, 2006 (Illustration 18.6). The %DV column (for percent of Daily Value) will not be used for trans fats because there is no recommended level of intake. Products labeled "trans fat–free" must contain less than 0.5 gram of both trans and saturated fats. It is expected that the requirement to label the trans fat content of food products will increase the number of foods labeled "trans fat–free" and "no trans fat" (Illustration 18.7) and decrease its use in foods. Food companies are busy developing other ways to produce foods without them.[25]

trans fats
Unsaturated fatty acids in fats that contain atoms of hydrogen attached to opposite sides of carbons joined by a double bond:

H
–C=C–
      H   H H
                        –C=C–
Trans fatty   Cis fatty
  acid   acid

Fats containing fatty acids in the trans form are generally referred to as trans fats. Cis fatty acids are the most common, naturally occurring form of unsaturated fatty acids. They contain hydrogens located on the same side of doubly bonded carbons.

Checking Out Cholesterol

Cholesterol is a lipid found *only* in animal products. It is tasteless and odorless and contained in both the lean and fat parts of animal products. Table 18.6 lists some sources of cholesterol. Plants don't contain cholesterol because they can't produce it and don't need it to function and grow normally.

Sources of Cholesterol

The cholesterol used by the body comes from two sources. Most (about two-thirds) of the cholesterol available to the body is produced by the liver. The rest comes from the diet (Illustration 18.8). Because the liver produces cholesterol from other substances in our diet, it does not qualify as an essential nutrient.

The Contributions of Cholesterol

Would you be surprised to learn that cholesterol:

- is found in every cell in your body?
- serves as the building block for estrogen, testosterone, and the vitamin D that is produced in your skin upon exposure to sunlight?

TABLE 18.5	
TRANS FATTY ACID CONTENT OF FOOD PRODUCTS. Values may change as companies lower the trans fat content of foods.	
FOOD	TRANS FATTY ACIDS (GRAMS)
Kentucky Fried Chicken pot pie	8
French fries, large serving	4–7
Cake doughnut	6
Breaded fish sticks, 3	5
Margarine, 1 tbs	4
Dutch apple pie	3–4
Wheat crackers	3
Chocolate chip cookie	2
Snack crackers,1 2 oz	2
Shortening, 1 tbs	1
Tub margarine, 1 tbs	1
Butter, 1 tbs	0
"No trans fat" margarine	0
Olive oil	0

Illustration 18.6

Trans fat: the newest addition to Nutrition Facts panels.

Nutrition Facts

Serving Size 1 Entree
Serving Per Container 1

Amount Per Serving

Calories 380 Calories from Fat 170

	%Daily Value
Total Fat 19g	**29%**
Saturated Fat 10g	**50%**
Trans Fat 2g	
Cholesterol 85g	**28%**
Sodium 810mg	**34%**
Total Carbohydrate 33g	**11%**
Dietary Fiber 3g	**12%**
Sugars 5g	
Protein 20g	

Vitamin A 10%	Vitamin C 0%
Calcium 10%	Iron 15%

Percent Daily Values are based on a 2000 calorie diet. Your daily values may be higher or lower depending on your calorie needs:

	Calories	2000	2500
Total Fat	Less Than	65g	80g
Sat Fat	Less Than	20g	25g
Cholesterol	Less Than	300mg	300mg
Sodium	Less Than	2400mg	2400mg
Total Carbohydrate		300g	375g
Dietary Fiber		25g	30g

Illustration 18.7

Products that feature "no trans fats" and "trans fat–free" labels.

Richard Anderson

- is a major component of nerves and the brain?
- cannot be used for energy (so it provides no calories)?

The body has many uses for cholesterol (Table 18.7). It doesn't just accumulate in arteries!

Fat Substitutes

The fat content of processed foods can be partially or fully replaced by fat substitutes. Fat substitutes attempt to imitate the taste, texture, and cooking properties of our favorite fats, but with fewer calories. Table 18.8 lists some of the more than 60 products that are already on the market; others are under development. The degree to which these products succeed in imitating the qualities of fat varies. Making the

TABLE 18.6

FOOD SOURCES OF CHOLESTEROL.

Note that all the sources are animal products. Cholesterol in foods is a clear, oily liquid found in the fat and lean portions of many animal products.

ANIMAL PRODUCT	AMOUNT	CHOLESTEROL (MILLIGRAMS)
Brain	3 oz	1746
Liver	3 oz	470
Egg	1	212
Veal	3 oz	128
Shrimp	3 oz	107
Prime rib	3 oz	80
Chicken (no skin)	3 oz	75
Turkey (no skin)	3 oz	65
Hamburger, regular	3 oz	64
Pork chop, lean	3 oz	60
Fish, baked (haddock, flounder)	3 oz	58
Ice cream	1 cup	56
Sausage	3 oz	55
Hamburger, lean	3 oz	50
Milk, whole	1 cup	34
Crab, boiled	3 oz	33
Lobster	3 oz	29
Cheese (cheddar)	1 oz	26
Milk, 2%	1 cup	22
Yogurt, low-fat	1 cup	17
Milk, 1%	1 cup	14
Butter	1 tsp	10
Milk, skim	1 cup	7

perfect fat substitute is not an easy task. Not all of the fat substitutes that have entered the marketplace have met the test of consumer acceptability.[28]

Are Fat Substitutes Safe?

Consumers sometimes wonder about the safety of fat substitutes. Because most fat substitutes are derived from ingredients of food such as carbohydrates, protein, and vegetable oils, they are often assumed to be safe. Fat replacers that are made from substances that do not occur naturally in foods must be tested for safety. Olean, a non-naturally occurring fat replacer that cannot be digested, was tested extensively before being approved for use. It can cause diarrhea and oily stools if consumed in large amounts.[29]

Can Fat Substitutes Benefit Health?

When substituted for foods containing saturated fats, fat substitutes may benefit health by lowering blood cholesterol levels. Studies suggest that the use of foods containing fat substitutes reduces total fat consumption, but whether they lead to weight loss is controversial. So, although they may not be the answer to weight control, fat substitutes may provide part of the solution to high fat intakes and the health problems related to them.[30]

Finding Out about the Fat Content of Food

Not all of the fat in food is visible. To avoid being fooled, it helps to use a reference on the fat composition of foods. Table 18.9 lists the fat content of common food sources of fat, including candy. Vegetables and fruits (except avocado and coconut) and grains are not listed because they contain relatively little fat. Other references can also be used, such as the food composition tables in Appendix A, the Diet Analysis Plus Program software, and the nutrition labels on food products.

TABLE 18.7

HOW THE BODY USES CHOLESTEROL.

- Cholesterol is a component of all cell membranes, the brain, and nerves.
- It is needed to produce estrogen, testosterone, and vitamin D.

Illustration 18.8
Food sources of cholesterol in the U.S. diet.[26]
Percentages indicate the proportion of cholesterol each type of food contributes to the diet.

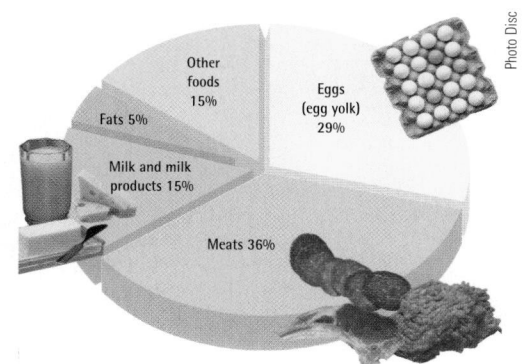

Other foods 15%
Eggs (egg yolk) 29%
Fats 5%
Milk and milk products 15%
Meats 36%
Photo Disc

TABLE 18.8

SOME FAT SUBSTITUTES.[27]

CARBOHYDRATE-BASED (0–4 CAL/G)	PROTEIN-BASED (1–2 CAL/G)	FAT-BASED (0–5 CAL/G)	COMBINATIONS
Amalean I and II	K-Blazer	Benefat	Fitesse
CrystaLean	Proplus	Caprenin	Nutrifat
Instant Stellar	Simplesse	Olean	Prolestra
Juguar			
Litesse			
Maltrin			
N-Lite			
Oat fiber			
Oatrim			
Opta Grade			
Pure-gel			
Sta-Slim			
Sta-Lite			

Knowledge of the caloric and fat content of a food can be used to calculate the percentage of calories provided by fat. For example, suppose that a slice of cherry pie provides 350 calories and 15 grams of fat. To calculate the percentage of fat calories, multiply 15 grams by 9 (the number of calories in each gram of fat), divide the result by 350 calories, and then multiply by 100:

$$15 \text{ grams fat} \times 9 \text{ calories/gram} = 135 \text{ calories}$$
$$135 \text{ calories}/350 = 0.39$$
$$0.39 \times 100 = 39\% \text{ of total calories from fat}$$

Fat Labeling

Nutrition labeling regulations for fat require food manufacturers to adhere to standard definitions of "low fat," "fat-free," and related terms used on food labels. Similarly, claims made about the cholesterol content of food products must comply with standard definitions (Table 18.10, on page 18-14). If a claim is made about the fat content of a food, the Nutrition Facts panel must specify the food's fat, saturated fat, trans fat, and cholesterol content. If a claim is made about cholesterol content (and claims can be made only for products that normally contain cholesterol), the nutrition panel must also reveal the product's fat and saturated fat content. To prevent the use of unrealistically small serving sizes as a way to appear to cut down on a product's fat content, standard serving sizes must be used on food labels.

TABLE 18.9

THE FAT CONTENT OF SOME FOODS.

FOOD	AMOUNT	GRAMS	PERCENTAGE OF TOTAL CALORIES FROM FAT
Fats and oils			
Butter	1 tsp	4.0	100%
Margarine	1 tsp	4.0	100
Oil	1 tsp	4.7	100
Mayonnaise	1 tbs	11.0	99
Heavy cream	1 tbs	5.5	93
Salad dressing	1 tbs	6.0	83
Meats and fast foods			
Hot dog	1 (2 oz)	17.0	83
Bologna	1 oz	8.0	80
Sausage	4 links	18.0	77
Bacon	3 pieces	9.0	74
Salami	2 oz	11.0	68
Pork steak	3 oz	18.0	62
Hamburger, regular (20% fat)	3 oz	16.5	62
Chicken, fried with skin	3 oz	14.0	53
Big Mac	6.6 oz	31.4	52
Quarter Pounder with cheese	6.8 oz	28.6	50
Whopper	8.9 oz	32.0	48
Steak (rib eye)	3 oz	9.9	47
Hamburger, lean (10% fat)	3 oz	9.5	45
Steak (T-bone), lean	3 oz	8.9	44
Rabbit	3 oz	7.0	38
Veggie pita	1	17.0	38
Ranch chicken pita	1	18.0	34

TABLE 18.9

THE FAT CONTENT OF SOME FOODS. (CONTINUED)

FOOD	AMOUNT	GRAMS	PERCENTAGE OF TOTAL CALORIES FROM FAT
Meats and fast foods—*continued*			
Steak (round), lean only	3 oz	5.2	29
Chicken, baked without skin	3 oz	4.0	25
Turkey wrap	1	9.0	24
Hamburger, extra lean (4% fat)	3 oz	2.3	23
Venison	3 oz	2.7	18
Subway, club	6"	5.0	14
Flounder, baked	3 oz	1.0	13
Subway, veggie	6"	3.0	11
Shrimp, boiled	3 oz	1.0	10
Milk and milk products			
Cheddar cheese	1 oz	9.5	74
American cheese	1 oz	6.0	66
Milk, whole	1 cup	8.5	49
Cottage cheese, regular	½ cup	5.1	39
Milk, 2%	1 cup	5.0	32
Milk, 1%	1 cup	2.7	24
Cottage cheese, 1% fat	½ cup	1.2	13
Milk, skim	1 cup	0.4	4
Yogurt, frozen	¾ cup	0.0–6.6	0–3
Other			
Olives	4 medium	1.5	90
Avocado	½	15.0	84
Almonds	1 oz	15.0	80
Sunflower seeds	¼ cup	17.0	77
Peanut butter	1 tbs	8.0	76
Peanuts	¼ cup	17.5	75
Cashews	1 oz	13.2	73
Egg	1	6.0	61
Potato chips	1 oz (13 chips)	11.0	61
Chocolate chip cookies	4	11.0	54
French fries	20 fries	20.0	49
Taco chips	1 oz (10 chips)	6.2	41
Candy			
Mr. Goodbar	1.7 oz	15.0	56
Peanut butter cups, 2 regular	1.6 oz	15.0	54
Milk chocolate	1.6 oz	14.0	53
Almond Joy	1.8 oz	14.0	50
Kit Kat	1.5 oz	12.0	47
M and M's, peanut	1.7 oz	13.0	47
Nestlé's Crunch	1.6 oz	11.0	45
Twix	2.0 oz	14.0	45
Baby Ruth	2.1 oz	14.0	43
Pay Day	1.9 oz	12.0	43
Snickers	2.1 oz	13.0	42
Butterfinger	2.1 oz	12.0	39
M and M's, plain	1.7 oz	10.0	39
Milky Way	2.2 oz	11.0	35
3 Musketeers	2.1 oz	9.0	31
Tootsie Roll	2.3 oz	6.0	21

TABLE 18.10

WHAT CLAIMS ABOUT THE CHOLESTEROL CONTENT OF FOODS THAT NORMALLY CONTAIN CHOLESTEROL MUST MEAN.

- No cholesterol or cholesterol-free: Contains less than 3 milligrams of cholesterol per serving.
- Low cholesterol: Contains 20 milligrams or less of cholesterol per serving.
- Reduced cholesterol: Contains at least 75% less cholesterol than normal.
- Less cholesterol: Contains at least 25% less cholesterol than normal. The percentage less must be stated on the label.

Reasons for the Revised Recommendations for Fat Intake

Several lines of evidence led to the recent changes in recommendations for fat intake. One line of evidence has taught us that although people tend to lose weight on low-fat diets, people also lose weight on high-fat, low- or high-carbohydrate, and low- or high-protein diets. Whether people lose weight and keep it off depends strongly on the acceptability of the dietary changes made, and not on the relative proportions of fat, carbohydrate, and protein content in the diet that lead to weight loss.[31]

Fat consumption among Americans decreased from 43% of total calories in 1970 to around 33% of calories in recent years. Yet, the incidence of overweight and obesity in the United States has increased since 1970. Consequently, it is difficult to argue that high fat diets are at the root of the obesity epidemic.[32] As fat intake has decreased, carbohydrate intake has increased. Scientists are asking the question about whether high intakes of carbohydrate are related to rising rates of obesity. Some evidence suggests that consumption of high-carbohydrate diets by people with low levels of physical activity and obesity may hasten the development of type 2 diabetes and other disorders related to insulin resistance.[33]

The experience of groups of people who have traditionally consumed high-fat diets yet have average or below-average rates of heart and other chronic diseases has taught us to view fat intake in the context of overall diet and lifestyle.[34] A classic example of the paradox between high-fat diets and low disease rates comes from Greece and the traditional Mediterranean diet. This diet is based on whole grain products, vegetables, fruits, nuts, olive oil, dried beans, wine, fish, and poultry (Illustration 18.9). Over 40% of the calories in the diet come from fat, most of which is provided in the form of monounsaturated fats. People consuming the traditional Mediterranean diet tend to be physically active and of normal weight or lean. Despite this diet's high fat content, populations consuming the diet and living the typical lifestyle have low rates of heart disease and cancer, and long life expectancy.[35]

A different example making the point that fat intake and health relationships should be evaluated in the context of the overall diet and lifestyles come from Nigeria. The Fulani pastoralists of Northern Nigeria subsist on a diet primarily composed of animal blood, meat, and dairy products. It is very high in fat and rich in saturated fat—yet the Fulani have normal levels of blood lipids and are not at increased risk for heart disease. The Fulani are highly physically active, lean, and tend not to smoke.[36]

Rates of heart disease, cancer, type 2 diabetes, and obesity tend to increase in populations as they move from traditional diets and high levels of physical activity to Western-type diets and sedentary lifestyles.[37] In the new context, the relationship between fat intake and health changes—it becomes similar to the situation in the United States and Canada. High-fat diets in these two countries are often high in calories and saturated fat and low in vegetables, fruits, and whole grain products. General levels of physical activity in these countries tend to be low, and rates of obesity high. A high-fat, high-saturated fat diet under these circumstances is related to increased blood cholesterol levels and the risk of heart disease.[38] New recommendations for fat intake take into consideration the effects on health of different sources and amounts of dietary fat, and the potentially problematic influence of low-fat diets on increasing carbohydrate intake in populations that tend to be inactive and obese. Recommendations for fat intake are changing in part because of evidence pointing to the fact that some types of fat are better for health than others.

Illustration 18.9
A look at the cuisine of the Mediterranean diet.

Good Fats, Bad Fats

Fats come in many types in foods, and with few exceptions, they serve as a source of energy and provide a number of essential functions in the body. With regard to raising or lowering the risk of heart disease and stroke, however, fats differ. Those that elevate total cholesterol and LDL-cholesterol levels are regarded as "bad" or "unhealthy" fats. Those that lower total cholesterol and LDL-cholesterol and raise blood levels of HDL-cholesterol (the one that helps the body get rid of cholesterol in the blood) are considered "good" or healthy."[39]

The list of unhealthy fats includes trans fats, saturated fats, and cholesterol. Fats labeled bad are generally solid at room temperature and are included in foods such as high-fat meats and dairy products, hard margarines, shortening, and crispy snack foods.[40] Monounsaturated fats, polyunsaturated fats, alpha-linolenic acid, DHA, and EPA are considered healthy fats and are present in food in the form of oils (Table 18.11).

REALITY CHECK

Good fats, bad fats
What foods provide "healthy" fats

Who gets thumbs up

?

Paprika:
How can I be wrong? Low-fat food products are best for healthy fat because they contain almost no fat!

Butch:
I'm thinking foods like fish, peanut butter, and trans fat-free margarine contain healthy fats.

Answers on next page

Paprika

ANSWERS TO **REALITY CHECK**

Good fats, bad fats

Low-fat foods contain less fat than the regular version of the foods. But that doesn't mean the products contain no fat, or only good fats. Food sources of fish oils, unsaturated fat, and trans fat-free products provide the healthy fats. As always, healthy diets aren't based on individual foods, they are based on overall diets. You can emphasize foods providing healthy fats without feeling bad about occasionally eating foods branded with the bad fat label.

Butch

Recommendations for Fat and Cholesterol Intake

Current recommendations for adults call for consumption of 20–35% of total calories from fat. The AIs (Adequate Intakes) for the essential fatty acid linoleic acid is set at 17 grams a day for men and 12 grams for women. AIs for the other essential fatty acid, alpha-linolenic acid, are 1.6 grams per day for men and 1.1 grams for women. It is recommended that intake of trans fats and saturated fats be as low as possible while consuming a nutritionally adequate diet. Only a small proportion of Americans consume too little linoleic acid, but intakes of alpha-linolenic acid tend to be low. Americans are being encouraged to increase consumption of EPA and DHA by eating fish more often. In addition, saturated fat intake averages 11–12% of calories, an amount that increases the risk of heart disease.[41]

There is no recommended level of cholesterol intake, because there is no evidence to indicate that cholesterol is required in the diet. The body is able to produce enough cholesterol, and people do not develop a cholesterol deficiency disease if it

TABLE 18.11

HEALTHY AND UNHEALTHY FATS AND EXAMPLES OF FOOD SOURCES.

HEALTHY FATS	UNHEALTHY FATS
DHA, EPA (omega-3 fatty acids) fish and seafood	**Trans fats** Snack and fried foods, bakery goods
Monounsaturated fats Olive and peanut oil, nuts, avocados	**Saturated fats** Animal fats
Polyunsaturated fats Vegetable oils	**Cholesterol** eggs, seafood, and meat
Alpha-linolenic acid Soybeans, walnuts, flaxseed	

is not consumed. Because blood cholesterol levels tend to increase somewhat as consumption of cholesterol increases, it is recommended that intake should be minimal. Although cholesterol intake averages around 250 mg per day in the United States, a more health-promoting level of intake would be less than 200 mg a day.[42]

Recent recommendations for fat intake represent an unusually large but necessary change in dietary intake guidance. The rationale for the changes has been developing for years as research results emerged, showing consistent results that supported the new recommendations. Much remains to be understood about the effects of dietary fats on health, and how other components of the diet and lifestyle and genetic traits modify relationships between fat intake and health. A beneficial by-product of the new recommendations for fat intake is that nutrient and health relationships are much more likely to be studied in the context of overall diets and lifestyles in the future.

Nutrition UP CLOSE

The Healthy Fats in Your Diet

FOCAL POINT: Identify your healthy fat food choices.

Are the fats in your diet the healthy type? Check it out by answering these questions:

How Often Do You Eat:

	Seldom or Never	1–2 Times per Week	3–5 Times per Week	Almost Daily
1. Sausage, hot dogs, ribs, and luncheon meats?	☐	☐	☐	☐
2. Heavily marbled steaks or roasts and chicken with the skin?	☐	☐	☐	☐
3. Soybean products such as tofu or soynuts?	☐	☐	☐	☐
4. Nuts or seeds?	☐	☐	☐	☐
5. Whole milk, cheese, or ice cream?	☐	☐	☐	☐
6. Soft margarine or olive oil?	☐	☐	☐	☐
7. French fries, snack crackers, commercial bakery products?	☐	☐	☐	☐
8. Rich sauces and gravies?	☐	☐	☐	☐
9. Fish or seafood?	☐	☐	☐	☐
10. Peanut butter?	☐	☐	☐	☐

FEEDBACK (including scoring) can be found at the end of Unit 18.

Key Terms

essential fatty acids, page 18-3   *lipids*, page 18-2   *trans fats*, page 18-9

hydrogenation, page 18-8

www links

www.nlm.nih.gov/medlineplus/
dietaryfat.html
Provides a menu that connects you to sites such as good and bad fats, interpreting blood lipid profiles, benefits of flaxseed, and fat substitutes.

www.nal.usda.gov/fnic
Find out more about fats and fat substitutes from the extensive list of topics covered under the search term "fats."

www.healthfinder.gov
Good source of information on fats, trans fat, and cholesterol through search terms such as dietary fat and healthy fats.

www.mayoclinic.org
Healthy fats, bad fats, know your fats, fats and heart disease, and other topics are intelligently covered in sites available through the Mayo Clinic's home page.

www.epa.gov/ost/fish
The Environmental Protection Agency's site for looking up national and local advisories on fish contamination and consumption.

Notes

1. Dietary Reference Intakes, Energy, carbohydrate, fiber, fat, fatty acids, cholesterol, protein, and amino acids. Institute of Medicine, National Academy of Sciences, Washington, DC: National Academies Press; 2002; and Mensink RP et al., Effects of dietary fatty acids and carbohydrates on the ratio of serum total to HDL cholesterol and serum lipids, Am J Clin Nutr 2003;77:1146–55.

2. Dietary Reference Intakes.

3. Fat in your diet. How low should you go? www.mayoclinic.com/invoke .cfm?id=HQ00670, accessed 6/03.

4. Dietary Reference Intakes.

5. Fats: the good and the bad. www .mayoclinic.com/invoke.cfm?id=NU00262, accessed 9/03.

6. Connor WE. Importance of n-3 fatty acids in health and disease. Am J Clin Nutr 2000;71(suppl):171S–5S.

7. Hepburn FN et al. Provisional tables on the content of omega-3-fatty acids and other fat components of selected foods. J Am Diet Assoc 1986;86:788–93.

8. Everything you always wanted to know about those newfangled eggs. Nutr Today 2003;38:75.

9. Dietary Reference Intakes.

10. Vanschoonbeek K et al. Fish oil, consumption and reduction of arterial disease. J Nutr 2003;133:657–60.

11. Krauss RM et al., Revision 2000: a statement for healthcare providers from the Nutrition Committee of the American Heart Association, J Nutr 2001; 131:132-46; and Kris-Etherton PM et al. New guidelines focus on fish, fish oil, omega-3 fatty acids. AHA Statement. Circulation, 2002;Oct. 18.

12. Kris-Etherton, New guidelines.

13. Connor, Importance of n-3 fatty acids; and Hepburn et al., Provisional tables.

14. Kris-Etherton, New guidelines.

15. Stone NJ. Fish consumption, fish oil, lipids, and coronary heart disease. Am J Clin Nutr 1997;65:1083–6.

16. Kris Etherton, New guidelines; and Clarkson TW, Strain JJ, Nutrition factors may modify the toxic action of methyl mercury in fish-eating populations, J Nutr 2003;133:1539S-43S.

17. Kris-Etherton, New guidelines.

18. Kris-Etherton, New guidelines.

19. Krauss et al., Revision 2000.

20. Kris-Etherton PM et al. Polyunsaturated fatty acids in the food chain in the United States. Am J Clin Nutr 2000;71 (suppl):179S–88S.

21. Walsh J. Low fat, no fat, some fat . . . high fat? Envir Nutr 1998;21(April):1, 6.

22. Krauss et al., Revision 2000.

23. Stuppy P, Transitioning away from trans fatty acids, Today's Dietitian 2003;Jan.: 12–14; and Dietary Reference Intakes.

24. Dietary Reference Intakes.

25. Vranico S. PepsiCo sets health-snack effort. Wall Street Journal 2003, Sept. 23, p. B6.

26. Sabar AF et al. Dietary sources of nutrients among U.S. adults. J Am Diet Assoc 1998;98:537–47.

27. Gershoff SN, Nutrition evaluation of dietary fat substitutes, Nutr Rev 1995;53:305–13; and Mattes RD, Position of the American Dietetic Association: fat replacers, J Am Diet Assoc 1998;98:463–8.

28. Mattes, Position of the ADA: fat replacers.

29. Bray GA et al. A 9-month randomized controlled trial comparing fat-substituted and fat-reduced diets in healthy obese men: the Ole Study. Am J Clin Nutr 2002;76:928–34.

30. Mattes, Position of the ADA: fat replacers; and Bray et al., A 9-month randomized controlled trial.

31. Olson RE. Dietary fats: friend or foe? Nutr Notes, American Society of Nutritional Sciences, 2000;Mar.:3.

32. Olson, Dietary fats: friend or foe?

33. Dietary Reference Intakes; and Schwartz J-M et al., Hepatic de novo lipogenesis in normoinsulinemic and hyperinsulinemic subjects consuming high-fat, low-carbohydrate and low-fat, high carbohydrate isoenergetic diets, Am J Clin Nutr 2003;77:43–50.

34. Dietary Reference Intakes.

35. Trichopoulou A et al. Adherence to a Mediterranean diet and survival in a Greek population. N Engl J Med 2003;348:2599–608.

36. Glew RH et al. Cardiovascular disease risk factors and diet of the Fulani pastoralists of Northern Nigeria. Am J Clin Nutr 2001;74:730–6.

37. Hu FB. The Mediterranean diet and mortality—olive oil and beyond. N Engl J Med 2003;348:2595–6.

38. Dietary Reference Intakes; and Krauss et al., Revision 2000.

39. Fats: the good and the bad (www .mayoclinic.com/invoke.cfm?id= NU00262); and Kris-Etherton et al., New guidelines.

40. Fats: the good and the bad (www .mayoclinic.com/invoke.cfm?id= NU00262).

41. Dietary Reference Intakes.

42. Krauss et al., Revision 2000.

Nutrition UP CLOSE

The Healthy Fats in Your Diet

Feedback for Unit 18

Give yourself a point for each time you checked the "3–5 Times per Week" or "Almost Daily" columns for numbers 3, 4, 6, 9, and 10. These foods are sources of healthy unsaturated fats or DHA and EPA. Take a point away for each time your answer ended up in the same columns for foods listed in numbers 1, 2, 5, 7, and 8. These foods provide saturated or trans fats. If you have any points left, your selection of food sources of fat regularly include healthy fats.

Nutrition and Heart Disease

Nutrition Scoreboard

	TRUE	FALSE
1 Heart disease is the leading cause of death among men and women in the United States.		
2 The most effective way to lower blood cholesterol is to exercise regularly.		
3 Inadequate intake of folate (a B vitamin) is a risk factor for heart disease.		
4 The risk factors for heart disease are the same for women as for men.		

Answers on next page

[KEY CONCEPTS AND FACTS]

- Heart disease is the leading cause of death in men and women in the United States.[1]

- Of the multiple contributors to heart disease, dietary and lifestyle factors are among the most important.[2]

- Moderate-fat diets that provide primarily "healthy fats" decrease heart disease risk to a greater extent than do low-fat, high-carbohydrate diets.[3]

- Lowering high blood cholesterol levels reduces the risk of heart disease.[4]

Answers to *Nutrition Scoreboard*

		TRUE	FALSE
1	About 30% of the deaths of men *and* women in the United States each year are related to heart disease.[5]	✔	
2	The most effective way to lower blood cholesterol levels is to reduce saturated and trans fat intake.[6]		✔
3	Increasing folate intake may substantially reduce the risk of heart disease in many adults.[7]	✔	
4	Risk factors for heart disease differ between women and men.[8]		✔

Photo Disc

The Diet–Heart Disease Connection

Suspicions that dietary fat may be related to heart disease were first raised over 200 years ago. During the late eighteenth century, physicians noted that people who died of heart attacks had fatty streaks and deposits in the arteries that led to the heart. Unlike many early theories about the causes of diseases, this one has basically held up over time. It is clearly established that diets high in saturated fat and trans fat (rather than total fat as originally suspected) are a major risk factor for heart disease in the United States.[9] Other risk factors for heart disease have been identified and, except for genetic tendencies and age, can be modified. That makes heart disease a highly preventable cause of death and disability.

Declining Rates of Heart Disease

No disease that can be treated by diet should be treated with any other means.
—Maimonides, 1135–1204

As Americans have become more aware of the relationship between diet and heart disease, food choices have changed, and the average cholesterol level in adults has dropped from 240 milligrams in the early 1970s to 203 in recent years.[10] As cholesterol levels have declined, so has the rate of heart disease in the United States (Illustration 19.1). The 50% drop in deaths from heart disease since 1972 appears to be primarily related to declines in blood cholesterol levels, reduced rates of smoking, improved blood pressure control, and advances in medical care. Rising rates of obesity are currently slowing down improvements in the occurrence of heart disease.[11]

Despite the general decline in cholesterol levels and rates of heart disease, the average cholesterol level of 203 milligrams among U.S. adults is still too high and further improvements are stalled.[12] Every 1% drop in average cholesterol level in countries with high rates of heart disease produces a 2% decline in heart disease rate. Raising levels of beneficial HDL cholesterol by 1 milligram per deciliter reduces rates by 2–3%.[13]

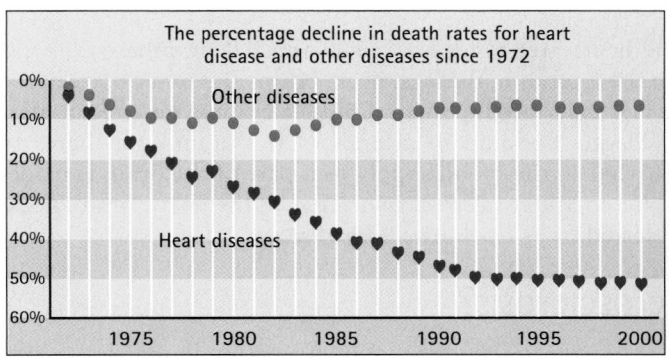

Illustration 19.1
The percentage decline in death rates for heart disease and other diseases since 1972.
Source: National Center for Health Statistics, 2003.

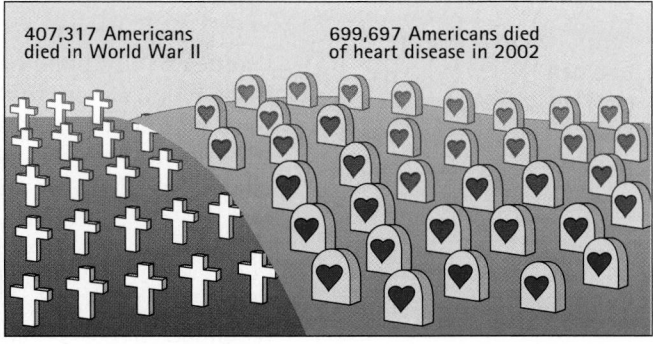

Illustration 19.2
The impact of heart disease. *If deaths from stroke and other diseases related to atherosclerosis are included, the annual death toll reaches 863,298.*
Source: National Center for Health Statistics, 2003.

A Primer on Heart Disease

There is no bigger health problem in the United States and other developed countries than heart disease (Illustration 19.2). Worldwide, heart disease accounts for one out of every four deaths.[14] It is an "equal opportunity" disease, striking as many women as men, although women on average die 10 years later from heart disease than do men. It is an *un*equal opportunity disease in that African Americans are more likely to die from it than are Caucasian Americans.[15]

What Is Heart Disease?

Heart disease, or more correctly "coronary" heart disease, is a term referring to several disorders that result from inadequate blood circulation to parts of the heart. Heart disease is almost always due to a narrowing of the arteries leading to the heart. Arteries become narrow due to a buildup of *plaque* (Illustration 19.3). People with narrowed arteries have *atherosclerosis,* or "hardening of the arteries" as it is often called. Heart disease develops silently over time (usually decades).

When arteries are narrowed by 50% or more, the shortage of blood to the heart can produce chest pain (called "angina"). If the blockage is extensive enough, a blood clot may form in the narrow opening and cut off blood supply to part of the heart. If that happens, a heart attack occurs (Illustration 19.4). Although heart disease primarily affects individuals over the age of 55, it's a progressive disease that may begin in childhood (Illustration 19.5).[16]

heart disease
One of a number of disorders that result when circulation of blood to parts of the heart is inadequate. Also called *coronary heart disease.* ("Coronary" refers to the blood vessels at the top of the heart. They look somewhat like a crown.)

plaque
Deposits of cholesterol, other fats, calcium, and cell materials in the lining of the inner wall of arteries.

atherosclerosis
"Hardening of the arteries" due to a build-up of plaque.

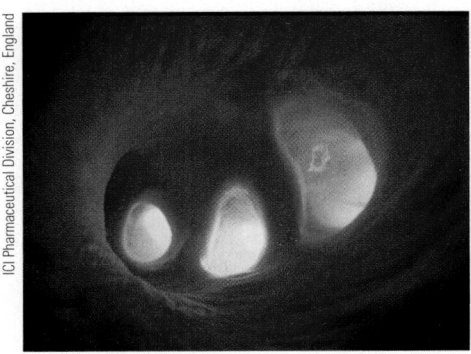

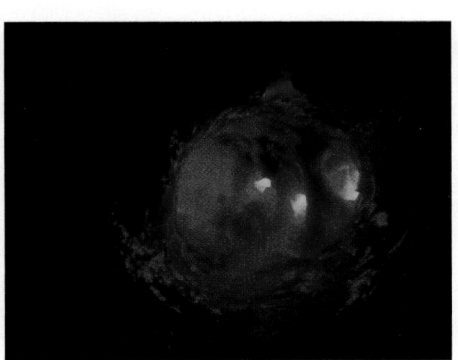

Illustration 19.3
The progression of atherosclerosis.
As plaque builds up, arteries narrow, reducing or stopping the supply of blood to the heart, brain, muscle, or other affected parts of the body.

cardiovascular disease
Disorders related to plaque build-up in arteries of the heart, brain, and other organs and tissues.

Arteries leading to the heart aren't the only ones affected by atherosclerosis. Plaque can also build up in the arteries in the legs, neck, brain, and other body parts. If the blood supply to the legs is reduced, pain and muscle cramps may result after brief periods of exercise. Plaque build-up in arteries of the brain contributes to stroke—an event that occurs when the blood supply to a part of the brain is inadequate.[17] Health problems due to atherosclerosis in arteries of the heart, brain, neck, and legs are collectively referred to as ***cardiovascular disease.***

What Causes Atherosclerosis?

A number of conditions are known to increase plaque formation in arteries; a very influential one is blood cholesterol level. (Some people with atherosclerosis don't have elevated blood cholesterol, however.[18]) In general, the higher the blood cholesterol level, the more likely it is that plaque will build up in arteries and that heart disease will occur (Illustration 19.6).[19]

What Raises Blood Cholesterol?

A person's cholesterol level is determined by a number of factors, including dietary intake, behaviors such as smoking and exercise, and heredity. Diets high in saturated fat elevate cholesterol levels in most people. Such diets are characterized by the regular consumption of high-fat milk and cheese, eggs, and beef. One type of unsaturated fat raises blood cholesterol levels more than saturated fat do, and that is trans fat.[20] This type of fat is produced when vegetable oils are hydrogenated—made solid by the addition of hydrogen. Blood cholesterol levels can also be raised by high cholesterol intakes. But blood cholesterol responds less to dietary cholesterol intake than to trans or saturated fat intake.[21] Depending on genetic background, some individuals "hyperrespond"; that is, they experience a sizable rise in blood cholesterol level when they consume large amounts of dietary cholesterol. Other people are "hyporesponders" and experience little change in blood cholesterol level when they consume additional cholesterol. For many individuals, the addition of an egg or two a day to a low-fat diet has little or no effect on blood cholesterol levels.[22]

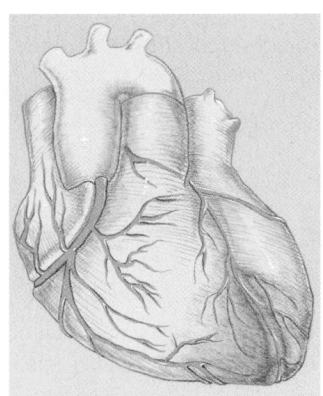

Illustration 19.4
The heart after a heart attack. *The dark portion at the base of the heart is affected by the blockage in blood flow.*

Dunagin's People / By Ralph Dunagin

"And stay away from saturated fats."

Illustration 19.5
Source: © Tribune Media Services. All rights reserved. Reprinted with permission.

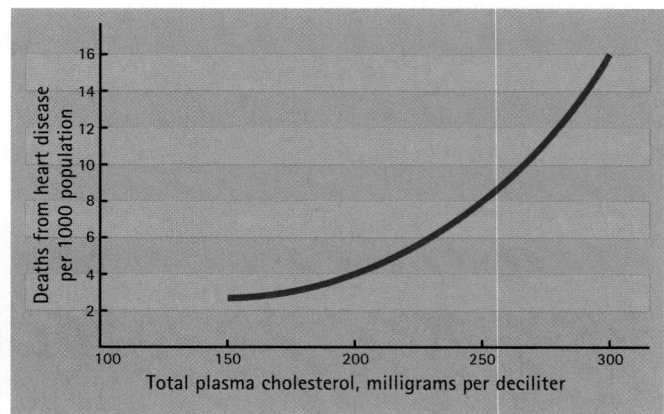

Illustration 19.6
The relationship between blood cholesterol level and death from heart disease.
Source: Grundy SM. Cholesterol and coronary heart disease: a new era. JAMA 1986;256:2849–58. Copyright 1986, American Medical Association.

REALITY CHECK

Is shrimp off the menu for people with high blood cholesterol levels?

Who gets thumbs up

?

Lupe:
Shrimp has some cholesterol, but it also has good fats like DHA. You can still eat it on a cholesterol-lowering diet.

Sharon:
You can afford shrimp? Save your money and bring your cholesterol down. Forget about eating shrimp.

Answers on next page

All Blood Cholesterol Is Not Equal

Cholesterol is soluble in fat, but blood is mostly water. Therefore, cholesterol must be bound to compounds that mix with water, or it would float in blood. For this reason, cholesterol present in blood is bound to protein, which is soluble in water. The resulting combination is called a "lipoprotein," and there are a number of different types. Two lipoproteins have gained notoriety by virtue of their coverage in the popular press and their role in the development of heart disease. One is HDL cholesterol (for "high-density lipoprotein" cholesterol—it could be nicknamed "Heart-Disease-Lowering" cholesterol). It is the "good" type of cholesterol, and you want high levels of it in your blood. LDL cholesterol (low-density lipoprotein cholesterol) is the villain. (See Illustration 19.7 for a lesson and a chuckle.) The composition and roles of both types of lipoproteins are summarized in Illustration 19.8.

Understanding HDL and LDL] HDL gets its reputation as the good cholesterol because it helps remove cholesterol from the blood. An avid cholesterol acceptor, HDL escorts cholesterol to the liver for its eventual excretion from the body. High HDL-cholesterol levels (over 60 milligrams per deciliter) are protective against heart disease. LDL cholesterol carries more cholesterol than does HDL, and its cholesterol can be incorporated into plaque. The higher the LDL-cholesterol level, the more likely it is that atherosclerosis will develop and progress into heart disease.[24]

Triglycerides and Heart Disease Risk] Triglycerides are transported in blood attached to VLDL (very low-density lipoprotein) cholesterol. Until recently, the risk to heart disease posed by elevated blood levels of triglycerides has taken a back seat to blood cholesterol. Newer study results confirm that high blood levels of triglycerides increase heart disease risk and that efforts to prevent and treat heart disease should include a focus on triglyceride levels.[25] In addition to increasing the risk of heart disease, elevated triglyceride levels may signal the presence of metabolic syndrome. This condition is characterized by elevated blood levels of triglycerides and glucose, insulin resistance, abdominal obesity, high blood pressure, and low levels

ON THE SIDE

Seven out of 10 Americans over the age of 18 do not know the difference between "good" cholesterol (HDL) and "bad" cholesterol (LDL).

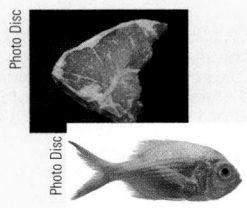

Lupe

Sharon

ANSWERS TO **REALITY CHECK**

Is shrimp off the menu for people with high blood cholesterol levels?

Shrimp contains cholesterol—about 100 mg in 3 ounces. But that doesn't mean you should not eat it if you're watching your cholesterol intake. Shrimp also contains DHA and EPA, which are heart healthy, and only a trace of saturated fat. Blood cholesterol level is much less responsive to dietary cholesterol than to saturated fat intake.[23]

Close to Home: *By John McPherson*

"To help you better understand this good cholesterol / bad cholesterol thing, Nurse Bowman and Nurse Strickling are going to do a little skit for you."

Illustration 19.7
Source: © 2001 John McPherson. Distributed by Universal Press Syndicate.

of HDL cholesterol. People with metabolic syndrome are at particularly high risk for heart disease.[26]

Other Risk Factors for Heart Disease

Plaque formation increases when LDL cholesterol becomes oxidized due to inflammation around the plaque area and when LDL is exposed to oxygen. Antioxidants such as vitamin E and C have been given to people with heart disease to reduce the oxidation of LDL and plaque build-up. Beneficial effects have been reported for both of these vitamins, but evidence that vitamin C decreases plaque formation is much more convincing than that for vitamin E.[27]

Ample intakes of vegetables and fruits have repeatedly been shown to lower the risk of heart disease. It is thought that naturally occurring antioxidants in these plant foods reduce the oxidation of LDL cholesterol. The link between beneficial components of plant foods and reduced heart-disease risk extends to include fiber, whole grain products, soy products, and dried beans.[28]

Illustration 19.8
The "bad" cholesterol (LDL) and the "good" cholesterol (HDL) found in blood.

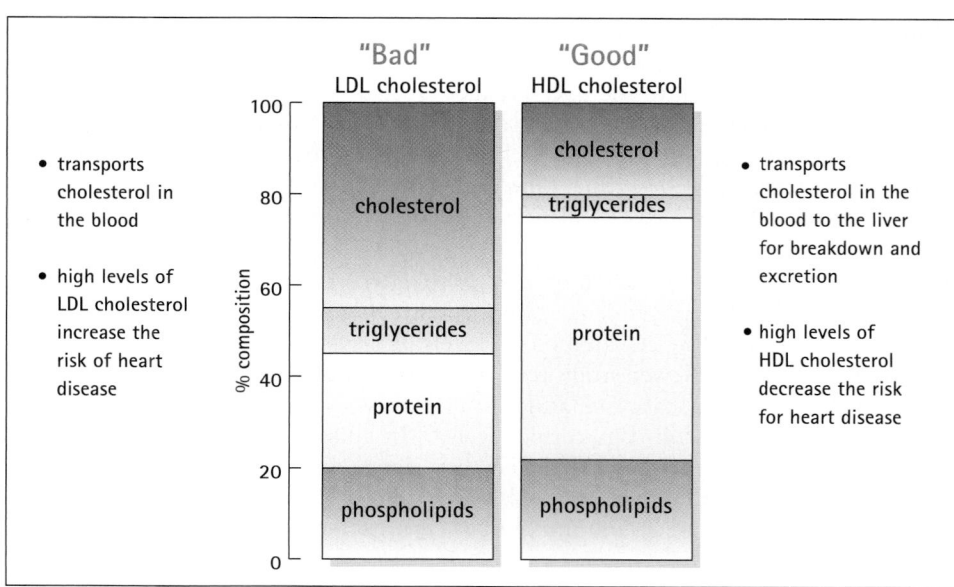

- transports cholesterol in the blood

- high levels of LDL cholesterol increase the risk of heart disease

- transports cholesterol in the blood to the liver for breakdown and excretion

- high levels of HDL cholesterol decrease the risk for heart disease

High levels of stored iron in the body have been reported to increase plaque formation. The effect, however, appears to be weak. High iron stores are not currently considered to be an important risk factor for heart disease.[29]

Folate and Atherosclerosis

A truly exciting development related to folate (a B vitamin) and atherosclerosis has emerged. Numerous studies have confirmed that inadequate folate is strongly related to the development of plaque-lined arteries.[30] This news is exciting because a practical solution is at hand—folate intakes can be increased.

Folate helps to prevent atherosclerosis by reducing blood levels of homocysteine, an amino acid that increases plaque formation if present in high amounts in the blood. Folate occurs primarily in two forms in our diets: naturally occurring folate in vegetables and fruits, and synthetic folic acid, the form of folate added to fortified foods and used in supplements. Naturally occurring folates are only about 50% absorbed, whereas folic acid is approximately 90% absorbed. Consequently, it takes more naturally occurring folates to reduce blood homocysteine levels than it takes folic acid. As shown in Table 19.1, very high intakes of green vegetables (spinach, green peas, asparagus, broccoli, and collard greens, for example) and citrus fruit decrease blood levels of homocysteine about 9%. Consumption of 250–400 micrograms of folic acid in fortified foods or supplements reduces it by over 20%. High homocysteine levels are rare among people with folic acid intakes of 400 micrograms per day.[32]

Since 1998, refined grain products such as bread, crackers, pasta, and rice have been fortified with folic acid. Most breakfast cereals are fortified with 100 micrograms of folic acid per serving, and some are highly fortified (see Table 19.2). The recommended intake of folate of 400 micrograms can be more than met by consuming 5 servings of vegetables and fruits and 6 servings of folic acid–fortified bread and cereal products daily.

Are you getting enough folate? You can estimate your intake by following the steps in the "Health Action" on p. 19–9.

TABLE 19.1

DECLINES IN BLOOD HOMOCYSTEINE LEVELS ASSOCIATED WITH INTAKE OF FOLATE FROM VEGETABLES AND CITRUS FRUITS, FOLIC ACID–FORTIFIED CEREALS, AND FOLIC ACID SUPPLEMENTS.[31]

FOLATE OR FOLIC ACID INTAKE/DAY	DECLINE IN BLOOD HOMOCYSTEINE LEVEL
A. 418 micrograms (mcg) folate from vegetables and fruits (about 8 servings/day)	 A. 9%  B. 7%
B. 70–120 mcg folic acid from fortified, refined grain products	
C. 300 mcg folic acid from fortified cereals	 C. 24% 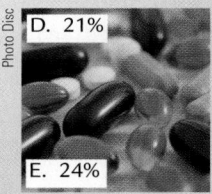 D. 21%
D. 250 mcg folic acid from supplements	
E. 400 mcg folic acid from supplements	E. 24%

TABLE 19.2

EXAMPLES OF BREAKFAST CEREALS FORTIFIED WITH 400 MICROGRAMS OF FOLIC ACID.

All-Bran

Complete

Healthy Choice

Just Right

Life

MultiGrain Cheerios

Product 19

Smart Start

Special K

Total

Walnuts are one of the oldest foods known to humans. Historical records related to walnut consumption date back to 7000 B.C.[36]

plant stanols or sterols Substances in corn, wheat, oats, rye, olives, wood, and some other plants that are similar in structure to cholesterol but that are not absorbed by the body. They decrease cholesterol absorption.

What about Fish?

Are fish good for the heart? Would Americans benefit if they ate more fish? The answers are yes and yes.

Fish, especially fatty fish from cold waters, are rich sources of DHA and EPA. These omega-3 fatty acids help protect the body against heart disease by decreasing the tendency of blood to clot and block arteries, and by decreasing plaque build-up, blood pressure, and blood triglyceride levels. As few as two fish meals a week can reduce heart-disease risk.[33] Pregnant and breastfeeding women and infants should not eat the following types of fish because they may potentially contain high levels of mercury: swordfish, shark, king mackerel, and tilefish.[34]

Nuts!

There is a direct relationship between nut consumption and a decreased risk of heart disease. Nuts are high in fat, but the fats they contain are the good ones that lower LDL cholesterol. Eating as little as 1 ounce of nuts a day can decrease LDL cholesterol in many people. Research results have convinced the FDA to approve a heart-healthy claim for nuts and products made from them.[35]

Special Spreads and Blood Cholesterol Reduction

Margarine spreads containing **plant stanols or sterols,** such as Take Control and Benecol, are widely available in grocery stores in many countries (Illustration 19.9). These spreads lower blood cholesterol levels by blocking cholesterol absorption.[37] Daily consumption of 2 tablespoons of a spread containing plant stanols or sterols is related to a 10% drop in total blood cholesterol level, and a 14% decline in LDL-cholesterol concentration. Blood levels of HDL cholesterol are not affected by consumption of plant stanols.[38] Although expensive, these spreads represent one more tool for lowering high blood cholesterol levels. By characterizing diets that lower risk, Table 19.3 pulls together information on dietary risk factors for heart disease.

Who's at Risk for Heart Disease?

Frankly, you may be. Take a look at Table 19.4 and see. Most Americans are at risk for heart disease because they have high blood cholesterol levels, hypertension, or a family history of heart attack before the age of 55; or because they smoke, are physically inactive, are obese, or have diabetes or high blood pressure (Illustration 19.10). High blood cholesterol levels are common. If 100 adults representing a cross section of Americans were to file past you, 20 of them would be at risk of heart disease due to high cholesterol levels.[41]

Some of the risk for heart disease may arise very early in life. Recent evidence suggests that infants who weigh less than 5.5 pounds at birth due to the mother's poor nutrition during pregnancy may be at increased risk for heart disease later in life. According to the "fetal origins hypothesis" of chronic disease development, it is theorized that adaptations made by a fetus in response to an inadequate supply of energy or nutrients may compromise long-term health. Small, malnourished infants may be at increased risk for heart disease as adults due to less than optimal development and growth of the liver.[42] Although researchers have long believed that heart disease begins in childhood, the risk for heart problems may begin even earlier than previously imagined.

HEALTH ACTION I *Are You Getting Enough Folate from Your Diet?*

To estimate whether you are consuming approximately 400 micrograms of absorbable folate per day, follow these steps:

	Number of Servings	Folate Content

1. Write down the number of servings of vegetables you usually consume in a day. Multiply that number by 20 to estimate the amount of folate that will be absorbed from vegetables.

  _____ × 20 = _____ mcg

2. Write down the number of servings of cold breakfast cereal you eat each day. (A serving is generally an ounce or a cup.) Multiply that number by 100. If you consume a highly fortified cereal (see Table 19.2), multiply the number of daily servings by 400.

  _____ × 100 = _____ mcg

3. Write down the number of servings of bread, pasta, rice, bagels, crackers, biscuits, or other refined grain products you tend to eat in one day. Multiply this figure by 40.

  _____ × 40 = _____ mcg

4. Write down the number of half-cup servings of cooked dried beans you eat in a usual day. Multiply that number by 70.

  _____ × 70 = _____ mcg

5. Add the figures in the Folate Content column.

  Total = _____ mcg

If the result is approximately 400 micrograms of folate, you're right on target! If less—well, you know what to do.

P.S. The Upper Limit for folic acid intake is 1000 micrograms per day.

TABLE 19.3

CHARACTERISTICS OF DIETS THAT LOWER HEART-DISEASE RISK.[39]

- Provide 20–35% of total calories from fat, 10–25% from protein, and 45–65% from carbohydrates
- Emphasize healthy fats
  > plant sources of mono- and polyunsaturated fats
  > DHA and EPA from fish
- Limit intake of unhealthy fats
  > trans fat in processed foods
  > saturated fat from animal products, palm and coconut oil
  > dietary cholesterol
- Are high in fiber; 35 grams per day for men and 25 grams per day for women
- Contain over 3 servings of vegetables and 2 servings of fruit daily
- Include whole grain products and nuts
- Include 400 mcg of folic acid daily
- Include spreads with plant stanols or sterols

Keith Weller/ARS/USDA

Illustration 19.9
Examples of spreads containing LDL-cholesterol-lowering plant stanols and sterols.

Are the Risks the Same for Women as for Men?

It is generally assumed that risk factors for heart disease identified in men apply to women as well. This assumption, along with the notion that heart disease represents a major health problem only for men, led to the exclusion of women from research studies. Both of these assumptions have been shown to be incorrect. Heart disease is a major health problem of women. Indeed, it is the leading cause of death among women in many countries.[43] Furthermore, risk factors for heart disease identified in studies of men do not always apply to women.

Results of recent studies involving women reveal that high blood cholesterol levels are a weak predictor of heart disease in older women and that low HDL-cholesterol levels are a stronger predictor than in men. Diabetes, obesity, high blood triglyceride levels, and increasing age are particularly strong risk factors for heart disease in women.[44]

The risk for heart disease in women increases substantially after menopause, which occurs around the age of 50 years. Menopause is characterized by declines in the levels of the hormone estrogen and HDL cholesterol and increases in the levels of LDL cholesterol. It is not clear, however, whether these increases in LDL cholesterol or other factors increase heart-disease risk, or whether lowering LDL-cholesterol levels reduces risk in women.[45] Compared to men, it appears that LDL-cholesterol levels are a much weaker predictor of heart disease in women.[46]

Women normally have higher levels of HDL cholesterol throughout their lives than men, and their total blood cholesterol level is a bit higher, too.[47] As a result, interpreting total blood cholesterol levels in women is difficult. Total cholesterol may exceed the level considered "too high," but that may be due to high levels of HDL cholesterol. Since high levels of HDL cholesterol (over 60 milligrams per deciliter) are highly protective against heart disease,[48] it's a mistake to allocate risk based on either total cholesterol levels or LDL-cholesterol levels alone. Yet many women are advised to modify their diets based on total blood cholesterol results. It has been suggested that weight loss by overweight women, substitution of monounsaturated fats for trans and saturated fats, increased fish intake, adequate folate, and high-fiber diets may be a more healthful way for women to reduce their risk of heart

TABLE 19.4

LEADING RISK FACTORS FOR HEART DISEASE.[40]

- High blood cholesterol level (particularly LDL-cholesterol level)
- Low HDL-cholesterol level
- Diet high in saturated fat, cholesterol, and trans fats
- Family history of early heart attack
- Inadequate folate intake
- Diet low in fruits, vegetables, and whole grains
- High blood triglyceride level
- Hypertension
- Smoking
- Physical inactivity
- Obesity (especially central fat)
- Diabetes
- Age over 45 for men, over 55 for women

Illustration 19.10
How to have a heart attack.

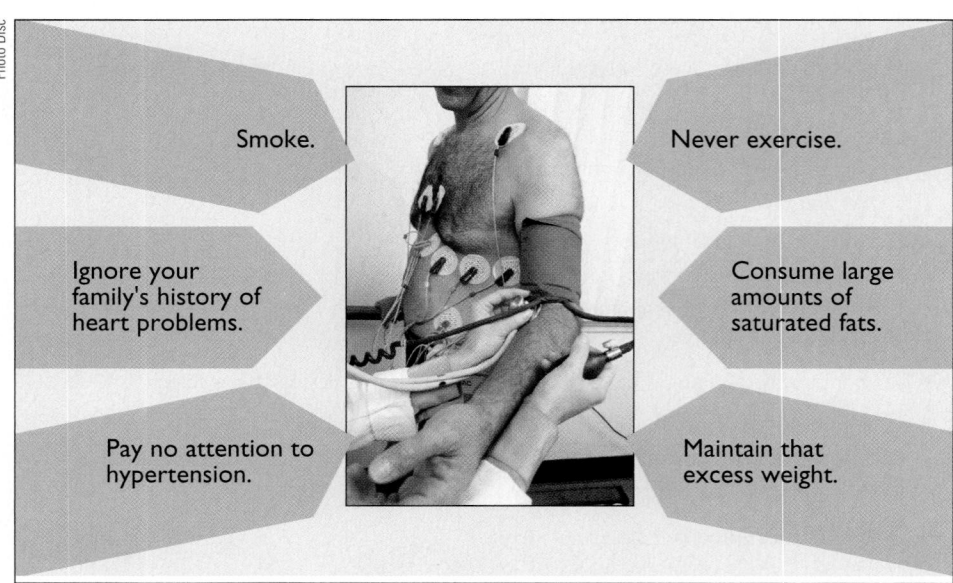

Smoke.

Never exercise.

Ignore your family's history of heart problems.

Consume large amounts of saturated fats.

Pay no attention to hypertension.

Maintain that excess weight.

disease than consumption of low-fat diets.[49] We have much more to learn about dietary risk factors for heart disease—and other conditions such as cancer—in women.

Estrogen Replacement Therapy and Heart-Disease Risk

The popularity of estrogen replacement therapy for heart-disease prevention in post-menopausal women has declined dramatically in recent years. Based on results of observation studies, it was assumed that raising estrogen levels in older women would reduce LDL cholesterol, raise HDL cholesterol, and reduce the risk of heart disease. Clinical trials have not, however, found that estrogen replacement therapy reduces the risk of heart disease in women. The use of estrogen for the prevention or treatment of heart disease is not recommended.[50]

Diet and Lifestyle in the Management of Heart Disease

Many factors are involved in the development of heart disease, so approaches to treatment need to be broad. Treatment begins and continues with dietary and lifestyle modifications, and includes reduction of high blood pressure and body weight (if needed), drugs in some cases, and smoking cessation. The goals of heart-disease treatment are improved overall health and blood lipid profiles.[51]

Modification of Blood Lipid Levels

Blood levels of total, LDL, and HDL cholesterol, and triglycerides considered desirable and abnormal, have been defined (Table 19.5) and are used to assess heart disease risk and treatment progress. Total cholesterol represents the sum of the levels of LDL, HDL, and VLDL cholesterol (the lipoprotein that transports triglycerides in blood).

Table 19.6 summarizes the effects of dietary components and lifestyle factors on blood levels of LDL cholesterol, HDL cholesterol, triglycerides, and homocysteine. Because some types of fat have both good and bad effects on blood lipid levels, dietary recommendations for heart disease focus on consumption of specific types of fat and foods that lower levels of LDL cholesterol and triglycerides and raise levels

---

TABLE 19.5

RISK CATEGORIES FOR TOTAL, LDL-, AND HDL-CHOLESTEROL LEVELS AND TRIGLYCERIDE LEVELS IN ADULTS.[52]

	TOTAL CHOLESTEROL (mg/dL)	LDL CHOLESTEROL (mg/dL)	HDL CHOLESTEROL (mg/dL)	TRIGLYCERIDE (mg/dL)
Desirable/optimal	<200	<100	60+	<150
Near optimal	–	100–129	<40	–
Borderline high	200–239	130–159	(low) <40	150–200
High	240+	160–189	–	200–499
Very high	–	190+	–	500+

*High cholesterol levels among women and people over the age of 70 may be a poor predictor of heart disease risk.

TABLE 19.6

EFFECTS OF DIETARY COMPONENTS AND LIFESTYLE FACTORS ON BLOOD LIPID AND HOMOCYSTEINE LEVELS.[53]

BLOOD LIPID	DIETARY COMPONENTS	LIFESTYLE FACTORS
LDL cholesterol (low levels are heart healthy)	**Increases levels** • trans fats • saturated fats • dietary cholesterol **Decreases levels** • mono- and polyunsaturated fats • whole grain products • fiber • vegetables and fruits • soy protein • plant stanols/sterols • nuts	 **Decreases levels** • weight loss (if needed) • physical activity
HDL cholesterol (high levels are heart healthy)	**Increases levels** • moderate fat intake • saturated fats • alcohol (1–2 drinks per day if appropriate) • moderate carbohydrate intake **Decreases levels** • polyunsaturated fats • high-carbohydrate diets • trans fats	**Increases levels** • physical activity • weight loss (if needed)
Triglycerides (low levels are heart healthy)	**Increases levels** • high-carbohydrate diets • high alcohol intake **Decreases levels** • moderate fat diets • omega-3 fatty acids (DHA and EPA)	**Decreases levels** • weight loss (if needed) • physical activity
Homocysteine (low levels are heart healthy)	**Decreases levels** • adequate intakes of folate, vitamins B_6 and B_{12} • soy protein	**Increases levels** • low intake of folate, vitamins B_6 and B_{12}

of HDL cholesterol. Adequate intake of folate and vitamins B_6 and B_{12} are recommended to decrease elevated homocysteine levels.

High levels of LDL cholesterol are reduced by limiting saturated fat intake to less than 7% of total calories, and by excluding processed foods that contain trans fats. These fats are replaced in the diet by unsaturated fats, and in particular by good food sources of monounsaturated fats like canola oil, olive oil, safflower oil, nuts, and avocados. Monounsaturated fats are preferred for lowering LDL cholesterol because they do not decrease HDL-cholesterol levels the way polyunsaturated fats do. Whole grain products, fiber, vegetables, fruits, soy protein foods, and plant stanols or sterols also lower LDL cholesterol without decreasing HDL levels. Dietary recommendations for LDL-cholesterol reduction include limiting cholesterol intake to less than 200 mg daily. Improvements in weight status and increased physical activity result in lower levels of this risky lipoprotein.[54]

HDL-cholesterol levels can be increased by exercise, weight loss, and moderate alcohol consumption (1–2 drinks per day if appropriate), and by including soy protein products and nuts in the diet. It's better to increase HDL cholesterol through these means than by consuming foods with saturated fats because saturated fats also increase LDL cholesterol.[55]

Low-fat, high-carbohydrate diets were a mainstay of heart-disease prevention and treatment approaches in the past, principally because this type of diet lowers LDL-cholesterol levels. However, because low-fat, high-carbohydrate diets also raise triglyceride levels and lower HDL-cholesterol levels, the recommendation has changed. Advice to consume low-fat, high-carbohydrate diets is being replaced by recommendations to consume healthy fats and foods that lower LDL cholesterol while maintaining triglyceride and HDL cholesterol at acceptable levels.[56] Food choices that are compatible with current recommendations for the dietary treatment of heart disease are shown in Table 19.7.

Cholesterol-lowering drugs are indicated for the treatment of heart disease if blood lipid changes achieved by diet and lifestyle improvements are insufficient.[58]

The Statins

Statins have been hailed as wonder drugs for treating heart disease. These drugs, known by names such as Lipitor, Zocor, and Mevacor, markedly reduce cholesterol production in the liver; they also combat the ill effects of heart disease in other ways. Their use is related to a 30% drop in LDL-cholesterol levels and a 30–40% reduction in heart attack and stroke in both women and men.[59] Statins are not a substitute for diet and lifestyle changes. They improve blood lipid levels more when combined with dietary and lifestyle changes than when used alone.[60]

Statins are used widely, but are expensive and have side effects such as muscle pain and weakness, liver disease, and kidney failure.[61] The cost and side effects of statins have prompted researchers to take a close look at alternatives—like extreme cholesterol-lowering diets. Table 19.8 shows an example of the extreme diet used to decrease LDL-cholesterol levels. A menu for the type of diet more typically used to lower LDL cholesterol is provided in the table for comparison. The extreme diet is vegetarian and based on soy milk and soy-based foods, fiber, oat bran, almonds, plant sterols, and vegetables and fruits. Far from representing usual food preferences, the diet may be a challenge to follow in the long run. The diet did, however, reduce LDL cholesterol to levels that are achieved by statins.[63] The results of one study showing dramatic effects of an extreme diet on blood lipids were widely covered as a *wonder diet* in national newspapers. The end result appears to be that therapeutic diets are being increasingly viewed as one way to dramatically lower LDL-cholesterol levels while keeping the dose of statin required as low as possible.[64]

TABLE 19.7
FOOD CHOICES THAT PROMOTE HEALTH AND LOWER HEART-DISEASE RISK.[57]

- Oils: canola, peanut, olive, safflower, flaxseed
- Fish
- Nuts
- Vegetables
- Fruits
- Whole grain products
- Soy products
- Lean, unfried, and unprocessed meats
- Low-fat dairy products
- Dried beans
- Spreads containing plant stanols or sterols

Heart healthy food choices are basic, unprocessed, and low in saturated fats.

TABLE 19.8

TYPE OF FOOD INCLUDED IN A HEART-DISEASE-PREVENTING DIET AND
AN EXTREME DIET FOR SUBSTANTIAL REDUCTION IN LDL CHOLESTEROL.[62]

PREVENTING HEART DISEASE	LDL CHOLESTEROL LOWERING
BREAKFAST	
breakfast cereal (fortified)	oat bran cereal
skim milk	with added psyllium fiber
fruit	soy milk
whole grain toast	oat bran toast
soft margarine	plant-sterol enriched margarine
coffee/tea	jam, strawberries
LUNCH	
turkey chili and onions	bean soup with rice
whole grain roll	oat bran bread with plant-sterol
soft margarine	enriched margarine
cole slaw	soy milk
skim milk	
DINNER	
salmon steak	spicy sautéed tofu
tartar sauce	ratatouille
green peas and corn	cooked barley
pasta, whole grain	broccoli, cauliflower
tossed salad with avocado and Italian dressing	plant-sterol enriched margarine
angel food cake with fruit	soy milk
wine/coffee/tea	almonds
SNACKS	
nuts, popcorn,	nuts, soy bar,
skim milk, soy milk,	psyllium in juice,
fruit, raw vegetables,	fruit, raw vegetables
peanut butter on whole grain bread	

Do low cholesterol levels cause cancer or aggressive behavior? The answer appears to be no.[65] Unusually low cholesterol levels are more likely to be due to the overly ambitious use of cholesterol-lowering drugs and certain diseases such as cancer that lower blood cholesterol than to diets low in saturated fat and cholesterol.[66] Total cholesterol levels below 160 mg/dL have been linked to an increased risk of suicide in people with depression. They have also been related to violence when combined with depression. The cause of these potential relationships is unkown.[67]

Looking toward the Future

Approaches to the prevention and treatment of heart disease have changed rather dramatically over recent years and will continue to evolve. Concerns about the cost of cholesterol-lowering drugs and their side effects, and the availability of low-cost preventive and treatment approaches, will factor into these changes. Approaches that utilize diet and lifestyle modification, changes in the quality of the food supply in stores and restaurants, and increased consumer involvement in risk reduction may lead the way to higher rates of decline in heart disease. An end to escalating rates of obesity and physical inactivity would serve our collective heart especially well.

Nutrition UP CLOSE

Score Your Diet for Fat

FOCAL POINT: Assess your chances for developing heart disease.

To estimate your risk for developing heart disease, circle the one number in each column that best describes you. Then, total the numbers from all categories to determine if your lifestyle encourages heart health or heart disease.

Heredity	Exercise	Age	Weight	Habits of Tobacco	Eating Fat
1 No known familial history of heart disease	**1** **Intensive** exercise at work and during recreation	**1** 15–24	**0** More than 5 lb below standard weight	**0** Nonuser	**1** No animal fat
2 **One** immediate family member who developed heart disease **over** age 55	**2** **Moderate** exercise at work and during recreation	**2** 24–34	**1** 65 lb standard weight	**1** Cigar or pipe	**2** Very little animal fat (<10% of total calories)
3 **Two** immediate family members who developed heart disease **over** age 55	**3** Sedentary work with **intensive** exercise during recreation	**3** 35–44	**2** 6–20 lb overweight	**2** 10 cigarettes or fewer per day	**3** Some animal fat (11–20% of total calories)
4 **One** immediate family member who developed heart disease **under** age 55	**5** Sedentary work with **moderate** exercise during recreation	**4** 45–54	**4** 21–35 lb overweight	**4** 20 cigarettes or more per day	**4** Moderate animal fat (21–30% of total calories)
6 **Two** immediate family members who developed heart disease **under** age 55	**6** Sedentary work with **light** exercise during recreation	**6** 55+ years	**6** 36–50 lb overweight	**6** 30 cigarettes or more per day	**6** Excessive animal fat (>30% of total calories)

Total of all categories _____

FEEDBACK (including scoring) can be found at the end of Unit 19.

Source: Adapted from Heart Health Quiz (Loma Linda University).

Key Terms

atherosclerosis, page 19-3

cardiovascular disease, page 19-4

heart disease, page 19-3

plant stanols or sterols, page 19-8

plaque, page 19-3

www links

www.nlm.nih.gov/medlineplus/
coronarydisease.html
Links to information on coronary heart disease, coronary artery disease, risk factor reduction, current news, and latest study results.

www.nlm.nih.gov/medlineplus/cholesterol
.html
Get the latest news about good and bad cholesterol, trans fats, statins, dietary and blood cholesterol, and heart disease at this site. Or, visit the Virtual Fitness Room.

www.nlm.nih.gov/medlineplus/
heartdiseasesprevention.html

A site from the National Institutes of Health dedicated to presenting information on heart-disease prevention.

www.deliciousdecisions.org
American Heart Association's nutrition Web site includes links to sites on nutrition basics, a cookbook tailored to people with heart disease or those who want to prevent it, and tips for grocery shopping and eating out.

www.americanheart.org
The American Heart Association's home page. Go there to find an array of infor-

mation for consumers on heart disease and stroke.

www.nhlbi.nih.gov/health/public/heart/
index.htm
Shows the index of postings on heart disease offered by the National Heart, Lung, and Blood Institute of the National Institutes of Health.

www.medscape.com
An excellent site for health, medical, and nutrition information searches related to heart disease, stroke, and arteriosclerosis.

Notes

1. Health, United States, 2002. National Center for Health Statistics, www.cdc.gov/nchs/Default.htm.

2. Topol EJ. Heart disease attributable to common risk factors. JAMA 2003;290: 891–904, 947–9.

3. Dietary Reference Intakes. Energy, carbohydrate, fiber, fat, fatty acids, cholesterol, protein, and amino acids. Institute of Medicine, National Academy of Sciences, Washington, DC: National Academies Press, 2002.

4. Detection, evaluation, and treatment of high blood cholesterol in adults. Adult Treatment Panel III. National Heart, Lung, and Blood Institute of the National Institutes of Health, 2001.

5. Health, United States, 2002 (www.cdc.gov/nchs/Default.htm).

6. Detection, evaluation, and treatment of high blood cholesterol in adults.

7. Detection, evaluation, and treatment of high blood cholesterol in adults.

8. Stampher MJ et al. Primary prevention of coronary heart disease in women through diet and lifestyle. N Engl J Med 2000;343:16–22.

9. Dietary Reference Intakes.

10. Ford ES et al. U.S. population levels of total cholesterol. Circulation 2003;107.

11. Lenfant C. Clinical research to clinical practice—lost in translation? N Engl J Med 2003;349:868–74.

12. Lenfant, Clinical research to clinical practice.

13. Jellinger PS et al. The American Association of Clinical Endocrinologists medical guidelines for the diagnosis and treatment of dyslipidemia and prevention of atherogenesis. Endocrin Prac 2000; 6:1–213.

14. Dwyer T et al. Differences in HDL cholesterol concentrations in Japanese, American, and Australian children. Circ 1997;76:2830–6.

15. Health, United States, 2002 (www.cdc .gov/nchs/Default.htm).

16. Dwyer et al., Differences in HDL cholesterol concentrations.

17. Mosca L et al. Serum cholesterol and risk of stroke death in women. Stroke: J Am Heart Assoc 2002;July.

18. Topol, Heart disease.

19. Detection, evaluation, and treatment of high blood cholesterol in adults.

20. Dietary Reference Intakes; Shea S et al., Age, sex, educational attainment, and race/ethnicity in relation to consumption of specific foods contributing to the atherogenic potential of diet, Prev Med 1993;22:203–18; and Mensink RP et al., Effects of dietary fatty acids and carbohydrates on the ratio of serum total to HDL cholesterol and serum lipids, Am J Clin Nutr 2003;77:1146–55.

21. Shea et al., Age, sex, educational attainment.

22. Herron KL et al., Men classified as hypo- and hyperresponders to dietary cholesterol feeding exhibit differences in lipoprotein metabolism, J Nutr 2003; 133:1036–42; and McNamara DJ, Eggs and heart disease: perpetuating the misconception, Am J Clin Nutr 2002;75: 333–5.

23. Herron et al., Men classified as hypo- and hyperresponders.

24. Detection, evaluation, and treatment of high blood cholesterol in adults; and Davidson MH, New tactics, new targets: the changing landscape of dyslipidemia management in coronary prevention, Medscape CME Activity, www .medscape.com, accessed 5/03.

25. Detection, evaluation, and treatment of high blood cholesterol in adults; and Eberly LE et al. Nonfasting triglyceride levels and cardiovascular risk. Arch Intern Med 2003;163:1077–83.

26. Detection, evaluation, and treatment of high blood cholesterol in adults.

27. Supplements for heart health. www .mayoclinic.com/invoke.cfm?id=HB00018, accessed 7/03.

28. Rissanen TH et al., Low intake of fruits and vegetables is associated with excess mortality in men, J Nutr 2003;133: 199–204; and Krauss RM et al., Revision 2000: a statement for healthcare providers from the Nutrition Committee of the American Heart Association, J Nutr 2001;131:132–46.

29. Sempos CT. Do body iron stores increase the risk of developing coronary heart disease? Am J Clin Nutr 2003;76: 501–3.

30. Stampfer MJ et al. Primary prevention of coronary heart disease in women through diet and lifestyle. N Engl J Med 2000;343:16–22.

31. Jacques PF et al., The effect of folic acid fortification on plasma folate and total homocysteine concentrations, N Engl J Med 1999;340:1449–54; and Kris-Etherton PM et al., New guidelines focus on fish, fish oil, omega-3 fatty acids, AHA Statement, Circulation 2002;Oct. 18; and Clarkson TW, Strain JJ, Nutritional factors may modify the toxic action of methyl mercury in fish-eating populations, J Nutr 2003;133:1539S–43S.

32. Jacques et al., The effect of folic acid fortification.

33. Kris-Etherton et al., New guidelines; Krauss et al., Revision 2000.

34. Clarkson and Strain, Nutritional factors.

35. Feldman EB. The scientific evidence for a beneficial health relationship between walnuts and coronary heart disease. J Nutr 2002;132:1062S–1101S.

36. Dreher ML et al. The traditional and emerging role of nuts in healthful diets. Nutr Rev 1996;54:241–5.

37. Levine BS. Plant stanol esters. Nutr Today 2000;35:61–6.

38. Kerckhoffs DAJM et al. Effects on the human serum lipoprotein profile of B-glucan, soy protein and isoflavinoids, plant sterols and stanols, garlic, and tocotrienols. J Nutr 2002;132:2494–2505.

39. Detection, evaluation, and treatment of high blood cholesterol in adults; and Krauss et al., Revision 2000.

40. Topol, Heart disease; Detection, evaluation and treatment of high blood cholesterol in adults; and Krauss et al., Revision 2000.

41. Berg JE, Høstmark AT, Coronary artery disease and lipid fractions, Epidemiol 1995;6:91; and Forman A, Heart disease handbook—part 2: deciphering blood cholesterol, Envir Nutr 1997;20:1, 4.

42. Yajnik C, Interactions of perturbations in intrauterine growth and growth during childhood on the risk of adult-onset disease, Proc Nutr Soc 2000;59:257–65; and Brown JE, Kahn ESB, Maternal nutrition and the outcome of pregnancy, Clin Perinatol 1997;24:433–49.

43. Kereiakes DJ et al., Managing cardiovascular disease in women. Medscape Women's Health, http://womenshealth .medscape.com/CMECircleWomensHealth/ 2000/CME02/pnt-CME02.html, accessed 1/01; and Stampfer et al., Primary prevention.

44. Sprecher DL et al. Metabolic coronary risk factors and mortality after bypass surgery. J Am Coll Cardiol 2000;36: 1159–65.

45. Jeppesen J et al. Effects of low-fat, high-carbohydrate diet on risk factors for ischemic heart disease in postmenopausal women. Am J Clin Nutr 1997;65:1027–33.

46. Stampfer et al., Primary prevention; and Hu FB et al. Dietary fat and the risk of coronary heart disease in women. N Engl J Med 1997;337:1491–9.

47. Ernst ND et al. Consistency between US dietary fat intake and serum total cholesterol concentrations: the National Health and Nutrition Examination Surveys. Am J Clin Nutr 1997;66(suppl): 965S–72S.

48. Detection, evaluation, and treatment of high blood cholesterol in adults.

49. Stampfer et al., Primary prevention; and Hegsted M, Kritchevsky D. Diet and serum lipid concentrations: where are we? Am J Clin Nutr 1997;65:1893–6.

50. Manson JE et al. Estrogen plus progestin and the risk of coronary heart disease. N Engl J Med 2003;349:523–34.

51. Detection, evaluation, and treatment of high blood cholesterol in adults.

52. Detection, evaluation, and treatment of high blood cholesterol in adults.

53. Dietary Reference Intakes; Detection, evaluation, and treatment of high blood cholesterol in adults; and Krauss et al., Revision 2000.

54. Detection, evaluation, and treatment of high blood cholesterol in adults; and Krauss et al., Revision 2000.

55. Dietary Reference Intakes; and Detection, evaluation, and treatment of high blood cholesterol in adults.

56. Dietary Reference Intakes.

57. Stampfer et al., Primary prevention; and Kerckhoffs et al., Effects on the human serum lipoprotein profile.

58. Detection, evaluation, and treatment of high blood cholesterol in adults; and Krauss et al., Revision 2000.

59. Schwartz RG et al. Benefits of statin therapy unrelated to lipoprotein levels. J Am Coll Cardiol 2003;42:600–13.

60. Davidson, New tactics, new targets (www.medscape.com).

61. Davidson, New tactics, new targets (www.medscape.com).

62. Stampfer et al., Primary prevention; Schwartz et al., Benefits of statin therapy; and Jenkins DJA et al., Effect of a vegetarian diet, a low fat, high fiber vegetarian dietary portfolio, and Mevacor on LDL-cholesterol levels on patients with high LDL-cholesterol levels (approximated), JAMA 290:502–513, 531–33.

63. Jenkins et al., Effect of a vegetarian diet.

64. Davidson, New tactics, new targets (www.medscape.com); and Vega C, Diet may lower cholesterol as much as statins, Medscape CME Activity, www.medscape.com, accessed 8/03.

65. Eichholzer M et al., Association of low plasma cholesterol with mortality for cancer at various sites in men: 17-y follow-up of the prospective basal study, Am J Clin Nutr 2000;72:569–74; and Wardle J et al., Randomized trial of the effects of cholesterol lowering dietary treatment on psychological function, Am J Med 2000;108:547–53.

66. Fagot-Campagna A. Serum cholesterol and mortality rates in a Native American population with low cholesterol levels. Circ 1997;95:1408–15.

67. Kim Y. Low cholesterol linked to suicide in depressed patients. Presentation at the 16th European College of Neuropsychopharmacology, Prague, Sept. 2003.

Nutrition UP CLOSE

Heart Health Quiz

Feedback for Unit 19

Find your score below to learn your estimated risk of developing heart disease:

4–9 very remote
10–15 less than average
16–20 average
21–25 moderately increased
26–30 excessive
31–35 too high—reduce score!

You have no control over certain risk factors for heart disease, such as heredity and age. Other conditions that detract from heart health, such as high blood pressure, diabetes, and high blood cholesterol, should be evaluated by your physician. Some risk factors are within your control, however. These lifestyle changes can help reduce your risk of developing heart disease:

- Keep your weight within a healthy range.
- Exercise regularly.
- Don't smoke.
- Reduce the amount of saturated fat in your diet if it is high.
- Eat at least five servings of vegetables and fruits each day.

Source: Adapted from Heart Health Quiz (Loma Linda University).

Vitamins and Your Health

Nutrition Scoreboard

	TRUE	FALSE
1 The only documented benefit of consuming sufficient amounts of vitamins is protection against deficiency diseases.		
2 Vitamins provide energy.		
3 Vitamin C is found only in citrus fruits.		
4 Nearly all cases of illness due to excessive intake of vitamins result from the overuse of vitamin supplements.		

Answers on next page

[KEY CONCEPTS AND FACTS]

- Vitamins are chemical substances found in food that are required for normal growth and health.

- Adequate intakes of vitamins protect people against deficiency diseases and help prevent a number of chronic diseases.

- Each vitamin has a range of intake in which it functions optimally. Intakes below and above the range impair health.

- Eating 5 or more servings of fruits and vegetables is a good way to get enough vitamins in your diet each day.

Answers to *Nutrition Scoreboard*

		TRUE	FALSE
1	For many vitamins, intake levels above those known to prevent deficiency diseases help protect humans from certain cancers, heart disease, and other disorders.		✔
2	Nope—only carbohydrates, proteins, and fats provide energy to the body. Vitamins are needed, however, to convert the energy in food into energy the body can use.		✔
3	Citrus fruits are good sources of vitamin C, but so are green peppers, collards, broccoli, strawberries, and a number of other fruits and vegetables.		✔
4	True. Nearly all cases of illness due to vitamin overdoses result from the excessive intake of vitamin supplements.	✔	

Vitamins: They're on Center Stage

These are exciting times for people interested in vitamins. Long relegated to the role of preventing deficiency diseases, vitamins are now being viewed from a very different vantage point. These essential nutrients clearly do more than the important job of protecting us from vitamin deficiency diseases. Vitamins are taking a preeminent position as protectors against a host of ills, ranging from certain birth defects and cataracts to heart disease and cancer.[1]

But first, what are vitamins? How much of them do we need? Where do we get them? And what's behind this new interest in them? Background information essential to understanding the current interest in the benefits of vitamins is next.

Vitamin Facts

Vitamins are chemical substances that perform specific functions in the body. They are essential nutrients because, in general, the body cannot produce them or produce sufficient amounts of them. If we fail to consume enough of any of the vitamins, specific deficiency diseases develop.

Thirteen vitamins have been discovered so far, and they are listed in Table 20.1. It is possible that a few more substances will be added to the list of vitamins in years to come.

Water- and Fat-Soluble Vitamins] Vitamins come in two basic types—those soluble in water (the B-complex vitamins and vitamin C) and those that dissolve in fat (vitamins D, E, K, and A, or the "deka" vitamins). Their key features are sum-

vitamins
Chemical substances that perform specific functions in the body.

TABLE 20.1

THIRTEEN VITAMINS ARE KNOWN TO BE ESSENTIAL FOR HEALTH. NINE ARE WATER SOLUBLE—THAT IS, THEY DISSOLVE COMPLETELY IN WATER—AND FOUR DISSOLVE ONLY IN FATS.

THE WATER-SOLUBLE VITAMINS*

B-complex vitamins
   Thiamin (B_1)
   Riboflavin (B_2)
   Niacin (B_3)
   Vitamin B_6 (pyridoxine)
   Folate (folacin, folic acid)
   Vitamin B_{12} (cyanocobalamin)
   Biotin
   Pantothenic acid (pantothenate)
Vitamin C (ascorbic acid)

THE FAT-SOLUBLE VITAMINS

Vitamin A (retinol) (provitamin is beta-carotene)
Vitamin D (1,25 dihydroxy-cholecalciferol)
Vitamin E (tocopherol)
Vitamin K (phylloquinone, menaquinone)

*Choline has been assigned Adequate Intake levels in the Dietary Reference Intakes. It is not included here as it has not been shown to be required in the diet during all stages of the life cycle.

marized in Table 20.2, which provides an "intensive course" on vitamins. With the exception of vitamin B_{12}, the water-soluble vitamins can be stored in the body only in small amounts. Consequently, deficiency symptoms generally develop within a few weeks to several months after the diet becomes deficient in water-soluble vitamins. Vitamin B_{12} is unique in that the body can build up stores that may last for a year or more after intake of the vitamin stops. Of the water-soluble vitamins, niacin, vitamin B_6, and vitamin C are known to produce ill effects if consumed in excessive amounts.

The fat-soluble vitamins are stored in body fat, the liver, and other parts of the body. Because the body is better able to store these vitamins, deficiencies of fat-soluble vitamins generally take longer to develop than deficiencies of water-soluble vitamins when intake from food is too low.

Bogus Vitamins]   Some substances that don't belong on the list of vitamins and won't end up there in years to come are listed in Table 20.3. This list includes some of the most common substances claimed to be vitamins by enterprising quacks, misdirected manufacturers of supplements, and some weight-loss and cosmetic product producers. Although these substances may help sales, they aren't essential and therefore cannot be considered vitamins. People do not develop deficiency diseases when they consume too little of the bogus "vitamins."

What Do Vitamins Do?

For starters, vitamins *don't* provide energy or serve as components of body tissues such as muscle and bone. A number of vitamins do play critical roles as *coenzymes* in the conversion of proteins, carbohydrates, and fats into energy. Coenzymes are also involved in reactions that build and maintain body tissues such as bone, muscle, and red blood cells. Thiamin, for example, is needed for reactions that convert glucose into energy. People who are thiamin deficient tire easily and feel weak (among other things). Folate, another B-complex vitamin, is required for reactions that build body proteins. Without enough folate, proteins such as those found in red blood cells form abnormally and function poorly. Vitamin A is needed for reactions

coenzymes
Chemical substances, including many vitamins, that activate specific enzymes. Activated enzymes increase the rate at which reactions take place in the body, such as the breakdown of fats or carbohydrates in the small intestine and the conversion of glucose and fatty acids into energy within cells.

text continued on page 20-10

TABLE 20.2

AN INTENSIVE COURSE ON VITAMINS.

THE WATER-SOLUBLE VITAMINS

	PRIMARY FUNCTIONS		CONSEQUENCES OF DEFICIENCY
Thiamin (vitamin B$_1$) AI[a] women: 1.1 mg men: 1.2 mg	• Helps body release energy from carbohydrates ingested • Facilitates growth and maintenance of nerve and muscle tissues • Promotes normal appetite	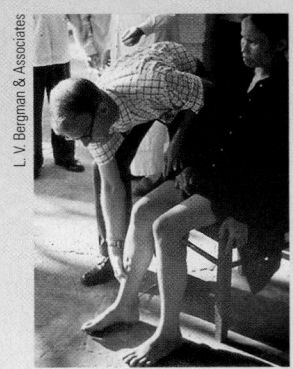 < A look at beriberi, a thiamin-deficiency disease.	• Fatigue, weakness • Nerve disorders, mental confusion, apathy • Impaired growth • Swelling • Heart irregularity and failure
Riboflavin (vitamin B$_2$) AI women: 1.1 mg men: 1.3 mg	• Helps body capture and use energy released from carbohydrates, proteins, and fats • Aids in cell division • Promotes growth and tissue repair • Promotes normal vision		• Reddened lips, cracks at both corners of the mouth • Fatigue
Niacin (vitamin B$_3$) RDA women: 14 mg men: 16 mg UL: 35 mg (from supplements and fortfied foods)	• Helps body capture and use energy released from carbohydrates, proteins, and fats • Assists in the manufacture of body fats • Helps maintain normal nervous system functions	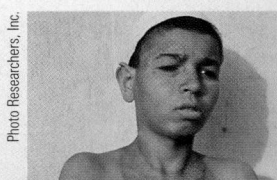 Pellagra: the niacin-deficiency disease.	• Skin disorders • Nervous and mental disorders • Diarrhea, indigestion • Fatigue

Vitamin B$_6$ (pyridoxine)
 	
AI   women: 1.3 mg
     men: 1.3 mg
    UL: 100 mg
	• Needed for reactions that build proteins and protein tissues
	• Assists in the conversion of tryptophan to niacin
	• Needed for normal red blood cell formation
	• Promotes normal functioning of the nervous system
			• Irritability, depression
			• Convulsions, twitching
			• Muscular weakness
			• Dermatitis near the eyes
			• Anemia
			• Kidney stones

[a]AI (Adequate Intakes) and RDAs (Recommended Dietary Allowances) are for 19–30 year olds; UL (Upper Limits) are for 19–70 year olds,

TABLE 20.2

AN INTENSIVE COURSE ON VITAMINS. (CONTINUED)

THE WATER-SOLUBLE VITAMINS

CONSEQUENCES OF OVERDOSE	PRIMARY FOOD SOURCES	HIGHLIGHTS AND COMMENTS
• High intakes of thiamin are rapidly excreted by the kidneys. Oral doses of 500 mg/day or less are considered safe.	• Grains and grain products (cereals, rice, pasta, bread) • Ready-to-eat cereals • Pork and ham, liver • Milk, cheese, yogurt • Dried beans and nuts	• Need increases with carbohydrate intake. • There is no "e" on the end of thiamin! • Deficiency rare in the U.S.; may occur in people with alcoholism. • Enriched grains and cereals prevent thiamin deficiency.
• None known. High doses are rapidly excreted by the kidneys.	• Milk, yogurt, cheese • Grains and grain products (cereals, rice, pasta, bread) • Liver, poultry, fish, beef • Eggs	• Destroyed by exposure to light
• Flushing, headache, cramps, rapid heartbeat, nausea, diarrhea, decreased liver function with doses above 0.5 g per day	• Meats (all types) • Grains and grain products (cereals, rice, pasta, bread) • Dried beans and nuts • Milk, cheese, yogurt • Ready-to-eat cereals • Coffee • Potatoes	• Niacin has a precursor—tryptophan. Tryptophan, an amino acid, is converted to niacin by the body. Much of our niacin intake comes from tryptophan. • High doses raise HDL-cholesterol levels.
• Bone pain, loss of feeling in fingers and toes, muscular weakness, numbness, loss of balance (mimicking multiple sclerosis)	• Oatmeal, bread, breakfast cereals • Bananas, avocados, prunes, tomatoes, potatoes • Chicken, liver • Dried beans • Meats (all types), milk • Green and leafy vegetables	• Vitamins go from B_3 to B_6 because B_4 and B_5 were found to be duplicates of vitamins already identified.

continued

TABLE 20.2

AN INTENSIVE COURSE ON VITAMINS. (CONTINUED)

THE WATER-SOLUBLE VITAMINS

	PRIMARY FUNCTIONS	CONSEQUENCES OF DEFICIENCY	
Folate (folacin, folic acid) RDA: women: 400 mcg men: 400 mcg UL: 1000 mcg (from supplements and fortified foods)	• Needed for reactions that utilize amino acids (the building blocks of protein) for protein tissue formation • Promotes the normal formation of red blood cells	• Megaloblastic anemia • Diarrhea • Red, sore tongue 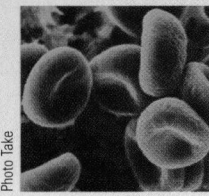 Normal red blood cells.	• Neural tube defects, low birth weight (in pregnancy); increased risk of heart disease and stroke • Elevated blood levels of homocysteine 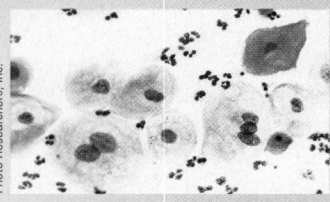 Red blood cells in megaloblastic anemia
Vitamin B₁₂ (cyanocobalamin) AI women: 2.4 mcg men: 2.4 mcg	• Helps maintain nerve tissues • Aids in reactions that build up protein tissues • Needed for normal red blood cell development	• Neurological disorders (nervousness, tingling sensations, brain degeneration) • Pernicious anemia • Fatigue • Deficiency reported in 39% of adults in one study • Elevated blood level of homocysteine	
Biotin AI women: 30 mcg men: 30 mcg	• Needed for the body's manufacture of fats, proteins, and glycogen	• Depression, fatigue, nausea • Hair loss, dry and scaly skin • Muscular pain	
Pantothenic acid (pantothenate) AI women: 5 mg men: 5 mg	• Needed for the release of energy from fat and carbohydrates	• Fatigue, sleep disturbances, impaired coordination • Vomiting, nausea	
Vitamin C (ascorbic acid) RDA women: 75 mg men: 90 mg UL: 2000 mg	• Needed for the manufacture of collagen • Helps the body fight infections, repair wounds • Acts as an antioxidant • Enhances iron absorption	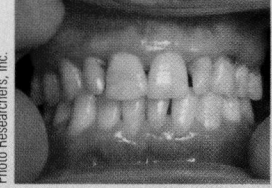 Gums that are swollen and bleed easily are signs of scurvy, the vitamin C-deficiency disease.	• Bleeding and bruising easily due to weakened blood vessels, cartilage, and other tissues containing collagen • Slow recovery from infections and poor wound healing • Fatigue, depression • Deficiency reported in 9–24% of adults in one study

TABLE 20.2

AN INTENSIVE COURSE ON VITAMINS. (CONTINUED)

THE WATER-SOLUBLE VITAMINS

CONSEQUENCES OF OVERDOSE	PRIMARY FOOD SOURCES	HIGHLIGHTS AND COMMENTS
• May cover up signs of vitamin B$_{12}$ deficiency (pernicious anemia)	• Fortified, refined grain products (bread, flour, pasta) • Ready-to-eat cereals • Dark green, leafy vegetables (spinach, collards, romaine) • Broccoli, brussels sprouts • Oranges, bananas, grapefruit • Milk, cheese, yogurt • Dried beans	• *Folate* means "foliage." It was first discovered in leafy green vegetables. • This vitamin is easily destroyed by heat. • Synthetic form added to fortified grain products is better absorbed than naturally occurring folates.
• None known. Excess vitamin B$_{12}$ is rapidly excreted by the kidneys or is not absorbed into the bloodstream. • Vitamin B$_{12}$ injections may cause a temporary feeling of heightened energy.	• Animal products: beef, lamb, liver, clams, crab, fish, poultry, eggs • Milk and milk products • Ready-to-eat cereals	• Older people and vegans are at risk for vitamin B$_{12}$ deficiency. • Some people become vitamin B$_{12}$ deficient because they are genetically unable to absorb it. • Vitamin B$_{12}$ is found in animal products and microorganisms only.
• None known. Excesses are rapidly excreted.	• Grain and cereal products • Meats, dried beans, cooked eggs • Vegetables	• Deficiency is extremely rare. May be induced by the overconsumption of raw eggs.
• None known. Excesses are rapidly excreted.	• Many foods, including meats, grains, vegetables, fruits, and milk • Required for the conversion of homocysteine to methionine	• Deficiency is very rare.
• Intakes of 1 g or more per day can cause nausea, cramps, and diarrhea and may increase the risk of kidney stones.	• Fruits: oranges, lemons, limes, strawberries, cantaloupe, honeydew melon, grapefruit, kiwi fruit, mango, papaya • Vegetables: broccoli, green and red peppers, collards, cabbage, tomato, asparagus, potatoes • Ready-to-eat cereals	• Need increases among smokers (to 110–125 mg per day). • Is fragile; easily destroyed by heat and exposure to air • Supplements may decrease duration and symptoms of colds. • Deficiency may develop within 3 weeks of very low intake. *continued*

TABLE 20.2

AN INTENSIVE COURSE ON VITAMINS. (CONTINUED)

THE FAT-SOLUBLE VITAMINS

	PRIMARY FUNCTIONS	CONSEQUENCES OF DEFICIENCY	
Vitamin A **I. Retinol** RDA women: 700 mcg 　　men: 900 mcg 　　UL: 3000 mcg	• Needed for the formation and maintenance of mucous membranes, skin, bone • Needed for vision in dim light	• Increased susceptibility to infection, increased incidence and severity of infection (including measles) • Impaired vision, blindness • Inability to see in dim light	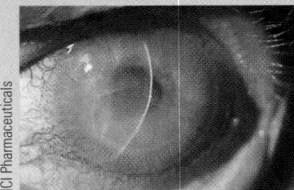 Xerophthalmia. Vitamin A deficiency is the leading cause of blindness in developing countries.
2. Beta-carotene (a vitamin A precursor or "provitamin") No RDA; suggested intake: 6 mg	• Acts as an antioxidant; prevents damage to cell membranes and the contents of cells by repairing damage caused by free radicals	• Deficiency disease related only to lack of vitamin A	
Vitamin E (alpha-tocopherol) RDA women:　15 mg 　　men: 15 mg 　　UL: 1000 mg	• Acts as an antioxidant, prevents damage to cell membranes in blood cells, lungs, and other tissues by repairing damage caused by free radicals • Reduces the ability of LDL cholesterol (the "bad" cholesterol) to form plaque in arteries	• Muscle loss, nerve damage • Anemia • Weakness • Many adults may have non-optimal blood levels.	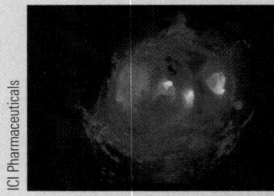 Vitamin E helps prevent plaque formation in arteries, may improve circulation in people with diabetes, and may decrease symptoms of Alzheimer's disease.

TABLE 20.2

AN INTENSIVE COURSE ON VITAMINS. (CONTINUED)

THE FAT-SOLUBLE VITAMINS

CONSEQUENCES OF OVERDOSE	PRIMARY FOOD SOURCES	HIGHLIGHTS AND COMMENTS
• Vitamin A toxicity (hypervitaminosis A) with acute doses of 500,000 IU, or long-term intake of 50,000 IU per day. Limit retinol use in pregnancy to 5000 IU daily. • Nausea, irritability, blurred vision, weakness • Increased pressure in the skull, headache • Liver damage • Hair loss, dry skin • Birth defects	• Vitamin A is found in animal products only. • Liver, butter, margarine, milk, cheese, eggs • Ready-to-eat cereals	• Symptoms of vitamin A toxicity may mimic those of brain tumors and liver disease. Vitamin A toxicity is sometimes misdiagnosed because of the similarities in symptoms. • 1 mcg retinol equivalent = 5 IU vitamin A or 6 mcg beta-carotene
• High intakes from supplements may increase lung damage. • With high intakes and supplemental doses (over 12 mg/day for months), skin may turn yellow-orange. • Possibly related to reversible loss of fertility in women	• Deep orange, yellow, and green vegetables and fruits are often good sources. • Carrots, sweet potatoes, pumpkin, spinach, collards, red peppers, broccoli, cantaloupe, apricots, tomatoes, vegetable juice	• The body converts beta-carotene to vitamin A. Other carotenes are also present in food, and some are converted to vitamin A. Beta-carotene and vitamin A perform different roles in the body, however. • May decrease sunburn.
• Intakes of up to 800 IU per day are unrelated to toxic side effects; over 800 IU per day may increase bleeding (blood-clotting time). • Avoid supplement use if aspirin, anticoagulants, or fish oil supplements are taken regularly.	• Oils and fats • Salad dressings, mayonnaise, margarine, shortening, butter • Whole grains, wheat germ • Leafy, green vegetables, tomatoes • Nuts and seeds • Eggs	• Vitamin E is destroyed by exposure to oxygen and heat. • Oils naturally contain vitamin E. It's there to protect the fat from breakdown due to free radicals. • Eight forms of vitamin E exist, and each has different antioxidant strength. • Natural form is better absorbed than synthetic form: 15 IU alpha-tocopherol = 22 IU d-alpha tocopherol (natural form) and 33 IU synthetic vitamin E.

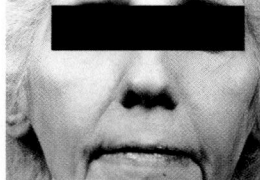

Brittle hair and dry, rough, scaly, and cracked skin from vitamin A overdose.

From: American Journal of Clinical Nutrition, Vol 71, No 4, 878–884. April 2000, Robert M. Russell.

continued

TABLE 20.2

AN INTENSIVE COURSE ON VITAMINS. (CONTINUED)

	THE FAT-SOLUBLE VITAMINS	
	PRIMARY FUNCTIONS	CONSEQUENCES OF DEFICIENCY
Vitamin D (1,25 dihydroxy-cholecalciferol) AI women: 5 mcg (200 IU) men: 5 mcg (200 IU) UL: 50 mcg (2000 IU)	• Needed for the absorption of calcium and phosphorus, and for their utilization in bone formation, nerve and muscle activity.	• Weak, deformed bones (children) • Loss of calcium from bones (adults), osteoporosis 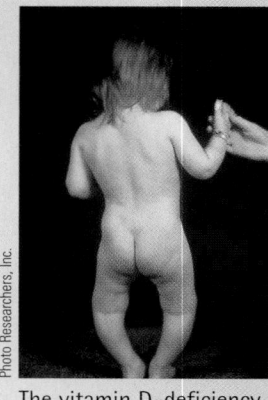 The vitamin D–deficiency disease: rickets
Vitamin K (phylloquinone, menaquinone) AI women: 90 mcg men: 120 mcg	• Is an essential component of mechanisms that cause blood to clot when bleeding occurs • Aids in the incorporation of calcium into bones 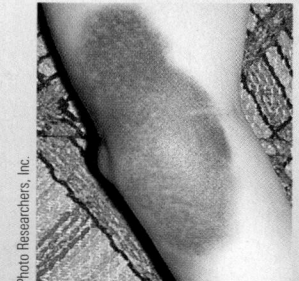 The long-term use of antibiotics can cause vitamin K deficiency. People with vitamin K deficiency bruise easily.	• Bleeding, bruises • Decreased calcium in bones • Deficiency is rare. May be induced by the long-term use (months or more) of antibiotics.

that generate new cells to replace worn-out cells lining the mouth, esophagus, intestines, and eyes. Without enough vitamin A, old cells aren't replaced and the affected tissues are damaged. And vitamin C is required for reactions that build and maintain collagen, a protein found in skin, bones, blood vessels, gums, ligaments, and cartilage. (Approximately 30% of the total amount of protein in the body is collagen.) With vitamin C deficiency, collagen becomes weak, causing tissues that contain collagen to weaken and bleed easily.

The examples just given all relate to the physical effects of vitamins. Vitamins participate in reactions that affect behaviors, too. Alterations in behaviors such as reduced attention span, poor appetite, irritability, depression, or paranoia often precede the physical signs of vitamin deficiency.[2] Vitamins are truly "vital" for health.

Protection from Vitamin Deficiencies and More

Current research on vitamins centers around their effects on disease prevention and treatment. It is a very active area of research and, no doubt, new and important

TABLE 20.2

AN INTENSIVE COURSE ON VITAMINS. (CONTINUED)

THE FAT-SOLUBLE VITAMINS

CONSEQUENCES OF OVERDOSE	PRIMARY FOOD SOURCES	HIGHLIGHTS AND COMMENTS
• Mental retardation in young children • Abnormal bone growth and formation • Nausea, diarrhea, irritability, weight loss • Deposition of calcium in organs such as the kidneys, liver, and heart	• Vitamin D–fortified milk and margarine • Butter • Fish • Eggs • Mushrooms • Milk products such as cheese, yogurt, and ice cream are generally not fortified with vitamin D.	• Vitamin D is manufactured from cholesterol in cells beneath the surface of the skin upon exposure of the skin to sunlight. • Deficiency may be common in ill, homebound, and elderly and hospitalized adults. • Breastfed infants with little sun exposure benefit from vitamin D supplements.
• Toxicity is only a problem when synthetic forms of vitamin K are taken in excessive amounts. That may cause liver disease.	• Leafy, green vegetables • Grain products	• Vitamin K is produced by bacteria in the gut. Part of our vitamin K supply comes from these bacteria. • Newborns are given a vitamin K injection because they have "sterile" guts and consequently no vitamin K–producing bacteria.

TABLE 20.3

THE "REAL" VITAMINS ARE LISTED IN TABLE 20.1. BUT A NUMBER OF BOGUS VITAMINS ARE ALSO ON THE MARKET. THEY TURN UP IN SUPPLEMENTS, SKIN CARE CREAMS, WEIGHT-LOSS AND HAIR CARE PRODUCTS, AND OTHER ITEMS. THESE ARE SOME OF THE MORE POPULAR NONVITAMINS.

NONVITAMINS

Bioflavonoids (vitamin P)

Coenzyme Q_{10}

Gerovital H-3

Hesperidin

Inositol

Laetrile (vitamin B_{17})

Lecithin

Lipoic acid

Nucleic acids

Pangamic acid (vitamin B_{15})

Para-amino benzoic acid (PABA)

Provitamin B_5 complex

Rutin

---

results will be announced after this text goes to press. (Nutrition textbooks should really be updated every few weeks. Stay tuned to your instructor for the latest developments.) Here are a few examples of recent developments in research on vitamins and disease prevention and treatment:

- *Folate, neural tube defects, and heart disease.* Daily consumption of 400 micrograms of folic acid (the synthetic form of folate added to refined grain products) before and early in pregnancy prevents about two-thirds of cases of neural tube defects in newborns.[3] Neural tube defects are abnormalities of the spinal cord and brain (Illustration 20.1). They are the most common type of malformation of newborns in the United States. Adequate intake of folate also reduces the risk of developing heart disease and other diseases associated with hardening of the arteries, such as stroke.[4]

    In 1998 manufacturers started fortifying refined grain products such as bread, pasta, and rice with folic acid. The addition of folic acid to these foods has produced substantial gains in people's folate status. Blood levels

Illustration 20.1
A baby with spina bifida, a form of neural tube defect associated with poor folate status early in pregnancy.

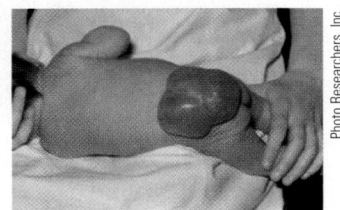

Photo Researchers, Inc.

homocysteine
A compound produced when the amino acid methionine is converted to another amino acid, cysteine. High blood levels of homocysteine increase the risk of hardening of the arteries, heart attack, and stroke.

antioxidants
Chemical substances that prevent or repair damage to cells caused by exposure to free radicals. Beta-carotene, vitamin E, and vitamin C function as antioxidants.

precursor
In nutrition, a nutrient that can be converted into another nutrient. (Also called *provitamin.*) Beta-carotene is a precursor of vitamin A.

free radicals
Chemical substances (usually oxygen or hydrogen) that are missing an electron. The absence of the electron makes the chemical substances reactive and prone to oxidizing nearby atoms or molecules by stealing an electron from them.

of folate have nearly doubled, and the prevalence of poor folate status has declined by 92%. Rates of neural tube defects appear to be declining substantially, as are blood levels of **homocysteine,** a risk factor for heart disease and stroke.[5]

- *Vitamin A—from measles to "liver spots."* Studies undertaken in both developing countries and the United States indicate that adequate vitamin A status decreases the severity of measles and other infectious diseases.[6] (It has been known for decades that adequate vitamin A intake also prevents blindness, an all-too-common consequence of vitamin A deficiency in developing nations.) This vitamin is showing promise as a treatment for skin and cervical cancer; and a vitamin A derivative (retinoic acid) is being used successfully in the treatment of serious cases of acne, skin wrinkles, and "liver" or "aging" spots on the skin due to overexposure to the sun.[7]

- *Vitamin D, osteoporosis, and rickets.* Vitamin D intakes of 800 to 1000 IU per day are associated with decreased bone loss and a reduced risk of osteoporosis.[8]

  Increased rates of breastfeeding in the United States have had the unintended side effect of increased vitamin D deficiency (rickets) in infants.[9] Breast milk contains a very low amount of vitamin D, so infants receiving human milk need to get it from exposure of their skin to sunlight or from supplements. Infants who develop vitamin D deficiency fail to get enough sunlight exposure due to winters in cold climates, or from being fully clothed or covered with sunscreen to protect their skin from the sun. Modest exposure to sunshine (about 30 minutes a week with diapers only) is sufficient to supply the required amount of vitamin D (Illustration 20.2). In cold climates, a vitamin D supplement is needed.[10]

- *Vitamin C and the common cold revisited.* Recent research confirms that vitamin C, the popular cold remedy, reduces the symptoms and duration of the common cold, but does *not* affect how often colds occur.[11]

Some of the most exciting recent findings on vitamins and disease concern vitamins that function as **antioxidants.**

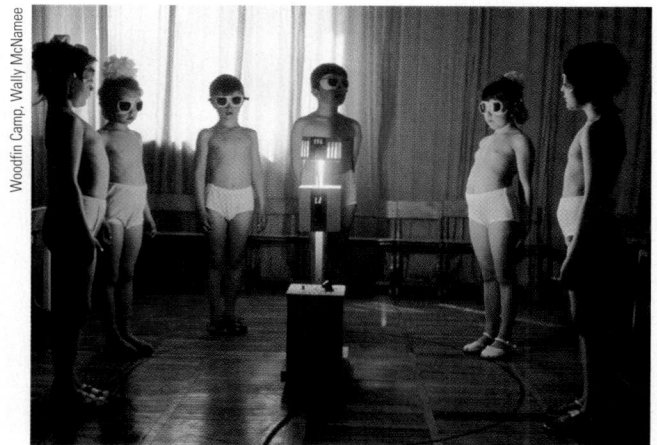

Illustration 20.2
Russian children are exposed to a quartz lamp to prevent vitamin D deficiency during the long winter.

The Antioxidant Vitamins

Beta-carotene (a **precursor** to vitamin A), vitamin E, and vitamin C function as antioxidants. This means they prevent or repair damage to components of cells caused by exposure to **free radicals.** A free radical is formed when an atom of hydrogen or oxygen loses an electron. Without the electron, there is an imbalance between the atom's positive and negative charges. This makes the atom reactive—it needs to steal an electron from a nearby atom or molecule in order to reestablish a balance between its positive and negative charges. Atoms and molecules that have lost electrons to free radicals are said to be "oxidized." These oxidized substances are reactive and can damage cell membranes, DNA, and other cell components. Antioxidants such as beta-carotene, vitamin E, and vitamin C donate electrons to stabilize oxidized molecules or repair them in other ways.

Why do free radicals exist in the body? Free radicals play a number of roles in the body, so they are always present. They are produced during energy formation, by breathing, and by the immune system to help destroy bacteria and viruses that enter the body. They can also be formed when the body is exposed to alcohol, radiation emitted by the sun, smoke, ozone, smog, and other environmental pollutants.

Cellular damage due to free radicals appears to play a role in the development of some types of cancer, bronchitis, emphysema, heart disease, cataracts, and premature aging.[12] Vitamin E specifically is associated with slowed progression of Alzheimer's disease, reduced plaque formation in arteries, and improved circulation in people with diabetes.[13] The antioxidant vitamins—along with the mineral selenium, substances found in plants, and other substances produced by the body that function as antioxidants—may help prevent or postpone these disorders by protecting cell components against free radical damage.

Vitamins: Getting Enough without Getting Too Much

Vitamins are widely present in basic foods (Table 20.4). Adequate amounts of them can be obtained from diets that include the variety of foods recommended in the Food Guide Pyramid. Most fruits and vegetables are good sources of vitamins, and eating 5 or more servings a day is one way for Americans to get their vitamins. In the United States, 80% of adults consume 2 or fewer servings of vegetables and fruits daily, and 10% consume none.[14]

Fortified foods such as ready-to-eat cereals, fruit juices and drinks, and snack bars are not listed in Table 20.4. They can increase vitamin intake substantially, and

text continued on page 20-17

TABLE 20.4

FOOD SOURCES OF VITAMINS.

THIAMIN

FOOD	SERVING SIZE	THIAMIN (Milligrams)
Meats:		
Pork roast	3 oz	0.8
Beef	3 oz	0.4
Ham	3 oz	0.4
Liver	3 oz	0.2
Nuts and seeds:		
Sunflower seeds	¼ cup	0.7
Peanuts	¼ cup	0.1
Almonds	¼ cup	0.1
Grains:		
Bran flakes	1 cup (1 oz)	0.6
Macaroni	½ cup	0.1
Rice	½ cup	0.1
Bread	1 slice	0.1
Vegetables:		
Peas	½ cup	0.3
Lima beans	½ cup	0.2
Corn	½ cup	0.1
Broccoli	½ cup	0.1
Potato	1 medium	0.1
Fruits:		
Orange juice	1 cup	0.2
Orange	1	0.1
Avocado	½	0.1

continued

Photo Disc

Beef and broccoli are good sources of thiamin.

TABLE 20.4

FOOD SOURCES OF VITAMINS. (CONTINUED)

RIBOFLAVIN

FOOD	SERVING SIZE	RIBOFLAVIN (Milligrams)
Milk and milk products:		
Milk	1 cup	0.5
2% milk	1 cup	0.5
Yogurt, low-fat	1 cup	0.5
Skim milk	1 cup	0.4
Yogurt	1 cup	0.1
American cheese	1 oz	0.1
Cheddar cheese	1 oz	0.1
Meats:		
Liver	3 oz	3.6
Pork chop	3 oz	0.3
Beef	3 oz	0.2
Tuna	3 oz	0.1
Vegetables:		
Collard greens	½ cup	0.3
Broccoli	½ cup	0.2
Spinach, cooked	½ cup	0.1
Eggs:		
Egg	1	0.2
Grains:		
Macaroni	½ cup	0.1
Bread	1 slice	0.1

NIACIN

FOOD	SERVING SIZE	NIACIN (Milligrams)
Meats:		
Liver	3 oz	14.0
Tuna	3 oz	10.3
Turkey	3 oz	9.5
Chicken	3 oz	7.9
Salmon	3 oz	6.9
Veal	3 oz	5.2
Beef (round steak)	3 oz	5.1
Pork	3 oz	4.5
Haddock	3 oz	2.7
Scallops	3 oz	1.1
Nuts and seeds:		
Peanuts	1 oz	4.9
Vegetables:		
Asparagus	½ cup	1.5
Grains:		
Wheat germ	1 oz	1.5
Brown rice	½ cup	1.2
Noodles, enriched	½ cup	1.0
Rice, white, enriched	½ cup	1.0
Bread, enriched	1 slice	0.7
Milk and milk products:		
Cottage cheese	½ cup	2.6
Milk	1 cup	1.9

Milk and milk products are good sources of riboflavin

Salmon is a good source of niacin.

TABLE 20.4

FOOD SOURCES OF VITAMINS. (CONTINUED)
VITAMIN B$_6$

FOOD	SERVING SIZE	VITAMIN B$_6$ (Milligrams)
Meats:		
Liver	3 oz	0.8
Salmon	3 oz	0.7
Other fish	3 oz	0.6
Chicken	3 oz	0.4
Ham	3 oz	0.4
Hamburger	3 oz	0.4
Veal	3 oz	0.4
Pork	3 oz	0.3
Beef	3 oz	0.2
Eggs:		
Egg	1	0.3
Legumes:		
Split peas	½ cup	0.6
Dried beans, cooked	½ cup	0.4
Fruits:		
Banana	1	0.6
Avocado	½	0.4
Watermelon	1 cup	0.3
Vegetables:		
Turnip greens	½ cup	0.7
Brussels sprouts	½ cup	0.4
Potato	1	0.2
Sweet potato	½ cup	0.2
Carrots	½ cup	0.2
Peas	½ cup	0.1

Bananas are a good source of vitamin B$_6$.

FOLATE

FOOD	SERVING SIZE	FOLATE (Micrograms)
Vegetables:		
Garbanzo beans	½ cup	141
Navy beans	½ cup	128
Asparagus	½ cup	120
Brussels sprouts	½ cup	116
Black-eyed peas	½ cup	102
Spinach, cooked	½ cup	99
Romaine lettuce	1 cup	86
Lima beans	½ cup	71
Peas	½ cup	70
Collard greens, cooked	½ cup	56
Sweet potato	½ cup	43
Broccoli	½ cup	43
Fruits:		
Cantaloupe	¼ whole	100
Orange juice	1 cup	87
Orange	1	59
Grains:[a]		
Ready-to-eat cereals	1 cup/1 oz	100–400
Oatmeal	½ cup	97
Noodles	½ cup	45
Wheat germ	2 tbs	40
Wild rice	½ cup	37

Beans are an especially rich source of folate.

[a]Fortified, refined grain products such as bread, rice, pasta, and crackers provide approximately 40 to 60 micrograms of folic acid per standard serving.

continued

TABLE 20.4

FOOD SOURCES OF VITAMINS. (CONTINUED)
VITAMIN B$_{12}$

FOOD	SERVING SIZE	VITAMIN B$_{12}$ (micrograms)
Meats:		
Liver	3 oz	6.8
Trout	3 oz	3.6
Beef	3 oz	2.2
Clams	3 oz	2.0
Crab	3 oz	1.8
Lamb	3 oz	1.8
Tuna	3 oz	1.8
Veal	3 oz	1.7
Hamburger, regular	3 oz	1.5
Milk and milk products:		
Skim milk	1 cup	1.0
Milk	1 cup	0.9
Yogurt	1 cup	0.8
Cottage cheese	½ cup	0.7
American cheese	1 oz	0.2
Cheddar cheese	1 oz	0.2
Eggs:		
Egg	1	0.6

VITAMIN C

FOOD	SERVING SIZE	VITAMIN C (milligrams)
Fruits:		
Orange juice, vitamin C–fortified	1 cup	108
Kiwi fruit	1 or ½ cup	108
Grapefruit juice, fresh	1 cup	94
Cranberry juice cocktail	1 cup	90
Orange	1	85
Strawberries, fresh	1 cup	84
Orange juice, fresh	1 cup	82
Cantaloupe	¼ whole	63
Grapefruit	1 medium	51
Raspberries, fresh	1 cup	31
Watermelon	1 cup	15
Vegetables:		
Green peppers	½ cup	95
Cauliflower, raw	½ cup	75
Broccoli	½ cup	70
Brussels sprouts	½ cup	65
Collard greens	½ cup	48
Vegetable (V-8) juice	¾ cup	45
Tomato juice	¾ cup	33
Cauliflower, cooked	½ cup	30
Potato	1 medium	29
Tomato	1 medium	23

Crab is one food source of vitamin B$_{12}$.

Citrus fruits are an excellent source of vitamin C.

TABLE 20.4

FOOD SOURCES OF VITAMINS. (CONTINUED)
VITAMIN A (RETINOL)

FOOD	SERVING SIZE	VITAMIN A (micrograms RE)[b]
Meats:		
Liver	3 oz	9124
Salmon	3 oz	53
Tuna	3 oz	14
Eggs:		
Egg	1 medium	84
Milk and milk products:		
Skim milk, fortified	1 cup	149
2% milk	1 cup	139
American cheese	1 oz	82
Whole milk	1 cup	76
Swiss cheese	1 oz	65
Fats:		
Margarine, fortified	1 tsp	46
Butter	1 tsp	38
Vitamin A and Beta-Carotene		

BETA-CAROTENE

FOOD	SERVING SIZE	VITAMIN A VALUE (micrograms RE)[b]
Vegetables		
Pumpkin, canned	½ cup	2712
Sweet potato, canned	½ cup	1935
Carrots, raw	½ cup	1913
Spinach, cooked	½ cup	739
Collard greens, cooked	½ cup	175
Broccoli, cooked	½ cup	109
Winter squash	½ cup	53
Green peppers	½ cup	40
Fruits:		
Cantaloupe	¼ whole	430
Apricots, canned	½ cup	210
Nectarine	1 medium	101
Watermelon	1 cup	59
Peaches, canned	½ cup	47
Papaya	½ cup	20

[b]RE (retinol equivalent) = 3.33 IU.

continued

Eggs are one food source of vitamin A (retinol).

Spinach and winter squash are two sources of vitamin A and beta–carotene.

concern has been expressed that fortified food consumption may increase intake too much. Although an important concern, overdose reactions from the new generation of vitamin-fortified foods have not yet been reported.

Updated recommendations for vitamin intakes associated with the prevention of deficiency and chronic diseases are represented by standards called Dietary Reference Intakes, or DRIs. DRIs include Recommended Dietary Allowances (RDAs) for vitamins for which convincing scientific data exist for establishing intake standards. Adequate Intakes, or AIs, are assigned to vitamins for which scientific information about levels of intake associated with chronic disease prevention is less convincing. Tolerable Upper Levels of Intake, abbreviated ULs, are also assigned to

TABLE 20.4

FOOD SOURCES OF VITAMINS. (CONTINUED)
VITAMIN E

FOOD	SERVING SIZE	VITAMIN E (IU)[c]
Oils:		
Oil	1 tbs	6.7
Mayonnaise	1 tbs	3.4
Margarine	1 tbs	2.7
Salad dressing	1 tbs	2.2
Nuts and seeds:		
Sunflower seeds	¼ cup	27.1
Almonds	¼ cup	12.7
Peanuts	¼ cup	4.9
Cashews	¼ cup	0.7
Vegetables:		
Sweet potato	½ cup	6.9
Collard greens	½ cup	3.1
Asparagus	½ cup	2.1
Spinach, raw	1 cup	1.5
Grains:		
Wheat germ	2 tbs	4.2
Bread, whole wheat	1 slice	2.5
Bread, white	1 slice	1.2
Seafood:		
Crab	3 oz	4.5
Shrimp	3 oz	3.7
Fish	3 oz	2.4

VITAMIN D

FOOD	SERVING SIZE	VITAMIN D (IU)[d]
Milk:		
Milk, whole, low-fat, or skim	1 cup	100
Fish and seafoods:		
Salmon	3 oz	340
Tuna	3 oz	150
Shrimp	3 oz	127
Organ meats:		
Beef liver	3 oz	42
Chicken liver	3 oz	40
Eggs:		
Egg yolk	1	27

[c]15 milligrams alpha-tocopherol = 22 IU d-alpha tocopherol (natural form) and 33 IU synthetic vitamin E.
[d]40 IU = 1 microgram.

Seafood and asparagus both provide vitamin E.

Shrimp and other seafoods are good sources of vitamin D.

vitamins and indicate levels of vitamin intake from foods, fortified foods, and supplements that should not be exceeded. The RDAs or AIs and the ULs for the vitamins are given in Table 20.2. (World Wide Web addresses listed at the end of this unit can lead you to the latest developments on the DRIs.)

Although people can get all the vitamins they need from supplements, it makes more sense to get them from basic foods. Foods offer fiber, minerals, and other healthful ingredients that don't come in supplements, and they certainly taste better on the way down!

Nutrition UP CLOSE

Antioxidant Vitamins—

How Adequate Is Your Diet?

FOCAL POINT: Determine if you eat enough antioxidant-rich foods.

Vitamin C, beta-carotene, and vitamin E, the antioxidant vitamins, help to maintain cellular integrity in the body. They may act to reduce the risk of heart disease, certain cancers, and other ailments. Check below to find out how frequently you consume foods containing these important, health-promoting nutrients.

How Often Do You Eat:	Seldom or Never	1–2 Times per Week	3–5 Times per Week	Almost Daily
Vitamin C-rich foods:				
1. Grapefruit, lemons, oranges, or pineapple?	☐	☐	☐	☐
2. Strawberries, kiwi, or honeydew melon?	☐	☐	☐	☐
3. Orange juice, cranberry juice cocktail, or tomato juice?	☐	☐	☐	☐
4. Green, red, or chili peppers?	☐	☐	☐	☐
5. Broccoli, Chinese cabbage, or cauliflower?	☐	☐	☐	☐
6. Asparagus, tomatoes, or potatoes?	☐	☐	☐	☐
Beta-carotene-rich foods:				
7. Carrots, sweet potatoes, pumpkin, or winter squash?	☐	☐	☐	☐
8. Spinach, collard greens, or chard?	☐	☐	☐	☐
9. Cantaloupe, papayas or mangoes?	☐	☐	☐	☐
10. Nectarines, peaches, or apricots?	☐	☐	☐	☐
Vitamin E-rich foods:				
11. Whole grain breads, whole grain cereals, or wheat germ?	☐	☐	☐	☐
12. Crab, shrimp, or fish?	☐	☐	☐	☐
13. Peanuts, almonds, or sunflower seeds?	☐	☐	☐	☐
14. Oils, margarine, butter, mayonnaise, or salad dressing?	☐	☐	☐	☐

FEEDBACK (including scoring) can be found at the end of Unit 20.

Key Terms

antioxidants, page 20-12

coenzymes, page 20-3

free radicals, page 20-12

homocysteine, page 20-12

precursor, page 20-12

vitamins, page 20-2

www links

www.healthfinder.gov
Search vitamins.

www.nal.usda.gov/fnic/foodcomp/
Look up the vitamin and other nutrient content of thousands of foods on this site.

www.iom.edu/fnb
Provides updated information on the DRIs.

www.merckhomeedition.com
The Merck Manual of Medical Information is free and searchable. It can be used to

find out more about health problems related to vitamins.

Notes

1. Hathcock JN. Vitamins and minerals: efficacy and safety. Am J Clin Nutr 1997;66:427–37.

2. Buzina R et al. Workshop on functional significance of mild-to-moderate malnutrition. Am J Clin Nutr 1989;50:172–6.

3. American Academy of Pediatrics. Folic acid and the prevention of neural tube defects. Pediatrics 1999;104:325–27.

4. Boushey CJ et al. A quantitative assessment of plasma homocysteine as a risk factor for vascular disease: probable benefits of increasing folic acid intakes. JAMA 1995;274:1049–57.

5. Shane B, Folate fortification: enough already? Am J Clin Nutr 2003;77-8–9; and Quinlivan EP, Gregory JF, Effect of food fortification on folic acid intake in the United States, Am J Clin Nutr 2003; 77:221–5.

6. Frieden TR et al. Vitamin A levels and severity of measles. Am J Dis Child 1992;146:182–6.

7. Soprano KJ, Soprano DR, Retinoic acid receptors and cancer, J Nutr 2002;132: 3809S–13S; and Rafal ES et al., Topical tretinoin (retinoic acid) treatment for liver spots associated with photodamage. N Engl J Med 1992;326:368–74.

8. Utiger RD. The need for more vitamin D. N Engl J Med 1998;338:828–9.

9. Welch TR et al. Vitamin D–deficient rickets: the reemergence of a once-conquered disease (editorial). J Pediatr 2000;1137:152–7.

10. Specker BL et al., Sunshine exposure and serum 25-hdyroxyvitamin D concentrations in exclusively breast-fed infants, J Pediatr 1985;107:372–6; and Welch et al., Vitamin D–deficient rickets.

11. Hemilä H, Herman ZS. A retrospective analysis of Chalmer's review. J Am Coll Nutr 1995;14:116–23.

12. Hu GZ, Cassano P. Antioxidant nutrients and pulmonary function. Am J Epidemiol 2000;151:975–81.

13. Grundman M, Vitamin E and Alzheimer disease: the basis for additional clinical trials, Am J Clin Nutr 2000;71(suppl): 630S–6S; Stephens NG et al., Randomized control trial of vitamin E in patients with coronary disease, Lancet 1996;347: 781–6; and Bursell S-E et al., High doses of vitamin E supplementation normalize retinal blood flow and creatinine clearance in patients with type 1 diabetes. Diabetes Care 1999;22:1245–51.

14. Johnston CS, Corte C. People with marginal vitamin C status are at high risk of developing vitamin C deficiency. J Am Diet Assoc 1999;99:854–56.

Nutrition UP CLOSE

Antioxidant Vitamins—
How Adequate Is Your Diet?

Feedback for Unit 20

Several responses in the last two columns indicate adequate antioxidant vitamin consumption. If you need to boost your intake, increase the overall amount of fruits, vegetables, and whole grains in your diet. Although nuts, seeds, oils, margarine, butter, mayonnaise, and salad dressing all contribute vitamin E, they are high-fat choices and should be consumed in moderation.

Phytochemicals and Genetically Modified Food

Nutrition Scoreboard

	TRUE	FALSE
1 Phytochemicals are found only in plants.		
2 Phytochemicals taken as supplements provide the same health benefits as do phytochemicals consumed in plant foods.		
3 Humans eat lots of DNA every day.		
4 Some chemical substances that occur naturally in food may be harmful to health.		

Answers on next page

[KEY CONCEPTS AND FACTS]

- Plants contain thousands of substances in addition to essential nutrients that affect body processes and health.

- Diets containing lots of vegetables, fruits, whole grains, and other plant foods are strongly associated with the prevention of chronic diseases such as heart disease and cancer.

- Biotechnology is rapidly changing the characteristics and types of foods available to consumers.

- Not all substances that occur naturally in foods are safe to eat.

Answers to *Nutrition Scoreboard*

		TRUE	FALSE
1	*Phytochemicals* means "plant chemicals."	✔	
2	Beneficial effects of phytochemicals on health appear to result primarily from complex interactions among them in foods. Supplements do not appear to provide these benefits.[1]		✔
3	The average meal contains approximately 250,000 miles (150,000 kilometers) of uncoiled DNA when it's digested.	✔	
4	Some foods contain "naturally occurring toxins" that can be harmful if consumed in excess.	✔	

Photo Disc

Phytochemicals: Nutrition Superstars

Things don't happen by accident in nature. If you observe it,
it has a reason for being there.
—Norman Krinsky, Tufts University Medical Center

As recently as 25 years ago, the science of nutrition focused on the study of the actions and health effects of protein, vitamins, minerals, and the other essential nutrients. Those days are gone. Now nutrition scientists are investigating the effects of thousands of other substances in food. Although we have barely scratched the surface of knowledge about these substances, current research shows that impressive benefits can be gained by diets high in plant foods. People who habitually consume lots of vegetables, fruits, whole grains, and other plant foods are less likely to develop heart disease, cancer, adult-onset diabetes, infections, eye disease, premature aging, and a number of other health problems than are people who do not.[3] Research results are sufficiently positive to change the way nutritionists and other scientists are thinking about food and health relationships.

At center stage in this new era in nutrition are the ***phytochemicals,*** or chemical substances found in plants (Illustration 21.1). The number of potentially beneficial substances in plants is mind-boggling. There are thousands of individual phytochemicals in plants, and some foods contain hundreds of them. A sampling of the phytochemicals and other substances in two foods is shown in Illustration 21.2. Meats, eggs, dairy products, and other foods of animal origin also contain chemical substances that are not considered essential nutrients but nevertheless affect body processes. Much less is known about these ***zoochemicals*** because their effects on health are not as clear as those of phytochemicals.

Phytochemicals are not considered to be essential nutrients for humans, because we do not develop a deficiency disease if we consume too little of them. They are similar to essential nutrients in that the body cannot make them—or enough of them. Consequently, they must be obtained from the diet. The chemical properties

phytochemicals
(phyto = *plant*)
Chemical substances in plants, some of which perform important functions in the body.

zoochemicals
Chemical substances in animal foods, some of which perform important functions in the body.

Ann Borman

Illustration 21.1
The foods shown here have star qualities.
They have been featured in recent research reports on the benefits of phytochemicals in food. Foods shown are grape juice, green tea, tomatoes, cabbage, carrots, tofu, crushed garlic; green, red, and yellow peppers; brown rice, whole grain pita bread, and peanut butter. (Food preparation courtesy of The Blue Moon Café, Menomonie, WI.)

of most phytochemicals tend to make them heat and light stable, so they are not easily destroyed by cooking or storage.[4]

Many phytochemicals are excreted within a day or two after ingestion, so intake of vegetables, legumes, nuts, fruits, and other food sources should be maintained.[5] Cooking vegetables, or consuming them with a small amount of fat, increases the body's absorption of some phytochemicals.

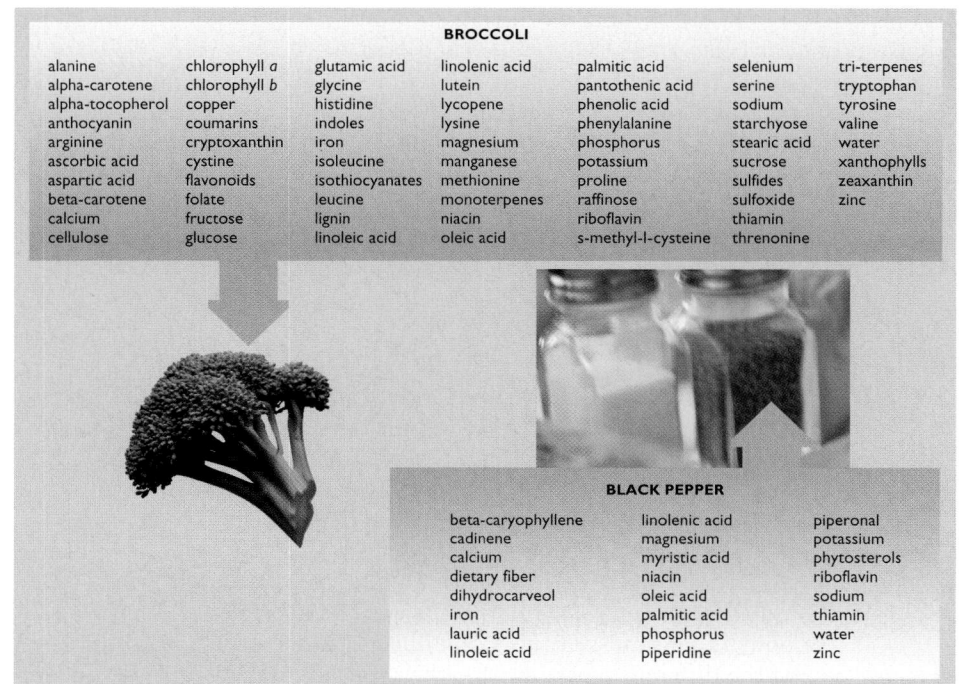

BROCCOLI

alanine	chlorophyll *a*	glutamic acid	linolenic acid	palmitic acid	selenium	tri-terpenes
alpha-carotene	chlorophyll *b*	glycine	lutein	pantothenic acid	serine	tryptophan
alpha-tocopherol	copper	histidine	lycopene	phenolic acid	sodium	tyrosine
anthocyanin	coumarins	indoles	lysine	phenylalanine	starchyose	valine
arginine	cryptoxanthin	iron	magnesium	phosphorus	stearic acid	water
ascorbic acid	cystine	isoleucine	manganese	potassium	sucrose	xanthophylls
aspartic acid	flavonoids	isothiocyanates	methionine	proline	sulfides	zeaxanthin
beta-carotene	folate	leucine	monoterpenes	raffinose	sulfoxide	zinc
calcium	fructose	lignin	niacin	riboflavin	thiamin	
cellulose	glucose	linoleic acid	oleic acid	s-methyl-l-cysteine	threnonine	

BLACK PEPPER

beta-caryophyllene	linolenic acid	piperonal
cadinene	magnesium	potassium
calcium	myristic acid	phytosterols
dietary fiber	niacin	riboflavin
dihydrocarveol	oleic acid	sodium
iron	palmitic acid	thiamin
lauric acid	phosphorus	water
linoleic acid	piperidine	zinc

Photo Disc

Illustration 21.2
A sampling of the chemical substances in two foods.

age-related macular degeneration
Eye damage caused by oxidation of the macula, the central portion of the eye that allows you to see details clearly. It is the leading cause of blindness in U.S. adults over the age of 65. Antioxidants provided by the carotenoids in dark green, leafy vegetables such as kale, collard greens, spinach, and Swiss chard may help prevent macular degeneration.[8]

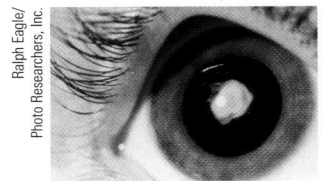

Ralph Eagle/
Photo Researchers, Inc.

cataracts
Complete or partial clouding over the lens of the eye (shown in illustration).

Illustration 21.3
People love it or hate it.
Arugula is, however, a rich source of phytochemicals.
Source: From the Wall Street Journal, permission CFS.

"I thought you were bringing the arugula."

Characteristics of Phytochemicals

Phytochemicals serve a wide variety of functions in plants. They provide color and flavor and protect plants from insects, microbes, and oxidation due to exposure to sunlight and oxygen. Some phytochemicals are components of a plant's energy-making processes, and others act as plant hormones. More than 2000 types of phytochemicals that act as pigments have been identified. There are at least 600 types of carotenoids, for example. These pigments give plants yellow, orange, and red color; they primarily function as antioxidants.

The amount and type of phytochemicals present in plants vary a good deal depending on the plant. Some plant foods, such as citrus fruits and dried beans, which are considered rich sources of vitamins and minerals, contain an abundance of phytochemicals. Apples, celery, green tea, and garlic, foods considered by nutrient composition tables to be relative "nutrient weaklings," also provide ample amounts of phytochemicals. Many of the richest sources of beneficial phytochemicals are foods that are infrequently consumed (see Illustration 21.3).

Not all of the phytochemicals in plants are beneficial to health, however. Some are "naturally occurring toxins" and can be harmful. This type of phytochemical is discussed later in the unit.

Phytochemicals and Health

It has been known for more than 30 years that diets rich in vegetables and fruits are protective against heart disease and certain types of cancer.[6] For most of this time, the benefits of such diets were attributed to the vitamin, mineral, or dietary fiber content of vegetables and fruits. More recently, scientists discovered that the essential nutrient content of the diet didn't explain all of the differences in disease incidence between people consuming diets low in plants and those eating diets rich in plants. As researchers looked for the reason for the difference, they discovered that other chemical substances in food might be responsible. Many chemically active substances that could contribute to disease prevention were identified and their actions in the body described.

Phytochemicals are associated with a reduced risk of developing heart disease, certain types of cancer (lung, breast, cervical, esophageal, stomach, and colon cancer, for example), *age-related macular degeneration, cataracts,* infectious diseases, diabetes in adults, stroke, hypertension, and other disorders.[7] Although evidence supports a primary role for phytochemicals in the prevention of these disorders, research studies required to demonstrate *cause-and-effect relationships* in humans have yet to be completed.

Phytochemicals Work in Groups

News about the potential benefits of phytochemicals has sent consumers in search of pills that contain them. The market is replete with such products (Illustration 21.4), even though there is no solid evidence that individual phytochemicals extracted from foods benefit health.[9]

The absorption of phytochemicals appears to depend on the presence of other phytochemicals and nutrients in foods. Most (if not all) phytochemicals act together, producing a desired effect in the body if consumed at the same time. Since the optimal combinations of the different types of phytochemicals are not yet known, it is recommended that foods rather than supplements provide them.[10]

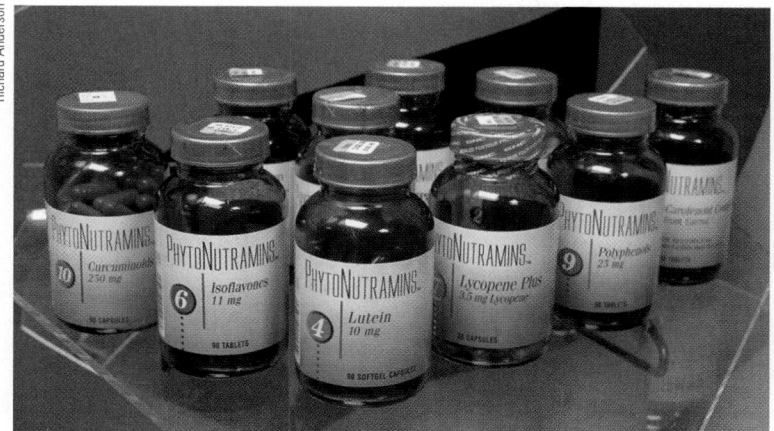

Illustration 21.4
Consumers have their choice of phytochemicals in pills.
The problem is that the supplements have not been shown to benefit health. Phytochemicals taken individually in supplements may actually harm health. They are best consumed in their natural packaging: vegetables, fruits, whole grains, and other plant foods.

The Case of Beta-Carotene Supplements] The lesson that phytochemicals act together was learned the hard way. Numerous studies had shown that smokers consuming diets high in vegetables and fruits and those with high levels of beta-carotene in their blood were much less likely to develop lung cancer than smokers who consumed few vegetables and fruits. Beta-carotene is a powerful antioxidant present in many vegetables and fruits, and researchers theorized that the beta-carotene provided by these plant foods might prevent lung cancer.

Three large and expensive clinical trials of beta-carotene supplementation among male smokers were conducted to test this theory. The results? None of the trials showed a benefit from the beta-carotene supplements, and two studies found a higher rate of lung cancer among the groups using the supplements.[11]

Scientists drew two major conclusions from these studies:

1. High blood levels of beta-carotene found in the group of smokers who did not develop cancer were likely a marker for a high intake of vegetables and fruits, and not the cause of the reduced risk.

2. Other phytochemicals in plants, or in a combination of plant foods, are likely responsible for the reduced risk of cancer.[12]

Vegetable Extracts and Essences] Dehydrated, powdered extracts of vegetables high in certain phytochemicals and of the parts of vegetables richest in phytochemicals are also widely available. As is the case for supplements of individual phytochemicals, there is no solid evidence that these extracts benefit health. One such product called "Vegetable Essence" claimed to have "200 pounds of vegetables in a bottle." That is, of course, impossible. "Broccoli Concentrate," another vegetable extract, was found to contain primarily sulforaphane, one of the many phytochemicals in vegetables of the ***cruciferous family*** (Illustration 21.5). The problem with this and other vegetable concentrates is that you can fit only so much of a vegetable into a capsule. You would have to consume approximately 100 of the pills to get the amount of sulforaphane present in one serving of broccoli.[13] Even if you did that,

cruciferous vegetables
Sulfur-containing vegetables whose outer leaves form a cross (or crucifix). Vegetables in this family include broccoli, cabbage, cauliflower, brussels sprouts, mustard and collard greens, kale, bok choy, kohlrabi, rutabaga, turnips, broccoflower, and watercress.

Illustration 21.5
Some examples of cruciferous vegetables.

Plants have been on this planet for about a billion years longer than humans. Although the total number of edible plants hasn't changed much over the centuries, most modern humans consume only 150–200 different types of plants, or about 0.2% of the total possibilities.[24]

you would still be missing out on the other phytochemicals and essential nutrients that are part of those little green trees.

Until scientific studies demonstrate that supplements of phytochemicals, individually or in combination, are safe and effective, it is best to forget about using them.

How Do Phytochemicals Work?

A variety of body processes involved in disease development are affected by phytochemicals. Some of these processes, and the phytochemicals responsible, are listed in Table 21.1. In general, phytochemicals can

1. act as hormone-inhibiting substances that prevent the initiation of cancer,
2. serve as antioxidants that prevent and repair damage to cells due to oxidation,
3. block or neutralize enzymes that promote the development of cancer and other diseases,
4. modify the absorption, production, or utilization of cholesterol, or
5. decrease formation of blood clots.

Some plant pigments are particularly powerful antioxidants. Zeaxanthin (pronounced ze-ah-zan-thun), which give plants a corn yellow color, anthocyanin (an-tho-sigh-ah-nin, the blue in blueberries and grapes), and lycopene (lie-co-peen), which helps make tomatoes, strawberries, and guava red, are all strong antioxidants. Dark chocolate contains flavonoids which act as antioxidants. Some of the phytochemicals that act as antioxidants help reduce plaque formation in arteries by preventing the oxidation of LDL cholesterol. Oxidized LDL is reactive and easily becomes a component of plaque.[15]

Diets High in Plant Foods

The discovery of a wide array of health-promoting phytochemicals in plants is one more very important reason why vegetables, fruits, whole grains, and other plant foods should be a major part of the diet. They are a minor part now for many Americans. In the United States, 80% of adults consume 2 or fewer servings of vegetables and fruits daily, and 10% consume none.[16] In reality, people would probably be healthier if they consumed more than the 5 servings of vegetables and fruits daily.

Naturally Occurring Toxins in Food

All that occurs naturally in foods is not necessarily good. There are many examples of foods containing phytochemicals that can have negative side effects. Spinach, collard greens, rhubarb, and other dark green, leafy vegetables contain oxalic acid. Eating too much of these foods can make your teeth feel as though they are covered with sand and give you a stomachache. Have you ever seen a potato that was partly colored green? (If not, take a look at Illustration 21.6.) The green area contains solanine, a bitter-tasting, insect-repelling phytochemical that is normally found only in the leaves and stalks of potato plants. Small amounts of solanine are harmless, but large quantities (an ounce or so) can interfere with the transmission of nerve impulses.

Phytates are an example of a naturally occurring substance that can have harmful effects. Phytates, which are found in whole grains, tightly bind zinc, iron, and other minerals, making them unavailable for absorption. Although not toxic, phytates do reduce the availability of some minerals in whole grain products. Cassava, a root con-

Illustration 21.6
Potatoes grown partly above ground develop a green color in the part exposed to the sun. *The green section contains solanine, a naturally occurring, toxic phytochemical.*

TABLE 21.1

EXAMPLES OF PHYTOCHEMICALS, THEIR FOOD SOURCES, AND POTENTIAL MECHANISMS OF ACTION IN DISEASE PREVENTION.[14]

PHYTOCHEMICAL	COMMENTS	FOOD SOURCES	PROPOSED ACTION
Indoles, isothiocynates (sulforaphanes)	These sulfur-containing compounds may be particularly protective against breast cancer.	Cruciferous vegetables (broccoli, brussels sprouts, cabbage, cauliflower)	Interfere with the cancer growth-promoting effects of genes on cells; increase the body's ability to neutralize cancer-causing substances.
Allicin	Health benefits are associated with as little as a half clove of garlic per day on average.	Garlic, onions, leeks, shallots, chives, scallions	Interferes with the replication of cancer cells; decreases the production of cholesterol by the liver, reduces blood clotting.
Terpenes (monoterpenes, limonene)	These substances give citrus fruits a slightly bitter taste. Limonene is from the same family of compounds as tamoxifen, a drug used to treat breast cancer.	Oranges, lemons, grapefruit, and their juices	Facilitate the excretion of cancer-causing substances; decrease tumor growth.
Phytoestrogens (plant estrogens; isoflavones, genistein, daidzein, lignans)	May decrease risk of some cancers, heart disease, and osteoporosis. Estrogen effects, if any, are weak. High doses of some individual phytoestrogens may have adverse effects on health.	Soybeans, soy food, chickpeas, other dried beans, peas, peanuts	Interfere with cancer growth-promoting effects of estrogen; block the action of cancer-causing substances; lower blood cholesterol; decrease menopausal symptoms and bone loss, act as antioxidants.
Lignans (phytoestrogens)	Flaxseed is an extremely rich source of lignans. In the gut, lignans are converted to substances that may help prevent breast cancer.	Flaxseed, flaxseed oil, seaweed, soybeans and other dried beans, bran	Interfere with the action of estrogen.
Saponins	Nearly every type of dried beans is an excellent source of saponins.	Dried beans, whole grains, apples, celery, strawberries, grapes, onions, green and black tea, red wine	Neutralize certain potentially cancer-causing enzymes in the gut.
Flavonoids (tannins, phenols)	There are over 4000 flavonoids, some of which are plant pigments. Originally called "vitamin P." Gives red wines and dark teas astringent or bitter taste.	Apples, celery, strawberries, grapes, onions, green and black tea, red wine, soy, purple grape juice, broccoli, dark chocolate	Protect cells from oxidation; decrease plaque formation and blood clotting; increase HDL cholesterol; decrease DNA damage related to cancer development.
Carotenoids (alpha-carotene, beta-carotene, lutein, zeaxanthin, beta-cryptoxanthin, lycopene)	There are more than 600 types of colorful carotenoids in plants. An orange contains at least 20 types. Dark green vegetables are often good sources; the green chlorophyll obscures the colors of carotenoids in these plants. Fat intake increases absorption.	Dark green vegetables, orange, yellow, and red vegetables and fruits	Neutralize oxidation reactions that can damage eyes and promote macular degeneration and cataracts; increase LDL cholesterol and cancer risk.
Plant stanols and sterols	Structurally similar to cholesterol; they block cholesterol absorption.	Edible and nonedible oils	Decrease blood LDL-cholesterol level.

Illustration 21.7
Ackee fruit and seeds.

sumed daily in many parts of tropical Africa, can be very toxic if not prepared properly because it contains cyanide. Soaking cassava roots in water for three nights gets rid of the cyanide, but soaking for shorter periods does not. When the soaking time is cut to one or two nights, as sometimes happens during periods of food shortage, enough of the toxin remains in the root to cause konzo, a disease caused by the cyanide overdose.[17] Konzo is characterized by permanent, spastic paralysis.

Beware of Ackee Fruit] Ackee fruit is another potential hazard to health. If you're from Jamaica or Africa, chances are excellent that you love the taste of the core of ackee—and know the fruit can be deadly. The national fruit of Jamaica, the yellow fleshy part around the seeds tastes like butter and looks like scrambled eggs (Illustration 21.7). The rest of the fruit is not edible. The ackee fruit—and the fruit of unopened, unripe ackee in particular—contains high concentrations of phytochemicals that cause severe vomiting and a drastic drop in blood glucose levels. Ingestion of the fruit has caused hundreds of deaths in Jamaica. Its sale was banned in the United States until 2000.[18]

Caffeine is another phytochemical in plants (primarily, coffee beans) that some people consider to be toxic. To others, it's a gift from the gods every morning. Does caffeine belong in the list of naturally occurring, harmful substances in food?

What's the Scoop on Caffeine?] Caffeine is one of many phytochemicals in coffee beans, cocoa beans, cola nuts, and tea leaves. Coffee, however, is far and away the leading source of caffeine in the diet. Table 21.2 displays the caffeine content of various beverages, chocolate, and nonprescription drugs. Although the effects of caffeine on health are widely debated, it has received a clean bill of health with a couple of exceptions. Too much caffeine causes sleeplessness and "caffeine jitters" in many people.[19] Large amounts of coffee (more than 4 cups a day) may somewhat delay the time it takes women to become pregnant and may slightly increase the risk of miscarriage.[20] Drinking unfiltered, boiled coffee (as is the custom in Turkey and some Scandinavian countries) raises LDL-cholesterol levels a little. However, two other phytochemicals in coffee, kahweol and cafestol, and not caffeine, may be responsible for this effect.[21] Suspicions that filtered coffee or caffeine is related to heart disease, cancer, or osteoporosis have not been confirmed.[22]

Regular coffee drinking can be habit forming. Cessation of coffee intakes that average two-and-a-half cups a day leads to headaches in about 50% of people.[23]

Genetically Modified Foods

Through *biotechnology,* researchers can modify the phytochemical and nutrient makeup of foods by altering the genetic makeup of plants. Resulting plants are called "genetically modified" or "GM" plants (Illustration 21.8). The process of biotechnology usually entails identifying a favorable genetic trait in one plant and transplanting that gene into another plant that lacks the characteristic.

Modifying the genetic makeup of plants has actually been practiced for centuries. Some 8000 years ago, Native Americans increased corn yields by cross-fertilizing corn plants. Over the centuries, thousands of plant hybrids derived from combining the genes of different plant species through cross-pollination have been developed. Advances related to genetic engineering in recent years have refined techniques so that single genes, rather than the full complements of a plant's genetic makeup, can be transferred to another plant.[25]

Genetic engineering is now used to transfer disease-resistant genes from one plant to another plant, conferring upon it an improved ability to resist disease. Fla-

biotechnology
As applied to food products, the process of modifying the composition of foods by biologically altering their genetic makeup. Also called *genetic engineering* of foods. The food products produced are sometimes referred to as "GM" and GMOs (genetically modified organisms).

vor genes can be transported from one plant to another, creating new taste sensations like garbanzo beans with a nutty, peppery taste and basil preseasoned with cinnamon. Watermelon and oranges have had their seeds removed through genetic engineering. Colors of vegetables and fruits can be modified by transferring the appropriate genes from one plant to another. Carrots have been modified to be dark red so they would match Texas A&M's school color. Tomatoes have been genetically altered to stay firm during shipment and to produce 10 times the normal amount of lycopene; and rice has been altered to be rich in beta-carotene, the precursor of vitamin A. High beta-carotene rice was produced to make a good source of vitamin A available to people in countries where vitamin A deficiency is widespread. Called "golden rice," it contains two genes extracted from daffodils and one bacterial gene that together lead to the production of beta-carotene within rice seeds.[26]

Genetic Modification of Animals

Food biotechnology applies to foods derived from animals as well as plants. Scientists have engineered Atlantic salmon that grow to market weight in 18 rather than the usual 24–30 months. Pigs that produce less smelly stools and gas, cattle with leaner muscles, and hens that lay more eggs are also products of animal genetic engineering. Genetically engineered animals are not, however, approved for entry into the food supply. Approval is being blocked by concerns that genetically engineered fish and other animals could escape, mate with wild animals, and introduce novel and potentially damaging genetic traits in animals.[27]

TABLE 21.2

CAFFEINE CONTENT OF FOODS, BEVERAGES, AND SOME DRUGS.

SOURCE	CAFFEINE (mg)
Coffee, one cup	
Drip	137–153
Decaffeinated (ground or instant)	0.5–4.0
Instant	61–70
Percolated	97–125
Espresso (2 ounces)	100
Tea, one cup	
Black, brewed 5 minutes, U.S. brands	32–144
Black, brewed 5 minutes, imported brands	40–176
Green, brewed 5 minutes	25
Instant	40–80
Soft drinks/bottled water	
Coca-Cola (8 ounces)	31
Cherry Coke (8 ounces)	31
Diet Coca-Cola (8 ounces)	31
Dr. Pepper (12 ounces)	40
Ginger ale (12 ounces)	0
Jolt (12 ounces)	108*
Kick (12 ounces)	56
Mountain Dew (12 ounces)	54
Pepsi-Cola (12 ounces)	38
Diet Pepsi-Cola (12 ounces)	37
7-Up (12 ounces)	0
Surge (8 ounces)	35
Water Joe (8 ounces)	125
Chocolate	
Cocoa, chocolate milk, one cup	10–17
Milk chocolate candy, one ounce	1–15
Chocolate syrup, one ounce (2 tablespoons)	4
Nonprescription drugs, one tablet	
Nodoz	200
Vibrin	200
Dexatrim	200

*Estimated from company Web site description of "twice the caffeine."

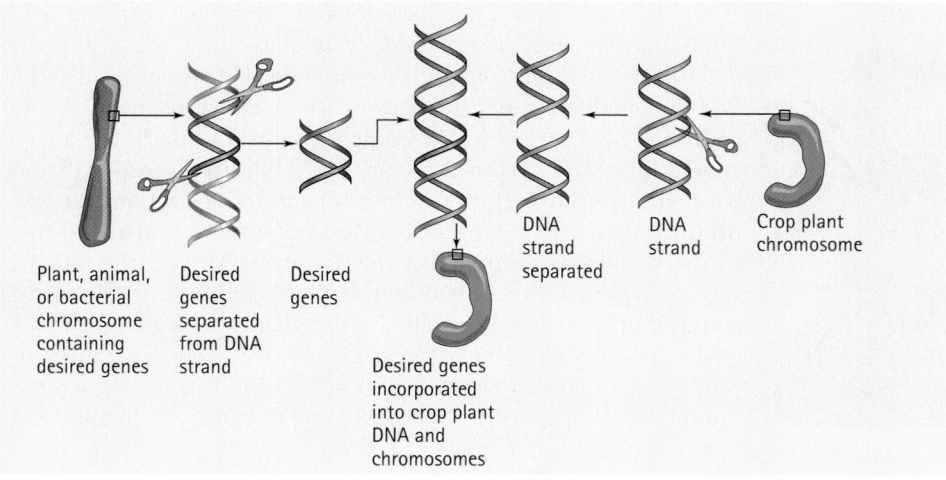

Illustration 21.8
Genetic modification of plants.

Plant, animal, or bacterial chromosome containing desired genes

Desired genes separated from DNA strand

Desired genes

Desired genes incorporated into crop plant DNA and chromosomes

DNA strand separated

DNA strand

Crop plant chromosome

Illustration 21.9

GM Foods: Are They Safe and Acceptable?

People have a fear of eating DNA. We eat 150,000 kilometers [250,000 miles] of DNA in an average meal.
—David Cove, PhD, Leeds University's Genetics Department

People generally don't think of DNA as a normal component of food, but since we eat cells when we consume foods, we get hefty helpings of DNA and chromosomes in every meal. People have been consuming foods with novel DNA in hybrid plants for many years without ill effects. Obviously, consuming DNA is not harmful by itself. Concerns about the safety of GM foods have less to do with consuming them and more to do with their potential impact on people's lives and the environment.

Most Americans are unaware of the broad presence of GM foods in the marketplace. Over 60% of processed foods contain GM ingredients, but only 19% of adults believe they have eaten a GM food.[28] Over half of all soybeans, and a third of corn grown in the United States are from genetically modified seeds. United States consumers regularly purchase and eat tomatoes, squash, cantaloupe, and potatoes that have been genetically modified (Illustration 21.9).[29] Although a relatively small percentage of U.S. consumers are concerned about the safety of GM foods, several baby and snack food producers have excluded them from their products. Individuals preferring GM-free foods in the United States and abroad are opting for organic products (Illustration 21.10) and locally grown produce.[30]

Foods derived from biotechnology are more heavily regulated than any other new foods. They are considered safe to eat, pose little threat of producing allergic reactions, and have real and potential benefits (which are listed in Table 21.3).[31]

Illustration 21.10
Some organic products in England sport a GM-free label.

Important concerns about GM crops and foods exist, and they include the inability of small farmers to afford seed license fees and the fate of locally grown crops (these concerns are also listed in Table 21.3). Serious concerns about GM foods are deeply rooted in some areas of the world. In 1998 the European Union banned GM foods, closing the European market to farmers who produce foods from GM seeds. Reflecting the sentiment of most of the people in countries of the European Union, the ban will stay in effect until products with GM ingredients are labeled as containing genetically modified ingredients and the origins of GM ingredients are stated on packages.[33] Leaders in Zimbabwe refused GM corn offered as food aid by the UN during a severe famine in 2002. Officials were worried that farmers would plant

REALITY CHECK

Eating genes

Is it safe to eat genes?

Who gets thumbs up

?

Hyde:
I have never eaten a gene
in my life, and I
never will.

Ejay:
I had a huge salad for
lunch yesterday and I
haven't turned into a
head of lettuce yet.

Answers on next page

the corn kernels or feed it to cattle, which would mean neither the crops nor the meat could be sold in Europe. Ground corn was eventually accepted to help avert the famine. The nearby countries of Zambia and Mozambique faced similar famine and refused to accept GM foods.[34]

The Future of GM Foods: The Plot Thickens

Genetically engineered foods remain a controversial subject, and some experts believe that the future of GM crops is at a crossroads. To date, most genetic modifications of crops have benefited producers and not consumers.[35] Sentiment against GM foods is growing in Europe, and food producers will discontinue using GM

TABLE 21.3

BENEFITS AND CONCERNS RELATED TO GM CROPS AND FOODS.[32]

REAL AND POTENTIAL BENEFITS	REAL AND POTENTIAL CONCERNS
• Reduced herbicide and pesticide use • Increased availability of crops that can grow in dry or salty soils, and in hot or cold climates • Increased crop yield • Decreased waste due to spoilage • Improved nutritional content of foods • Improved food flavors • Source of antigens that could help people become immune to food allergens	• GM crop traits may cross-pollinate with local crops and reduce the market for locally grown foods. • GM pollen may harm local butterfly larvae and other insects near GM crop areas. • GM crop seeds are relatively expensive and perhaps too expensive for small farmers. • Europe does not allow the importation of GM foods. • Nutritionally superior GM foods have yet to reach consumers. • Insects may become resistant to genetic modifications of plants intended to keep them away. • Foods produced may be unacceptable to the consumers they are intended to benefit. • Long-term effects of GM organisms on local plants and crops, insects, and animals are unknown. • A belief by some that humans should not mess with mother nature by changing the genetic makeup of life-forms.

Photo Disc

ANSWERS TO REALITY CHECK

Eating genes

Genes in food are primarily protein and don't make it through the digestive tract. Characteristics transferred to foods by genes are not acquired by humans.

Hyde

Ejay

ingredients in foods if they are not acceptable to consumers.[36] It is not clear if consumers will accept foods genetically modified to deliver specific nutrients. Populations in need of additional vitamin A may reject yellow rice, preferring short-grained, long-grained, sticky, or dry white rice.[37] GM foods that offer nutritional advantages are not expected to reach the market for several years.[38] The future of GM foods appears to depend on resolution of issues related to cross-pollination and alteration of existing crops, the viability of small farms, consumer benefits, and other concerns. It will take years of experience with GM crops and food to adequately address these issues.[39]

Key Terms

age-related macular degeneration,
    page 21-4
biotechnology, page 21-8

cataracts, page 21-4
cruciferous vegetables, page 21-5

phytochemicals, page 21-2
zoochemicals, page 21-2

www links

www.ers.usda.gov
Receive updates on biotechnology and GM foods.

http://5aday.gov/
The National Cancer Institute's 5-a-day program site offers encouragement and tips on selecting colorful vegetables and fruit daily for better health.

www.citizens.org
Presents opponents' view of GM foods.

www.nas.edu
Select "biotechnology" to get reports on results from the National Academy of Science's subcommittee on genetically modified plants.

www.easynet.co.uk/ifst
Check for food biotechnology and related information from Britain's Institute for Food Science and Technology.

Nutrition UP CLOSE

Have You Had Your Phytochemicals Today?

FOCAL POINT: Consuming good sources of the "other" beneficial components of food.

Good food sources of beneficial phytochemicals are listed below. Indicate foods you consumed at least twice last week, foods you like, and those you have never tried.

	Foods Eaten at Least Twice Last Week	Foods You Like	Foods You've Never Tried
Broccoli			
Cabbage			
Carrots			
Cauliflower			
Celery			
Collard greens			
Garlic			
Onions			
Tomatoes			
Turnips			
Apple/juice			
Orange/juice			
Grapefruit/juice			
Grapes/juice			
Strawberries			
Brown rice			
Dried beans			
Tofu			
Whole grain bread, cereal			

FEEDBACK (answers to these questions) can be found at the end of Unit 21.

Notes

1. Gingsburg J, Prelevic GM, Lack of significant hormonal effects and controlled trials of phytoestrogens, N Engl J Med 2000;355:163–4; and Hollon T, Isoflavonoids: always healthy? The Scientist 2000;14(Sept. 4):21.

2. Bunk S. Researchers feel threatened by disease genes. The Scientist 1999;13 (Oct. 11):7–8.

3. Arab L, Steck S, Lycopene and cardiovascular disease, Am J Clin Nutr 2000; 71(suppl):1691S–5S; Michaud DS et al., Intake of specific carotenoids and risk of lung cancer in 2 prospective US cohorts, Am J Clin Nutr 2000;72:990–7; Kelloff GJ et al., Progress in cancer chemoprevention: development of diet-derived chemopreventive agents, J Nutr 2000;130:467S–71S; and Potter SM et al., Soy protein and isoflavones: their effect on blood lipids and bone density on postmenopausal women, Am J Clin Nutr 1998; 68(suppl):1375S–9S.

4. Gross M, Flavonoids, platelets, and the risk of cardiovascular disease. Epidemiology Seminar, Minneapolis, 1999 Nov. 10; and Forman A, As beta-carotene promise fades, focus turns to other carotenoids. Environmental Nutrition Newsletter 1997 (Aug.):1, 4.

5. Gross, Flavonoids, platelets.

6. Craig WT. Phytochemicals: guardians of our health. J Am Diet Assoc 1997;97 (suppl 2):S199–S204.

7. Bunk, Researchers feel threatened by disease genes; Arab and Steck, Lycopene and cardiovascular disease; Michaud et al., Intake of specific carotenoids; Kelloff et al., Progress in cancer chemoprevention; Potter et al., Soy protein and isoflavones; Gross, Flavonoids, platelets; Forman, As beta-carotene promise

fades; Craig, Phytochemicals; and Bloch A, Thomson CA. Position of the American Dietetic Association: phytochemicals and functional foods. J Am Diet Assoc 1995;95:493–6.

8. Bloch and Thomson, Position of the ADA.

9. Gingsburg and Prelevic, Lack of significant hormonal effects.

10. Hollon, Isoflavonoids.

11. Knekt P, Jarvinen R et al. Dietary flavonoids and the risk of lung cancer and other malignant neoplasms. Am J Epidemiol 1997;146:223–30.

12. Michaud et al., Intake of specific carotenoids.

13. Johnston CS, Corte C. People with marginal vitamin C status are at high risk of developing vitamin C deficiency. J Am Diet Assoc 1999;99:854–6.

14. Gingsburg and Prelevic, Lack of significant hormonal effects; Bunk, Researchers feel threatened by disease genes; Arab and Steck, Lycopene and cardiovascular disease; Michaud et al., Intake of specific carotenoids; Kelloff et al., Progress in cancer chemoprevention; Potter et al., Soy protein and isoflavones; Gross, Flavonoids, platelets; Forman, As beta-carotene promise fades; Craig, Phytochemicals; Bloch and Thoomson, Position of the ADA; and Knekt and Jarvinen et al., Dietary flavonoids.

15. Gingsburg and Prelevic, Lack of significant hormonal effects

16. Johnston and Corte, People with marginal vitamin C status.

17. Tylleskar T, Banea M et al. Dietary determinants of a non-progressive spastic paparesis (konzo). Int'l J Epidemiol 1995;24:949–56.

18. Holson D. Toxicity, plants: ackee fruit. www.emedicine.com, accessed 9/02.

19. Shirlow MJ, Mathers CD. A study of caffeine consumption and symptoms: indigestion, palpitations, tremor, headache, and insomnia. Int'l J Epidemiol 1985;14:239–48.

20. Wei C, Shu X et al. Caffeine consumption and reproductive outcomes. (in press) 1998.

21. Tol AV, Urgert R et al., The cholesterol-raising diterpenes from coffee beans, Atherosclerosis 1997;132: 251–4; and Grubben MJ et al., Unfiltered coffee increases plasma homocysteine concentrations in healthy volunteers: a randomized trial, Am J Clin Nutr 2000;71:480–4.

22. Lloyd T, Rollings N et al., Dietary caffeine intake and bone status of postmenopausal women, Am J Clin Nutr 1997;65:1826–30; and Lloyd T et al., Caffeine not related to bone loss in postmenopausal women, Am Coll Nutr 2000;19:256–61.

23. Silverman K, Evans SM et al. Withdrawal syndrome after the double-blind cessation of caffeine consumption. N Engl J Med 1992;327:1109–14.

24. Herber D. The stinking rose: organosulfur compounds and cancer. M J Clin Nutr 1997;66:425–6.

25. Babcock BC, Francis CA. Solving global nutrition challenges requires more than new biotechnologies. J Am Diet Assoc 2000;100:1308–11.

26. Tylleskar and Banea, Dietary determinants.

27. Vanderbergh JG et al., Potential environmental problems with animal biotech raises concerns, National Research Council, National Academies of Science. www.nas.edu, accessed 8/02; and Lewis C, A new kind of fish story: the coming of biotech animals. FDA Consumer, Jan.–Feb. 2001.

28. Falk MC et al. Food biotechnology: benefits and concerns. J Nutr 2002; 132:1384–90.

29. Holson, Toxicity, plants.

30. Falk et al., Food biotechnology.

31. Falk et al., Food biotechnology; and Merrigan KA et al. Executive summary. Nutr Rev 2003;61:S95–S100.

32. Babcock and Francis, Solving global nutrition challenges; Falk et al., Food biotechnology; Genetic food fight, Wall Street Journal, 5/15/03, p. A16; and Merrigan et al., Executive summary.

33. Genetic food fight, Wall Street Journal.

34. Genetic food fight, Wall Street Journal; and Chege W. Africa mulls gene modified food as hunger claws. World Food Program press release, www.medscape.com, accessed 7/02.

35. Falk et al., Food biotechnology.

36. Merrigan et al., Executive summary.

37. Merrigan et al., Executive summary.

38. Falk et al., Food biotechnology.

39. Babcock and Francis, Solving global nutrition challenges; and Falk et al., Food biotechnology.

Nutrition UP CLOSE

Have You Had Your Phytochemicals Today?

Feedback for Unit 21

Did you eat four or more of the foods listed at least twice last week? If yes, listen carefully and you'll hear your cells say "thank you."

If you didn't eat four or more twice last week, go for the foods you didn't eat but like. Be adventurous! Try some of the phytochemical-rich foods you have never eaten before.

Diet and Cancer

Nutrition Scoreboard

		TRUE	FALSE
1	Some types of cancer are contagious.		
2	Cancer is primarily an inherited disease.		
3	People who eat plenty of fruits and vegetables are less likely to get cancer than people who don't.		
4	High levels of body fat contribute to the development of many types of cancer.		
5	Two out of three Americans never get cancer.		

Answers on next page

- Cancer has many different causes. Diet is a major factor that influences the development of most types of cancer.

- Diets primarily based on plant foods that include lean meats, fish, and low-fat dairy products; regular physical activity; and normal levels of body fat reduce cancer risk.

- Cancer is largely preventable, but there are no absolute guarantees that an individual will not develop cancer.

Answers to *Nutrition Scoreboard*

		TRUE	FALSE
1	Cancer doesn't spread from person to person.		✔
2	Cancer is primarily related to environmental factors such as diet, smoking, and overexposure to the sun. Genetic background plays a role by placing some individuals at increased risk for developing cancer.		✔
3	Among the arsenal of cancer-protective measures, the consumption of 5 or more servings of fruits and vegetables a day stands out.	✔	
4	High levels of body fat *are* related to the development of many types of cancer.	✔	
5	Statistics from the National Cancer Institute demonstrate that most people don't get cancer.	✔	

What Is Cancer?

cancer
A group of diseases in which abnormal cells grow out of control and can spread throughout the body. Cancer is not contagious and has many causes.

Cancer, the second leading cause of death in the United States, is a group of conditions that result from the uncontrolled growth of abnormal cells. Although these cells can begin to grow in any tissue in the body, the lungs, colon, *prostate,* and breasts are the most common sites for cancer development (Illustration 22.1).[1] Some forms of cancer are highly curable.

How Does Cancer Develop?

prostate
A gland located above the testicles in males. The prostate secretes a fluid that surrounds sperm.

Cancer develops by complex processes that are not yet fully understood. Adding to the complexity is the fact that cancer development often does not proceed in a straight line—cancer can progress two steps forward and then take a step or two back.

initiation
The start of the cancer process; it begins with the alteration of DNA within cells.

Illustration 22.2 summarizes the processes involved in the development of cancer, as we currently understand them. Cancer begins when something goes wrong within cells that modifies cell division. Every minute, 10 million cells in the body divide. Usually, the cells divide the right way and on schedule due to a set of regulatory mechanisms that control the replication (duplication) of DNA, the genetic material that becomes part of new cells.

promotion
The period in cancer development when the number of cells with altered DNA increases.

During the *initiation* phase of cancer development, something alters the DNA in certain cells. During the *promotion* phase of cancer, the cells with altered DNA divide, eventually producing large numbers of abnormal cells. This phase of cancer development commonly takes place over a span of 10 to 30 years. Unless they are hindered by the body or corrected by some other means, the abnormal cells continue to divide, leading to the *progression* phase of cancer development.

progression
The uncontrolled growth of abnormal cells.

During the progression phase, the body loses control over the abnormal cells, and their numbers increase rapidly. Eventually, the cells become so numerous that they erode the normal functions of the tissue where they are growing. During this phase, the abnormal cells may migrate to other tissues and cause DNA damage and abnormal cell development in these tissues, too.[4]

What Causes Cancer?

Explaining what causes cancer turns out to be very difficult. It appears, however, that approximately 80–90% of all cancers are related to environmental factors such as smoking, exposure to asbestos, chemical pollutants, and radiation. Diet is one of the major environmental factors, and it accounts for 40% of cancer risk.[5] Some of the most convincing evidence of the robust relationship between environmental factors and cancer come from studies of cancer rates in people migrating to another country. Rates of breast cancer, for example, are low in rural Asia. When individuals from rural parts of Asia immigrate to the United States, rates of breast cancer become the same or higher than the U.S. rate by the third generation. Rates of prostate cancer similarly increase as people move from countries with low rates to countries with high rates.[6]

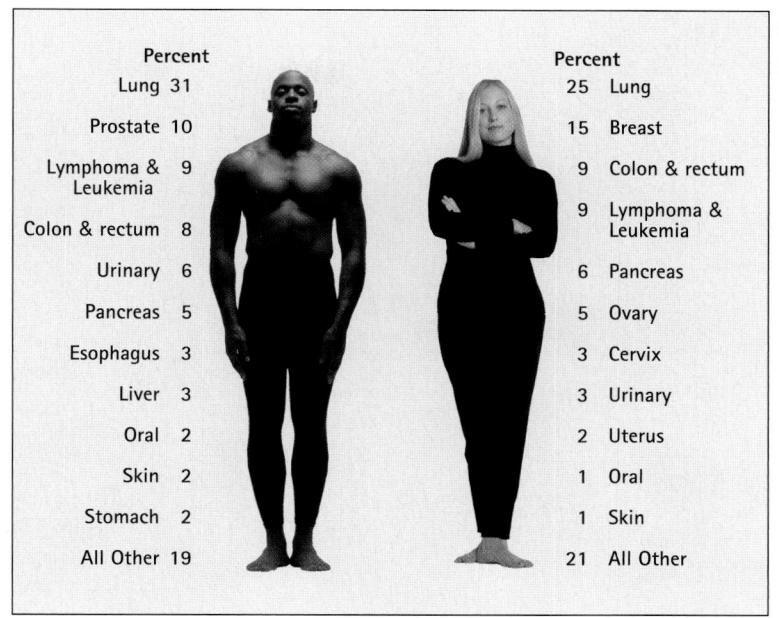

Illustration 22.1
Percentage of cancer deaths by selected sites and sex, 2002. *In 2002, 23% of deaths in the United States were due to cancer.[2]*

Westernization of dietary intake and lifestyles increases the risk of many types of cancer. Rates of breast cancer in Japanese and Alaskan Eskimo women have increased substantially as they have adopted Westernized diets and lifestyles.[7]

Some people have a genetic tendency toward cancer, which means they have a tendency to develop cancer if regularly exposed to certain substances in the diet or environment. Genetic factors appear to account for 42% of the risk for prostate cancer, 5 to 27% of the risk for breast cancer, and 36% of the risk for pancreas cancer, for example. Endometrial cancer (cancer of the lining of the uterus), oral, thyroid, and bone cancer do not appear to be related to inherited factors.[8]

Given the high percentage of cancers related to diet and other environmental factors, cancer is considered a largely preventable disease. Increasing rates of new

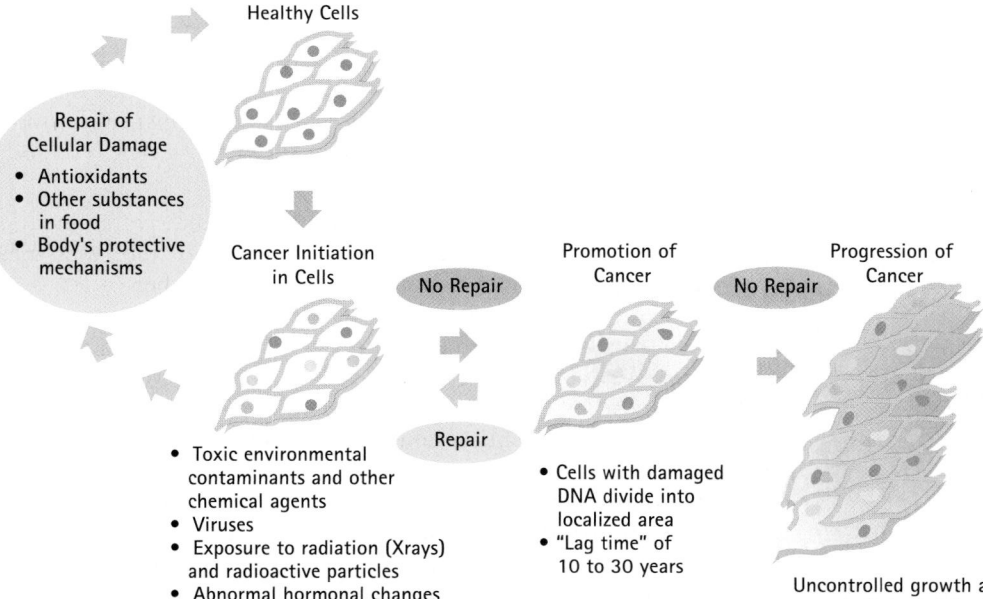

Illustration 22.2
Steps in the development of cancer.[3]
Antioxidants, other substances in food, and the body's protective mechanisms may repair damage to cells and halt the progression of cancer.

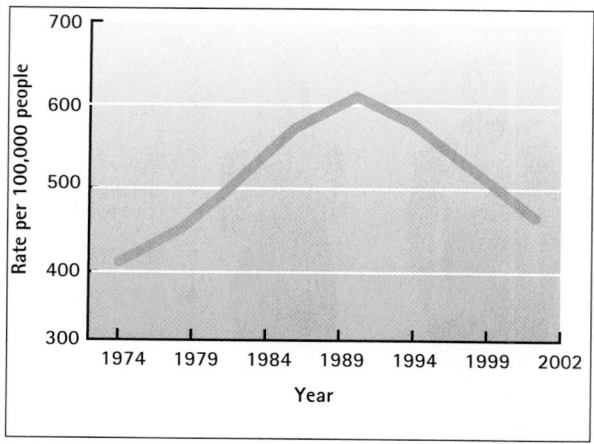

Illustration 22.3

Rates of new cases of cancer in the United States have declined since 1992.[9]

cases of cancer took a turn for the better after 1992 and correspond to declines in rates of tobacco use (Illustration 22.3)[10] It is anticipated that other improvements in lifestyles and diets will lead to further declines in cancer rates.[11]

Fighting Cancer with a Fork

Leading the list of cancer-promoting diets are those low in vegetables and fruits.[12] High intakes of saturated fat from meats and dairy products, regular intake of charred and nitrate-cured meats, and excessive alcohol consumption are associated with the development of cancer. Low intakes of plant foods such as whole grains, dried beans, nuts, and seeds have been found to increase cancer risk. Other, major risk factors for many types of cancer include smoking, physical inactivity, and excess body fat.[13] Table 22.1 summarizes characteristics of diets and lifestyle that are related to a reduced risk of cancer.

Frequent consumption of certain types of food is sometimes more strongly related to particular cancers than to other types. For example, regular consumption of tomato products is related in particular to decreased risk of prostate cancer, and regular intake of green tea appears to contribute to breast cancer reduction.[15] Diets and lifestyles that best prevent cancer are represented by a set of characteristics and not by hard rules about specific foods, dietary restrictions, or types of physical activities.[16]

Dietary Risk Factors for Cancer: A Closer Look

Foods contain a variety of vitamins and minerals, as well as fiber and phytochemicals that participate in protecting the body against cancer. These substances in food, particularly plant foods, appear to work together in ways that confer the protection. Attempts to prevent cancer by giving large groups of people individual components of plants that may biologically explain the plant's beneficial effects have not been successful.[17] Particular types of food clearly provide greater levels of protection against cancer than extracts of specific substances in food or other types of food.

Fruits and Vegetables and Cancer] People who regularly consume plenty of vegetables and fruits (5 or more servings daily) have a lower risk of developing a number of types of cancer than people who eat fewer servings.[19] Because cancer incidence continues to decline as intake of vegetables and fruits increases, some experts are advising people to consume 5 to 9 servings daily.[20] It is not yet clear why fruits and vegetables exert this effect, but a number of possibilities exist.

One possibility is the contribution of the antioxidants vitamin C, beta-carotene, vitamin E, and selenium in vegetables and fruits. Much of the damage to DNA related to cancer initiation is thought to be caused by exposure to oxygen and other oxidizing substances that disrupt molecules within DNA; antioxidants are able to repair damaged DNA.[21] Phy-

TABLE 22.1

DIETARY PATTERNS AND LIFESTYLES RELATED TO REDUCED RISK OF CANCER AND HEART DISEASE.[14]

1. Consume a plant-based diet.
   - 5+ servings of vegetables and fruits daily
   - 3+ whole grains/products daily
   - regular consumption of dried beans, nuts, and seeds
2. Eat foods that are low in saturated fat.
   - focus on lean meats, low-fat dairy products, fish
3. Exclude charred meats, nitrate-preserved meats.
4. Exclude smoking.
5. Exclude excess alcohol consumption.
6. Include physical activity.
   - 30 minutes 5+ days a week
7. Lose excess weight.

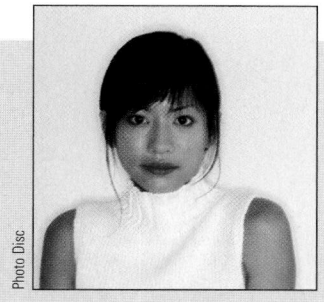
tochemicals in vegetables and fruits appear to participate in cancer prevention by protecting cells from damage due to oxidation and by inhibiting the multiplication of abnormal cells.

Consuming 3 servings per week of vegetables from the cruciferous family (broccoli, cabbage, cauliflower, and brussels sprouts) substantially reduces the risk of lung, bladder, and prostate cancer (Illustration 22.4).[22] If you enjoy eating these vegetables, consider yourself lucky.

Color-Coding Vegetable and Fruit Choices] Many of the phytochemicals participating in mechanisms that likely help to prevent cancer can be identified by their color. This feature of some phytochemicals, and the advantages of eating a variety of vegetables and fruits, has led to the advice to select and consume colorful vegetables and fruits daily.[23] Table 22.2 lists phytochemicals, the color they impart to mature vegetables and fruits, and their sources. The phytochemicals listed all act as antioxidants.

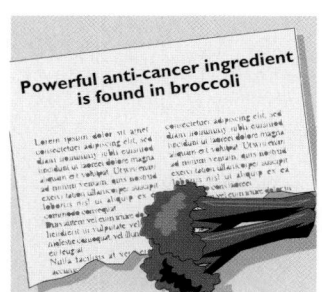

Illustration 22.4
An example of headline coverage of the cancer-protective effects of vegetables.

TABLE 22.2

COLOR-CODING YOUR VEGETABLES AND FRUITS.[24]

COLOR	PHYTOCHEMICAL ANTIOXIDANT	VEGETABLE AND FRUIT SOURCES
Red	lycopene (lie-co-peen)	tomatoes, red raspberries, watermelon, strawberries, red peppers
Yellow-green	lutein zeaxanthin (lou-te-in, ze-ah-zan-thun)	leafy greens, avocado, honeydew melon, kiwi fruit
Red-purple	anthocyanins (antho-sigh-ah-nin)	grapes, berries, wine, red apples, plums, prunes
Orange	beta-carotene	carrots, mangos, papayas, apricots, pumpkins, yams
Orange-yellow	flavonoids (fla-von-oids)	oranges, tangerines, lemons, plums, peaches, cantaloupe
Green	glucosinolates (glu-co-sin-oh-lates)	broccoli, brussels sprouts, kale, cabbage
White-green	allyl sulfides (al-lill sulf-ides)	onions, leeks, garlic

Sebastian

Jasmine

Whole Grains and Cancer]   Whole grains contain vitamins, minerals, unsaturated fatty acids, fiber, and phytochemicals that play roles in cancer prevention.[25] Effects of whole grains and whole grain products on cancer risk are related to the combined action of these substances. If you isolate a single substance, you in effect are destroying its ability to function in cancer prevention. Americans are advised to include three or more whole grain products in their daily diet.[26]

Saturated Fat and Cancer Promotion]   High intake of saturated fats from meats and dairy products increases the risk of cancer.[27] Consequently, recommendations for diet and cancer prevention include the use of plant sources of protein because they provide unsaturated fats. There are additional advantages to increased reliance on plant foods in the diet. Plant foods such as dried beans, soy products, nuts, and seeds also provide vitamins, minerals, fiber, and phytochemicals that help ward off cancer progression. Fish are a recommended source of protein, too. Regular consumption of fish is related to lower rates of cancer.[28]

Nitrate-Preserved and Grilled Meats and Cancer]   Cancer of the stomach and liver appears to be related to the regular consumption of foods such as hot dogs, luncheon meats, bacon, and pickled eggs and vegetables preserved with nitrates.[29] (You can identify foods that contain nitrates by examining food ingredient labels.) Most cases of cancer associated with nitrate use in smoked, salted, and pickled foods now occur in China, parts of the former Soviet Union, and Central and South America, where such foods are commonly consumed.[30]

Certain nitrogen-containing and other substances in beef, chicken, fish, and other meats may become cancer promoting if heated to a high temperature.[31] Such high temperatures can be reached by broiling and grilling food. The potentially cancer-promoting substances formed by meats are found in the charred portions of the meat and the fatty coating that accumulates on meat when fat drips into the heat source and smokes (Illustration 22.5).[32]

The Department of Agriculture concludes that eating broiled or grilled foods several times a week poses no threat to health. People should be careful not to eat the charred portions of meat and to wipe or rinse off any coating on meat that results from fatty flare-ups on the grill.

Illustration 22.5
The charred and black, oily coating on grilled or broiled meats is the part you shouldn't eat.

22-6

Diet and Cancer Guidelines

Dietary patterns and lifestyle changes that reduce the risk of cancer (Table 22.1) are similar to, and compatible with, dietary recommendations to reduce the risk of heart disease.[33] Considered together, dietary recommendations for cancer prevention can be transferred to dietary intake by selection of the types of foods shown in the example menu in Table 22.3.

Alcohol Intake and Cancer]   Consumption of excessive amounts of alcoholic beverages has been linked to cancers of the breast, mouth, throat, and liver.[34] The risk of developing cancers of the mouth and throat increases for people who smoke cigarettes or chew tobacco as well as drink.[35]

Excess Body Fat and Cancer]   Obesity, and in particular obesity characterized by excess stores of central body fat, increases the risk of cancer at several sites. High levels of central fat appear to alter metabolism of hormones such as estrogen, testosterone, and insulin in ways that promote the growth of abnormal cells. Calorie-reduced diets combined with at least 30 minutes of moderate or vigorous activity five or more days a week are recommended for cancer prevention.[36]

Bogus Cancer Treatments

Unorthodox, purported cancer cures such as macrobiotic diets; hydrogen peroxide ingestion; laetrile tablets; vitamin, mineral, herbal, and phytochemical supplements; and animal gland therapy have *not* been shown to be effective treatments for cancer. Such remedies have been promoted since the early 1900s. They still exist because, although not proven to work, they offer some cancer patients their last ray of hope. They should *not* be used as a substitute for conventional cancer treatments.[37]

Eating to Beat the Odds

Two out of every three people in the United States do *not* develop cancer. You can likely improve your odds of being among the two by not smoking or drinking excessively, by regularly consuming 5 or more servings of fruits and vegetables each day, by sticking to a low-saturated-fat diet, and by being physically active and achieving or maintaining a normal level of body fat. Although there are no guarantees, people can help themselves prevent cancer by adhering to a good diet and a healthy lifestyle.

TABLE 22.3

TYPES OF FOODS INCLUDED IN DIETARY PATTERNS THAT LOWER THE RISK OF CANCER AND HEART DISEASE.

BREAKFAST

Oat flakes with banana and skim/low-fat milk
Cracked wheat toast with soft margarine and grape jam
Orange juice
Tea/coffee

LUNCH

Skinless chicken sandwich on whole grain bread with mozzarella cheese
Broccoli, cauliflower, raisin salad
Carrot sticks
Skim/low-fat milk

DINNER

Tomato soup
Broiled tuna with olive oil-lemon sauce
Baked beans
Coleslaw
Ice milk with strawberries
Skim/low-fat milk, coffee, tea, 1 glass wine

SNACKS

Nuts, seeds
Yogurt
Raw vegetables
Fruits
Skim/low-fat milk

Nutrition UP CLOSE

A Cancer Risk Checkup

FOCAL POINT: Reducing cancer risk.

A number of behaviors that help protect people from developing cancer are listed below. Check those that apply to you.

	Yes	No	Don't Know
1. I eat 5 or more servings of vegetables and fruits daily.	_____	_____	_____
2. I consume whole grain products.	_____	_____	_____
3. I eat a high-fat diet.	_____	_____	_____
4. I smoke or chew tobacco.	_____	_____	_____
5. I exercise regularly.	_____	_____	_____
6. I have too much body fat.	_____	_____	_____

FEEDBACK (including scoring) can be found at the end of Unit 22.

Key Terms

cancer, page 22-2

initiation, page 22-2

progression, page 22-2

promotion, page 22-2

prostate, page 22-2

www links

www.nlm.nih.gov/medlineplus/cancergeneral.html
This address sends you to summarized, scientific information about cancer. Changing the term "cancergeneral" to "breastcancer," "coloncancer," "alternativemedicine" and so on takes you to related topics.

www.cancer.gov
The U.S. government's gateway to information about cancer.

www.cdc.gov/nchs/hus.htm
Contains the report "Health, United States, 2003"—or 2004, 2005. Updated yearly, this report provides statistics on cancer and other disease rates.

www.nci.nih.gov
Web site for the National Cancer Institute; provides evidence-based information about cancer prevention and treatment, and research updates.

www.cancer.org
The American Cancer Society's site provides a wealth of information on cancer clinics, treatments, prevention, and statistics.

www.healthfinder.gov
Search engine for reliable sources of information on cancer and other health concerns.

www.nal.usda.gov/fnic/
Click on "cancer" and find recent results related to diet, nutrition, and cancer development.

Notes

1. Cancer facts and figures, 2002. American Cancer Society, Inc. www.cancer.org/downloads/STT/CancerFacts&Figures2002TM.pdf, accessed 10/03;

2. Cancer statistics, http://progressreport.cancer.gov, accessed 10/03; and Weir HK et al., Annual report to the nation on the status of cancer, J Natl Cancer Inst 2003;95:1258–61.

3. Lichtenstein P et al. Environmental and heritable factors in the causation of cancer. N Engl J Med 2000;343:78–85.

4. Lichtenstein et al., Environmental and heritable factors.

5. Hoover RN, Cancer—nature, nurture, or both, N Eng J Med 2000;343:135–6; and Finley JW, The antioxidant responsive element (ARE) may explain the protective effects of cruciferous vegetables on cancer, Nutr Rev 2003;61:250–54.

6. Hoover, Cancer—nature, nurture, or both.

7. Cowing BC, Saker KE. Polyunsaturated fatty acids and epidermal growth factor receptor. J Nutr 2001;131:1125–8.

8. Lichtenstein et al., Environmental and heritable factors; and Hoover, Cancer—nature, nurture, or both.

9. Cancer facts (www.cancer.org/downloads/STT/CancerFacts&Figures2002TM.pdf); and Cancer statistics (http://progressreport.cancer.gov).

10. Cancer facts (www.cancer.org/downloads/STT/CancerFacts&Figures2002TM.pdf); and Cancer statistics (http://progressreport.cancer.gov).

11. Byers T et al. Behavior change could prevent 60,000 cancer deaths a year: IOM report. www.medscape.com, accessed 7/03.

12. Byers T et al. Guidelines on nutrition and physical activity for cancer prevention: reducing the risk of cancer with healthy food choices and physical activity. CA Cancer J 2002;52:92–119.

13. Byers et al., Guidelines on nutrition; Go VLW et al., Diet, nutrition and cancer prevention: where do we go from here? J Nutr 2001;131:3121S–26S; and NCI fact sheet. www.cancer.gov, accessed 10/03.

14. Byers et al., Guidelines on nutrition; Go et al., Diet, nutrition and cancer prevention; and NCI fact sheet (www.cancer.gov).

15. NCI fact sheet (www.cancer.gov); and Wu AH, Green tea intake and breast cancer, Int J Cancer 2003;106:574–9.

16. Byers et al., Guidelines on nutrition; Rock C, Cancer symposium: observational vs. intervention studies, Exp Biol, San Diego, 4/11/03; and Campbell TC, Critique of Report on "Food, nutrition and the prevention of cancer: a global perspective," Nutr Today 2001;36:80–4.

17. Gerber M. The comprehensive approach to diet: a critical review. J Nutr 2001;131:3051S–5S.

18. Rock, Cancer symposium.

19. Byers et al., Guidelines on nutrition.

20. Herber D, Bowerman S. Applying science to changing dietary patterns. J Nutr 2001;131:3078S–81S.

21. Rock, Cancer symposium; and Seifried HE et al., Antioxidants and cancer, Cancer Res 2003;63:4295–8.

22. Kelloff GJ et al. Progress in cancer chemoprevention: development of diet-derived chemopreventive agents. J Nutr 2000;130:467S–71S.

23. Gerber, The comprehensive approach to diet; and Herber and Bowerman, Applying Science.

24. Gerber, The comprehensive approach to diet; and Herber and Bowerman, Applying science.

25. Slavin JL. Epidemiological and clinical studies on whole grains. Nutr Today 2001;36:61–6.

26. Slavin, Epidemiological and clinical studies.

27. Cho E et al., Relationship of fat intake to breast cancer in premenopausal women, J Natl Cancer Inst 2003;95:1079–85; and Bingham S et al., Dietary fat intake and the risk of breast cancer, Lancet 2003;362:212–14.

28. Byers et al., Guidelines on nutrition.

29. Byers et al., Guidelines on nutrition.

30. Food, nutrition and the prevention of cancer: a global perspective. Washington, DC: World Cancer Research Fund, American Institute for Cancer Research; 1997.

31. Engel LS et al. Risk factors for esophageal and gastric cancer. J Natl Cancer Inst 2003;95:1404–13.

32. de Verdier MG et al. Meat, cooking methods and colorectal cancer. Int'l J Cancer 1991;49:520–5.

33. Byers et al., Guidelines on nutrition.

34. Cancer statistics (http://progressreport.cancer.gov); and Steinmetz KA, Potter JD, Vegetables, fruit, and cancer prevention: a review, J Am Diet Assoc 1996;96:1027–39.

35. Lichtenstein et al., Environmental and heritable factors.

36. Bianchini F et al., Overweight, obesity, and cancer risk, Lancet, Oncology 2002;3:565–74; and Friedenreich CM et al., Physical activity and cancer prevention: etiologic evidence and biological mechanisms, J Nutr 2002;132:3456S–64S.

37. Straus SE. Complementary and alternative medicine and cancer: what you should know. www.peoplelivingwithcancer.org, accessed 10/03.

Nutrition UP CLOSE

A Cancer Risk Checkup

Feedback for Unit 22

Best answers:

1. Yes   4. No
2. Yes   5. Yes
3. No   6. No

If you responded "Don't Know" to statements 1, 2, or 3, analyze your diet using the Diet Analysis Plus program. "Regular exercise" in statement 5 means exercising 20 to 30 minutes three times per week. If you didn't know the answer to statement 6, can you pinch an inch of fat above your ribs while you are standing? If yes, you probably have too much body fat.

Good Things to Know about Minerals

Nutrition Scoreboard

	TRUE	FALSE
1 The sole function of minerals is to serve as a component of body structures such as bone, teeth, and hair.		
2 Bones continue to grow and mineralize through the first 30 years of life.		
3 Ounce for ounce, spinach provides more iron than beef does.		
4 Worldwide, the most common nutritional deficiency is iron deficiency.		
5 More than 1 in 4 Americans have hypertension.		

Answers on next page

[KEY CONCEPTS AND FACTS]

- Minerals are single atoms that cannot be created or destroyed by the human body or by any other ordinary means.

- Minerals serve as components of body structures and play key roles in the regulation of body processes.

- Deficiency diseases occur when too little of any of the 15 essential minerals are provided to the body, and overdose reactions occur when too much is provided.

- Inadequate intakes of certain minerals are associated with the development of chronic disorders, including osteoporosis, iron deficiency, and hypertension.

Answers to *Nutrition Scoreboard*

		TRUE	FALSE
1	Minerals serve as structural components of the body, but they also play important roles in stimulating muscle and nerve activity and in other functions.		✔
2	Bones continue to grow and mineralize well after we reach adult height. Bone growth and development can continue to about age 30.	✔	
3	Spinach is a nutritious food, providing 0.6 milligram iron per ounce. It contains less iron than beef (1 milligram iron per ounce). The iron in spinach is poorly absorbed, but the iron in meat is well absorbed by the body. (When several hundred college students were asked this question, 82% voted for spinach.)[1]		✔
4	Iron deficiency is the most common nutritional deficiency in both developed and developing countries. Approximately one-third of the world's population is iron deficient.[2]	✔	
5	An estimated 9% of Americans have hypertension (high blood pressure).[3]	✔	

Mineral Facts

What substances are neither animal nor vegetable in origin, cannot be created or destroyed by living organisms (or by any other ordinary means), and provide the raw materials from which all things on Earth are made? The answer is the **mineral** elements, and they are displayed in full in the periodic table presented in Illustration 23.1. Minerals considered "essential," or required in the diet, are highlighted.

The body contains 40 or more minerals. Only 15 are an essential part of our diets; we obtain the others through the air we breathe or from other essential nutrients in the diet such as protein and vitamins.

Minerals are unlike the other essential nutrients in that they consist of single atoms. A single atom of a mineral typically does not have an equal number of protons (particles that carry a positive charge) and electrons (particles that carry a negative charge), and it therefore carries a charge. The charge makes minerals reactive. Many of the functions of minerals in the body are related to this property.

minerals

In the context of nutrition, minerals are specific, single atoms that perform particular functions in the body. There are 15 essential minerals—or minerals required in the diet.

Getting a Charge out of Minerals

The charge carried by minerals allows them to combine with other minerals of the opposite charge and form fairly stable compounds that become part of bones, teeth, cartilage, and other tissues. In body fluids, charged minerals serve as a source of electrical power that stimulates muscles to contract and nerves to react. The electrical current generated by charged minerals when performing these functions can be

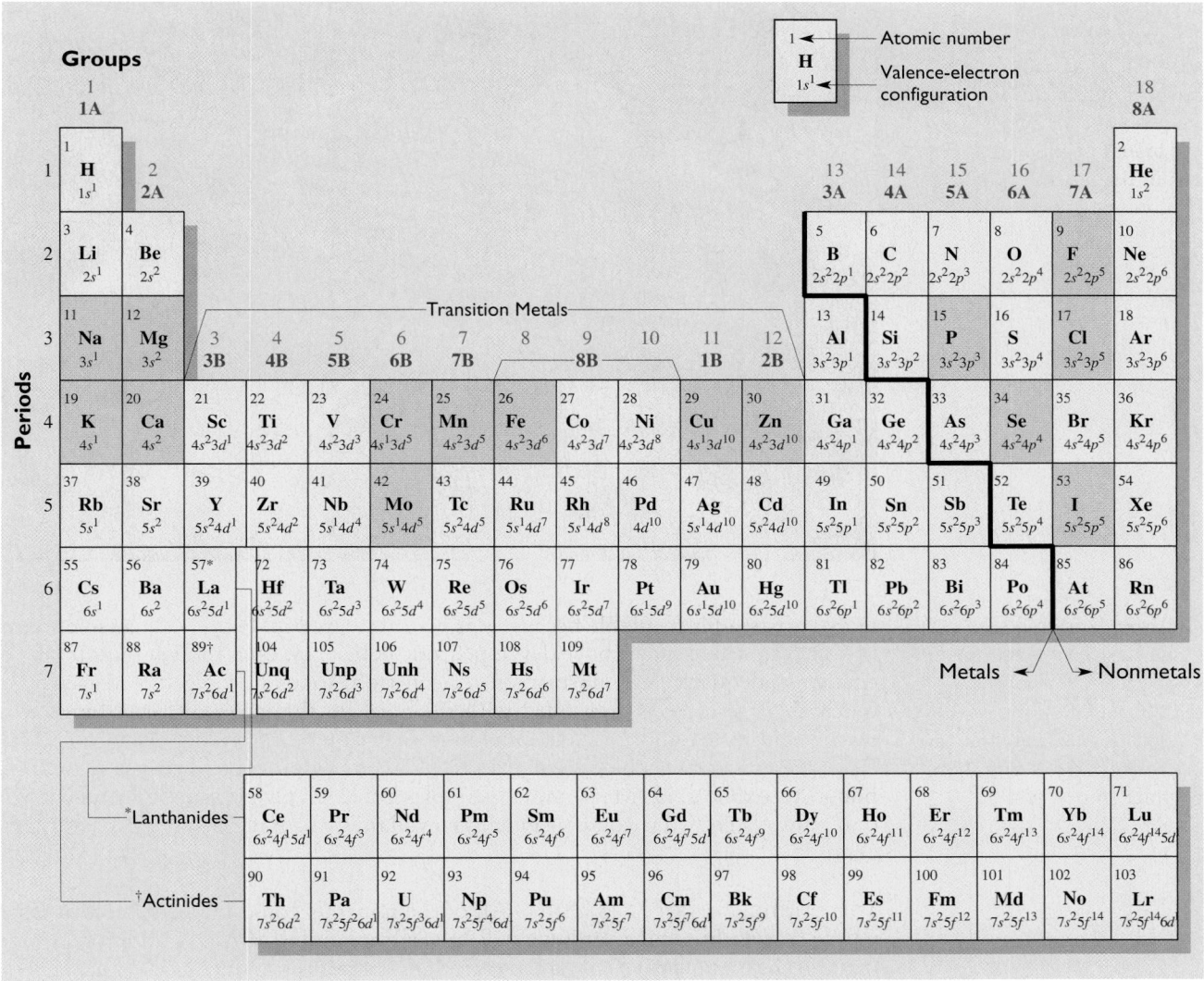

Illustration 23.1
The periodic table lists all known minerals.
The highlighted minerals are required in the human diet.

recorded by an electrocardiogram (abbreviated EKG or ECG) and an electroencephalogram (EEG). Abnormalities in the pattern of electrical activity in EKGs signal pending or past problems in the heart muscle. An EKG recording is shown in Illustration 23.2. Electroencephalograms similarly record electrical activity in the brain.

The charge minerals carry is related to many other functions. It helps maintain an adequate amount of water in the body and assists in neutralizing body fluids when they become too acidic or basic. Minerals that perform the roles of *cofactors* are components of proteins and enzymes, and they provide the "spark" that initiates enzyme activity.

Charge Problems] Because minerals tend to be reactive, they may combine with other substances in food and form highly stable compounds that are not easily absorbed. Absorption of zinc from foods, for example, can vary from zero to 100%, depending on what is attached to it. Zinc in whole grain products is very poorly absorbed because it is bound tightly to a substance called phytate. In contrast, zinc

cofactors
Individual minerals required for the activity of certain proteins. For example:

• iron is needed for hemoglobin's function in oxygen and carbon dioxide transport,

(continued next page)

Illustration 23.2
The electrical current measured by an EKG results from the movement of charged minerals across membranes of the muscle cells in the heart.

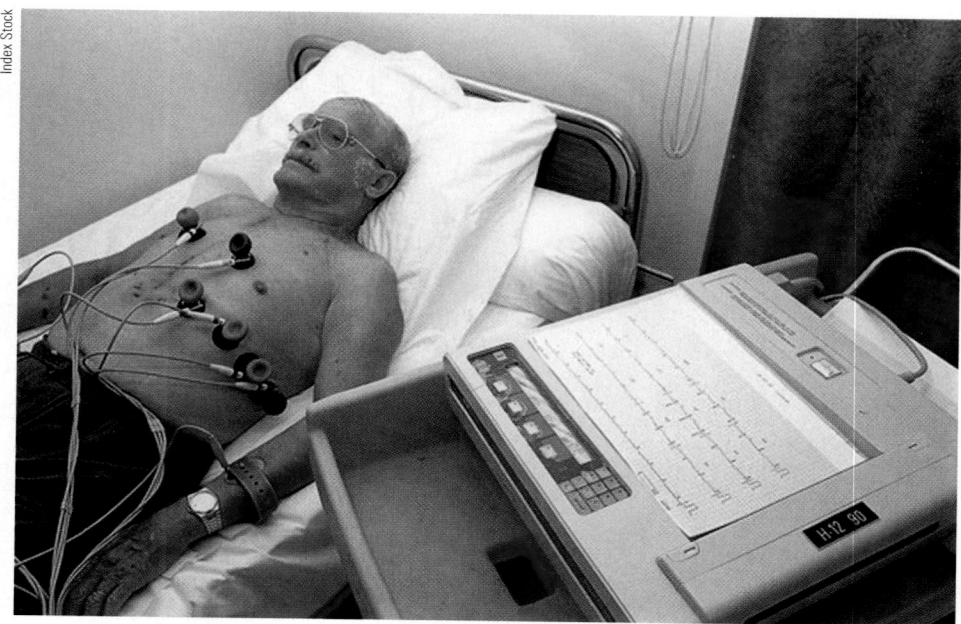

Index Stock

cofactors, continued.
• zinc is needed to activate or is a structural component of over 200 enzymes, and
• magnesium activates over 300 enzymes involved in the formation of energy and proteins.

in meats is readily available because it is bound to protein. People whose sole source of zinc was whole grains have developed zinc deficiency, even though their intake of zinc was adequate.[4] The absorption of iron from foods in a meal decreases by as much as 50% if tea is consumed with the meal. In the intestines, iron binds with tannic acid in tea and forms a compound that cannot be broken down.[5] The calcium present in spinach and collard greens is poorly absorbed because it is firmly bound to oxalic acid. Many more examples could be given. The point is that you don't always get what you consume; the availability of minerals in food can vary a great deal.

The Boundaries of This Unit] All of the minerals could be the subject of fascinating stories, but in this unit we will concentrate on just three. In addition, a summary of the main features of all the essential minerals is provided in Table 23.1. This table lists the recommended intake level, functions, consequences of deficiency and overdose, and food sources of each mineral. Table 23.12 at the end of the unit lists food sources for many of the essential minerals.

Minerals highlighted are calcium, iron, and sodium. They have been selected primarily because they play important roles in the development of osteoporosis, iron-deficiency anemia, and hypertension, respectively. These disorders are widespread in the United States and many other countries, and improved diets offer a key to their prevention and treatment.

Selected Minerals: Calcium

What you've heard about calcium is true: it's good for bones and teeth. About 99% of the 3 pounds of calcium in the body is located in bones and teeth. The remaining 1% is found in blood and other body fluids. We don't hear so much about this 1%, but it's very active. Every time a muscle contracts, a nerve sends out a signal, or blood clots to stop a bleeding wound, calcium in body fluids is involved. Calcium's most publicized function, however, is its role in bone formation and the prevention of osteoporosis.

A Short Primer on Bones

Most of the bones we see or study are hard and dead. As a result, people often have the impression that bones in living bodies are that way. Nothing could be further from the truth. The 206 bones in our bodies are slightly flexible living tissues infiltrated by blood vessels, nerves, and cells.

The solid parts of bones consist of networks of strong protein fibers (called the "protein matrix") embedded with mineral crystals (Illustration 23.3). Calcium is the most abundant mineral found in bone, but many other minerals such as phosphorus, magnesium, and carbon are also incorporated into the protein matrix. The combination of water, the tough protein matrix, and the mineral crystals makes bone very strong yet slightly flexible and capable of absorbing shocks.

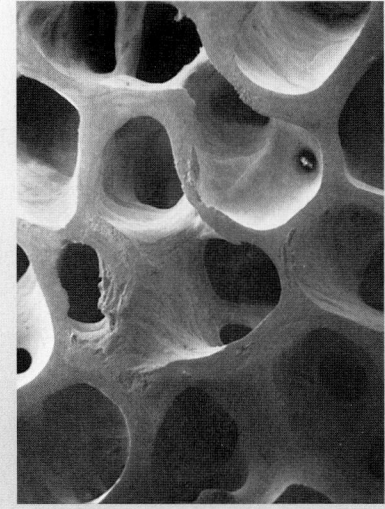 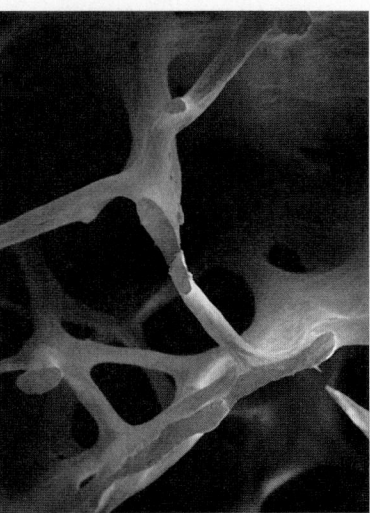

Illustration 23.3
(left) Electron micrograph of healthy bone. (right) Electron micrograph of bone affected by osteoporosis.
Reproduced with permission from Dempster et al, J Bone Min. Res 1, 15–21, 1986.

Teeth Are a Type of Bone] Teeth have the same properties as other bone plus a hard outer covering called enamel, which is not infiltrated by blood vessels or nerves. Enamel serves to protect the teeth from destruction by bacteria and mechanical wear and tear.

Remodeling Your Bones] Bones slowly and continually go through a repair and replacement process known, appropriately enough, as ***remodeling***. During remodeling, the old protein matrix is replaced and remineralized. If insufficient calcium is available to complete the remineralization, or if other conditions prevent calcium from being incorporated into the protein matrix, ***osteoporosis*** results.

Osteoporosis

If you are female, you have a one in four chance of developing osteoporosis in your lifetime. If you are a Caucasian, Hispanic, or Asian female, you have a higher risk of developing osteoporosis than if you are an African American woman or a male. If you are male, your chance of developing osteoporosis is one in eight. Approximately 44 million adults in the United States have osteoporosis, and 1.5 million suffer a broken wrist or hip or crushed spinal vertebrae each year because of the disease.[6]

Osteoporosis is a disabling disease that reduces the quality of life and dramatically increases the need for health care.[7] Because the incidence of osteoporosis increases with age, its importance as a personal and public health problem is intensifying as the U.S. population ages. It currently appears, however, that a large percentage of the cases of osteoporosis can be prevented. The key to prevention is to build dense bones during childhood and the early adult years and then keep bones dense as you age.[8]

The Timing of Bone Formation] Bones develop and mineralize throughout the first three decades of life. Even after the growth spurt occurs during adolescence and

remodeling
The breakdown and buildup of bone tissue.

osteoporosis
(*osteo* = bones; *poro* = porous, *osis* = abnormal condition) A condition characterized by porous bones; it is due to the loss of minerals from the bones.

text continued on page 23-10

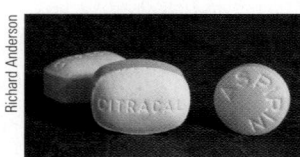

Calcium supplements are large because our daily need for calcium is high (1000 milligrams/day). Four pills provide 800 milligrams calcium, the amount in 2⅔ cups of milk. An aspirin is shown for comparison.

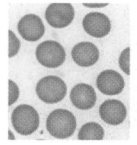 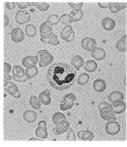

Iron-deficiency anemia is characterized by microcytic anemia and small, pale red blood cells (right photo). Normal red blood cells are shown on the left.
CNRI - Martin M. Rotker

TABLE 23.1

AN INTENSIVE COURSE ON ESSENTIAL MINERALS.

	PRIMARY FUNCTIONS	CONSEQUENCES OF DEFICIENCY
CALCIUM AI* women: 1000 mg men: 1000 mg UL: 2500 mg	• Component of bones and teeth • Needed for muscle and nerve activity, blood clotting	• Poorly mineralized, weak bones (osteoporosis) • Rickets in children • Osteomalacia (rickets in adults) • Stunted growth in children • Convulsions, muscle spasms
PHOSPHORUS RDA women: 700 mg men: 700 mg UL: 4000 mg	• Component of bones and teeth • Component of certain enzymes and other substances involved in energy formation • Needed to maintain the right acid-base balance of body fluids	• Loss of appetite • Nausea, vomiting • Weakness • Confusion • Loss of calcium from bones
MAGNESIUM RDA women: 310 mg men: 400 mg UL: 350 mg (from supplements only)	• Component of bones and teeth • Needed for nerve activity • Activates enzymes involved in energy and protein formation	• Stunted growth in children • Weakness • Muscle spasms • Personality changes
IRON RDA women: 18 mg men: 8 mg UL: 45 mg	• Transports oxygen as a component of hemoglobin in red blood cells • Component of myoglobin (a muscle protein) • Needed for certain reactions involving energy formation	• Iron deficiency • Iron-deficiency anemia • Weakness, fatigue • Pale appearance • Reduced attention span and resistance to infection • Mental retardation, developmental delay in children
ZINC RDA women: 8 mg men: 11 mg UL: 40 mg	• Required for the activation of many enzymes involved in the reproduction of proteins • Component of insulin, many enzymes	• Growth failure • Delayed sexual maturation • Slow wound healing • Loss of taste and appetite • In pregnancy, low-birth-weight infants and preterm delivery

*AI (Adequate Intakes) and RDA's (Recommended Dietary Allowances) are for 19–30 year olds; UL (Upper Limits) are for 19–70 year olds, 1997–2004.

TABLE 23.1

AN INTENSIVE COURSE ON ESSENTIAL MINERALS. CONTINUED

CONSEQUENCES OF OVERDOSE	PRIMARY FOOD SOURCES	HIGHLIGHTS AND COMMENTS
• Drowsiness • Calcium deposits in kidneys, liver, and other tissues • Suppression of bone remodeling • Decreased zinc absorption	• Milk and milk products (cheese, yogurt) • Broccoli • Dried beans • Calcium-fortified foods (some juices, breakfast cereals, bread, for example)	• The average intake of calcium among U.S. women is approximately 60% of the DRI. • One in four women and one in eight men in the U.S. develop osteoporosis. • Adequate calcium and vitamin D status must be maintained to prevent bone loss.
• Muscle spasms	• Milk and milk products (cheese, yogurt) • Meats • Seeds, nuts • Phosphates added to foods	• Deficiency is generally related to disease processes.
• Diarrhea • Dehydration • Impaired nerve activity due to disrupted utilization of calcium	• Plant foods (dried beans, tofu, peanuts, potatoes, green vegetables) • Milk • Bread • Ready-to-eat cereals • Coffee	• Magnesium is primarily found in plant foods where it is attached to chlorophyll. • Average intake among U.S. adults is below the RDA.
• Hemochromatosis ("iron poisoning") • Vomiting, abdominal pain • Blue coloration of skin • Liver and heart damage, diabetes • Decreased zinc absorption • Atherosclerosis (plaque buildup) in older adults	• Liver, beef, pork • Dried beans • Iron-fortified cereals • Prunes, apricots, raisins • Spinach • Bread • Pasta	• Cooking foods in iron and stainless steel pans increases the iron content of the foods. • Vitamin C, meat, and alcohol increase iron absorption. • Iron deficiency is the most common nutritional deficiency in the world. • Average iron intake of young children and women in the U.S. is low.
• Over 25 mg/day is associated with nausea, vomiting, weakness, fatigue, susceptibility to infection, copper deficiency, and metallic taste in mouth. • Increased blood lipids	• Meats (all kinds) • Grains • Nuts • Milk and milk products (cheese, yogurt) • Ready-to-eat cereals • Bread	• Like iron, zinc is better absorbed from meats than from plants. • Marginal zinc deficiency may be common, especially in children. • Zinc supplements may decrease duration and severity of the common cold.

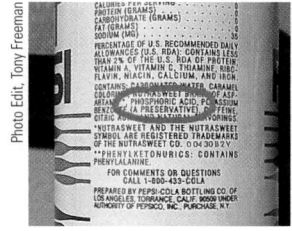

Phosphates are a common, multipurpose food additive.

Magnesium is to chlorophyll in plants as iron is to hemoglobin in humans.

continued

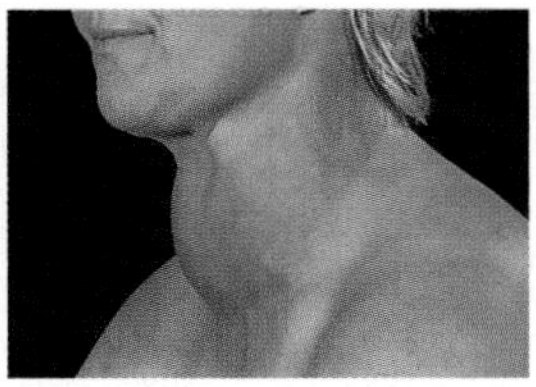

Goiter is a highly visible sign of iodine deficiency.

TABLE 23.1

AN INTENSIVE COURSE ON ESSENTIAL MINERALS. CONTINUED

	PRIMARY FUNCTIONS	CONSEQUENCES OF DEFICIENCY
FLUORIDE AI women: 3 mg men: 4 mg UL: 10 mg	• Component of bones and teeth (enamel)	• Tooth decay and other dental diseases
IODINE RDA women: 150 mcg men: 150 mcg UL: 1100 mcg	• Component of thyroid hormones that help regulate energy production and growth	• Goiter • Cretinism (mental retardation, hearing loss, growth failure)
SELENIUM RDA women: 55 mcg men: 55 mcg UL: 400 mcg	• Acts as an antioxidant in conjunction with vitamin E (protects cells from damage due to exposure to oxygen) • Needed for thyroid hormone production	• Anemia • Muscle pain and tenderness • Keshan disease (heart failure), Kashin-Beck disease (joint disease)
COPPER RDA women: 900 mcg men: 900 mcg UL: 10,000 mcg	• Component of enzymes involved in the body's utilization of iron and oxygen • Functions in growth, immunity, cholesterol and glucose utilization, brain development	• Anemia • Seizures • Nerve and bone abnormalities in children • Growth retardation
MANGANESE AI women: 2.3 mg men: 1.8 mg	• Needed for the formation of body fat and bone	• Weight loss • Rash • Nausea and vomiting
CHROMIUM AI women: 35 mcg men: 25 mcg	• Required for the normal utilization of glucose and fat	• Elevated blood glucose and triglyceride levels • Weight loss

Custom Medical Stock

TABLE 23.1

AN INTENSIVE COURSE ON ESSENTIAL MINERALS. CONTINUED

CONSEQUENCES OF OVERDOSE	PRIMARY FOOD SOURCES	HIGHLIGHTS AND COMMENTS
• Fluorosis • Brittle bones • Mottled teeth • Nerve abnormalities	• Fluoridated water and foods and beverages made with it • Tea • Shrimp, crab	• Toothpastes, mouth rinses, and other dental care products may provide fluoride. • Fluoride overdose has been caused by ingestion of fluoridated toothpaste.
• Over 1 mg/day may produce pimples, goiter, and decreased thyroid function.	• Iodized salt • Milk and milk products • Seaweed, seafoods • Bread from commercial bakeries	• Iodine deficiency was a major problem in the U.S. in the 1920s and 1930s. Deficiency remains a major health problem in some developing countries. • Amount of iodine in plants depends on iodine content of soil. • Most of the iodine in our diet comes from the incidental addition of iodine to foods from cleaning compounds used by food manufacturers.
• "Selenosis." Symptoms of selenosis are hair and fingernail loss, weakness, liver damage, irritability, and "garlic" or "metallic" breath.	• Meats and seafoods • Eggs • Whole grains	• Content of foods depends on amount of selenium in soil, water, and animal feeds. • May play a role in the prevention of some types of cancer.
• Wilson's disease (excessive accumulation of copper in the liver and kidneys) • Vomiting, diarrhea • Tremors • Liver disease	• Bread • Potatoes • Grains • Dried beans • Nuts and seeds • Seafood • Ready-to-eat cereals	• Toxicity can result from copper pipes and cooking pans. • Average intake in the U.S. is below the RDA.
• Infertility in men • Disruptions in the nervous system (psychotic symptoms) • Muscle spasms	• Whole grains • Coffee, tea • Dried beans • Nuts	• Toxicity is related to overexposure to manganese dust in miners.
• Kidney and skin damage	• Whole grains • Wheat germ • Liver, meat • Beer, wine • Oysters	• Toxicity usually results from exposure in chrome-making industries or overuse of supplements. • Supplements do not build muscle mass or increase endurance.

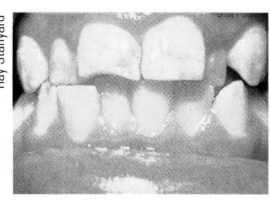

"Mottled teeth" result from an excessive intake of fluoride in children.

Iodine deficiency during pregnancy produces cretinism in the offspring.

continued

TABLE 23.1

AN INTENSIVE COURSE ON ESSENTIAL MINERALS. CONTINUED

	PRIMARY FUNCTIONS	CONSEQUENCES OF DEFICIENCY
MOLYBDENUM RDA women: 45 mcg men: 45 mcg UL: 2000 mcg	• Component of enzymes involved in the transfer of oxygen from one molecule to another	• Rapid heartbeat and breathing • Nausea, vomiting • Coma
SODIUM AI women: 1.5g men: 1.5g UL: 2.3g	• Needed to maintain the right acid-base balance in body fluids • Helps maintain an appropriate amount of water in blood and body tissues • Needed for muscle and nerve activity	• Weakness • Apathy • Poor appetite • Muscle cramps • Headache • Swelling
POTASSIUM AI women: 4.7g men: 4.7g	• Same as for sodium	• Weakness • Irritability, mental confusion • Irregular heartbeat • Paralysis
CHLORIDE AI women: 2.3g men: 2.3g UL: 3.5g	• Component of hydrochloric acid secreted by the stomach (used in digestion) • Needed to maintain the right acid-base balance of body fluids • Helps maintain an appropriate water balance in the body	• Muscle cramps • Apathy • Poor appetite • Long-term mental retardation in infants

Illustration 23.4
This woman's stooped appearance is due to osteoporosis.

PhotoEdit - Robert Brenner

people think they are as tall as they will ever be, bones continue to increase in width and mineral content for 10 to 15 more years. Peak bone density, or the maximal level of mineral content in bones, is reached somewhere between the age of 30 and 40 years. After that, bone mineral content no longer increases. The higher the peak bone mass, the less likely it is that osteoporosis will develop. People with higher peak bone mass simply have more calcium to lose before bones become weak and fracture easily. That also is the reason males experience osteoporosis less often than females do: they have more bone mass to lose.[9]

Bone size and density often remain fairly stable from age 30 to the mid-40s, but then bones tend to demineralize with increasing age. By the time women are 70, for example, their bones are 30 to 40% less dense on average than they once were.[10] A woman may lose an inch or more in height with age and develop the "dowager's hump" that is characteristic of osteoporosis in the spine (Illustration 23.4).

TABLE 23.1

AN INTENSIVE COURSE ON ESSENTIAL MINERALS. CONTINUED

CONSEQUENCES OF OVERDOSE	PRIMARY FOOD SOURCES	HIGHLIGHTS AND COMMENTS
• Loss of copper from the body • Joint pain • Growth failure • Anemia • Gout	• Dried beans • Grains • Dark green vegetables • Liver • Milk and milk products	• Deficiency is extraordinarily rare.
• High blood pressure in susceptible people • Kidney disease • Heart problems	• Foods processed with salt • Cured foods (corned beef, ham, bacon, pickles, sauerkraut) • Table and sea salt • Bread • Milk, cheese • Salad dressing	• Very few foods naturally contain much sodium. • Processed foods are the leading source of dietary sodium. • High-sodium diets are associated with the development of hypertension in "salt-sensitive" people.
• Irregular heartbeat, heart attack	• Plant foods (potatoes, squash, lima beans, tomatoes, plantains, bananas, oranges, avocados) • Meats • Milk and milk products • Coffee	• Content of vegetables is often reduced in processed foods. • Diuretics (water pills) and other antihypertension drugs may deplete potassium. • Salt substitutes often contain potassium.
• Vomiting	• Same as for sodium. (Most of the chloride in our diets comes from salt.)	• Excessive vomiting and diarrhea may cause chloride deficiency. • Legislation regulating the composition of infant formulas was enacted in response to formula-related chloride deficiency and subsequent mental retardation in infants.

What has just been described is the way things are in the United States, but this situation is neither normal nor inevitable. The incidence of osteoporosis is far less in many other countries than in the United States.[11] Although genetic and other factors play a role, the causes of osteoporosis are mainly related to diet and other lifestyle behaviors. That means that osteoporosis can largely be prevented.

How Do You Build and Maintain Dense Bones?] Since the late 1980s, a rash of studies have examined the relationships among dietary calcium intake, vitamin D status, and bone density. Results have been quite consistent: bone mass before the age of 30 is increased by calcium intakes of approximately 1200 milligrams per day along with adequate vitamin D status. In adults over the age of 50 years, bone density tends to be preserved with calcium intakes around 1200 to 1500 milligrams per day and 400–800 IU of vitamin D.[12] In addition, calcium intakes of 1200

milligrams or more per day with adequate vitamin D status decrease the incidence of stress fractures in athletes.[13] Low bone mineral density can be increased somewhat by switching from a low to an adequate intake of calcium and improved vitamin D status. The earlier in life this occurs, the more improvement in bone density is noted. High levels of bone density once achieved do not last forever; bone must be continually renewed by an adequate supply of calcium and vitamin D.[14]

Vitamin D is important for building dense bones because it increases calcium absorption and the deposition of calcium into bone. Our need for vitamin D can be met from vitamin D–fortified milk, vitamin D–fortified breakfast cereals, and seafoods or by exposing the skin to the sun. A precursor of vitamin D lies beneath our skin. When skin is exposed to ultraviolet rays from sunlight, this precursor is converted to vitamin D. Exposing the hands and face to sunshine for 5 to 15 minutes produces a day's supply of vitamin D. (Windows and sunscreen block ultraviolet rays and prevent the production of vitamin D in the skin.) The "Health Action" on the next page summarizes behaviors that help build dense bones and prevent osteoporosis.

Illustration 23.5

The wide availability of soft-drink vending machines in some high schools is thought to contribute to excessive consumption of these beverages at the expense of milk. *Some school systems are taking the machines out.*

We Are Consuming Too Little Calcium]

Unfortunately, about half of U.S. women belong to the "600 Club"; they consume 600 milligrams of calcium per day or less.[15] Only 14% of girls and 36% of boys aged 12 to 19 years consume the recommended amount of calcium during these critical years for building bone density. Low calcium intake during the growing years increases the probability that fractures and osteoporosis will occur at younger ages than is the norm. Men tend to consume more calcium than women but on average still fall short of the recommended intake.[16]

Why Do Women Consume Too Little Calcium?} Women consume too little calcium for several reasons. Although dairy products such as milk, cheese, and yogurt are our richest sources of calcium, some women consider them "fattening" and give them up during the teen years. Skipping breakfast and eating out are common among teens and tend to reduce calcium intake. Many think that milk is a food that only children need, while others stop drinking milk because it "gives them gas." (Such people may be lactose intolerant.)[17]

Illustration 23.6

Would you be more likely to drink milk at fast-food restaurants if it was included as part of a bargain special? *Some college women think this is a good idea—one they would go for.*

Increasing intakes of soft drinks, combined with reduced consumption of milk, also appear to be contributing to bone fractures and future osteoporosis in today's youth and women.[18] The wide availability of soft-drink machines in schools (Illustration 23.5) is thought to be a factor related to increased consumption by teens. According to some college students, one reason their intake of calcium is low is that they frequently eat at fast-food restaurants that feature soft drinks rather than milk on the menu. But when asked if they would choose specials that included fresh-tasting milk (not the kind that tastes like the milk container) as part of the meal, some students reported they would select the specials (Illustration 23.6).[19]

MEAL DEALS

ONLY $3.59 — HAMBURGER, FRIES AND MEDIUM SOFT DRINK

ONLY $2.19 — APPLE-CINNAMON MUFFIN AND FRESH MILK

ONLY $3.95 — CHICKEN FILET ON A BUN, COLESLAW AND FRESH MILK

HEALTH ACTION I *Behaviors That Help Prevent Osteoporosis*

- Consuming adequate amounts of Vitamin D (400 IU); exposing the hands and face to direct sunlight for 5 to 15 minutes daily

- Moderating alcohol intake
- Consuming enough dietary calcium (1000 to 1200 milligrams daily)

- Getting regular physical activity
- Not smoking

What Else Influences Osteoporosis?} Table 23.2 summarizes the risk factors identified for osteoporosis. If you were to find all of the various risk factors combined in one person, that individual would be a thin woman with light skin who has consumed too little calcium and vitamin D and is physically inactive, an excessive alcohol drinker, a smoker, and genetically "small-boned." Additionally, she would have had her ovaries surgically removed for medical reasons before the age of 45.[21] Few people meet every aspect of that description, but people who have several of these characteristics are more likely to develop osteoporosis than those who don't.

At one time, researchers thought that a relatively high intake of phosphorus or protein might intensify the effects of low intakes of calcium on the development of osteoporosis. Studies in humans, however, have failed to demonstrate harmful effects of high phosphorus or protein intake on the development of osteoporosis given an average level of calcium intake.[22]

How Is Osteoporosis Treated?} Calcium supplements or adequate calcium intakes (around 1000–1500 milligrams per day), vitamin D supplements (800 IU), and weight-bearing exercise like walking and tennis decrease the progression of osteoporosis in many people.[23] Table 23.3 summarizes current recommendations for the treatment of osteoporosis.

Calcium: Where to Find It

Over half of the calcium supplied by the diets of Americans comes from milk and milk products.[25] Milk (including chocolate milk), cheese, and yogurt are all good sources of calcium (see Table 23.12). Some plants such as kale, broccoli, and bok choy provide appreciable amounts of calcium, too. On average, 32% of the calcium content of milk and milk products, calcium-fortified orange juice, and calcium supplements is absorbed, compared to approximately 5 to 60% of the calcium from different plants (Table 23.4)[26] Calcium absorption decreases with age, vitamin D inadequacy, and ingestion of supplemental iron and zinc.[27] The best way to meet the recommended level of intake (1000 milligrams per day for most adults) is to choose rich sources of absorbable calcium.

Many foods rich in calcium aren't loaded with calories or fat. As Table 23.4 shows, low-fat yogurt, skim and soy milk, kale, and broccoli, for example, provide good amounts of calcium at a low cost in calories.

Calcium is increasingly appearing in unexpected places such as in candy, snack bars, waffles, and bread (Illustration 23.7). The array of calcium-fortified foods is increasing dramatically in the United States in response to publicity about our need for more of it. Although development of such products is making it easier to consume more calcium, the availability of calcium in some of these foods has not been established.[28]

TABLE 23.2

RISK FACTORS FOR OSTEOPOROSIS.[20]

- Female
- Menopause
- Deficient calcium intake
- Caucasian or Asian heritage
- Thinness ("small bones")
- Cigarette smoking
- Excessive alcohol intake
- Ovarectomy (ovaries removed) before age 45
- Physical inactivity
- Deficient vitamin D status
- Genetic factors

TABLE 23.3

APPROACHES TO THE TREATMENT OF OSTEOPOROSIS.[24]

- Daily intake of 1000–1500 milligrams of calcium
- Regular physical activity
- Adequate vitamin D intake (800 IU daily)
- Medication

TABLE 23.4

CALORIC CONTENT, CALCIUM LEVEL, AND PERCENTAGE OF AVAILABLE CALCIUM FROM DIFFERENT FOODS.

FOOD	AMOUNT	CALORIES	CALCIUM (mg)	CALCIUM ABSORBED (%)	AVAILABLE CALCIUM (mg)
Yogurt, low fat	1 cup	143	413	32	132
Skim milk	1 cup	85	301	32	96
Soy milk (fortified)	1 cup	79	300	31	93
1% milk	1 cup	163	274	32	88
Tofu	1 cup	188	260	31	81
Cheese	1 ounce	114	204	32	65
Kale, cooked	1 cup	42	94	49	46
Broccoli, cooked	1 cup	44	72	61	44
Bok choy, raw	1 cup	9	73	54	39
Dried beans	1 cup	209	120	24	29
Spinach, cooked	1 cup	42	244	5	12

Watch for Calcium on the Label] Health claims relating to the benefits of calcium in reducing the risk of osteoporosis may appear on the labels of food products that qualify as good sources of calcium. Look for health claims on foods such as milk, yogurt, and calcium-fortified foods such as orange juice, breakfast cereals, and grain products (Illustration 23.8).

Illustration 23.7
The number of calcium-fortified foods is increasing.

Illustration 23.8
Food products that are good sources of calcium can be labeled with a health claim.

"Regular exercise and a healthy diet with enough calcium help maintain good bone health and may reduce the risk of osteoporosis later in life."

Nutrition Facts

Serving Size: 1 cup (240ml)
Servings per Container: 16

Amount per Serving

Calories 110   Calories from Fat 20

	% Daily Value*
Total Fat 2.5g	4%
Saturated Fat 1.5g	8%
Cholesterol 15mg	4%
Sodium 135mg	6%
Total Carbohydrate 13g	4%
Dietary Fiber 0g	0%
Sugars 12g	
Protein 8g	

Vitamin A 10%   •   Vitamin C 4%

Calcium 30% Iron 0% Vitamin D 25%

Phosphorus 10%

*Percent Daily Values are based on a 2,000 calorie diet.

23-14

Can You Consume Too Much Calcium?] Yes, indeed, you can consume too much calcium. Supplemental calcium in doses exceeding 2.5 grams per day may produce drowsiness, lead to constipation, and cause calcium to deposit in tissues such as the liver and kidneys.[29]

Selected Minerals: Iron

Most of the body's iron supply is found in *hemoglobin*. Small amounts are present in *myoglobin*, and free iron is involved in processes that capture energy released during the breakdown of proteins and fats.

The Role of Iron in Hemoglobin and Myoglobin

What happens to a car when its paint gets scratched? After a while, the exposed metal rusts. The iron in the metal combines with oxygen in the air, and the result is iron oxide or "rust." Iron readily combines with oxygen, and that property is put to good use in the body. From its location in hemoglobin in red blood cells, iron loosely attaches to oxygen when blood passes near the inner surface of the lungs. The bright red, oxygenated blood is then delivered to cells throughout the body. When the oxygenated blood passes near cells that need oxygen for energy formation or other reasons, oxygen is released from the iron and diffuses into cells (Illustration 23.9). The free iron in hemoglobin then picks up carbon dioxide, a waste product of energy formation. When carbon dioxide attaches to iron, blood turns from bright red to dark bluish red. Blood then circulates back to the lungs, where carbon dioxide is released from the iron and exhaled into the air. The free iron attaches again to oxygen that enters the lungs, and the cycle continues.

Iron in myoglobin traps oxygen delivered by hemoglobin, stores it, and releases it as needed for energy formation for muscle activity. In effect, myoglobin boosts the supply of oxygen available to muscles.

The functions of iron just described operate smoothly when the body's supply of iron is sufficient. Unfortunately, that is often not the case.

hemoglobin
The iron-containing protein in red blood cells.

myoglobin
The iron-containing protein in muscle cells.

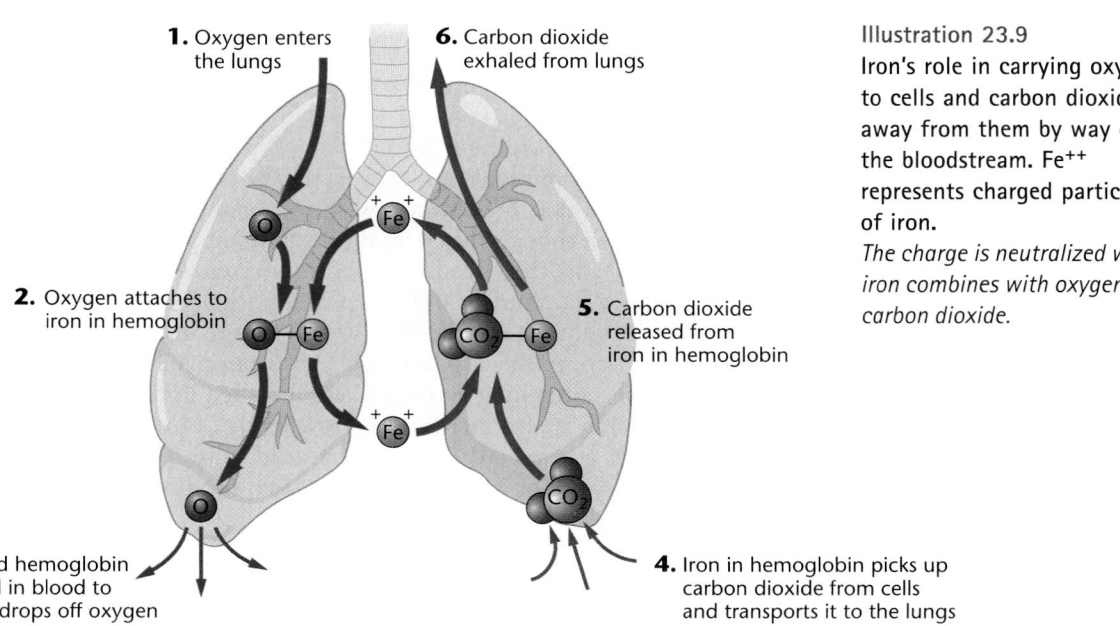

1. Oxygen enters the lungs

6. Carbon dioxide exhaled from lungs

2. Oxygen attaches to iron in hemoglobin

5. Carbon dioxide released from iron in hemoglobin

3. Oxygenated hemoglobin transported in blood to body cells, drops off oxygen

4. Iron in hemoglobin picks up carbon dioxide from cells and transports it to the lungs

Illustration 23.9
Iron's role in carrying oxygen to cells and carbon dioxide away from them by way of the bloodstream. Fe^{++} represents charged particles of iron.
The charge is neutralized when iron combines with oxygen or carbon dioxide.

TABLE 23.5

INCIDENCE OF IRON DEFICIENCY.[30]

	POPULATION WITH IRON DEFICIENCY (%)
Worldwide (children under 5 years):	
Developing countries	51
Developed countries	12
United States:	
Children, 1 to 2 years	9
Pregnant women	12
Females, 16 to 19 years	11
Females, 20 to 49 years	11
Males, 12 to 49 years	1 or less

iron deficiency
A disorder that results from a depletion of iron stores in the body. It is characterized by weakness, fatigue, short attention span, poor appetite, increased susceptibility to infection, and irritability.

iron-deficiency anemia
A condition that results when the content of hemoglobin in red blood cells is reduced due to a lack of iron. It is characterized by the signs of iron deficiency plus paleness, exhaustion, and a rapid heart rate.

TABLE 23.6

IRON INTAKE OF CHILDREN, WOMEN, AND MEN IN THE UNITED STATES.[32]

	PERCENTAGE OF RECOMMENDED INTAKE
Children	67%
Women	41
Men	84

Iron Deficiency Is a Big Problem

The most widespread nutritional deficiency in both developing and developed countries is ***iron deficiency*** (Table 23.5). It is estimated that one out of every three people in the world is iron deficient.[31] For the most part, iron deficiency affects children and women of childbearing age who have a high need for iron and frequently consume too little of it (Table 23.6). Iron deficiency may develop in people who have lost blood due to injury, surgery, or ulcers. Donating blood more than three times a year can also precipitate iron deficiency.[33]

Consequences of Iron Deficiency] Many body processes sputter without sufficient oxygen. People with iron deficiency usually feel weak and tired. They have a shortened attention span and a poor appetite, are susceptible to infection, and become irritable easily. If the deficiency is serious enough, ***iron-deficiency anemia*** develops and additional symptoms occur. People with iron-deficiency anemia look pale, are easily exhausted, and have rapid heart rates. Iron-deficiency anemia is a particular problem for infants and young children because it is related to lasting retardation in mental development.[34]

Getting Enough Iron in Your Diet

"Enough" iron according to the RDAs is 8 milligrams for men and 18 milligrams per day for women aged 19 to 50 years. Consuming that much iron can be difficult for women. On average, 1000 calories' worth of food provides about 6 milligrams of iron. Women would have to consume around 2500 calories per day to obtain even 15 milligrams of iron on average. Selection of good sources of iron has to be done on a better-than-average basis if women are to get enough.

Iron is found in small amounts in many foods, but only a few foods such as liver, beef, and prune juice are rich sources. (A list of food sources of iron is given in Table 23.12.) Foods cooked in iron and stainless steel pans, however, can be a significant source of iron because some of the iron in the pan leaches out during cooking. On average, approximately 1 milligram of iron is added to each 3-ounce serving of food cooked in these pans.[35]

Most of the iron in plants and eggs is tightly bound to substances such as phytates or oxalic acid that limit iron absorption, making these foods relatively poor sources of iron even though they contain a fair amount of it. (Differences in the proportions of iron absorbed from various food sources are shown in Illustration 23.10.) A 3-ounce hamburger and a cup of asparagus both contain approximately 3 milligrams of iron, for example. But 20 times more iron can be absorbed from the hamburger than from the asparagus.[37] Absorption of iron from plants can be increased if foods containing vitamin C are included in the same meal.

Iron absorption is also increased by low levels of iron stores in the body (Illustration 23.11). In other words, if you're in need of iron, your body sets off mechanisms that allow more of it to be absorbed from foods or supplements. When iron stores are high, less iron is absorbed.

The body's ability to regulate iron absorption provides considerable protection against iron deficiency and overdose. The protection is not complete, however, as evidenced by widespread iron deficiency and the occurrence of iron toxicity.

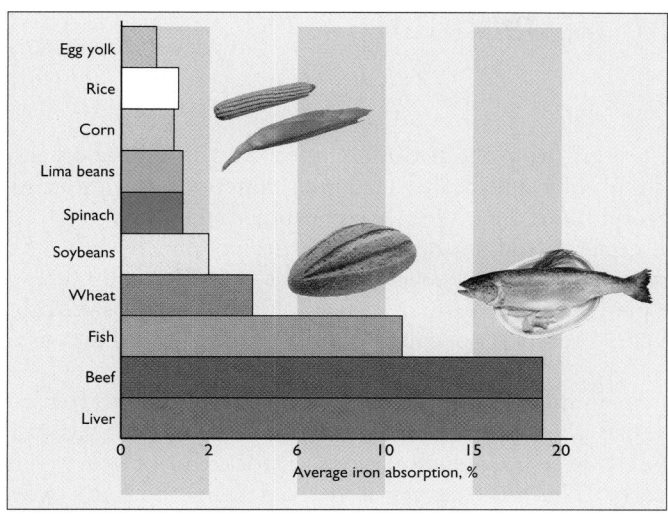

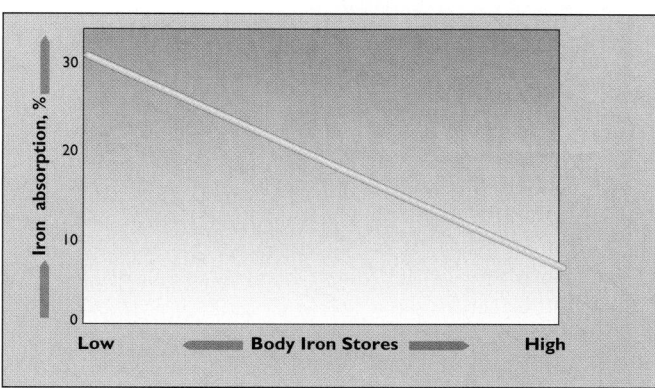

Illustration 23.11
People absorb more iron from foods and supplements when body stores of iron are low than when stores of iron are high.

Illustration 23.10
Average percentage of iron absorbed from selected foods by healthy adults.[36]

Overdosing on Iron]   Excess iron absorbed into the body cannot be easily excreted. Consequently, it is deposited in various tissues such as the liver, pancreas, and heart. There, the iron reacts with cells, causing damage that can result in liver disease, diabetes, and heart failure. One in 200 people in the United States has an inherited tendency to absorb too much iron (a disorder called hemochromatosis and pronounced hem-oh-chrom-ah-toe-sis). Other people develop iron toxicity from consuming large amounts of iron with alcohol (alcohol increases iron absorption) or from very high iron intakes, usually due to overdoses of iron supplements.

Each year in the United States, more than 10,000 people accidentally overdose on iron supplements.[38] Commonly, victims are young children who mistakenly think iron pills are candy (Table 23.7 and Illustration 23.12). The lethal dose of iron for a 2-year-old child is about 3 grams,[40] the amount of iron present in 25 pills containing 120 milligrams of iron each. Iron supplements that are not being used should be thrown away or stored in a place where toddlers cannot get to them.

TABLE 23.7

ACCIDENTAL OVERDOSES OF CHEMICAL SUBSTANCES BY CHILDREN UP TO AGE 4 IN THE UNITED STATES.[39]

SUBSTANCE	CASES PER 100,000 CHILDREN
Aspirin and substitutes	12.7
Solvents/petroleum products	11.2
Tranquilizers	9.9
Iron supplements	8.7
Corrosives and caustics	7.4

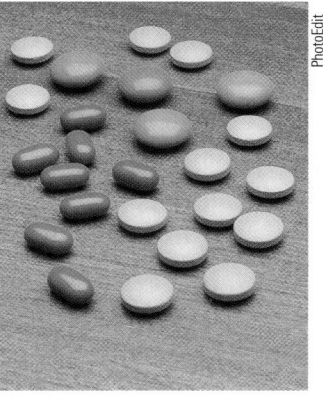

Illustration 23.12
If you were three years old, could you tell which "pills" are candy and which are the iron supplements?*
*Overdoses of iron supplements are a leading cause of accidental poisoning in young children.
*The iron supplements are the lightest green in color.

23–17

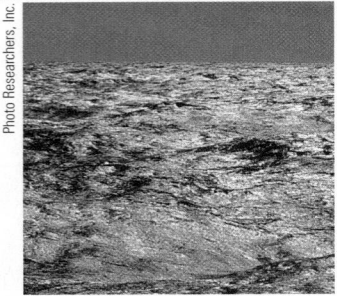

Illustration 23.13
The oceans contain enough salt to cover the surface of the Earth to a depth of more than 400 feet.

water balance
The ratio of the amount of water outside cells to the amount inside cells; this balance is needed for normal cell functioning.

hypertension
High blood pressure. It is defined as blood pressure exerted inside blood vessel walls that typically exceeds 140/90 millimeters of mercury.

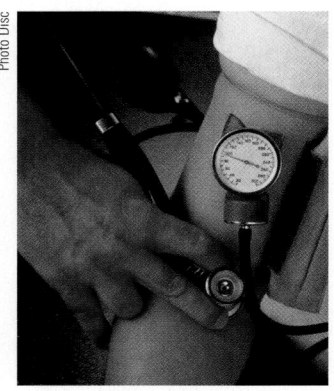

Illustration 23.14
A blood pressure test.
Blood pressure is expressed as two numbers:

Systolic pressure measures the force of the blood when the heart contracts.

110
——
70

Diastolic pressure measures the force of the blood when the heart is at rest.

Selected Minerals: Sodium

Excess of salty flavor hardens the pulse.
—Abernathy, 1000 B.C.

On the one hand, sodium is a gift from the sea (Illustration 23.13); on the other, it is a hazard to health. Sodium's life- and health-sustaining functions are frequently overshadowed by its effects on blood pressure when consumed in excess.

An extremely reactive mineral, sodium occurs in nature combined with other elements. The most common chemical partner of sodium is chloride, and much of the sodium present on this planet is in the form of sodium chloride—table salt. Table salt is 40% sodium by weight; 1 teaspoon of salt contains about 2400 milligrams of sodium.

Although salt is found in abundance in the oceans, humans have not always had access to enough salt to satisfy their taste for it. During times when salt has been scarce, wars have been fought over it. In such periods, a pocketful of salt was as good as a pocketful of cash. (The word *salary* is derived from the Latin word *salarium*, "salt money.") People were said to be "worth their salt" if they put in a full day's work. Today, salt is widely available in most parts of the world, and the problem is limiting human intakes to levels that do not interfere with health.

What Does Sodium Do in the Body?

Sodium appears directly above potassium in the periodic table, and the two work closely together in the body to maintain normal *water balance*. Both sodium and potassium chemically attract water, and under normal circumstances, each draws sufficient water to the outside or inside of cells to maintain an optimal level of water in both places.

Water balance and cell functions are upset when there's an imbalance in the body's supplies of sodium and potassium. You have probably noticed that you become thirsty when you eat a large amount of salted potato chips or popcorn. Salty foods make you thirsty because your body loses water when the high load of dietary sodium is excreted. The thirst signal indicates that you need water to replace what you have lost.

The loss of body water that accompanies ingestion of large amounts of salt explains why seawater neither quenches thirst nor satisfies the body's need for water. When a person drinks seawater, its high concentration of sodium causes the body to excrete more water than it retains. Rather than increasing the body's supply of water, the ingestion of seawater increases the need for water.

In healthy people, the body's adaptive mechanisms provide a buffer against upsets in water balance due to high sodium intakes. It appears that many people are overwhelming the body's ability to cope with high sodium loads, however. High dietary intakes of sodium appear to play an important role in the development of *hypertension* in many people.[41]

A Bit about Blood Pressure]   In order to circulate through the body, blood must exist under pressure in the blood vessels. The amount of pressure exerted on the walls of blood vessels is greatest when pulses of blood are passing through them (that's when "systolic" blood pressure is measured) and least between pulses (that's when "diastolic" blood pressure is taken). Blood pressure measurements note the highest and lowest pressure in blood vessels (Illustration 23.14).

Blood pressure levels less than 120/80 millimeters of mercury (mmHg) are considered normal, whereas levels between 120/80 and 139/89 mmHg are classified as "prehypertension." Values of 140/90 mm Hg and higher qualify as hypertension (Table 23.8). Several blood pressure measurements, taken while a person is relaxed, are needed to obtain an accurate result. (Even going to a clinic or doctor's office to have blood pressure measured can raise it for some people.)

Hypertension is a major public health problem in the United States and other countries. Although not considered a disease by itself, the presence of hypertension substantially increases the risk that a person will develop heart disease or kidney failure or will experience a heart attack or stroke. Hypertension occurs in 29% of U.S. adults and more than 20% of adults worldwide. It is more likely to develop as people become older. About 40% of all cases of hypertension are classified as mild.[43]

What Causes Hypertension?]

About 10% of all cases of hypertension can be directly linked to a cause. People who have hypertension with no identifiable cause (over 90% of all cases) are said to have *essential hypertension.*[44]

A number of risk factors for hypertension have been identified, and dietary factors are among the most important (Table 23.9). Foremost among the evidence linking diet to hypertension are population studies that show a relationship between average salt intake and hypertension. As average salt intake rises, so do rates of hypertension. When people with hypertension reduce their salt intake, their blood pressure tends to drop. Further scrutiny of the data on salt and blood pressure, however, reveals that not everyone is equally susceptible to high-salt diets. Some people are genetically susceptible to salt, or have subtle forms of kidney disorders that raise their blood pressure in response to high-sodium diets.[46] This condition is known as *salt sensitivity.*

Salt Sensitivity]

Approximately 51% of people with hypertension and 26% of people with normal blood pressure are salt sensitive. Reduction in salt intake by people who are salt sensitive, along with weight loss if overweight, substantially improves blood pressure in most cases. Because it is currently difficult to identify who in the population is salt sensitive and who is not; because most Americans consume much more sodium than needed; and because high-sodium foods are often high in fat, official advice for Americans is to limit their sodium intake to 2400 milligrams per day (the equivalent of approximately one level teaspoon of salt from all sources).[47]

Other Risk Factors for Hypertension]

Obesity is a major risk factor for hypertension. For obese people with hypertension, the most effective treatment is weight

essential hypertension
Hypertension of no known cause; also called *primary* or *idiopathic hypertension,* it accounts for 90–95% of all cases of hypertension.

salt sensitivity
A genetically influenced condition in which a person's blood pressure rises when high amounts of salt or sodium are consumed. Such individuals are sometimes identified by blood pressure increases of 10% or more when switched from a low-salt to a high-salt diet.

TABLE 23.8

HYPERTENSION CATEGORIES.[42]

CATEGORY	SYSTOLIC (mm Hg)*	DIASTOLIC (mm Hg)
Normal	<120	80
Prehypertension	120–139	80–89
Hypertension		
Stage 1	140–159	90–99
Stage 2	160+	100+

mm Hg = millimeters of mercury.

TABLE 23.9

RISK FACTORS FOR HYPERTENSION.[45]

- Age
- Family history
- High-sodium diet
- Obesity
- Physical inactivity
- Excessive alcohol consumption
- Low vegetable and fruit consumption

HEALTH ACTION I *Dietary and Lifestyle Recommendations for Preventing Hypertension*

- Consume a diet rich in vegetables and fruits.
- Consume low-fat dairy products.
- Reduce saturated fat intake.
- Include whole grains, poultry, fish, nuts, and seeds in your diet.

- Reduce sodium intake to about 2400 mg (1 tsp total salt) or less per day.
- Limit alcohol intake to 1 to 2 drinks per day, if any.
- Exercise 30 minutes or more almost every day.

- Lose weight if overweight; maintain a healthy weight.
- Don't smoke.

From the National Heart, Lung, and Blood Institute, 2003.

loss. Excessive alcohol intake can prompt the development of hypertension, and moderation of intake (to two or fewer alcoholic drinks per day) can bring blood pressure back down.[48] Low intakes of calcium, vegetables, fruits, and potassium and high intakes of saturated fat all contribute to the development of hypertension. Physically inactive lifestyles also foster the development of hypertension.[49]

What Is the Best Way to Prevent Hypertension?

The "Health Action" above highlights recommendations for decreasing the chances that hypertension will develop. These recommendations contribute to an overall healthy lifestyle whose benefits extend beyond the prevention of hypertension.

How Is Hypertension Treated?] The recommended approach to treatment of all cases of hypertension consists of dietary and lifestyle changes and the use of medications if necessary.[50] Weight loss and smoking cessation (if needed), a moderate-sodium diet, regular exercise, moderate alcohol consumption (if any), and the DASH diet are basic components of the approach to treatment (Table 23.10). The DASH diet (Illustration 23.15) is based on vegetables, fruits, low-fat dairy products, whole grains, and poultry and fish. Its composition is similar to that of diets recommended for heart disease and cancer prevention. People who adhere to the DASH diet often bring their blood pressure levels back into the normal range. If blood pressure remains elevated after dietary and lifestyle changes have been implemented, or if blood pressure is quite high when diagnosed, antihypertension drugs are prescribed.[54]

Cutting Back on Salt] The leading sources of salt (and therefore of sodium) in the U.S. diet are processed foods (Table 23.11). Restricting the use of foods to which

TABLE 23.10

APPROACHES TO THE TREATMENT OF HYPERTENSION.[51]

- Weight loss (if needed)
- Moderate salt intake (2400 mg of sodium per day or less)
- Moderate alcohol consumption (if any)
- Regular physical activity
- The DASH diet
- Antihypertension drugs (if hypertension not controlled by above measures)

TABLE 23.11

MAJOR SOURCES OF SODIUM IN THE U.S. DIET.[55]

SODIUM SOURCE	CONTRIBUTION TO SODIUM INTAKE (%)
Processed foods	77
Fresh foods	12
Salt added at the table	6
Salt added during cooking	5

The DASH diet

In the mid-1990s a revolutionary approach to the control of mild and moderate hypertension was tested, and the results have changed health professionals' thinking about high blood pressure prevention and management. Called the DASH diet (Dietary Approaches to Stop Hypertension), it didn't focus on salt restriction and was related to significant reductions in blood pressure within 2 weeks in most people tested. In some people, reductions in blood pressure were sufficient to erase the need for antihypertension medications, and for others the diet reduced the amount or variety of medications needed. A subsequent study showed that a low-sodium diet boosts the blood pressure lowering effects of the DASH diet, especially in African Americans.[52]

The DASH diet consists of eating patterns made up of the following food groups:

	DAILY SERVINGS
Vegetables	4–5
Fruits	4–5
Grain products	7–8
Low-fat milk and dairy products	2–3
Lean meats, fish, poultry	2 or less
Nuts, seeds, dried beans	about 1

Although it does not work for all individuals with hypertension, the DASH diet is a mainstay in clinical efforts to control hypertension. Benefits of the diet last as long as the diet does.[53] A very useful Web site is available that describes refinements in the diet based on research studies, and that gives practical tips on how to follow the diet.

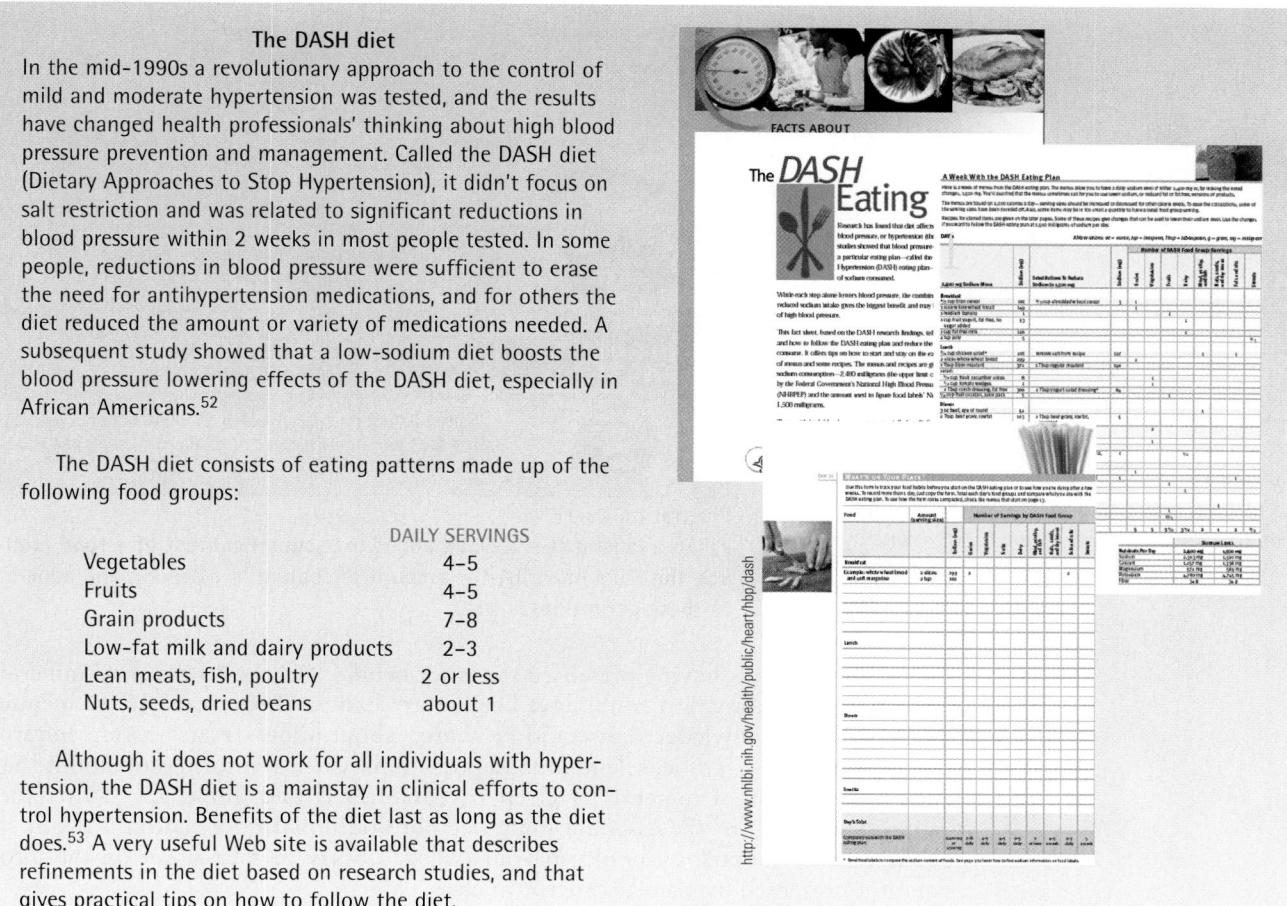

http://www.nhlbi.nih.gov/health/public/heart/hbp/dash

salt has been added during processing is the most effective way to lower salt intake and achieve a moderate- to low-salt diet.[56] High-salt processed foods include salad dressings, pickles, canned soups, corned beef, sausages and other luncheon meats, and snack foods such as potato chips and cheese twists.

Only a small proportion of our total sodium intake enters our diet from fresh foods. Very few foods naturally contain much sodium—at least not until they are processed (Illustration 23.16).

Using spices and lemon juice to flavor foods, consuming fresh vegetables (rather than processed ones) and fresh fruits, and checking out the sodium content of foods and then selecting low-sodium ones can also help reduce sodium intake.

Label Watch] Not all processed foods with added sodium taste salty. To find out which processed foods are high in sodium, you have to examine the label. Increasingly, low-salt processed foods are entering the market and can be easily identified by the "low-salt" message on the label (Illustration 23.17). Terms used to identify low-salt (or low-sodium) foods are defined by the Food and Drug Administration. To be considered low-sodium, foods must contain 140 milligrams or less of sodium per serving. Food manufacturers must adhere to the definitions when they make claims about the salt or sodium content of a food on the label.

Illustration 23.15
Recommendations for Preventing Hypertension*

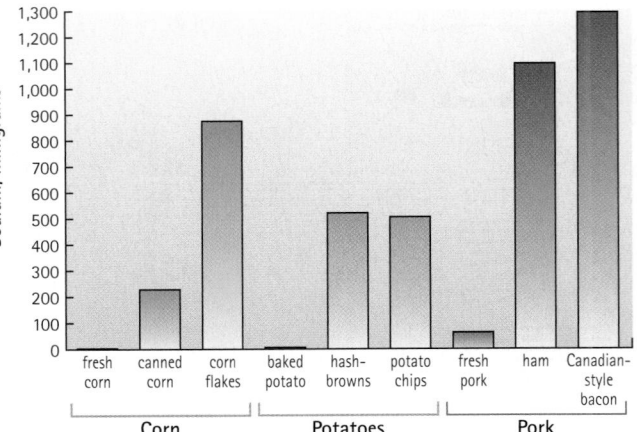

Illustration 23.16
Examples of how processing increases the sodium content of foods. *Sodium values are for a 3-ounce serving of each food shown.*

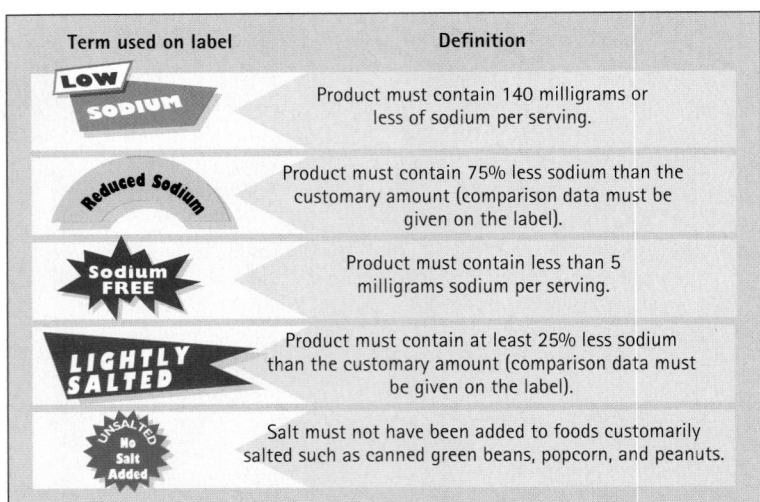

Illustration 23.17
When a label makes a claim about the sodium content of a food product, the label must list the amount of sodium in a serving and adhere to these definitions.

This unit ends having presented the story behind 3 of the 15 essential minerals. Much more information could have been relayed about these three, not to mention the wealth of knowledge that could be shared about minerals such as phosphorus, magnesium, zinc, iodine, selenium, and potassium. Critical information about these and other essential minerals is given in Table 23.1, and Table 23.12 lists food sources of many of the essential minerals. For additional information, look at the comprehensive nutrition books in your college library or take notes on the information presented by your instructor in class.

TABLE 23.12

FOOD SOURCES OF MINERALS.

MAGNESIUM

FOOD	AMOUNT	MAGNESIUM (mg)	FOOD	AMOUNT	MAGNESIUM (mg)
Legumes:			**Vegetables:**		
Lentils, cooked	½ cup	134	Bean sprouts	½ cup	98
Split peas, cooked	½ cup	134	Black-eyed peas	½ cup	58
Tofu	½ cup	130	Spinach, cooked	½ cup	48
			Lima beans	½ cup	32
Nuts:					
Peanuts	¼ cup	247	**Milk and milk products:**		
Cashews	¼ cup	93	Milk	1 cup	30
Almonds	¼ cup	80	Cheddar cheese	1 oz	8
			American cheese	1 oz	6
Grains:					
Bran buds	1 cup	240	**Meats:**		
Wild rice, cooked	½ cup	119	Chicken	3 oz	25
Breakfast cereal, fortified	1 cup	85	Beef	3 oz	20
Wheat germ	2 tbs	45	Pork	3 oz	20

TABLE 23.12

FOOD SOURCES OF MINERALS. (CONTINUED)

CALCIUM*

FOOD	AMOUNT	CALCIUM (mg)
Milk and milk products:		
Yogurt, low-fat	1 cup	413
Milk shake		
(low-fat frozen yogurt)	1¼ cup	352
Yogurt with fruit, low-fat	1 cup	315
Skim milk	1 cup	301
1% milk	1 cup	300
2% milk	1 cup	298
3.25% milk (whole)	1 cup	288
Swiss cheese	1 oz	270
Milk shake (whole milk)	1¼ cup	250
Frozen yogurt, low-fat	1 cup	248
Frappuccino	1 cup	220
Cheddar cheese	1 oz	204
Frozen yogurt	1 cup	200
Cream soup	1 cup	186
Pudding	½ cup	185
Ice cream	1 cup	180
Ice milk	1 cup	180
American cheese	1 oz	175
Custard	½ cup	150
Cottage cheese	½ cup	70
Cottage cheese, low-fat	½ cup	69
Vegetables:		
Spinach, cooked	½ cup	122
Kale	½ cup	47
Broccoli	½ cup	36
Legumes:		
Tofu	½ cup	260
Dried beans, cooked	½ cup	60
Foods fortified with calcium:		
Orange juice	1 cup	350
Frozen waffles	2	300
Soy milk	1 cup	200–400
Breakfast cereals	1 cup	150–1000

Actually, the richest source of calcium is alligator meat; 3½ ounces contain about 1231 milligrams of calcium; but just try to find it on your grocer's shelf!

SELENIUM

FOOD	AMOUNT	SELENIUM (mg)
Seafood:		
Lobster	3 oz	66
Tuna	3 oz	60
Shrimp	3 oz	54
Oysters	3 oz	48
Fish	3 oz	40
Meats/Eggs:		
Liver	3 oz	56
Egg	1 medium	37
Ham	3 oz	29
Beef	3 oz	22
Bacon	3 oz	21
Chicken	3 oz	18
Lamb	3 oz	14
Veal	3 oz	10

ZINC

FOOD	AMOUNT	ZINC (mg)
Meats:		
Liver	3 oz	4.6
Beef	3 oz	4.0
Crab	½ cup	3.5
Lamb	3 oz	3.5
Turkey ham	3 oz	2.5
Pork	3 oz	2.4
Chicken	3 oz	2.0
Legumes:		
Dried beans, cooked	½ cup	1.0
Split peas, cooked	½ cup	0.9
Grains:		
Breakfast cereal, fortified	1 cup	1.5–4.0
Wheat germ	2 tbs	2.4
Oatmeal, cooked	1 cup	1.2
Bran flakes	1 cup	1.0
Brown rice, cooked	½ cup	0.6
White rice	½ cup	0.4
Nuts and seeds:		
Pecans	¼ cup	2.0
Cashews	¼ cup	1.8
Sunflower seeds	¼ cup	1.7
Peanut butter	2 tbs	0.9
Milk and milk products:		
Cheddar cheese	1 oz	1.1
Whole milk	1 cup	0.9
American cheese	1 oz	0.8

continued

TABLE 23.12

FOOD SOURCES OF MINERALS. (CONTINUED)

SODIUM*

FOOD	AMOUNT	SODIUM (mg)
Miscellaneous:		
Salt	1 tsp	2132
Dill pickle	1 (4½ oz)	1930
Sea salt	1 tsp	1716
Chicken broth	1 cup	1571
Ravioli, canned	1 cup	1065
Spaghetti with sauce, canned	1 cup	955
Baking soda	1 tsp	821
Beef broth	1 cup	782
Gravy	¼ cup	720
Italian dressing	2 tbs	720
Pretzels	5 (1 oz)	500
Green olives	5	465
Pizza with cheese	1 wedge	455
Soy sauce	1 tsp	444
Cheese twists	1 cup	329
Bacon	3 slices	303
French dressing	2 tbs	220
Potato chips	1 oz (10 pieces)	200
Catsup	1 tbs	155
Meats:		
Corned beef	3 oz	808
Ham	3 oz	800
Fish, canned	3 oz	735
Meat loaf	3 oz	555
Sausage	3 oz	483
Hot dog	1	477
Fish, smoked	3 oz	444
Bologna	1 oz	370
Milk and milk products:		
Cream soup	1 cup	1070
Cottage cheese	½ cup	455
American cheese	1 oz	405
Cheese spread	1 oz	274
Parmesan cheese	1 oz	247
Gouda cheese	1 oz	232
Cheddar cheese	1 oz	175
Skim milk	1 cup	125
Whole milk	1 cup	120
Grains:		
Bran flakes	1 cup	363
Cornflakes	1 cup	325
Croissant	1 medium	270
Bagel	1	260
English muffin	1	203
White bread	1 slice	130
Whole-wheat bread	1 slice	130
Saltine crackers	4 squares	125

IRON

FOOD	AMOUNT	IRON (mg)
Meat and meat alternates:		
Liver	3 oz	7.5
Round steak	3 oz	3.0
Hamburger, lean	3 oz	3.0
Baked beans	½ cup	3.0
Pork	3 oz	2.7
White beans	½ cup	2.7
Soybeans	½ cup	2.5
Pork and beans	½ cup	2.3
Fish	3 oz	1.0
Chicken	3 oz	1.0
Grains:		
Breakfast cereal, iron-fortified	1 cup	8.0 (4–18)
Oatmeal, fortified, cooked	1 cup	8.0
Bagel	1	1.7
English muffin	1	1.6
Rye bread	1 slice	1.0
Whole-wheat bread	1 slice	0.8
White bread	1 slice	0.6
Fruits:		
Prune juice	1 cup	9.0
Apricots, dried	½ cup	2.5
Prunes	5 medium	2.0
Raisins	¼ cup	1.3
Plums	3 medium	1.1
Vegetables:		
Spinach, cooked	½ cup	2.3
Lima beans	½ cup	2.2
Black-eyed peas	½ cup	1.7
Peas	½ cup	1.6
Asparagus	½ cup	1.5

TABLE 23.12

FOOD SOURCES OF MINERALS. (CONTINUED)

PHOSPHOROUS

FOOD	AMOUNT	PHOSPHOROUS (mg)
Milk and milk products:		
Yogurt	1 cup	327
Skim milk	1 cup	250
Whole milk	1 cup	250
Cottage cheese	½ cup	150
American cheese	1 oz	130
Meats:		
Pork	3 oz	275
Hamburger	3 oz	165
Tuna	3 oz	162
Lobster	3 oz	125
Chicken	3 oz	120
Nuts and seeds:		
Sunflower seeds	¼ cup	319
Peanuts	¼ cup	141
Pine nuts	¼ cup	106
Peanut butter	1 tbs	61
Grains:		
Bran flakes	1 cup	180
Shredded wheat	2 large biscuits	81
Whole-wheat bread	1 slice	52
Noodles, cooked	½ cup	47
Rice, cooked	½ cup	29
White bread	1 slice	24
Vegetables:		
Potato	1 medium	101
Corn	½ cup	73
Peas	½ cup	70
French fries	½ cup	61
Broccoli	½ cup	54
Other:		
Milk chocolate	1 oz	66
Cola	12 oz	51
Diet cola	12 oz	45

POTASSIUM

FOOD	AMOUNT	POTASSIUM (mg)
Vegetables:		
Potato	1 medium	780
Winter squash	½ cup	327
Tomato	1 medium	300
Celery	1 stalk	270
Carrots	1 medium	245
Broccoli	½ cup	205
Fruits:		
Avocado	½ medium	680
Orange juice	1 cup	469
Banana	1 medium	440
Raisins	¼ cup	370
Prunes	4 large	300
Watermelon	1 cup	158
Meats:		
Fish	3 oz	500
Hamburger	3 oz	480
Lamb	3 oz	382
Pork	3 oz	335
Chicken	3 oz	208
Grains:		
Bran buds	1 cup	1080
Bran flakes	1 cup	248
Raisin bran	1 cup	242
Wheat flakes	1 cup	96
Milk and milk products:		
Yogurt	1 cup	531
Skim milk	1 cup	400
Whole milk	1 cup	370
Other:		
Salt substitutes	1 tsp	1300–2378

Key Terms

cofactors, page 23-3

essential hypertension, page 23-19

hemoglobin, page 23-15

hypertension, page 23-18

iron deficiency, page 23-16

iron-deficiency anemia, page 23-16

minerals, page 23-2

myoglobin, page 23-15

osteoporosis, page 23-5

remodeling, page 23-5

salt sensitivity, page 23-19

water balance, page 23-18

www links

www.nichd.nih.gov/milk/
Information on the benefits of calcium, and a "Kids and Teens" page with interactive tools for learning about calcium are available from this site.

www.nal.usda.gov/fnic
It's a quick trip from this site to information on minerals, osteoporosis, hyperten-

sion, iron deficiency, and other topics covered in this unit.

www.nhlbi.nih.gov/
NIH's Web site on hypertension includes current research and recommendations on prevention and treatment.

www.nof.org
Web address for the National Osteoporosis Foundation.

Notes

1. Mitchell SJ. Changes after taking a college basic nutrition course. J Am Diet Assoc 1990;90:955–61.

2. Parvanta I. The hidden hunger: micronutrient malnutrition. Centers for Disease Control, www.cdc.gov/epo/mmwr/preview/mmwrhtml/00051880.htm, March 2000.

3. Hajjar I et al. Incidence of hypertension in the United States. JAMA 2003; 290:199–206.

4. Prasad AS. Discovery and importance of zinc in human nutrition. Federation Proc 1984;43:2829–34.

5. Monsen ER, Balintfy JL. Calculating dietary iron bioavailability: refinement and computerization. J Am Diet Assoc 1982;80:307–11.

6. Follin SL, Hansen LB, Current approaches to the prevention and treatment of postmenopausal osteoporosis, Am J Health—Syst Pharm 2003;60: 883–901; and Prestwood KM, Raisz LG, Prevention and treatment of osteoporosis, Clin Cornerstone 2002; 4:31–41.

7. Follin and Hansen, Current approaches.

8. Follin and Hansen, Current approaches.

9. Follin and Hansen, Current approaches.

10. Maximizing peak bone mass: calcium supplementation increases bone mineral density in children. Nutr Rev 1992;50: 335–7.

11. Committee on Diet and Health (Food and Nutrition Board; National Research Council). Diet and health: implications for reducing chronic disease risk. Washington, DC: National Academy Press; 1989.

12. Follin and Hansen, Current approaches; and Prestwood and Raisz, Prevention and treatment.

13. Harris S. Calcium trials. Experimental Biology Annual Meeting, San Diego, Calif. 2000 Apr. 4.

14. Looker AC. Interaction of science, consumer practices, and policy: calcium and bone health as a case study. J Nutr 2003;133:1987S–91S; and Prestwood and Raisz, Prevention and treatment.

15. Follin and Hansen, Current approaches.

16. Milk matters, J Am Diet Assoc 2002; 102:471; and What we eat in America, Nutr Today, 1997;Jan/Feb:37–40.

17. Bone builders. FDA Consumer, 1997; Sept/Oct.

18. Wyshak G. Consumption of carbonated drinks by teenage girls associated with bone fractures, Arch Pediatr Adol Med 2000;154:542–3, 610–13; and Tucker K et al, Cola consumption and bone mineral density in women, Medscape Medical News. 2003, www.medscape.com, accessed 10/03.

19. Lewis NM, Hollingsworth M. Food choices of young college women consuming low- or moderate-calcium diets. Nutr Res 1992;12:843–8.

20. Follin and Hansen, Current approaches; Prestwood and Raisz, Prevention and treatment.

21. Prestwood and Raisz, Prevention and treatment.

22. Rapuri PB et al. Protein intake: effects on bone mineral density and the rate of bone loss in elderly women. Am J Clin Nutr 2003;77:1517–25.

23. New SA. Osteoporosis—ask the experts: dairy and bone health. Medscape Ob/Gyn & Women's Health 2003;8(2).

24. New, Osteoporosis—Ask the experts.

25. New, Osteoporosis—Ask the experts.

26. Rapuri et al., Protein intake; and Martini L, Wood RJ, Relative bioavailability of calcium-rich dietary sources in the elderly, Am J Clin Nutr 2002;76: 1345–50.

27. Follin and Hansen, Current approaches.

28. Klausner A. EN's guide to calcium in unexpected places. Envir Nutr 1999; May:5.

29. Looker, Interaction of science.

30. Parvanta, The hidden hunger; and Committee on Diet and Health, Diet and health.

31. Parvanta, The hidden hunger.

32. Parvanta, The hidden hunger.

33. Fairbanks VF. Iron in medicine and nutrition. In: Modern nutrition in health and disease, Shils ME et al, eds. 9th ed.

Nutrition UP CLOSE

The Salt of Your Diet

FOCAL POINT: Estimate your sodium intake.

Salt added to foods during processing, during home cooking, and at the table is the major source of sodium in the American diet. To determine if you consume too much sodium, answer the following questions.

How Often Do You Usually:	Daily	4–6 Times per Week	1–3 Times per Week	Less than Weekly
1. Eat processed foods such as luncheon meats, bacon, hot dogs, sausage, canned soups, broths, gravy, TV dinners, or smoked fish?	☐	☐	☐	☐
2. Eat pickles, green olives, potato or tortilla chips, pretzels, or salted crackers?	☐	☐	☐	☐
3. Use soy sauce, garlic salt, steak sauce, or catsup in recipes or on foods?	☐	☐	☐	☐
4. Salt foods at the table?	☐	☐	☐	☐
5. Add salt to cooking water or to foods you are preparing?	☐	☐	☐	☐

FEEDBACK (including scoring) can be found at the end of Unit 23.

Source: Adapted from Limit use of sodium. Washington, DC: USDA Home and Garden Publication 232–4; 1986.

Philadelphia: Lippincott Williams & Wilkins, 1999: pp. 193–221.

34. Fairbanks, Iron in medicine and nutrition; and Hurtado EK et al. Early childhood anemia and mild or moderate mental retardation. Am J Clin Nutr 1999; 69:115–9.

35. Park J, Brittin HC. Increased iron content of food due to stainless steel cookware. J Am Diet Assoc 1997;97: 659–61.

36. Monsen ER et al. Estimation of available dietary iron. Am J Clin Nutr 1978;31:134–41.

37. Monsen et al., Estimation of available dietary iron.

38. Shannon M. Ingestion of toxic sub-

stances by children. N Engl J Med 2000;342:186–89.

39. Trinkoff AM, Baker SP. Poisoning hospitalizations and deaths from solids and liquids among children and teenagers. Am J Public Health 1986;76:657–60.

40. National Research Council Subcommittee on Iron (Committee on Medical and Biological Effects of Environmental Pollutants). Iron. Washington, DC: Division of Medical Sciences, Assembly of Life Sciences, National Academy of Sciences; 1979.

41. Messerli RE et al. Salt: a perpetrator of hypertensive target organ disease? Arch Int Med 1997;157:2449–52.

42. NHLBI issues new high blood pressure clinical practice guidelines. NHLBI Communications Office, 5/14/03, www.nhlbi.nih.gov/index.htm.

43. Messerli et al., Salt; August P, Initial treatment of hypertension, N Engl J Med 2003;348:610–7; Chobanian AV, Control of hypertension—an important national priority, N Engl J Med 2001; 345:534–5; and Committee on Diet and Health, Diet and health.

44. August, Initial treatment of hypertension.

45. NHLBI issues new high blood pressure clinical practice guidelines (www.nhlbi. nih.gov.index.htm).

46. Law MR et al., By how much does dietary salt reduction lower blood

pressure? I. Analysis of observational data among populations. Br Med J 1991;302:811–15; Korhonen MH et al., Effects of a salt restricted diet on the intake of other nutrients, Am J Clin Nutr 2000;72:414–20; and Johnson RJ et al. Subtle acquired renal injury as a mechanism of salt-sensitive hypertension. N Engl J Med 2002;346:913–23.

47. Kaplan NM. The dietary guideline for sodium: should we shake it up? No. Am J Clin Nutr 2000;71:1020–6; NHLBI Statement on sodium intake and high blood pressure. National Heart, Lung, and Blood Institute, NIH, www.nhlbi.nih.gov/nhlbi, Aug. 1998; and NHLBI issues new high blood pressure clinical practice guidelines (www.nhlbi.nih.gov.index.htm).

48. Kaplan, The dietary guideline; Committee on Diet and Health, Diet and health; and The Trials of Hypertension

Prevention Collaborative Research Group: the effects of nonpharmacologic intervention on blood pressure of persons with high normal levels, Results of the trials of hypertension prevention, phase 1. JAMA 1992; 267:1213–20.

49. NHLBI issues new high blood pressure clinical practice guidelines (www.nhlbi .nih.gov.index.htm); and Caulin-Glaser T, Primary prevention of hypertension in women, J Clin Hypertens 2000;2: 204–9, 214.

50. NHLBI issues new high blood pressure clinical practice guidelines (www.nhlbi .nih.gov.index.htm).

51. NHLBI issues new high blood pressure clinical practice guidelines (www.nhlbi .nih.gov.index.htm).

52. Conlin PR et al., DASH diet can control stage 1 hypertension, Am J Hyper-

tens 2000;13:949–55; and Sacks FM et al., Effects on blood pressure of reduced dietary sodium and the Dietary Approaches to Stop Hypertension (DASH) diet, N Engl J Med 2001; 344:3–10.

53. Karanja NM et al. Descriptive characteristics of the dietary patterns used in the Dietary Approaches to Stop Hypertension trial. J Am Diet Assoc 1999; 99(suppl):S19–S27.

54. Sacks et al., Effects on blood pressure; and NHLBI issues new high blood pressure clinical practice guidelines (www.nhlbi.nih.gov.index.htm).

55. Mattes RD, Donnelley D. Relative contribution of dietary sodium sources. J Am Coll Nutr 1991;10:383–93.

56. Law et al., By how much does dietary salt reduction lower blood pressure?

Nutrition UP CLOSE

The Salt of Your Diet

Feedback for Unit 23

People in the United States generally consume two to three times more sodium than they need, and you may be doing so too, if you answered "daily" or "4–6 times per week" to several of the questions. If you wish to cut back on sodium, here are some suggestions:

- Choose fresh instead of processed foods more often.
- Read labels to check for sodium content.
- Reduce salt in cooking and at the table.
- When you do choose processed foods, try the lower-sodium versions available.
- To enhance the flavor of foods, try herbs and spices instead of salt.

Dietary Supplements and Functional Foods

Nutrition Scoreboard

		TRUE	FALSE
1	Products classified as "dietary supplements" consist of herbs and vitamins and mineral supplements.		
2	Dietary supplements must be tested for safety and effectiveness before they can be sold.		
3	Herbal remedies have been used for over 100 years in Germany so they must be safe and effective.		
4	"Probiotics" are "friendly bacteria" that have positive effects on health.		

Answers on next page

[KEY CONCEPTS AND FACTS]

- Dietary supplements include vitamin and mineral pills, herbal remedies, proteins and amino acids, fish oils, and other products.

- Dietary supplements do not have to be shown to be safe or effective prior to being sold.

- Although food is the preferred source of vitamins and minerals, certain people benefit from judiciously selected vitamin or mineral supplements.

Answers to *Nutrition Scoreboard*

		TRUE	FALSE
1	Dietary supplements include herbs, vitamin and mineral supplements, protein powders, amino acid and enzyme pills, fish oils and fatty acids, hormone extracts, and other products.		✔
2	Dietary supplements can be sold without proof of their safety or effectiveness.	✔	
3	Results of clinical trials, and not historical use, are the gold standard for determining the safety and effectiveness of herbal remedies and other dietary supplements.	✔	
4	Yes! There are healthful bacteria in a number of foods.	✔	

dietary supplements

Any products intended to supplement the diet, including vitamin and mineral supplements; proteins, enzymes, and amino acids; fish oils and fatty acids; hormones and hormone precursors; and herbs and other plant extracts. Such products must be labeled "Dietary Supplement."

What do vitamin E supplements, amino acid pills, and herbal remedies all have in common? They are members of the increasingly popular group of products called *dietary supplements*—and they are all discussed in this unit. Types of dietary supplements available to consumers are presented in Table 24.1. Over half of U.S. adults use one or more of these products.[1] Dietary supplements are supposed to "supplement the diet." Many do that and have beneficial health effects in some people, but others neither supplement the diet nor provide health benefits. These disparate products are grouped together because they are regulated by a common set of rules.

TABLE 24.1

TYPES OF DIETARY SUPPLEMENTS.

TYPE	EXAMPLE
1. Vitamins and minerals	Vitamins C and E, selenium
2. Herbs (botanicals)	Dong quai, ginseng, saw palmetto
3. Proteins and amino acids	Shark cartilage, chondroitin, creatine
4. Hormones, hormone precursors	DHEA, "Andro"
5. Fats	Fish oils, DHA, lecithin
6. Other plant extracts	Garlic capsules, fiber, cranberry concentrate, bee pollen

Photo Disc

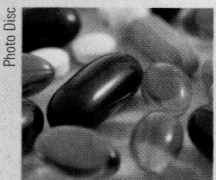

Vitamin, mineral, protein, and amino acid supplements.

Botanical supplements, such as dong quai, often come in gel caps.

Plant extracts can also be taken in liquid form, as tinctures.

Herbal teas are a familiar source of dietary supplements.

You will learn from this unit that taking some of the dietary supplements available on the market can be a gamble. This "buyer beware" situation exists because of the loose rules that govern dietary supplements. You will also learn about "functional foods" and some exciting, new developments related to intestinal fertilizers and friendly bacteria (no kidding).

Regulation of Dietary Supplements

You can call anything a dietary supplement,
even something you grow in your back yard.
—Donna Porter, RD, PhD, Congressional Research Service

In 1994 Congress passed the Dietary Supplement Health and Education Act, which started the explosion in the availability of dietary supplements. Under the act, dietary supplements are minimally regulated by the Food and Drug Administration (FDA); they do not have to be tested prior to marketing or shown to be safe or effective.[2] Although often advertised to relieve certain ailments, they are not considered to be drugs. Consequently, dietary supplements are not subjected to vigorous testing to prove safety and effectiveness, like drugs must be. Responsibility for evaluating the safety of dietary supplements lies with manufacturers and not the FDA. Supplements are deemed unsafe when the FDA has proof they are harmful. Since few dietary supplements have been adequately tested, and because results of studies showing negative effects may never see the light of day,[3] it is difficult to prove them to be unsafe. The FDA largely relies on reports of ill effects from manufacturers, health professionals, and consumers to assess supplement safety. Between 1993 and 2000, the FDA received over 2800 reports of adverse effects of supplements (primarily for herbs), including 105 deaths.[4]

According to FDA regulations (Table 24.2), dietary supplements must be labeled with a "Supplemental Facts" panel that lists serving size, ingredients, and percent

---

TABLE 24.2

FDA REGULATIONS FOR DIETARY SUPPLEMENT LABELING.

1. Product must be labeled "Dietary Supplement."

2. Product must have a "Supplemental Facts" label that includes serving size, amount of the product per serving, % Daily Value of essential nutrients, a list of other ingredients, and the manufacturer's name and address.

3. Health claims made on the label (e.g., "high fiber" or "low salt") must be justified based on nutrition labeling regulations established for foods.

4. Structure claims (such as those listing the parts of a plant used or a particular form of a vitamin) and function claims (e.g., "promotes normal bowel function") can be made on product labels. If a function claim is made, this FDA disclaimer must appear:

This statement has not been evaluated by the FDA. This product is not intended to diagnose, treat, cure, or prevent any disease.

Nutrition Facts
Serving size 1 Tablet

Amount Per Serving	% DV
Melatonin 3 mg	*

*Daily Value (DV) not established

Other Ingredients: Dicalcium Phosphate, Cellulose (Plant Origin), Vegetable Stearic Acid, Vegetable Magnesium Stearate, Silica, Croscarmellose.

GUARANTEED FREE OF: wheat, yeast, soy, corn, sugar, starch, milk, eggs. No artificial colors, flavors. No chemical additives. No preservatives. No animal derivatives.

KEEP OUT OF REACH OF CHILDREN

Directions: As a dietary supplement for adults, take one (1) tablet, under the direction of a physician, only at bedtime as Melatonin may produce drowsiness. **DO NOT EXCEED 3 MG IN A 24 HOUR PERIOD.**

Warning: For Adults. Use only at bedtime. This product is not to be taken by pregnant or lactating women. If you are taking medication or have a medical condition such as an auto-immune condition or a depressive disorder, consult your physician before using this product. **NOT FOR USE BY CHILDREN 16 YEARS OF AGE OR YOUNGER.** Do not take this product when driving a motor vehicle, operating machinery or consuming alcoholic beverages.

In case of accidental overdose, seek professional assistance or contact a Poison Control Center immediately.

Daily Value (%DV) of essential nutrient ingredients. Products can be labeled with a health claim, such as "high in calcium" or "low fat," if the product qualifies according to the nutrition labeling regulations. Supplements can also be labeled with "structure/function" claims. These are not subject to regulation, with the exception that they cannot refer to disease prevention or treatment effects. Claims such as "improves circulation," "supports the immune system," and "helps maintain mental health" can be used, whereas "prevents heart disease" or "cures depression" cannot be. If a function claim is made on the label or package inserts, the label or insert must include the FDA disclaimer that states the FDA does not support the claim. (This is done to reduce the FDA's liability for problems that may be caused by supplements.) Nonetheless, many people wholeheartedly believe health claims made for supplements.[5]

The Federal Trade Commission (FTC) regulates claims for dietary supplements made in print and broadcast advertisements, including direct marketing, Internet sites, and infomercials. Claims made for dietary supplements in advertisements are supposed to be truthful, but often are not.[6] Although some companies have been prosecuted for making false and misleading claims, neither the FDA nor the FTC has sufficient resources to fully monitor products and enforce laws related to dietary supplements.[7] The FDA has recently developed an "Adverse Events Reporting System" Web site (www.cfsan.fda.gov) that simplifies recording and tracking of adverse effects of dietary supplements.

Vitamin and Mineral Supplements

Multivitamin and mineral supplements—such as vitamins E or C, calcium, or magnesium—are among the wide variety of vitamin and mineral supplements used by consumers. They represent the most popular type of dietary supplement, with approximately one-third of U.S. adults using them daily.[8] Intake levels of vitamins and minerals below the "Tolerable Upper Limits" of the Dietary Reference Intake values (given on the inside front cover of this book) are safe for the vast majority of people, although lower amounts are related to their optimal functioning in the body.[9] Certain concerns related to vitamin and mineral supplements exist. The *bioavailability* of nutrients contained in some supplements remains uncertain. In general, the wider the assortment of vitamins and minerals in a supplement, the lower the absorption of each.[10] Minerals are particularly prone to forming unabsorbable complexes with each other, reducing the bioavailability of multiple minerals in the supplement.

bioavailability
The amount of a nutrient consumed that is available for absorption and use by the body.

Vitamin and Mineral Supplements: Who Benefits?

Supplements can have positive effects on health, as shown by the examples given in Table 24.3. Vitamin and mineral supplements have come a long way from primarily being used to treat vitamin and mineral deficiency diseases.

Using Vitamin and Mineral Supplements for the Wrong Reason] People often take supplements as a sort of insurance policy against problems caused by poor diets. Although multivitamin and mineral supplements may help fill in some of the nutrient gaps caused by poor food habits, they can't make a bad diet good. Whether a diet is good or bad is determined by more than its vitamin and mineral content. The "goodness" of a diet also depends on its content of essential fatty acid, fiber, water, and other nutrients. Additionally, supplements do not provide phytochemicals found in food such as flavones and antioxidant pigments that have positive influences on health.

One of the most serious consequences of supplements results when they are used as a remedy for health problems that can be treated, but not by vitamins or minerals. Vitamin and mineral supplements have not been found to be an effective treatment for behavioral problems, diabetes, autism, chronic fatigue syndrome, obesity, or stress, for example.

The Rational Use of Vitamin and Mineral Supplements] Like all medications, vitamin and mineral supplements should be taken only if there is a need for them. If they are taken, dosages should not be excessive. Guidelines for the selection and use of vitamin and mineral supplements can be found in the "Health Action" below.

Herbal Remedies

A weed is what we call a plant whose virtues have not yet been discovered.
—Ralph Waldo Emerson

Herbal remedies have a long history, are used worldwide, and are moving from alternative to mainstream health care in North America. Approximately 49% of U.S. adults use herbal supplements each year.[12]

The herb pharmacopoeia includes over 550 primary herbs known by at least 1800 names. Approximately 30% of all modern drugs are derived from plants,[13] and there's no doubt that many additional plants and plant ingredients benefit health. Plant products known to treat disease are considered drugs, however. Those that have not passed the scientific tests needed to demonstrate safety and effectiveness in disease treatment are often considered herbs. Yet, the truth is that some products sold as herbs in the United States have drug-like effects on body functions. An example of an herb that acts powerfully like a drug is given in Illustration 24.1.

TABLE 24.3
WHO MAY BENEFIT FROM VITAMIN AND MINERAL SUPPLEMENTS? HERE ARE SOME EXAMPLES:[11]

- People with vitamin and mineral deficiency diseases
- Newborns (vitamin K)
- People living in areas without a fluoridated water supply (fluoride)
- Vegans (vitamins B_{12} and D)
- Pregnant women (iron and folate)
- People experiencing blood loss (iron)
- Elderly persons on limited diets (multiple vitamins and minerals)
- People on restricted diets (multiple vitamins and minerals)
- Adults at risk for heart disease and cancer (folic acid, vitamin C)
- People at risk for osteoporosis due to low calcium intake and poor vitamin D status (calcium, vitamin D)
- People with alcoholism
- People being treated for depression (folic acid)

HEALTH ACTION I *Guidelines for Choosing and Using Vitamin and Mineral Supplements*

1. Purchase supplements labeled "USP."
   - Terms such as "release assured," "laboratory tested," "quality tested," and "scientifically blended" on supplement labels guarantee nothing.
2. Check the expiration date on supplements. Use unexpired supplements.

3. Choose supplements containing 100% of the Daily Value or less.
4. Take supplements with meals.
5. If you have a diagnosed need for a specific vitamin or mineral, take that individual vitamin or mineral and not a multiple supplement.
   - Avoid calcium supplements made from oyster shells or

bone. They may contain lead or aluminum.
6. Store supplements where small children cannot get at them.

REMEMBER! Consult your health care provider about health problems before you start taking supplements to try to treat the problems.

Illustration 24.1 Herb or drug?

The line between food, supplements, and drugs is so blurry these days that it's hard to know what any product is.
—Donna Porter, Congressional Research Service

Statins are a popular type of prescription drug that effectively lower LDL cholesterol and raise HDL cholesterol (the good one), and reduce the risk of heart disease. It turns out that statins are also available in herbal form.

Cholestin is an "herb" made from rice grown with red yeast. If subjected to specific growing conditions, the concentration of naturally occurring statins in the rice-yeast mixture can be increased greatly. Cholestin works like prescription statins, costs less, and is widely available to the public. Because cholestin is marketed as an herb rather than a drug, there is more room for the product to be mislabeled, contaminated, or unsafe, and to contain less active ingredients than claimed on the label.[14]

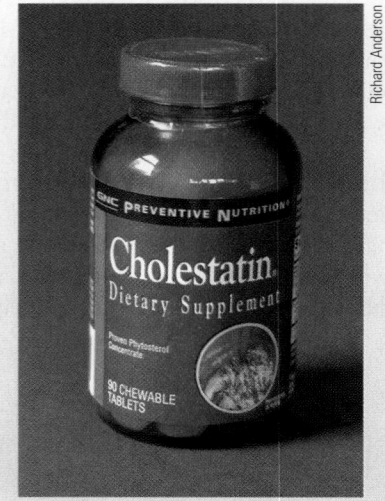

Effects of Herbal Remedies

Purported effects and side effects of some herbal and similar remedies are listed in Table 24.4. Herbal remedies, like drugs, have biologically active ingredients that can have positive, negative, and neutral effects on body processes. Basically, an herbal remedy (or a drug) is considered valuable if it has beneficial effects on body processes, and if the benefits are not outweighed by the risks. Knowledge of the risks and benefits of many herbal supplements remains incomplete. However, available evidence suggests that some herbal remedies are safe and effective while others appear to be neither.

Which herbal remedies are likely ineffective or unsafe? Human experimentation with various botanicals over the centuries to the present time has helped to identify herbs that lack beneficial effects or have negative side effects. Table 24.5 lists some of these herbs. The extent to which the herbs included in the table pose a risk to health depends on the amount taken and the duration of use, the age and health status of the user, and other factors.

**Illustration 24.2
Rose hips in Tuscany.**

Which Herbal Supplements Are Potentially Beneficial?] Results of many studies on the safety and benefits of specific herbal remedies have been mixed. The bulk of existing evidence, however, indicates that some may be beneficial:

- Ginkgo biloba may decrease symptoms in patients with Alzheimer's disease and increase blood flow to the brain.
- Echinacea may decrease the duration of colds.
- Cranberry may relieve urinary tract infection.
- Garlic can reduce elevated blood cholesterol levels and blood pressure.
- Saint John's wort may relieve mild and moderate depression.
- Rose hips have mild laxative and diuretic effects (Illustration 24.2).
- SAMe may relieve mild depression and arthritis pain.
- Saw palmetto may reduce frequent bouts of a need to urinate and improve urine flow in older men.
- Senna may relieve constipation.
- Ginger may prevent motion sickness.[17]

TABLE 24.4

PROPOSED EFFECTS AND POTENTIAL SIDE EFFECTS OF HERBAL AND SIMILAR REMEDIES.[15]

HERB/OTHER REMEDY	PROPOSED EFFECTS	POTENTIAL SIDE EFFECTS
Glucosamine-chondroitin sulfate	Slows progression of osteoarthritis and its pain	Gastrointestinal upset, fatigue, headache
Ginseng	Increases energy, normalizes blood glucose, stimulates immune function, relieves impotence in males, cancer prevention	Insomnia, hyperactivity, hypertension, diarrhea, menstrual dysfunction Interacts with blood thinners*
SAMe	Relieves mild depression, arthritis pain	May trigger manic excitement, nausea
Garlic	Lowers blood cholesterol, relieves colds and other infections	Heartburn, gas, blood thinner
Cholestin	Maintains desirable blood cholesterol levels	Safety of some ingredients unknown
Echinacea	Prevents and treats colds and sore throat	Allergies to plant components
DHEA	Improves memory, mood, physical well-being	Increases risk of breast cancer
Creatine	Sport supplement (increased performance in short, high-intensity events)	Kidney disease
Saw palmetto	Improves urine flow, reduces urgency of urination in men with prostate enlargement	Nausea, abdominal pain
Ginkgo biloba	Increases mental skills, delays progression of Alzheimer's disease, increases blood flow and sexual performance, decreases depression	Nervousness, headache, diarrhea, nausea. Interacts with blood thinners.
Shark cartilage	Treats lung cancer	Safety unknown
Saint John's wort	Relieves mild-to-moderate depression	Dry mouth, dizziness, sensitivity to light Interacts with many drugs and chemotherapy
Ephedra (Ma Huang)	Promotes weight loss, improves respiratory illnesses	Insomnia, headaches, nervousness, seizures, death
Kava (kava-kava)	Relaxation, stress relief, sleep aid, mood enhancer	Liver injury
Black cohosh	Improves menopausal and PMS symptoms	Gastric upset, dizziness, headache, low blood pressure. May increase risk of breast cancer. Interacts with anti-hypertension drugs
Coenzyme Q10 (ubiquinone)	Remedy for heart disease, cancer, Parkinson's disease	Nausea, diarrhea, rash, low blood glucose Interacts with some drugs

*Blood thinners include aspirin, warfarin, coumarin.

TABLE 24.5

EXAMPLES OF DIETARY SUPPLEMENTS THAT MAY NOT BE EFFECTIVE OR SAFE.[16]

Apricot pits (laetrile)	Eyebright	Poke weed
Belladonna	Ginkgo seed	Sassafras
Blue cohosh	Licorice root	Skullcap
Borage	Liferoot	Star anise
Broom	Lily of the valley	Vinca
Chaparral	Ma huang/ephedra	Wild yam
Chinese yew	Mandrake	Wormwood
Comfrey	Mistletoe	Yohimbe
Dong quai	Pennyroyal oil	

Illustration 24.3
These symbols on the labels of dietary supplements certify quality ingredients and accurate labeling, but do not address product safety or effectiveness.

The list of effective herbal remedies may grow, or shrink, with time as additional studies are completed.

Ephedra (Ma Huang)] Ephedra, or Ma huang, is chemically and functionally similar to adrenaline, the "flight or fight" hormone. It is traditionally used in China to treat respiratory problems, but has gained popularity in the United States as a weight-loss product and athletic performance enhancer. Ephedra delivers limited benefits for these purposes, however, and can produce life-threatening side effects on the heart and nervous system. Side effects and deaths related to ephedra use led to its being banned from the market early in 2004.[18]

A Measure of Quality Assurance for Dietary Supplements

Not all dietary supplements contain the amounts of herbal and other ingredients declared on the label, and some contain contaminants such as bacteria, mold, and lead.[19] Analyses of the composition of 25 ginseng products, for example, found that concentrations of ginseng compounds in the supplements were up to 36 times different than labeled amounts.[20] Similar studies of echinacea products found that 10% of samples contained no echinacea, and half contained the labeled amount.[21] Some male "enlargement" supplements have been found to be contaminated with *E. coli*, mold, lead, and pesticide residues.[22] Oddly enough, dietary supplements are often labeled as "pure," "natural," or "quality assured."

There is no government body that monitors the contents of herbal supplements. Private groups, such as the U.S. Pharmacopeia (USP), the National Formulary (NF), and Consumer Laboratories (CL), offer testing services to ensure that dietary supplements meet standards for disintegration, purity, potency, and labeling (Illustration 24.3). Products that pass these tests can display "USP," "NF," or the CL symbol boldly on product labels. These letters represent quality ingredients and labeling, but do not address product safety or effectiveness. New regulations for dietary supplements are being considered by the FDA and other groups.[23]

Due to the lack of studies and potential dangers, the FDA has advised dietary supplement manufacturers not to make claims related to pregnancy for herbs and other products.[25] Considerations for the use of herbal supplements are summarized in the "Health Action" below.

HEALTH ACTION | *Considerations for the Use of Herbal Remedies*

1. Don't use herbal remedies for serious, self-diagnosed conditions such as depression, persistent headaches, and memory loss. (You might benefit more from a different treatment.)

2. Let your doctor know what herbal remedies you take.

3. If you take prescription medications, clear the use of herbal remedies with your doctor.

4. Don't use herbal remedies without medical advice if you are attempting to become pregnant or if you are pregnant or breastfeeding.

5. Don't mix herbal remedies.

6. If you are allergic to certain plants, make sure herbal remedies are not going to be a problem before you use them.

7. If you have a bad reaction to an

herbal remedy, stop using it and report the reaction to the FDA from the site www.cfsan.fda.gov.

8. Buy herbs labeled with "USP," "NF," or the CL in a beaker symbol.

9. Investigate brands, herb safety, and effectiveness by checking into one or more of the Web sites and resources listed at the end of this unit.

REALITY CHECK

Herbals on the Web

Can you trust information on herbal products you see on the Web?

Who gets thumbs up

?

Candy:
When I'm sick, the first place I go is to the Web to find an herb that will make me feel better.

Photo Disc

Yaroslav:
Herbal products I see advertised on the Web look like they'll work for my problem. But, I'm conflicted about buying them, because I'm not sure I can trust the information.

Answers on next page

Functional Foods

Leave no aisle unfortified.
—Headline for article in a food industry magazine

Also known as "neutraceuticals," **functional foods** include a variety of foods and products that have in reality or theoretically been modified to enhance their contribution to a healthy diet (Table 24.6).[26] All foods are functional in that they provide nutrients. Foods considered "functional," however, are generally specifically formulated to supply one or more dietary ingredients that may improve health, or they are foods containing high amounts of substances that tend to prevent certain diseases.[27] Because there is no statutory definition for what constitutes functional foods, there are no specific regulations that apply to them. Health claims can be made for functional foods given approval by the FDA.[28] Functional foods containing food additives not on the Generally Accepted As Safe (GRAS) list must be tested and approved by the FDA before being sold.

Functional foods that appear to benefit health are listed in Table 24.7. Increasingly, however, the list of functional foods is becoming infiltrated with sports bars, soups, beverages, and cereals spiked with vitamins, minerals, and herbs.

Some of these products carry labels with unsubstantiated health claims and may be of no benefit or are potentially unsafe.[30] For these products, the label "functional food" is a marketing term.

Prebiotics and Probiotics: From "Pharm" to Table

The terms *prebiotics* and *probiotics* were derived from "antibiotics" due to their probable effects on increasing resistance to various diseases. They are in a class of functional

functional foods
Generally taken to mean foods, fortified foods, and enhanced food products that may benefit health beyond the effects of essential nutrients they contain.

TABLE 24.6

HOW ARE FOODS MADE TO BE "FUNCTIONAL"?

Foods are made to be "functional" by:
1. Taking out potentially harmful components (e.g., cholesterol in egg yolk and lactose in milk)
2. Increasing the amount of nutrients and beneficial nonnutrients (e.g., fiber-fortified liquid meals, calcium-, and vitamin C–fortified orange juice)
3. Using beneficial substances in food production or products (e.g., using "friendly" bacteria in fermented milk and soy products.

Candy

Yaroslav

prebiotics

"Intestinal fertilizer." Certain fiber-like forms of nondigestible carbohydrates that support the growth of beneficial bacteria in the gut.

probiotics

"Pro-life." Strains of *lactobacillus* (lac-toe-bah-sil-us) and bifidobacteria (bif-id-dough bacteria) that have beneficial effects on the body. Also called "friendly bacteria."

foods by themselves. **Prebiotics** are fiber-like, nondigestible carbohydrates that are broken down by bacteria in the colon. The breakdown products foster the growth of beneficial bacteria. For this reason they are considered "intestinal fertilizer." **Probiotics** is the term for live, beneficial—or "friendly"—bacteria that enter food through fermentation and aging processes.[31] Table 24.8 lists food and other sources of pre- and probiotics. Availability of foods and other products containing prebiotics and probiotics is much more common in Japan and European countries than in Canada or the United States.[33] However, availability of such products is increasing as research results shed light on their safety and effectiveness.

The digestive tract, particularly the colon, is home to over 500 species of microorganisms representing 100 trillion bacteria (and billions of viruses and fungi, too). Some species of bacteria such as *E. coli* may cause disease, whereas others such as lactobacillus and bifidobacteria prevent various diseases.[34] Pre- and probiotics have been credited with important health effects (Table 24.9); the right combination of each fosters the proliferation of healthful bacteria in the colon, nose, and some other internal canals of the body. The concept of the combined benefits of pre- and

TABLE 24.7

EXAMPLE OF FUNCTIONAL FOODS WITH APPARENT HEALTH BENEFITS.[29]

FUNCTIONAL FOOD	BENEFIT
Stanol and sterol fortified margarine, psyllium fiber, soy protein, whole oat products, garlic, nuts	Reduced blood levels of LDL cholesterol
Omega-3 fatty acids	Reduced heart disease risk
Grapes, grape juice	Decreased blood clots in blood vessels
Cranberry juice	Decreased urinary tract infections
Green tea, cooked tomato products, cruciferous vegetables, conjugated linoleic acid	Decreased risk of certain types of cancer
Folic acid-fortified breads and cereals	Decreased risk of neural tube defects, heart disease
Probiotics	Decreased risk of infection, lactose intolerance, food allergies, other disorders

TABLE 24.8

FOOD AND OTHER SOURCES OF PREBIOTICS AND PROBIOTICS.[32]

PREBIOTICS	PROBIOTICS
Chicory	Fermented or aged milk and milk products:
Jerusalem artichokes	• Yogurt with live culture
Wheat	• Buttermilk
Barley	• Kefir
Rye	• Cottage cheese
Onions	• Dairy spreads with added inulin
Garlic	Other fermented products
Leeks	• Soy sauce
Prebiotics tablets and powders and	• Tempeh
nutritional beverages	• Fresh sauerkraut
	• Miso
	Breast milk

A lactobacillus species (blue) taking over harmful E. coli bacteria (red).
Source: Probiotics: Their tiny worlds . . . The Scientist July 22, 2002, pp. 20–22; author: Bob Beale.

probiotics has been termed "symbiotics."[36] Because it is difficult to recolonize gut bacteria, the benefits of pre- and probiotics last only as long as dietary intake does.[37]

Prebiotics and probiotics are assumed to be safe because they have been part of the human diet for centuries. The primary, negative side effect of their ingestion is flatulence.[38]

Final Thoughts

From dietary supplements to bacteria: the universe of substances considered dietary ingredients is expanding. Knowledge about potential benefits of pre- and probiotics is charging ahead, and advances are catching the attention of consumers and health care professionals. Perhaps you never thought that "intestinal fertilizer" or "friendly bacteria" would ever intentionally pass through your lips. But, that may well be the nature of dietary ingredients to come. If you didn't know before, you do now.

TABLE 24.9

APPARENT AND POTENTIAL BENEFITS OF PREBIOTICS AND PROBIOTICS.[35]

PREBIOTICS	PROBIOTICS
• Prevention and treatment of diarrhea and constipation	• Treatment for chronic diarrhea and traveler's diarrhea
• Prevention of colon cancer	• Treatment of lactose intolerance and some food allergies
• Increased mineral absorption	• Prevention of infections in gastrointestinal tract, ear canals
• Decreased blood triglyceride, glucose, and insulin levels	• Recolonization of colon after antibiotic therapy
	• Increased bacterial production of some B vitamins
	• Decreased blood cholesterol levels
	• Decreased risk of certain types of cancer
	• Treatment of irritable bowel syndrome
	• Decreased dental caries
	• Decreased high blood pressure
	• Prevention of wound infections

Nutrition UP CLOSE

Supplement Use and Misuse

FOCAL POINT: Decide if a dietary supplement is warranted in these situations.

People take dietary supplements for many reasons, but is their use justified?
Apply the information from this chapter to determine if you agree with the
decisions made in each of the following scenarios.

1. Martha works part-time and takes a full load of classes. Like many college students, she is always on the go, often grabbing something quick to eat at fast-food restaurants or skipping meals altogether. Nevertheless, Martha feels confident her health will not suffer, because she takes a daily vitamin and mineral supplement.

   Is a supplement warranted in this case? _____

   Why or why not? _____

2. Sylvia is a 23-year-old student diagnosed with iron-deficiency anemia. She has learned in her nutrition class that it is preferable to get vitamins and minerals from food instead of supplements. Therefore, instead of taking the iron pills her doctor has prescribed, Sylvia has decided to counteract the anemia by increasing her consumption of iron-rich foods.

   Is a supplement warranted in this case? _____

   Why or why not? _____

3. John is a 21-year-old physical education major involved in collegiate sports. He is very aware that nutrition plays an important role in the way he feels, so he is careful to eat well-balanced meals. In addition, John takes megadoses of vitamins and minerals daily. He is convinced they enhance his physical performance.

   Is a supplement warranted in this case? _____

   Why or why not? _____

4. Roberto, a native Californian, is backpacking through Europe when he is slowed down by constipation. He visits a pharmacy where English is spoken and is given senna by the clerk rather than a fast-acting medicine he expected. Roberto has never taken an herb before and is not sure how his body will react to it, or if it will work.

   Should Roberto try the senna, ask for a nonherbal drug, or take another action? (Assume they cost the same.) _____

   What's the rationale for this decision?

   _____

5. While shopping at the mall, Yuen notices a kiosk selling "Hypermetabolite," a weight-loss product that guarantees you'll lose 5 pounds a week without dieting. Having gained 10 pounds since she started working full-time, Yuen decides to try it. Her examination of the product's label reveals that an ephedra-derivative and Asian ginseng are major ingredients.

   Should Yuen take Hypermetabolite for weight loss? _____

   Why or why not? _____

FEEDBACK (including answers) can be found at the end of Unit 24.

Key Terms

bioavailability, page 24-4

dietary supplements, page 24-2

functional foods, page 24-9

prebiotics, page 24-10

probiotics, page 24-10

www links

www.nccam.nih.gov
Home page for the National Center for Complementary and Alternative Medicine.

http://dietary-supplements.info.nih.gov
Gain access to state-of-the-science information on vitamin, mineral, herbal, and other supplements from the National Institutes of Health.

http://ods.od.nih.gov/index.aspx
The National Institutes of Health, Office of Dietary Supplements home page. Provides

descriptions for more than 80 vitamin, mineral, and herbal supplements.

www.cfsan.fda.gov
Provides news on latest developments, warnings, and health effects of dietary supplements. Access the Adverse Events Reporting System from this address.

http://vm.cfsan.fda.gov/~dms/dietsupp.html
Contains the provisions of the Dietary Supplement Health and Education Act of 1994.

www.mayoclinic.org
Search "herbs" and get the skinny on the helpful and harmful ones.

Notes

1. Halsted CH. Dietary supplements and functional foods: 2 sides of a coin? Am J Clin Nutr 2003;77(suppl);1001S–7S.

2. ADA/APhA Special Report from the Joint Working Group on Dietary Supplements. A healthcare professional's guide to evaluating dietary supplements. American Dietetic Association, American Pharmaceutical Association, 2000: 47;1–40.

3. De Smet P, Herbal remedies, N Engl J Med 2002;347:2046–56; and Rennie D, DeAngelis CD. Stricter regulations of dietary supplements needed. JAMA 2003;289:1568–70.

4. ADA/APhA Special Report; and Community Nutrition Institute Newsletter, 2000 July 21:2.

5. ADA/APhA Special Report.

6. FDA seizes dietary supplements bearing drug claims, www.medscape.com, accessed 2/03; and Morris C et al., Herbal remedy sellers on the Web break the rules, www.nlm.nih.gov/medlineplus/news/fullstory_14069.html, accessed 10/03.

7. Community Nutrition Institute Newsletter, 2000 July 21.

8. Willett WC, Stampfer MJ. What vitamins should I be taking, doctor? N Engl J Med 2001;345:1819–24.

9. Fletcher RH, Fairfield KM, Vitamins for chronic disease prevention in adults: clinical applications, JAMA 2002;287:

3127–9; and Osganian SK et al., Vitamin C and the risk of heart disease, J Am Coll Cardiol 2003;42:246–55.

10. Lind T et al. A community-based randomized trial of iron and zinc supplementation. Am J Clin Nutr 2003;77:883–90.

11. Willett and Stampfer, What vitamins should I be taking? Fletcher and Fairfield, Vitamins for chronic disease prevention in adults; Osganian et al., Vitamin C and the risk of heart disease; Cogswell ME et al., Iron supplementation during pregnancy, anemia, and birth weight: a randomized controlled trial, Am J Clin Nutr 2003;78:773–81; Taylor MJ et al., Folate for depressive disorders, Cochrane Database Syst Rev 2003;2:CD003390; and Barringer TA et al., Effect of multivitamin and mineral supplement on infection and quality of life: a randomized, double-blind, placebo-controlled trial, Ann Intern Med 2003;93:365–71.

12. Morris et al., Herbal remedy sellers (www.nlm.nih.gov/medlineplus/news/fullstory_14069.html).

13. Belew C. Herbs and the childbearing woman. Guidelines for midwives. J Nurse Midwifery 1999;44:231–52.

14. Fairfield KM, Fletcher RH. Vitamins for chronic disease prevention in adults: scientific review. JAMA 2002;9:288:1720–4.

15. Halsted, Dietary supplements and functional foods; De Smet, Herbal remedies; and Angwin J, Some "enlargement" pills pack impurities. Wall Street Journal 2003;Aug.8:B1; Ishii H, Herbal weight-loss product and liver injury, Ann Intern Med 2003;139:488–92; and Kava-containing dietary supplements may be associated with severe liver injury, FDA Consumer Advisory, www.cfsan.fda.gov/~dms/supplmnt.html, accessed 10/03.

16. Halsted, Dietary supplements and functional foods; De Smet, Herbal remedies; and FDA advises people not to drink star anise teas, www.fda.gov, accessed 9/03; and Kay LK, Herbal weight-loss products: effective and appropriate? Today's Dietitian, Aug 2003:12–16.

17. Halsted, Dietary supplements and functional foods; and De Smet, Herbal remedies.

18. Key, Herbal weight-loss products; and Rennie and DeAngelis, Stricter regulations.

19. Community Nutrition Institute Newsletter, 2000 July 21; Harkey MR et al., Variability in commercial ginseng products: an analysis of 25 preparations, Am J Clin Nutr 2001;73:1101–6; and Gilroy CM et al., Echinacea and truth in labeling, Arch Intern Med 2003;163:699–704.

20. Harkey et al., Variability in commercial ginseng products.

21. Gilroy et al., Echinacea and truth in labeling.

22. Angwin, Some "enlargement" pills pack impurities.

23. Rennie and DeAngelis, Stricter regulations.

24. Morris et al., Herbal remedy sellers (www.nlm.nih.gov/medlineplus/news/fullstory_14069.html).

25. HHS Statement: FDA statement concerning structure/function rule and pregnancy claims. 2000 Jan. 6.

26. Hyman P, Claims for functional foods under the current food regulatory scheme, Nutr Today 2002;37:217–9; and Hasler CM, Functional foods: benefits, concerns, and challenges, J Nutr 2002;132:3772–81.

27. Halsted, Dietary supplements and functional foods; and Hasler, Functional foods.

28. Hyman, Claims for functional foods.

29. Hyman, Claims for functional foods; Hasler, Functional foods; Beale B, Probiotics: their tiny worlds are under scrutiny, The Scientist 2002;July:20–2; and Bailey LB et al., Folic acid supplements and fortification affect the risk for neural tube defects, vascular disease, and cancer, J Nutr 2003;133:1961S–8S.

30. Hasler, Functional foods.

31. Sanders ME, Probiotics: considerations for human health, Nutr Rev 2003;61:91–7; and Brannon CB. Prebiotics: feeding friendly bacteria. Today's Dietitian 2003;Sept:12–16.

32. Halsted, Dietary supplements and functional foods; and Brannon, Prebiotics.

33. Brannon, Prebiotics.

34. Beale, Probiotics; and Sanders, Probiotics.

35. Sanders, Probiotics; and Brannon, Prebiotics.

36. Hasler, Functional foods.

37. Brannon, Prebiotics.

38. Brannon, Prebiotics.

Nutrition UP CLOSE

Supplement Use and Misuse

Feedback for Unit 24

1. *Is a supplement warranted in this case?* No

Why or why not? Martha is deceiving herself! She may be getting the vitamins and minerals she needs by taking a supplement, but this cannot make up for her poor food habits. She needs to improve the overall quality of her diet to ensure optimal health. If Martha follows the Food Guide Pyramid recommendations, she should be able to get all the nutrients she needs from what she eats.

2. *Is a supplement warranted in this case?* Yes

Why or why not? This is one time when a supplement is in order. Sylvia needs to follow her doctor's advice and take the prescribed iron preparation to increase her hemoglobin and replete her iron stores. However, she is correct in consuming more iron-rich foods, too, so that after the anemia has been treated, she will not experience a relapse.

3. *Is a supplement warranted in this case?* No

Why or why not? There is no scientific evidence that vitamin and mineral megadoses enhance physical performance. In fact, John may be setting himself up for toxicity reactions with prolonged intake of supplements at extremely high dosages.

4. *Should Roberto try the senna, ask for a nonherbal drug, or take another action?* He should take another action—get more information from the pharmacist so he can make a better-informed decision.

5. *Should Yuen take Hypermetabolite for weight loss?* No

Why or why not? She shouldn't take the product because the ephedra-derivative may have serious side effects and might interact with ginseng, and because quick weight-loss strategies don't work in the longer run. Plus, there's no guarantee that the product will work, or that it isn't mislabeled.

Water Is an Essential Nutrient

 *Nutrition Scoreboard*

	TRUE	FALSE
1 Bottled water is better for health than municipal water from your faucet.		
2 Drinking lots of water helps moisturize the skin.		
3 Adequate water intake reduces the risk of some types of cancer and kidney stones.		
4 You can't drink too much water.		

 Answers on next page

[KEY CONCEPTS AND FACTS]

- Water is an essential nutrient. It is a required part of the diet. Deficiency symptoms develop when too little is consumed, and toxicity symptoms occur when too much is ingested.

- Functions of water include maintenance of body hydration and temperature, removal of waste products,

and participation in energy formation. It is our major source of fluoride.

- Water is a precious resource whose availability and quality are threatened by wasteful use and pollution.

Answers to *Nutrition Scoreboard*

		TRUE	FALSE
1	Bottled waters have not been shown to be better for health or to be "purer" than tap water obtained from city water supplies.		✔
2	Excess water intake is excreted—it doesn't go to the skin.		✔
3	Adequate intake of water from fluids and food (9 to 12 cups a day) reduces the risk of breast, colon, and bladder cancer and of kidney stone formation.[1]	✔	
4	Yes, you can drink too much water, and it results in water intoxication.[2]		✔

Photo Disc

Water: Where Would We Be without It?

Ask any three people you know to name as many essential nutrients as they can. If they mention water, give them a prize. Our need for water is so obvious that it is often taken for granted. Well, that's not going to happen in this text!

Water differs from other essential nutrients in that it is liquid, and our need for it is measured in cups rather than grams or milligrams. Without it, our days are limited to about six. Water is the largest single component of our diet and body. It is a basic requirement of all living things. Now, how could something this important be so easily forgotten?

Water's Roles as an Essential Nutrient

Water qualifies in all respects as an essential nutrient. It is a required part of our diet; it performs specific functions in the body; and deficiency and toxicity signs develop when too little or too much is consumed.

Water is our body's main source of fluoride, an essential mineral needed for the formation and maintenance of enamel and resistance to tooth decay. Fluoride is naturally present in some water supplies, and is added to virtually all municipal water supplies in the United States and Canada. Because it is a gas that dissolves in water, fluoride is not removed by water pitcher or faucet filters.

Water is the medium in which many chemical reactions take place within our body. Water plays key roles in energy formation—it is produced as an end-product of energy formation from carbohydrates, proteins, and fats. We continue to produce and excrete water even if we quit drinking it for awhile, because energy production is an ongoing process. Water is needed to "carry" nutrients to cells and waste products away from them. Additionally, water acts as the body's cooling system. When our internal temperature gets too high, water transfers heat to the skin and releases it in perspiration. When we're too cool, less water—and heat—is released through the skin. Water's functions are summarized in Table 25.1.

TABLE 25.1

KEY FUNCTIONS OF WATER IN THE BODY.

- Provides a medium for chemical reactions
- Participates in energy formation
- Transports nutrients and waste products
- Helps regulate body temperature

Water has been given credit for other functions in the body, but undeservedly. Drinking more water than is normally needed does not prevent dry, wrinkled skin, lead to weight loss, or flush toxins out of the body.[3] Nor will it cure chronic fatigue, arthritis, migraines, or hypertension.

Water, Water, Everywhere . . .] The body of a 160-pound person contains about 12 gallons of water (Illustration 25.1). Adults are approximately 60 to 70% water by weight. Water is distributed in the body in blood, the spaces in between cells, and in all cells. The proportion of water in body tissues varies: blood is 83% water, muscle 75%, and bone 22%. Even fat cells are 10% water.[4]

Most Foods Contain Lots of Water, Too} Most beverages are more than 85% water, and fruits and vegetables are 75 to 90% (Illustration 25.2). Meats, depending upon their type and how well-done they are, contain between 50 and 70% water. Although it is nearly impossible to meet your need for water from solid foods alone, the water content of foods makes an important contribution to our daily intakes. On average, about 31% of water intake comes from plain water, 44% from other beverages, and 25% from foods.[5]

Health Benefits of Water] Adequate water consumption may benefit long-term as well as day-to-day health. Consumption of over 10 cups of fluid each day is associated with a decreased risk of bladder, breast, and colon cancer as well as of kidney stone formation. People feel and perform better when they are adequately hydrated.[6]

Illustration 25.1
The body of a 160-pound person contains approximately 12 gallons of water.
That's 96 pounds of water!

Illustration 25.2
The water content of some foods.

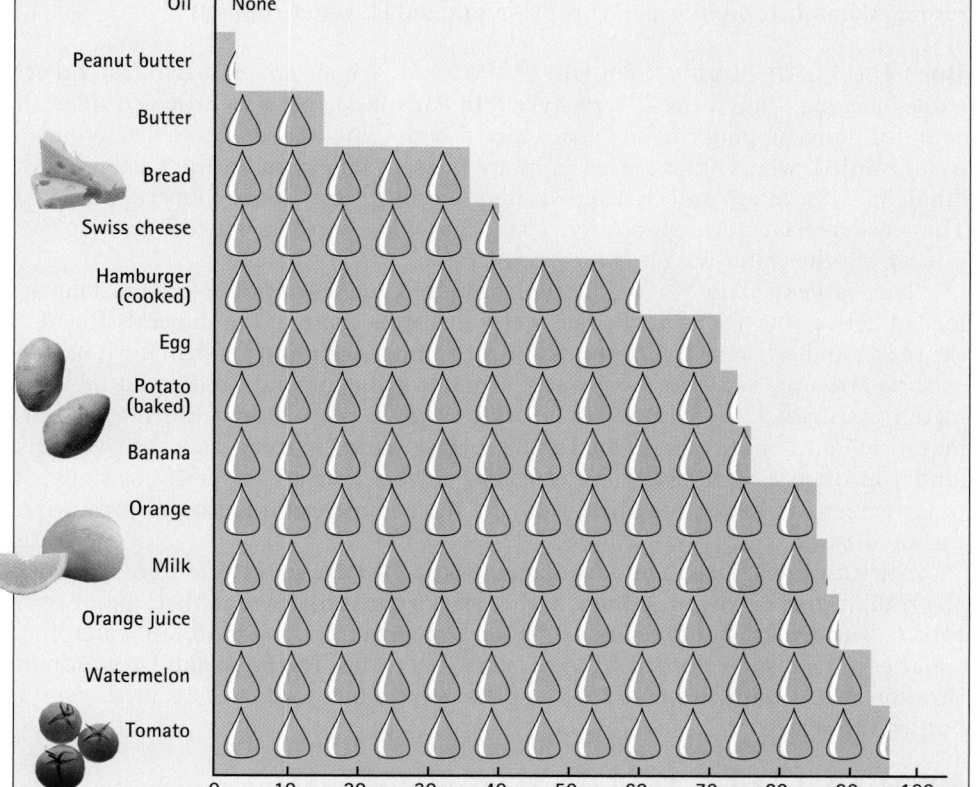

Oil — None
Peanut butter
Butter
Bread
Swiss cheese
Hamburger (cooked)
Egg
Potato (baked)
Banana
Orange
Milk
Orange juice
Watermelon
Tomato

0 10 20 30 40 50 60 70 80 90 100

% Water

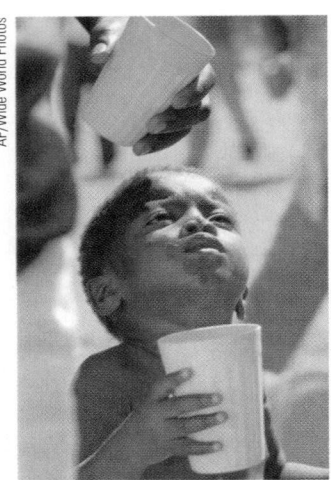

AP/Wide World Photos

Illustration 25.3
A child waits for water at a refugee camp.

Illustration 25.4
Contaminated water supplies are a threat to the public's health.

DANGER BEACH CLOSED

Photo Disc

The Nature of Our Water Supply

Water covers about three-quarters of the Earth's surface, yet very little of it is drinkable. Nearly 97% of the total supply is salt water, and only 3% is fresh. Of the freshwater supply, only a fourth of the total amount is available for use. The rest is located in polar and glacier ice. Although fresh water is readily available in most locations in the United States, this is not the case in a number of other countries.

Water is sufficiently scarce in parts of Russia that drinking-water dispensers are coin-operated. Fresh water is so highly prized in sections of the arid Middle East that having a decorative fountain in one's home is considered a sign of affluence. Drinking water is sampled, judged, and celebrated in Middle Eastern countries as ceremoniously as fine wines are in France. Although water is not a subject of envy or a cause for celebration in most parts of America, it is being increasingly viewed as a precious resource that must be protected.

Water scarcity may become a primary concern of many more nations in the twenty-first century (Illustration 25.3). The demand for water is increasing by 100 to 500% per year in countries in Central and South America, Africa, and Europe; but the supply is dwindling. Wasteful use of water, groundwater depletion, pollution, and leaky public water systems are contributing to water shortages and safety concerns (Illustration 25.4).[7] Without more effective protection of the world's water, the quantity of clean water available for agriculture and consumption may be insufficient, limiting food production and increasing the incidence of water-borne illnesses.[8]

The Environmental Protection Agency is responsible for the safety of public water supplies and has set maximum allowable levels of contaminants. Water quality is monitored by local water utilities, and the results are reported to state and federal officials. Problems identified are remedied and attempts are made to prevent future contamination in order to provide safe public water supplies.[9]

Does the Earth Supply "Gourmet" Water?] About 35 years ago, the French made drinking "fine waters" very stylish. In Paris, boulevardiers crowded sidewalk cafes for hours sipping chilled Perrier served with thinly sliced lemon. The popularity of bottled waters skyrocketed in many Western countries. In the United States, "mineral," "spring," and "seltzer" waters are now best-sellers (Illustration 25.5). These waters have a strong, positive image. But what *are* these waters? How do they differ from the water we get from the faucet?

True *mineral water* is taken from protected underground reservoirs that are lodged between layers of rock. The water dissolves some of the minerals found in the rocks, and as a result, it contains a higher amount of minerals than most sources of surface water. Actually, most water contains some minerals and could be legitimately considered "mineral water." *Spring water* is taken from freshwater springs that form pools or streams on the Earth's surface. True "seltzers" (not the sweetened kind you often find for sale) are *sparkling waters* that are naturally carbonated. Most seltzers, however, become bubbly by the commercial addition of pressurized carbon dioxide.

Bottled waters have their advantages. Bottled waters are calorie-free, are generally sodium-free or low in sodium, and quench a thirst better than their major competitor, soft drinks. But they are no "purer" or better for you than tap water.[10] As a matter of fact, some bottled waters contain tap water. The Food and Drug Administration (FDA) estimates that 25% of the bottled water sold in the United States is bottled tap water.

Is Bottled Water Safe?} Every now and then, a story about contaminated bottled water makes the news, raising concerns about the safety of bottled water and the

Illustration 25.5

Is bottled water better for you than tap water? *Consumers often perceive that bottled waters have fewer impurities and are better for health than tap water. Such beliefs are unfounded.*

way the industry is monitored. The FDA regulates the bottled water industry. Domestic bottlers must conform to specified standards of water safety (such as allowable levels of chemical contaminants) and labeling requirements. "Mineral" or "spring" water must be just that under FDA regulations. Bottled tap water must be labeled as such.

Bottlers are also subject to unannounced inspections by the FDA. Bottled water is classified as a very low-risk product, however, so domestic bottling plants are inspected, on average, every 5 years.[11] The FDA supplements its monitoring role by requiring that bottlers periodically test their water. Foreign bottling plants are not under the FDA's jurisdiction, but imported waters may be tested when they enter the United States.

Bottling plants that adhere to FDA regulations produce safe water. If you have questions about your favorite brand or want more information, call the Environmental Protection Agency's Safe Drinking Water Hotline (1-800-426-4791) between 8:30 a.m. and 5:00 p.m. Eastern Standard Time.

Water Gimmicks} New waves of bottled water products are finding their way to grocery shelves throughout America. Tapping into consumer interest in vitamins, minerals, herbs, and fitness, producers are marketing "enhanced" bottled water, fortified with everything from ginseng to oxygen (Illustration 25.6). Although mineral waters provide absorbable calcium and magnesium that contribute to nutrient status and potentially to health,[12] the rationale for many of the products (other than their commercial appeal) is elusive. It makes little sense to fortify water with herbs, oxygen, or with nutrients better obtained from foods. Safety of some of the combinations of herbs put into bottled water is unclear, and they do not appear to supplement the diet in meaningful ways.[13] Adding oxygen to beverages and foods is total nonsense. Oxygen our bodies use for energy formation and metabolic processes comes from the air we breathe, and not the beverages we drink or foods we consume.

Illustration 25.6
"Specialty" bottled water products are of uncertain value to health or fitness.

Richard Anderson

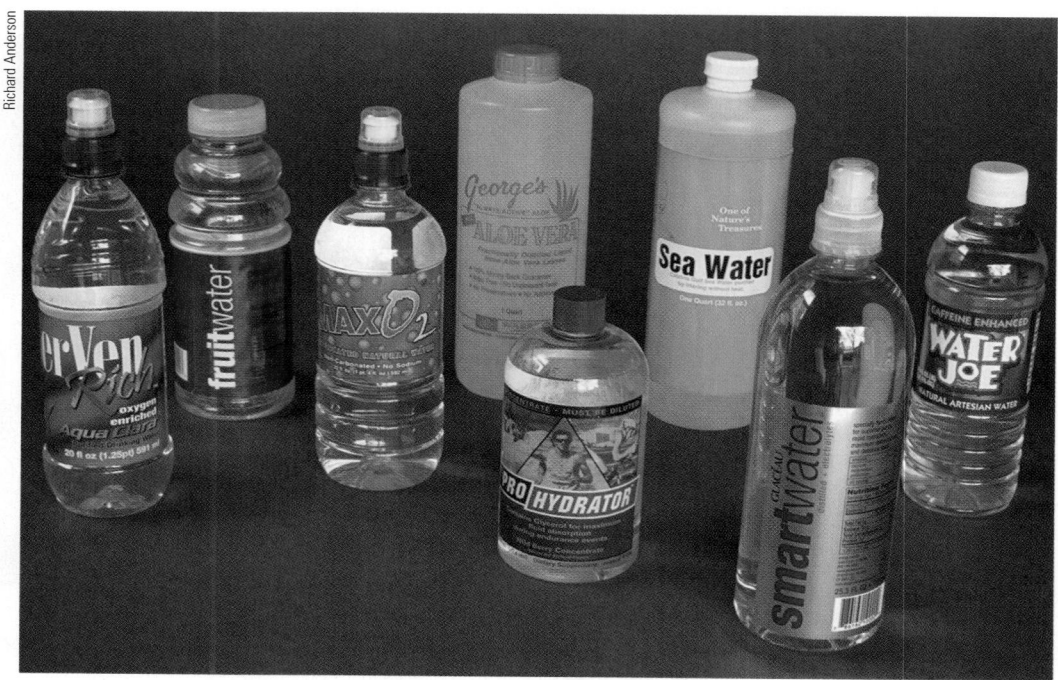

Meeting Our Need for Water

In general, individuals require enough water each day to replace water lost in urine, perspiration, stools, and exhaled air. Adequate Intakes (AIs) of water from fluids and food are set at 11 cups per day for women, and 15 cups per day for men.[14]

Built-in mechanisms that trigger thirst generally protect people from consuming too much or too little water.[15] People who do strenuous work, athletic or otherwise, in hot and humid weather need to consume enough water to replace the amount that is lost in sweat, urine, respiration, and evaporation of water from the skin's surface. How much extra water this takes varies from person to person, but it is often within the ballpark of a 50% increase. You know you are drinking enough water if you haven't lost weight after the physical activity and if your urine is pale yellow and produced in normal volume.[16]

Prolonged bouts of vomiting, diarrhea, and fever increase water need, and that is why you should drink plenty of fluids when you experience these conditions. High-protein and high-fiber diets and alcohol increase your need for water. Losses in body water that accompany high levels of protein consumption are the reason people on high-protein weight-loss diets are encouraged to "drink a lot of water." Adding fiber to your diet augments water need because fiber increases water loss in stools. The increased loss of water that goes along with alcohol intake explains why people get very thirsty after overindulging in spirits.

Is Caffeine Hydrating?] In the past it was widely believed that beverages containing caffeine were not hydrating because caffeine acted as a diuretic. Recent and better research has demonstrated that this conclusion is incorrect. Caffeine does not increase urine output in people accustomed to drinking coffee, tea, and other beverages that contain caffeine.[17] So, count the coffee or tea you consume as contributing to your overall water intake.

Water Deficiency

A deficiency of water can lead to dehydration. Dehydrated people feel very sick. They are generally nauseated, have a fast heart rate and increased body temperature, feel dizzy, and may find it hard to move. The ingestion of fluids produces quick recovery in all but the most serious cases of dehydration. If it is not resolved, however, dehydration can lead to kidney failure and death.[18]

Water Toxicity

People can overdose on water if they drink too much of it. High intake of water can lead to a condition known as hyponatremia—or low blood sodium level, and excessive water accumulation in the brain and lungs. The consequences can be devastating and include confusion, severe headache, nausea, vomiting, and even seizure, coma, and death.[19]

Water intoxication is rare, but it has occurred in marathon runners who consumed too much water during an event, infants given too much water or overdiluted formula (Illustration 25.7), and psychotic patients taking medications that produce cravings for water. The drive for water created by antipsychotic medications can be so strong that access to water (even when showering) has to be limited.[20]

Illustration 25.7
This headline is accurate—overdilution of infant formula can lead to water intoxication in infants.

Nutrition UP CLOSE

Foods as a Source of Water

FOCAL POINT: Water is a primary component of many foods.

Our requirement for water is met by fluids and solid foods. But how much water is in foods? You may be surprised. Circle the food in each set that you think contains the highest percentage of water by weight.

1. avocado   potato, boiled   corn, cooked
2. egg   almonds   ripe (black) olives
3. watermelon   celery   pineapple, fresh
4. 2% milk   coke   cranberry juice, low calorie
5. cheddar cheese   banana   refried beans
6. hot dog   pork sausage,   cooked ham, extra lean
7. apple   mushrooms, raw   orange
8. onions, cooked   lettuce   okra, cooked
9. peanut butter   butter   margarine
10. cake with frosting   bagel   Italian bread

FEEDBACK (answers to questions) can be found at the end of Unit 25.

www links

www.epa.gov
Concerned about the safety of your water supply? Check it out by entering this Environmental Protection Agency Web site and linking with water safety.

www.healthfinder.gov
A gateway to reliable information on water.

www.cdc.gov
The address for the Centers for Disease Control; provides access to water safety information and advice for travelers.

www.iom.edu/fnb
You'll be able to access the 2004 DRI Report on water and recommended intake levels from this site.

www.science.gov
Find out more about water and water quality at this Web site.

Notes

1. Kleiner SM. Water: an essential but overlooked nutrient. J Am Diet Assoc 1999;99:200–6.

2. Kleiner, Water.

3. Fiske H. Measuring water's benefits and optimal intake recommendations. Today's Dietitian 2003;Jan:22–4.

4. Askew EW. Water. In: Ziegler EE, Filer LJ Jr., editors. Present knowledge in nutrition. Washington, DC: ILSI Press; 1996: pp. 98–108.

5. Grandjean AC et al. Hydration: issues for the 21st century. Nutr Rev 2003;61: 261–71.

6. Askew, Water; and Grandjean et al., Hydration.

7. Water. www.intelihealth.com, accessed 6/03.

8. Water resources in the twenty-first century. Washington, DC: Food Policy Research Institute; 1997.

9. Water (www.intelihealth.com)

10. Water (www.intelihealth.com)

11. How safe is bottled water? American Institute for Cancer Research Newsletter 1992;36:8.

12. Sabatier M et al. Meal effect on magnesium bioavailability from mineral water in healthy women. Am J Clin Nutr 2002;75:65–71.

13. Welland D. Drink to good health, especially water: Here's why and how much. Environ Nutr 1999;22(Oct):1,6.

14. Dietary Reference Intakes for Water, Potassium, Sodium Chloride, and Sulfate. Food and Nutrition Board, National Academy of Science. http://books.nap.edu, accessed 5/04.

15. Grandjean, Hydration.

16. Fiske, Measuring water's benefits; and Noakes TD. Too many fluids as bad as too few. BMJ 2003;327:113–4.

17. Grandjean, Hydration.

18. Grandjean, Hydration.

19. Noakes, Too many fluids; and Grandjean, Hydration.

20. Noakes, Too many fluids; Keating JP, Schears GJ, Dodge PR, Oral water intoxication in infants: an American epidemic, Am J Dis Child 1991;145: 985–90; and Goldman MB, Luchins DJ, Robertson GL. Mechanisms of altered water metabolism in psychotic patients with polydipsia and hyponatremia. N Engl J Med 1988;318:397–403.

Nutrition UP CLOSE

Foods as a Source of Water

Feedback for Unit 25

The percentage of water is listed after each food.

1. **avocado (80%)** potato, boiled (77%) corn, cooked (73%)
2. egg (75%) almonds (4%) **ripe (black) olives (80%)**
3. watermelon (91%) **celery (95%)** pineapple, fresh (86%)
4. 2% milk (89%) coke (89%) **cranberry juice, low calorie (95%)**
5. cheddar cheese (37%) **banana (74%)** refried beans (72%)
6. hot dog (53%) pork sausage, cooked (45%) **ham, extra lean (74%)**
7. apple (84%) **mushrooms, raw (92%)** orange (87%)
8. onions, cooked (88%) **lettuce (96%)** okra, cooked (90%)
9. peanut butter (1%) **butter (16%)** margarine (16%)
10. cake with frosting (22%) bagel (33%) **Italian bread (36%)**

Nutrient-Gene Interactions in Health and Disease

Nutrition Scoreboard

	TRUE	FALSE
1 Most diseases are genetically caused.		
2 Individual differences in genetic traits are responsible for large differences in nutrient needs between individuals.		
3 Heart disease, cancer, obesity, and hypertension primarily result from interactions among environmental and genetic factors.		

Answers on next page

[KEY CONCEPTS AND FACTS]

- Nutrients interact in important ways with gene functions and thereby affect health status. Nutrients can turn genes on or off, and nutrient intake can compensate for abnormally functioning genes.[1]

- Health problems related to nutrient-gene interactions originate within cells.

- Advances in knowledge of nutrient-gene interactions are dramatically changing nutritional approaches to disease prevention and treatment.

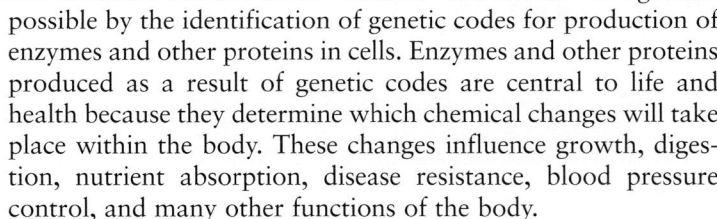

Answers to *Nutrition Scoreboard*

		TRUE	FALSE
1	That's incorrect. Only a very small proportion of disease states are caused by genetic traits.[2]		✔
2	Individual genetic traits alter nutrient needs to a small extent in many people, and to a large extent in very few. The Dietary Reference Intakes cover the nutrient needs of nearly all healthy people.		✔
3	How true. Common diseases result from interactions between multiple genetic traits and environmental factors such as dietary intake.[3]	✔	

The science of nutritional genomics is in its infancy, but it has the potential to transform the science of nutrition.
—D. Shattuck, Journal of the American Dietetic Association

Nutrition and Genomics

In the history of the relatively young science of nutrition, at least two breakthroughs have defined the future. The first resulted from experiments in the late 1800s demonstrating that some constituents of foods are essential for life. The second is represented by completion of the draft of the human *genome*. (Find definitions and explanations of many of the genetic terms used in this unit in Table 26.1.) That breakthrough marked the beginning of a new era in the discovery of contributing factors to health and disease, and the roles of nutritional and other "nurture" factors. It is being made possible by the identification of genetic codes for production of enzymes and other proteins in cells. Enzymes and other proteins produced as a result of genetic codes are central to life and health because they determine which chemical changes will take place within the body. These changes influence growth, digestion, nutrient absorption, disease resistance, blood pressure control, and many other functions of the body.

Diet-Gene Interactions

Research on pinpointing dietary-gene interactions is revolutionizing nutritional approaches to health promotion and to disease prevention and treatment.[4] A number of nutrient-gene interactions identified to date have gotten the revolution well under way. The following are examples of the effects gene types (or genotypes) have on the body's response to nutritional factors. In some people:

- Whole oats lower blood cholesterol levels.

- High folate intake decreases the risk of heart disease.

TABLE 26.1

DEFINITIONS AND EXPLANATIONS OF GENETIC TERMS.

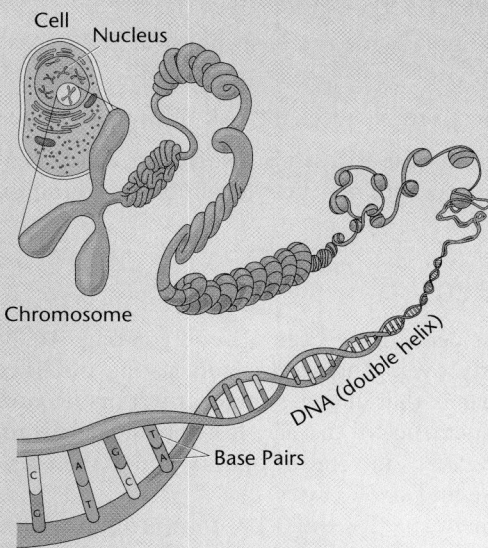

DNA is packed in chromosomes located in cell nuclei

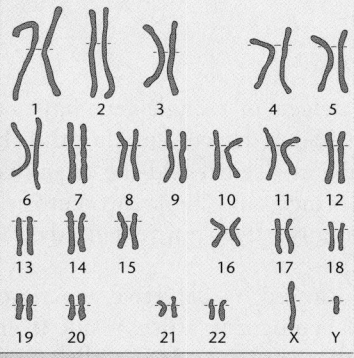

Human chromosomes

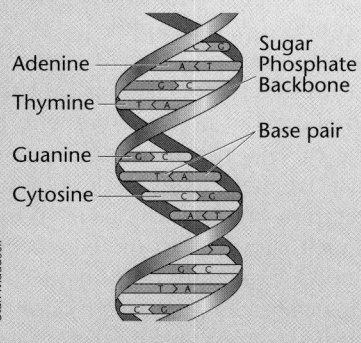

DNA

- **Genome** Combined term for "genes" and "chromosomes." It represents all the genes and DNA contained in an organism, which are principally located in chromosomes. The human genome consists of about 40,000 genes.

- **Genes** The basic units of heredity that occupy specific places (loci) on chromosomes. Genes consist of large DNA molecules, each of which contains the code for the manufacture of a specific enzyme or other protein.

- **Genomics** The study of the functions and interactions of all genes in the genome. Unlike genetics, it includes the study of genes related to common conditions and their interaction with environmental factors.

- **Chromosomes** Structures in the nuclei of cells that contain genes. Humans have 23 pairs of chromosomes (shown below); half of each pair comes from the mother and half from the father.

- **DNA (deoxyribonucleic acid)** Segments of genes that provide instructions for the manufacture of enzymes and other proteins by cells. DNA looks something like an immensely long ladder twisted into a helix, or coil. The sides of the ladder structure of DNA are formed by a backbone of sugar-phosphate molecules, and the "rungs" consist of pairs of bases joined by weak chemical bonds. Bases are the "letters" that spell out the genetic code, and there are over 3 billion of them in human DNA. There are two types of base pairs: adenine-guanine and cytosine-thymine. Each sequence of three base pairs on DNA codes for a specific amino acid. A specific enzyme or other protein is formed when coded amino acids are collected and strung together in the sequence dictated by DNA.

Some sections of DNA do not transmit genetic information because they don't code for protein production. They appear to signal which genes will turn on and how long they will be activated. Characteristics of these segments of DNA vary among individuals, making it possible to identify individuals based on "DNA fingerprinting."

- **DNA fingerprinting** The process of identifying specific individuals by their DNA. This is possible because no two individuals have the same genetic makeup. Differences among individuals are due to variations in the sections of DNA molecules that do not transmit genetic information.

- High polyunsaturated fat, low dietary cholesterol, or low-saturated-fat diets lower blood cholesterol levels.
- High-carbohydrate diets increase the risk of type 2 diabetes.
- High alcohol intake during pregnancy produces physical and behavioral abnormalities in the fetus.
- Regular consumption of green tea reduces the risk of prostate cancer.[5]

Research into the genetic bases of disease risk is also clarifying which health characteristics have little or nothing to do with genetic traits and everything to do with environmental factors.[6]

Genetic Secrets Unfolded

Our bodies may be new, but our genes have been around for over 40,000 years. Genes replicate themselves exactly over generations, and lasting modifications in them almost never occur. This means that changes in genetic traits do not account for increases or decreases in the incidence of disease. It is why genetic traits cannot be given as the cause of recent increases in rates of obesity and diabetes, nor credit for major declines in heart disease and stroke rates.

Humans are 99.9% alike genetically. It's the 0.1% that is different that makes every individual distinct.[7] Some of the 0.1% of genetic variation represents genes that contribute to disease resistance, disease development, and the way people respond to particular drugs.[8] Many questions about the roles of genetic traits in health and disease remain to be answered. Researchers know the most about disorders related to **single-gene defects.**

single-gene defects
Disorders resulting from one abnormal gene. Also called "inborn errors of metabolism." Over 800 single-gene defects have been cataloged, and most are very rare.

Single-Gene Defects

Hundreds of diseases related to one or more defects in a single gene have been identified, and many of these affect nutrient needs. Such defects can alter the absorption or utilization of nutrients such as amino acids, iron, zinc, and the vitamins B_{12}, B_6, or folate.[9] PKU, celiac disease, lactose intolerance, and hemochromatosis are four examples of single-gene defects that substantially affect nutrient needs. (These are described in Table 26.2.)

PKU and lactose intolerance are caused by defective genetic codes for enzymes, whereas celiac disease and hemochromatosis result from genetic abnormalities in the formation of other proteins. PKU, celiac disease, and lactose intolerance are treated by diets that limit phenylalanine, gluten, or lactose, respectively. Hemochromatosis is treated by a low iron diet.[11]

Not all single-gene abnormalities produce ill effects. For instance, lack of the gene that codes for an enzyme that helps the body excrete a specific sulfur-containing chemical unique to cruciferous vegetables (such as cabbage, broccoli, and bok choy) may be good. Smokers and nonsmokers who have the gene for the enzyme that helps excrete this beneficial sulfur compound and who consume these vegetables regularly are more likely to develop lung cancer than are people who eat the vegetables but lack the gene. Without the gene for the enzyme that causes its quick elimination, the beneficial sulfur-containing compound lingers in the body, extending the time it can play a role in cancer prevention.[12] Risk of lung cancer is decreased by not smoking.

Most diseases related to genetic traits are not as straightforward as are single-gene defects. They are more likely to represent an interwoven mesh of genetic and environmental risk factors.

Are you reminded an hour or two later that you ate asparagus? If you said yes, you are among the 1 out of 10 people born with a gene for detecting the odor of a sulfur-containing compound excreted in urine by 4 in 10 people after consumption of asparagus.

Photo Disc

TABLE 26.2

EXAMPLES OF SINGLE-GENE DISORDERS THAT AFFECT NUTRIENT NEED.[10]

PKU (phenylketonuria)	A very rare disorder caused by the lack of the enzyme phenylalanine hydroxylase. Lack of this enzyme causes phenylalanine, an essential amino acid, to build up in the blood. High blood levels of phenylalanine during growth lead to mental retardation, poor growth, and other problems. PKU is treated by low-phenylalanine diets.
Celiac disease	An intestinal malabsorption disorder caused by an inherited intolerance to gluten in wheat, rye, and barley. It causes multiple nutrient deficiencies and is treated with gluten-free diets. Celiac disease is also called "nontropical sprue" and "gluten enteropathy."
Lactose intolerance	A common disorder in adults in many countries resulting from lack of the enzyme lactase. Ingestion of lactose in dairy products causes gas, cramps, and nausea due to the presence of undigested lactose in the gut.
Hemochromatosis	A disorder affecting 1 in 200 people that occurs due to a genetic deficiency of a protein that helps regulate iron absorption. Individuals with hemochromatosis absorb more iron than normal and have excessive levels of body iron. High levels of body iron have toxic effects on tissues such as the liver and heart. The disorder can also be produced by excessive levels of iron intake over time and frequent iron injections or blood transfusions.

Chronic Disease: Nurture *and* Nature

> *A very large part of our health story is more genetically influenced than genetically caused.*
> —Elbert Branscomb, Director of the Joint Human Genome Institute

Major health problems of our day, including heart disease, cancer, hypertension, obesity, diabetes, and disorders associated with aging, are not due to single-gene defects but result from interactions among multiple genetic traits and environmental factors. Nutrients, a prominent environmental factor, can turn certain genes on or off and can compensate for some of the ill effects certain genotypes have on body processes.[13]

Heart disease primarily stems from plaque build-up in arteries near the heart (Illustration 26.1). Plaque build-up "hardens the arteries," narrows artery openings, and increases the risk of artery blockage and subsequent heart attack.

Some people who consume high-saturated fat diets build up plaque in their arteries, whereas others don't. For others, low folate and vegetable diets increase the risk that plaque will accumulate. These differences are due to variations in genotypes. Heart disease is also influenced by blood pressure, body weight, and levels of triglycerides and clotting factors in the blood. Each of these is influenced by environmental factors such as high salt or alcohol intake in people with specific genetic

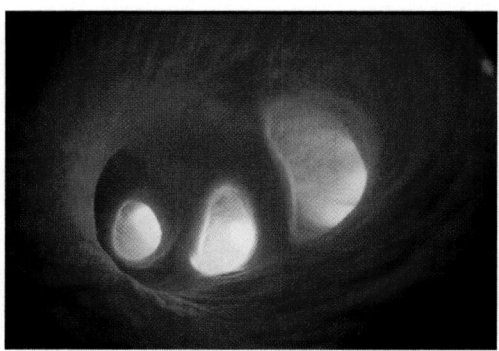
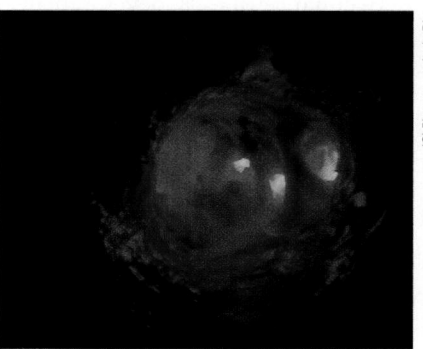

Illustration 26.1
Healthy (left) versus clogged (right) arteries.
How much plaque builds up in arteries is influenced by multiple environmental and genetic factors.

ICI Pharmaceuticals Div.

26-5

TABLE 26.3

ESTIMATES OF THE CONTRIBUTION OF ENVIRONMENTAL AND GENETIC FACTORS TO DEVELOPMENT OF SOME TYPES OF CANCER.[17]

CANCER SITE	ENVIRONMENTAL FACTORS (%)	GENETIC FACTORS (%)
Endometrial (uterine wall)	100	0
Ovary	78	22
Lung	74	26
Breast	73–95	5–27
Stomach	72	28
Colon	65	35
Pancreas	64	36
Prostate	58	42

salt sensitivity
A genetically determined condition in which a person's blood pressure rises when high amounts of salt or sodium are consumed. Such individuals are sometimes identified by blood pressure increases of 10% or more when switched from a lower-salt (1–3 grams) to a higher-salt (12–15 grams) diet.[19]

Illustration 26.2
Obesity is related to complex interactions among nutritional, other environmental and genetic factors.

Photo Disc

traits. Risk for heart disease is the sum of all the pro and con influences of environmental and genetic factors.[15]

Cancer

Most types of cancer are primarily related to environmental factors such as high fat and alcohol intakes, low vegetable and fruit diets, high levels of body fat, smoking, and toxins in the environment.[16] Some cancers have genetic components that interact with environmental exposures. Studies of the roles played by genetic and environmental factors, to the extent they can currently be separated, show differences by cancer site (Table 26.3). As can be seen by the large contributions of environmental factors to cancer development, the hope for cancer prevention primarily lies in our ability to modify harmful environmental exposures.

Hypertension

The primary causes of most cases of hypertension are unknown, but factors such as high body fat, high alcohol intake, physical inactivity, and genetic traits play a role.[18] High salt (sodium, really) intakes are also related, but not in all studies. Differences in results between studies may be related to the proportion of people in study samples who are genetically predisposed to the effects of salt intake on blood pressure. Such people are *salt sensitive;* their blood pressure increases when they consume high amounts of salt. Approximately 51% of people with hypertension, and 26% of people with normal blood pressure, are salt sensitive.[20]

Obesity

All children inherit—along with their parents' genes—
their parents, peers, and the community in which they live.
—L. Eisenberg, 1999

Obesity is primarily influenced by excessive-calorie diets and inadequate physical activity. However, over 200 genetic traits have been associated with obesity development, and many of these appear to interact strongly with environmental factors (Illustration 26.2). High-carbohydrate diets may influence gene expression and obesity development in some people, while a genetic predisposition toward inactivity may enhance obesity development in others.[21]

When people in developing countries who have subsisted on low-calorie diets for generations switch to high-calorie intakes, rates of obesity tend to increase dramatically. It appears that genetic traits expressed as a result of high food intakes, or body processes that have adapted to survive on low caloric intakes, promote the development of obesity. Diabetes and heart-disease rates in population groups also rise with increasing rates of obesity.[22] Many other nurture and nature interactions on obesity are possible.

Diseases Related to Aging

The incidence of disorders such as diabetes, cancer, heart disease, memory impairment, hypertension, and stroke usually increases with age. Such declines in health

REALITY CHECK

Changing family history

Elena and Alfredo both come from families with a number of relatives who have died from heart disease.

Who gets thumbs up

?

Answers on next page

Elena:
Heart disease is in my genes. There's nothing I can do about it.

Alfredo:
I already know my LDL cholesterol is high and my HDL cholesterol is low. I'm eating and exercising to beat the odds.

are neither inevitable nor largely determined by genetic factors. They are heavily influenced by diet, exercise, and other components of lifestyle (Illustration 26.3).[23]

Genetics of Food Selection

Food preferences are largely learned, but which vegetables we like or don't like may be influenced by genetic traits. There are over 80 genes that help people taste bitter foods, and some people get the set of genes that make them highly sensitive to bitter-tasting foods (which are mainly vegetables). People born with a high sensitivity to bitter tastes tend to dislike cooked cabbage, collard greens, spinach, brussels sprouts, or other vegetables that taste bitter to them. People who tend to like these vegetables generally don't perceive them to be bitter-tasting. A genetic tendency to reject these vegetables is likely to limit intake and therefore may be linked to diseases associated with low vegetable intake.[24]

Looking Ahead

In our rush to fit medicine with the genetic mantle, we are losing sight of other possibilities for improving public health.
—N. A. Holtzman, Johns Hopkins Medical Institute

The promise of advances in knowledge of the genetic bases of disease is longer, healthier lives. No doubt drugs that counter ill effects of genetic traits will continue to be developed, and attempts to fix abnormal genes by gene therapy will broaden. Some of the most meaningful breakthroughs in upcoming decades will be in the area of disease prevention and treatment through nutritional changes.[25]

Although used to some extent now, personalized modifications of dietary intake based on genotypes will become standard practice in clinical dietetics and medicine (Illustration 26.4).

Illustration 26.3
Diseases related to aging are neither inevitable nor largely determined by genetic factors.
They are influenced more by diet, exercise, and other components of lifestyle.

Illustration 26.4
In the future, advice given by registered dietitians may be tailored to the specific genetic traits of individuals.

ANSWERS TO **REALITY CHECK**

Changing family history

Alfredo recognizes that a family history of heart disease doesn't mean he is destined to die from it. A small minority of diseases are solely due to genetic factors; many are primarily due to interactions between genetic traits and environmental factors. Disease-promoting traits can be diminished by the right environmental changes.[14]

Elena

Alfredo

Illustration 26.5
A menu of the future.

Welcome to the Future Café
Today's Specials
Prime Rib au jus
or, for our saturated-fat sensitive guests, try our
Mediterranean Sole
with almonds

Are Gene-Based Designer Diets in Your Future?

In less than 10 years, you'll be able to go to a lab and complete a set of genetic tests to identify your personal disease susceptibilities. When you leave you'll be armed with a list of foods to eat and foods to avoid and recommendations of dietary supplements to help prevent your disease.
—D. Evans, CEO of Wellgen, Inc.

Sound far-fetched? It isn't.[23] In the future, people will have the option of, for example, reducing cancer risk by regularly consuming cruciferous vegetables (cabbage, broccoli, cauliflower, and the like), diabetes risk by switching to a high-protein diet, or heart disease by limiting dietary cholesterol or saturated fat intake (Illustration 26.5). In truth, we have enough knowledge today to implement healthful, disease-preventing dietary changes. The kinds of advances described above will not bring an end to disease or guarantee health, but they will take us a long way down that road.

Key Terms

salt sensitivity, page 26-6

single-gene defects, page 26-4

www links

www.nhgri.nih.gov
This great site from the National Human Genome Research Institute provides news of the latest developments in genetic research, educational resources, and a written and spoken glossary of genetic terms and useful illustrations.

www.healthfinder.gov
Search any disease of interest to find out current information about it.

www.ornl.gov/TechResources/
Human_Genome/medicine/genetherapy.
html
Forgive the long address, but it will send you directly to information about the

Human Genome Project and the current status of research.

www.cdc.gov/genomics
This is CDC's Office of Genomics and Disease Prevention home page. You'll find information on genomic discoveries and how they can be used to improve health and prevent disease.

Photo Disc

Nutrition UP CLOSE

Nature and Nurture

FOCAL POINT: Gaining insight into your family's and your own "dietary" health history.

Families often share dietary factors that interact with genetic traits to influence chronic disease development. Check out the dietary health history of you and your family by completing the following family food tree activity. If you don't know all of your relatives included in the activity or their dietary behaviors, fill out the parts based on what you do know.

Check each dietary behavior that applies:

Relative	Excess Calorie Consumption		Low Vegetable and Fruit Intake		Low Whole Grains Intake		Low Fiber Intake		Low Fish Intake		High Alcohol Intake	
	yes	no	yes	no	yes	no	yes	no	yes	no	yes	no
A. On your mother's side:												
Grandmother	—	—	—	—	—	—	—	—	—	—	—	—
Grandfather	—	—	—	—	—	—	—	—	—	—	—	—
B. On your father's side:												
Grandmother	—	—	—	—	—	—	—	—	—	—	—	—
Grandfather	—	—	—	—	—	—	—	—	—	—	—	—
C. Your mother	—	—	—	—	—	—	—	—	—	—	—	—
D. Your father	—	—	—	—	—	—	—	—	—	—	—	—
E. Yourself	—	—	—	—	—	—	—	—	—	—	—	—

FEEDBACK (answers to these questions) can be found at the end of Unit 26.

Notes

1. Guttmacher AE et al. Welcome to the genomic era. N Engl J Med 2003;349: 996–8.

2. Guttmacher et al., Welcome to the genomic era.

3. Murray RF, Nutrigenetics/nutrigenomics. Experimental Biology Annual Meeting, San Diego, CA, 4/12/03.

4. Shattuck D. Nutritional genomics. J Am Diet Assoc 2003;103:16–18.

5. German JB et al., Genomics and metabolomics as markers for the inter-action of diet and health: lessons from lipids, J Nutr 2003;133:2078S-83S; and Simopoulos AP, Genetic variation and nutrition, Experimental Biology Annual Meeting, San Diego, CA, 4/12/03.

6. German et al., Genomics and metabolomics.

7. Murray, Nutrigenetics/nutrigenomics; and Shattuck, Nutritional genomics.

8. Murray, Nutrigenetics/nutrigenomics; and Peters RJG et al., Gene polymor-phisms and the risk of myocardial infarction—an emerging relation, N Engl J Med 2002;347:1963–5.

9. Shils ME et al., eds. Modern nutrition in health and disease. 9th ed. Philadelphia: Lippincott, Williams & Wilkins; 1998.

10. Khoury MJ et al., Population screening in the age of genomic medicine, N Engl J Med 2003;348:50–8; and Guttmacher AE et al., Genomic medicine—a primer, N Engl J Med 2002; 347:1512–20.

11. Guttmacher et al., Welcome to the genomic era.

12. London SJ, Yuan JM, Coetzee GA et al. CYP1A1 1462V genetic polymorphism and lung cancer risk in a cohort of men in Shanghai, China. Cancer Epidemiol Biomarkers Prev 2000;9:987–91.

13. Guttmacher et al., Welcome to the genomic era; and Shattuck, Nutritional genomics.

14. Guttmacher et al., Welcome to the genomic era; and Peters et al., Gene polymorphisms.

15. Simopoulos, Genetic variation and nutrition; and Peters et al., Gene polymorphisms.

16. Lichtenstein P, Holm NV, Verkasalo PK et al. Environmental and heritable fac-tors in the causation of cancer—analyses of cohorts of twins from Sweden, Den-mark, and Finland. N Engl J Med 2000;343:78–85.

17. Lichtenstein et al., Environmental and heritable factors; Terry P, Baron JA, Weiderpass E et al., Lifestyle and endometrial cancer risk: a cohort study from the Swedish Twin Registry, Int'l J Cancer 1999;82:38–42; and Hunter DJ, Willett WC, Nutrition and breast cancer, Cancer Causes and Control 1996; 7:56–68.

18. Wong ZY, Stebbing M, Ellis JA et al. Genetic linkage of beta and gamma sub-units of epithelial sodium channel to systolic blood pressure. Lancet 1999; 353:1222–5.

19. Morimoto A, Uzu T, Fujii T et al. Sodium sensitivity and cardiovascular events in patients with essential hyper-tension. Lancet 1997;350:1734–7.

20. Kaplan NM. The dietary guideline for sodium: should we shake it up? No. Am J Clin Nutr 2000;71:1020–6.

21. Allison DB, Genetic and environmental influences on human body weight, Nutr Today 2000;35:18–21; and Martinez JA et al., Obesity risk associated with car-bohydrate intake in women carrying the Gln27Glu B_2-adrenoceptor polymor-phism, J Nutr 2003;133:2549–54.

22. Berdanier CD. Nutrient-gene interac-tions. Nutr Today 2000;35:8–17.

23. Rowe JW. Geriatrics, prevention, and the remodeling of Medicare [editorial]. N Engl J Med 1999;340:720–1.

24. Drewnowski A, Henderson SA, Hann CS et al. Genetic taste markers and pref-erences for vegetables and fruit of female breast care patients. J Am Diet Assoc 2000;100:191–7.

25. Guttmacher et al., Welcome to the genomic era.

22. Begley S. Poisons aren't toxic to every-one, creating a dilemma. Wall Street Journal 2003;Jan. 24:B1.

23. Dave Evans is president and CEO of Wellgen, Inc., a company commercializ-ing Rutgers University technology. It appeared in Peregrin, T. The new fron-tier of Nutritional Science: Nutrige-nomics. J Am Diet Assoc 2001;101: 1306–7.

Nutrition UP CLOSE

Nature and Nuture

Feedback for Unit 26

The more yes responses, the better the odds that some of your family's shared dietary traits may influence disease risk. The most important responses are those you gave yourself.

--- UNIT 27 ---

Nutrition and Physical Fitness for Everyone

Nutrition Scoreboard

	TRUE	FALSE
1 Physical fitness means being very muscular.		
2 Overweight people can be physically fit.		
3 Exercise is more effective than diet in preventing heart attacks.		
4 Physical fitness can be achieved only by exercising intensively for at least an hour every day of the week.		

Answers on next page

[KEY CONCEPTS AND FACTS]

- Physical fitness, along with a good diet, confers a number of physical and mental health benefits.

- You don't have to be an athlete or be lean to be physically fit. Fitness depends primarily on muscular strength, endurance, and flexibility.

- People who are physically fit have respiratory and circulatory systems capable of delivering large amounts of oxygen to muscles, and muscular systems that can utilize large amounts of oxygen for prolonged periods of time.

- Physical fitness can be achieved by resistance training, aerobic exercises, and stretching.

- The fitness level of most people in the United States is poor.

Answers to *Nutrition Scoreboard*

		TRUE	FALSE
1	Muscle strength is one part of physical fitness. The other primary components are endurance and flexibility.		✔
2	You can be fit and fat. Physical fitness is not measured by body fat content.	✔	
3	Regular exercise does help reduce the risk of heart attack[1]—but not as much as a combined program of exercise, weight loss and smoking cessation (if needed), and consumption of a healthy diet.[2]		✔
4	You don't have to exercise to that extent to become physically fit.[3]		✔

AP Photo/The Denver Post, Cyrus McCrimmon

Physical Fitness: It Offers Something for Everyone

I would not wish to imagine a world in which there were no games to play and no chance to satisfy the natural human impulse to run, to jump, to throw, to swim, to dance. Sport and recreation are, in themselves, eminently worthwhile and desirable. . . . Our aim should be, I suggest, to inspire everyone to become involved in sport and recreation by making the choice irresistible in its scope and variety. . . . I see health as a happy consequence of sporting activity.
—Sir Roger Bannister, 1989

Physical activity has much to offer the athlete and nonathlete alike. It provides recreation good for the body and soul, it doesn't have to cost anything, and it benefits almost everyone. At its best, physical activity is play with happy consequences for health and well-being. As long as there are activities people enjoy doing, there's an "athlete" in everyone.

The "Happy Consequences" of Physical Activity

Ask people who are in good physical condition what they get from exercise, and you're likely to hear a variety of responses. Someone who is 20 years old may say he wants to stay in shape and improve his stamina. A 40-year-old individual may say that exercise helps keep her cholesterol and weight from getting too high. People who are 80 might tell you they exercise so that they won't have to use a walker and will be able to maintain their independence. Ask children why they exercise, however, and they may not understand the question. Children engage in active play; they don't exercise. For them, fitness is truly an unintended consequence of play.

TABLE 27.1

BENEFITS OF REGULAR PHYSICAL ACTIVITY.[4]

REDUCED RISK OF CERTAIN DISEASES AND DISORDERS	IMPROVED SENSE OF WELL-BEING
• Heart disease	• Increases feeling of well-being
• Colon and breast cancer	• Decreases depression and anxiety
• Hypertension	• Helps relieve stress
• Stroke	
• Osteoporosis	
• Back and other injuries	
• Obesity, excess abdominal fat	
• Diabetes (non-insulin dependent)	
• Bone and joint diseases	

Photo Disc

Regular physical activity benefits both physical and psychological health in people of all ages (Table 27.1). By improving a person's physical health, regular exercise may help ward off heart disease, some types of cancer, hypertension and stroke, osteoporosis, back injury, and diabetes. It tends to increase a person's feeling of well-being and helps relieve depression, anxiety, and stress.[5]

The Bonus Pack: Exercise plus a Good Diet]   Exercise benefits health most when combined with a good diet and other healthy behaviors. The risk of developing heart disease, for example, is meaningfully lower when people combine a diet low in animal fat and high in vegetables, fruits, whole grains, and fiber with regular exercise than when exercise alone is relied upon for protection against heart disease.[6] Regular physical activity helps build bone mass and reduces the risk of osteoporosis. But the risk is lowered to a greater extent if exercise is combined with a diet that supplies adequate amounts of calcium and vitamin D. Some other benefits of exercise, such as a reduced chance of developing colon and breast cancer, may be related to the effect of exercise on body fat content. Regular exercise is one of the few factors yet identified that helps people achieve long-term weight control.[7]

Exercise and Body Weight}   Combined with a moderate decrease in usual caloric intake—on the order of 200 calories per day—exercise helps people lose fat, build muscle mass, and become physically fit.[8] Because the body needs more calories to maintain muscle than fat, exercise that results in an increase in muscle mass leads to an increase in caloric requirements. For some people, this increase in caloric requirement makes it easier to maintain weight within the normal range and to lose weight and keep it off.[9]

Exercise combined with a stable caloric intake can lead to weight loss, but the loss is generally smaller than can be achieved by reducing caloric intake.[10] Even if weight loss is not intended or is achieved at a slow pace, the happy consequences of exercise come to all basically healthy individuals who undertake it on a regular basis.

Physical Activity and Fitness

Many of the benefits of physical activity are related to the *physical fitness* it can produce. Physical fitness is not defined by bulging muscles, thin waistlines, or amount of physical activity. Obese as well as thin people can be physically fit or not. According to the American College of Sports Medicine, physical fitness is a state of health measured by strength, endurance, and flexibility.[11]

physical fitness
The health of the body as measured by muscular strength, endurance, and flexibility in the conduct of physical activity.

The strength component of physical fitness relates to the level of maximum force that muscles can produce. Endurance refers to the length of time muscles can perform physical activities, and flexibility refers to a person's range of motion. Physical fitness results when all three exist.

Muscle Strength] In adults, muscles grow in size and strength when muscle cells increase in size. Muscle cell size—and correspondingly, the strength of muscles—increases in response to weight-bearing or resistance exercise. Activities that require muscles to work harder than usual increase muscular strength.[12]

To increase muscle strength, people should lift or push against a heavy weight; to increase muscle endurance, they should lift lighter weights and push against them repeatedly. Resistance training should include exercises that build both muscle strength and endurance.

Endurance: A Measure of Aerobic Fitness] How long a person can perform an activity depends on inherited traits and conditioning.[13] Both factors contribute to a person's stamina—or the extent of one's ability to deliver oxygen to muscles and the ability of muscles to use the oxygen for work. The amount of oxygen an individual is able to deliver to muscles that can be used by muscles for physical activity corresponds to the level of *aerobic fitness*. The more aerobically fit a person is, the longer and harder she or he is able to exercise.

Aerobic activities include jogging, basketball, swimming, soccer, and other low- and moderate-intensity activities. They give the whole body, or most of it, a continuous workout.

How Is Aerobic Fitness Determined?} Aerobic fitness is classically assessed by measuring *maximal oxygen consumption* (abbreviated as VO_2 max) in a specially equipped laboratory (Illustration 27.1). In the lab, individuals are exercised at increasingly higher intensities, for example, by elevating the grade or speed of a treadmill. The individual performing the exercise breathes through a tube that delivers room air. Monitoring equipment measures the amount of oxygen from the air that is breathed and utilized during the exercise. The maximal amount of oxygen a person delivers to working muscles is the amount used when the intensity of exercise can no longer be increased. The higher the level of oxygen utilized at the peak level of activity, the higher the level of aerobic fitness and the longer physical activity can be performed.

aerobic fitness
A state of respiratory and circulatory health as measured by the ability to deliver oxygen to muscles, and the capacity of muscles to use the oxygen for physical activity.

maximal oxygen consumption
The highest amount of oxygen that can be delivered to, and utilized by, muscles for physical activity. Also called VO_2 *max* and *maximal volume of oxygen*.

Illustration 27.1
Determining VO_2 max.
This man is running on a treadmill. His nose is plugged, so he must breathe through a tube in his mouth. The tube is attached to an apparatus that measures the total amount of air breathed in and out, and the difference between the amount of oxygen inhaled and exhaled. Every few minutes the slope or speed of the treadmill is increased, making the exercise more intense. The amount of oxygen he utilizes at the point when his exercise intensity can go no higher is considered "maximal oxygen consumption," or VO_2 max.

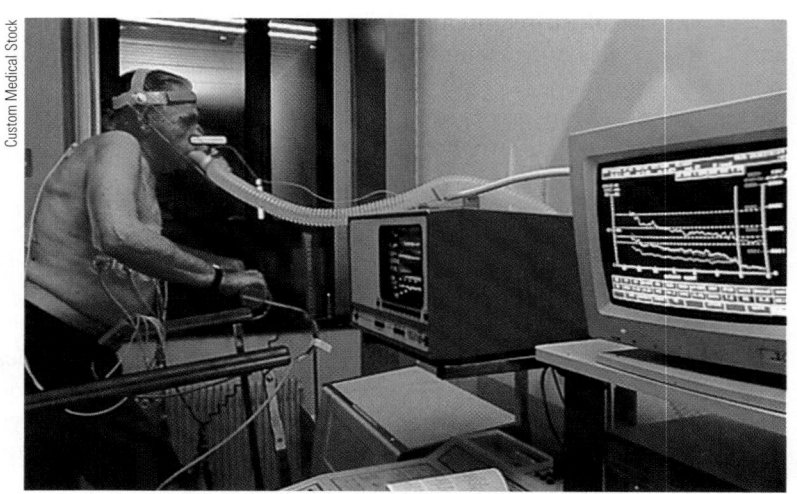

Custom Medical Stock

People can perform physical activity at 100% of VO_2 max for only a few minutes. Consequently, aerobic fitness goals are set below that level. In general, it is recommended that beginners start an aerobic fitness program with a goal of exercising at 40 to 60% of VO_2 max and working up to a higher level.[14] Aerobically fit people may train at 70 to 85% of VO_2 max.

How Do You Know Your VO_2 Max?} There are several ways to estimate VO_2 max that don't require a supervised laboratory test. One commonly employed method uses the percentage of maximal heart rate. This formula uses heart rate as a proxy measure of oxygen consumption, and it provides a fairly good estimate of VO_2 max because the amount of blood circulated to muscles is roughly equivalent to the number of times the heart sends out a pulse of blood each minute. The more oxygen muscles need to perform an activity, the more oxygen the lungs have to deliver and the faster the heart has to beat (that is, up to the point where the lungs and heart can no longer send an additional supply of oxygen to muscles). The heart rate at which the highest level of oxygen consumption by the body occurs is considered "maximal heart rate." It roughly corresponds to VO_2 max.

The formula for estimating maximal heart rate is quite simple: subtract your age from 220. To obtain a target heart rate for exercise, multiply the result by the desired percentage of maximal heart rate. (Table 27.2 shows an example.) The resulting figure is your "target heart rate" for aerobic exercise. Table 27.3 lists maximal heart rate, 60% of the maximum, and 75% of the maximum rate for different ages.

How Do You Know Your Heart Rate?} You can determine if you are exercising at the appropriate level for aerobic fitness by measuring your pulse rate during a break or as soon as you're done exercising. To take your pulse, place a fingertip gently on an artery in your wrist or neck (Illustration 27.2). Count the number of pulses you feel in a 10-second period and multiply that number by 6. That's your heart rate, measured as heartbeats (or pulses) per minute.

TABLE 27.2

APPLYING THE PERCENTAGE OF MAXIMAL HEART RATE (MHR) FORMULA TO A 22-YEAR-OLD INDIVIDUAL WHO WILL EXERCISE AT 60% OF MHR.

- Target heart rate = (220 − age) × %MHR
- Target heart rate = (220 − 22) × 0.60
   = 198 × 0.60
   = 119

TABLE 27.3

TARGET HEART RATES ESTIMATED AT 60% AND 75% OF MAXIMUM FOR PEOPLE AT DIFFERENT AGES USING THE PERCENTAGE OF MAXIMAL HEART RATE FORMULA.

AGE (YEARS)	AVERAGE MAXIMAL HEART RATE (MHR)	60% OF MHR	75% OF MHR
20	200	120	150
25	195	117	146
30	190	114	142
35	185	111	138
40	180	108	135
45	175	105	131
50	170	102	127
55	165	99	123
60	160	96	120
65	155	93	116
70	150	90	113

Photo Disc

Illustration 27.2
To find out if you're exercising at the right level, gently take your pulse during a break or right after you finish exercising.

Flexibility] Stretching major muscle groups to the point of mild discomfort increases the range of motion of joints and muscles. It helps protect muscles and their connective tissues from injury and shields you from resulting stiffness and pain. Stretching also helps your muscles and joints warm up prior to exercise, and to cool down after exercise. Appropriate stretching involves slow stretches of 15 to 20 seconds, with no bouncing, for about 10 minutes three to seven days a week. Breathe rhythmically during stretches and other exercises. Your body needs the oxygen.

Nutrition and Fitness

Your diet can make a strong contribution to your level of physical fitness. That's because good diets contribute to health, and it is easier to become physically fit if you are in good health. Diet also makes a difference to physical fitness because the foods you eat serve as the source of energy for physical activity. For some activities, the body needs glucose as its primary energy source. For others, the fuel of choice is fat.

Muscle Fuel

Muscles use fat, glucose, and amino acids for energy. The proportion of each that is used, as well as the amount, depends on the intensity of activity (Illustration 27.3). When we're inactive, fat supplies between 85 and 90% of the total amount of energy needed by muscles. The rest is provided by glucose (about 10%) and amino acids (5% at most).[15] Fat is also the primary source of fuel for activities of low-to-

Illustration 27.3
Schematic representation showing the proportionate use of sources of energy for physical activities of various intensities.
Most activities are fueled by both fat and glucose.

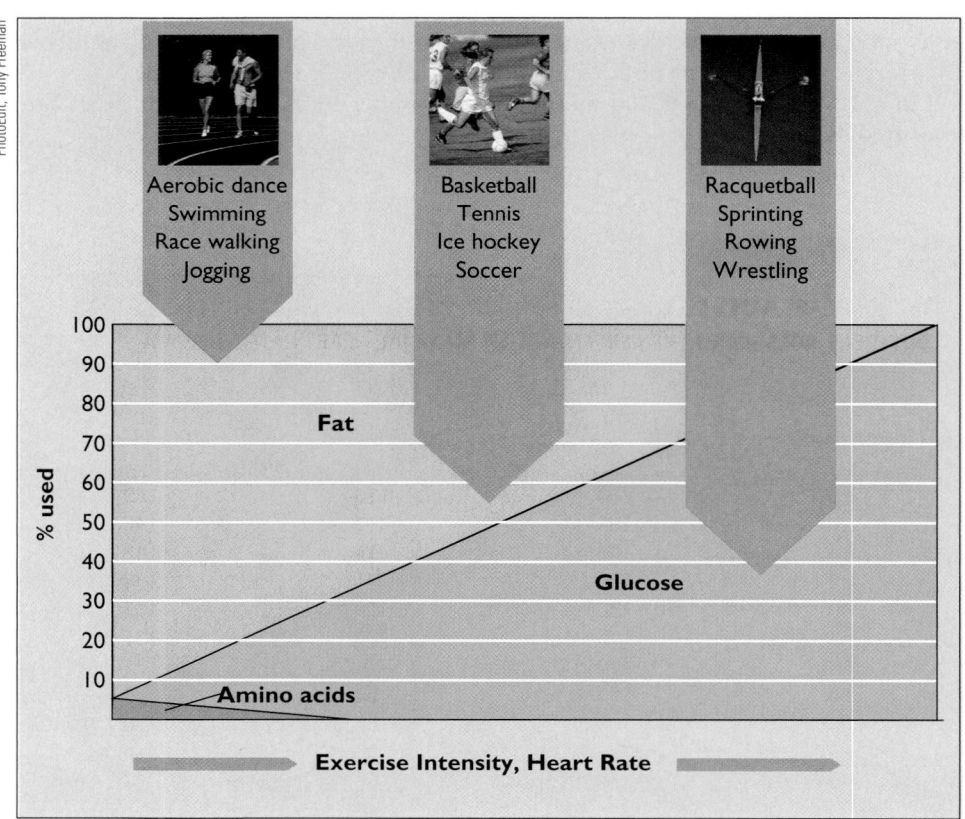

moderate intensity such as walking, running, and swimming. As can be seen from Illustration 27.3, amino acids make a relatively small contribution to meeting energy needs for physical activity.

Oxygen is required to convert fat into energy. That makes activities of low-to-moderate intensity "aerobic"—or "oxygen requiring." Consequently, fat-burning, oxygen-requiring exercises are the type used to increase aerobic fitness. Jogging, long-distance running, swimming, and aerobic dance are all aerobic activities.

High-intensity, short-duration activities like spiking a volleyball, running down a deep fly ball, or sprinting down the block to catch a bus are fueled primarily by glucose. Our supply of glucose for intense activities comes principally from glycogen, the storage form of glucose. Glycogen is stored in muscles and the liver and can be rapidly converted to glucose when needed by working muscles. The conversion of glucose to energy for intense activity doesn't require oxygen (it's *an*aerobic). People can undertake very intensive activity only as long as their stores of glucose last.

Activities such as basketball, hockey, tennis, and soccer that involve walking, running, sprinting, and the occasional high-intensity, quick move use mainly fat or glucose as the energy source depending on the intensity of the physical effort.

Stiff Muscles in the Out of Shape] People who are out of shape and exercise intensely, or who change the type of exercise performed, may experience stiff muscles 24 to 48 hours after exercise (Illustration 27.4).[16] Although the cause is not clear, the stiffness usually lasts a day or two. The probability of this happening decreases as people get into shape.

Diet and Aerobic Fitness

The best type of diet for physical fitness is the same one that is recommended for people in general. It includes a variety of vegetables and fruits, whole grain products, lean meats and fish, and low-fat dairy products. Healthy diets promote physical fitness because they facilitate maintenance of normal weight; reduce plaque build-up in arteries; and supply adequate amounts of vitamins, minerals, essential fatty acids, fiber, and other substances that keep the body running like a well-tuned

Illustration 27.4
She's going to feel this tomorrow. People may get stiff muscles if they exercise too hard when they're out of shape.

Photo Disc

Tex:
When I played basketball in school, the coach would never let us drink until half-time. Now I know why he did that.

REALITY CHECK

To drink or not to drink?

Drinking water during exercise upsets your stomach and slows you down. That's what some people say. Are they right?

Who gets thumbs up

?

Photo Disc

Rose:
I get thirsty when I exercise, so I drink water. It never gave me a stomachache.

Answers on next page

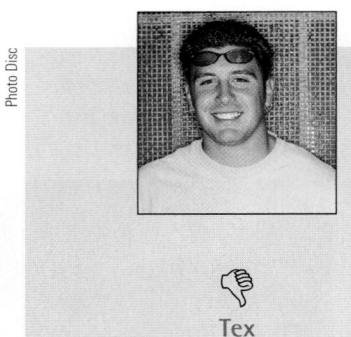

Tex

ANSWERS TO **REALITY CHECK**

To drink or not to drink?

Rose is onto something. Thirst in healthy people signals the need for water. Drinking enough water or other dilute fluids during exercise keeps you hydrated. Not drinking enough slows you down.[21]

Rose

machine. Diets that promote physical fitness most certainly include well-planned vegetarian diets.[17]

Actually, people who exercise tend to have better diets than people who don't. They are often health-conscious, select foods with care, and, because they generally eat more than sedentary people, have a better chance of getting all the nutrients they need.[18]

A Reminder about Water

Physical activity increases the body's need for water, and if the climate is hot and humid, this need increases even more. In general, people should drink in response to thirst and, overall, consume enough water to replace the amount lost in sweat, respiration, and urine during exercise. [19] (The amount of water lost during exercise is equivalent to the amount of weight that is lost during the exercise.) You are consuming the right amount of water if your urine is pale yellow and normal in volume. For exercise that lasts over an hour, alternating water with a sports drink or diluted fruit juice may improve hydration.[20] It's important to keep up fluid intake when exercising in cold weather. Cold air holds less water vapor than hot air, so you lose more water through breathing in cold weather.

A Personal Fitness Program

Commenting on physical activity for the long run, Sir Roger Bannister has said:

> The best advice to those who are unfit is to take exercise unobtrusively. I am not an enthusiast for fitness and exercise schemes that are boring because they tend to fizzle out. . . . The vast majority of people, I think, can only be attracted for any length of time towards recreational activities if these are rewarding, enjoyable, and satisfying in themselves.

Physical fitness is not something that, once achieved, lasts a lifetime. The beneficial effects of training diminish within 2 weeks after training stops and disappear altogether in 2 to 8 months.[22] That makes it critical to undertake physical activities that "wear well"—those that a person finds rewarding and will reserve a place for in a busy schedule. Too many well-intentioned attempts to get into shape have been abandoned because they were overly ambitious or no fun. As the guidelines for becoming physically fit are presented in the next section, keep firmly in mind that

ON THE SIDE

People in Washington, DC, and Wyoming tend to be the most physically active, whereas physical activity levels are lowest in Kentucky and Louisiana.[30]

TABLE 27.4

AN OVERVIEW OF THE COMPONENTS OF PHYSICAL FITNESS TRAINING.

(Top) Moving muscles against resistance for strength.
(Middle) Aerobic exercise for oxygen delivery.
(Bottom) S-T-R-E-T-C-H-I-N-G for flexibility.

A. **Resistance training:**
- Goal: To improve muscle strength, flexibility, and endurance.
- Components: Two to three resistance training sessions per week.
  → Each session includes 10 exercises repeated in sets of 8 to 12.
  → Exercises should involve all major muscle groups.

B. **Aerobic training:**
- Goal: To increase the ability of the respiratory and circulatory systems to deliver oxygen to working muscles (to build "cardiorespiratory" fitness).
- Components: Three exercise sessions per week (lasting 20 to 30 minutes each) at 60 to 75% of maximal heart rate.
  → Aerobic exercise should be a continuous activity (not stop-and-go) and involve all or most major muscle groups.

C. **Flexibility:**
- Goal: Increase range of motion of muscles and joints.
- Components: Stretching of all major muscle groups.
  → Stretching about 10 minutes three to seven times a week.
  → Stretch arm, legs, back, and stomach area slowly in multiple directions to the point of mild discomfort. Hold each stretch 15 to 20 seconds. Don't forget to breathe during the stretches, and don't bounce.
  → Warm up prior to exercise, and cool down afterwards, by stretching your muscles and joints.

you should seek a realistic and achievable plan—one that feels right and will last because it's good to you and for you.

Becoming Physically Fit: What It Takes

Physical fitness results from regular physical activity that leads to muscle strength, endurance, and flexibility (Table 27.4). These components of physical fitness can be achieved by resistance training and aerobic exercise.

A Resistance Training Plan] The resistance training program for physical fitness doesn't prepare people for bodybuilding contests. It increases muscle strength, flexibility, and endurance and improves the performance of aerobic exercises.

The American College of Sports Medicine recommends two to three resistance training sessions per week. Two sessions are enough for those who are not interested in "bulking up." Strength training twice weekly will give you 70 to 80% of the benefit of training three times a week.

Each workout session includes about 10 different exercises, repeated 8 to 12 times each. Each set of exercises should leave the muscles exercised and feeling a bit tired. Exercises selected should involve each of the major muscle groups in the legs and buttocks, arms and shoulders, and abdomen and back. Table 27.5 lists examples of exercises for the different muscle groups.

Resistance training can also be used for additional gains in muscular strength. Muscle strength is increased by the use of progressively heavier weights in workouts.

TABLE 27.5

THE MATCH BETWEEN RESISTANCE EXERCISES AND THE BODY'S MAJOR MUSCLE GROUPS.

MUSCLE GROUP EXERCISED	ACTIVITY
Legs and buttocks	Stair climbing, resistance leg pushing, cycling
Arms and shoulders	Free weight lifts, pull-ups, push-ups
Abdomen and back	Abdominal curls, sit-ups, leg lifts (front and side)

The Aerobic Fitness Plan

Aerobic fitness can be achieved by exercising at 60 to 75% of maximal heart rate for at least 30 minutes most of the week (Illustration 27.5).[23] The best kind of aerobic exercise is one that involves all, or most, of the body's major muscle groups in the workout. Beginners should start at 40 to 60% of maximum with low-intensity activities, such as alternating walking with jogging, and increase to a higher level as exercise gets easier. Exercises such as swimming, jogging, running, aerobic dance, and cycling are good choices for subsequent aerobic exercise. Activities such as mowing the lawn, shoveling dirt or snow, and household tasks that elevate heart rate to the target level for 30 minutes are also aerobic exercise. Actually, the most common physical activities of American adults are walking, gardening, and yard work.[24]

Population-Based Physical Activity Recommendations

The American College of Sports Medicine and the Centers for Disease Control and Prevention recommend levels of physical activity for Americans that would benefit health and contribute to a decline in the rising incidence of obesity in the United States.[25] Specifically, it is recommended that moderate-intensity activities be undertaken for at least 30 minutes most days of the week. Moderate-intensity activities include brisk walking, lifting, jogging, tennis playing, lawn mowing (with a push mower), carpet cleaning, child chasing, and fast dancing. Higher, daily levels of moderate-intensity physical activity (60 minutes or more daily) increase weight loss and improve weight control to a greater extent than does less exercise.[26] To complement these aerobic activities, the surgeon general also suggests that people engage in strength-building exercises twice a week.

Some Exercise Is Much Better than None] Sedentary people who take up walking, golfing, biking, gardening, or other similar activities improve their fitness levels. Duration of low- and moderate-intensity exercises counts toward improved blood cholesterol concentrations, decreased body fat, and aerobic fitness.[27]

What's the largest muscle in your body? If your body is like most, its largest muscle is the gluteus maximus (buttocks).

Photo Disc

Illustration 27.5
Source: © 2001 United Feature Syndicate, Inc. Reprinted by permission.

Dilbert / By Scott Adams

TABLE 27.6

AVERAGE ENERGY OUTPUT PER POUND OF BODY WEIGHT FOR SELECTED TYPES OF EXERCISE.

EXERCISE	INTENSITY	CALORIES (POUND/HOUR)	EXERCISE	INTENSITY	CALORIES (POUND/HOUR)
Walking	3 mph (20 min/mi)	1.6	Weight lifting		2.9
	3½ mph (17 min/mi)	1.8	Wrestling		6.2
	4 mph (15 min/mi)	2.7	Handball	Moderate	4.8
	4½ mph (13 min/mi)	2.9		Vigorous	6.2
Jogging	5 mph (12 min/mi)	4.1	Swimming	Resting strokes	1.4
	5½ mph (11 min/mi)	4.5		20 yd/min (mod.)	2.9
	6 mph (10 min/mi)	4.9		40 yd/min (vig.)	4.8
	6½ mph (9 min/mi)	5.2	Rowing	(sculling or machine)	4.8
	7 mph (8½ min/mi)	5.6	Downhill skiing		3.8
	7½ mph (8 min/mi)	6.0	Cross-country	4 mph (15 min/mi)	4.3
Running	8 mph (7½ min/mi)	6.3	skiing (level)	6 mph (10 min/mi)	5.7
	8½ mph (7 min/mi)	6.7		8 mph (7½ min/mi)	6.7
	9 mph (6⅔ min/mi)	7.1		10 mph (6 min/mi)	7.6
	9½ mph (6⅓ min/mi)	7.4	Aerobic dancing	Moderate	3.4
	10 mph (6 min/mi)	7.8		Vigorous	4.3
	11 mph (5½ min/mi)	8.5	Racquetball/		
	12 mph (5 min/mi)	9.5	squash	Moderate	4.3
Rebound				Vigorous	4.8
trampoline	50–60 steps/min	4.1	Tennis	Moderate	3.4
Cycling				Vigorous	4.3
(stationary)	Mild effort	2.9	Volleyball	Moderate	3.4
	Moderate effort	3.4		Vigorous	3.8
	Vigorous effort	4.3	Basketball	Moderate	3.8
Cycling (level)	6 mph (10 min/mi)	1.5		Vigorous	4.8
	8 mph (7½ min/mi)	1.8	Football	Moderate	3.8
	10 mph (6 min/mi)	2.0		Vigorous	4.3
	12 mph (5 min/mi)	2.8	Baseball/golf/woodcutting/horseback riding/		
	15 mph (4 min/mi)	3.9	badminton/canoeing		2.4
	20 mph (3 min/mi)	5.7	Soccer/hill climbing/fencing/judo/snowshoeing		5.3
Skating		2.9	Bowling/archery/pool		1.2
Calisthenics	Moderate	2.4			
	Vigorous	2.9			
Rope skipping	Moderate	4.8			
Bench stepping	12" high, 24 steps/min	3.2			

Source: Values calculated from Guidelines for graded exercise testing and exercise prescription, 2d ed. Philadelphia: American College of Sports Medicine, Lea & Febiger; 1984.

The Caloric Value of Exercise] Some people prefer to base their aerobic exercise workout on a particular level of calories (such as 200), while others simply like to know how many calories they're burning off. Table 27.6 lists the caloric value of various exercises in case you would like to know.

A Cautionary Note] People who have a chronic disease such as diabetes or hypertension, or are out of shape and over 40, should consult a physician before starting

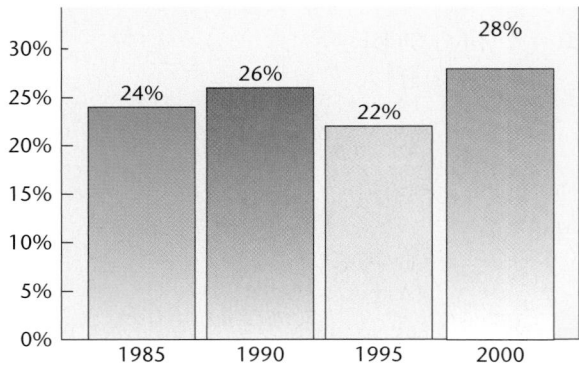

Illustration 27.6

The proportion of people ages 6 and older who engage in no leisure-time physical activity.

Sources: Centers for Disease Control and Prevention, npps.nccd.cdc.gov

an exercise program. (It may well turn out that exercise will be just what the doctor orders.) Moderate exercise programs (those undertaken at 40 to 60% of VO_2 max) are beneficial and safe for almost everyone.[28]

The Paradox of Death during Exercise} "Famous athlete dies while jogging" reads the headline. Why would this happen if exercise is good for you? That's the exercise paradox.

Most people who die while shoveling snow, playing basketball, or jogging are those with existing heart disease either due to hardened arteries (atherosclerosis) or an inborn heart problem. It is a very rare event (because exercise really is good for the heart), and the risk can be reduced by regular exercise versus sporadic and intense physical activity.[29]

U.S. Fitness: America Needs to Shape Up

You don't have to be a high-performance athlete to benefit from exercise. Do the things you enjoy—walk, ride a bike, garden, swim, play. Just get out and move at least 5 times per week for 30 minutes or more.

The benefits of regular exercise are sufficient to make it a matter of public health policy in the United States. Americans have much to gain by getting in better shape.

Studies are undertaken periodically to see how well Americans are doing at increasing their physical activity level. Not all of the results are encouraging (Illustration 27.6). Approximately 55% of adults do not achieve recommended levels of physical activity, and 27% are inactive. The encouraging news is that recent data indicate that rates of physical activity are beginning to increase. If maintained, this trend would help lower rates of overweight, obesity, type 2 diabetes, and other disorders.[31]

Photo Disc

A Focus on Fitness in Children

Physical fitness advocates are especially concerned about the lack of opportunity for physical activity in schools. They point out that children who are not physically active are more likely to be overweight and that a lifelong love for physical activity and a physically active lifestyle often develop during childhood. Regular exercise benefits obese children by lowering body fat, increasing bone density, and improving aerobic fitness. Fewer than one in 10 public schools require students to participate in physical activity daily—and, on average, children undertake vigorous physical activity for only 25 minutes per week.[32]

Schools are being encouraged to include physical activity as a daily part of the curriculum for all students in all grades. It is recommended that students engage in moderate-to-vigorous activities for about an hour each day. Physical fitness goals should be stressed above competitive or sport performance goals.[33] Tests of physical fitness, such as the muscle strength and endurance assessment shown in Illustration 27.7, should be performed periodically on all children and the results used to modify instruction and activities.

Physical activity programs for schoolchildren are most likely to be successful when they offer recreational and sporting activities that are irresistible and fulfill the human desire to run, jump, and play. The happy consequences of physical activity come easiest to those who do it for play.

Illustration 27.7

This child is performing modified pull-ups as part of a fitness assessment.
Source: Greg Merhar, reprinted with permission of the Journal of Physical Education, Recreation and Dance, 59 (9), 1987.

Nutrition UP CLOSE

Exercise: Your Options

FOCAL POINT: Assess your level of physical activity.

More than one-quarter of U.S. adults are physically inactive, and most Americans do not exercise regularly. How much physical activity are you getting?

Physical Activity

Check the category that best describes your usual overall daily activity level. *Do not* consider activities that are part of an exercise program.

- *Inactive:* Sitting most of the day, with less than 2 hours moving about slowly or standing. ☐

- *Average:* Sitting most of the day, walking or standing 2 to 4 hours each day, but not engaging in strenuous activity. ☐

- *Active:* Physically active 4 or more hours each day. Little sitting or standing, engaging in some physically strenuous activities. ☐

Exercise

Do you exercise four or more days a week for at least 30 minutes? ☐ ☐
                                                                    Yes No

FEEDBACK (including scoring) can be found at the end of Unit 27.

Key Terms

aerobic fitness, page 27-4

maximal oxygen consumption, page 27-4

physical fitness, page 27-3

www links

www.nlm.nih.gov/medlineplus/exercisephysicalfitness.html
Covers what's new, exercise recommendations, fitness tips, and public health policies related to physical activity.

www.mayoclinic.org
Head to the Mayo Clinic's Health Oasis on Fitness and Exercise to learn about the whys and hows of increasing your physical activity level.

http://presidentschallenge.org
A fun Web site for kids that include a fitness calculator, activity logs, and awards for meeting goals.

www.kidnetic.com
A popular, highly interactive Web site for children that focuses on healthy eating and regular physical activity.

Notes

1. Dwyer JD et al. Leisure time physical activity and atherosclerosis. Am J Med 2003;115:19–25.

2. Risk factors and coronary heart disease. AHA Scientific Position. www.medscape.com, accessed 7/03.

3. McCaffree J. Physical activity: how much is enough? J Am Diet Assoc 2003; 103:153–4.

4. Physical activity and health: a report of the surgeon general. Washington, DC: U.S. Department of Health and Human Services; 1996.

5. Dwyer et al., Leisure time physical activity; and Balady GJ, Survival of the fittest—more evidence. N Engl J Med 2002;346:852–3; and Atkinson ML, Panama Canal's locomotive "mules" lock ships dead-center in waterway, Newhouse News Service, Sept. 1995.

6. Risk factors and coronary heart disease (www.medscape.com); and Krauss RM et al, AHA Dietary Guidelines, Revision 2000: a statement for healthcare providers from the Nutrition Committee of the American Heart Association, Circulation 2000;102:2296–2311.

7. Dwyer et al., Leisure time physical activity; and Jeffery RW et al., Physical activity and weight loss: does prescribing higher physical activity goals improve outcomes? Am J Clin Nutr 2003;78: 684–9; Physical activity levels in children aged 9–13 years, United States, 2002, MMWR 2003;52:785–8; and Sharma AM, Effects of exercise on plasma lipoproteins, N Engl J Med 2003;348:15.

8. Broeder CE et al., The effects of either high-intensity resistance or endurance training on resting metabolic rate, Am J Clin Nutr 1992;55:802–10; and Broeder C et al., The effects of aerobic fitness on resting metabolic rate, Am J Clin Nutr 1992;55:795–801.

9. Jeffery et al., Physical activity and weight loss; Physical activity levels in children (MMWR); and Sharma, Effects of exercise.

10. Saris WHM. Physiological aspects of exercise in weight cycling. Am J Clin Nutr 1989;49(suppl):1099–1104.

11. Wei M et al., Relationship between low cardiorespiratory fitness and mortality in normal weight, overweight, and obese men, JAMA 1999;278:1547–53; and ACSM fitness book, 2nd ed. American College of Sports Medicine, Champaign (IL): Human Kinetics Publishers, 1998.

12. Coleman E. Eating for endurance. Palo Alto (CA): Bull Publishing Co., 1988.

13. Bouchard C et al., Heredity and trainability of aerobic and anaerobic performances. Sports Med 1988;5:69–73.

14. Position of the American Dietetic Association, Dietitians of Canada, and the American College of Sports Medicine: nutrition and athletic performance. J Am Diet Assoc 2000;100:1543–56.

15. Horton ES. Metabolic fuels, utilization, and exercise. Am J Clin Nutr 1989; 49(suppl):931–2.

16. Satlin B, Astrand P-O. Free fatty acids and exercise. Am J Clin Nutr 1993; 57(suppl):752S–8S.

17. Barrett S. Don't buy phoney "ergogenic aids." Nutrition Forum 1997;May/June: 17, 19–21, 24.

18. Nieman DC et al., Nutrient intake of marathon runners. J Am Diet Assoc 1989;89:1273–8.

19. Noakes TD. Too many fluids as bad as too few. BMJ 2003;327:113–4.

20. Position of the ADA et al.: nutrition and athletic performance.

21. Noakes, Too many fluids.

22. Surgeon general, Physical activity and health: a report, 1966.

23. Lie D. Moderate intensity exercise effective for weight loss. www.medscape.com, accessed 10/03.

24. Surgeon general, Physical activity and health: a report, 1966.

25. Lie, Moderate intensity exercise.

26. Jeffery et al., Physical activity and weight loss; Physical activity levels in children (MMWR); Sharma, Effects of exercise; Weinsier RL et al., Free-living activity energy expenditure in women successful and unsuccessful at maintaining a normal body weight, Am J Clin Nutr 2002;75:499–504; and Dietary Reference Intakes, chapter 12: Physical activity, Washington, DC: National Academies Press; 2002.

27. McCaffree, Physical activity.

28. Surgeon general, Physical activity and health: a report, 1966.

29. Maron BJ. The paradox of exercise. N Engl J Med 2000;343:1409–11.

30. Most US adults not meeting physical activity recommendations. MMWR 2003;53:764–8.

31. Most US adults not meeting physical activity recommendations (MMWR).

32. Health Objectives for the Nation, 2010. Physical Activity. www.cdc.gov/nchs; accessed 8/00; Barbeau P et al., Correlates of individual differences in body composition resulting from physical training in obese children, Am J Clin Nutr 1999;69:705–11; School study seeks to promote effective interventions, Chronic Disease News and Notes 2002;15:9–10; and New report released on Americans' overall physical activity levels, Today's Dietitian 2003;June:5A.

33. Study suggests schools lacking in exercise programs for children, Today's Dietitian 2003;April:12; and Kuntzleman CT, Childhood fitness: What's happening? What needs to be done? Prev Med 1993;22:520–32.

Nutrition **UP CLOSE**

Exercise: Your Options

Feedback for Unit 27

For those of you who answered "Active" to the first question and "Yes" to the second question, congratulations! You qualify as a mover and shaker. Keep up the good work!

If you answered "Inactive" or "Average" to the first question or "No" to the second, now is a good time to think about increasing your level of physical fitness. Besides promoting a feeling of well-being, regular exercise can help beat stress, weight gain, heart disease, and high blood pressure. Maybe it's time you exercised your options by incorporating more physical activity into your daily routine.

Nutrition and Physical Performance

Nutrition Scoreboard

	TRUE	FALSE
1 The diet an athlete consumes affects energy substrate availability to muscles during exercise.		
2 Carbohydrate loading is a waste of time.		
3 Foods and beverages high in carbohydrate and protein increase muscle protein synthesis after strenuous physical activity.		
4 Scientists can't explain why, but bee pollen actually improves physical performance.		

Answers on next page

- Endurance is affected by genetics, training, and nutrition.

- Glycogen stores and endurance can be increased by diet and training.

- Abnormal or absent menstrual cycles in female athletes should not be dismissed, but corrected with increased caloric intake.

- Glucose and fat are used as energy sources during exercise, but in different proportions. Intense physical activity is primarily fueled by glucose, and fats are the main source of energy for low to moderate-intensity exercise.

- A few ergogenic aids that claim to improve performance work to some extent, but most do not.

Answers to *Nutrition Scoreboard*

		TRUE	FALSE
1	Carbohydrate and fat availability to muscles during exercise varies based on usual diet.	✔	
2	Short-term carbohydrate loading increases endurance in trained athletes.		✔
3	Protein powders or bars have not been found to increase muscle mass whether consumed during training or not.		✔
4	In bees, maybe—but not in humans.		✔

Sports Nutrition

BAM! Nobody heard it, but Lou felt it. She had "hit the wall." She was ahead of her planned pace, but now her legs felt like lead. She would have to finish the last two miles of the marathon at the slow pace her legs would allow.

From her carefully crafted and scrupulously followed training program to her refined shaping of mental attitude, Lou thought she had done everything right. She had left one thing out, however, and that may have cost her the race. Lou failed to pay attention to her diet while training and ran out of *glycogen* too soon (Illustration 28.1).

Three major factors affect physical performance: genetics, training, and nutrition.[1] The first gives some people an innate edge in sprinting or endurance, and nothing can be done about it. The second is acknowledged as a basic truth. Most athletes know a good bit about proper training, and the trick is to follow the right plan. The third is often ignored or, when taken seriously, misunderstood.

Nutrition has important effects on physical performance, but the legitimate role of nutrition is often poorly understood by athletes and coaches alike.[2] Athletes' lack of knowledge may make them vulnerable to phony nutrition claims and bad dietary advice. It leaves some coaches wishing they had taken a college course on nutrition. Only 1 in 10 coaches has had a course in nutrition, but nearly all of them regularly dispense nutritional advice.[3] (To the coaches and future coaches who are taking this nutrition course: Bravo!)

Basic Components of Energy Formation during Exercise

Understanding the role of nutrition and performance, and the potential effects of some *ergogenic aids,* can be fostered by basic knowledge of how energy is formed within muscle cells. Illustration 28.2 summarizes the processes by which energy for muscle movement is formed.

glycogen
The storage form of glucose. Glycogen is stored in muscles and the liver.

ergogenic aids
(*ergo* = work; *genic* = producing) Substances that increase the capacity for muscular work.

There are two main substrates for energy formation in muscles: glucose from muscle and liver glycogen, and fatty acids released from fat stores. How much of each is used depends on the intensity and duration of the exercise, as well as the body's ability to deliver each along with oxygen to muscle cells. Each substrate is used to form *ATP* from *ADP*. ATP serves as the source of energy for muscle contraction.

Anaerobic Energy Formation] Glucose from the liver and muscle glycogen (which is converted to glucose) form ATP without oxygen. This route of energy formation is "anaerobic," or "without oxygen," and it generates most of the energy used for intense muscular work (70% VO$_2$ max).[4] Creatine phosphate (abbreviated CrP in Illustration 28.2), an amino acid containing a high-energy phosphate molecule, also converts ADP to ATP to a limited extent. Creatine phosphate stores are limited and decrease rapidly during intensive exercise.

Glucose is converted to "pyruvate" during anaerobic energy formation. In the absence of oxygen, pyruvate is converted to lactate. Lactate can build up in muscles and blood if not reconverted to pyruvate by the addition of oxygen. Pyruvate yields additional energy when it enters "aerobic" energy formation pathways along with fatty acids from fat stores.

Aerobic Energy Formation] The conversion of pyruvate and fatty acids to ATP requires oxygen. Much more ATP is delivered by the breakdown of fatty acids than glucose (fats have 9 calories per gram; glucose has only 4). The rate of energy formation from fatty acids is four times slower than that from glucose, however. It's the reason fatty acids are used to fuel low- and moderate-intensity exercise, or those below 60% VO$_2$ max.[5] Unlike glucose, energy formation from fatty acids is not limited by availability. Muscle cells can continue to produce energy from fatty acids as long as delivery of oxygen from the lungs and the circulation is sufficient.[6]

PhotoEdit - Tony Freeman

Illustration 28.1
Hitting the wall. *This runner ran out of muscle glycogen before the finish line.*

ATP, ADP
Adenosine triphosphate (ah-den-o-scene tri-phos-fate) and adenosine diphosphate. Molecules containing a form of phosphorous that can trap energy obtained from the macronutrients. ADP becomes ATP when it traps energy, and returns to being ADP when it releases energy for muscular and other work.

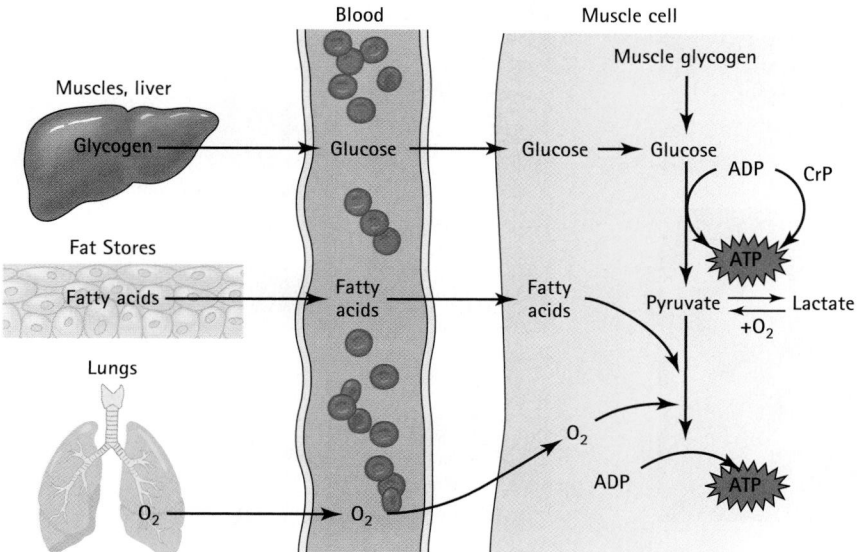

Illustration 28.2
Schematic representation of how ATP is formed for muscular movement.

Nutrition and Physical Performance

When everything else is equal, nutrition can make the difference between winning and losing.
—R Maughan, "The Athlete's Diet"[7]

Glycogen stores; foods eaten before, between, and after exercise events; water intake; and *electrolyte* balance are all related to energy formation and physical performance.

Glycogen Stores and Performance

The vast majority of energy for muscle activity comes from fat and glucose. For most activities, both types of fuel are used; and the more intense the activity, the higher the amount of fat and glucose used for it. Fats, however, are the principal source of fuel for activities of low to moderate intensity, and glucose is the main source of energy for activities of high intensity (Illustration 28.3).[8]

Glycogen stores in muscles and the liver can deliver about 2000 calories worth of energy, whereas adults have access to over 100,000 calories from fat. Consequently, a person's ability to perform continuous, intense physical activity is limited by the amount of glycogen stored.[9] People who run out of muscle glycogen during an event "hit the wall"—they have to slow down their pace substantially because they can no longer use muscle glycogen as a fuel. The pace they are able to maintain will be dictated by the body's ability to use fat and liver glycogen as fuel for muscular work. If athletes keep pushing themselves after muscle glycogen runs out, they may end up "bonking"—using up the liver's supply of glycogen. That's worse than hitting the wall because hypoglycemia (low blood sugar) develops, and the person becomes dizzy and shaky and may pass out.[10] Obviously, endurance athletes do not want to exhaust their glycogen stores before an event is over.

Illustration 28.3
Fat is the main source of energy for low- and moderate-intensity activities, whereas glycogen is the primary fuel for high-intensity events.

Activities primarily fueled by fat.

Activities primarily fueled by glycogen.

TABLE 28.1

EFFECTS OF DIET ON MUSCLE GLYCOGEN STORES DURING TRAINING.[12]

CARBOHYDRATE INTAKE LEVEL FOR PREVIOUS 3 DAYS	MUSCLE GLYCOGEN (GRAMS PER 100 GRAMS OF MUSCLE)	TIME TO EXHAUSTION AT 75% VO$_2$ MAX
Low	0.6	60 minutes
Average	1.8	126 minutes
High (65–70% of total calories)	3.5	189 minutes

TABLE 28.2

HOW MANY GRAMS OF CARBOHYDRATE ARE IN A DIET THAT PROVIDES 60 OR 70% OF TOTAL CALORIES FROM CARBOHYDRATES? IT DEPENDS ON YOUR CALORIC INTAKE.

USUAL CALORIC INTAKE	GRAMS OF CARBOHYDRATE	
	60% CARB DIET	70% CARB DIET
1600	240 g	280 g
1800	270 g	315 g
2000	300 g	350 g
2200	330 g	385 g
2400	360 g	420 g
2600	390 g	455 g
2800	420 g	490 g
3000	450 g	525 g
3200*	480 g	560 g

*For each additional 200 calories of usual intake, add 30 grams carbohydrate for a 60% carbohydrate diet and 35 grams for a 70% carbohydrate diet.

Athletes consuming a typical U.S. diet normally have enough glycogen stores to fuel continuous, intense exercise for about an hour or two.[11] Both glycogen stores and endurance can be increased, however, by loading up muscles with glycogen for three days prior to an event (Table 28.1).

Carbohydrate Loading] The recommended method for increasing glycogen stores for endurance events is much less stringent than in years past. It basically calls for increasing carbohydrate intake to 60 to 70% of calories for the 24-hour period preceding high-intensity exercise. Loading up on carbohydrates in this 24-hour period can increase glycogen stores to the same extent as previous regimes that lasted up to 6 days.[13] Table 28.2 shows how many grams of carbohydrate correspond to 60 or 70% of total calories for people with different caloric needs. Average diets provide about 50% of calories from carbohydrates. See Illustration 28.4 for an example of a 1-day diet that provides 68% of total calories from carbohydrate.

Illustration 28.4

A high-carbohydrate diet. *Carbohydrates provide 68% of the 2420 calories in this day's menus, protein provides 11%, and fats provide 21%. Grams of carbohydrate are shown in parentheses.*

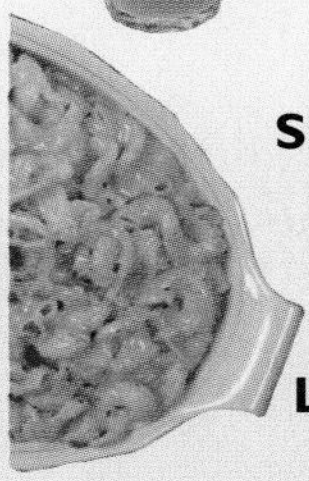

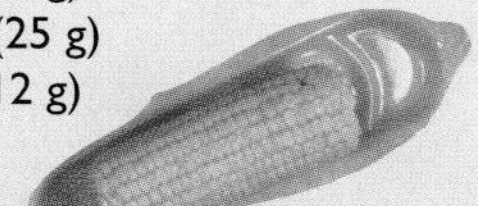

BREAKFAST
Orange juice — 1 cup (25 g)
Oatmeal — 1 cup (25 g)
Brown sugar — 1 tablespoon (13 g)
Skim milk — 1 cup (12 g)
Banana — 1 (27 g)

SNACK
Granola — 1/2 cup (29 g)
Apple juice — 1 cup (30 g)

LUNCH
Lettuce salad — 2 cups (10 g)
Salad dressing — 2 tablespoons (4 g)
Macaroni & cheese — 1 cup (40 g)
Crackers — 4 squares (9 g)
Sliced tomato — 1/2 (3 g)
Skim milk — 1 cup (12 g)

SNACK
Cornflakes — 1 cup (20 g)
Canned peaches — 1/2 cup (25 g)
Skim milk — 1 cup (12 g)

DINNER
Spanish rice — 1 cup (41 g)
Baked potato — 1 (35 g)
Corn — 1/2 cup (17 g)
Apple crisp — 1/2 cup (43 g)
Iced tea — 2 cups (0 g)

TABLE 28.3	
HIGH AND MODERATE GLYCEMIC INDEX FOODS AND INGREDIENTS.[14]	
HIGH GLYCEMIX INDEX	MODERATE GLYCEMIC INDEX
Glucose (dextrose, dextrin)	Popcorn
Maltose (malt)	Carrots
Coco Pops	Honey
Corn Chex	Sucrose
Cornflakes	Bread
Corn Pops	Oatmeal
Gatorade	Soft drinks
Doughnuts	Orange juice
Jelly beans	Muffins
Potatoes	Bananas

Glycemic Index and Carbohydrate Loading} Food sources of carbohydrate that have high or moderate glycemic index (GI) values (Table 28.3) appear to maximize glycogen storage. These foods increase blood glucose levels more than do low-GI carbohydrates such as dried beans, fructose, apple juice, and oranges.[15] Regular intake of high-GI carbohydrates is generally discouraged because it may increase the risk of insulin resistance and type 2 diabetes. Athletes tend to be very sensitive to the action of insulin, however, and exercise in general lowers the risk of diabetes.[16] Until the appropriate studies are undertaken, effects of chronically high intake of high-GI foods on the health status of athletes will remain unknown.

Carbohydrate-Loading Caveats} There is a limit on glycogen storage in muscle. Each gram of glycogen binds with 2 grams of water, limiting the total amount that can "fit" into muscle tissue. Once the capacity of muscles to store glycogen is reached, any additional carbohydrate is largely converted to fat and stored.[17]

Not all athletes tolerate the relatively high-carbohydrate, low-fat, carbohydrate-loading diet equally well. Some athletes dislike the stiff, heavy feeling that may occur when muscles become "loaded" with glycogen. In addition, it should be remembered that carbohydrate loading does not improve performance in athletes participating in physical activities that last less than an hour or two. The body normally has enough glycogen to meet the needs of athletes participating in intense but short-duration activities.

Protein Need

Many athletes require no more than their RDA for protein. Individuals undertaking strength training may need up to 15 grams more protein daily, however. Even then, most people (including athletes) consume far more than their RDA for protein, so they are unlikely to need extra protein in their diets.[18]

Preexercise and Recovery Foods

The types of foods eaten before and after hard physical activity influence performance. Generally, at least one meal and snack are consumed prior to exercise, and both should be high in carbohydrates. A high-carbohydrate meal helps top off glycogen stores, and a high-carbohydrate snack consumed within an hour of exercise increases glucose availability for energy formation while reducing reliance on glycogen stores.[19] The snack consumed postexercise should provide carbohydrate

In 1928 Richard Halliburton paid 36 cents as the toll for swimming the length of the Panama Canal. It's the lowest toll ever paid.[23]

AP Photo

AP

REALITY CHECK

Sweet energy burst?

Can you get a burst of energy by eating honey, candy, or another sweet just before a competition?

Who gets thumbs up

?

Toro:
That's what I do! It's why they call me El Toro!

Lakisha:
That's not what I heard. My coach says it will just make you lose steam sooner.

Answers on next page

and protein. Carbohydrates raise blood insulin levels, and that increases transfer of amino acids (provided by the protein) into muscle cells. Amino acid entry into muscle cells increases protein synthesis and hastens muscle recovery.[20] Illustration 28.5 provides examples of food and beverages that qualify as preexercise meals and snacks, and recovery snacks.

Athletes should make sure to drink sufficient water and electrolyte-replacement fluids before, during, and after exercise to maintain the body's fluid balance.

Fat Loading

Fat is the best fuel for endurance exercise; trials have been run to see if high-fat diets (60–70% of calories) increase muscle fat content and spare glycogen. Results of the trials show that fat loading does increase muscle content of fat and fatty acid utilization during endurance exercise. It also decreases glycogen utilization to an extent. However, gains are small, and muscle utilization of fat for energy plateaus within 2 weeks after fat loading begins.[22]

Hydration

Muscular activity produces heat that must be eliminated to prevent the body from becoming overheated. To keep the body cool internally, the heat produced by muscles is collected in the blood and then released both through blood that circulates near the surface of the skin and in sweat. Sweating cools the body because heat is

Illustration 28.5
Examples of a preexercise meal and snacks, and recovery snacks.

1 PREEXERCISE MEAL
Orange juice
Oatmeal
Honey
Muffin
Coffee, milk

2 PREEXERCISE SNACKS
Bananas
Crackers
Jell-O
Sherbet
Fruit juice
Bagel

3 RECOVERY SNACKS
Low-fat yogurt
Frozen yogurt
Chocolate skim milk
High-carb/protein bar

Toro

Lakisha

Photo Disc

ANSWERS TO **REALITY CHECK**

Sweet energy burst?

Bravo, El Toro! Yours is the better answer. It won't make you start a race like you were shot out of a cannon, but it helps. Consumption of foods that increase blood glucose levels within an hour of exercise helps preserve glycogen stores, decreases mental fatigue, and increases endurance somewhat in trained athletes.[21]

released when water evaporates on the skin. People sweat more during physical activity in hot, humid weather because it is harder to add moisture and heat to warm, moist air than to dry, cool air. To stay cool during exercise in hot, humid conditions, the body must release more heat and water than when exercise is undertaken in drier, cooler conditions.

Athletes engaged in short events (less than an hour) can generally stay appropriately hydrated by drinking 2 cups of water 15 to 20 minutes before the event. When events are longer or conditions are hot and humid, athletes should consume enough fluid from water and electrolyte-replacement beverages to maintain fluid balance. Athletes should drink when they are thirsty—and more than that if they lose weight during exercise. Fluid balance is maintained when athletes do not lose weight during exercise and when their urine remains pale yellow and is normal in volume.[24]

Illustration 28.6
Effects of dehydration.[26]
Loss of 9 to 12% of body water can be fatal. Even a low level of dehydration impairs physical performance.

Bob Winslett/Index Stock

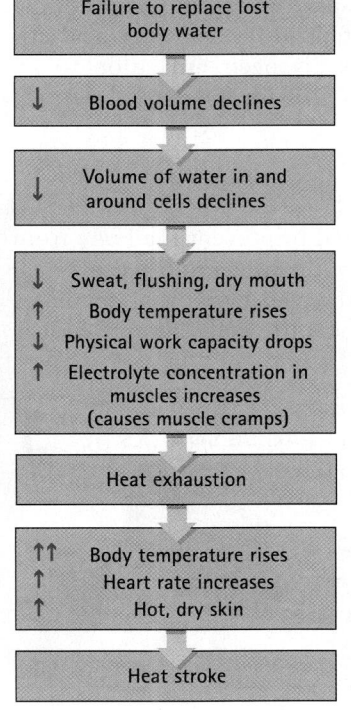

Failure to replace lost body water

↓ Blood volume declines

↓ Volume of water in and around cells declines

↓ Sweat, flushing, dry mouth
↑ Body temperature rises
↓ Physical work capacity drops
↑ Electrolyte concentration in muscles increases (causes muscle cramps)

Heat exhaustion

↑↑ Body temperature rises
↑ Heart rate increases
↑ Hot, dry skin

Heat stroke

Water is usually sufficient for exercise lasting less than an hour. For bouts of exercise longer than an hour, drinking water alternated with a 4 to 8% carbohydrate and electrolyte sports drink improves hydration and performance.[25]

Dehydration: The Consequences] Loss of more than 2% of body weight (2–4 pounds) during an event indicates that the body is becoming dehydrated. Effects of dehydration range from mild to severe, depending on how much body water is lost (Illustration 28.6). Any amount of dehydration impairs physical performance. At the extreme, dehydration can lead to heat exhaustion or heat stroke (Table 28.4). People who overexercise in hot weather when they are out of condition are most likely to suffer heat exhaustion or heat stroke, but these conditions occasionally occur among seasoned athletes, too. Heat exhaustion can be remedied by fluids and electrolytes, but heat stroke requires emergency medical care.[28]

Hyponatremia and Excess Water] A few cases of hyponatremia—or sodium deficiency— have occurred in athletes competing in marathons over the past few years. The news hit the sports headlines, causing renewed concern about replacing sodium during exercise.

Hyponatremia is a rare but serious condition that produces some of the same symptoms as heat stroke. It can occur when only water is consumed during long events such as marathons. Low sodium status can be prevented by maintaining fluid balance with water and sports drinks containing sodium, potassium, and other electrolytes.[29]

Fluids high in sugar, such as soft drinks, don't quench a thirst and should not be consumed for fluid replacement. Their high-sugar content may draw fluid from the blood into the intestines, thereby increasing the risk of dehydration, nausea, and bloating. Alcohol-containing beverages such as beer, wine, and gin and tonics are not hydrating, either.[30]

Body Fat and Weight: Heavy Issues for Athletes

As a group, athletes tend to be very concerned about their weight and body shape, and they are leaner than the U.S. population in general (Table 28.5).[31] But the advantages of low body weight and body fat disappear when they become too low.[32] Body fat levels of less than 3% in men and less than 12% in women can seriously interfere with health. These percentages represent the obligatory body fat content that people need for the functions of fat unrelated to energy production.[33] Everybody needs some fat to use for hormone production, to maintain normal body temperature in cold weather, and to cushion internal organs.

Exercise, Body Fat, and Health in Women] It is not uncommon for female athletes to experience irregular or absent menstrual periods or the late onset of periods during adolescence (Table 28.6).[35] These aberrations in menstrual period appear to be related to specific effects of high levels of physical activity or caloric intakes deficits on hormone production or the body's metabolic function.[36] Female athletes

TABLE 28.4

A PRIMER ON HEAT EXHAUSTION AND HEAT STROKE.[27]

Heat exhaustion: A condition caused by low body water and sodium content due to excessive loss of water through sweat in hot weather. Symptoms include intense thirst, weakness, paleness, dizziness, nausea, fainting, and confusion. Fluids with electrolytes and a cool place are the remedy. Also called "heat prostration" and "heat collapse."

Heat stroke: A condition requiring emergency medical care. It is characterized by hot, dry skin, labored and rapid breathing, a rapid pulse, nausea, blurred vision, irrational behavior, and, often, coma. Internal body temperature exceeds 105°F due to a breakdown of the mechanisms for regulating body temperature. Heat stroke is caused by prolonged exposure to environmental heat or strenuous physical activity. The person affected by heat stroke should be kept cool by any means possible, such as removing clothing and soaking the person in ice-cold water. If conscious, the person should be given fluids. Also called "sun stroke."

TABLE 28.5

AVERAGE BODY FAT CONTENT OF VARIOUS ATHLETES.

	BODY FAT CONTENT (AS PERCENTAGE OF BODY WEIGHT)	
	WOMEN	MEN
Bodybuilders	10%	6%
Long-distance runners	17	6–13
Baseball players	–	12–14
Basketball players	21–27	7–10
Football players	–	9–19
Gymnasts	10	5
Soccer players	–	10
Tennis players	20	16
Wrestlers	–	9

TABLE 28.6

INCIDENCE OF IRREGULAR OR ABSENT MENSTRUAL CYCLES IN FEMALE ATHLETES AND SEDENTARY WOMEN.[34]

Joggers (5 to 30 miles per week)	23%
Runners (over 30 miles per week)	34
Long-distance runners (over 70 miles per week)	43
Competitive bodybuilders	86
Noncompetitive bodybuilders	30
Volleyball players	48
Ballet dancers	44
Sedentary women	13

Illustration 28.7

Dropping weight quickly in the days before a match can wreck a wrestler's chances and harm his health.

with missing and irregular periods are at risk for developing low bone density, osteoporosis, and bone fractures. Women at particular risk of bone fractures are those with the "female-athlete triad" of disordered eating, amenorrhea (pronounced "a-men-or-re-ah" and meaning "no menstrual periods"), and osteoporosis. Abnormal menstrual cycles should not be dismissed as a normal part of training. Normal menstrual periods should be reinstated through increased caloric intake.[37] Suppressed testosterone levels or other metabolic changes due to caloric deficits in male athletes may be the parallel to menstrual irregularities in female athletes.[38]

Wrestling: The Sport of Weight Cycling]   "Making weight" is a common and recurring practice among wrestlers (Illustration 28.7). Most wrestlers will "cut" 1 to 20 pounds over a period of days between 50 and 100 times during a high school or college career.[39] That makes wrestling more than a test of strength and agility. It makes it a contest of rapid weight loss.

For competitive reasons, wrestlers often want to stay in the lowest weight class possible and may go to great lengths to achieve it. They may fast, "sweat the weight off" in saunas or rubber suits, or vomit after eating to lose weight before the weigh-in. These practices can be dangerous, even life-threatening, if taken too far. After the match is over, the wrestlers may binge and regain the weight they lost.

Wrestlers, like other athletes involved in intense exercise, perform better if they have a good supply of glycogen and a normal amount of body water. Fasting before a weigh-in dramatically reduces glycogen stores, and withholding fluids or losing water by sweating puts the wrestler at risk of becoming dehydrated. Trying to stay within a particular weight class too long may also stunt or delay a young wrestler's growth.

The American Medical Association and the Association for Sports Medicine recommend that wrestling weight be determined after 6 weeks of training and normal eating. In addition, a minimum of 7% body fat should be used as a qualifier for assigning wrestlers to a particular weight class.

Wisconsin high school wrestling programs adopted the medical association's recommendations and found that wrestlers, coaches, and parents supported the changes. Participation in wrestling increased because the athletes didn't have to deal with weight cutting.[40]

Iron Status of Athletes

Iron status is an important topic in sports nutrition because iron-deficiency (or low iron stores) and iron-deficiency anemia (or low blood hemoglobin level) decreases endurance. Iron is a component of hemoglobin, a protein in blood that carries oxygen to cells throughout the body, and it works with enzymes involved in energy production. When iron stores or hemoglobin levels are low, less oxygen is delivered to cells and less energy is produced than normal.[41]

Female athletes are at higher risk of iron-deficiency anemia than other females. Consequently, it is recommended that female athletes especially pay attention to the amount of iron consumed.[42]

TABLE 28.7

FASTER, HIGHER, STRONGER, LONGER. A BRIEF HISTORY OF PERFORMANCE-ENHANCING SUBSTANCES USED BY ATHLETES.[43]

B.C.	Large quantities of beef consumed by athletes in Greece to obtain "the strength of 10 men." Deer liver and lion heart were consumed for stamina.
1880s	Morphine used to increase performance in (painful) endurance events.
1910s	Strychnine consumed for the same reason as morphine.
1930s	Amphetamines used to increase energy levels and endurance; testosterone taken to increase muscle mass.
1980s	Blood doping, EPO used to increase endurance.

Ergogenic Aids: The Athlete's Dilemma

The quest for ownership of a competitive edge has drawn athletes to ergogenic aids throughout much of history. (See Table 28.7 for a historical review of ergogenic aids use.) Relatively few of the hundreds of products available work, and most are sold as "dietary supplements" so they do not have to be tested for safety. Those found to increase muscle mass, strength, or endurance are usually banned for use by competitive athletes. Several of the aids clearly pose serious risks to health, and others simply represent misguided hopes and misspent money. Table 28.8 summarizes claims made for a variety of ergogenic aids, research results, and what is known about adverse effects.

The Path to Improved Performance] Genetics, training, and nutrition: these are the real keys to physical performance. Although other aids will be sought, those that exceed the boundaries of what is considered fair and safe will not be approved for use by athletes. After all, athletic competition is not a test of drugs or performance aids. It's a test of an individual's ability to excel. Anything less wouldn't be sporting.

TABLE 28.8

ERGOGENIC AIDS: CLAIMS AND EVIDENCE.[44]

ERGOGENIC AID— NORMAL FUNCTIONS	CLAIMS	EVIDENCE
Amphetamines • No normal function.	Reduce fatigue, improve concentration, increase aggressiveness, decrease appetite and weight.	Reduce fatigue, appetite, and weight; increase aggressiveness. **Adverse effects:** Many, including dizziness, tremors, confusion, paranoia, increased heart rate, chest pain, cardiovascular collapse.
Androgenic anabolic steroids • Testosterone and related substances. • Promote protein synthesis, decrease protein tissue breakdown, increase blood volume and red blood cell production.	Gradually increasing doses of several forms of steroids increase muscle mass.	Increase strength, muscle mass; decrease recovery time. **Adverse effects:** Liver, heart, psychiatric disorders, testicle atrophy, reduced HDL and elevated LDL cholesterol.
Bee pollen • Pollen collected by bees from flowers and mixed with bee secretions. Provides food for bee larva.	Detoxification of blood, improves performance.	No such effects found. **Adverse effects:** Allergic reaction in people who are allergic to bee stings and honey, and who have asthma.
Boron • A mineral that influences calcium, magnesium, and phosphorus utilization.	Increases testosterone and muscle mass.	No such effects found. May improve bone density, arthritis.
Branched chain amino acids • Isoleucine, leucine, valine. Are essential amino acids. May lower brain serotonin levels.	Improve performance, delay fatigue, build muscle, increase mental concentration.	May delay muscle breakdown, spare glycogen, increase fat utilization and time to exhaustion. **Adverse effects:** High doses (>20 g) may impair performance.

continued

TABLE 28.8

ERGOGENIC AIDS: CLAIMS AND EVIDENCE. CONTINUED

ERGOGENIC AID—NORMAL FUNCTIONS	CLAIMS	EVIDENCE
Caffeine • Central nervous system stimulant, diuretic, circulatory and respiratory stimulant.	Decreases fatigue in endurance events.	May reduce fatigue and improve endurance. **Adverse effects:** Nervousness, sleeplessness, irregular heartbeat; fatal dose 3–10 g. Olympic illegal dose equivalent to 6–8 cups coffee 2–3 hours pre-event.
Carnitine • A compound required for energy formation from fatty acids.	Enhances fatty acids utilization for energy.	Most studies show no effect on fatty acid utilization during exercise in healthy adults.
Cocaine • No normal functions.	Same as for stimulants but much shorter acting.	
Chromium Picolinate • Involved in energy metabolism and insulin utilization.	Increases lean body mass, decreases fat mass.	No effect on body composition. **Adverse effects:** May increase oxidative damage to muscles.
Creatine • Component of muscle; generates ATP from ADP in muscle.	Increases performance in high-intensity, short-duration exercises.	Increases peak muscle force in short-term, high-intensity exercises, no increase in strength but may increase energy available for training. **Adverse effects:** Nausea, vomiting, diarrhea, increased blood pressure, muscle cramps.
DHEA, Andro **(Dehydroepinandrosterone, Androstendione)** • Precursors of testosterone.	Promote testosterone production and increase muscle mass.	Promote testosterone production and increase muscle mass. **Adverse effects:** The same as for anabolic steroids.
Ephedra (ma huang) • No normal function.	Central nervous system stimulant; enhances endurance, strength, and body fat loss.	Increases aerobic capacity when combined with caffeine. **Adverse effects:** Nervousness, anxiety, rapid heart rate, headache, hypertension, death. Not recommended for use; no longer sold.
EPO (Erythropoietin) • A hormone that increases red blood cell production by bone marrow.	Replaces blood doping, or injection of red blood cells for increasing oxygen availability and endurance.	Increases endurance. **Adverse effects:** Increases blood viscosity, blood pressure, blood clot formation; headache.
GBL (Gammabutyrolactone, furanodehydro) • No normal function.	Increases muscle mass and muscle recovery from exercise.	No such effects **Adverse effects:** Vomiting, seizures, death. FDA has called for its voluntary removal from the market.

TABLE 28.8

ERGOGENIC AIDS: CLAIMS AND EVIDENCE. CONTINUED

ERGOGENIC AID— NORMAL FUNCTIONS	CLAIMS	EVIDENCE
Ginseng, Siberian • No normal function.	Enhances endurance and strength.	Increases muscular strength and aerobic work capacity.
Glycerol • A clear, sweet fluid attached to fatty acids in fat molecules.	Increases hydration.	Conflicting results. **Adverse effects:** Bloating, nausea.
HGH (Human growth hormone) • Stimulates muscle and bone growth; converted to glucose.	Increases muscle mass and strength; similar effects as anabolic steroids but safer.	Increases muscle mass but probably not strength. **Adverse effects:** Growth of facial bones, heart disease, stroke, hypertension, diabetes. Effects generally irreversible.
HMB (β-hydroxy-β-methylobutyrate) • Derivative of the amino acid leucine.	Improve muscle size and strength.	No performance improvement, may help muscle recovery. **Adverse effects:** Appears safe at doses of 3 grams per day.
Insulin • Speeds entry of glucose into cells, increases production of glycogen.	Increases muscle mass.	Increases glucose passage into cells and glycogen and protein synthesis. **Adverse effects:** Low blood sugar, seizure, brain damage, death.
Oxygen, oxygen drinks and bars	Increase oxygen availability for muscular work.	No effect on oxygen availability in the body. **Adverse effects:** Burping if consumed.
Protein powders • Can contribute to meeting the body's need for protein.	Increase muscle mass and strength.	No such effects found. **Adverse effects:** Dehydration, liver and kidney problems.
Vitamin and mineral supplements • May contribute to meeting need for vitamins and minerals.	Improve energy and stamina.	Such supplements improve energy and stamina in deficient individuals only. **Adverse effects:** Toxicity reactions.

Nutrition UP CLOSE

Testing Performance Aids

FOCAL POINT: The critical examination of studies on performance aids.

Read the following summary of a study (imaginary) of the effects of a phosphorous supplement on strength, and then answer the critical thinking questions.

- *Purpose:* To assess the effect of a phosphorous supplement on strength.
- *Methods:* Twenty volunteers from the crew team were given the phosphorous supplement for a week. Strength, assessed as the maximum number of push-ups a study participant could do in one session, was assessed before and after supplementation. Participants recorded any supplement side effects.
- *Results:* The number of push-ups increased by an average of 5% after supplementation. Diarrhea was the only side effect consistently noted by participants.

Critical Thinking Questions

1. Does the study demonstrate that the supplement is safe?

   Yes _____ No _____ Give reasons for your answer.

   _____

2. Does the study demonstrate that the supplement increases strength?

   Yes _____ No _____ Give reasons for your answer.

   _____

FEEDBACK (answers to these questions) can be found at the end of Unit 28.

Key Terms

ATP, ADP, page 28-3

electrolytes, page 28-4

ergogenic aids, page 28-2

glycogen, page 28-2

www links

www.mayohealth.org
The Mayo Clinic's Web site offers reliable advice on exercise and sports nutrition.

www.acsm.org
The American College of Sports Medicine Web site offers information on sports nutrition and the opportunity to join the national Coalition for Promoting Physical Activity.

www.nal.usda.gov/fnic/
Select the "sport nutrition" link.

www.nutrition.gov
Search "physical performance" and find a wealth of information on study results, evaluation systems, and guidelines.

www.runnersworld.com
This site isn't just for runners! Athletes

looking for a competitive edge should check this site out.

www.gssiweb.com
Connects you to the Gatorade Sports Science Institute. Site provides valuable resources for sports nutrition professionals through "Science Center" and "Topics" buttons.

Notes

1. Bannister R. Special presentation. Health, fitness, and sport. Am J Clin Nutr 1989;49(suppl):927–30.

2. Barrett S. Don't buy phony "ergogenic aids." Nutrition Forum 1997;May/June: 17, 19–21, 24.

3. Corley G et al. Nutrition knowledge and dietary practices of college coaches. J Am Diet Assoc 1990;90:705–9.

4. Coleman E. Fat loading for endurance sports. Today's Dietitian. 2003;Mar.: 12–16.

5. Coleman, Fat loading.

6. Brown RC. Nutrition for optimal performance during exercise: carbohydrate and fat. Curr Sports Med 2002;1:222–9.

7. Maughan R. The athlete's diet: Nutritional goals and dietary strategies. Proc Nutr Soc 2002;61:87–96.

8. Coleman, Fat loading.

9. Brown, Nutrition for optimal performance.

10. Coleman E. Eating for endurance. Palo Alto (CA): Bull Publishing Co., 1988.

11. Maughan, The athlete's diet.

12. Foster C et al. Effects of preexercise feeding on endurance performance. Med Sci Sports Exer 1979;11:1–5.

13. Fairchild TJ et al. Rapid carbohydrate loading after a short bout of near maximal-intensity exercise. Med Sci Sports Exercise 2002;34:980–6.

14. Noakes TD, Too many fluids as bad as too few, BMJ2003;327–113–4; and Foster-Powell K et al., International tables of glycemic index and glycemic load, Am J Clin Nutr 2002;76:5–56; and Brand-Miller J et al., Low glycemic index foods and glycemic control in diabetes, Diabetes Care 2003;26:2261–7.

15. Coleman, Fat loading; and Suzuki M, Glycemic carbohydrates consumed with amino acids or protein right after exercise enhance muscle formation, Nutr Rev 2003;61:S88–S94.

16. Jenkins DJ et al, Glycemic index: overview of implications in health and disease, Am J Clin Nutr 2002;76(suppl): 266S–73S; and Schrauwen-Hinderling VB, The increase in intra-myocellular lipid content is a very early response to training, J Clin Endocrin Metab 2003; 88:1610–6.

17. Brouns F et al. Utilization of lipids during exercise in human subjects: metabolic and dietary constraints. Brit J Nutr 1998;79:117–28.

18. Kleiner SM. Healthy muscle gains. Phys Sports Med 1995;23:21–22.

19. Maughan, The athlete's diet.

20. Suzuki, Glycemic carbohydrates.

21. Lieberman HR et al. Carbohydrate administration during a day of sustained aerobic activity improves vigilance, as assessed by a novel ambulatory monitoring device, and mood. Am J Clin Nutr 2002;76:120–7.

22. Coleman, Fat loading; Schrauwen-Hinderling, The increase in intramyocellular lipid content.

23. Atkinson ML. Panama Canal's locomotive "mules" lock ships dead-center in waterway. Newhouse News Service, Sept. 1955.

24. Noakes TD. Too many fluids as bad as too few. BMJ 2003;327:113–4.

25. Kleiner SM. Water: an essential but overlooked nutrient. J Am Diet Assoc 1999;99:200–6.

26. Noakes, Too many fluids; and Gisolfi CV, Fluid balance for optimal performance, Nutr Rev 1996;54:S159–S68.

27. Noakes, Too many fluids; and Bouchama A, Knochel JP, Heatstroke, N Engl J Med 2002;346:1978–88.

28. Bouchama and Knochel, Heatstroke.

29. Barr SI, Costill DL, Water: can the endurance athlete get too much of a good thing? J Am Diet Assoc 1989;89:1629–35; Noakes, Too many fluids; and Gisolfi, Fluid balance.

30. Coleman, Eating for endurance; and Gisolfi, Fluid balance.

31. Wilmore JH. Body composition in sports and exercise: directions for future research. Med Sci Sports Exer 1983;15: 21–31.

32. Manson JE, Lee I-M. Exercise for women: how much pain for optimal gain? N Engl J Med 1996;334:1325–6.

33. Elliot DL, Goldberg L. Nutrition and exercise. Med Clin North Am 1985;69: 71–81.

34. Beals KA, Eating behaviors, nutritional status, and menstrual function in elite female adolescent volleyball players, J Am Diet Assoc 2002;102:1293–6; Warren MP et al., Hormone therapy and bone loss in dancers with amenorrhea, Fertil Steril 2003;80:398–404; Ireland ML, Nattiv A, The female athlete, Philadelphia: W.B. Saunders; 2003; Lloyd T et al., Interrelationships of diet, athletic activity, menstrual status, and bone density in collegiate women, Am J Clin Nutr 1987;46:681–4; Walberg JL, Johnston CS, Menstrual function and eating behavior in female recreational weight lifters and competitive body builders, Med Sci Sports Exer 1991; 23:30–36; Loucks AB, Effects of exercise training on the menstrual cycle: existence and mechanisms, Med Sci Sports Exer 1990;22:275–80.

35. Beals, Eating behaviors; Warren et al., Hormone therapy; and Ireland and Nattiv, The female athlete.

36. Lloyd et al., Interrelationships of diet.

37. Ireland and Nattiv, The female athlete.

38. Brownell KP et al. Weight regulation practices in athletes: analysis of metabolic and health effects. Med Sci Sports Exer 1987;19:546–56.

39. Steen SN et al. Metabolic effects of repeated weight loss and regain in adolescent wrestlers. JAMA 1988;260: 47–50.

40. Oppliger RA et al. The Wisconsin wrestling minimum weight project: a model for weight control among high school wrestlers. Med Sci Sports Exer 1995;27:1220–24.

41. Brownlie T et al. Marginal iron deficiency without anemia impairs aerobic adaptation among previously untrained women. Am J Clin Nutr 2002;75: 734–42.

42. Manson and Lee, Exercise for women.

43. Icledon T, Sports supplements: where we were, are, and where we're going. Annual Meeting of the American Dietetic Association (FNCE), Philadelphia, PA, Oct. 2002; Lynch R et al., Dangers of abusing insulin, Br J Sports Med 2003;37:356–7; Birchard K, Past, present, and future of drug abuse at the Olympics, Lancet 2000;356:1008.

44. Icledon, Sports supplements; Vincent JB, The potential value and toxicity of chromium picolinate as a nutritional

supplement, weight loss agent and muscle development agent, Sports Med 2003;33:213–30; Lund BC, Perry PJ, Androgenic anabolic steroids: an overview for clinicians, Medscape 1999;69:705–11; Lund BC, Perry PJ, Nonsteroid performance-enhancing agents in athletic competition: an overview for clinicians. Medscape Pharmacotherapy: www.medscape.com/ Medscape/pharmacology/journal/ 2000/v02,n05; Sept. 25, 2000; and Bren L, Oxygen and sports, FDA Consumer Magazine 2002; Nov.–Dec.:1–2.

Nutrition **UP CLOSE**

Testing Performance Aids

Feedback for Unit 28

1. Answer: No. Side effects, such as blood pressure changes, diarrhea, abnormalities in kidney function, and weight loss, were not assessed. Safety of long-term use was not addressed.

2. Answer: No. A 5% increase in strength is small and may have been due to usual differences in the number of push-ups a person can do from time to time. Any increase in strength may have been due to increased training during the week the supplement was taken. There was no control group. All participants knew they were taking phosphorous, a supplement that some may believe increases strength. This belief may have changed performance on the push-up test.

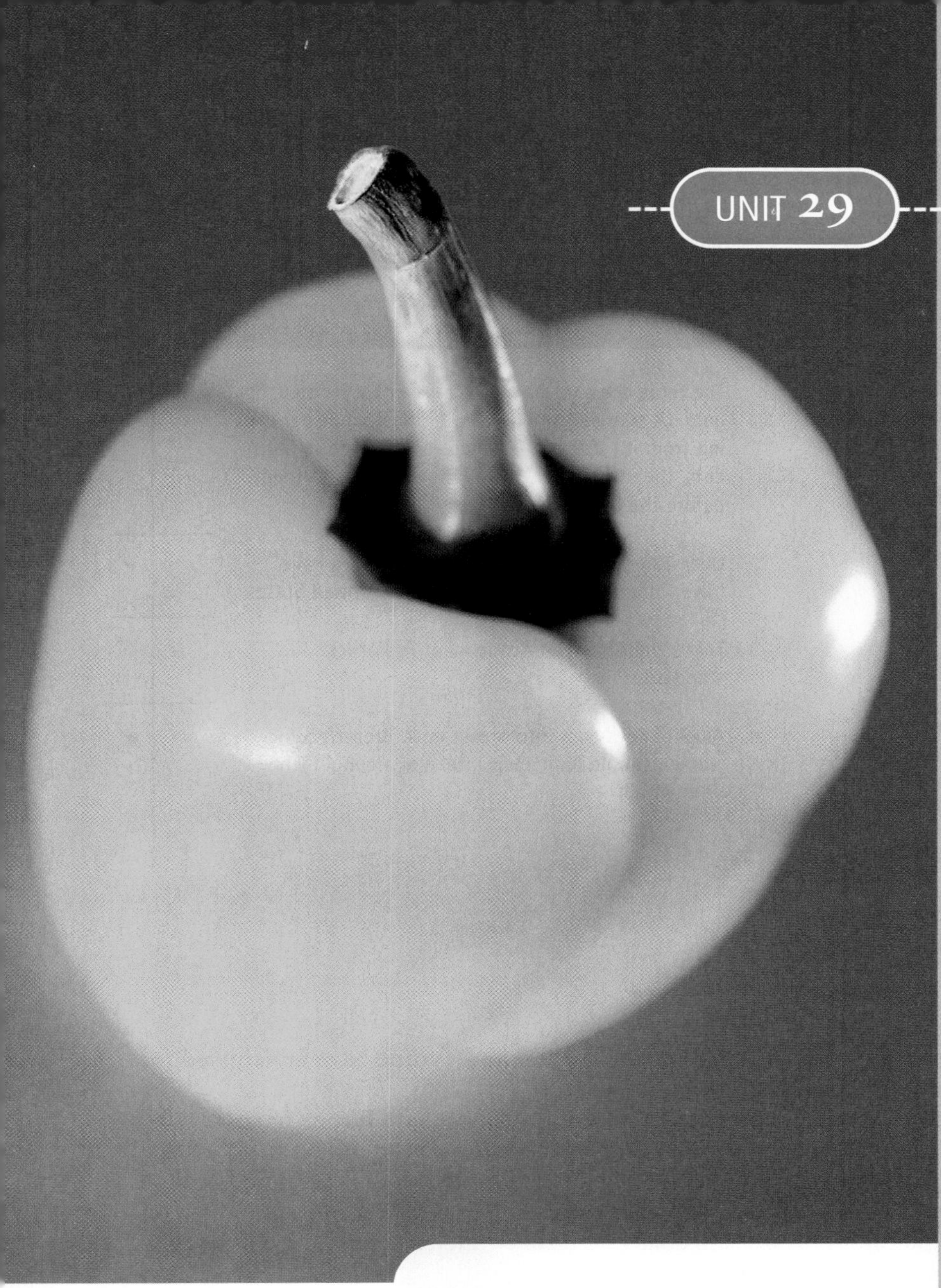

Good Nutrition for Life: Pregnancy, Breastfeeding, and Infancy

Nutrition Scoreboard

		TRUE	FALSE
1	The fetus is a parasite.		
2	The United States has the lowest rate of infant mortality in the world.		
3	Normal health, growth, and development of infants are defined by the experience of breastfed infants.		
4	Drinking alcohol-containing beverages is safe during breastfeeding because the alcohol doesn't pass into the milk.		

Answers on next page

[KEY CONCEPTS AND FACTS]

- There are no "maternal instincts" that draw women to select and consume a good diet during pregnancy.

- The fetus is not a parasite.

- Fetal and infant growth are characterized by "critical periods" during which all essential nutrients needed for growth and development must be available or growth and development will not proceed normally.

- An adult's risk of certain chronic diseases may be partially determined by maternal nutrition during pregnancy and the person's own nutrition early in life.

- Breastfeeding is the optimal method of feeding infants.

- Rates of growth and development are higher during infancy than at any other time in life.

Answers to *Nutrition Scoreboard*

		TRUE	FALSE
1	The fetus does not harm the mother for its own gain. (A true parasite injures the host while benefiting from it.) If there is a shortage of essential nutrients, the mother generally will get access to them before the fetus does.		✔
2	Over 20 economically developed countries have lower infant mortality rates than the United States.		✔
3	That's the American Academy of Pediatrics's standard.	✔	
4	Alcohol does pass into breast milk. Breastfeeding women should limit their intake of alcohol if they drink at all.		✔

Photo Disc

Healthy Start

The day has finally arrived, the one Crystal and Tyrone have anticipated for a year. Crystal's pregnancy test is positive!

This is a planned pregnancy, and both Crystal and Tyrone have done all they could to prepare. Crystal's diet has been the picture of perfection, she has kept up her regular exercise schedule, and she has quit drinking alcohol entirely. Now they are ready to have the baby of their dreams. They are convinced that not only will this baby be healthy and strong, but it will be above average in every respect.

All parents want to give their children every advantage in life that they can. Although no one can guarantee that a baby will be born healthy and strong—no matter what the parents do—there are steps parents can take to make the best baby possible.

Unhealthy Starts on Life

Unfortunately, many infants born in the United States do not have the advantage of good health at birth. Among countries with populations over 2.5 million, the United States ranks twenty-fourth in terms of infant mortality rate (Table 29.1). The relatively high rate of infant deaths in the United States is due primarily to the proportion of infants born with low birthweights (or a weight of less than 5½ pounds) or preterm (delivered before 37 weeks of pregnancy); see Illustration 29.1. One in 14 U.S. infants is born too small, and over 1 in 10 is born too early. These infants are at particular risk for requiring intensive and continuing care and of dying within the first year of life. In contrast, infants weighing between 7 pounds, 11 ounces and 8 pounds, 13 ounces (or 3500 to 4000 grams) are least likely to die within the first year. Currently, the average weight of U.S. infants is 7 pounds, 7 ounces (3350 grams).[3]

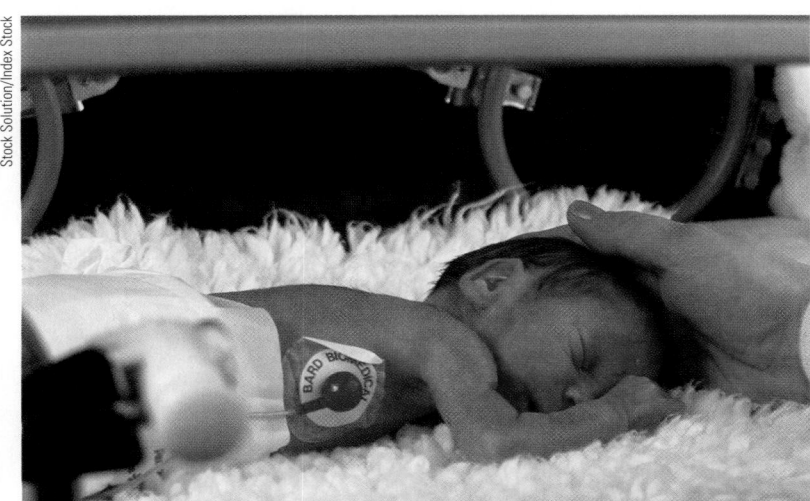

Stock Solution/Index Stock

Illustration 29.1
In 2001, 6.8% of the infants born in the United States had a low birthweight, and 11.6% were born preterm.[2]

TABLE 29.1	
INFANT MORTALITY PER 1000 LIVEBIRTHS IN 1999 IN THE 24 COUNTRIES WITH THE LOWEST RATES AND POPULATIONS OVER 2.5 MILLION.[1]	
COUNTRY	INFANT MORTALITY PER 1000 LIVEBIRTHS
Hong Kong	3.1
Japan	3.2
Sweden	3.4
Singapore	3.5
Norway	3.9
Finland	4.2
Denmark	4.2
Austria	4.4
France	4.4
Germany	4.5
Switzerland	4.6
Czech Republic	4.6
Netherlands	5.2
Canada	5.3
Belgium	5.5
Greece	5.5
Ireland	5.5
Australia	5.6
Portugal	5.6
Israel	5.8
United Kingdom	5.8
New Zealand	6.1
Cuba	6.4
United States	7.1

Improving the Health of U.S. Infants

A high proportion of poor infant outcomes in the United States is attributed to a combination of factors including poverty, poor nutrition, limited access to health care, and a maternal lifestyle that includes the use of illicit drugs, cigarettes, and excessive amounts of alcohol.[4] Many infant deaths, preterm infants, and low-birthweight infants can be prevented, however, by optimizing maternal behaviors that influence health. Of the various behaviors that affect maternal health, none offers potentially greater advantage to both pregnant women and their infants than good nutrition.

Nutrition and Pregnancy

Nutrition is important during pregnancy because the *fetus* depends on it. Energy and every nutrient the fetus needs for *growth* and *development* must be supplied by the mother's diet (preferably) or by supplements. If either insufficient or excessive amounts of nutrients are supplied, fetal growth and development may be compromised. The nature and extent of the impairment depend on which organ or tissue is growing most rapidly when the nutritional deficiencies or overdoses occur.[5]

Critical Periods

Fetal growth and development proceed in a series of critical periods. A *critical period* is an interval of time during which cells of a tissue or organ are genetically programmed to multiply. The period is considered critical because if the cells do not multiply as programmed during this set time interval, they cannot make up the deficiency later. The level of nutrients required for cell multiplication to occur normally must be available during this specific time interval. If the nutrients are not available, the developing tissue or organ will contain fewer cells than normal, will form abnormally, or will function less than optimally.

fetus
A baby in the womb from the eighth week of pregnancy until birth. (Before then, it is referred to as an *embryo*.)

growth
A process characterized by increases in cell number and size.

development
Processes involved in enhancing functional capabilities. For example, the brain *grows*, but the ability to reason *develops*.

Illustration 29.2
Baby's first picture.
An ultrasound image of a rapidly growing and developing 16-week-old fetus.

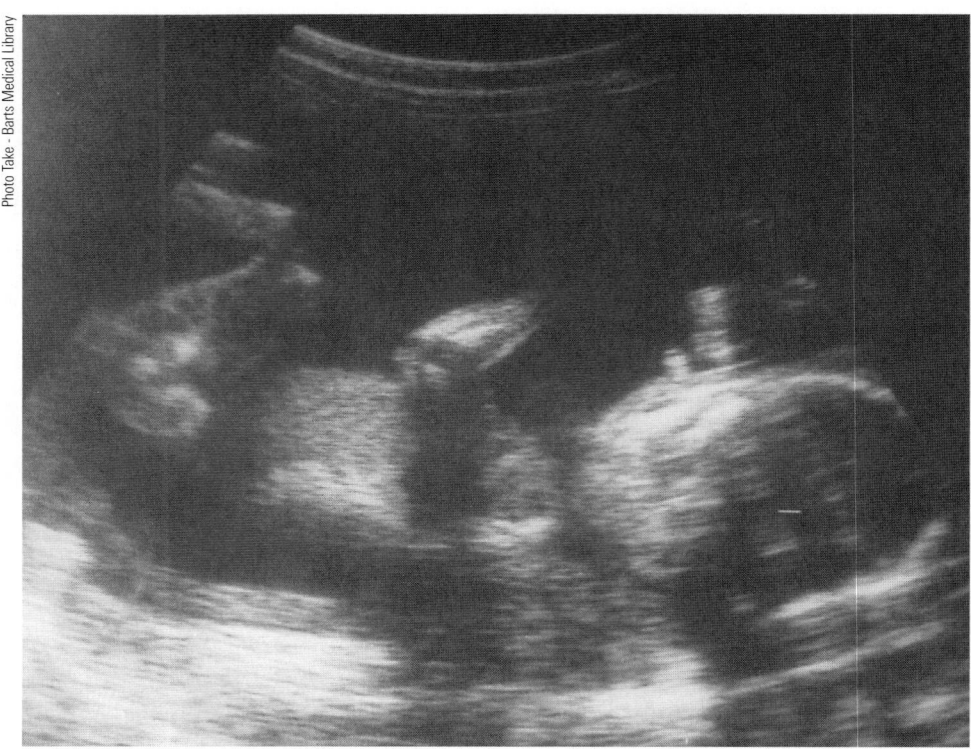

Photo Take - Barts Medical Library

critical period
A specific interval of time during which cells of a tissue or organ are genetically programmed to multiply. If the supply of nutrients needed for cell multiplication is not available during the specific time interval, the growth and development of the tissue or organ are permanently impaired.

The roof of the mouth (or the hard palate), for example, is formed early in the third month of pregnancy when two developing plates fuse together. This process can only occur early in the third month. If excessive amounts of vitamin A are present in fetal tissues during this period, the two plates may fail to combine, resulting in a cleft palate. (The divided hard palate can be surgically corrected after birth.)

Critical periods of cell multiplication are most intensive in the first few months of pregnancy, when fetal tissues and organs are forming rapidly (Illustration 29.2)—hence the importance of the nutritional status of women at the time of conception and very early in pregnancy. For some organs and tissues, cell multiplication even continues through the first two years after birth. Most of the growth that occurs in the fetus late in pregnancy and throughout the rest of the growing years, however, is due to increases in the size of cells within tissues and organs.

The Fetal Origins Hypothesis

One of the most striking advances in research on pregnancy concerns the potential effects of maternal nutrition on the baby's risk of developing certain chronic diseases later in life. It appears that an important element of increased susceptibility to heart disease, stroke, diabetes, obesity, hypertension, and other disorders may be "programmed" by inadequate or excessive supplies of energy or nutrients during pregnancy.[6]

Given optimal conditions, fetal growth and development proceed according to the genetic blueprint established at conception. Organs and tissues are well developed and ready to function optimally after birth. Under less than optimal growing conditions, such as those introduced by maternal weight loss, poor nutrient intakes, or diseases in the mother that alter her ability to supply the fetus with energy or nutrients, fetal growth and development are modified. Fetal tissues undergoing critical phases of development at that time have to make adaptations to cope with the under- or oversupply of nutrients. These adaptations may produce long-term changes in the structure and function of tissues. Low maternal energy intake in the last few months of pregnancy, for example, may hinder the development of cells that produce insulin and limit the body's ability to control blood glucose levels later in life. This could increase the risk of developing diabetes later in life.[7]

A large body of evidence supports the fetal origins hypothesis. Nevertheless, much remains to be discovered about the effects of diet during pregnancy on the risk of later disease.

The Fetus Is Not a Parasite]   From early history to current times, every culture has had myths about diet in pregnancy. According to one of the more common pregnancy myths, the fetus is a parasite, capable of drawing whatever nutrients it needs from the mother at the expense of her health. If this were true, there would be no such thing as a small or poorly nourished newborn. Obviously, it is not true.

The fetus is not a parasite because, with few exceptions, it receives an adequate supply of nutrients only if the mother's intake is sufficient to maintain her own health.[8] The situation makes sense in relation to survival of the species. By meeting the mother's needs first, nature protects the reproducer.

Rather striking evidence that the fetus is not a parasite comes from studies that have identified deficiencies of vitamin B_{12}, thiamin, iodine, folate, zinc, and other nutrients in newborns but not in their mothers.[9] Similarly, infants born to women who consumed excessive levels of vitamin or mineral supplements during pregnancy are more likely to display signs of nutrient overdose than are the mothers.[10]

Prepregnancy Weight Status and Prenatal Weight Gain Are Important

Women who enter pregnancy underweight or who fail to gain a certain minimum of weight during pregnancy are much more likely to deliver low-birthweight and preterm infants than are women who enter pregnancy at normal weight or above and gain an appropriate amount of weight.[11] The risk of low birthweight can be reduced substantially by good diets that lead to a desired rate and amount of weight gain during pregnancy. Along with the duration of pregnancy and smoking, prepregnancy weight status and weight gain in pregnancy are the major factors known to influence an infant's birthweight (Table 29.2).

What's the Right Amount of Weight to Gain during Pregnancy?]   Weight-gain recommendations for pregnancy vary depending on whether a woman enters pregnancy underweight, normal weight, overweight, or obese. Thus, a woman who is obese when she becomes pregnant will need to gain less weight than a woman who enters pregnancy underweight. It is recommended that underweight women gain 28 to 40 pounds, normal-weight women 25 to 35 pounds, overweight women 15 to 25 pounds, and obese women at least 15 pounds during pregnancy. Women carrying twins should gain 35 to 45 pounds during pregnancy.[13] Table 29.3 can be used to identify prepregnancy weight status by a woman's weight and height or body mass index (BMI) category, and the appropriate recommended weight gain during pregnancy. After the goal is identified, a woman may want to plot her prenatal weight gain on a chart like the one in Illustration 29.3. The weight gained should be the result of a high-quality diet that leads to gradual and consistent gains in weight throughout pregnancy.

TABLE 29,2

MAJOR FACTORS THAT DIRECTLY INFLUENCE BIRTHWEIGHT.[12]

- Duration of pregnancy
- Prenatal weight gain
- Prepregnancy weight status
- Smoking

TABLE 29.3

IDENTIFYING PREPREGNANCY WEIGHT STATUS
AND THE RECOMMENDED PREGNANCY WEIGHT-GAIN GOAL.

HEIGHT		WEIGHT STATUS (pounds)			
FEET	INCHES	UNDERWEIGHT	NORMAL WEIGHT	OVERWEIGHT	OBESE
4	9	92 or less	93–113	114–134	135 or more
4	10	94 or less	95–117	118–138	139 or more
4	11	97 or less	98–120	121–142	143 or more
5	0	100 or less	101–123	124–146	147 or more
5	1	103 or less	104–127	128–150	151 or more
5	2	106 or less	107–131	132–155	156 or more
5	3	109 or less	110–134	135–159	160 or more
5	4	113 or less	114–140	141–165	166 or more
5	5	117 or less	118–144	145–170	171 or more
5	6	121 or less	122–149	150–176	177 or more
5	7	124 or less	125–153	154–181	182 or more
5	8	128 or less	129–157	158–186	187 or more
5	9	131 or less	132–162	163–191	192 or more
5	10	135 or less	136–166	167–196	197 or more
5	11	139 or less	140–171	172–202	203 or more
6	0	142 or less	143–175	176–207	208 or more
Body mass index		<19.8	19.8–26.0	>26.0–29.0	>29.0
Prenatal weight gain goal (pounds)		28–40	25–35	15–25	15+

For twin pregnancies, the weight-gain goal is 35–45 pounds,

Note: Height is measured without shoes; weight includes light indoor clothing. Weight-for-height ranges are calculated from the 1959 Metropolitan Height and Weight Tables for Women over the age of 25 years. A midpoint value was determined from the range of weight for height for women of "medium frame." The cut point for underweight women is designated as a weight for height that is more than 10% below the midpoint. The normal-weight range is calculated as plus or minus 10% of the midpoint for each height. The overweight range is calculated as greater than 10% through 30% above the midpoint of weight for height. The cut point in weight for the obese category is calculated as a weight for height that is more than 30% above the midpoint of weight for height.

Source: Courtesy of the Healthy Infant Outcome Program, University of Minnesota, 1989; updated 1998.

Where Does the Weight Gain Go?] If healthy infants tend to weigh about 8 pounds at birth, why do women need to gain more than that? Where does the rest of the weight go? Many pregnant women ask themselves these questions, especially when they don't relish the thought of gaining weight. Illustration 29.4 shows where the weight gain goes.

Fetal growth is accompanied by marked increases in maternal blood volume, fat stores, and breast and uterus size, all of which contribute to weight gain. In addition, water accumulates in the amniotic fluid, which cushions and protects the fetus, and the volume of fluid that exists outside cells increases. The placenta (the tissue that transfers nutrients in the mother's blood supply to the fetus) also accounts for some of the weight that is gained during pregnancy.

Where Does the Weight Gain Go—after Pregnancy?] Delivery is one of the greatest weight-loss plans known to humankind. On average, women lose 15 pounds within the first week after delivery, and weight loss generally continues for up to a year.[14] Women who gain more than the recommended amount of weight, however, and those who gain weight after delivery may have extra pounds to lose to get back to prepregnancy weight.

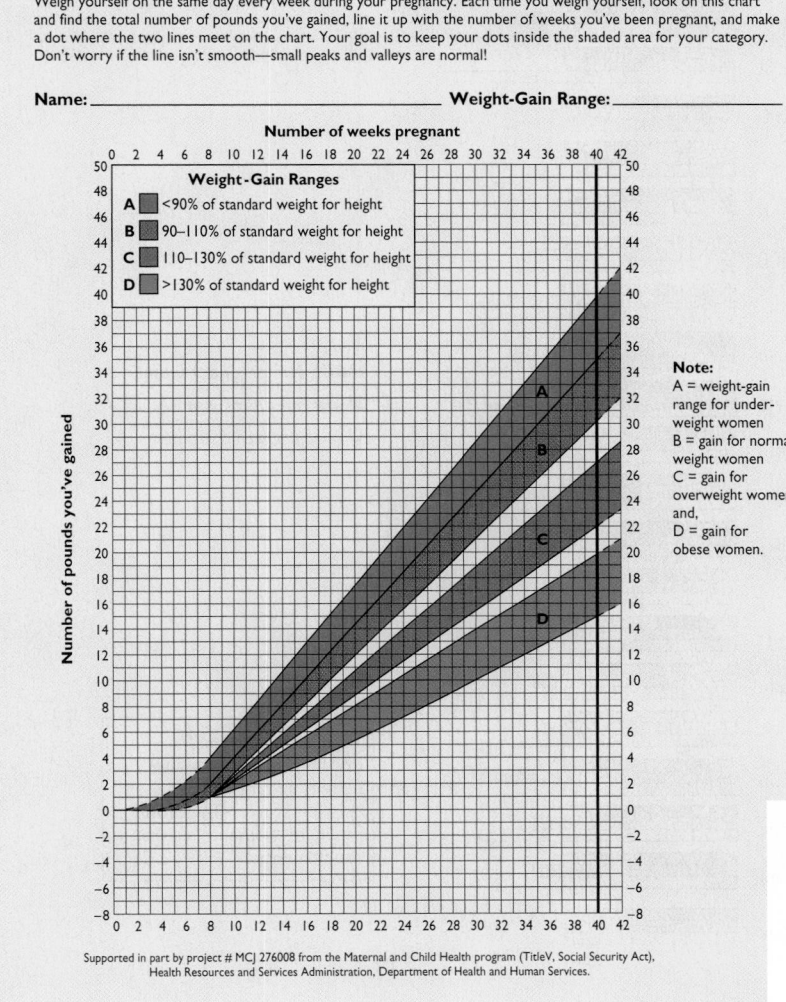

Your Weight-Gain Chart

Weigh yourself on the same day every week during your pregnancy. Each time you weigh yourself, look on this chart and find the total number of pounds you've gained, line it up with the number of weeks you've been pregnant, and make a dot where the two lines meet on the chart. Your goal is to keep your dots inside the shaded area for your category. Don't worry if the line isn't smooth—small peaks and valleys are normal!

Name: _____   Weight-Gain Range: _____

Number of weeks pregnant

Weight-Gain Ranges

A <90% of standard weight for height
B 90–110% of standard weight for height
C 110–130% of standard weight for height
D >130% of standard weight for height

Note:
A = weight-gain range for under-weight women
B = gain for normal-weight women
C = gain for overweight women and,
D = gain for obese women.

Number of pounds you've gained

Supported in part by project # MCJ 276008 from the Maternal and Child Health program (TitleV, Social Security Act), Health Resources and Services Administration, Department of Health and Human Services.

Illustration 29.3
An example of a prenatal weight–gain chart for pregnant women.

Illustration 29.4
Where does all the weight go?

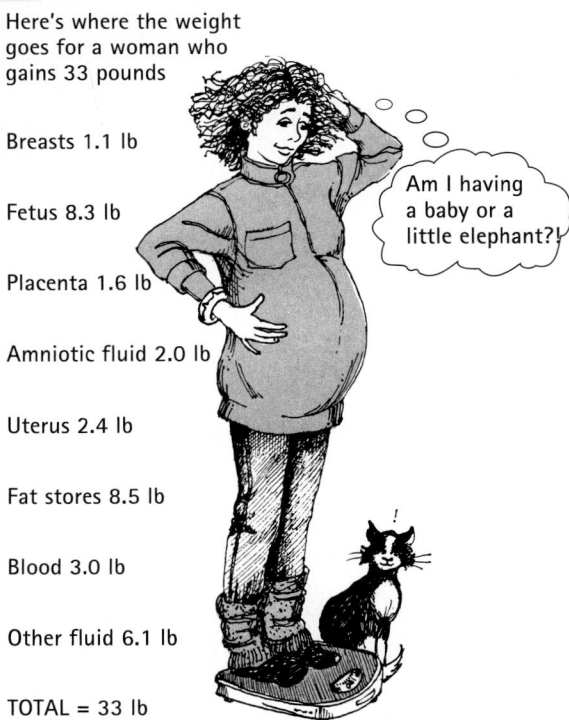

Here's where the weight goes for a woman who gains 33 pounds

Breasts 1.1 lb

Fetus 8.3 lb

Placenta 1.6 lb

Amniotic fluid 2.0 lb

Uterus 2.4 lb

Fat stores 8.5 lb

Blood 3.0 lb

Other fluid 6.1 lb

TOTAL = 33 lb

Am I having a baby or a little elephant?!

The Need for Calories and Key Nutrients during Pregnancy

Pregnant women need more calories, protein, and essential nutrients than non-pregnant women need. All of these are important; but calories, folate, vitamin B_6, vitamin A, calcium, iron, and zinc are of paramount concern. Although many pregnant women are concerned about protein, low protein intakes are rarely a problem in pregnancy. Most women consume 10 to 30 grams more protein daily than the recommended intake level of 71 grams.[15]

According to dietary intake recommendations, pregnant women need approximately 15% more calories and up to 50% more of various nutrients than do nonpregnant women (Illustration 29.5).[16] Relatively high nutrient requirements mean that pregnant women should increase their intake of nutrient-dense foods more than their consumption of calorie-rich foods.

Illustration 29.5
Percentage increases in the RDAs or AIs for pregnant and breastfeeding women compared to other women.

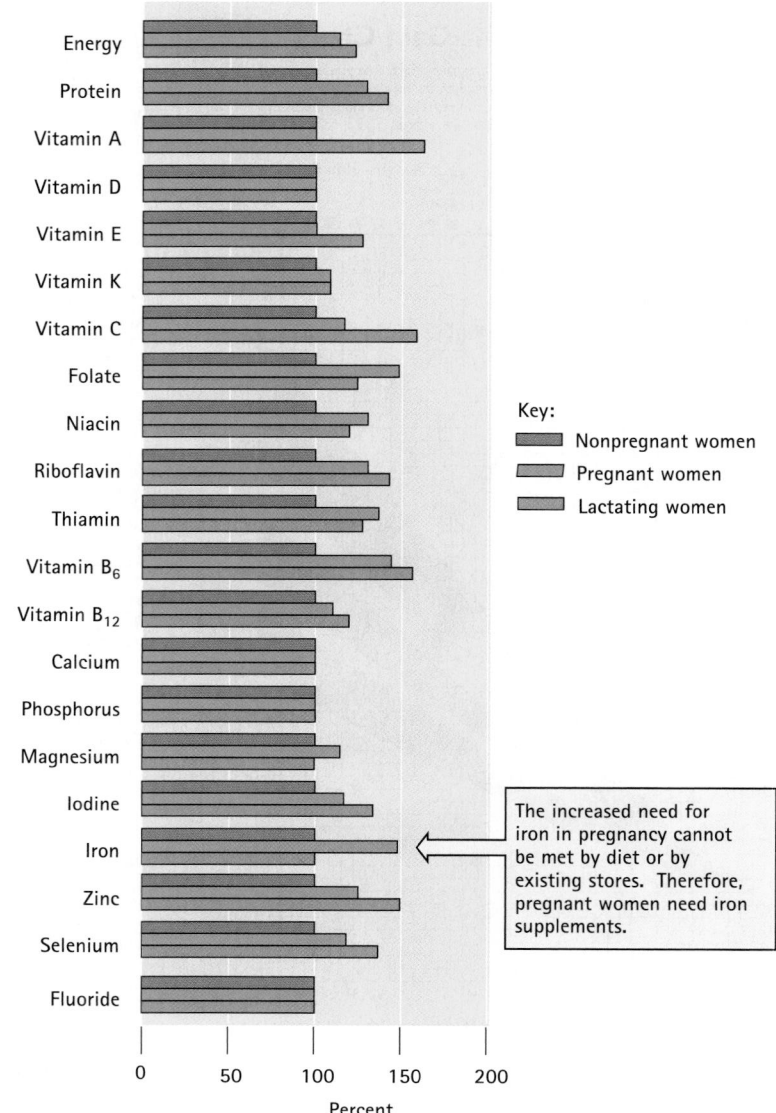

Key:
- Nonpregnant women
- Pregnant women
- Lactating women

The increased need for iron in pregnancy cannot be met by diet or by existing stores. Therefore, pregnant women need iron supplements.

trimester
One-third of the normal duration of pregnancy. The first trimester is 0 to 13 weeks, the second is 13 to 26 weeks, and the third is 26 to 40 weeks.

neural tube defects
Malformations of the spinal cord and brain. They are among the most common and severe fetal malformations, occurring in approximately 1 in every 1000 pregnancies. Neural tube defects include spina bifida (spinal cord fluid protrudes through a gap in the spinal cord; shown in Illustration 29.6), anencephaly (absence of the brain or spinal cord), and encephalocele (protrusion of the brain through the skull).

Calories] On average, women need an additional 340 calories a day in the second *trimester,* and 450 calories in the third trimester of pregnancy.[17] Women entering pregnancy underweight will need more calories than this, and those entering overweight will need fewer. In addition, physically active pregnant women require higher caloric intakes than average. Rather than counting calories, however, it is generally easier and more accurate to monitor the adequacy of caloric intake by tracking weight gain.

Folate] Folate is required for protein tissue construction and therefore is in high demand during pregnancy. It is considered an "at risk" nutrient because many pregnant women do not consume enough to meet their daily need for folate.[18]

Folate deficiency has long been associated with fetal growth failure and malformations, but its link to *neural tube defects* such as spina bifida (Illustration 29.6) is fairly recent. Adequate folate status very early in pregnancy is associated with a

50–70% reduction in the occurrence of neural tube defects.[19] This finding, along with the knowledge that many women fail to consume enough folate, has led to the fortification of refined grain products with folic acid. (Examples of fortified grain products are shown in Illustration 29.7.) Folic acid is a form of folate used in fortified foods and supplements. It is absorbed from foods about twice as completely as folate is. The vast majority of ready-to-eat breakfast cereals are fortified with at least 100 micrograms (0.1 milligram) of folic acid per serving.

Neural tube defects form before 30 days after conception, so it is important that women are consuming enough folate when they enter pregnancy. "Enough" is 600 micrograms (or 0.6 milligram) daily before and during pregnancy. Women can generally obtain this level of folate by consuming 2 servings of ready-to-eat breakfast cereal, 6 servings of grain products, and 3 servings of vegetables daily.[20] If the need for folate is not met by a diet that includes fortified foods, a 400-microgram folic acid supplement should be taken. Fortification of refined grain products with folic acid is improving folate status in Americans, and is decreasing the incidence of neural tube defects.[21]

Vitamin B$_6$] Vitamin B$_6$ is of concern not so much due to the likelihood of deficiency during pregnancy, but because it is often used to treat nausea and vomiting of early pregnancy, or morning sickness. Doses of 75 milligrams of vitamin B$_6$ a day (a dose well above the RDA of 1.9 milligrams) effectively resolve morning sickness in many women. This dose of vitamin B$_6$ appears to be safe for pregnant women and their babies.[22]

Vitamin A] Both low and high intakes of vitamin A may cause problems during pregnancy. Too little vitamin A is associated with poor fetal growth. Too much vitamin A in the form of retinol from supplements can cause fetal malformations.[23]

The effects of vitamin A overdoses during pregnancy came to public attention in the early 1990s, when women taking accutane or retinoic acid for acne were found to deliver far more than the expected number of infants with malformations. Use of these drugs before and very early in pregnancy increases the risk that babies will be born with malformations of facial features and the heart; the intake of more than 10,000 to 15,000 IU of retinol daily during the same period has the same effect. As a precaution, the American College of Obstetrics and Gynecology recommends that

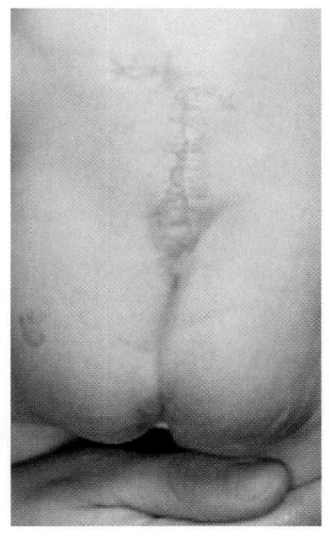

Illustration 29.6
Spina bifida—a neural tube defect.
The interruption in the spinal cord results in paralysis below the injury. This photo shows the back after surgery to close the interruption.

Illustration 29.7
Foods fortified with folic acid.

women who are or could become pregnant should limit their intake of retinol to less than 5000 IU per day and should not use vitamin A–containing medications. Beta-carotene, a precursor to vitamin A, is not harmful, however.[24]

Calcium] Calcium used to support the mineralization of bones in the fetus is supplied by the mother's diet and, if needed, by the calcium contained in the long bones of the mother's body. Consequently, the fetus has access to as much calcium as it needs. Uptake of calcium by the fetus is especially high during the third trimester, when the fetus's bones are mineralizing.

Pregnant women who regularly consume low-calcium diets lose calcium from their bones during pregnancy, but usually regain it after delivery. Because any calcium losses are from the long bones—not from the teeth—the old saying, "for every baby a tooth" has no basis in fact. Several studies have shown that teeth do not demineralize as a result of low-calcium diets during pregnancy.[25]

Iron] Iron deficiency is the most common nutrient deficiency in pregnant women.[26] It develops when a woman enters pregnancy with low iron stores and fails to consume enough iron during her pregnancy. Iron requirements increase during pregnancy due to increases in hemoglobin production and storage of iron by the fetus.

Because of the large increase in need, it is difficult for pregnant women to get enough iron from foods. Consequently, 30 milligrams of supplemental iron per day are recommended for the second and third trimesters of pregnancy. Women who do not take supplemental iron are more likely to develop iron deficiency and to deliver infants that are small and at risk of developing iron deficiency in their first year than are women who do take supplements.[27]

The importance of iron to maternal and infant health appears to have led some health professionals to overreact and routinely prescribe iron in amounts that are too high. It is not uncommon for pregnant women to be given supplements that provide over 100 milligrams of iron daily. Excessive amounts of supplemental iron are poorly absorbed and can produce heartburn, gas, and constipation.[28] Rather than endure the side effects, some women simply put the supplements in the medicine cabinet or on a shelf. Later, when the pregnancy is over and there's a toddler in the house, the child may find the iron pills and eat them like candy.[29] The result can be iron toxicity that requires emergency care. Iron pills that aren't used should be discarded or stored in a toddler-proof place.

Zinc] On average, women in the United States consume around 9 milligrams of zinc per day, a level below the RDA of 11 milligrams for pregnancy.[30] Although the average intake of zinc is sufficient to prevent serious deficiency, marginal deficiency may be quite common. Zinc deficiency during pregnancy is associated with long labors and the delivery of small and malformed infants.[31]

One important way to reduce the risk of zinc deficiency is to limit iron supplements to 30 to 60 milligrams per day. Zinc absorption is reduced substantially when it is consumed with large doses of supplemental iron.[32] Another way is to consume at least 2 servings of meats and meat alternates each day.

What's a Good Diet for Pregnancy?

Contrary to folklore, women do not instinctively select and consume a healthy diet during pregnancy. A good diet takes planning.

Regardless of a pregnant woman's age and whether or not she is a vegetarian, a healthy diet provides all the nutrients she needs with the possible exception of iron. Foods are recommended over supplements because foods provide fiber, antioxidants, and other beneficial substances that supplements do not. Pregnancy diets should include sufficient fluid and fiber and should consist of regular meals and snacks. A pregnant woman should not consume alcohol and should drink coffee only in moderation. Table 29.4 lists these and other recommendations for the diet during pregnancy. Planning a diet for pregnancy around the basic food groups is the most straightforward approach to meeting nutrient needs (Illustration 29.8).

Why Alcohol and Pregnancy Don't Mix] As early as the 1800s, maternal consumption of alcohol during pregnancy was said to cause the birth of "sickly" infants. The ill effects of alcohol on babies were not fully acknowledged, however, until the 1970s, when several research reports described a condition called fetal alcohol syndrome (Illustration 29.9). Women who drank heavily or frequently binged on alcohol during pregnancy were found to be at high risk of delivering infants with specific malformations and retarded physical and mental development.[33] The effect of maternal alcohol consumption on the fetus worsens as intake increases. Heavy drinking in the first half of pregnancy is closely associated with the birth of malformed, small, mentally impaired infants. When excessive drinking occurs only in the second half of the pregnancy, infants are less likely to be malformed, but are still likely to be small and to suffer abnormal mental development.[34] These conditions are permanent; they cannot be fully corrected with special treatment, and the child does not outgrow them.

TABLE 29.4

DIETARY RECOMMENDATIONS FOR PREGNANCY.

- Consume sufficient calories for adequate weight gain.
- Eat a variety of foods from each food group.
- Eat regular meals and snacks.
- Consume sufficient dietary fiber (about 28 grams per day).
- Consume 10 or more cups of water each day from fluids and foods.
- Use salt to taste (within reason).
- Do not drink alcoholic beverages.
- Limit coffee to 4 or fewer cups per day.

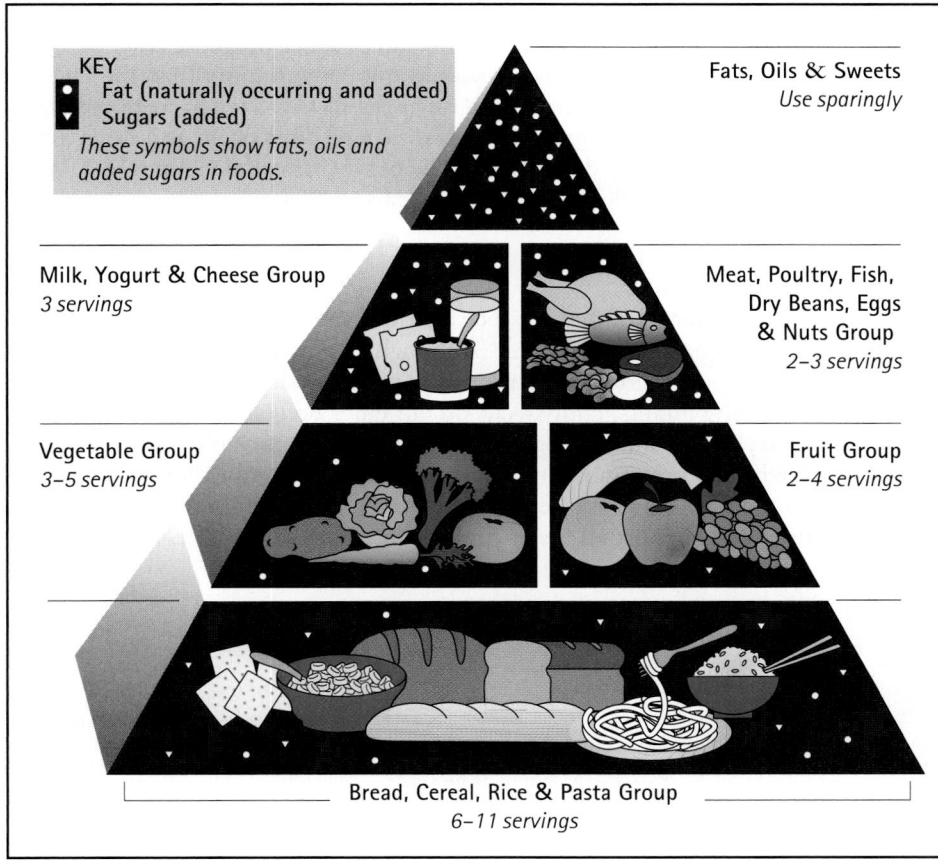

Illustration 29.8
Food group recommendations for pregnant and breastfeeding women based on the Food Guide Pyramid.

KEY
○ Fat (naturally occurring and added)
▽ Sugars (added)
These symbols show fats, oils and added sugars in foods.

Fats, Oils & Sweets
Use sparingly

Milk, Yogurt & Cheese Group
3 servings

Meat, Poultry, Fish, Dry Beans, Eggs & Nuts Group
2–3 servings

Vegetable Group
3–5 servings

Fruit Group
2–4 servings

Bread, Cereal, Rice & Pasta Group
6–11 servings

PhotoEdit - David Young-Wolff

Illustration 29.9
Children with fetal alcohol syndrome (FAS) have characteristic facial features, mental retardation, heart defects, and other problems. *Less severe cases can include small size and slow development.*

Illustration 29.10

Photo Disc

No amount of alcohol has been found to be absolutely safe during pregnancy. When only an occasional drink is consumed, however, the adverse effects of alcohol on fetal development appear to be small and rare. To exclude the possibility of even small impairments in fetal growth and development, it is recommended that women not drink alcohol at all during pregnancy or when they are attempting to become pregnant.[35]

Vitamin and Mineral Supplements] Iron is the only supplement recommended for all pregnant women. About 83% of pregnant women take multiple vitamin and mineral supplements, however. Apparently, supplements are being prescribed as insurance against the possibility of poor diets. Women should be given supplements the same way they are given medications—when they are indicated. Under certain conditions, supplementation with nutrients besides iron is indicated.[36]

A multivitamin-mineral preparation is recommended for pregnant women who do not ordinarily consume an adequate diet and for those in high-risk categories, such as women carrying more than one fetus, heavy cigarette smokers, and alcohol and drug abusers. The preparation should be taken daily beginning in the second trimester and should contain the following nutrients:

Iron	30 milligrams	Vitamin B_6	2 milligrams
Zinc	15 milligrams	Folate	300 micrograms
Copper	2 milligrams	Vitamin C	50 milligrams
Calcium	250 milligrams	Vitamin D	5 micrograms

Vegans (those who consume no animal products at all) should supplement their diet daily with 10 micrograms (400 IU) of vitamin D and 2 micrograms of vitamin B_{12} or consume plant foods fortified with these vitamins.[37]

Because a number of vitamins and minerals can produce overdose reactions when taken in excessive amounts, supplements should not contain more than the DRI levels. As little as 15,000 IU of vitamin A (five times the RDA) taken early in pregnancy appears to be linked to the development of central nervous system and bone abnormalities, for example, and daily doses of 2000 IU of supplemental vitamin D (10 times the DRI) are related to mental retardation and heart defects in infants.[38]

Teen Pregnancy

Approximately 3 of every 100 females between the ages of 15 and 17 years in the United States deliver babies each year.[39] Although teens who do not smoke, drink alcohol, or use drugs and who enter pregnancy in good physical and nutritional health tend to deliver healthy infants, many teens do not enter pregnancy in such good condition. Poor lifestyle habits are thought to be primarily responsible for the high rate of low-birthweight and preterm infants born to teens. Few teens (1%) meet all of the Food Guide Pyramid's food group recommendations. Individualized attention, nutrition intervention, and support all foster healthy outcomes of teen pregnancy.[40]

Breastfeeding

Food is the first enjoyment of life.
—Lin Yutang

A woman's capacity to nourish a growing infant does not end at birth; it continues in the form of breastfeeding (Illustration 29.10). Breast milk from healthy, well-nourished women is ideally suited for infant nutrition and health.[41]

What's So Special about Breast Milk?

The breastfed infant is the reference model against which all alternative feeding methods must be assessed with regard to growth, health, and development; and other short- and long-term health outcomes.

—HHS Blueprint for Action on Breastfeeding, 2000

Breast milk is like a bonus pack. In addition to serving as a complete source of nutrition for infants for the first 4 to 6 months of life, breast milk contains substances that convey a significant degree of protection against a variety of illnesses (Table 29.5)—including infectious diseases such as polio; the "flu"; and ear, respiratory tract, and gastrointestinal tract infections. Evidence indicates that breast milk may confer a degree of protection against the development of cancer of the lymph system (lymphoma), asthma, and diabetes during childhood. In addition, breast milk contains essential and nonessential fats and other substances that appear to promote optimal growth and development of the nervous system and eyes. Intelligence, as measured by IQ, tends to be higher in babies receiving breast milk than in those given formula.[43] The disease-preventing and development-promoting components of human milk are lifesaving assets in many developing countries, where a safe water supply and medical care may be unavailable. Although breastfeeding doesn't protect infants from all infectious diseases and food allergies, it's the ounce of prevention that's worth a pound of cure.

Breastfeeding also offers other benefits—ones that may mean a lot to parents. Table 29.6 discusses them.

Is Breastfeeding Best for All New Mothers and Infants?

Over 96% of women are biologically capable of breastfeeding, and the vast majority of infants thrive on breast milk.[44] But breastfeeding is not best for every woman and infant. Successful breastfeeding involves more than biology; it is heavily influenced by environmental and psychological conditions.

In the United States, about 70% of new mothers breastfeed their infants to some extent (Illustration 29.11).[46] Because rates of breastfeeding were relatively low throughout the 1960s and 1970s, many women giving birth now grew up in an environment where formula feeding was preferred to breastfeeding (Illustration 29.12).

The increase in women returning to work soon after delivery, a lack of health care provider and emotional support for breastfeeding, embarrassment, early hospital discharge, free formula samples, and inadequate knowledge about how to breastfeed all appear to be deterring U.S. women from breastfeeding.[47] If more women are to have the opportunity to breastfeed, these and other barriers must be broken down. Perhaps a good place

TABLE 29.5

A SAMPLE OF U.S. INFANTS SHOWS FEWER SIGNIFICANT EPISODES OF ILLNESS AMONG THOSE WHO WERE BREASTFED.[42]

ILLNESS	PERCENTAGE OF INFANTS FED:	
	BREAST MILK	FORMULA/OTHER
Ear infections	3.7%	9.1%
Respiratory illness to age 7	17.0	32.0
Diarrhea, vomiting	3.5	6.9
Hospital admissions	1.0	3.0
Average number of episodes of illness per infant	8.2	21.1

TABLE 29.6

FIFTEEN REASONS TO BREASTFEED.

1. The milk container is easy to clean.
2. Breast milk is a renewable resource.
3. There's no packaging to discard.
4. Breast milk comes in an attractive container.
5. The temperature of breast milk is always perfect right out of the container.
6. Breast milk tastes real good.
7. There are no leftovers.
8. You don't have to go to the kitchen in the middle of the night to get breast milk ready.
9. It takes just seconds to get a meal ready.
10. There's no bottle to repeatedly pick up off the floor.
11. The price is right.
12. The meal comes in a perfect serving size.
13. Meals and snacks are easy to bring along on a trip or outing.
14. Feeding units come in an assortment of beautiful colors and sizes.
15. One food makes a complete meal.

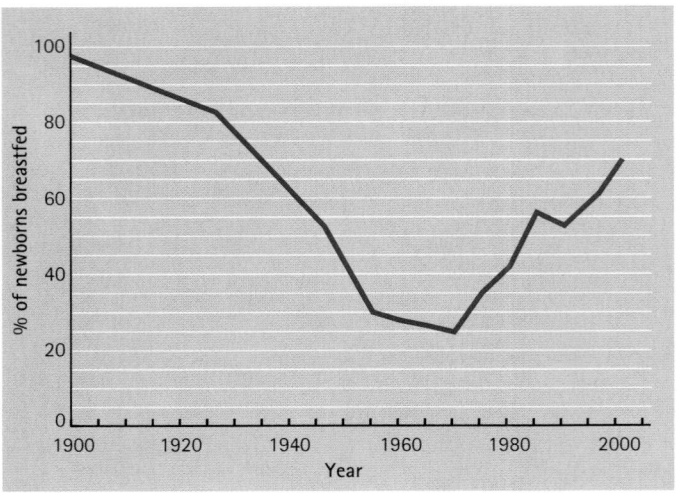

Illustration 29.11
Percentage of breastfed newborns in the United States.[45]

Illustration 29.12
Said a friend when she saw this photo in 1951:
"This is the best advertisement for breastfeeding
I've ever seen!"

Courtesy of Hans and Libby Schapiro

ON THE SIDE

Can adopting moms breastfeed their infants? Yes, they can. It takes hormones and motivation, however. Breastfeeding success is higher the younger infants are when they begin to breastfeed with their adoptive moms.[50]

Photo Disc

to start is with the value our society places on breastfeeding. A number of European countries that actively promote breastfeeding allow women to stay in the hospital after delivery until breastfeeding is going well. The usual practice in some African countries is to relieve a breastfeeding mother's workload so that she can devote nearly full time to feeding and caring for her young infant. A relative may move in with the family and take over household chores, or the mother and baby may live with her parents for a time.

Public health initiatives are also in place to facilitate breastfeeding. In order to meet national health objectives for improving rates of breastfeeding in the United States by 2010 (see Table 29.7), it is recommended that:

• Health care workers encourage and facilitate breastfeeding.
• "How to" advice and problem solving guidance be available for breastfeeding women from qualified health care staff.
• Breastfeeding women returning to work have access to on-site child care facilities and private rooms for breastfeeding.

TABLE 29.7

ACTUAL VERSUS NATIONAL GOALS FOR BREASTFEEDING IN THE UNITED STATES.[48]

	NEWBORNS	6 MONTHS	12 MONTHS
National goal	75%	50%	25%
Actual (all women)	70	33	25

Breastfeeding is at its best when both the mother and infant benefit from the experience. If mutual benefit is not possible, then formula feeding may be necessary. Infant growth and development are well supported by commercially available infant formulas.[49]

How Breastfeeding Works

The mother's body prepares for breastfeeding during pregnancy. Fat is deposited in breast tissue, and networks of blood vessels and nerves infiltrate the breasts. Ducts that will channel milk from the milk-producing cells forward to the nipple—the milk collection ducts—also mature (Illustration 29.13). Hormonal changes that occur at delivery signal milk production to begin.

Breast milk produced during the first three days or so after delivery is different from the milk produced later. Called *colostrum,* this early milk contains higher levels of protein, minerals, and antibodies than does "mature" milk.

Breast Milk Production] While an infant is consuming one meal, she or he is "ordering" the next. The pressure produced inside the breast by the infant's sucking and the emptying of the breasts during a feeding cause a hormone to be released from specific cells in the brain. The hormone stimulates the production of milk so that more milk is produced for the next feeding. It generally takes about 2 hours for the milk-producing cells to manufacture enough milk for the next feeding. An important exception to this occurs when an infant enters a growth spurt and consumes more milk than usual. Then milk production takes longer, perhaps a day, to catch up with demand.

Only very rarely is a breastfeeding woman unable to produce enough milk. As long as an infant is allowed to satisfy her or his appetite by breastfeeding as often as desired, milk production will catch up with the baby's need.

colostrum
The milk produced during the first few days after delivery. It contains more antibodies, protein, and certain minerals than the mature milk that is produced later. It is thicker than mature milk and has a yellowish color.

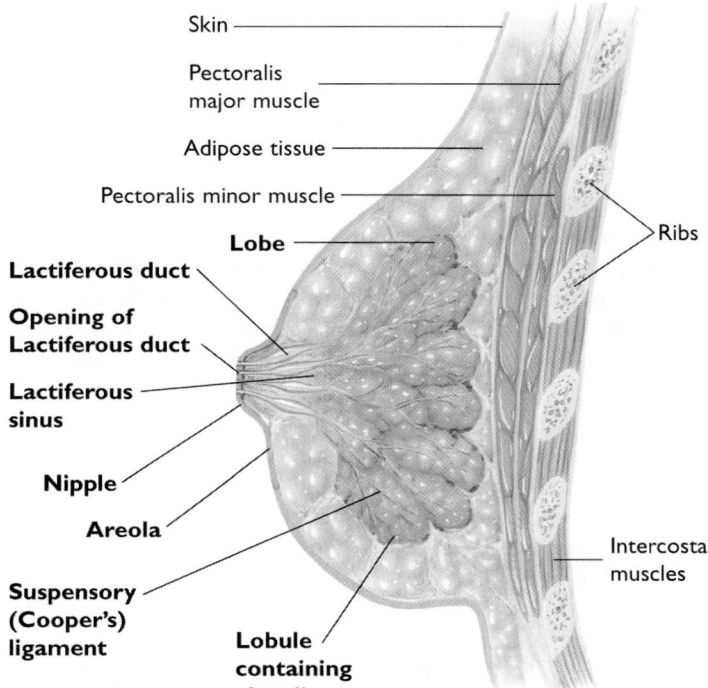

Illustration 29.13
A view of the interior of the breast.

Skin

Pectoralis major muscle

Adipose tissue

Pectoralis minor muscle

Lobe

Lactiferous duct

Opening of Lactiferous duct

Lactiferous sinus

Nipple

Areola

Suspensory (Cooper's) ligament

Lobule containing alveoli

Ribs

Intercostal muscles

Nutrition and Breastfeeding

A breastfeeding woman needs an adequate and balanced diet to replenish her body's nutrient stores, maintain her health, and produce sufficient milk for her baby. Increases in recommended dietary intakes for breastfeeding are generally higher than those for pregnancy. As during pregnancy, proportionately higher amounts of nutrients than calories are required, indicating the need for a nutrient-dense diet (see Illustration 29.5 on p. 29–8).

Calorie and Nutrient Needs] The RDA for calories is about 15% higher for breastfeeding women than for other women. Actually, a breastfeeding woman needs about 30% more calories than the RDA for women who are not breastfeeding, but she does not have to consume that level of calories from food. Energy supplied from fat stores that normally accumulate during pregnancy contributes to meeting these needs during breastfeeding, so not all of the calories must come from the mother's diet.[51]

What's a Good Diet for Breastfeeding Women?] The calories and nutrients needed by the breastfeeding woman can be obtained from a varied diet that includes foods from each of the basic food groups. The recommended numbers of servings from each food group are shown in Illustration 29.8; they are the same as for pregnant women. Dietary recommendations for breastfeeding women are summarized in Table 29.8.

Increases in hunger and food intake that accompany breastfeeding generally take care of meeting caloric needs. Failure to consume enough calories from food can decrease milk production, however. Low-calorie diets (those providing less than 1500 calories per day) and weight loss that exceeds 1.5 to 2 pounds per week—even in women with a good supply of fat stores—can reduce the amount of milk women produce. Weight loss of about a pound a week starting a month after delivery appears to be safe and helpful in women's attempts to return to prepregnancy weight.[53]

Are Supplements Recommended for Breastfeeding Women?] Supplements are not recommended for breastfeeding women. Instead, breastfeeding women should try to get all of the nutrients they need from food.[54] Supplements, if needed, should be prescribed on an individual basis.

TABLE 29.8

DIETARY RECOMMENDATIONS FOR BREASTFEEDING WOMEN.[52]

- Diets should supply all of the nutrients breastfeeding women need. (The routine use of vitamin and mineral supplements is not recommended.)
- Fluids should be consumed to thirst (10 cups of fluid each day).
- Weight loss should not exceed 6 to 8 pounds per month after the first month after delivery; caloric intakes should not fall below 1500.
- Alcohol intake should be avoided.

Dietary Cautions for Breastfeeding Women] Almost anything a woman consumes may end up in her breast milk. When a breastfeeding woman drinks coffee, her infant receives a small dose of caffeine. Breastfed infants of women who are heavy coffee drinkers (10 or more cups per day) may develop "caffeine jitters." Alcohol also is transferred from a woman's body to breast milk. The development of the brain and nervous system of infants breastfed by chronic, heavy drinkers appears to be retarded. There is also concern that alcohol ingestion by breastfeeding women may decrease milk production.[55]

Environmental contaminants, such as lead, DDT, chlordane, PCBs, and PBBs, for example, are transferred into breast milk. Many environmental contaminants may be stored in a woman's fat or bone tissues. When the fat stores are later broken down for use in breast milk, or the calcium in bone is mobilized, the contaminants stored in the fat or bone may enter the breast milk. The ingestion of fish from contaminated waters in Lake Ontario and Lake Michigan was linked to abnormally high levels of PCBs in breast milk. Infants exposed to PCBs can develop rashes, digestive upsets, and nervous system problems. With the exception of breast-feeding women known to be exposed to excess levels of environmental toxins, how-ever, it is concluded that the benefits of breastfeeding outweigh the risks to infants from harmful substances in the environment.[56]

Infants are much smaller than women, and it takes a smaller dose of caffeine, alcohol, drugs, or environmental contaminants to have an effect on them than on an adult. Whereas breastfeeding mothers may show no adverse effects from these substances, their infants may.

Breastfeeding women are advised to limit their consumption of regular coffee to 4 or fewer cups a day and to avoid alcohol or limit it to one drink with a meal per day.[57] Drugs or medications should be taken only on the advice of a health care provider.

Illustration 29.14
Infants develop at a truly amazing rate.

Infant Nutrition

> *He who possesses virtue in abundance may be compared to an infant.*
> —Lao Tzu, sixteenth century B.C.

At no other time during life outside the womb do growth and development proceed at a faster pace than during infancy. Infants grow out of clothes within weeks, long before they wear them out. Each day infants learn new behaviors, and their minds absorb large chunks of information that will serve them well in the future (Illustration 29.14). The rate at which growth and development proceed in the first year of life is truly amazing. You can get an idea of just how rapidly infants grow and develop by reviewing Illustration 29.15. Just as infants need security, love, and attention to flourish, so too do they need calories and essential nutrients.

Infant Growth

During the first week or two of life, infants generally lose 5 to 10% of their birth-weight while adjusting to the new surroundings. After that, infants grow rapidly. Most infants double their birthweight by 4 months and triple it by 1 year. Length usually increases by 50% during the first year. If this rate of growth were to con-tinue, 10-year-old children would be about 10 stories high and weigh over 220 tons! After infancy, the growth rate declines and remains at a fairly low level until the adolescent growth spurt begins.

Growth Charts for Infants]   In 2000 the Centers for Disease Control released new growth charts for female and male infants, children, and adolescents. Separate charts were not developed by ethnic group or race, because growth potential is a shared human trait that varies primarily due to environmental factors such as nutri-tional status and disease. An example of the growth charts that include infants is shown in Illustration 29.16. The charts should be carefully plotted based on accu-rate, periodic measures of an infant's size. They are best used for screening growth problems; confirmation of underweight or obesity requires assessment of body fat by skinfold thickness and other measures.[58]

Body Composition Changes with Growth]   Humans grow up and out, and their body composition and proportions change as growth progresses. Although generally measured by gains in pounds and inches, growth also reflects changes in bone mass, organ size, body proportions, and composition (the proportionate amounts of water, muscle, and fat). Infants' heads are very large in relation to the rest of their bodies, for example. Brain growth takes precedence over trunk and limb

Illustration 29.15
Growth and developmental characteristics from birth through one year.

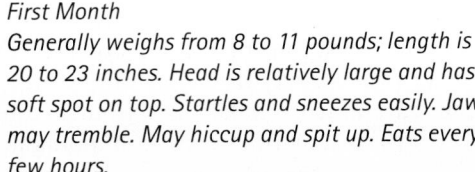

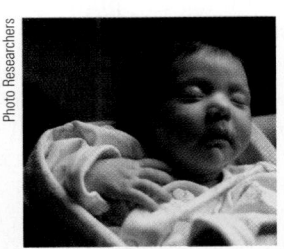

First Month
Generally weighs from 8 to 11 pounds; length is 20 to 23 inches. Head is relatively large and has soft spot on top. Startles and sneezes easily. Jaw may tremble. May hiccup and spit up. Eats every few hours.

One to Three Months
Lifts head briefly when placed on stomach; smiles, coos, and gurgles. Whole body moves when infant is touched or lifted. Eats every three to four hours.

Four to Six Months
Weight nearly doubled. Has grown three to four inches. Follows objects with eyes. Reaches toward objects with both hands; puts fingers and objects into mouth. Turns over, sits unassisted. Awake longer at feeding time. Eats six to seven times per day. Sleeps six to seven hours at night.

Seven to Eleven Months
Gains in weight and height are less rapid, appetite has decreased. Stands up with help, hitches self along the floor. Reaches for, grasps, and examines objects with hands, eyes, and mouth. Has one or two teeth. Takes two naps a day.

Twelve Months
Usually has tripled birthweight and increased length by 50%. Grasps and releases objects with fingers. Holds spoon, but uses it poorly. Begins to walk unassisted.

growth early in life, and the body eventually grows to "fit" the head. During the first 10 years of life, a child who initially has the shape of a loaf of bread may normally come to resemble a string bean.

Nutrition and Mental Development

Malnutrition has the greatest impact on mental development when it is severe and occurs during the critical period for brain cell multiplication. For humans, this vulnerable period begins during pregnancy and ends after the first year of life. Impair-

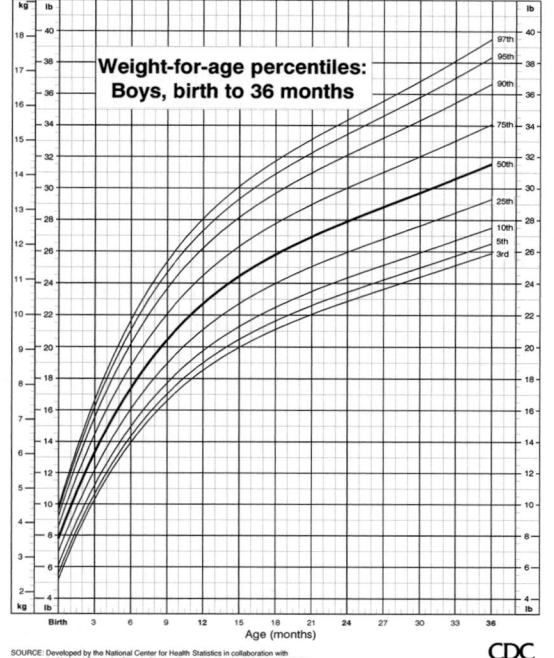

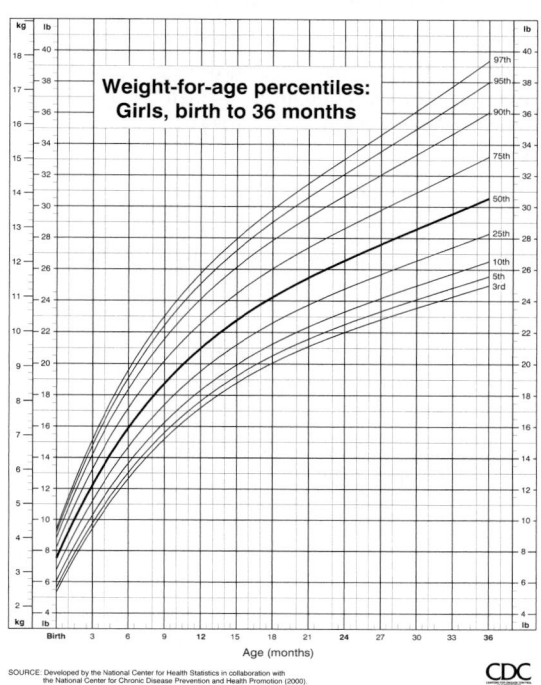

Illustration 29.16

An example of the CDC's growth charts.

The charts can be copied from the Internet using the address www.cdc.gov/growthcharts.
Source: National Center for Chronic Disease Prevention and Health Promotion.

ment in mental development is less severe when malnutrition occurs only during pregnancy or only during infancy than if it occurs throughout both pregnancy and infancy.[59]

Children's mental development is also greatly influenced by the social and psychological environment in which they are raised. Because malnutrition is generally accompanied by both social and psychological deprivation, these factors often contribute jointly to poor mental development. For children in the United States, poverty, neglect, illnesses, and psychological problems appear to be the main cause of undernutrition and poor mental development.[60]

Infant Feeding Recommendations

Current dietary recommendations for infants call, preferably, for breastfeeding or, alternatively, for formula feeding for the first 12 months of life (Table 29.9).[62] Infants should be fed "on demand," that is, when they indicate they are hungry, rather than on a rigid schedule set by the clock. In the first few months of life, most infants will get hungry every 3 or 4 hours.

Soft, solid foods should be added to an infant's diet of breast milk or infant formula between the ages of 4 and 6 months. At one time, it was thought that infants should be offered solid foods within the first few months of life. Although such young infants are unable to swallow much of the food (most of it ends up on their face and bib) or to digest completely what they do swallow, solid foods were thought to help the baby grow and sleep through the night. This belief was incorrect. Infants

TABLE 29.9

OVERVIEW OF INFANT FEEDING RECOMMENDATIONS.[61]

- Exclusive breastfeeding for the first 4 to 6 months of life and continuation of breastfeeding through the first year of life are preferred. Iron-fortified infant formulas may be used as a secondary option to breastfeeding.

- Introduce basic, specially prepared solid foods at 4 to 6 months.

- Infants should be fed "on demand" and not by a set schedule. Feeding should stop when the baby loses interest in eating.

- Breastfed infants receiving little sun exposure should be given a vitamin D supplement (200 IU or 5 micrograms per day).

- Infants not receiving fluoridated water should be given fluoride drops (dose depends on age).

who receive solids before the age of 4 months are no more likely to grow normally or sleep through the night than are infants who start to receive solid foods between 4 and 6 months of age.[63] The age at which an infant begins to sleep for 6 or more hours during the night depends on other factors, including the infant's developmental level and how much he or she has slept during the day. Neither the infant nor the infant's parents are likely to get a full night's sleep for at least 4 months.

Introducing Solid Foods] Recommendations for the timing and sequence of offering solid foods to infants are shown in Illustration 29.17. Rice cereal should be offered first. It is easily digested by infants of this age, and allergies to rice are extremely rare. It is recommended that iron-fortified rice cereal be given to all breastfed infants and to bottle-fed infants who are not receiving iron-fortified formula. The iron provided by iron-fortified rice cereal or formula helps to restock the infant's iron stores, which have been drawn upon since birth. Generally, solid foods prepared for infants should consist of single, basic foods such as strained vegetables, fruits, or meats.[64]

Solid foods for infants can be purchased as commercial baby food or prepared at home. When baby foods are prepared at home, care should be taken to avoid contamination and to achieve the right consistency. Infants should be offered new foods one at a time. The new offerings should be separated by several days, so that any allergic reactions to a food can be identified. Variety is the key to achieving a healthy diet for infants in their second 6 months of life, and an assortment of basic foods should be given.

Illustration 29.17
Introduction of "solid" foods to infants.

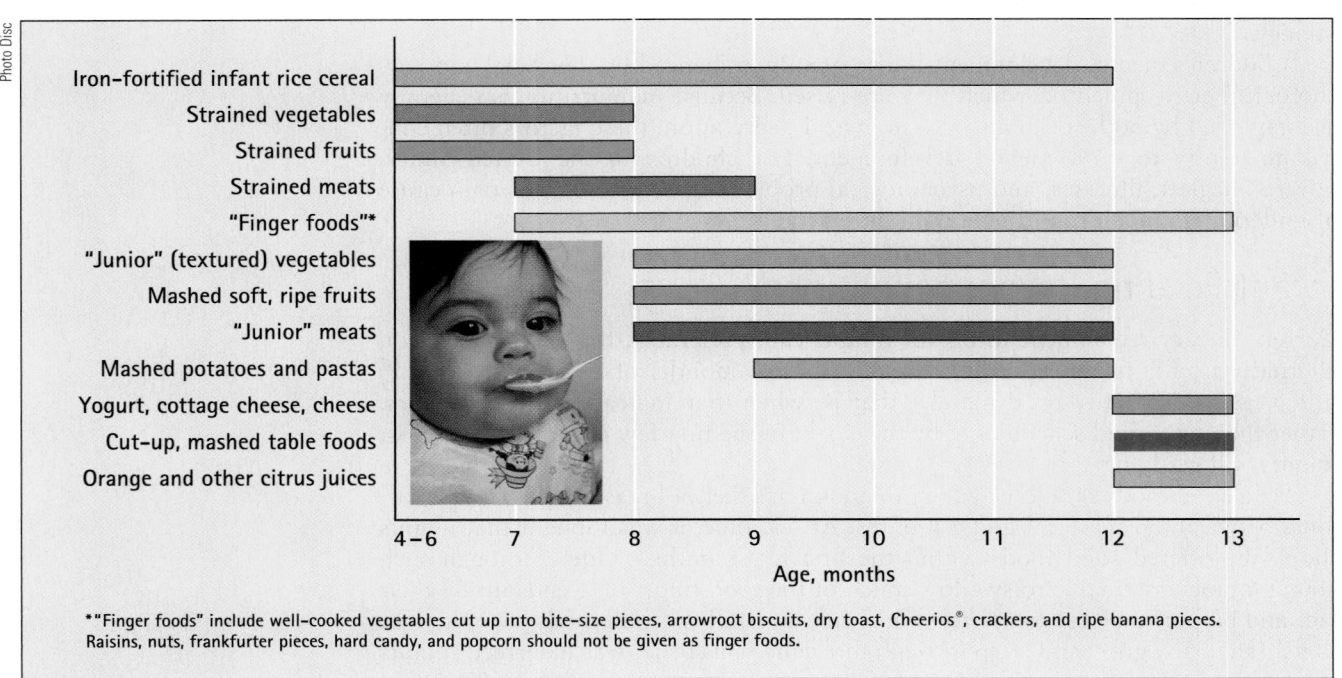

*"Finger foods" include well-cooked vegetables cut up into bite-size pieces, arrowroot biscuits, dry toast, Cheerios®, crackers, and ripe banana pieces. Raisins, nuts, frankfurter pieces, hard candy, and popcorn should not be given as finger foods.

TABLE 29.10

FOODS TO AVOID IN THE FIRST YEAR OF LIFE.

FOODS THAT MAY CAUSE ALLERGIC REACTIONS	FOODS THAT MAY CAUSE OTHER PROBLEMS	FOODS THAT MAY LEAD TO CHOKING
Cow's milk	Blueberries	Grapes
Egg white	Coffee	Frankfurter pieces
Fish, seafood	Corn	Hard candy
Nuts	Fruit drinks	Hard pieces of vegetables
Peanut butter	Honey (unpasteurized)	Meat chunks
Soy protein	Prune juice	Nuts and seeds
Wheat products	Tea	Popcorn
		Raw vegetables

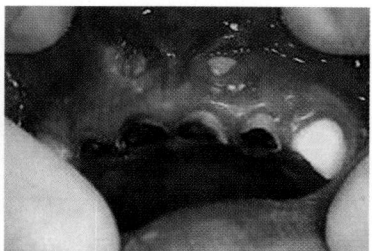
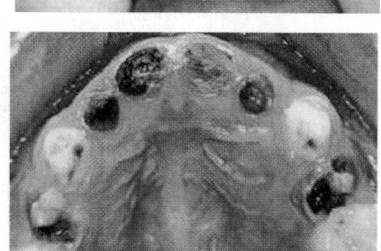

Illustration 29.18
Baby-bottle tooth decay.
© Dietmar A. J. Kennel

By 9 months of age, infants are ready for mashed foods and foods such as yogurt, applesauce, ripe banana pieces, and grits. Most infants have several teeth by this time, and they are able to bite into and chew soft foods.

Infants graduate to adult-type foods after the age of 12 months. Although most foods still need to be mashed or cut up into small pieces for them, 1-year-olds are able to eat the same types of food as the rest of the family. They can drink from a cup and nearly feed themselves with a spoon. Infants have come a long way in 12 months.

Foods to Avoid] Not all foods can be considered "baby foods." Some foods should be omitted from an infant's diet because they are apt to cause allergic reactions or are too difficult for infants to chew into small pieces and swallow. Table 29.10 provides a list of foods that should not be offered to infants.

Reduced-fat products are not recommended for infants. Infants need a relatively high-fat diet for brain and nervous system development. Unpasteurized honey is not recommended either, because it may cause botulism in infants due to their still maturing gastrointestinal defenses against bacteria. Beverages containing a lot of sugar, such as soft drinks and sweetened fruit juices, should not be given to infants because they promote tooth decay. To help prevent "baby-bottle tooth decay," shown in Illustration 29.18, infants should not be put to sleep with a bottle containing sweet fluids or formula.

Do Infants Need Supplements?] Two situations call for the use of supplements during infancy (see Table 29.11). Breastfed infants and infants receiving formula from a concentrate that is not diluted with fluoridated water need fluoride supplements after 6 months of age. Since breast milk contains a low amount of vitamin D, and since cases of rickets are increasing among breastfed infants in the United States, breastfed infants not exposed regularly to sunshine should receive a vitamin D supplement.[68]

The Development of Healthy Eating Habits Begins in Infancy

Older infants and young children should be offered a wide variety of nutritious foods in a positive eating environment. They alone should make the decision about how much to eat of any food offered.

TABLE 29.11

INDICATIONS FOR SUPPLEMENTING INFANTS.

1. Fluoride supplementation is recommended for infants older than 6 months who live in areas with no or low fluoride in the household water supply (0.25 mg of fluoride/day from 6 months to 3 years of age).[65]

2. Vitamin D supplements (200 IU/day) should be given to breastfed infants receiving little exposure to sunshine.[66]

- Breastfed infants exposed to about 30 minutes of sunshine each week with only diapers on, or those receiving 2 hours of sun exposure while wearing clothes but no hat, are likely meeting their needs for vitamin D by its manufacture in their skin.[67]

Illustration 29.19
Infant food preferences.

Clara Davis conducted her most quoted feeding experiment with older infants (7 to 9 months old) living in an orphanage connected to a large hospital in Chicago.

Three times a day a nurse would offer the infants a tray of various foods. The infants would pick up or grab for the foods they wanted, and the nurse would bring the food selected to the infant's hands or mouth.

Altogether 32 different foods were offered, consisting of fresh, unprocessed, unseasoned, and simply prepared basic foods. Foods included milk, beef, kidney, bone marrow, liver, brain,

thymus, fish, whole-wheat cereals, raw eggs, sea salt, 15 fresh fruits, and 10 fresh vegetables. Desserts, sugars, syrups, and other sweetened foods were not offered.

Infants quickly formed preferences and narrowed their choices to 14 of the 32 foods. Vegetables were the least liked, whereas milk, bone marrow, eggs, bananas, apples, oranges, and oatmeal were the best liked. They ate enough to grow normally and to remain in good health.

Dr. Davis concluded that appetite is the best guide to how much food an infant or young child needs, and that self-selection will only have doubtful value if the diet is chosen from inferior foods.[70]

TABLE 29.12

TIPS FOR TEACHING INFANTS THE RIGHT LESSONS ABOUT FOOD AND EATING.

1. Infants learn to eat a variety of healthy foods by being offered an assortment of nutritious food choices. There are *no* inborn mechanisms that direct babies to select a nutritious diet.

2. Infants must be allowed to eat when they are hungry and to stop eating when they are full. Infants, not parents, know when they are hungry and have had enough to eat.

3. Food should be offered in a pleasant environment with positive adult attention.

4. Food should not be used as a reward, punishment, or pacifier.

5. Infants or children should never be coerced into eating anything.

6. Food preferences change throughout infancy. Because an infant rejects a food one time doesn't mean she or he will not accept the food if offered later. Offering a food on a number of occasions often improves acceptance of the food. Babies still may not like strong-flavored vegetables until they are older, however.

Food preferences are primarily learned and are unique to every individual. The learning process begins early in life and is affected by a variety of influences. "Instincts" that draw infants to select certain, nutritious foods are not one of the influences, however. Infants do not instinctively know what to eat. Rather, they are born with mechanisms that help regulate how much they eat. These basic characteristics of infant food intake were beautifully demonstrated by Clara Davis in experiments with 7- to 9-month-old infants in the 1920s and 1930s.[69] One of her most famous experiments is described in Illustration 29.19.

Subsequent research reinforced earlier conclusions that infants and young children should be offered a wide variety of nutritious foods, and that decisions about how much to eat should be left up to the infant or child.[71] Beginning at birth, infants who are hungry eat enthusiastically. They stop eating when they are full. Coaxing, cajoling, and pleading by parents or caregivers can override infants' decisions about how much food to eat. That, unfortunately, may upset infants' built-in food intake regulating mechanisms and lead to over- or undereating, and to "fussy" eaters.

Teaching Infants the Right Lessons about Food] Recommendations for feeding infants are primarily based on energy and nutrient needs, the developmental readiness of infants for solid foods, and the prevention of food allergies. But infant feeding recommendations also include a large educational component. Many of the lessons infants learn about food and eating make an impression that lasts a lifetime. Later food habits and preferences, appetite, and food intake regulation are all influenced by early learning experiences.[72] Table 29.12 provides a lesson plan for teaching your infant the right lessons about food and eating.

Making Feeding Time Pleasurable]

More than any other area of development, feeding is an arena in which parents and baby work out the continuing struggle between dependence (being fed) and independence (feeding oneself). Independence must win. Pushing a child to eat is the surest way to create problems. Feeding has got to be pleasurable.

—T. Berry Brazelton, 1993

As Dr. Brazelton, an expert on child development, points out, there is much more to the feeding of infants than meets the eye. Infants learn about communication, relationships, and independence during pleasurable eating experiences. If undertaken in positive and supportive circumstances, mealtimes for infants can provide lessons that will support their physical and psychological health well into the future.

Key Terms

colostrum, page 29-15

critical period, page 29-3

development, page 29-2

fetus, page 29-2

growth, page 29-2

trimester, page 29-8

neural tube defects, page 29-8

www links

www.lalecheleague.org
A comprehensive online resource for breastfeeding information and support resources.

www.aap.org
The American Academy of Pediatrics's Web site is loaded with information on breastfeeding and infant and child nutrition.

www.4woman.gov
A Web site from the Public Health Service that focuses on breastfeeding and other health topics of interest to women.

www.cdc.gov/breastfeeding/
The Centers for Disease Control Web site on breastfeeding.

www.cdc.gov/growthcharts
Need more information about the growth charts, or want copies of them? Get it here.

www.hc-sc.gc.ca/english/
Health Canada's Web site. Search pregnancy, breastfeeding, infants, or other topics.

www.nlm.nih.gov/medlineplus/breastfeeding.html
Good source for updates and advice on breastfeeding, problem solving, alternative therapies, and nutrition.

http://mchlibrary.info
The Maternal and Child Health Bureau's site for information on resources, hot topics, and announcements of new publications.

www.babycenter.com
A favorite Web site of many consumers, it offers articles, bulletin boards, and features (and advertisements!) related to fertility, pregnancy, and infant feeding.

Notes

1. MacDorman MF et al. Annual summary of vital statistics. Pediatrics 2002;110: 1037–50.

2. MacDorman et al., Annual summary of vital statistics.

3. MacDorman et al., Annual summary of vital statistics; Li C et al., Birthweight and risk of overall and cause-specific childhood mortality, Pediatr Perinat Epidemiol 2003;17:164–70; and Health, United States 2003. www.cdc.gov/nchs/, accessed 10/03.

4. Healthy people 2000: national health promotion and disease prevention objectives. Washington, DC: U.S. Department of Health and Human Services, DHHS publication No. (PHS) 91-50212; 1991.

5. Rosso P. Nutrition and metabolism in pregnancy. New York: Oxford University Press; 1990.

6. Lewis R. New light on fetal origins of adult disease. The Scientist 2000(Oct 30);14:1, 16.

Nutrition UP CLOSE

You Be the Judge!

FOCAL POINT: Apply your nutrition knowledge to the diet of a pregnant woman.

Gloria's baby is due in 5 months. Before becoming pregnant, she never gave much thought to what she ate, but now she is trying to eat healthy foods for her baby and herself. Compare her choices with the Food Guide Pyramid recommendations for a pregnant woman to find out how well she is doing. Assume that each food item listed is the equivalent of 1 serving.

- *Breakfast:* Bran muffin, cereal with strawberries, and a glass of milk.
- *Lunch:* Vegetable soup, carrot sticks, an apple, and a glass of orange juice.
- *Afternoon snack:* Yogurt.
- *Dinner:* Chicken, turnip greens, a tossed salad, and iced tea.

Categorize Gloria's choices into the following food groups to determine if she ate the recommended servings from the Food Guide Pyramid. The breads and cereals group has been done for you.

Breads and Cereals	**Vegetables**
bran muffin	_____
cereal	_____
Gloria's servings: __2__	Gloria's servings: _____
Recommended servings: __6-11__	Recommended servings: __3-5__

Fruits	**Milk & Milk Products**
_____	_____
_____	_____
Gloria's servings: _____	Gloria's servings: _____
Recommended servings: __2-4__	Recommended servings: __3__

Meat and Alternates	**Miscellaneous (fats, sugars)**
_____	_____
_____	_____
Gloria's servings: _____	Gloria's servings: _____
Recommended servings: __2__	Recommended servings: __in moderation__

FEEDBACK (including answers) can be found at the end of Unit 29.

Source: Adapted from Healthy Infant Outcome Program "You Be the Judge," University of Minnesota, 1989.

7. Brown JE, Kahn ESB, Maternal nutrition and the outcome of pregnancy, Clin Perinatol 1997;24:433–49; and Clark PM, Programming of the hypothalamo-pituitary-adrenal axis and the fetal origins of adult disease hypothesis, Eur J Pediatrics 1998;157(suppl):S7–S10.

8. Rosso, Nutrition and metabolism in pregnancy.

9. Doyle JJ et al., Nutritional vitamin B_{12} deficiency in infancy: three case reports and a review of the literature, Pediatr Hematol Oncol 1989;6:161–72; and Winick M, Nutrition and pregnancy, White Plains (NY): March of Dimes Birth Defects Foundation; 1986.

10. Miller DR, Hayes KC, Vitamin excess and toxicity, In: Hathcock JN, editor. Nutritional toxicology, New York: Academic Press; 1982: pp 81–131; and Cochrane WA, Overnutrition in prenatal and neonatal life: a problem? Can Med Assoc J 1965;93:893–9.

11. Brown JE et al., Prenatal weight gains related to the birth of healthy-sized infants to low-income women, J Am Diet Assoc 1986;86:1679–83; and National Academy of Science (Institute of Medicine), Nutrition during pregnancy: I. Weight gain, II. Nutrient supplements, Washington, DC: National Academy Press; 1990: p 48.

12. National Academy of Science, Nutrition during pregnancy.

13. National Academy of Science, Nutrition during pregnancy.

14. Ohlin A, Rossner S. Maternal body weight development after pregnancy. Int J Obesity 1990;14:159–73.

15. Dietary Reference Intakes: Energy, carbohydrate, fiber, fatty acids, cholesterol, protein, and amino acids. Washington, DC: National Academies Press; 2002.

16. Dietary Reference Intakes, 2002.

17. Dietary Reference Intakes, 2002.

18. Brown JE et al. Predictors of red cell folate level in women attempting pregnancy. JAMA 1997;277:548–52.

19. Daly LE et al. Folate levels and neural tube defects. JAMA 1995;274: 1698–1702.

20. Brown et al., Predictors of red cell folate level.

21. Folate levels among American women, Morbidity and Mortality Weekly Report 2000;49:962–65; and Palomaki GE et al., Prenatal screening for open neural tube defects in Maine (letter), N Engl J Med 1999;340:1049–50.

22. Sahakian V et al. Vitamin B_6 is effective therapy for nausea and vomiting of pregnancy: a randomized, double-blind placebo-controlled study. Obstet Gynecol 1991;78:33–36.

23. National Academy of Science, Nutrition during pregnancy; and Brown and Kahn, Maternal nutrition.

24. Brown and Kahn, Maternal nutrition.

25. National Academy of Science, Nutrition during pregnancy; and Rosso, Nutrition and metabolism in pregnancy.

26. Recommendations to prevent and control iron deficiency in the United States. MMWR 1998;47:1–28.

27. National Academy of Science, Nutrition during pregnancy; and Preziosi P et al. Effect of iron supplementation on the iron status of pregnant women: consequences for newborns. Am J Clin Nutr 1997;66:1178–82.

28. National Academy of Science, Nutrition during pregnancy.

29. Litovitz T, Manoguerra A. Comparison of pediatric poisoning hazards: an analysis of 3.8 million exposure incidents. Pediatrics 1992;89:999–1006.

30. Dietary Reference Intakes, 2002.

31. Taper LJ et al., Zinc and copper retention during pregnancy: the adequacy of prenatal diets with and without dietary supplementation, Am J Clin Nutr 1985;41:1184–92; and Swanson CA, King JC. Zinc and pregnancy outcome, Am J Clin Nutr 1987;46:763–71.

32. Herbert V. Recommended dietary intake (RDI) of iron in humans. Am J Clin Nutr 1987;45:679–86.

33. Jones KL. Pattern of malformation in offspring of chronic alcoholic mothers. Lancet 1973;i:1267–71; and Spohr H-L et al. Prenatal alcohol exposure and long-term developmental consequences. Lancet 1993;341:907–10.

34. Mattson SN et al., Heavy prenatal alcohol exposure with or without physical features of fetal alcohol syndrome leads to IQ deficits. J Pediatr 1997;131: 718–21.

35. National Academy of Science, Nutrition during pregnancy.

36. National Academy of Science, Nutrition during pregnancy.

37. National Academy of Science, Nutrition during pregnancy.

38. Hathcock JN et al., Evaluation of vitamin A toxicity, Am J Clin Nutr 1990; 52:183–202; Brown and Kahn, Maternal nutrition; and National Academy of Science, Nutrition during pregnancy.

39. Births: Final Data, 1998. Washington, DC: National Center for Health Statistics, 2000; www.cdc.gov/nchs.

40. Da Silva AAM et al., Young maternal age and preterm birth, Pediatr Perinat Epidemiol 2003;17:332-9; Muñoz KA et al., Food intakes of US children and adolescents compared with recommendations, Pediatrics 1997;100:323–9; and Dubois S et al., Ability of the Higgins Nutrition Intervention Program to improve adolescent pregnancy outcome, J Am Diet Assoc 1997;97:871–8.

41. HHS Blueprint for Action on Breastfeeding. www.4woman.gov, accessed 11/8/00.

42. Cummings AS, Morbidity in breastfed and artificially fed infants, J Pediatr 1979;90:726–9; Cunningham AS et al., Breastfeeding and health in the 1980s: A global epidemiologic review, J Pediatr 1991;118:659–65; and Wilson AC et al., Relation of infant diet to childhood health, Brit Med J 1998;316:21–25.

43. HHS Blueprint (www.4woman.gov); Wilson et al., Relation of infant diet; and U.S. Public Health Service. A review of the medical benefits and contraindications to breastfeeding in the United States. Washington, DC: Maternal and Child Health Bureau; 1997: pp 9–10.

44. Simopoulos AP, Grave GD. Factors associated with the choice and duration of infant-feeding practices. Pediatrics 1984;74(suppl):603–14.

45. Health, U.S. 2003 (www.cdc.gov/nchs/); HHS Blueprint (www.4woman .gov); and Ryan AS et al. A comparison of breast-feeding data from the National Surveys of Family Growth and the Ross Laboratories Mothers Surveys. Am J Public Health 1991;81:1049–52.

46. Health, U.S. 2003 (www.cdc.gov/nchs/).

47. Health, U.S. 2003 (www.cdc.gov/nchs/).

48. Health, U.S. 2003 (www.cdc.gov/nchs/).

49. Committee on Nutrition, American Academy of Pediatrics. Breastfeeding and the use of human milk. Pediatrics 1997;100:1035–9.

50. Lakhkar BB. Breastfeeding in adoptive babies. Indian Pediatr 2000;37:1114–16.

51. National Academy of Sciences (Institute of Medicine). Nutrition during lactation. Washington, DC: National Academy Press; 1991.

52. National Academy of Sciences, Nutrition during lactation.

53. National Academy of Sciences, Nutrition during lactation; and Lovelady CA et al. The effect of weight loss in overweight, lactating women on the growth of their infants. N Engl J Med 2000; 342:449–53.

54. National Academy of Sciences, Nutrition during lactation.

55. Watkinson B, Fried PA, Maternal caffeine use before, during and after pregnancy and effects upon offspring, Neurobehavioral Toxicology and Teratology 1985;7:9–17; Fetal alcohol syndrome and alcohol-related neurodevelopmental disorders, American Academy of Pediatrics. Pediatrics 2000;106:358–61; and Menella JA, Beauchamp GK, The transfer of alcohol to human milk, N Engl J Med 1991; 325:981–85.

56. Schwartz PM et al., Lake Michigan fish consumption as a source of polychlorinated biphenyls in human cord serum, maternal serum, and milk, Am J Public Health 1983;73:293–6; Miller SA, Chopra JG, Problems with human milk and infant formulas, Pediatrics 1984; 74(suppl):639–47; and Rogan WJ, Pollutants in breast milk, Arch Pediatr Adol Med 1996;150:981–90.

57. National Academy of Sciences, Nutrition during lactation.

58. Thompson D et al. New CDC growth charts. WIC National Meeting, Salt Lake City, Utah, 9/8/00.

59. Winick M. Malnutrition and brain development. New York: Oxford University Press; 1976.

60. Lloyd-Still JD et al. Intellectual development after severe malnutrition in infancy. Pediatrics 1974;54:306–11.

61. HHS Blueprint (www4woman.gov); and Rappo PD et al. Pediatrician's responsibility for infant nutrition. Pediatrics 1997;99:749–50.

62. Rappo et al., Pediatrician's responsibility.

63. Beal VA. Termination of night feeding in infancy. J Pediatr 1969;75:690–2.

64. Krebs NK. Dietary zinc and iron sources, physical growth and cognitive development of breastfed infants. J Nutr 2000;130:358S–60S.

65. ADA Council on Access, Prevention, and Interpersonal Relations. J Am Dent Assoc 1995;126 (suppl):19S.

66. Vitamin D supplementation for breast-fed infants. Committee on Nutrition, American Academy of Pediatrics. Pediatrics 2003;111:908-10.

67. Specker BL et al. Sunshine exposure and serum 25-hydroxyvitamin D concentrations in exclusively breast-fed infants. J Pediatr 1985;107:372–6.

68. Specker et al., Sunshine exposure.

69. Marianne BM et al., Genetic and environmental influences on eating patterns of twins ages >50 years, Am J Clin Nutr 1999;70:456–65; Story M, Brown JE, Do young children instinctively know what to eat? The studies of Clara Davis revisited, N Engl J Med 1987;316:103–6; and Davis C. Self-selection of diets by newly weaned infants: an experimental study. Am J Dis Child 1928;36:651–79.

70. Davis, Self-selection of diets.

71. Birch LL, Deysher M. Caloric compensation and sensory specific satiety: evidence for self-regulation of food intake by young children. Appetite 1986;7:323–31.

72. Skinner JD et al. Do food-related experiences in the first 2 years of life predict variety in school-aged children? J Nutr Ed Behav 2002;34:310–5.

Nutrition UP CLOSE

You Be the Judge!

Feedback for Unit 29

Gloria's choices:

Breads and cereals	*2 servings* (bran muffin, cereal)
Vegetables	*4 servings* (vegetable soup, carrot sticks, turnip greens, tossed salad)
Fruits	*3 servings* (strawberries, apple, orange juice)
Milk and milk products	*2 servings* (milk, yogurt)
Meat and alternates	*1 serving* (chicken)
Miscellaneous (fats, sugars)	*0 servings!*

All of Gloria's choices are healthy ones, but she is missing some important foods. While concentrating hard on eating enough fruits and vegetables, she has neglected to consume enough high-protein and carbohydrate-rich foods from the remaining food groups. She needs to include in her diet more dairy products, more meat or protein alternates, and more starchy foods like breads, pasta, rice, tortillas, corn, and legumes. With a few additions to her existing choices, the quality of her diet can be greatly improved. Here is a more balanced version of Gloria's menu for a day:

- *Breakfast:* Bran muffin, cereal with strawberries, and a glass of milk.
- *Lunch:* Vegetable soup, carrot sticks, an apple, **burrito**, and a glass of orange juice.
- *Afternoon snack:* Yogurt, **cheese**, and **whole-grain crackers.**
- *Dinner:* Chicken, turnip greens, tossed salad, **grits, pita bread,** and iced tea.

Nutrition for the Growing Years: Childhood through Adolescence

Nutrition Scoreboard

	TRUE	FALSE
1 Parents should not allow their children to eat "junk" foods.		
2 It is up to parents and caretakers to decide how much children should eat.		
3 The incidence of overweight among 6- to 19-year-olds has more than tripled since the 1960s.		
4 Childhood and adolescence represent "grace periods" during which the diet consumed does not influence future health.		

Answers on next page

[KEY CONCEPTS AND FACTS]

- There is no evidence to support the notion that children are born knowing what foods they should eat.

- Children are born with regulatory processes that help them decide how much to eat.

- Parents and caretakers should decide *what* foods to offer children. Children should decide *how much* to eat.

- Diet and other behaviors of children and adolescents affect health before and during the adult years.

Answers to *Nutrition Scoreboard*

		TRUE	FALSE
1	Withholding sweets and other favorite foods makes kids value and want them more. Such foods should not be given special emphasis.		✔
2	Children, not parents or caregivers, should decide how much food is consumed.		✔
3	About 5% of 6- to 19-year-olds were overweight in the 1960s. Now over 15% are overweight.[1]	✔	
4	Childhood and adolescent diets can affect future health. Heart disease and diabetes, for example, can have "pediatric" origins that are related to dietary intake patterns.	✔	

Photo Researchers - Juneberg Clark

The Span of Growth and Development

Physical and mental development proceed at a high rate from infancy through adolescence (Illustration 30.1). Young children generally enter this phase of life able to take a step or two and to guide a spoon into their mouths sideways. They will likely leave adolescence with the ability to drive, work for pay, and solve complex problems. These are the formative years that lay the foundation for the rest of life.

The Nutritional Foundation

As the twig is bent, so grows the bough.

Good nutrition takes on particular importance during the growing years, for many reasons. Growth is an energy- and nutrient-requiring process. It will not proceed normally unless the diet supplies enough of both.

Children learn about food and its importance to health and well-being during these early years of life, and they also establish food preferences and physical activity patterns that may endure into the adult years. Food intake regulatory mechanisms are affected by the lessons children learn early in life. Given control over decisions about how much to eat, children generally become responsive to internal cues that signal when they should eat and when they should stop eating.[2] If the lessons go well, they learn to eat for the right reason.

Eating well and learning the right lessons about food and health have implications that transcend the growing years. Early diets may have long-term effects on the risk of developing a number of diseases later in life.

Characteristics of Growth in Children

Growth slows substantially after the first year of life. Between the ages of 2 and 10 years, children normally gain somewhere around 5 pounds and 2 to 3 inches in

One to Two Years

Gains in height and weight continue at a lower rate; appetite is less. Uses finger and thumb to pick up things. Soft spot grows smaller and then disappears. Baby teeth continue to appear. Usually takes one long nap a day. Drinks from a cup, attempts to feed self with a spoon. Likes to eat with hands. Pulls self up to standing position. Walks alone.

Two to Three Years

Slower and more irregular gains in height and weight. Has all 20 teeth. Runs and climbs, pushes, pulls, lugs, walks upstairs one step at a time. Feeds self using fingers, spoon, and cup; spills a lot. At times has one favorite food. Associates the sensation of hunger with a need for food.

Three to Four Years

Gains 4–6 pounds and grows about 2–3 inches. Feeds self and drinks from a cup quite neatly, carries things without spilling. May give up sleep at nap time, substituting quiet play.

Four to Five Years

Gains in height and weight about same as previous year. Hops and skips, throws ball. Has increasingly good coordination, masters buttons and shoelaces. Can use knife and fork and is a good self-feeder.

Five to Six Years

Growth continues at about the same rate. Legs lengthen. Six-year permanent molars usually appear (new teeth, not replacing baby teeth). Begins to lose front baby teeth. Prefers plain, bland, and unmixed foods.

Illustration 30.1

Growth and development characteristics from 1 through 16 years.

continued

Illustration 30.1—continued
Growth and development characteristics from 1 through 16 years.

Six to Nine Years
Slow gains in height and weight (2–3 inches and 4–6 pounds a year). Some additional permanent teeth appear. Likely to have the childhood communicable diseases. Sleeps 11 to 13 hours.

Nine to Twelve Years
May be long legged and rangy, but health is generally sturdy. Permanent teeth continue to appear. Appetite good. Needs about 10 hours sleep. Growth spurt in girls usually begins. May get very irritable when hungry.

Twelve to Fourteen Years
Wide differences in height and weight in children of either sex of same age. Menstruation usually begins (sometimes earlier). Girls develop breasts, are usually taller and heavier than boys of the same age. Growth spurt of boys begins. Muscle growth rapid. Appetite increases. Likes to spend time with friends.

Fourteen to Sixteen Years
Boys are in period of growth spurt. Voice deepens. Girls have usually achieved maximum growth. Menstruation is established though may still be irregular. Enormous appetite. Pimples a common problem. Sleep reaches its adult pattern. All permanent teeth except wisdom teeth.

height per year (Illustration 30.2). Gains in weight and height occur in "spurts," rather than continuously and gradually. Prior to a growth spurt, appetite and food intake increase (given an adequate food supply), and the child puts on a few pounds of fat stores (Illustration 30.3). During the growth spurt, these fat stores are used to supply energy needed for growth in height.

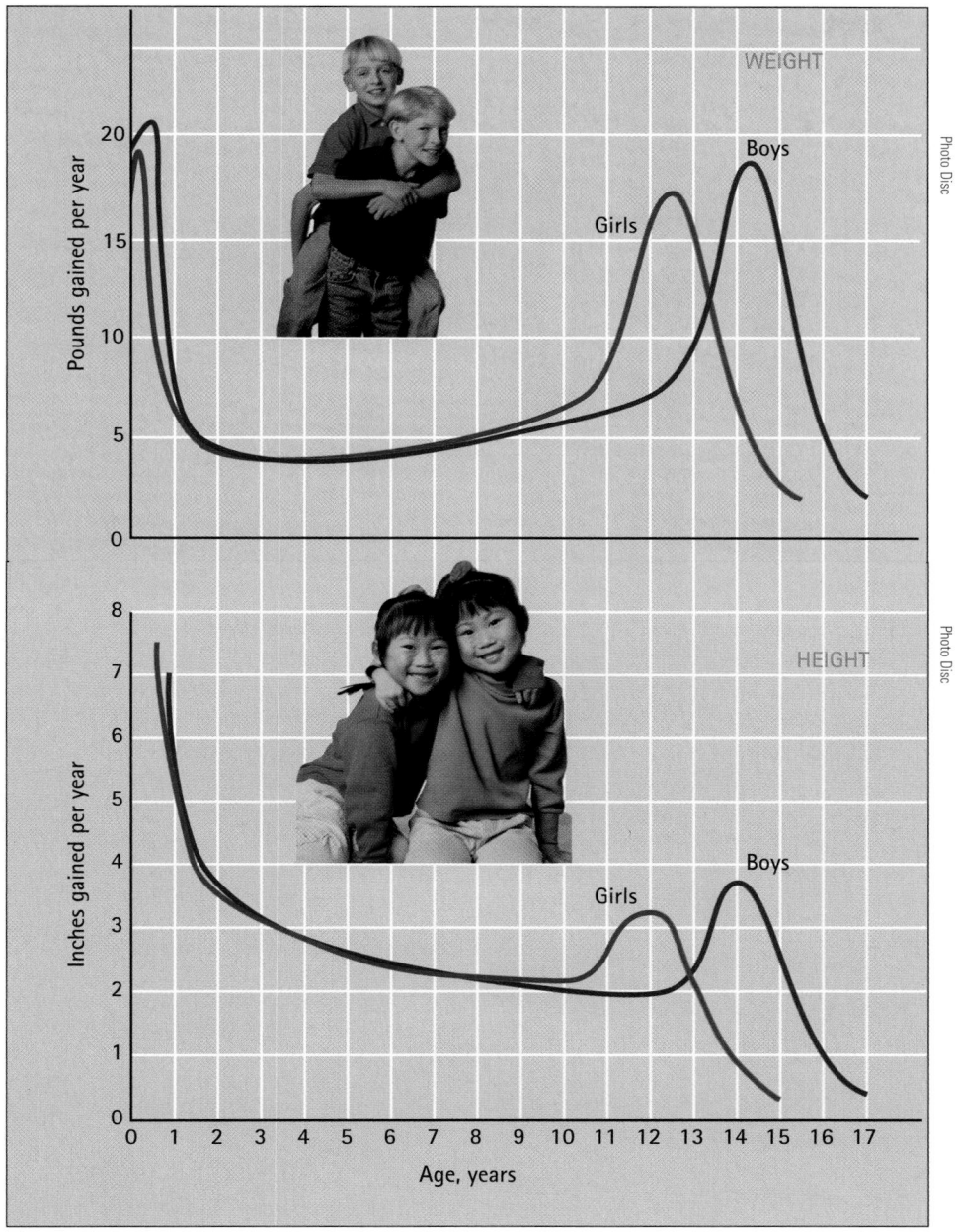

Illustration 30.2
Average yearly growth in weight and height during childhood and adolescence.

Photo Disc

Photo Disc

CDC's Growth Charts for Children and Adolescents] Growth progress during childhood and adolescence is generally monitored with the use of the Centers for Disease Control (CDC) growth charts. Charts are available for females and males from 0 to 36 months old, and from 2 to 20 years. Growth charts for 2- to 20-year-olds consist of graphs of:

- Weight for age (see Illustration 30.4)
- Height (stature) for age
- Weight for height
- Body mass index (BMI) for age

Illustration 30.3
Growth occurs in a series of "spurts." *Each spurt follows this sequence.*

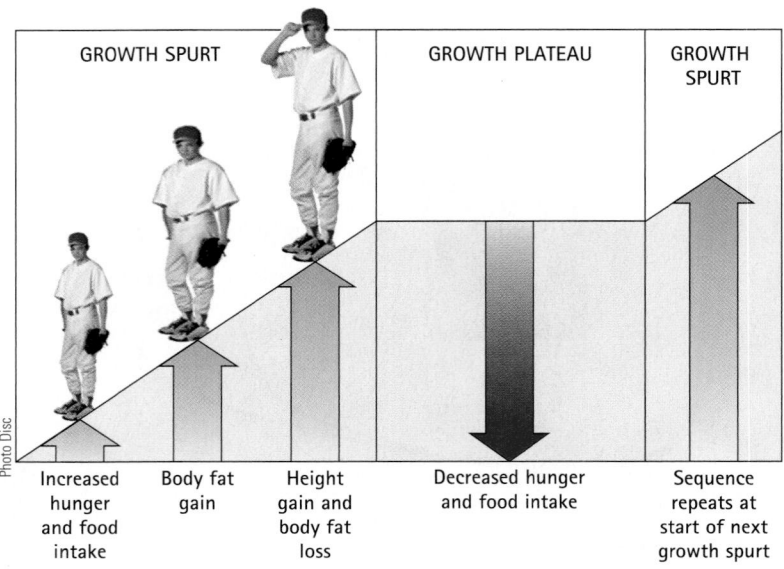

| GROWTH SPURT | GROWTH PLATEAU | GROWTH SPURT |

Increased hunger and food intake | Body fat gain | Height gain and body fat loss | Decreased hunger and food intake | Sequence repeats at start of next growth spurt

Illustration 30.4
An example of the CDC's growth charts: Weight for age for boys and girls. *The charts can be printed off the Internet using the address www.cdc.gov/growthcharts.*

Each graph provides percentiles that reflect the distribution of these measures in a representative sample of 2- to 20-year-olds in the United States. So, for example, if a child's weight for age is between the 50th and 75th percentiles, it means that this child's weight is somewhat higher than that of most children and similar to that of 25% of children who are represented in the 50th to 75th percentile range. Children and adolescents whose weight status measurements place them in the highest and lowest percentile ranges should be evaluated further for potential underlying nutrition and health problems.

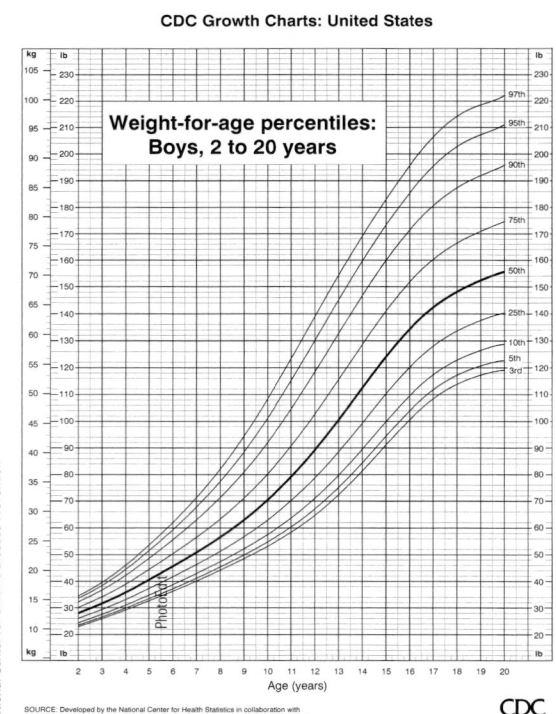

CDC Growth Charts: United States

Weight-for-age percentiles: Boys, 2 to 20 years

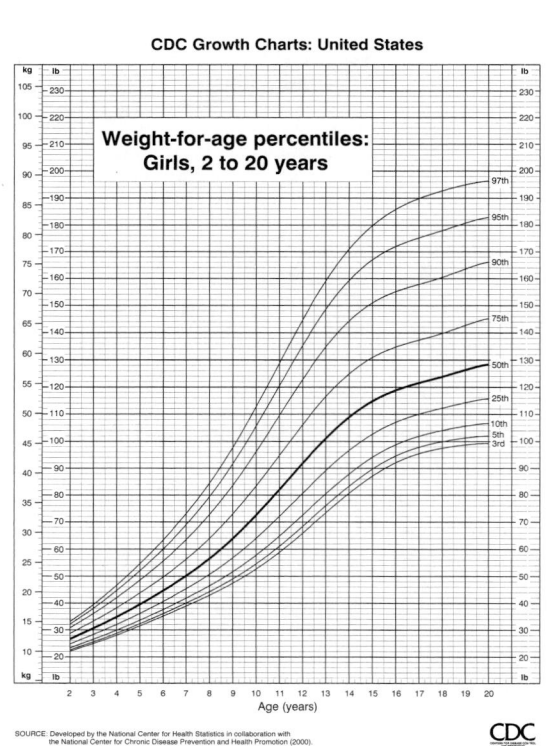

CDC Growth Charts: United States

Weight-for-age percentiles: Girls, 2 to 20 years

BMI charts for age are a feature of the growth charts. Since BMI increases with age in children and adolescents, ranges of BMI used to classify weight status in adults cannot be used. BMIs for age that fall between the 85th and 95th percentiles are considered "at risk" for overweight, and those above the 95th percentile at risk for obesity.[3] Although now recommended for determining weight status in children and adolescents, use of the CDC's new BMI for age charts requires that BMI be calculated. Illustration 30.5 shows you how to calculate BMI.

Food Jags and Normal Appetite Changes]

When growth is not occurring, children are often disinterested in food and eat very little at times. They may go on "food jags," insisting on eating only a few favorite foods like peanut butter and jelly sandwiches or breakfast cereal.

The ups and downs of children's food intake can be unnerving for parents. But as long as growth continues normally and children remain in good health, there's little reason to worry about food jags and fluctuations in appetite and food intake. Only children know when they are hungry or full.[4] (Actually, children appear to regulate caloric intake meal by meal better than adults do. Read about it in Illustration 30.6.)

Hunger Can Make Kids Irritable!]

It was a hot summer day, and 10-year old Kim and her friends had spent it in the pool. The fun went on and on . . . until five o'clock. That's when the argument started and Kim left the pool in a huff.

When Kim arrived home, still in a huff, her mom knew what was going on. Kim was cranky because she was hungry. She was so cranky, in fact, that suggesting she'd feel better if she ate something would only make her more

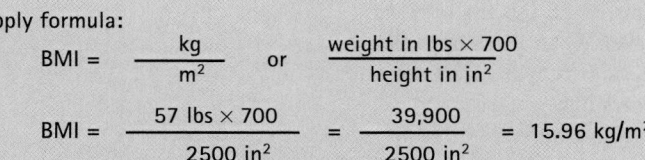

How to calculate BMI

Lucy just turned 7 years old and has had her weight and height carefully measured. She weighs 57 pounds (lb), and her height (stature) is 4 feet (ft), 2 inches (in). What's her BMI?

1. Convert her height of 4 ft 2 in to inches:

$$4 \text{ ft} \times 12 \text{ in per foot} = 48 \text{ in}$$
$$+2 \text{ in}$$
$$\overline{50 \text{ in}}$$

2. Square the inches:

$$50 \text{ in} \times 50 \text{ in} = 2500 \text{ in}^2$$

3. Apply formula:

$$BMI = \frac{kg}{m^2} \quad \text{or} \quad \frac{\text{weight in lbs} \times 700}{\text{height in in}^2}$$

$$BMI = \frac{57 \text{ lbs} \times 700}{2500 \text{ in}^2} = \frac{39,900}{2500 \text{ in}^2} = 15.96 \text{ kg/m}^2$$

In this example, Lucy's BMI for age of 15.96 kg/m² falls between the 50th and 75th percentiles, or well within the normal range.

Illustration 30.5
How to calculate BMI.

Illustration 30.6
The pudding study.

Lunch at the day care center the day of the study was a bit different than usual—it started with pudding. Preschoolers were offered a serving of pudding that contained 150 calories or another with 40 calories. Both looked and tasted the same. The amount of pudding consumed was secretly recorded, as was the children's food intake during the rest of lunch. The experiment was repeated on adults.

Children compensated almost perfectly for the calories in the different puddings. Those given the higher-calorie pudding ate less at lunch, and those receiving the low-calorie pudding ate more. The adults, however, didn't fare so well. Their caloric intakes for the rest of lunch bore no relationship to calories consumed in the pudding. The study suggests that children are more sensitive to caloric intake on a short-term basis than adults.[5]

irritated! Kim was ravenously hungry, but she didn't want anybody telling her what her problem was!

Kim's mom had the solution to her daughter's grumpiness on the table. "Dinner's ready," her mom called softly. Kim stomped to the table and grudgingly started to eat. Within minutes, she did feel better, but it wasn't because she was no longer hungry!

Only children know when they are hungry, but sometimes when they stay hungry too long, the signal to eat is overpowered by fun or the effects of not eating. Gently offering food to an irritable child who has skipped a meal or played too long can provide the cure for irritability.

The Adolescent Growth Spurt

It dawned on us that Max was entering his growth spurt when we had to wait for him to finish eating before we could all go out to dinner.
—Max's parents

The adolescent growth spurt usually occurs in girls between the ages of 9 and 12 years. For boys, this period of growth generally begins around the age of 12 to 14. Actually, though, the age when adolescents start their growth spurt normally varies considerably. Pictured in Illustration 30.7 are three friends, all aged 12. It's not hard to tell which one of them has experienced a growth spurt! The difference was also

Illustration 30.7
This photo (on the left) was taken when these boys, born 2 months apart, were 12 years old.
Ben (on the left) started his growth spurt at age 11, while Max (in the middle) began his at age 13. David (on the right) began to spring up in height after he turned 14. The photo (on the right) shows the same boys at age 19.

clear at the dinner table. By the time all three were 19 years old, Max (the one in the middle) had caught up with Ben (on the left), and David had grown taller but less than Max or Ben.

During these years of growth, teenagers gain approximately 50% of their adult weight, 20 to 25% of adult height, and 45% of their total bone mass. In the year of peak growth, girls gain 18 pounds on average, and boys 20 pounds.[6]

Can You Predict or Influence Adult Height?] Ultimate height is difficult to predict. On average, children tend to achieve adult heights that are between the heights of their biological parents.[7] There are many exceptions to this general finding, however, and that means that heredity is not the only influence on height. The dramatic increases in height of Japanese youth since World War II provide clear evidence that nongenetic factors have the strongest influence on height. Since the late 1940s, Japanese youth have grown an average of 2 inches taller each generation. (The increase in height has meant that everything from shoes to beds must be produced in larger sizes.) The increase in size of Japanese people is largely attributed to higher calorie and protein intakes and more nutrient-dense diets. Indeed, people in most economically developed countries continue to grow taller; a maximal genetically determined height has not yet been reached. If children are less well nourished than their parents, though, they tend to be shorter than their parents as adults.[8]

A healthy diet during the growing years—and, likely, exercise and freedom from frequent bouts of illness as well—support growth in height. There are no supplements, powders, or special diets that can be used to increase growth rate. Growth hormone injections can increase height somewhat, but the side effects of growth hormone are numerous and its use is limited.

Overweight and Type 2 Diabetes: Growing Problems

Rates of overweight in children and adolescents in the United States have tripled since the 1960s (Table 30.1), and overweight-related disorders are also on the rise. Conditions such as type 2 diabetes, bone and joint disorders, and elevated blood pressure that were very rarely observed in children and adolescents in the past are being diagnosed with increasing frequency.[10] The emergence of type 2 diabetes as a problem of the early years is of particular concern because diabetes generally worsens with time and causes long-term health impairments. It is estimated that 4% of children and adolescents have impaired glucose tolerance (a strong risk factor for type 2 diabetes), and that 6–17% of overweight and obese children and adolescents have type 2 diabetes.[11]

Causes of Overweight in Young People] Jean Mayer, a noted professor of nutrition, said we shouldn't be surprised that so many children are overweight: "A society where most children sit and watch TV, where nobody walks, and where no domestic chores are required is a society where we may expect to have obese children."[12]

A number of "obesigenic" trends have developed over the past several decades that likely account for rising rates of overweight. Compared to 20 or so years ago, children and adolescents now have fewer opportunities for physical activity and are exposed to a generally plentiful supply of energy-dense foods. Schools offer few physical activity classes and opportunities for exercise, while access to energy-dense,

TABLE 30.1

CHANGES IN THE PREVALENCE OF OVERWEIGHT IN US CHILDREN AND ADOLESCENTS.[9]

	PERCENT OVERWEIGHT		
	AGE 2–5 YEARS	AGE 6–11 YEARS	AGE 12–19 YEARS
1960s	–	4%	5%
1992	7	11	11
2000	10	15	16

empty-calorie foods in vending machines and cafeteria lines is increasing.[13] Children and adolescents are now exposed to portion sizes of energy-dense fast foods that are up to five times larger than in the past.[14] Genetically based, inborn processes that regulate appetite and satiety are being overwhelmed by environmental conditions that influence decisions about food intake.

Medicalization of Obesity and Treatment] Stomach stapling and diet drugs are being increasingly used to treat overweight and obesity in children and adolescents, but this trend is not viewed by some as a healthy one. Stomach stapling is forever, and little is known about the psychological and health effects of this type of surgery in children and adolescents. Diet drug use in adults is accompanied by potentially serious side effects and the requirements for long-term use for continued effectiveness.[15] Given that such drastic measures as surgery and diet drugs may not be attractive, how is the problem of the obesity epidemic in children and adolescents to be solved? That's right. The answer is prevention.

Prevention of Overweight] If environmental factors play the predominant role in the overweight epidemic, then the path to prevention is paved with environmental changes. Children and adolescents need more opportunities for physical activity, and a wider array of healthy food options at schools, in fast-food restaurants, as well as at home. Children who regulate their food intake based on appetite and satiety rather than on environmental cues will have a jump start on overweight prevention.

How Do Food Preferences Develop?

I do not like broccoli, and I haven't liked it since I was a little kid and my mother made me eat it. And I'm President of the United States, and I'm not going to eat any more broccoli.
—George Bush, 1990

Food preferences are a highly individual matter (Illustration 30.8). It's hard to find two people who share the same food likes and dislikes, and many people are very picky about how food is prepared and served. How *do* food preferences form? Are they "shaped" by early learning experiences, or are they inborn and beyond anyone's control?

With the exception of genetically determined sensitivity to bitter-tasting foods and an inborn preference for sweet-tasting foods, there is no direct evidence that food preferences are inborn.[16] The belief that young children instinctively know what to eat can be hazardous to a child's health. Parents may take an overly relaxed attitude toward poor food habits and inadvertently contribute to the development of health problems later in life. Informed parents should decide what types of food their child should be offered, but the decision about how much to eat should be left to the child.[17]

Food likes and dislikes appear to be almost totally shaped by the environment in which children learn about food. Which foods are offered, the way they are offered, and how frequently particular foods are offered all influence whether a child will like a given food or not.[18]

Humans are born with a tendency to be cautious about accepting new things, including foods. They may need to get used to a new food before they trust it. Parents may have to offer a new food on five or more occasions before the child makes a decision.[19] Sometimes children decide they like the food, and sometimes they

Illustration 30.8

Why select the apple?

Food likes and dislikes appear to be almost totally shaped by the environment in which children learn about food. Which foods are offered, the way they are offered, and how frequently particular foods are offered all influence whether a child will like a given food or not.

really don't like it. Often infants and young children do not like strong-flavored vegetables, spicy foods, and mixed foods (Illustration 30.9).

Forcing a young child to eat foods she or he does not like, or restricting access to favorite foods, can have lifelong negative effects on food preferences and health. When attempts to get a child to eat a particular food turn the dinner table into a battleground for control, nobody wins. Foods should be offered in an objective, nonthreatening way so that the child has a fair chance to try the food and make a decision about it. Restricting access to, or prohibiting intake of, children's favorite "junk" foods tends to strengthen their interest in the foods and consumption of those foods when they get a chance. Such prohibitions have the opposite effect of that intended because they make kids want the foods even more.[20]

What's a Good Diet for Children and Adolescents?

> *Children today are tyrants. They contradict their parents, gobble their food, and tyrannize their teachers.*
> —Socrates, 470–399 B.C.

Good diets for the growing years can be achieved by following the food group recommendations shown in Illustration 30.10. The type of foods selected from the food groups is important. Whole grain breads and

Illustration 30.9

USDA's Food Guide Pyramid for young children.

Illustration 30.10

Food group recommendations for children aged 2 to 6 and 7 to 15 years, based on the Food Guide Pyramid.*

*Children between the ages of 1 and 2 need the same variety of foods as older children but smaller serving sizes.

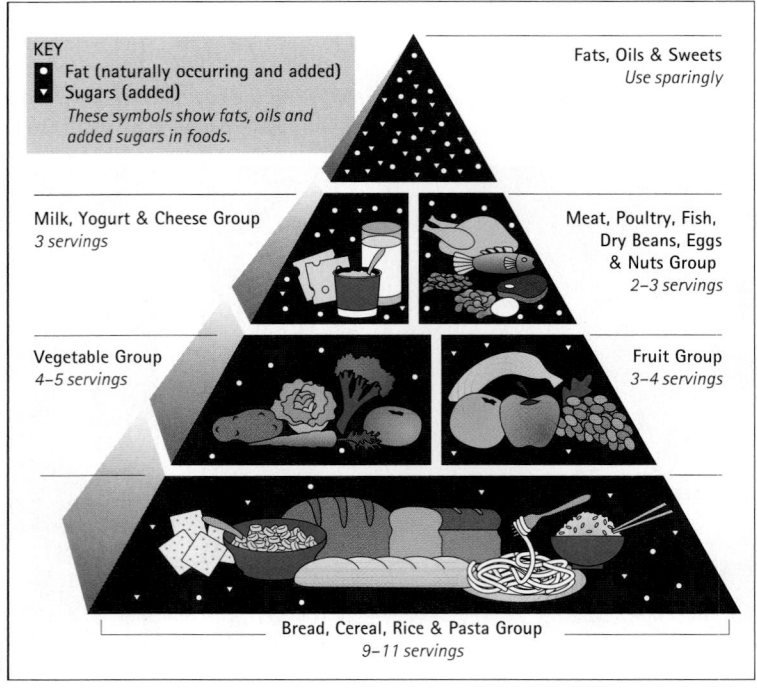

KEY
○ Fat (naturally occurring and added)
▽ Sugars (added)
These symbols show fats, oils and added sugars in foods.

Fats, Oils & Sweets
Use sparingly

Milk, Yogurt & Cheese Group
3 servings

Meat, Poultry, Fish, Dry Beans, Eggs & Nuts Group
2–3 servings

Vegetable Group
4–5 servings

Fruit Group
3–4 servings

Bread, Cereal, Rice & Pasta Group
9–11 servings

USDA

cereals should be represented in diets; and boiled, broiled, and baked foods should be consumed more often than fried foods.

Foods from the "miscellaneous" group that contribute mainly calories may be needed by children and teenagers. Although cakes, candy, cookies, potato chips, and so on are generally considered "sometimes" foods rather than "always" foods, they can serve as sources of needed calories. Snacks also are an important source of calories in the diets of children and teenagers (Illustration 30.11). Good choices for snacks include

Yogurt	Bananas	Carrots
Cheese	Oranges	Cucumbers
Low-fat milk	Apples	Popcorn
Ice milk	Apple juice	Peanuts
Pears	Mangos	Cherry tomatoes
Melons	Raisins	
Peanut butter		

Illustration 30.11

Snacks are an important source of calories and nutrition in the diets of children and teenagers.

Blair Seitz/Photo Researchers

Recommendation for Fat Intake] Normal growth and health have been demonstrated in children and adolescents consuming 21 to 35% of total calories from fat. Problems are noted when diets reach 35% or more calories from fat, however. High fat intakes are associated with excessive caloric intake, very low intakes of folate and vitamin C, and high saturated fat intake. Consequently, it is recommended that children older than 3 years and adolescents consume diets that provide 25 to 35% of total calories from fat. Additionally, it is recommended that 3- to 18-year-olds consume as little saturated fat as possible as while consuming a nutritionally adequate diet.[21]

Should Children Drink Milk?] Yes! Children certainly should drink milk (Illustration 30.12). If this sounds like a silly question, good. It is, but the notion that children (and adults, for that matter) should not drink milk or consume other animal products erupts in the media from time to time.

Children don't have to consume milk in order to have an adequate diet, because a variety of other foods could provide the nutrients in milk. However, children who fail to drink milk tend to have low calcium intakes, lower bone density, and more fractures than children who drink milk regularly.[22]

Illustration 30.12
Children should drink milk in order to meet their needs for calcium and vitamin D.
Virtually all school lunch programs include milk as a beverage.

Status of Children's and Adolescents' Diets

How well are children and teenagers doing at meeting the Food Guide Pyramid recommendations? Children over the age of 2 in the United States tend to consume too few vegetables and fruits; too little calcium, zinc, and vitamins E, D, C, and B_6; and too little dietary fiber.[23] Few (9%) children consume 5 servings of vegetables and fruits a day, and most consume 2.5 servings or less. Adolescents are even less likely to consume their "five a day"; only 1% of them meet the vegetable and fruit intake goal.[24] Low vegetable and fruit intakes contribute to less than optimal intakes of vitamins and minerals by children and teens.

Approximately half of all children fail to consume the recommended intake of calcium daily. Low calcium and vitamin D intakes during childhood and adolescence decrease peak bone mass and increase the risk of osteoporosis later in life. A part of the reason children and teens fail to get enough calcium and vitamin D is that they tend to substitute soft drinks for milk during these years.[25]

Diets of children and adolescents generally provide about half of the recommended intake of dietary fiber of 19–38 grams per day.[26] This makes constipation a problem for many children and teens.

Early Diet and Later Disease

Children and adolescents generally feel vigorous and healthy even when they are accumulating diet-related risk factors that will influence disease development later in life. Many studies indicate that in populations with a low incidence of heart disease, blood cholesterol values of both children and adults are lower than in countries with high rates of heart disease. Typical cholesterol levels of children in the United States are higher than those of children in countries where the rate of heart disease is lower. In addition, obesity during adolescence is associated with the presence of risk factors for subsequent heart disease: high LDL cholesterol, low HDL cholesterol, and increased blood pressure.[27] High-sodium diets tend to increase blood pressure in children and adolescents who have a hypertensive parent. U.S. children tend to consume far more sodium than is recommended. Adolescents who fail to consume sufficient calcium and to obtain adequate levels of vitamin D are at later risk of osteoporosis.[28] Children and adolescents who become obese are at higher risk not only for obesity later in life but also for disorders related to obesity such as diabetes, heart disease, some cancers, and hypertension. Obesity may also lead to lower self-esteem, a poor body image, and eating disorders.[29]

Nutrition is related to both current and future health and quality of life for people of all ages.

Nutrition **UP CLOSE**

Eating on the Run

FOCAL POINT: Using the Diet Analysis Plus program to check out a teenager's diet.

Tanya is 16 and her life is hectic. She has skating practice at 6:30 a.m., attends school from 7:45 a.m. to 2:30 p.m., and works after school at a super- market. Here's a list of the foods she eats in a typical day:

Breakfast:	Orange juice, 12 oz
Lunch:	Bagel, 2 oz
	M&M's candy, 2 oz
	Diet cola, 12 oz
Snack at work:	Apple, 1
	Potato chips, 1 oz
Dinner:	French fries, 1 cup
	Hamburger, 3 oz
	Cheddar cheese, 1½ oz
	Bun, 2 oz
	Diet cola, 12 oz
Evening snack:	Banana-peanut butter shake (banana, 1; peanut butter, 2 tbs; 2% milk, 1 cup)

Checking It Out Use the Diet Analysis Plus program's Food Guide Pyramid option to determine how Tanya's typical diet compares to that recommended. Fill in this table with the number of servings of foods from each food group in Tanya's diet. Then list two strengths and two weaknesses of her diet.

Food Groups

	Bread, Cereal, Rice, Pasta	Vegetables	Fruits	Milk Yogurt, Cheese	Meat, Poultry, etc.	Fats, Sweets
Recommended number of servings	9–11	4–5	3–4	3	2–3	limit
Tanya's number of servings	_____	_____	_____	_____	_____	_____

List two strengths of Tanya's diet:

1. _____

2. _____

List two weaknesses:

1. _____

2. _____

FEEDBACK (including answers) can be found at the end of Unit 30.

www links

www.cdc.gov/nccdphp/dnpa/kidswalk .htm or 1-800-CDC-4NRG
"KidsWalk" program kit available from the CDC provides guides for encouraging increased physical activity in children.

www.aap.org
The American Academy of Pediatrics's Web site is loaded with information on child nutrition.

www.cdc.gov/growthcharts
Need more information about the growth charts or printouts of them? Get it here.

www.hc-sc.gc.ca/english/
Health Canada's Web site. Search infant and child nutrition for information and resources.

www.eatright.org and www.dietitians.ca
The American Dietetic Association and Dietitians of Canada offer sound nutrition advice for parents and children.

www.usda.gov/cnpp/
Provides access to "Tips for Using the Food Guide Pyramid for Young Children 2 to 6 Years Old," as well as the Food Guide Pyra-

mid booklet and the Dietary Guidelines for Americans. Obtain revised Food Guide Pyramids from this site in 2005 and later.

www.navigator.tufts.edu
Click on "lifecycle" and check out recommended Web sites for child and adolescent nutrition.

Notes

1. Satter E. The feeding relationship: problems and interventions. J Pediatr 1990; 117:S181–S9.

2. Satter, The feeding relationship.

3. Thompson D et al. New CDC growth charts. WIC National Meeting, Salt Lake City, Utah, 9/8/00.

4. Birch LL. Presentation at the National Conference on Nutrition Education Research, Chicago, September 1986.

5. Birch, Presentation at National Conference.

6. Tanner JM. Fetus into man: physical growth from conception to maturity. Cambridge (MA): Harvard University Press; 1978.

7. Smith GD. Growth and its disorders. Philadelphia: W. B. Saunders; 1979.

8. Bronner F. Adaptation and nutritional needs. (letter) Am J Clin Nutr 1997; 65:1570.

9. Ogden CL et al. Prevalence and trends in overweight among US children and adolescents, 1999–2000;288:1728–32.

10. Ogden et al., Prevalence and trends; and Gottesman MM, Healthy eating and activity together (HEAT): weapons against obesity, J Pediatr Health Care 2003;17:210–5.

11. Goran MI, Impaired glucose tolerance in obese children and adolescents (letter), N Engl J Med 2002;347:290; and Uwaifo GI et al., Impaired glucose tolerance in obese children and adolescents (letter), N Engl J Med 2002;347:290.

12. Mayer J. Obesity during childhood. In: Winick M, ed. Childhood obesity. New York: John Wiley & Sons; 1975: pp 73–80.

13. Unhealthy à la carte choices in schools bad for kids' overall eating habits, Nutr Week 2003;July 14:1; and Fisher JO et al., Children's bite size and intake of an entrée are greater with large portions than with age-appropriate or self-selected portions, Am J Clin Nutr 2003; 77:1164–70.

14. Young LR, Nestle M. Expanding portion sizes in the US marketplace: implications for nutrition counseling. J Am Diet Assoc 2003;103:231–4.

15. Winslow R, Rundle RL. For obese teens, a radical solution: stomach surgery. Wall Street Journal 2003;Oct.7:A1.

16. Story M, Brown JE. Do young children instinctively know what to eat? The studies of Clara Davis revisited. N Engl J Med 1987;316:103–6.

17. Satter, The feeding relationship.

18. Birch LL. The role of experience in children's food acceptance patterns. J Am Diet Assoc 1987;87(suppl):S36–S40.

19. Birch, The role of experience.

20. Fisher JO, Birch LL. Restricting access to palatable foods affects children's behavioral response, food selection, and intake. Am J Clin Nutr 1999;69:1264–72.

21. Dietary Reference Intakes: Energy, carbohydrate, fiber, fat, fatty acids, cholesterol, protein, and amino acids, Ch. 8, 11. Washington, DC: National Academies Press, 2002.

22. Black RE et al. Children who avoid drinking cow's milk have low dietary calcium intakes and poor bone health. Am J Clin Nutr 2002;76:675–80.

23. Dietary guidance for healthy children aged 2 to 11 years: Position of the

American Dietetic Association, J Am Diet Assoc 1999;99:93–103; and Stang J et al., Relationship between vitamin and mineral supplement, dietary intake, and dietary adequacy among adolescents, J Am Diet Assoc 2000;100:905–10.

24. Dietary guidance for healthy children; and Muñoz KA et al., Food intakes of US children and adolescents compared with recommendations, Pediatrics 1997;100:323–9.

25. Dietary guidance for healthy children; and Harnack L et al., Soft drink consumption among U.S. children and adolescents: nutritional implications, J Am Diet Assoc 1999;99:436–41.

26. Dietary Reference Intakes.

27. American Academy of Pediatrics (Committee on Nutrition), Prudent life-style for children: dietary fat and cholesterol, Pediatrics 1986;78:521–5; and Bergstrom E et al., Insulin resistance syndrome in adolescents, Meta Clin Exp 1996;28:908–14.

28. Roy CC, Galeano N, Childhood antecedents of adult degeneration disease, Pediatr Clin North Am 1985; 32:517–33; Frank GC et al., Sodium, potassium, calcium, magnesium, and phosphorus intakes of infants and children: Bogalusa Heart Study, J Am Diet Assoc 1988;88:801–7; and Stear SJ et al., Effect of calcium and exercise interventions on the bone mineral status of 16–18-y-old adolescent girls, Am J Clin Nutr 2003;77:985–92.

29. Dietary Reference Intakes.

Nutrition UP CLOSE

Eating on the Run

Feedback for Unit 30

	Bread	Vegetables	Fruits	Milk	Meat	Fats
Tanya's number of servings	4	2	4	2	1⅓	3

Two strengths of Tanya's diet:

1. Six (!) vegetables and fruits
2. Fruit for snacks

Two weaknesses:

1. Includes a number of foods high in saturated fat (hamburger, cheddar cheese)
2. Includes too many fats and sweets (potato chips, candy)

Other: Diet is low in milk products, breads and cereals, and meat and meat alternates.

Nutrition and Health Maintenance for Adults of All Ages

Nutrition Scoreboard

	TRUE	FALSE

1 Healthy eating contributes to longevity in part by delaying the age at which chronic disease develops.

2 Aging is a kind of disease, and people can do little to prevent their health and quality of life from declining as they age.

3 Calorie restriction and underweight increase longevity.

Answers on next page

[KEY CONCEPTS AND FACTS]

- Age does not necessarily predict health status. Healthy adults come in all ages.

- Dietary intake, body weight, and physical activity influence changes in health status with age.

- Aging processes begin at the cellular level.

- Medications, diseases, and biological processes associated with aging influence adults' requirements for certain essential nutrients.

Answers to *Nutrition Scoreboard*

		TRUE	FALSE
1	Healthy eating pays off.	✔	
2	Aging is not a disease process! Many things can be done to extend life and its quality. (This unit discusses some of those things.)		✔
3	Calorie restriction has not been shown to increase longevity in humans; underweight is associated with shorter life expectancy than normal weight.[1]		✔

Photo Disc

You Never Outgrow Your Need for a Good Diet

One of the most important types of health changes that occurs with age is no change at all.
—Daphne Roe, 1985

Eating right during the adult years is a wise practice. Good nutrition helps adults feel healthy and vigorous as they age and improves their overall sense of well-being.[2] Adults who eat healthfully tend to develop heart disease, cancer, hypertension, and diabetes at older ages and have more years of life than adults who do not.[3]

Aging is a normal process, not a disease. Although the incidence of many diseases increases with age, the causes of the diseases are often unrelated to aging. They may fully or partially result from the cumulative effects of diets high in saturated fat and low in vegetables and fruits, obesity, smoking, physical inactivity, excessive stress, or other habits that insidiously influence health on a day-to-day basis.[4] Aging cannot be prevented, but how healthy we are during aging can be influenced by what we do to our bodies.

Maintaining health as we age is becoming an increasingly important concern. There are more older adults in the United States than ever before, and the numbers are growing.

The Age Wave

life expectancy
The average length of life of people of a given age.

The proportion of people in the United States aged 65 years and older is steadily increasing (Illustration 31.1). By the year 2050, approximately 20% of the U.S. population will be 65 years of age or older. Since 1900, **life expectancy** at birth has increased by 63%, from 47.3 to 77.2 years.[5] Although advances in life expectancy

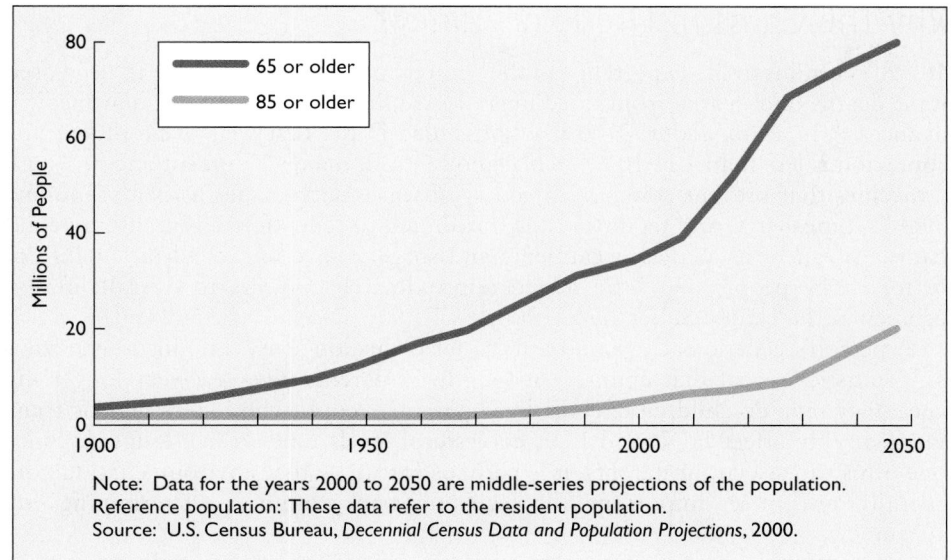

Illustration 31.1
Expected growth in the U.S. population of people aged 65 and 85 years and older.

Note: Data for the years 2000 to 2050 are middle-series projections of the population.
Reference population: These data refer to the resident population.
Source: U.S. Census Bureau, *Decennial Census Data and Population Projections*, 2000.

are welcomed, there is more progress to be made. The United States ranks fourteenth in life expectancy worldwide, behind countries such as Japan, Sweden, Canada, and Greece. Theoretically, the human life span could reach 130–135 years.[6]

All groups in the U.S. population are not benefiting equally from the increase in life expectancy. Life expectancy varies a good deal by sex and race. Table 31.1 shows overall life expectancy for Caucasian and African American men and women in 1900, 1940, and 2001. African American men have the shortest life expectancy, followed by Caucasian men. As in other industrialized countries, life expectancy for females exceeds that for males.[8] In the United States, the gender difference in average life span appears to be largely related to behaviors. U.S. males tend to smoke more, consume more alcoholic beverages, pay less attention to what they eat, and seek medical care less often than do females.[9]

TABLE 31.1

AVERAGE LENGTH OF LIFE FOR CAUCASIAN AND AFRICAN AMERICAN MALES AND FEMALES IN THE UNITED STATES IN 1900, 1940, AND 2001.[7]

	LIFE EXPECTANCY (YEARS)		
	1900	1940	2001
Caucasian Americans			
Males	46.6	62.1	75.0
Females	48.7	66.6	80.2
African Americans			
Males	32.5	51.5	68.6
Females	33.5	54.9	75.5

(Data for other ethnic groups were unavailable.)

Why the Gains in Life Expectancy?

Most of the gains in life expectancy in the United States are attributable to decreased infant deaths and deaths from infectious disease, improved nutrition, and medical advances.[10] In 1900, 1 out of 10 newborns died in the first year of life; today the proportion is less than 1 in 100.[11] Since about 1930, the development and mass use of vaccines that prevent common infectious diseases such as pertussis ("whooping cough"), diphtheria, and tetanus have contributed to reductions in infant and child deaths. Advances in medical treatments and surgery have improved the quality of life for many people and have added approximately 5 years to overall life expectancy in the United States since 1900.[12]

A person's genetic background affects her or his longevity. Life insurance company statistics reveal that children of long-lived parents have a 3-year longer life expectancy than do children of short-lived parents. People who inherit genetic traits that favorably affect HDL and LDL cholesterol levels tend to live longer.[13] Since none of us can select our parents, it is perhaps fortunate that environmental factors generally are more important than genetic background in determining life expectancy.

Living in the Bonus Round: Diet and Life Expectancy

Characteristics of diets of adults and older people who experience low disease rates and increased longevity include

- Regular consumption of vegetables and fruits
- Above-average intake of whole grain products
- Lower consumption of saturated fats
- Alcohol in moderation
- Eating breakfast[15]

Taking time to socialize and celebrate with friends and engaging in regular physical activity also appear to contribute bonus years to life (Illustration 31.2).[16]

Jeanne-Louise Calment of France had the longest documented life span: 122.5 years. Giant tortoises, by the way, can live up to 177 years.[14]

Photo Disc

Illustration 31.2
Fun with friends may increase longevity.

Photo Disc

Calorie Restriction and Longevity

For decades it has been known that laboratory animals fed diets providing 30% or less than normal levels of calories and adequate levels of essential nutrients have increased life expectancies. Why this happens isn't clear, but the reasons could be related to modifications in nutrient utilization and gene expression. To date, no human studies have tested the theory that caloric restriction increases longevity. Health effects of starvation have been examined, but those results don't apply. Diets of people who starve are inadequate in many nutrients in addition to calories.[17]

There are some who take the leap and extrapolate results of animal studies to themselves. The number of people practicing calorie restriction is increasing, although their numbers are still small. It's too soon to be adopting a calorie-restricted diet for the purpose of longevity. People who tend to live the longest have normal weights. Life expectancy decreases as body weight decreases below, or increases above, normal.[18]

Nutrition Issues for Adults of All Ages

Researchers and others interested in nutrition and health during middle age and beyond have tended to focus their attention on the cumulative effects of diet on chronic disease. Nutrition exerts its effects on chronic disease development over time, and therefore diseases related to poor diets are most likely to express themselves during the older adult years. The occurrence of diseases related to behavioral traits such as smoking and physical inactivity also increases among adults as they age.

In addition to the health effects of behaviors, people may age biologically. The combined effects of poor diets, other risky lifestyle behaviors, and biological aging increase the rates of serious illness during adulthood. How soon a disease develops largely depends on the intensity and duration of exposure to behavioral risks that contribute to disease development. Behavioral factors also affect the progress of biological aging processes.

Breaking the Chains of Chronic Disease Development

For the most part, the development of chronic disease in middle-aged and older adults can be viewed as a chain that represents the accumulation over time of problems that impair the functions of cells. Each link that is added to the chain, or each additional insult to cellular function, increases the risk that a chronic disease will develop. The presence of a disease indicates that the chain has gotten too long—that the accumulation of problems is sufficient to interfere noticeably with the normal functions of cells and tissues.

It appears that the chain can be shortened by healthful dietary and other behaviors. For example:

- Correcting obesity and stabilizing weight during the adult years may lengthen life expectancy.[19]

- Dietary intakes that correspond to the Dietary Guidelines for Americans are related to decreased death rates in women around the age of 60.[20]

- Regular intake of vitamin C and vitamin E from plant foods or supplements, and adequate folic acid intake, may be protective against declines in cognitive function and Alzheimer's disease.[21]

TABLE 31.2

EXAMPLES OF BIOLOGICAL CHANGES DURING AGING AND NUTRITIONAL CONSEQUENCES.[25]

BIOLOGICAL CHANGE	NUTRITIONAL CONSEQUENCES
• Lowered stomach acidity	• Decreased absorption of vitamin B_{12} and C
• Decreased lean muscle mass	• Reduced caloric need
• Reduced production of vitamin D in the skin	• Increased dietary requirement for vitamin D, increased risk for inadequate intake of vitamin D and calcium
• Decreased thirst	• Dehydration risk

- Maintaining adequate calcium and vitamin D intake and engaging in regular physical activity during the adult years may prevent, postpone, or lessen the severity of osteoporosis.[22]

- Above-average intakes of fruit and vegetables may delay or prevent the development of a number of types of cancer, heart disease, and stroke, and the formation of cataracts that "cloud over" the lens of the eye.[23]

The health status of adults is not necessarily "fixed" by age; it can change for the better or the worse, or not much at all.[24]

Nutrient Needs of Middle-Aged and Older Adults

Biological processes and lifestyle changes that generally accompany aging affect caloric and nutrient needs (Table 31.2). A person's need for calories generally declines with age as physical activity, muscle mass, and basal metabolic rate decrease. However, people who remain physically active into their older years maintain muscle mass and have a higher need for calories than people who are inactive.[26]

While caloric need decreases, the requirements for certain nutrients such as protein, vitamin C, vitamin D, vitamin B_{12}, and calcium may increase with aging.[27] The increased need for protein appears to be a result of decreased efficiency of protein utilization. Dietary requirements for vitamin C and calcium may increase due to lower levels of stomach acidity that often occur with advancing age. Decreased stomach acidity reduces the body's ability to absorb vitamins C and B_{12} and calcium. Vitamin D status may become a problem for older adults for two reasons: their intake of milk and milk products is often low, and exposure of the skin to sunlight produces about half of the amount of vitamin D as it does in young adults. An estimated 15% of older adults in the United States may be deficient in vitamin D.[28]

A Food Guide Pyramid constructed to reflect the nutrient needs of people over 70 years is presented in Illustration 31.3. This pyramid takes into account the lower food

Illustration 31.3
A Food Guide Pyramid for adults over 70.
Source: Tufts University, © 1999.

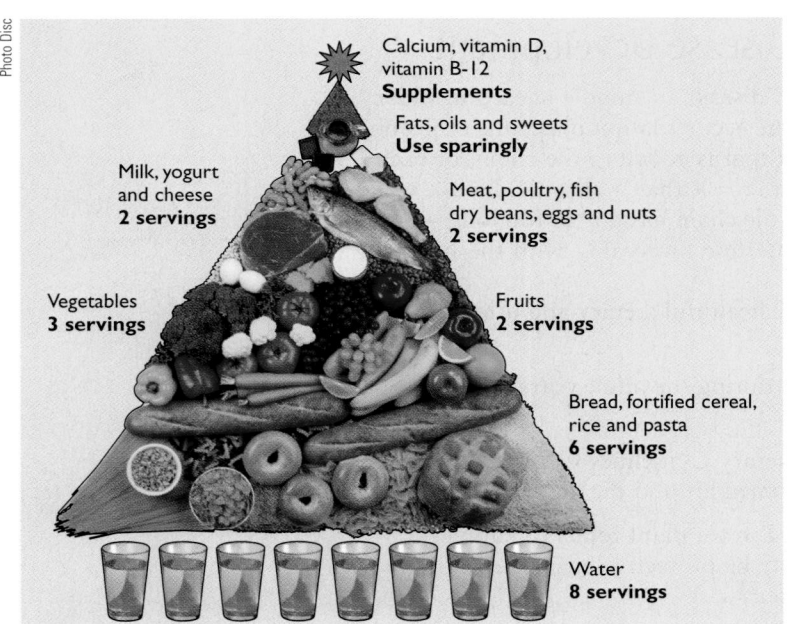

intakes of most older people by recommending 6 servings of grain products daily rather than the 6–11 recommended in the Food Guide Pyramid. It also highlights nutrient-dense foods that are most likely to provide adequate nutrient intakes in the relatively low-calorie diets of many elderly.

Fluid Needs]   For reasons that aren't entirely clear, many older people don't get thirsty when their bodies are running low on water. The implication of this change is that older people may become dehydrated and need medical assistance. Approximately 1 million elderly men and women are admitted to hospitals each year due to dehydration.[29] Fluid needs can be met through consumption of water, juices, teas, and other beverages. The food guide for the elderly shown in Illustration 31.3 emphasizes the importance of adequate fluid consumption by placing 8 glasses of water at the base of the pyramid.

Does Taste Change with Age?]   Poor diets observed in some middle-aged and older adults have been ascribed to "declining taste" with age. Besides being a dreadful thought, it's not true that taste declines with age to the extent that it makes eating less pleasurable than before. Although taste sensitivity does diminish somewhat with age, chances are your favorite foods will taste as good to you in your older years as they did in your youth (Illustration 31.4). Sight, smell, and hearing senses usually decline with age a good deal more than taste does.[30]

The senses of taste and smell are affected by medications such as antibiotics, antihistamines, some lipid-lowering drugs, and cancer treatments; diseases including Alzheimer's disease, cancer, and allergies; and surgeries that affect parts of the brain and nasal passages. These treatments and conditions are more prevalent in older adults than in middle-aged populations and are primarily responsible for declines in the senses of taste and smell. Changes in these senses increase the likelihood that the person will experience food poisoning or consume an inadequate diet and a low intake of food.[31]

Psychological and Social Aspects of Nutrition in Older Adults

Preparing meals and eating right may not be as simple as it sounds for many older adults. Consuming an adequate and balanced diet may not be easy when you depend on someone else to take you shopping, when mealtimes involve little social life, or when you "don't feel up" to making a meal. Isolation, loneliness, depression, and poor health can be major contributors to poor diets in older adults. The diets of older people are often lacking in nutrients, and because many older adults do not, or cannot, consume enough nutrients to meet their increased need for them, supplementation may be required.

Eating Right during Middle Age and the Older Years

The best diet for middle-aged adults is one that contains a wide variety of basic foods. A balanced and adequate diet can be obtained by judiciously selecting foods from the basic food groups. It is important to select *judiciously* because not all foods within the respective food groups are equally desirable. Food choices should emphasize the members of the food groups that are low in saturated fat, trans fat, and sodium, and

Illustration 31.4
Taste is less affected by age than are some other senses.

PhotoEdit - Jeff Greenberg

high in fiber. Diets should highlight vegetables, fruits, low-fat dairy products, and whole grains due to their health benefits.

Altering food habits in favor of healthy diets appears to be a common practice among adults of all ages. A study of 100 women and men over the age of 60 found that 100% had made some change in their food choices in the recent past. The changes had been for health, taste, convenience, and social reasons.[32] Clearly, a person's food choices and intake change throughout life, and the changes can be for the better.

Nutrition UP CLOSE

Does He Who Laughs, Last?

FOCAL POINT: Critically thinking about factors that influence longevity.

The following is an adaptation of a letter printed in the "Dear Abby" newspaper column:

> Dear Abby,
> Since I've reached my 80s, my mail is full of ads for health products to help me live longer.
> I once had many friends, all of whom were health vigilantes. They shook their heads knowingly as I avoided all health food fads and exercise. They made liquid out of good vegetables and spent fortunes buying all the latest supplements. They argued that "organic" was better and "natural" was best. I would tell them that snake venom, poison ivy and manure were "natural". But they wouldn't listen and they didn't laugh.
> Now my friends are all dead and I have no one left to argue with.

Critically think about the contents of this letter, and identify three alternate explanations for why this individual outlived his friends:

_____

_____

_____

FEEDBACK can be found at the end of Unit 31.

Key Terms

life expectancy, page 31-2

www links

www.4woman.gov
The National Women's Health Information Center provides information on diabetes, diet and heart health, and other conditions of concern to women. Or call 1-800-994-WOMAN, TDD: 1-888-220-5446.

www.agingwell.state.ny.us
This site provides a "health and wellness village" for mature adults that includes

nutrition quizzes, healthy recipes, and diet and exercise tips.

www.healthfinder.gov
A search engine for middle-aged and older adult nutrition and health topics.

www.nia.nih.gov/
The National Institute on Aging provides health information, exercise videos, and news related to senior health topics.

www.aarp.org
You can obtain nutrition and health educational materials from the American Association of Retired Person's site.

Notes

1. Schneider EL. Weighing your longevity part II: should you restrict your calories? www.healthandage.com, accessed 8/03.

2. Folsom AR et al., Waist-hip ratio and morbidity and mortality risk in older women, Arch Intern Med 2000;160: 2177–228.

3. Daviglus ML et al., Eating behavior in midlife and health care costs, www.nlm.nih.gov/medlineplus/heartdiseases.html, accessed 7/03; and Dausch JG, Aging issues moving mainstream. J Am Diet Assoc 2003;103:683–4.

4. Rowe JW, Kahn RL, Human aging: usual and successful, Clin Nutr 1990; 9:26–33; and Hughes VA et al., Longitudinal changes in body composition in older men and women: role of body weight change and physical activity, Am J Clin Nutr 2002;76:473–81.

5. US Census Bureau's International Data Base. www.about.com, accessed 10/03.

6. Health United States, 2003. www.cdc.gov/nchs/, accessed 10/03.

7. US Census Bureau (www.about.com).

8. US Census Bureau (www.about.com).

9. Women's health, Report of the Public Health Service Task Force on Women's Health Issues, Vol. 1, Public Health Rep 1985;100:73–106; and Women's Health USE 2003, www.hrsa.gov/womenshealth/, accessed 8/03.

10. Bernarducci MP, Owens NJ. Is there a fountain of youth? A review of current life extension strategies. Pharmacotherapy 1996;16:183–200.

11. US Census Bureau (www.about.com).

12. Guyer B et al., Annual summary of vital statistics: trends in the health of Americans during the 20th century, Pediatrics 2000;106:307–17; and Gold MR, Senior Adviser, Office of Disease Prevention and Health Promotion, Public Health Service, U.S. Department of Health and Human Services, 1997.

13. Cooper R, Sempos C, The price of progress, J National Med Assoc 1984;76:163–6; and Study finds gene mutation linked to long life, www.nlm.nih.gov/medlineplus/heartdiseases.html, accessed 10/03.

14. Mariani SM. Magic potions and the quest for eternal youth. Medscape Molecular Medicine 2003;5:1–7.

15. Kant AK et al., Consumption of recommended foods and mortality risk in women, JAMA 2000;283:2109–15; Lasheras C et al., Mediterranean diet and age with respect to overall survival in institutionalized, nonsmoking elderly people, Am J Clin Nutr 2000;71: 987–92; and Johnson MA et al., Nutritional patterns of centenarians, Internatl J Aging Hum Dev 1992;34:57–76.

16. Leaf A. Observations of a peripatetic gerontologist. Nutr Today 1973; (Sept./Oct.):4–12.

17. Schneider, Weighing your longevity part II.

18. Johannes L, The surprising rise of the radical diet: "calorie restriction." Wall Street Journal 2002;June 3:A1, A10; and Schneider, Weighing your longevity part II.

19. Stevens J et al. The effect of age on the association between body-mass index and mortality. New Engl J Med 1998; 338:1–7.

20. Kant et al., Consumption of recommended foods.

21. Seshadri S et al., Plasma homocysteine as a risk factor for dementia and Alzheimer's disease, N Engl J Med 2002;346:476–83; and Haan MN, Can vitamin supplements prevent cognitive decline and dementia in old age? Am J Clin Nutr 2003;77:62–3.

22. Dawson-Hughes B et al. Effects of withdrawal of calcium and vitamin D supplements on bone mass in elderly men and women, Am J Clin Nutr 2000; 72:745–50.

23. Steinmetz KA, Potter JD, Vegetables, fruit, and cancer prevention: a review, J Am Diet Assoc 1996;96:1027–39; Daviglus et al., Eating behavior in midlife; and Brunce GE, Antioxidant nutrition and cataract in women: a prospective study, Nutr Rev 1993;51: 84–86.

24. Rowe and Kahn, Human aging.

25. Garry PJ, Vellas BJ, Aging and nutrition. In: Present knowledge in nutrition, Ziegler EE, Filer, LJ eds., ISLI Press: Washington, DC; 1998:pp. 404–19; Guo S et al., Aging, body composition, and lifestyle: the Fels Longitudinal Study, Am J Clin Nutr 1999;70:405–11; and Drinka PJ, Goodwin JS, Prevalence and consequences of vitamin deficiency in the nursing home: a critical review, J Am Geriatr Soc 1991;39:1008–17.

26. Garry and Vellas, Aging and nutrition; Guo et al., Aging, body composition, and lifestyle.

27. Blumberg J. Nutritional needs of seniors. J Am Coll Nutr 1997;16:517–23.

28. Blumberg, Nutritional needs of seniors; and Drinka and Goodwin, Prevalence and consequences of vitamin deficiency.

29. Vogelzang JL et al. Overview of fluid maintenance/prevention of dehydration. J Am Diet Assoc 1999;99:605–9.

30. Shamburek RD, Farrar JT, Disorders of the digestive system in the elderly, N Engl J Med 1990;322:438; and Rolls BJ, Do chemosensory changes influence food intake in the elderly? Physiol Behav 1999;66:193–7.

31. Schiffman SS, Taste and smell losses in normal aging and disease. JAMA 1997;278:1357–62; and Bilderbeck N et al., Changing food habits among 100 elderly men and women in the United Kingdom. J Hum Nutr 1981;35:448–55.

32. Schiffman, Taste and smell losses; and Bilderbeck et al., changing food habits.

Nutrition UP CLOSE

Does He Who Laughs, Last?

Feedback for Unit 31

Alternate explanations:

1. The letter writer's friends may have switched to perceived health foods when they found out they were sick.

2. The letter writer may have consumed a lifelong healthy diet without choosing organic or "natural" foods.

3. The letter writer may come from a family whose members tend to have long lives.

Other possible explanations:

• A sense of humor may help extend life.

• Many factors in addition to diet, exercise, and genetic background influence longevity.

• One individual's experience may not apply to the masses.

The Multiple Dimensions of Food Safety

Nutrition Scoreboard

	TRUE	FALSE
1 The ingestion of pesticides on foods is a major contributor to the high incidence of cancer in humans.		
2 Freezing kills bacteria in food.		
3 Alaska has the highest incidence of botulism in the United States.		

Answers on next page

[KEY CONCEPTS AND FACTS]

- The leading food safety problem in America is contamination of food by bacteria and viruses.

- Foodborne illnesses primarily result from unsafe methods of producing, storing, and handling food.

- Foodborne illnesses are linked to hundreds of foods but most commonly to raw and undercooked meats and eggs, shellfish, and unpasteurized milk.

- Most cases of foodborne illness are preventable.

Answers to *Nutrition Scoreboard*

		TRUE	FALSE
1	Consumers in general are at low risk of developing cancer due to pesticides on food.[1] People at highest risk for pesticide-related illnesses are farmers and others who work with pesticides and do not follow appropriate safety procedures.		✔
2	Freezing causes most bacteria to cease multiplying, but does not kill them. If a food is contaminated with bacteria before you freeze it, bacteria will be present when you thaw it.	✔	
3	Here's a question for your nutrition trivia file. Alaska does have the highest incidence of botulism of all 50 states.[2] The reason is explained in this unit. Any guesses?	✔	

ON THE SIDE

More than 50% of consumers do not realize that foodborne diseases can be transmitted by fruits and vegetables.[5]

Photo Disc

foodborne illnesses
An illness related to consumption of foods or beverages containing disease-causing bacteria, viruses, parasites, toxins, or other contaminants.

Threats to the Safety of the Food Supply

Hamburgers contaminated with *E. coli*, chicken tainted with *Salmonella*, fruit coated with organophosphate, mad cow disease—it's enough to give you food fright. Actually, Americans have good reason to be concerned about the safety of some foods (Illustration 32.1). The Centers for Disease Control and Prevention (CDC) estimates that each year in the United States foodborne illnesses cause:

- sickness in 76 million people,

- 325,000 hospitalizations, and

- over 5000 deaths.[3]

Foodborne illnesses can be caused by bacteria, viruses, marine organisms, fungi, and the toxins they produce, as well as by chemical contaminants in foods or water. They are spread by a wide assortment of foods, including meats, eggs, unpasteurized milk, shellfish, raspberries, sprouts, breakfast cereals, and hummus. Foods most commonly associated with foodborne illness, however, are raw and undercooked meat and eggs, shellfish, and unpasteurized milk.[4]

How Good Foods Go Bad

Bacteria and viruses are the most common causes of **foodborne illnesses** and largely enter the food supply during food processing, storage, or preparation (Illustration 32.2).[6] They are transferred to humans in foods through many different routes, a major one being the contamination of food with animal feces. The lower intestines of many healthy farm animals are colonized by bacteria that may be harmful to humans. These bacteria contaminate food when unsanitary practices are used to prepare and process meats, and when vegetables and fruits are fertilized with animal manure. They can also be transferred to foods by humans who have colonies of harmful bacteria in their gastrointestinal tracts through the use of human sewage on crops, and food handling by people carrying the bacteria on their hands. Humans also transfer certain harmful microorganisms to food when fluids from infected injuries or body secretions contact food.[7] These types of foodborne illnesses are mainly caused by bacteria that are "on" rather than "in" foods.

Although less common, bacteria can be present on the inside of foods. They can enter vegetables and fruits, for example, if the protective skin coatings are broken, allowing an entrance for bacteria. *Salmonella* bacteria can infect the ovaries of hens and cause them to lay normal-looking but infected eggs. Shellfish can concentrate microorganisms present in surrounding water, and although the bacteria may be harmless to the shellfish, they can provide large doses of harmful bacteria and viruses to humans.[8]

Illustration 32.1
Problems with food safety often make the headlines.

Cross-Contamination of Foods

The source of many cases of foodborne illness is food that has come into contact with contaminated food. This situation, referred to as "cross-contamination," increases the reach of foodborne illnesses.

Illustration 32.2
Can you find three potential opportunities for the spread of foodborne illness in this photograph?

Illustration 32.3
Gathering broccoli samples for pesticide testing.
Pesticide residues do not appear to be a major cause of foodborne illnesses.

Illustration 32.4
Protective gear—including a chemical-resistant suit, respirator, and goggles—is required for use with some of the most toxic pesticides, but is not always worn.

Microorganisms including bacteria, viruses, toxins, and other harmful substances can contaminate safe foods during processing, shipping, preparation, or storage. Opportunities for cross-contamination of foods during processing, for example, are plentiful. The hamburger you eat may contain meat from hundreds of different cows, an omelet in a restaurant may contain the eggs of hundreds of different chickens, and the chicken you baked may have bathed with hundreds of others when they were all washed in the same vat of water at the meat processing plant. Cross-contamination also occurs on cutting boards across America. The failure to routinely wash cutting boards in between the preparation of different raw foods is a major route to the spread of foodborne illness.[9]

Antibiotics, Hormones, and Other Substances in Foods

Potential sources of food contamination include antibiotics and hormones given to animals, pesticides, and industrial pollutants. Effects of these substances on health vary from potentially life threatening to uncertain.

Antibiotic Resistance]   Chickens, cattle, and other farm-raised animals are commonly given antibiotics in feed to prevent infectious disease. In many cases, the antibiotics used are the same ones given to humans to treat infections. Unfortunately, microorganisms are very clever: they can transform themselves to become resistant to antibiotics used to kill them. People can become infected with these new strains of antibiotic-resistant microorganisms when they are consumed in foods. These infections can be difficult to treat because the disease-causing microorganisms are not sensitive to the antibiotics used. Many common forms of foodborne illnesses today are represented by these new forms of bacteria. The FDA is taking action to limit the use of potentially hazardous antibiotics in animals.[10]

Hormones]   Hormones are commonly given to farm-raised animals to promote growth or improve milk production. Many consumers are concerned about the effects, however, and would prefer such hormones not be present in food. Hormones given to animals do show up in foods along with other hormones naturally present in animal tissues. Whether animal products free of additional hormones are safer than those without them is unknown.[11]

Pesticides and PCBs]   Pesticides containing organophosphates, mercury-containing fungicides, and DDT remain causes of foodborne illnesses, as do PCBs.[12] Use of DDT on insects and PCBs in transformers was phased out over 20 years ago due to links with cancer, but these long-lasting chemicals still contaminate some land, lakes, and streams. Fewer than 1 in 10,000 foods on the market contains an excessive level of pesticide, and two out of three contain no trace of agricultural chemicals (Illustration 32.3). Farmers and food processing employees are most likely to experience illnesses associated with exposure to agricultural chemicals (Illustration 32.4). Health problems due to agricultural chemicals appear to be very rare in other groups of people.[13]

Causes and Consequences of Foodborne Illness

Over 250 types of foodborne illnesses caused by infectious agents (bacteria, viruses, and parasites) and noninfectious agents (toxins and chemical contaminants) have been identified. Their impact on health ranges from a day or two of nausea and diarrhea to death within minutes. Effects of foodborne illnesses are generally most severe in people with weakened immune systems or certain chronic illnesses, pregnant women, young children, and older persons (Table 32.1).[15] Symptoms of foodborne illness most commonly consist of nausea, vomiting, abdominal cramps, and diarrhea. Many cases go unreported, making it difficult to identify the most common causes of foodborne illness. Estimates of the causes are usually based on cases reported by health care professionals and health departments.[16]

The most prevalent reported causes of foodborne illness result from *Salmonella*, *Campylobacter*, and *E. coli* 0157:H7 bacteria and Norwalk-like viruses.[17] Table 32.2 summarizes facts about these top four leading causes of foodborne illness.

Salmonella] Over 37,000 cases of *Salmonella* infection are reported to the CDC yearly, and it is estimated that 38 times that number, or 1,412,498 cases, actually occurred.[19] (A photograph of *Salmonella* bacteria is presented in Illustration 32.5). An outbreak of *Salmonella* infection that affected approximately 224,000 people

TABLE 32.1

HIGH-RISK GROUPS FOR SEVERE EFFECTS OF FOODBORNE ILLNESS.[14]

- People with weakened immune systems due to HIV/AIDS and others with weakened defenses against infections
- People with certain chronic illnesses such as diabetes and cancer
- Pregnant women
- Young children
- Older persons

TABLE 32.2

THE TOP FOUR CAUSES OF FOODBORNE ILLNESS.[18]

BACTERIA	ONSET	ILLNESS DURATION	SYMPTOMS	FOODS MOST COMMONLY AFFECTED	USUAL SOURCE OF CONTAMINATION
Salmonella	1–3 days	4–7 days	Diarrhea, abdominal pain, chills, fever, vomiting, dehydration	Uncooked or undercooked eggs, unpasteurized milk, raw meat and poultry, vegetables and fruits	Infected animals, human feces on food, contaminated water
Campylobacter	2–5 days	2–5 days	Diarrhea (may be bloody), abdominal cramps, fever, vomiting	Undercooked poultry, unpasteurized milk	Infected poultry and other animals
E. coli 0157:H7	1–8 days	5–10 days	Watery, bloody diarrhea, abdominal cramps, little or no fever	Raw or undercooked beef, unpasteurized milk, raw vegetables and fruits, contaminated water	Infected cattle
Norwalk-like viruses	1–2 days	1–3 days	Nausea, vomiting, diarrhea	Undercooked seafood	Human feces contamination of oysters and other shellfish beds

Illustration 32.5
A magnified view of *Salmonella* bacteria.

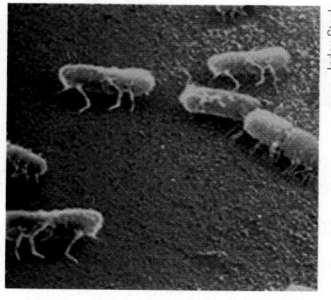

Index Stock

was linked to the transport of an ice cream mixture in a tanker that had previously transported *Salmonella*-infected eggs. The infected ice cream was distributed in 41 states.[20] It is estimated that 5% of the U.S. population experiences a *Salmonella* infection every year.[21]

Campylobacter] This bacterium is a common contaminant on chicken and is the cause of the greatest number of cases of bacteria-related foodborne illness each year. That figure is a startling 2,453,926.[22] The bacteria are found on up to 80% of chickens, and 20% represent a strain of *Campylobacter* resistant to antibiotics. In addition to poultry, *Campylobacter* infection has been traced to unpasteurized milk and contaminated water.[23]

E. Coli] *E. coli* 0157:H7 is a new strain of *E. coli* that has evolved into a potential killer. As few as 10 of these bacteria can lead to death by kidney failure in vulnerable people. Nearly 80,000 cases of *E. coli* 0157:H7 occur each year in the United States and 8% of infected people die from it.[24] *E. coli* infections are particularly associated with consumption of undercooked ground beef. A major outbreak of *E. coli* 0157:H7 infection occurred when contaminated ground beef leftover from one day's production was added to the next day's batch of beef. Use of the leftover, contaminated meat kept recontaminating subsequent batches.[25] Some 25 million pounds of ground beef had to be recalled as a result.

Norwalk-Like Viruses] Norwalk-like viruses are an extremely common but underreported cause of foodborne illnesses. The viruses go underreported because laboratory tests required for diagnosis are not widely performed. Norwalk-like viruses usually cause an acute bout of vomiting that resolves within 2 days. They are thought to be primarily spread by infected kitchen workers and fishermen who have dumped sewage waste into waters above oyster beds. An estimated 23,000,000 cases of illnesses related to Norwalk-like viruses occur in the United States each year.[26]

Other Causes of Foodborne Illnesses

Of the hundreds of other causes of foodborne illness, seven have been selected for brief review here. They represent a sampling of foodborne illnesses stemming from seafood consumption, and others that can result from a bacterial toxin, a parasite, and prions (a type of protein).

Illustration 32.6
Seafoods are an important potential cause of foodborne illness.

Photo Disc

Foodborne Illnesses Related to Seafoods] Many types of seafood (Illustration 32.6) may become contaminated due to industrial or human pollution of fresh and ocean waters. Examples presented relate to mercury contamination, ciguatera fish poisoning, and the neurotoxin (or nervous system toxin) responsible for "red tide."

Mercury Contamination] Seafoods have come under fire as a potential source of foodborne illnesses due to mercury contamination of waters and fish by fungicides, fossil fuel exhaust, smelting plants, pulp and paper mills, leather-tanning facilities, and chemical manufacturing plants. High levels of mercury are most likely to be present in large, long-lived fish such as shark, swordfish, and tuna. Because mercury can interfere with fetal brain development, pregnant women should limit their consumption of these types of fish. The Food and Drug Administration (FDA) advises that women who are pregnant or may become pregnant not eat shark, swordfish, king mackerel, or tilefish. Pregnant women should limit consumption of fish caught by family or friends to 6 ounces a week, and they may eat an average of

12 ounces of fish from stores and restaurants weekly. Consumption of fish by adults in general does not appear to pose a health risk.[27]

Ciguatera] Ciguatera poisoning from fish is caused by a neurotoxin (ciguatoxin) present in microorganisms called dinoflagellates that live in reefs (Illustration 32.7). The toxin is transferred through herbivorous reef fish to carnivorous fish, and then to humans who eat these fish. Over 200 types of fish may cause ciguatera poisoning, the most common being grouper, red snapper, and barracuda. Primary areas affected by ciguatera poisoning include the Caribbean and South Pacific Islands. Symptoms develop within 1 to 30 hours after ingestion of poisoned fish and cause nausea, vomiting, abdominal cramps, watery diarrhea, and then numbness, shooting pains in the legs, and other symptoms. The toxin causing the poisoning is not destroyed by cooking, freezing, or digestive enzymes, and there is no effective treatment.[28]

Red Tide] "Red tide" may occur between June and October on Pacific and Atlantic coasts. It is due to the accumulation of a microorganism that produces a nerve toxin. Oysters and other shellfish that consume the microorganism become contaminated, as do humans who ingest the shellfish. Resistant to cooking, the toxin produced causes a burning or prickling sensation in the mouth from 5 to 30 minutes after contaminated shellfish are consumed. This symptom is followed by nausea, vomiting, muscle weakness, and a loss of feeling in the hands and feet. Recovery is usually complete, but it's easy to understand that the warning not to eat mussels, clams, oysters, scallops, or other shellfish from red-tide waters is for real.[29]

Botulism] *C. botulinum* bacteria produce a toxin that is one of the deadliest known. It can cause nerve damage and respiratory failure. (An antidote is available but must be given soon after the infection begins.) These bacteria are commonly present in soil and ocean and lake sediment, and they may contaminate crops, honey, animals, and seafood. In humans, botulism usually results from eating underheated, contaminated foods stored in airtight containers. The bacteria thrive without oxygen and produce gases as they grow. The gases expand the food container. (For a real example of a can that exploded due to gas produced by the growth of bacteria, see Illustration 32.8.) Consequently, foods in cans, plastic bags and wraps, and other airtight containers that have bulged-out areas should *not* be eaten.

Alaska has the highest incidence of botulism in the United States. Why does the 49th state rank first? The reason is that some Native Alaskans place uncooked or partially cooked salmon eggs, whale blubber, and other seafoods that harbor *C. botulinum* in plastic bags to ferment. The bags are squeezed to expel the air before sealing, and the no-oxygen environment fosters the growth of botulinum.[30]

Parasites] Various parasitic worms, such as tapeworms, flatworms, and roundworms, may enter food and water through fecal material and soil. They are generally killed by freezing and always killed by high temperature. Once consumed, roundworms may attach to the lining of the intestine and feed on the person's blood. This can lead to anemia. One type of roundworm can bore a hole through the stomach within an hour after the worm's source—raw fish—is eaten (Illustration 32.9). The severe pain that results sends people to their doctors on the double. Hundreds of Japanese people experience this foodborne illness every year.[31]

It's a good thing most parasites are destroyed by freezing. One study in Seattle assessed the parasite content of raw fish used in sushi. About 40% of the raw fish samples contained roundworms, but since the fish had been frozen, all the worms were dead.[32]

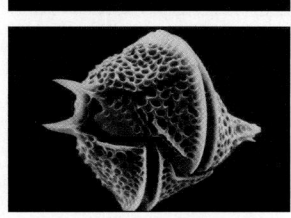
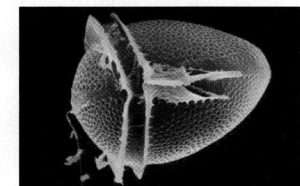

Illustration 32.7
Magnified view of dinoflagellates that cause ciguatera poisoning.

Biophoto Associates/Photo Researchers (All)

Illustration 32.8
Pressure caused by bacterial gases in this can of glucose drink made the can explode.

Richard Anderson

HERMAN

"I need 148 get-well cards."

Illustration 32.9
Source: Jim Unger/dist. by Laugh-ingStock Licensing Inc. Used by permission.

Mad Cow Disease] Technically called bovine spongiform encephalopathy, this rare disease in cattle is suspected of causing at least 144 human deaths in Europe. Only one case of the disease in humans—caused by consumption of affected beef—has been diagnosed in the United States, and it originated in England. The disease appears to have started when cows, who are herbivores by nature, were given sheep intestines and parts of the spinal cord in their feed. Some of the sheep harbored a protein, called a prion, that caused a deadly disease when consumed by cows. A prion is not a bacterium or other microorganism; it's a small protein that can transmit disease when consumed by a similar species. Only one other known foodborne illness is spread by the consumption of otherwise healthy body parts. That disease is kuru, and it is transmitted by cannibalism.[33]

Researchers concluded that mad cow disease was transferred to humans who ate the meat of prion-infected cows. The disease represents a new form of Creutzfeldt-Jakob disease previously identified in humans. It inevitably leads to death in humans, after many years, due to brain damage. Needless to say, it is no longer legal to feed cows animal parts that may transmit the disease. Although the risk of consuming beef from cattle with mad cow disease is extremely small, the possibility that animals may develop this or a similar disease exists. The possibility that mad cow disease may affect cattle in the United States became a reality late in 2003. Stringent monitoring efforts are in place in the United States and many European countries to prevent the occurrence and spread of prions.[34]

Because the disease was first recognized in 1994 and may take 20 or more years to develop, we may not know the full impact of the exposure to contaminated beef for years to come.

Preventing Foodborne Illnesses

There are two major approaches to the prevention of foodborne illnesses. The first relies on food safety regulations that control food processing and handling practices, and the second involves consumer behaviors that lessen the risk of consuming contaminated foods. (Read what can happen when both of these principles are violated in the nearby "Health Action.")

Food Safety Regulations

According to the Federal Food, Drug, and Cosmetic Act, it is illegal to produce or dispense foods that are contaminated with substances that cause illness in humans. Foods are considered "safe" if there is a reasonable certainty that no harm will result from repeated exposure to any substance added to foods. The act governs all substances that are added intentionally or accidentally to foods—except pesticides. According to legislation passed in 1996, pesticides are permitted if their consumption is associated with a "negligible risk" of cancer or other health problems.[35]

Irradiation of Foods] Increasing rates of foodborne illness have led some health officials to recommend the expanded use of food irradiation. Food irradiation is a safe process when conducted under specified conditions. It destroys bacteria, parasites, and viruses present in or on foods. Availability of irradiated foods is growing but is still rather limited due to lingering, low consumer acceptance of the process and of the foods.[36]

Although the proposal has merits, food irradiation is not the silver bullet that will prevent all cases of foodborne illness. Prions, toxins, pesticides, mercury, and PCBs are resistant to radiation, so radiation is unlikely to affect the safety of foods with respect to these contaminants. Once irradiated, foods can later become con-

HEALTH ACTION | *A Double Whammy*

From the true story department of *Nutrition Now* comes the tale of a man who took a night off with his wife to dine at their favorite restaurant. The man, a physician, ordered his favorite item: the luscious chicken burritos. He was served more than he could eat, so he took the leftovers home and promptly put them in the refrigerator. About 8 hours later, he felt like his intestines were exploding; he thought he'd rather die than be so sick. Then the light bulb went off—he had probably consumed *C. perfingens* toxin somehow.

By the third day he began to feel better and noticed he was ravenously hungry. Then he remembered there was a leftover burrito in the refrigerator. . . . Happily, the second round of symptoms wasn't as bad as the first.

The message? It's difficult for consumers to prevent every case of foodborne illness, but we can prevent more cases than we do. Throw out spoiled food.

---

taminated in such places as a packing plant, grocery store, restaurant, or home. The absence of all microorganisms in irradiated food may enable individual types of bacteria that contact the food after it is sterilized to grow at unusually high rates. Irradiation does not work well on all foods—some fruits and vegetables develop a poor texture and change flavor when irradiated.[37]

The Consumer's Role in Preventing Foodborne Illnesses

> *Food safety education will never be a substitute for a safe food supply.*
> —Caroline Smith DeWall, 1997

Due in large measure to the demands of individuals and consumer groups, many mechanisms are in place to help ensure the safety of the food supply (Illustration 32.10). However, existing safeguards in food production and processing, and government regulations and enforcement efforts, are insufficient to guarantee that all foods purchased by consumers will be free from contamination. This situation, along with the possibility that food can become contaminated in unanticipated ways, means that responsibility for food safety is shared by consumers.

Food Safety Basics]   The first rule of food safety is to wash your hands thoroughly with soap and water before and after handling food (Illustration 32.11). According to the Centers for Disease Control, this is the single most important means of preventing the spread of foodborne illness caused by bacteria.

There's a Right Way to Wash Your Hands} Ever watch a TV dramatic series about doctors and hospitals and wondered why they show surgeons scrubbing their hands and upper arms before surgery for so long? TV got it right. It takes about 20 seconds to sanitize your hands. First you need to lather up with soap and very warm water. The soap makes germs lose their grip on your skin. Next you have to make sure to scrub all the crevices between your fingers and under your fingernails. Rinse your hands thoroughly with really warm water and dry them with a paper towel. (If you use a dishcloth, you may reinfect your hands.)[38]

Illustration 32.10
Source: © L. Trepel.

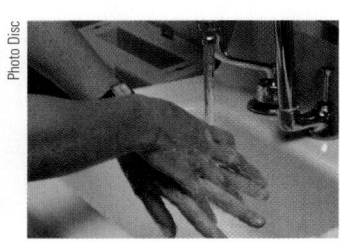

Illustration 32.11
There's a right way to wash your hands.

32-9

Keep Hot Foods Hot and Cold Foods Cold} Good foods can go bad if stored improperly. One of the most effective ways to prevent foods from spoiling is to store them at, and heat them to, the right temperatures (see Table 32.3 and Illustrations 32.12 and 32.13). When holding prepared foods before serving, hot foods should be kept hot, and cold foods cold. Freezing foods halts the growth of all of the main types of bacteria that contaminate foods. Once the foods are thawed, however, bacteria growth resumes, and the foods may become contaminated with new bacteria while thawing.

Bacteria grow best at temperatures between 40° and 135°F (see Illustration 32.12). The room temperature of homes is within this range, and that's why foods that spoil should not be left outside the refrigerator for more than an hour. If a food has been improperly stored or kept past the expiration date on the label, it should not be eaten even if it tastes all right. This general rule should apply: when in doubt, throw it out!

TABLE 32.3

A GUIDE FOR COOKING FOODS TO SAFE TEMPERATURES.

RAW FOOD	INTERNAL TEMPERATURE
Ground Meats	
Hamburger	160°F
Beef, veal, lamb, pork	160°F
Chicken, turkey	165°F
Beef, Veal, Lamb	
Roasts and steaks	
Medium rare	145°F
Medium	160°F
Well done	170°F
Pork	
Chops, roasts, ribs	
Medium	160°F
Well done	170°F
Ham, fresh	160°F
Sausage, fresh	160°F
Poultry	
Chicken, whole and pieces	180°F
Duck	180°F
Turkey (unstuffed)	180°F
Whole	180°F
Breast	170°F
Dark meat	180°F
Stuffing (cooked separately)	165°F
Eggs	
Fried, poached	Yolk and white are firm
Casseroles	160°F
Sauces, custards	160°F

*Fish should be cooked until the flesh is not clear-looking and flakes easily with a fork; shrimp, lobster, and crab until shells are red and the flesh is opaque and not clear- or raw-looking inside; and clams, oysters, and mussels until the shells open.

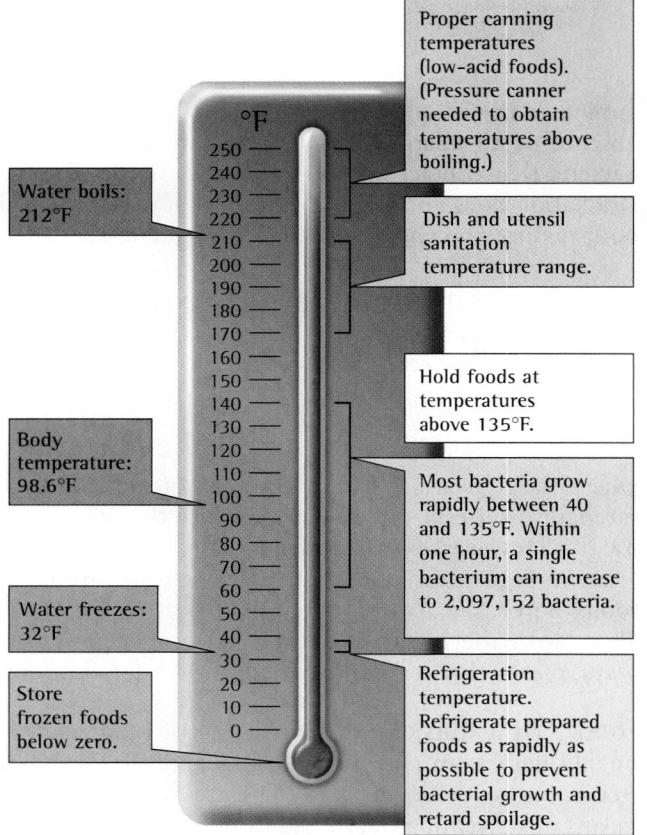

Illustration 32.12

Temperature guide for safe handling of food in the home.
Source: Adapted from Minnesota Extension Service, University of Minnesota; updated in 2004, based on data from USDA.

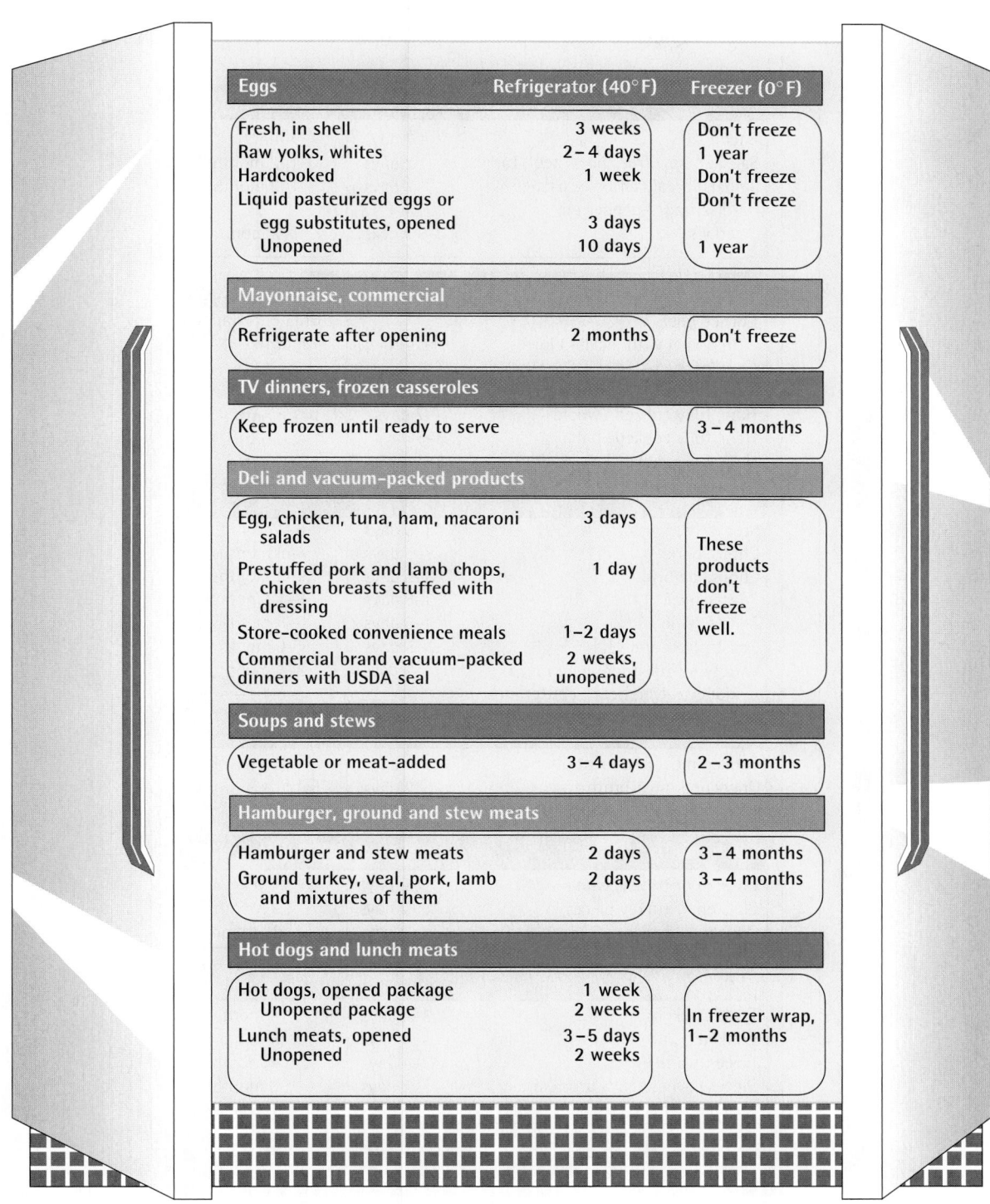

Eggs	Refrigerator (40°F)	Freezer (0°F)
Fresh, in shell	3 weeks	Don't freeze
Raw yolks, whites	2–4 days	1 year
Hardcooked	1 week	Don't freeze
Liquid pasteurized eggs or egg substitutes, opened	3 days	Don't freeze
Unopened	10 days	1 year

Mayonnaise, commercial		
Refrigerate after opening	2 months	Don't freeze

TV dinners, frozen casseroles		
Keep frozen until ready to serve		3–4 months

Deli and vacuum-packed products		
Egg, chicken, tuna, ham, macaroni salads	3 days	These products don't freeze well.
Prestuffed pork and lamb chops, chicken breasts stuffed with dressing	1 day	
Store-cooked convenience meals	1–2 days	
Commercial brand vacuum-packed dinners with USDA seal	2 weeks, unopened	

Soups and stews		
Vegetable or meat-added	3–4 days	2–3 months

Hamburger, ground and stew meats		
Hamburger and stew meats	2 days	3–4 months
Ground turkey, veal, pork, lamb and mixtures of them	2 days	3–4 months

Hot dogs and lunch meats		
Hot dogs, opened package	1 week	In freezer wrap, 1–2 months
Unopened package	2 weeks	
Lunch meats, opened	3–5 days	
Unopened	2 weeks	

(Continued on next page)

Illustration 32.13

Cold storage: these safe time limits will help keep refrigerated food from spoiling or becoming dangerous to eat.

These time limits will keep frozen food at top quality.

Source: U.S. Department of Agriculture, Food Safety and Inspection Service, September 1990 and 1997.

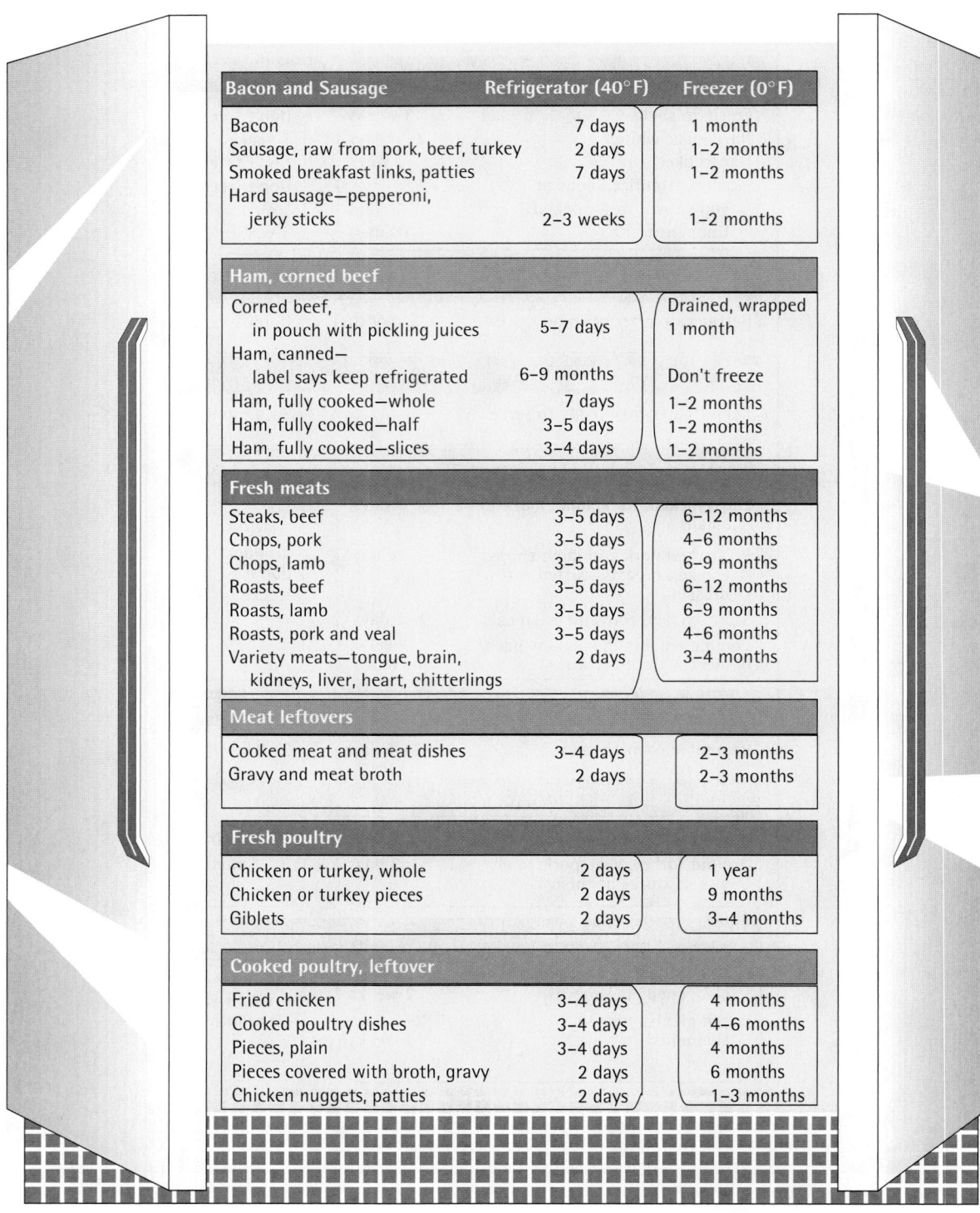

Bacon and Sausage	Refrigerator (40°F)	Freezer (0°F)
Bacon	7 days	1 month
Sausage, raw from pork, beef, turkey	2 days	1–2 months
Smoked breakfast links, patties	7 days	1–2 months
Hard sausage—pepperoni, jerky sticks	2–3 weeks	1–2 months

Ham, corned beef		
Corned beef, in pouch with pickling juices	5–7 days	Drained, wrapped 1 month
Ham, canned— label says keep refrigerated	6–9 months	Don't freeze
Ham, fully cooked—whole	7 days	1–2 months
Ham, fully cooked—half	3–5 days	1–2 months
Ham, fully cooked—slices	3–4 days	1–2 months

Fresh meats		
Steaks, beef	3–5 days	6–12 months
Chops, pork	3–5 days	4–6 months
Chops, lamb	3–5 days	6–9 months
Roasts, beef	3–5 days	6–12 months
Roasts, lamb	3–5 days	6–9 months
Roasts, pork and veal	3–5 days	4–6 months
Variety meats—tongue, brain, kidneys, liver, heart, chitterlings	2 days	3–4 months

Meat leftovers		
Cooked meat and meat dishes	3–4 days	2–3 months
Gravy and meat broth	2 days	2–3 months

Fresh poultry		
Chicken or turkey, whole	2 days	1 year
Chicken or turkey pieces	2 days	9 months
Giblets	2 days	3–4 months

Cooked poultry, leftover		
Fried chicken	3–4 days	4 months
Cooked poultry dishes	3–4 days	4–6 months
Pieces, plain	3–4 days	4 months
Pieces covered with broth, gravy	2 days	6 months
Chicken nuggets, patties	2 days	1–3 months

Illustration 32.13—continued
Cold storage : safe time limits

Food Handling and Storage] Particular care needs to be taken when handling and storing foods. Dairy products such as milk and cheese may spoil in about a week or sooner if contaminated with bacteria from the air or hands. If you hold cheese by the wrapper when you take it out of the refrigerator to cut a slice off, it will likely last longer than if you touch it with your hands. Handling foods with clean hands or plastic gloves and using clean utensils on washed surface areas helps to reduce the number of bacteria that come in contact with the food. Raw meats should be separated from other foods, and utensils and surface areas used to prepare meats should be thoroughly cleaned after each use.

The Safety of Canned Foods] Commercially canned foods are heated to the point where all bacteria are killed after the can is sealed. Consequently, the contents of canned foods are sterile and will not spoil if left on the shelf for years. Nevertheless, canned foods may spoil if the can develops a pinhole leak or if errors made during the canning process allow bacteria to enter the food. A cardinal sign that canned foods have spoiled is a build-up of pressure inside the can. The pressure will cause the top of the can to bulge out instead of curving in. Because a build-up of pressure inside a can may be due to bacteria that cause botulism, the food should be thrown away. These and other tips for safe food handling are listed on the USDA's Safe Handling Instructions label (Illustration 32.14) and in the "Health Action" below.

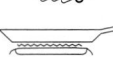

Safe Handling Instructions

THIS PRODUCT WAS PREPARED FROM INSPECTED AND PASSED MEAT AND/OR POULTRY. SOME FOOD PRODUCTS MAY CONTAIN BACTERIA THAT CAN CAUSE ILLNESS IF THE PRODUCT IS MISHANDLED OR COOKED IMPROPERLY. FOR YOUR PROTECTION, FOLLOW THESE SAFE HANDLING INSTRUCTIONS.

KEEP REFRIGERATED OR FROZEN. THAW IN REFRIGERATOR OR MICROWAVE.

KEEP RAW MEAT AND POULTRY SEPARATE FROM OTHER FOODS. WASH WORKING SURFACES (INCLUDING CUTTING BOARDS), UTENSILS, AND HANDS AFTER TOUCHING RAW MEAT OR POULTRY.

COOK THOROUGHLY.

KEEP HOT FOODS HOT. REFRIGERATE LEFTOVERS IMMEDIATELY OR DISCARD.

Illustration 32.14
The USDA Safe Handling Instructions label.

HEALTH ACTION | *Tips for Safe Food Handling*

Tip 1 Wash hands with soap and water before handling foods.

Tip 2 Wash hands, surfaces, and utensils with soap and water after contact with raw meat and meat juices

Tip 3 Avoid contact between raw and cooked foods. Sanitize cutting boards and utensils in contact with raw meats by washing them in water that contains 1 tablespoon of chlorine (i.e., Clorox) per quart.

Tip 4 Cook meats to and store foods at safe temperatures (see Table 32.3 and Illustration 32.12). Eat thoroughly cooked hamburger only.

Tip 5 Never leave perishable foods such as potato and egg salad, meats, and dairy products at room temperature for more than an hour. Keep hot foods hot (over 135°F) and cold foods cold (under 40°F).

Tip 6 Freeze or refrigerate leftovers promptly in sealed containers.

Tip 7 Use flat containers that will allow food to cool quickly if storing large amounts of food.

Tip 8 Do not buy or eat food in cans or jars with bulging tops or loose lids.

Tip 9 Do not consume foods that are past their expiration date.

Tip 10 Follow label instructions for food storage (for example: "Refrigerate after opening").

Tip 11 Thaw foods such as turkey and roasts in the refrigerator to reduce bacterial growth.

Tip 12 When in doubt, throw it out. If a food has a strange color or odor, throw it away.

Tip 13 Wash fresh vegetables and fruits thoroughly with water before eating. Peel vegetables and fruits covered with a waxy film.

Tip 14 Avoid "traveler's diarrhea" by getting safety tips from the Centers for Disease Control before you travel (http://www.cdc.gov).

Because so many consumers have questions about food safety, the U.S. Department of Agriculture (USDA) has established a 24-hour hotline for meat and poultry. When in doubt, call the hotline at 1-800-535-4555. If you are planning to visit a foreign country and want to avoid "traveler's diarrhea," contact the Centers for Disease Control for safety tips (www.cdc.gov).

There Is a Limit to What the Consumer Can Do

In the end, it is up to consumers to demand a safe food supply, up to industry to produce it, up to researchers to develop better ways of doing so, and up to government to see that it happens, to make sure it works, and to identify problems still in need of resolution.

—CDC's Report on Foodborne Illnesses (www.cdc.gov)

Keeping food safe from the farm, lake, or ocean to the table represents a major challenge to industry, government, and consumers. Until contamination of food is prevented, consumers will continue to play an important role in ensuring food's safety.

Nutrition UP CLOSE

Food Safety Detective

FOCAL POINT: Investigating paths to foodborne illness.

Assume an outbreak of foodborne illness occurred among people who ate foods from the onion and lettuce platter shown being prepared in Illustration 32.2. Review the food safety tips in the "Health Action" on page 32–13, and identify three potential causes of the outbreak related to the preparation of the foods as shown in the illustration:

1. _____

2. _____

3. _____

FEEDBACK (answers to these questions) can be found at the end of Unit 32.

Key Terms

foodborne illnesses, page 32-2

www links

www.fightbac.org
Provides materials for use by the media and consumers. The site gives you access to workbooks, posters about food safety, and "mug shots" of the 10 least-wanted foodborne pathogens.

www.foodsafety.gov
The national gateway to food safety information, resources, outbreak alerts, and consumer advice. Or call the Food Safety 24-Hour Hotline at 1-888-SAFEFOOD.

www.cdc.gov
Search foodborne illnesses, food safety, or international travel, and find a plethora of information.

http://vm.cfsan.fda.gov
The FDA's Center for Food Safety and Applied Nutrition offers press releases, fact sheets, food safety warnings, and consumer advice.

www.epa.gov/ost/fish
The place to go for local fish advisories.

www.cdc.gov/mmwr/
Sends you to the Centers for Disease Control's Morbidity and Mortality Monthly Report. The site includes updates on foodborne illnesses.

Notes

1. CDC issues new 5-year report of foodborne illness outbreaks. Food Safety Educator 1997;2:1–6.

2. Mead PS et al. Food-related illness and death in the United States. www.cdc.gov, accessed 10/03.

3. Mead et al., Food-related illness.

4. Mead et al., Food-related illness; and Foodborne illnesses. www.cdc.gov, accessed 1/26/01.

5. www.fightbac.org, accessed 1/27/01.

6. Mead et al., Food-related illness.

7. Mead et al., Food-related illness.

8. Foodborne illnesses (www.cdc.gov).

9. Foodborne illnesses (www.cdc.gov); and Zhao P et al. Development of a model for evaluation of microbial cross-contamination in the kitchen. J Food Protection 1998;61:960–3.

10. Peregrin T, Limiting the use of antibiotics in livestock: helping patients understand the science behind this issue, J Am Diet Assoc 2002;102:768; Wegener HC, The consequences for food safety of the use of fluoroquinolones in food animals, N Engl J Med 1999;340:1581–2; and Mathews AW, FDA announces policy designed to curb animal-antibiotic use, Wall Street Journal 2003;Oct. 24:A6.

11. FDA warns milk manufacturers on hormone label. www.medscape.com, accessed 10/03.

12. Anderson ER et al. Diagnosis and management of foodborne illness: a primer for physicians. MMWR, 2001;Jan. 26.

13. Position of the American Dietetic Association: food and water safety. J Am Diet Assoc 1997;97:184–9.

14. Mead et al., Food-related illness.

15. Dietary Guideline for Americans. U.S. Dept. of HHS and USDA, 2000.

16. Mead et al., Food-related illness.

17. Mead et al., Food-related illness.

18. Food safety: a challenge for everyone in public health. Webcast from the University of North Carolina, 1/26/01.

19. Mead et al., Food-related illness.

20. Hennessy TW et al. A national outbreak of *salmonella enteritidis* infection from ice cream. N Engl J Med 1996;334:1281–6.

21. Mead et al., Food-related illness.

22. Mead et al., Food-related illness.

23. Wegener, The consequences for food safety.

24. Mead et al., Food-related illness.

25. Welland D. *E. coli . . . salmonella . . . Listeria . . .* unwelcome house guests. Envir Nutr 1997;20(Oct.):1, 6.

26. Mead et al., Food-related illness.

27. EPA national advice on mercury in freshwater fish for women who are or may become pregnant, nursing mothers, and young children, www.epd.gov, 1/2001; and Clarkson TW et al., The toxicology of mercury—current exposures and clinical manifestations, N Engl J Med 2003;349:1731–8.

28. Treatment of ciguatera poisoning with gabapentin. N Engl J Med 2001;344:692–3.

29. The Merck Manual, 17th ed. Whitehouse Station (NJ): Merck Research Laboratories; 1999, p. 262.

30. Lewis R. Botulism from blubber. The Scientist 2003;Mar. 10:11.

31. Tighter reins sought for imported produce. Envir Nutr 1997;20(Nov.):1, 7.

32. Tighter reins sought for imported produce.

33. Mariani SM, Prions—are they viruses, nucleic acids, or "infectious" proteins? Medscape General Medicine 2003;5:1–8; and Madigan D, After Delaney: how will the new pesticide law work? CNI Weekly Report 1996;Oct. 11:4–5.

34. Mariani, Prions; and Madigan, After Delaney.

35. Mariani, Prions; and Madigan, After Delaney.

36. Shea KM. Technical report: irradiation of food. Pediatrics 2000;106:1505–9.

37. Shea, Technical report.

38. Henneman A. Don't mess with food safety myths. Nutr Today 1999;34:23–8.

39. Henneman, Don't mess with food safety myths.

Nutrition **UP CLOSE**

Food Safety Detective

Feedback for Unit 32

Potential causes of the foodborne illness outbreak:

1. Disease spread from hands to foods.

2. Use of contaminated cutting board, knife, or food platter.

3. Cross-contamination between onions and lettuce or between other foods stored together.

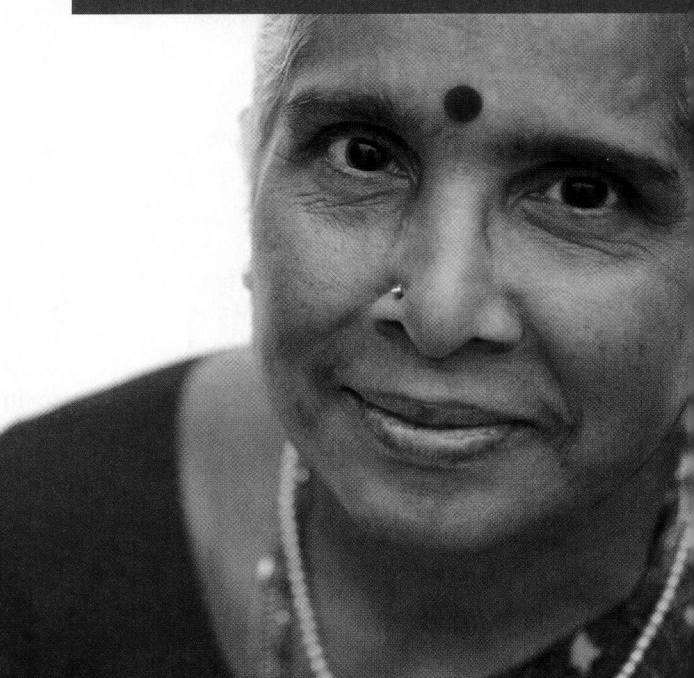

Aspects of Global Nutrition

Nutrition Scoreboard

		TRUE	FALSE
1	Approximately 30% of the world's population does not have access to a safe supply of water.		
2	At most, 25% of deaths of young children in developing countries are related to malnutrition and infection.		
3	The main cause of famine in developing countries is that too little food is available.		

Answers on next page

[KEY CONCEPTS AND FACTS]

- The world produces enough food for all its people.

- Poverty, corrupt governments, the HIV/AIDS epidemic, low rates of breastfeeding, unsafe water supplies, and discrimination against females all contribute to malnutrition.

- Malnutrition early in life has long-term effects on mental and physical development.

- Rates of diseases such as heart disease, diabetes, and hypertension increase in developing countries that adopt Western lifestyles and eating habits.

Answers to *Nutrition Scoreboard*

		TRUE	FALSE
1	Contaminated water supplies are a major source of infection in many developing countries.	✔	
2	Nearly 50% of deaths of children in developing countries are associated with malnutrition and infection.[1]		✔
3	Poverty is the primary cause of famine. There are many other important causes as well.		✔

State of the World's Health

The global community consists of over 100 countries grouped into the categories of "industrialized nations," "developing nations," and "least developed nations." (Countries within each category are listed in Table 33.1.) Only 31 countries are considered industrialized, most are developing (113), and, of the developing countries, 49 are least developed.[2] All countries in the global community have interests and issues in common, such as the adequacy of the food supply, availability of health care and education, and safe water. But the countries of the world are also dissimilar in many ways. Differences in financial resources, population growth, and political and other systems translate into large disparities in health status and life expectancy (Table 33.2). People in the least developed and developing countries are more likely to have substantially shorter life expectancies, to die from infectious diseases, and to experience malnutrition than individuals living in industrialized countries.[4]

The general state of health of populations in various countries is monitored by tracking key environmental, health, and behavioral characteristics. Pregnancy outcome and child growth, rates of breastfeeding, and access to safe drinking water, for example, are key indicators of the health status of a population.[5] As can be seen from Illustration 33.1 and Table 33.3, rates of these key indicators vary substantially worldwide. When percentages of low birthweight in newborns and underweight in young children are high, and rates of breastfeeding and access to a safe water supply are low, one can rightfully assume that malnutrition, infection, and other health problems are common. When these key indicators show improvement, the health status of the population and longevity improve.[8]

The health and nutritional status of populations in developing countries is monitored by the World Health Organization (WHO), the Food

TABLE 33.1

COUNTRIES CLASSIFIED BY THE UNITED NATIONS INTERNATIONAL CHILDREN'S EMERGENCY FUND (UNICEF) AS INDUSTRIALIZED, DEVELOPING, AND LEAST DEVELOPED.*

INDUSTRIALIZED COUNTRIES

Andorra; Australia; Austria; Belgium; Canada; Denmark; Finland; France; Germany; Greece; Holy See; Iceland; Ireland; Israel; Italy; Japan; Liechtenstein; Luxembourg; Malta; Monaco; Netherlands; New Zealand; Norway; Portugal; San Marino; Slovenia; Spain; Sweden; Switzerland; United Kingdom; United States.

DEVELOPING COUNTRIES

Afghanistan; Algeria; Angola; Antigua and Barbuda; Argentina; Armenia; Azerbaijan; Bahamas; Bahrain; Bangladesh; Barbados; Belize; Benin; Bhutan; Bolivia; Botswana; Brazil; Brunei Darussalam; Burkina Faso; Burundi; Cambodia; Cameroon; Cape Verde; Central African Rep.; Chad; Chile; China; Colombia; Comoros; Congo; Congo, Dem. Rep.; Cook Islands; Costa Rica; Côte d'Ivoire; Cuba; Cyprus; Djibouti; Dominica; Dominican Rep.; Ecuador; Egypt; El Salvador; Equatorial Guinea; Eritrea; Ethiopia; Fiji; Gabon; Gambia; Georgia; Ghana; Grenada; Guatemala; Guinea; Guinea-Bissau; Guyana; Haiti; Honduras; India; Indonesia; Iran; Iraq; Jamaica; Jordan; Kazakhstan; Kenya; Kiribati; Korea, Dem. People's Rep.; Korea, Rep. of; Kuwait; Kyrgyzstan; Lao People's Dem. Rep.; Lebanon; Lesotho; Liberia; Libya; Madagascar; Malawi; Malaysia; Maldives; Mali; Marshall Islands; Mauritania; Mauritius; Mexico; Micronesia, Fed. States of; Mongolia; Morocco; Mozambique; Myanmar; Namibia; Nauru; Nepal; Nicaragua; Niger; Nigeria; Niue; Oman; Pakistan; Palau; Panama; Papua New Guinea; Paraguay; Peru; Philippines; Qatar; Rwanda; Saint Kitts and Nevis; Saint Lucia; Saint Vincent/Grenadines; Samoa; Sao Tome and Principe; Saudi Arabia; Senegal; Seychelles; Sierra Leone; Singapore; Solomon Islands; Somalia; South Africa; Sri Lanka; Sudan; Suriname; Swaziland; Syria; Tajikistan; Tanzania; Thailand; Togo; Tonga; Trinidad and Tobago; Tunisia; Turkey; Turkmenistan; Tuvalu; Uganda; United Arab Emirates; Uruguay; Uzbekistan; Vanuatu; Venezuela; Viet Nam; Yemen; Zambia; Zimbabwe.

LEAST DEVELOPED COUNTRIES

Afghanistan; Angola; Bangladesh; Benin; Bhutan; Burkina Faso; Burundi; Cambodia; Cape Verde; Central African Rep.; Chad; Comoros; Congo, Dem. Rep.; Djibouti; Equatorial Guinea; Eritrea; Ethiopia; Gambia; Guinea; Guinea-Bissau; Haiti; Kiribati; Lao People's Dem. Rep.; Lesotho; Liberia; Madagascar; Malawi; Maldives; Mali; Mauritania; Mozambique; Myanmar; Nepal; Niger; Rwanda; Samoa; Sao Tome and Principe; Sierra Leone; Solomon Islands; Somalia; Sudan; Tanzania; Togo; Tuvalu; Uganda; Vanuatu; Yemen; Zambia.

*Data for some eastern European countries, former Soviet republics, and the Baltic states are not available. Least developed countries represent a subgroup of the developing countries.

TABLE 33.2

LIFE EXPECTANCY IN SELECTED COUNTRIES FROM LOWEST TO HIGHEST.[3]

	LIFE EXPECTANCY (YEARS)
Zambia	37.2
Mozambique	37.5
Malawi	37.6
Botswana	39.3
Ethiopia	45.2
Afghanistan	45.9
Chad	50.5
Sudan	56.6
Russia	67.2
Philippines	67.5
Iran	69.7
Costa Rica	75.8
Cuba	76.2
Ireland	76.8
United States	77.2
Greece	78.4
France	78.8
Spain	78.8
Italy	79.0
Sweden	79.6
Australia	79.8
Singapore	80.1
Japan	80.7

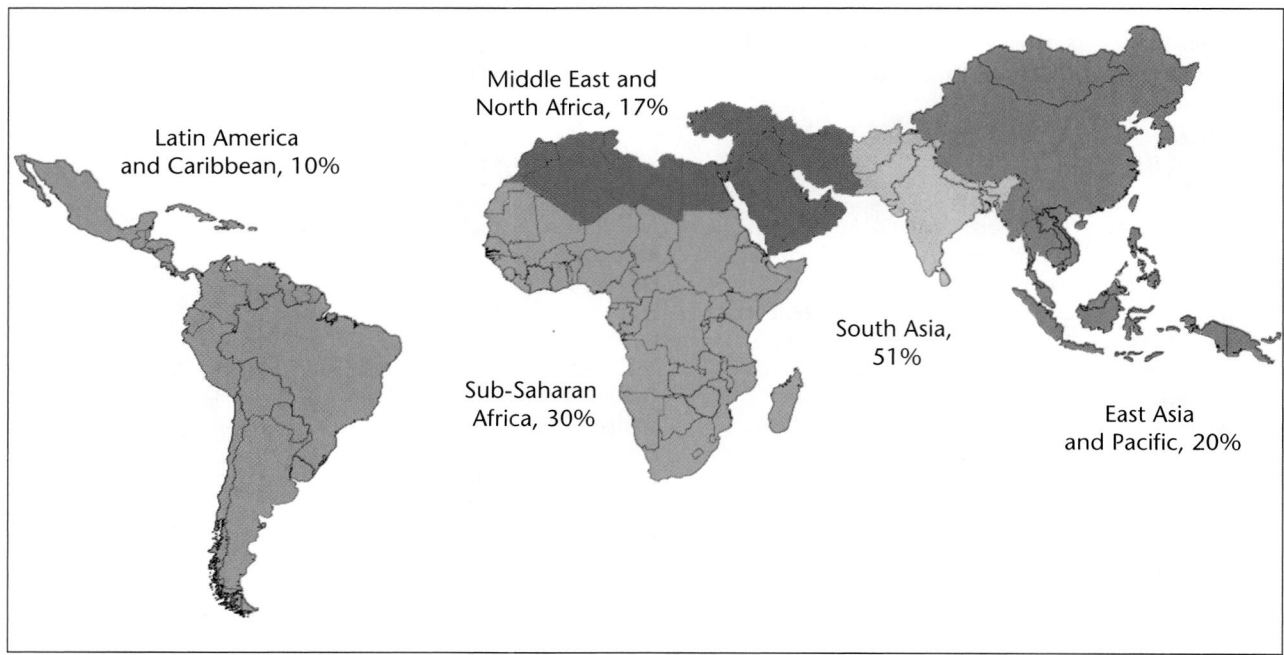

Latin America
and Caribbean, 10%

Middle East and
North Africa, 17%

South Asia,
51%

Sub-Saharan
Africa, 30%

East Asia
and Pacific, 20%

Illustration 33.1

Underweight children are an indicator of malnutrition.

This map shows the percentage of children who were underweight in developing regions of the world in 1996–1998.[6]

and Agriculture Organization (FAO), and the United Nations International Children's Emergency Fund (UNICEF). Recent reports have identified leading problem areas related to malnutrition that must be addressed as part of global strategies aimed at improving health and well-being (Table 33.4). Important progress is being made, and rates of malnutrition, poor growth, and infectious disease are decreasing slowly worldwide.[10] Much progress remains to be made, however.

TABLE 33.3

PERCENTAGE OF LOW-BIRTHWEIGHT INFANTS, EXCLUSIVE BREASTFEEDING, AND ACCESS TO SAFE WATER IN DIFFERENT AREAS OF THE WORLD.[7]

REGION[a]	LOW-BIRTHWEIGHT INFANTS (%)	EXCLUSIVE BREASTFEEDING, 0–3 MONTHS (%)	ACCESS TO SAFE WATER (%)
Sub-Saharan Africa	16	32	49
Middle East and North Africa	11	43	81
South Asia	33	46	80
East Asia and Pacific	11	56	67
Latin America and Caribbean	10	38	77
Developing countries	18	44	70
Least developed countries	23	46	54
Industrialized countries[b]	6	—	100
World	17	44	71

[a]Data for the former Soviet republics and the Baltic states are not available.
[b]Data for exclusive breastfeeding are not available; the figure is approximately 50%.

TABLE 33.4

PRIORITY PROBLEM AREAS RELATED TO MALNUTRITION IN DEVELOPING COUNTRIES.[9]

1. Childhood protein-calorie malnutrition
2. Vitamin deficiencies: vitamin A, folate
3. Mineral deficiencies: iodine, iron, zinc
4. Lack of breastfeeding
5. Alcohol abuse
6. Overweight and obesity
7. Insufficient physical activity
8. Low vegetable and fruit intake
9. Malnutrition and increased complications from HIV/AIDS
10. Poor nutritional status of women of child-bearing age

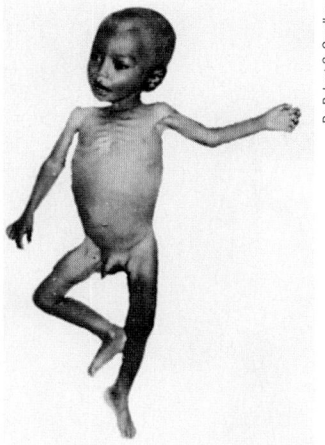

Illustration 33.2
The look of marasmus.
Children with marasmus look like "skin and bones." They don't get enough calories and protein in their diet. This child is suffering from the extreme emaciation of marasmus.

Dr. Robert S. Goodhard

Food and Nutrition: The Global Challenge

He who is healthy has hope; and he who has hope has everything.
—Arab proverb

We have all heard about starvation in Somalia, Ethiopia, Sudan, and Bangladesh. The pictures of starving children with desperation in their big eyes and heads too large for their frail bodies burn an image in our minds. What we don't hear about in the news, however, is the extent of starvation and malnutrition in the world. These are not just problems of isolated areas experiencing civil war or crop failures. In the developing regions, malnutrition is an ongoing problem for 10 to 51% of children under the age of 5, depending on the region. Approximately 50 to 70% of women worldwide are iron deficient, and 2 billion children suffer from iodine deficiency. Vitamin A deficiency exists among 21% of children in developing countries. These children are at risk of frequent and severe infectious disease, decreased growth, and blindness because of the deficiency. It is estimated that one-third of the world's population is mildly to moderately zinc deficient.[11]

marasmus
Malnutrition caused by a lack of calories and protein. Also called *protein-calorie malnutrition.*

kwashiorkor
Malnutrition resulting from inadequate protein intake in children.

The children we usually see in pictures from famine areas are victims of *marasmus,* a disease caused primarily by a lack of calories and protein. These children look starved (Illustration 33.2) and consume far fewer calories and far less protein and other essential nutrients than they need. In victims of marasmus, the body uses its own muscle and other tissues as an energy source.

Sometimes malnutrition is primarily due to an inadequate protein intake rather than a deficiency of calories. Children who lack protein develop *kwashiorkor.* These children may appear fat due to massive swelling that occurs with the disease (Illustration 33.3). Most victims of kwashiorkor are somewhat marasmic as well. Children with either type of severe malnutrition are generally deficient in multiple vitamins and minerals and at high risk of dying from infectious diseases due to weakened immune systems.[12]

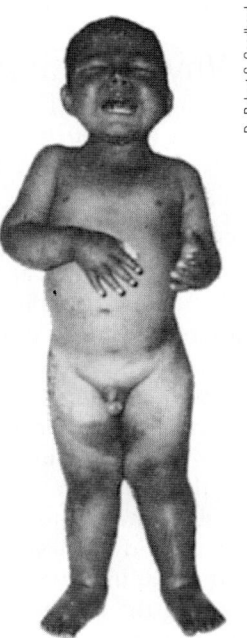

Dr. Robert S. Goodhard

Survivors of Malnutrition

Malnutrition that occurs before a child's brain has completely grown has lasting effects on development.[13] Brain growth occurs during pregnancy through the age of 2

Illustration 33.3
The look of kwashiorkor.
This child has the characteristic "moon face" (edema), swollen belly, and patchy dermatitis often seen with kwashiorkor, a protein deficiency. Children with kwashiorkor may not look starved because of the massive swelling.

years. If children suffer severe malnutrition during these years, they will experience permanent delays in mental development even if refeeding improves their physical growth and health. The severity of the mental delays depends on the timing and duration of the malnutrition. The longer malnutrition exists, the harder it becomes for young children to achieve a brighter future.[14] In addition, the state of hunger and starvation leads people to become self-centered and to lose any sense of well-being. The need for food for survival may prompt unethical behaviors such as stealing and injuring others to obtain food. The devastating psychological effects of starvation may follow children into adulthood and have a lasting effect on behavior throughout life. Persistent malnutrition saps the physical and mental energy of people and, ultimately, compromises the economic and social progress of nations.[15]

Malnutrition and Infection] There is a very close relationship between malnutrition and infection: malnutrition weakens the immune system and increases the likelihood and severity of infection. Simultaneously, repeated infection and the bouts of diarrhea that often accompany it can produce undernutrition. Poor sanitary conditions, contaminated water supplies, and lack of refrigeration contribute to the spread of infectious diseases and thus to the malnutrition the diseases promote. Most deaths of children under the age of 5 years in developing countries are related to the presence of both malnutrition and infection.[16]

Low rates of breastfeeding promote the spread of infection, especially in countries where contaminated water is used to prepare formula. In addition to providing optimal nutrition, breast milk contains a number of substances that protect babies from infection.[17]

Vitamin A deficiency and malnutrition in children play important roles in the spread and severity of infections. Children who are deficient in vitamin A are much more likely to die from a measles infection, for example, than are children whose vitamin A status is sufficient.[18] HIV infection progresses more rapidly into AIDS in malnourished individuals and shortens the time parents can work and support their families.[19] Life expectancy in Botswana, for example, has decreased from 65 years in 1995 to 39 years in 2001. Over 42 million people in the world have HIV infection, and the incidence is increasing.[20]

Why Do Starvation and Malnutrition Happen?

> It is very much an oversimplification to assert, without qualification, that people starve because they do not have enough to eat.
> —M. Pyke, 1972

Malnutrition and starvation don't have to happen anywhere. The world's agricultural systems have the capacity to produce enough food to feed everyone on Earth.[21] People become malnourished or starve primarily because of poverty. Human-made disasters, including discrimination against women, the HIV/AIDS epidemic, racism, corrupt governance, and other factors (Table 33.5) heavily contribute to malnutrition. In some instances, starvation and malnutrition are due to natural disasters that lead to crop failures or the inability to distribute food to those in need.[23]

The famine in Somalia that led to the death of over 1.5 million people was due to the collapse of the government, fighting among rival clans, and general lawlessness. Farmers and their families were driven off their land by bandits who stole their crops and possessions. Many had nowhere to go for help. Mass starvation in Ethiopia and Sudan was similarly initiated by violent conflict within the country. Seizing the opposing side's food and preventing them from obtaining food aid are among the primary weapons used to fight these civil wars. In Bangladesh, where

TABLE 33.5

ROOT CAUSES OF MALNUTRITION IN DEVELOPING COUNTRIES.[22]

- Poverty
- Discrimination against females
- HIV/AIDS epidemic
- Racism, ethnocentrism
- Poor and corrupt governance
- Unsafe water
- Low levels of education
- Unequitable distribution of the food supply
- Lack of economic opportunities
- Low agricultural productivity

poor people have no choice but to live on floodplains, cyclones wipe out crops and regularly result in the death of thousands due to floods, starvation, and infectious disease. None of these cases of mass famine was due primarily to the ravages of nature. Starvation and malnutrition are largely initiated by humans, and it is within the power of humans to end them.

Women and female children are at particular risk for malnutrition in some societies because cultural practices call for food to be allocated to men and boys first (Illustration 33.4). If too little food is available, what remains after meals for women and daughters may be too little to support health and growth. In many developing countries, discrimination against women in education and employment, sanctions against the use of birth control, and violence toward women place women at high risk of developing malnutrition and having a low quality of life.[24]

Ending Malnutrition

Adequate food is the cradle of normal resistance, the playground of normal immunity, the workshop of good health, and the laboratory of long life.
—Charles Mayo

The long-term solution to malnutrition will depend on the ability of humans to work together to achieve educational and economic development, peace, population growth control, improved sanitation, social equity for women and children, and environmentally sound and productive agricultural policies and practices in developing countries (Illustration 33.5). Bottom-up approaches to problem solving work better than top-down approaches. The people most affected by malnutrition must be active participants in the planning and implementation of improvement programs. Efforts to improve the nutritional status of populations can and have paid off.

Illustration 33.4
Women and children in some developing countries are particularly vulnerable to malnutrition because they, and their nutritional needs, may be considered less important than men and their needs. *Similarly, boys' needs may take precedence over girls' needs. These children are 2-year-old twins: the child on the left is a girl, and the child on the right is a boy.*

Success Stories] Examples of strikingly successful efforts to reduce malnutrition in various countries around the world are available for telling. A repeated theme is the reduction in malnutrition as economic, educational, nutritional, and sanitary conditions improve. Improvements in these same areas were

Illustration 33.5
Safe water supplies represent a major advance in quality of life for people in developing countries.

Photo Researchers - Porterfield/Chickering

responsible for the dramatic declines in malnutrition and infectious diseases and the increases in life expectancy experienced in the United States, Canada, and many European countries around 100 years ago.[25] In other instances, specific nutrient deficiencies have been greatly reduced or eliminated by food fortification and nutrition education programs. Here are several specific examples:

- Vitamin A supplements and education on vitamin A–rich foods were related to a dramatic decrease in cases of severe and moderate vitamin A deficiency and infection in children in Indonesia.

- Food supplements given to groups of infants in Russia, Brazil, South Africa, and China were associated with higher IQ scores at age 8.

- Iodization of salt has eliminated iodine deficiency in Bolivia and Ecuador. Worldwide, iodization of salt is associated with a drop in the number of iodine-deficient children from 48 million in 1990 to 28 million in 1997.

- Fortification of flour with iron has led to a decrease in iron-deficiency anemia in the Philippines.[26]

In Thailand, a country with a serious iodine-deficiency problem, a highly creative approach to increasing iodine intake was implemented by the king. To celebrate his birthday, salt producers gave the king 10,000 tons of iodized salt. The king had the salt packaged in small plastic bags and, with the help of the Red Cross and the army, delivered a bag to every household in Thailand. A message from the king about the importance of using iodized salt was attached to each bag.[27]

Deaths of young children from malnutrition and its related diseases have dropped substantially in countries experiencing a resurgence of breastfeeding (Illustration 33.6). Breast milk protects infants and young children from a variety of infectious diseases, supports the growth and health of infants, and protects them from the hazards of formulas reconstituted with contaminated water. Worldwide health initiatives have led to the development of the International Code of Marketing Breast

Milk Substitutes and to "Baby Friendly Hospitals." The international marketing code calls for the prohibition of free formula samples as well as the promotion of infant formulas by health care professionals. More than 12,700 hospitals have adopted Baby Friendly Hospital policies that promote and facilitate breastfeeding.[28]

Health status of people in developing countries has also been improved by the widespread use of oral rehydration fluids that protect children with diarrhea from dehydration. Broadly based vaccination programs have protected many from various diseases.[29]

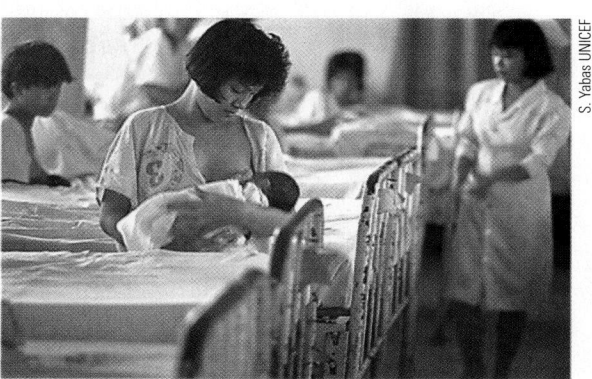

The World Food Summit] The 1996 World Food Summit ended with a pledge that by 2015, countries would reduce by half the number of people in the world who are hungry and undernourished.[30] Progress is being made toward meeting the goal. Access to food and rates of undernutrition are improving, but at a slower pace than envisioned. It is anticipated that it will take until the year 2030 to cut rates of hunger and undernutrition by half.[31]

Countries in "Nutrition Transition"] As countries develop economically and food supplies increase, the incidence of undernutrition usually decreases, and chronic health problems such as heart disease, diabetes, and hypertension tend to increase. Elevated rates of chronic disease appear to be closely related to high saturated fat and low vegetable and fruit intakes, low levels of physical activity, and other characteristics of Western lifestyles and dietary habits.[32]

Countries undergoing the nutrition transition often experience rising rates of obesity in children.[34] Children born to poorly nourished women may be biologically programmed while in the womb to conserve energy. When exposed to a generous food supply later in life, they may gain fat more readily than height.[35] Excess accumulation of fat later in life may place the children at increased risk for hypertension, diabetes, heart disease, and a number of other "diseases of civilization."[36] Although infections such as malaria and tuberculosis will predominate through the year 2025, the incidence of diseases of "Western civilization" will continue to expand.[37]

The Future

> Knowing is not enough; we must apply.
> Willingness is not enough; we must do.
> —Goethe

Benefits derived from improving the nutritional status of the world's population are immense. Adequate nutrition is the bedrock on which present generations secure a future for themselves and the next generation. People who are well nourished are more productive, happier, require less medical care, and are more likely to be self-sufficient than malnourished people. When the malnutrition problem is solved, the world will likely wonder why it didn't happen sooner.[38]

If you would like to be part of the solution to food insecurity, hunger, or malnutrition, consider volunteering at a food or meals program, in community courses for language or self-sufficiency skills, or in other public service programs. The experience will likely enrich your life as well as the lives of those you serve.

Illustration 33.6
After a rooming-in policy that enabled mothers to breastfeed their infants on demand was introduced in one hospital in the Philippines, the incidence of early infant deaths due to infection dropped by 95%.
Source: First Call for Children, A UNICEF Quarterly 1992 (January/ March):1.

Approximately 14% of the U.S. population lives in poverty, but there are sizable differences in poverty rates by age and racial group. Twenty percent of children, 30% of Hispanics, 29% of African Americans, and 11% of whites live in poverty.[33]

ON THE SIDE

Nutrition UP CLOSE

Ethnic Foods Treasure Hunt

FOCAL POINT: Values placed on foods are culturally specific.

The following questions deal with ethnic food preferences. Answer any *three* of the questions. Resources that will help you find the answers include knowledgeable individuals, the Internet, travel and anthropology books, and encyclopedias.

1. Name three foods commonly served at celebrations by the Igbo in Nigeria (Igbos were formerly referred to as Biafrans):

   a. _____

   b. _____

   c. _____

2. List two special foods given to Hmong women during pregnancy:

   a. _____

   b. _____

3. List the five foods that make up the traditional Costa Rican breakfast:

   a. _____

   b. _____

   c. _____

   d. _____

   e. _____

4. List three foods commonly sold at soccer matches in Italy:

   a. _____

   b. _____

   c. _____

5. Name two traditional foods of the Zulu in South Africa:

   a. _____

   b. _____

6. Name one of the staple foods of people in Nentsi (it's near Siberia):

7. List two foods considered to be "yin" and two considered to be "yang" in Chinese culture:

Yin	Yang
a. _____	a. _____
b. _____	b. _____

FEEDBACK (answers to these questions) can be found at the end of Unit 33.

Key Terms

kwashiorkor, page 33-5

marasmus, page 33-5

www links

www.who.org
Leads to causes of illness and death in industrialized and developing countries, information about disease outbreaks, WHO programs, and traveler's health.

www.nutrition.gov
Provides updates on government-sponsored nutrition programs.

www.fao.org
The United Nations' Food and Agriculture Organization address. Provides updates on

worldwide malnutrition and meeting the World Food Summit's goal of reducing hunger by half by 2015.

www.travel.state.gov
Traveling abroad? This site offers travel alerts on health conditions, political instability; crime statistics, and public announcements.

www.cdc.gov
Check out international foodborne illness and other infectious disease warnings here.

www.wfp.org
The address for the international World Food Programme and information about food aid, antihunger programs, and school feeding projects.

Notes

1. Darton-Hill I, Coyne ET. Feast and famine: socioeconomic disparities in global nutrition and health. Public Health Nutr 1998;1:23–31.

2. Darton-Hill and Coyne, Feast and famine.

3. Health, United States 2003, www.cdc.gov/nchs, accessed 10/03; and International life expectancy statistics. www.who.org, accessed 10/03.

4. The state of food insecurity in the world in 2000. FAO, United Nations Report, Nov. 2000, www.fao.org, accessed 2/01.

5. Darton-Hill and Coyne, Feast and famine.

6. United Nations International Children's Emergency Fund. The state of the world's children, 1998. New York: United Nations; 1998.

7. Darton-Hill and Coyne, Feast and famine.

8. Position of American Dietetic Association: addressing world hunger, malnutrition, and food insecurity. J Am Diet Assoc 2003;103:1046–56.

9. Helping other countries identify and eliminate micronutrient deficiencies, Chronic Dis Notes & Reports, Centers for Disease Control and Prevention 2002; Winter:8; Food as a front line defense against HIV/AIDS, www.wfp.org, accessed 10/03; Increasing fruit and vegetable consumption a global priority, www.fao.org, accessed 10/03; Global Nutrition Strategy,

Dietetics in Practice 2003; Fall:3; and Quantifying selected major risks to health, World Health Organization, www.who.int/whr/2002/chapter4/en/print.html, accessed 11/02.

10. Christan P. Micronutrients and reproductive health issues: an international perspective. J Nutr 2003;133:1969S–73S.

11. Quantifying selected major risks to health (www.who.int/whr/2002/chapter4/en/print.html).

12. Scrimshaw NS. Historical concepts of interactions, synergism and antagonism between nutrition and infection. J Nutr 2003;133:316S–21S.

13. Position of ADA: addressing world hunger.

14. UNICEF, The state of the world's children, 1998; and Global Nurtiion Strategy.

15. The state of food insecurity in the world in 2000.

16. Underwood BA. Scientific research: essential, but is it enough to combat world food insecurity? J Nutr 2003;133:1434S–7S.

17. Quantifying selected major risks to health (www.who.int/whr/2002/chapter4/en/print.html).

18. UNICEF, The state of the world's children, 1998.

19. Thurow R. In Africa, AIDS and famine now go hand in hand. Wall Street Journal 2003;July 7:A1, A6.

20. International life expectancty statistics (www.who.org); Naik G, HIV's impact is seen by UN as even worse, Wall Street Journal 2003; Feb. 27:B4; and Clinton WJ, Turning the tide on the AIDS pandemic, N Engl J Med 2003;348:1800–05.

21. Haddad L, Martorell R. Feeding the world in the coming decades requires improvements in investment, technology and institutions. J Nutr 2002;132:3435S–36S.

22. Position of ADA: addressing world hunger; Global Nutrition Strategy; and Haddad and Martorell, Feeding the world.

23. UNICEF, The state of the world's children, 1998.

24. Position of ADA: addressing world hunger; and Kumar A et al, Mortality in female and male infants from treatable causes in India, BMJ 2003;327:126–8.

25. Gold MR. Presentation at the Tenth Chronic Disease Conference, Washington, DC; 1996.

26. Decimosexta Conferencia Internacional de Nutrición, Montreal. Nutr View 1997; fall:1–36.

27. Food fortification to end micronutrient malnutrition: state of the art. Nutr View 1997; special issue:1–8.

28. UNICEF, The state of the world's children, 1998.

29. Pelletier DL, Frongillo EA. Changes in child survival are strongly associated

with changes in malnutrition in developing countries. J Nutr 2003; 133:107–19.

30. UNICEF, The state of the world's children, 1998.

31. Underwood, Scientific research; and Haddad and Martorell, Feeding the world.

32. Increasing fruit and vegetable consumption (www.fao.org); and Global Nutrition Strategy.

33. Health, United States, 2000. www.cdc.gov/nchs, accessed 8/00.

34. De Onis M et al. Prevalence and trends of overweight among preschool children in developing countries. Am J Clin Nutr 2000;72:1032–9.

35. Popkin BM. Stunting is associated with overweight in children of four nations that are undergoing the nutrition transition. J Nutr 1996;126:3009–16.

36. Mi J et al. Effects of infant birthweight and maternal body mass index in pregnancy on components of the insulin resistance syndrome in China. Ann Int Med 2000;132:253–60.

37. UNICEF, The state of the world's children, 1998.

38. Position of ADA: addressing world hunger.

Nutrition UP CLOSE

Ethnic Foods Treasure Hunt

Feedback for Unit 33

1. a. yams; b. kola nuts; c. breadfruit with corn and yams; d. cassava.

2. a. chicken; b. rice; c. ripe mango; d. grapes; e. ginger; f. milk; g. many more.

3. "Gallo pinto" includes the following:

   a. rice and black beans; b. eggs; c. tortillas; d. coffee; e. fruit; f. sometimes steak.

4. a. Italian ices; b. calzones; c. Italian sodas; d. pizza; e. pasta; f. hot dogs.

5. a. beef; b. cassava; c. milk; d. yams; e. pumpkin; f. millet; g. corn; h. dried beans; i. beer; j. honey; k. yogurt; l. cow's blood.

6. a. reindeer; b. polar bear; c. sour milk; d. caribou; e. tea.

7. Yin: a. raw kelp (seaweed); b. boiled rice; c. raw vegetables; d. fruit; e. dairy products.

   Yang: a. cooked kelp; b. cooked vegetables; c. fried rice; d. fish; e. eggs; f. chicken.

Appendixes

Appendix A
Table of Food Composition

Appendix B
Reliable Sources of Nutrition Information

Appendix C
The U.S. Food Exchange System

Appendix D
Table of Intentional Food Additives

Appendix E
Cells

Appendix F
Canadian Choice System

Appendix **A**

Table of Food Composition

This edition of the table of food composition includes a wide variety of foods from all food groups. It is updated with each edition to reflect nutrient changes for current foods, remove outdated foods, and add foods that are new to the marketplace.*

The nutrient database for this appendix is compiled from a variety of sources, including the USDA Standard Reference database (Release 16), literature sources, and manufacturers' data. The USDA database provides data for a wider variety of foods and nutrients than other sources. Because laboratory analysis for each nutrient can be quite costly, manufacturers tend to provide data only for those nutrients mandated on food labels. Consequently, data for their foods are often incomplete; any missing information is designated in this table as a blank space. Keep in mind that a blank space means only that the information is unknown and should not be interpreted as a zero.

Whenever using nutrient data, remember that many factors influence the nutrient contents of foods, including the mineral content of the soil, the diet of the animal or the fertilizer of the plant, the season of harvest, the method of processing, the length and method of storage, the method of cooking, the method of analysis, and the moisture content of the sample analyzed. With so many factors involved, users must view nutrient data as a close approximation of the actual amount.

For updates, corrections, and a list of 6000 foods and codes found in the diet analysis software that accompanies this text, visit **www.wadsworth.com/nutrition** and click on *Diet Analysis*.

- *Fats* Total fats, as well as the breakdown of total fats to saturated, monounsaturated, and polyunsaturated fats, are listed in the table. The fatty acids seldom add up to the total due to rounding and to other fatty acid components that are not included in these basic categories, such as *trans*-fatty acids and glycerol. *Trans*-fatty acids can comprise a large share of the total fat in margarine and shortening (hydrogenated oils) and in any foods that include them as ingredients.

- *Vitamin A and Vitamin E* In keeping with the 2001 RDA for vitamin A, which established a new measure of vitamin A activity—retinol activity equivalents (RAE)—this appendix presents data for vitamin A in micrograms (μg) RAE. Similarly, because the 2000 RDA for vitamin E is based only on the alpha-tocopherol form of vitamin E, this appendix reports vitamin E data in milligrams (mg) alpha-tocopherol (listed in the table as mg α).

- *Bioavailability* Keep in mind that the availability of nutrients from foods depends not only on the quantity provided by a food, but also on the amount absorbed and used by the body—the bioavailability. The bioavailability of folate from fortified foods, for example, is greater than from naturally occurring sources. Similarly, the body can make niacin from the amino acid tryptophan, but niacin values in this table (and most databases) report preformed niacin only. Chapter 10 provides conversion factors and additional details.

- *Using the Table* The items in this table have been organized into several categories, which are listed at the head of each right-hand page. Page numbers have been provided, and each group has been color-coded to make it easier to find individual items.

  In an effort to conserve space, the following abbreviations have been used in the food descriptions and nutrient breakdowns:

- diam = diameter
- ea = each

---

*This food composition table has been prepared for Wadsworth Publishing Company and is copyrighted by ESHA Research in Salem, Oregon—the developer and publisher of the Food Processor and Genesis nutritional software programs. The nutritional data are supported by over 1300 references. Because the list of sources is so extensive, it is not provided here, but is available from the publisher.

- enr = enriched
- f/ = from
- frzn = frozen
- g = grams
- liq = liquid
- pce = piece
- pkg = package
- w/ = with
- w/o = without
- t = trace
- 0 = zero (no nutrient value)

- blank space = information not available

- *Caffeine Sources* Caffeine occurs in several plants, including the familiar coffee bean, the tea leaf, and the cocoa bean from which chocolate is made. Most human societies use caffeine regularly, most often in beverages, for its stimulant effect and flavor. Caffeine contents of beverages vary depending on the plants they are made from, the climates and soils where the plants are grown, the grind or cut size, the method and duration of brewing, and the amounts served. The accompanying table shows that in general, a cup of coffee contains the most caffeine; a cup of tea, less than half as much; and cocoa or chocolate, less still. As for cola beverages, they are made from kola nuts, which contain caffeine, but most of their caffeine is added, using the purified compound obtained from decaffeinated coffee beans.

    The FDA lists caffeine as a multipurpose GRAS substance■ that may be added to foods and beverages. Drug manufacturers use caffeine in many kinds of drugs: stimulants, pain relievers, cold remedies, diuretics, and weight-loss aids.

■ Reminder: A GRAS substance is one that is "generally recognized as safe."

TABLE — Caffeine Content of Beverages, Foods, and Over-the-Counter Drugs

Beverages and Foods	Average (mg)	Range (mg)	Drugs[a]	Average (mg)
Coffee (5-oz cup)			Cold remedies (standard dose)	
Brewed, drip method	130	110–150	Dristan	0
Brewed, percolator	94	64–124	Coryban-D, Triaminicin	30
Instant	74	40–108	Diuretics (standard dose)	
Decaffeinated, brewed or instant	3	1–5	Aqua-ban, Permathene H$_2$Off	200
Tea (5-oz cup)			Pre-Mens Forte	100
Brewed, major U.S. brand	40	20–90	Pain relievers (standard dose)	
Brewed, imported brands	60	25–110	Excedrin	130
Instant	30	25–50	Midol, Anacin	65
Iced (12-oz can)	70	67–76	Aspirin, plain (any brand)	0
Soft drinks (12-oz can)			Stimulants	
Dr. Pepper	40		Caffedrin, NoDoz, Vivarin	200
Colas and cherry cola			Weight-control aids (daily dose)	
Regular		30–46	Prolamine	280
Diet		2–58	Dexatrim, Dietac	200
Caffeine-free		0–trace		
Jolt	72			
Mountain Dew, Mello Yello	52			
Fresca, Hires Root Beer, 7-Up, Sprite, Squirt, Sunkist Orange	0			
Cocoa beverage (5-oz cup)	4	2–20		
Chocolate milk beverage (8 oz)	5	2–7		
Milk chocolate candy (1 oz)	6	1–15		
Dark chocolate, semisweet (1 oz)	20	5–35		
Baker's chocolate (1 oz)	26			
Chocolate flavored syrup (1 oz)	4			

NOTE: A pharmacologically active dose of caffeine is defined as 200 milligrams.

[a]Because products change, contact the manufacturer for an update on products you use regularly.

Table A–1

Food Composition   (Computer code number is for West Diet Analysis program)   (For purposes of calculations, use "0" for t, <1, <.1, <.01, etc.)

Computer Code Number	Food Description	Measure	Wt (g)	H₂O (%)	Ener (kcal)	Prot (g)	Carb (g)	Dietary Fiber (g)	Fat (g)	Fat Breakdown (g) Sat	Mono	Poly
	BEVERAGES											
	Alcoholic:											
	Beer:											
22500	Regular (12 fl oz)	1½ c	356	94	117	1	6	<1	<1	0	0	0
22512	Light (12 fl oz)	1½ c	354	95	99	1	5	0	0	0	0	0
22679	Budweiser (12 fl oz)	1½ c	357		145	1	11		0	0	0	0
22616	Miller (12 fl oz)	1½ c	356	96	143	1	13		0	0	0	0
22617	Miller Light (12 fl oz)	1½ c	356		110	1	7		0	0	0	0
	Gin, rum, vodka, whiskey:											
22670	80 proof	1½ fl oz	42	67	97	0	0	0	0	0	0	0
22654	86 proof	1½ fl oz	42	64	105	0	<1	0	0	0	0	0
22661	90 proof	1½ fl oz	42	62	110	0	0	0	0	0	0	0
	Liqueur:											
22519	Coffee liqueur, 53 proof	1½ fl oz	52	31	175	<1	24	0	<1	t	t	t
22520	Coffee & cream liqueur, 34 proof	1½ fl oz	47	46	154	1	10	0	7	4.5	2.1	0.3
22521	Crème de menthe, 72 proof	1½ fl oz	50	28	186	0	21	0	<1	t	t	t
22551	Kahlua	1½ fl oz	30	30	106	<1	13	0	<1	t	t	t
	Wine, 4 fl oz:											
22673	Dessert, sweet	½ c	118	71	189	<1	16	0	0	0	0	0
22501	Red	½ c	118	88	85	<1	2	0	0	0	0	0
22502	Rosé	½ c	118	89	84	<1	2	0	0	0	0	0
22504	White medium	½ c	118	90	80	<1	1	0	0	0	0	0
20077	Nonalcoholic light	1 c	232	98	14	1	3	0	0	0	0	0
22681	Wine cooler	1 c	227	90	113	<1	13	<1	<1	t	t	t
	Carbonated:											
20006	Club soda (12 fl oz)	1½ c	355	100	0	0	0	0	0	0	0	0
20005	Cola beverage (12 fl oz)	1½ c	372	89	156	<1	40	0	0	0	0	0
20030	Diet cola w/aspartame (12 fl oz)	1½ c	355	100	4	<1	<1	0	0	0	0	0
20007	Diet soda pop w/saccharin (12 fl oz)	1½ c	355	100	0	0	<1	0	0	0	0	0
20008	Ginger ale (12 fl oz)	1½ c	366	91	124	0	32	0	0	0	0	0
20031	Grape soda (12 fl oz)	1½ c	372	89	160	0	42	0	0	0	0	0
20032	Lemon-lime (12 fl oz)	1½ c	368	90	147	0	38	0	0	0	0	0
20027	Pepper-type soda (12 fl oz)	1½ c	368	89	151	0	38	0	<1	0.3	0	0
20009	Root beer (12 fl oz)	1½ c	370	89	152	0	39	0	0	0	0	0
20149	Cherry Coke (12 fl oz)	1½ c	375		156	0	42	0	0	0	0	0
20148	Coca Cola Classic (12 fl oz)	1½ c	373		145	0	40	0	0	0	0	0
20150	Diet Coke (12 fl oz)	1½ c	359		1	0	<1	0	0	0	0	0
20207	Diet 7 UP (12 fl oz)	1½ c	360		0	0	0	0	0	0	0	0
20167	Diet Pepsi (12 fl oz)	1½ c	360		0	0	0	0	0	0	0	0
20166	Pepsi cola (12 fl oz)	1½ c	360	88	150	0	41	0	0	0	0	0
20163	Sprite (12 fl oz)	1½ c	373		144	0	39	0	0	0	0	0
	Coffees:											
20012	Brewed	1 c	237	99	9	<1	0	0	2	0	0	0
20592	Cappuccino w/lowfat milk	1½ c	244		110	8	11	0	4	2.5		
20639	Cappuccino w/whole milk	1½ c	244		140	7	11	0	7	4.5		
20668	Latte w/lowfat milk	1½ c	366		170	12	17	0	6	4		
20023	Prepared from instant	1 c	238	99	5	<1	1	0	0			
	Fruit drinks:											
20024	Fruit punch drink, canned	1 c	248	88	117	0	30	<1	0	0	0	0
20142	Gatorade	1 c	241	94	60	0	15	0	0	0	0	0
20026	Grape drink, canned	1 c	250	87	125	<1	32	0	0	0	0	0
20016	Koolade sweetened with sugar	1 c	262	90	97	0	25	0	<1	t	t	t
20017	Koolade sweetened with nutrasweet	1 c	240	95	43	0	11	0	0	0	0	0
20001	Lemonade, frzn concentrate (6-oz can)	¾ c	219	52	396	1	103	<1	<1	t	t	0.1
20000	Lemonade, from concentrate	1 c	248	86	131	<1	34	<1	<1	t	t	t
20003	Limeade, frzn concentrate (6-oz can)	¾ c	218	50	427	<1	106	<1	<1	0	0	0
20002	Limeade, from concentrate	1 c	247	89	104	<1	26	0	<1	0	0	0
20059	Pineapple grapefruit, canned	1 c	250	88	118	<1	29	0	<1	t	t	t
20025	Pineapple orange, canned	1 c	250	87	125	3	30	<1	0	0	0	0
20559	Powerade	1 c	247		72	0	19	0	0	0	0	0
20737	Snapple, fruit punch	1 c	252	88	110	0	29		0	0	0	0
20761	Snapple, tropical	1 c	252	89	110	0	27		0	0	0	0

PAGE KEY: A–4 = Beverages   A–6 = Dairy   A–10 = Eggs   A–12 = Fat/Oil   A–14 = Fruit   A–20 = Bakery   A–26 = Grain   A–32 = Fish   A–32 = Meats   A–36 = Poultry   A–38 = Sausage   A–40 = Mixed/Fast   A–44 = Nuts/Seeds   A–46 = Sweets   A–48 = Vegetables/Legumes   A–58 = Vegetarian   A–62 = Misc   A–62 = Soups/Sauces   A–64 = Fast   A–78 = Convenience   A–80 = Baby foods

Chol (mg)	Calc (mg)	Iron (mg)	Magn (mg)	Pota (mg)	Sodi (mg)	Zinc (mg)	VT-A (RAE)	Thia (mg)	VT-E (mg α)	Ribo (mg)	Niac (mg)	V-B$_6$ (mg)	Fola (µg)	VT-C (mg)
0	18	0.07	21	89	14	0.04	0	0.02	0	0.09	1.61	0.18	21	0
0	18	0.14	18	64	11	0.11	0	0.03	0	0.11	1.39	0.12	14	0
0	16	<.01	22	111	9	<.01		<.01		0.14	1.4	0.16		0
0					7									
0					6									
0	0	0.02	0	1	<1	0.02	0	<.01	0	<.01	<.01	<.01	0	0
0	0	0.02	0	1	<1	0.02	0	<.01	0	<.01	<.01	<.01	0	0
0	0	0.02	0	1	<1	0.02	0	<.01	0	<.01	<.01	<.01	0	0
0	1	0.03	2	16	4	0.02	0	<.01	0	<.01	0.07	0	0	0
27	8	0.06	1	15	43	0.08	81	<.01	0.21	0.03	0.04	<.01	1	0
0	0	0.04	0	0	2	0.02	0	0	0	0	<.01	0	0	0
0	0	0.02	0	4	2	0.01	0	<.01	0	<.01	0.02	0	0	0
0	9	0.28	11	109	11	0.08	0	0.02	0	0.02	0.25	0	0	0
0	9	0.51	15	132	6	0.11	0	<.01	0	0.03	0.1	0.04	2	0
0	9	0.45	12	117	6	0.07	0	<.01	0	0.02	0.09	0.03	1	0
0	11	0.38	12	94	6	0.08	0	<.01	0	<.01	0.08	0.02	0	0
0	21	0.93	23	204	16	0.19	0	0	0	0.02	0.23	0.05	2	0
0	13	0.62	12	102	19	0.13	<1	0.01	0.01	0.02	0.1	0.03	3	4
0	18	0.04	4	7	75	0.36	0	0	0	0	0	0	0	0
0	11	0.07	4	4	15	0.04	0	0	0	0	0	0	0	0
0	11	0.11	4	21	18	0	0	0.02	0	0.08	0	0	0	0
0	14	0.07	4	14	57	0.11	0	0	0	0	0	0	0	0
0	11	0.66	4	4	26	0.18	0	0	0	0	0	0	0	0
0	11	0.3	4	4	56	0.26	0	0	0	0	0	0	0	0
0	7	0.26	4	4	40	0.18	0	0	0	0	0.06	0	0	0
0	11	0.15	0	4	37	0.15	0	0	0	0	0	0	0	0
0	18	0.18	4	4	48	0.26	0	0	0	0	0	0	0	0
0				0	6	0	0	0	0	0	0	0	0	0
0				0	13	0								0
0				18	6	0		0						0
0					52	0	0		0		0			0
0	0	0		8	35	0								0
0	0	0			35	0								0
0				0	34	0								
0	2	0.02	5	114	2	0.02	0	0	0.05	0.12	0	<.01	5	0
15	250	0			110									2
30	250	0			105									2
25	400	0			170									4
0	10	0.1	7	71	5	0.02	0	0	0	<.01	0.56	0	0	0
0	20	0.22	7	62	94	0.02	5	0.05	0.05	0.06	0.05	0.03	10	73
0	0	0.12	2	27	96	0.05	0	0.01	0	<.01	0	0	0	0
0	8	0.25	8	82	2	0.05	<1	0.02	0	0.03	0.17	0.04	2	40
0	42	0.13	3	3	37	0.08	0	0	0	<.01	<.01	0	0	31
0	17	0.65	5	50	50	0.26	1	0.02	0	0.05	0.05	0	5	78
0	15	1.58	11	147	9	0.18	1	0.06	0.09	0.21	0.16	0.05	11	39
0	10	0.52	5	50	7	0.07	<1	0.02	0.02	0.07	0.05	0.02	2	13
0	28	0.07	11	94	20	0.09	<1	0.02	0	0.03	0.08	0.04	7	25
0	7	0.02	2	22	5	0.02	<1	<.01	0	<.01	0.02	<.01	2	6
0	18	0.78	15	152	35	0.15	<1	0.08	0.02	0.04	0.67	0.1	22	115
0	12	0.68	15	115	8	0.15	2	0.08	0.08	0.05	0.52	0.12	22	56
0	0	0		32	28	0								0
0					10									0
0					10									

A

Appendix

Table A–1

Food Composition

(Computer code number is for West Diet Analysis program)   (For purposes of calculations, use "0" for t, <1, <.1, <.01, etc.)

Computer Code Number	Food Description	Measure	Wt (g)	H₂O (%)	Ener (kcal)	Prot (g)	Carb (g)	Dietary Fiber (g)	Fat (g)	Fat Breakdown (g)		
										Sat	Mono	Poly
	BEVERAGES—Continued											
	Fruit and vegetable juices: see Fruit and Vegetable sections											
	Ultra Slim Fast, ready to drink, can:											
	Chocolate Royale	1 ea	350	83	220	10	40	5	3	1	1.5	0.5
	French Vanilla	1 ea	350	84	220	10	40	5	3	0.5	1.5	0.5
	Strawberries n' cream	1 ea	350	84	220	10	40	5	2	0.5	1.5	0.5
20041	Water, municiple	1 c	237	100	0	0	0	0	0	0	0	0
792	La Croix	1 c	236	100	0	0	0	0	0	0	0	0
20010	Tonic water	1½ c	366	91	124	0	32	0	0	0	0	0
	Tea:											
20014	Brewed, regular	1 c	237	100	2	0	1	0	0	0	0	0
20036	Brewed, herbal	1 c	237	100	2	0	<1	0	0			
20022	From instant, sweetened	1 c	259	91	88	<1	22	0	<1	t	t	t
20020	From instant, unsweetened	1 c	237	100	2	<1	<1	0	0	0	0	0
20443	Green tea bag	1 ea	2		0	0	0	0	0	0	0	0
	DAIRY											
	Butter: see Fats and Oils											
	Cheese:											
	Natural:											
1003	Blue	1 oz	28	42	99	6	1	0	8	5.2	2.2	0.2
1037	Brick	1 oz	28	41	104	7	1	0	8	5.3	2.4	0.2
1004	Brie	1 oz	28	48	94	6	<1	0	8	4.9	2.2	0.2
1006	Camembert	1 oz	28	52	84	6	<1	0	7	4.3	2	0.2
1007	Cheddar:	1 oz	28	37	113	7	<1	0	9	5.9	2.6	0.3
1008	Shredded	1 c	113	37	455	28	1	0	37	23.8	10.6	1.1
1088	Shredded, low fat, sodium	1 oz	28	65	48	7	1	0	2	1.2	0.6	t
1050	Edam	1 oz	28	42	100	7	<1	0	8	4.9	2.3	0.2
1016	Feta	1 oz	28	55	74	4	1	0	6	4.2	1.3	0.2
1054	Gouda	1 oz	28	41	100	7	1	0	8	4.9	2.2	0.2
1073	Gruyere	1 oz	28	33	116	8	<1	0	9	5.3	2.8	0.5
1538	Gorgonzola	1 oz	28	42	97	6	1	0	8	5		
1055	Limburger	1 oz	28	48	92	6	<1	0	8	4.7	2.4	0.1
1017	Monterey Jack	1 oz	28	41	104	7	<1	0	8	5.3	2.5	0.3
1056	Mozzarella whole milk	1 oz	28	50	84	6	1	0	6	3.7	1.8	0.2
1019	Part-skim, low moisture	1 oz	28	46	85	7	1	0	6	3.5	1.6	0.2
1021	Muenster	1 oz	28	42	103	7	<1	0	8	5.4	2.4	0.2
1060	Neufchatel	1 oz	28	62	73	3	1	0	7	4.1	1.9	0.2
1075	Parmesan, grated:	1 oz	28	21	121	11	1	0	8	4.8	2.3	0.3
57	Cup	1 c	100	21	431	38	4	0	29	17.3	8.4	1.2
1023	Provolone	1 oz	28	41	98	7	1	0	7	4.8	2.1	0.2
1064	Ricotta, whole milk	1 c	246	72	428	28	7	0	32	20.4	8.9	0.9
1024	Part-skim milk	1 c	246	74	339	28	13	0	19	12.1	5.7	0.6
1027	Swiss	1 oz	28	37	106	8	2	0	8	5	2	0.3
	Substitute:											
47947	Vegan cheese substitute, slice	1 ea	21		20	2	3		0	0	0	0
47946	Vegetarian cheese substitute topping	2 tsp	5		15	2	<1		<1	0		
	Cottage:											
1099	Low sodium, low fat	1 c	225	84	162	28	6	0	2	1.4	0.6	t
1013	Creamed, large curd	1 c	225	79	232	28	6	0	10	6.4	2.9	0.3
1012	Creamed, small curd	1 c	210	79	216	26	6	0	9	6	2.7	0.3
1049	With fruit	1 c	226	80	219	24	10	<1	9	5.2	2.3	0.3
1014	Low fat 2%	1 c	226	79	203	31	8	0	4	2.8	1.2	0.1
1047	Low fat 1%	1 c	226	82	163	28	6	0	2	1.5	0.7	t
	Cream:											
1015	Regular	1 tbs	15	54	52	1	<1	0	5	3.3	1.5	0.2
47949	Substitute	1 oz	28		49	3	3		3	0		
	Pasteurized processed:											
1456	American	1 oz	28	39	105	6	<1	0	9	5.5	2.5	0.3
1458	Swiss	1 oz	28	42	94	7	1	0	7	4.5	2	0.2
1437	American cheese food, jar	½ c	57	43	188	10	4	0	14	8.5	4.1	0.6

Chol (mg)	Calc (mg)	Iron (mg)	Magn (mg)	Pota (mg)	Sodi (mg)	Zinc (mg)	VT-A (RAE)	Thia (mg)	VT-E (mg α)	Ribo (mg)	Niac (mg)	V-B$_6$ (mg)	Fola (µg)	VT-C (mg)
5	400	2.7	140	600	220	2.25		0.52	13.64	0.6	7	0.7	120	60
5	400	2.7	140	600	220	2.25		0.52	13.64	0.6	7	0.7	120	60
5	400	2.7	140	600	220	2.25		0.52	13.64	0.6	7	0.7	120	60
0	5	0	2	0	5	0	0	0	0	0	0	0	0	0
0					5									
0	4	0.04	0	0	15	0.37	0	0	0	0	0	0	0	0
0	0	0.05	7	88	7	0.05	0	0	0	0.03	0	0	12	0
0	5	0.19	2	21	2	0.09	0	0.02	0	<.01	0	0	2	0
0	5	0.05	5	49	8	0.03	0	0	0	0.04	0.09	<.01	0	0
0	7	0.05	5	47	7	0.02	0	0	0	<.01	0.09	<.01	0	0
0	0	0			0		0							0
21	148	0.09	6	72	391	0.74	55	<.01	0.07	0.11	0.28	0.05	10	0
26	189	0.12	7	38	157	0.73	82	<.01	0.07	0.1	0.03	0.02	6	0
28	52	0.14	6	43	176	0.67	49	0.02	0.07	0.15	0.11	0.07	18	0
20	109	0.09	6	52	236	0.67	67	<.01	0.06	0.14	0.18	0.06	17	0
29	202	0.19	8	27	174	0.87	74	<.01	0.08	0.1	0.02	0.02	5	0
119	815	0.77	32	111	702	3.51	299	0.03	0.33	0.42	0.09	0.08	20	0
6	197	0.2	8	31	6	0.87		<.01	0.05	<.01	0.03	0.02	5	0
25	205	0.12	8	53	270	1.05	68	0.01	0.07	0.11	0.02	0.02	4	0
25	138	0.18	5	17	312	0.81	35	0.04	0.05	0.24	0.28	0.12	9	0
32	196	0.07	8	34	229	1.09	46	<.01	0.07	0.09	0.02	0.02	6	0
31	283	0.05	10	23	94	1.09	76	0.02	0.08	0.08	0.03	0.02	3	0
30	170	0.18			280									0
25	139	0.04	6	36	224	0.59	95	0.02	0.06	0.14	0.04	0.02	16	0
25	209	0.2	8	23	150	0.84	55	<.01	0.07	0.11	0.03	0.02	5	0
22	141	0.12	6	21	176	0.82	50	<.01	0.05	0.08	0.03	0.01	2	0
15	205	0.07	7	27	148	0.88	38	0.03	0.1	0.09	0.03	0.02	3	0
27	201	0.11	8	38	176	0.79	83	<.01	0.07	0.09	0.03	0.02	3	0
21	21	0.08	2	32	112	0.15	83	<.01	0.26	0.05	0.04	0.01	3	0
25	311	0.25	11	35	428	1.08	34	<.01	0.07	0.14	0.03	0.01	3	0
88	1109	0.9	38	125	1529	3.87	120	0.03	0.26	0.49	0.11	0.05	10	0
19	212	0.15	8	39	245	0.9	66	<.01	0.06	0.09	0.04	0.02	3	0
125	509	0.93	27	258	207	2.85	295	0.03	0.27	0.48	0.26	0.11	30	0
76	669	1.08	37	308	308	3.3	263	0.05	0.17	0.46	0.19	0.05	32	0
26	221	0.06	11	22	54	1.22	62	0.02	0.11	0.08	0.03	0.02	2	0
0	100			10	220									
0	60			50	80									
9	137	0.32	11	194	29	0.86	25	0.04	0.25	0.36	0.29	0.16	27	0
34	135	0.32	11	189	911	0.83	99	0.05	0.09	0.37	0.28	0.15	27	0
32	126	0.29	10	176	850	0.78	92	0.04	0.08	0.34	0.26	0.14	25	0
29	120	0.36	16	203	777	0.75	86	0.07	0.09	0.32	0.34	0.15	25	3
18	156	0.36	14	217	918	0.95	47	0.05	0.05	0.42	0.33	0.17	29	0
9	138	0.32	11	194	918	0.86	25	0.05	0.02	0.37	0.29	0.15	27	0
16	12	0.18	1	18	44	0.08	55	<.01	0.04	0.03	0.02	<.01	2	0
0	99			15	89									
26	155	0.05	8	47	417	0.8	71	<.01	0.08	0.1	0.02	0.02	2	0
24	216	0.17	8	60	384	1.01	55	<.01	0.1	0.08	0.01	0.01	2	0
46	325	0.32	18	166	721	1.82	115	0.04	0.13	0.29	0.1	0.04	4	0

A Appendix

Table A–1

Food Composition (Computer code number is for West Diet Analysis program)   (For purposes of calculations, use "0" for t, <1, <.1, <.01, etc.)

Computer Code Number	Food Description	Measure	Wt (g)	H₂O (%)	Ener (kcal)	Prot (g)	Carb (g)	Dietary Fiber (g)	Fat (g)	Fat Breakdown (g) Sat	Mono	Poly
	DAIRY—Continued											
1002	American cheese spread	1 tbs	15	48	44	2	1	0	3	2	0.9	t
1081	Nonfat cheese (Kraft Singles)	1 oz	28	61	44	6	4	0	0	0	0	0
1094	Velveeta cheese spread, low fat, low sodium, slice	1 pce	34	62	61	8	1	0	2	1.5	0.7	t
	Cream:											
	Sweet:											
69	Half & half (cream & milk)	1 c	242	81	315	7	10	0	28	17.3	8	1
500	Tablespoon	1 tbs	15	81	20	<1	1	0	2	1.1	0.5	t
501	Light, coffee or table	1 tbs	15	74	29	<1	1	0	3	1.8	0.8	0.1
511	Light, whipping cream, liquid	1 tbs	15	64	44	<1	<1	0	5	2.9	1.4	0.1
502	Heavy whipping cream, liquid	1 tbs	15	58	52	<1	<1	0	6	3.5	1.6	0.2
510	Whipped cream, pressurized	1 tbs	4	61	10	<1	<1	0	1	0.6	0.3	t
	Sour, cultured:											
79	Regular	1 c	230	71	492	7	10	0	48	30	13.9	1.8
504	Tablespoon	1 tbs	14	71	30	<1	1	0	3	1.8	0.8	0.1
577	Fat free	1 tbs	15	79	12	1	2	0	0	0	0	0
	Imitation and part-dairy:											
507	Coffee whitener, frozen or liquid	1 tbs	15	77	20	<1	2	0	1	1.4	t	t
506	Coffee whitener, powdered	1 tsp	2	2	11	<1	1	0	1	0.7	t	t
508	Dessert topping, frozen, nondairy	1 tbs	5	50	16	<1	1	0	1	1.1	t	t
509	Dessert topping, mix with whole milk	1 tbs	5	67	9	<1	1	0	1	0.5	t	t
514	Dessert topping, pressurized	1 c	70	60	185	1	11	0	16	13.2	1.3	0.2
505	Sour cream, imitation	1 tbs	14	71	29	<1	1	0	3	2.5	t	t
516	Sour dressing, part dairy	1 tbs	15	75	27	<1	1	0	2	2	0.3	t
	Milk: Fluid											
1	Whole milk	1 c	244	88	146	8	11	0	8	4.6	2	0.5
2	2% lowfat milk	1 c	244	89	122	8	11		5	2.3	2	0.2
147	2% milk solids added	1 c	245	89	125	9	12	0	5	2.9	1.4	0.2
4	1% lowfat milk	1 c	244	90	102	8	12		2	1.5	0.7	0.2
148	1% milk solids added	1 c	245	90	105	9	12	0	2	1.5	0.7	t
6	Nonfat milk, vitamin A added	1 c	245	91	83	8	12		<1	0.1	t	t
149	Nonfat milk solids added	1 c	245	90	91	9	12	0	1	0.4	0.2	t
7	Buttermilk, skim	1 c	245	90	98	8	12	0	2	1.3	0.6	t
	Canned:											
11	Sweetened condensed	1 c	306	27	982	24	166	0	27	16.8	7.4	1
10	Evaporated, nonfat	1 c	256	79	200	19	29	0	1	0.3	0.2	t
	Dried:											
32	Buttermilk, sweet	1 c	120	3	464	41	59	0	7	4.3	2	0.3
9	Instant, nonfat, vit A added (makes 1 qt)	1 ea	91	4	326	32	47	0	1	0.4	0.2	t
8	Instant nonfat, vit A added, cup	1 c	68	4	243	24	35	0	<1	0.3	0.1	t
23	Goat milk	1 c	244	87	168	9	11	0	10	6.5	2.7	0.4
51	Kefir	1 c	233	88	149	8	11	0	8			
	Milk beverages and powdered mixes:											
	Chocolate:											
20	Whole	1 c	250	82	208	8	26	2	8	5.3	2.5	0.3
18	2% fat	1 c	250	84	180	8	26	1	5	3.1	1.5	0.2
19	1% fat	1 c	250	84	158	8	26	1	2	1.5	0.8	t
	Chocolate-flavored beverages:											
12	Powder containing nonfat dry milk	1 oz	28	2	111	2	24	1	1	0.7	0.4	t
48	Prepared with water	1 c	275	86	151	2	32	1	2	0.9	0.5	t
14	Powder without nonfat dry milk:	1 oz	28	1	98	1	25	1	1	0.5	0.3	t
39	Prepared with whole milk	1 c	266	81	226	9	32	1	9	4.9	2.2	0.5
	Eggnog:											
17	Commercial	1 c	254	74	343	10	34	0	19	11.3	5.7	0.9
98	2% low-fat	1 c	254	85	191	12	17	0	8	3.7	2.7	0.7
	Instant breakfast:											
24	Envelope, pwd only	1 ea	37	7	131	7	24		1	0.2	0.1	0.1
25	Prepared with whole milk	1 c	281	78	253	15	36	<1	5	3.1		
26	Prepared with 2% milk	1 c	281	77	280	15	36	<1	9	5.3		

PAGE KEY: A–4 = Beverages   A–6 = Dairy   A–10 = Eggs   A–12 = Fat/Oil   A–14 = Fruit   A–20 = Bakery
A–26 = Grain   A–32 = Fish   A–32 = Meats   A–36 = Poultry   A–38 = Sausage   A–40 = Mixed/Fast   A–44 = Nuts/Seeds   A–46 = Sweets
A–48 = Vegetables/Legumes   A–58 = Vegetarian   A–62 = Misc   A–62 = Soups/Sauces   A–64 = Fast   A–78 = Convenience   A–80 = Baby foods

Chol (mg)	Calc (mg)	Iron (mg)	Magn (mg)	Pota (mg)	Sodi (mg)	Zinc (mg)	VT-A (RAE)	Thia (mg)	VT-E (mg α)	Ribo (mg)	Niac (mg)	V-B$_6$ (mg)	Fola (µg)	VT-C (mg)
8	84	0.05	4	36	202	0.39	26	<.01	0.03	0.06	0.02	0.02	1	0
7	221	0		88	398	0.88				0.15				0
12	233	0.15	8	61	2	1.13		0.01	0.17	0.13	0.03	0.03	3	0
90	254	0.17	24	315	99	1.23	235	0.08	0.8	0.36	0.19	0.09	7	2
6	16	0.01	2	20	6	0.08	15	<.01	0.05	0.02	0.01	<.01	<1	0
10	14	<.01	1	18	6	0.04	27	<.01	0.08	0.02	<.01	<.01	<1	0
17	10	<.01	1	15	5	0.04	42	<.01	0.13	0.02	<.01	<.01	1	0
21	10	<.01	1	11	6	0.03	62	<.01	0.16	0.02	<.01	<.01	1	0
3	4	<.01	0	6	5	0.01	8	<.01	0.03	<.01	<.01	<.01	<1	0
101	267	0.14	25	331	122	0.62	407	0.08	1.38	0.34	0.15	0.04	25	2
6	16	<.01	2	20	7	0.04	25	<.01	0.08	0.02	<.01	<.01	2	0
2	28	0			23									0
0	1	<.01	0	29	12	<.01	1	0	0.24	0	0	0	0	0
0	0	0.02	0	16	4	0.01	<1	0	0.01	<.01	0	0	0	0
0	0	<.01	0	1	1	<.01	<1	0	0.05	0	0	0	0	0
0	4	<.01	0	8	3	0.01	1	<.01	0.02	<.01	<.01	<.01	<1	0
0	4	0.01	1	13	43	<.01	3	0	0.6	0	0	0	0	0
0	0	0.05	1	23	14	0.17	0	0	0.1	0	0	0	0	0
1	17	<.01	2	24	7	0.06	<1	<.01	0.2	0.02	0.01	<.01	2	0
24	246	0.07	24	325	105	0.93	68	0.11	0.15	0.45	0.26	0.09	12	0
20	271	0.24	27	342	115	1.17	134	0.1	0.07	0.45	0.22	0.09	12	0
20	314	0.12	34	397	127	0.98	137	0.1	0.1	0.42	0.22	0.11	12	2
12	264	0.85	27	290	122	2.12	142	0.05	0.02	0.45	0.23	0.09	12	0
10	314	0.12	34	397	127	0.98	145	0.1	0.1	0.42	0.22	0.11	12	2
5	223	1.23	22	238	108	2.08	149	0.11	0.02	0.45	0.23	0.09	12	0
5	316	0.12	37	419	130	1	149	0.1	0	0.43	0.22	0.11	12	2
10	284	0.12	27	370	257	1.03	17	0.08	0.12	0.38	0.14	0.08	12	2
104	869	0.58	80	1135	389	2.88	226	0.28	0.49	1.27	0.64	0.16	34	8
10	742	0.74	69	850	294	2.3	302	0.12	0	0.79	0.45	0.14	23	3
83	1421	0.36	132	1910	620	4.82	59	0.47	0.12	1.89	1.05	0.41	56	7
16	1120	0.28	106	1552	500	4.01	645	0.38	<.01	1.59	0.81	0.31	46	5
12	837	0.21	80	1159	373	3	482	0.28	<.01	1.19	0.61	0.23	34	4
27	327	0.12	34	498	122	0.73	139	0.12	0.17	0.34	0.68	0.11	2	3
		0.3	33	373	107									
30	280	0.6	32	418	150	1.02	65	0.09	0.15	0.4	0.31	0.1	12	2
18	285	0.6	32	422	150	1.02	138	0.09	0.1	0.41	0.32	0.1	12	2
8	288	0.6	32	425	152	1.02	145	0.1	0.05	0.42	0.32	0.1	12	2
2	39	0.33	23	199	141	0.41	<1	0.03	0.15	0.16	0.16	0.03	0	1
3	60	0.47	33	270	195	0.58	<1	0.04	0.19	0.21	0.22	0.04	0	1
0	10	0.88	27	165	59	0.43	<1	<.01	0.01	0.04	0.14	<.01	2	0
24	253	0.8	48	458	154	1.28	69	0.11	0.16	0.48	0.38	0.09	13	0
150	330	0.51	48	419	137	1.17	114	0.09	0.51	0.48	0.27	0.13	3	4
194	270	0.71	32	369	155	1.26		0.11	0.58	0.55	0.21	0.15	30	2
4	105	4.74	84	350	142	3.16		0.31	5.31	0.07	5.27	0.42	105	28
24	403	4.87	119	726	264	4.12		0.41	5.48	0.48	5.48	0.53	118	31
38	396	4.87	117	721	262	4.09	0	0.41	5.55	0.47	5.47	0.52	118	31

A Appendix

Table A–1

Food Composition

(Computer code number is for West Diet Analysis program)   (For purposes of calculations, use "0" for t, <1, <.1, <.01, etc.)

Computer Code Number	Food Description	Measure	Wt (g)	H₂O (%)	Ener (kcal)	Prot (g)	Carb (g)	Dietary Fiber (g)	Fat (g)	Fat Breakdown (g) Sat	Mono	Poly
	DAIRY—Continued											
101	Prepared with 1% milk	1 c	281	79	233	15	36	<1	3	1.8		
27	Prepared with nonfat milk	1 c	282	80	216	16	36		1	0.7		
	Malted Milk:											
30	Chocolate Powder	3 tsp	21	1	79	1	18	1	1	0.5	0.2	t
34	Prepared with whole milk	1 c	265	81	225	9	30	1	9	5	2.2	0.6
38	Ovaltine with whole milk	1 c	265	81	223	9	29	1	9	5	2.2	0.5
28	Natural Powder:	3 tsp	21	2	87	2	16	<1	2	0.9	0.4	0.3
29	Prepared with whole milk	1 c	265	81	233	10	27	<1	10	5.4	2.4	0.7
	Milk Shakes:											
2020	Chocolate	1 c	166	72	211	6	34	3	6	3.8	1.8	0.2
2024	Vanilla	1 c	166	75	184	6	30	<1	5	3.1	1.4	0.2
	Milk Desserts:											
2062	Low-fat frozen dessert bars	1 ea	81	72	88	2	19	0	1	0.2	0.1	0.4
	Ice cream, vanilla (about 10% fat):											
2004	Hardened	1 c	132	61	265	5	31	1	15	9	3.9	0.6
2008	Soft serve	1 c	172	60	382	7	38	1	22	12.9	6	0.8
	Ice cream, rich vanilla (16% fat):											
2006	Hardened	1 c	148	57	369	5	33	0	24	15.3	6.6	1
2220	Ben & Jerry's	½ c	108		250	4	26	0	16	11		
	Ice milk, vanilla (about 4% fat):											
2009	Hardened	1 c	132	63	218	7	35	<1	5	3.4	1.1	0.2
2010	Soft serve (about 3.3% fat)	1 c	176	70	222	9	38	0	5	2.9	1.3	0.2
	Pudding, canned (5 oz can = .55 cup):											
2610	Chocolate	1 ea	142	69	197	4	33	1	6	1	2.4	2
2611	Tapioca	1 ea	142	74	169	3	28	<1	5	0.9	2.2	1.9
2612	Vanilla	1 ea	142	71	183	3	31	0	5	0.8	2.2	1.9
	Puddings, dry mix with whole milk:											
2604	Chocolate, regular, cooked	1 c	284	73	338	9	55	2	9	5.1	2.3	0.5
2606	Rice, cooked	1 c	288	72	348	9	60	<1	8	4.8	2.2	0.3
2607	Tapioca, cooked	1 c	282	74	324	8	55	0	8	4.9	2.2	0.3
2609	Vanilla, regular, cooked	1 c	280	75	314	8	52	<1	8	4.6	2	0.6
2011	Sherbet (2% fat)	1 c	198	66	285	2	60	7	4	2.3	1	0.2
	Frozen Yogurt, Low-Fat:											
2064	Cup	1 c	144	65	235	6	35	0	8	4.9	2.3	0.3
2082	Scoop	1 ea	79	74	78	4	16	0	<1	t	t	t
	Milk Substitutes:											
20440	Rice Milk	1 c	245	89	120	<1	25	0	2	0.2	1.3	0.3
20590	Rice/Soy Milk, blend	1 c	241	88	120	7	18	0	3	0.5		
20033	Soy Milk	1 c	245	89	120	9	11	3	5	0.5	0.8	2
7785	Edensoy	1 c	244	89	130	10	13	0	4	0.5	0.8	2.7
7804	Vanilla	1 c	244	87	150	6	23	0	3	0		
20693	Soy Dream	1 c	244		128	7	17	0	4	0.5		
483	Enriched	1 c	244		128	7	17	0	4	0.5		
20404	Veggie original	1 c	227		110	9	13		3	0		
20405	Chocolate	1 c	227		150	9	26		2	0		
	Yogurt:											
	Fat Free:											
2852	Strawberry, container	1 ea	227	86	120	8	22	0	0	0	0	0
2574	Vanilla	1 c	245	76	223	13	43	0	<1	0.3	0.1	t
2012	Plain	1 c	245	85	137	14	19	0	<1	0.3	0.1	t
	Low-Fat:											
2034	Fruit added with low-calorie sweetener	1 c	241	86	122	11	19	1	<1	0.2	t	t
2001	Fruit added	1 c	245	74	250	11	47	0	3	1.7	0.7	t
2000	Plain	1 c	245	85	154	13	17	0	4	2.4	1	0.1
2015	Vanilla or coffee flavor	1 c	245	79	208	12	34	0	3	2	0.8	t
2013	Whole Milk:	1 c	245	88	149	9	11	0	8	5.1	2.2	0.2
	EGGS											
	Raw, Large:											
19501	Whole, without shell	1 ea	50	76	74	6	<1	0	5	1.5	1.9	0.7
19506	White	1 ea	33	88	17	4	<1	0	<1	0	0	0

Appendix **A**

PAGE KEY: A–4 = Beverages   A–6 = Dairy   A–10 = Eggs   A–12 = Fat/Oil   A–14 = Fruit   A–20 = Bakery
A–26 = Grain   A–32 = Fish   A–32 = Meats   A–36 = Poultry   A–38 = Sausage   A–40 = Mixed/Fast   A–44 = Nuts/Seeds   A–46 = Sweets
A–48 = Vegetables/Legumes   A–58 = Vegetarian   A–62 = Misc   A–62 = Soups/Sauces   A–64 = Fast   A–78 = Convenience   A–80 = Baby foods

Chol (mg)	Calc (mg)	Iron (mg)	Magn (mg)	Pota (mg)	Sodi (mg)	Zinc (mg)	VT-A (RAE)	Thia (mg)	VT-E (mg α)	Ribo (mg)	Niac (mg)	V-B$_6$ (mg)	Fola (µg)	VT-C (mg)
14	406	4.87	119	731	267	4.12		0.41	5.4	0.48	5.48	0.53	118	31
9	407	4.83	112	755	268	4.14		0.4	5.3	0.42	5.47	0.52	118	31
0	13	0.48	15	130	53	0.17	1	0.04	0.02	0.04	0.42	0.03	11	0
26	260	0.56	40	456	159	1.09	69	0.14	0.16	0.49	0.69	0.12	24	0
26	339	3.76	45	578	231	1.17	904	0.76	0.16	1.32	11.08	1.01	19	32
7	63	0.15	20	159	104	0.21	17	0.11	0.07	0.19	1.1	0.09	10	1
32	310	0.24	45	485	209	1.14	87	0.21	0.32	0.64	1.38	0.17	21	1
22	188	0.51	28	332	161	0.68	43	0.1	0.18	0.41	0.27	0.08	8	1
18	203	0.15	20	289	136	0.6	61	0.07	0.1	0.3	0.31	0.09	8	1
1	81	0.04	9	107	44	0.26	38	0.03	0.07	0.11	0.06	0.03	3	1
58	169	0.12	18	263	106	0.91	156	0.05	0.4	0.32	0.15	0.06	7	1
157	225	0.36	21	304	105	0.89	279	0.08	1.05	0.31	0.16	0.08	15	1
136	173	0.5	16	232	90	0.7	269	0.06	0.75	0.25	0.12	0.07	12	0
75	100	0.36			60									0
33	153	0.11	18	275	98	0.96	182	0.05	0.12	0.23	0.13	0.04	7	0
21	276	0.11	25	389	123	0.93	51	0.09	0.11	0.35	0.21	0.08	11	2
4	128	0.72	30	256	183	0.6	14	0.04	0.41	0.22	0.49	0.04	4	3
1	119	0.33	11	136	226	0.38	<1	0.03	0.43	0.14	0.44	0.03	4	1
10	125	0.18	11	160	192	0.36	9	0.03	0	0.2	0.36	0.02	0	0
26	273	0.99	57	426	278	1.36	68	0.1	0.14	0.55	0.42	0.09	11	0
32	297	1.07	37	369	311	1.09	84	0.21	0.24	0.4	1.27	0.1	12	2
34	290	0.17	34	369	338	0.96	70	0.08	0.23	0.39	0.21	0.09	11	2
25	249	0.11	25	333	437	0.95	70	0.1	0.14	0.45	0.26	0.08	11	0
0	107	0.28	16	190	91	0.95	20	0.06	0.06	0.18	0.15	0.05	14	11
3	206	0.43	20	304	125	0.6	85	0.05	0.16	0.32	0.41	0.12	9	1
1	137	0.07	13	175	53	0.67	1	0.03	<.01	0.16	0.09	0.04	8	1
0	20	0.2	10	69	86	0.24	<1	0.08	1.76	0.01	1.91	0.04	91	1
0	13	1.08	40	270	85	0.9	0	0.09		0.1	0.4	0.08	26	
0	10	1.42	47	345	29	0.56	5	0.39	0.02	0.17	0.36	0.1	5	0
0	80	1.44	60	440	105	0.9	0	0.15		0.07	0.8	0.16	40	0
0	60	0.72	40	290	90	0.6	0	0.12		0.07	1.2	0.12	40	0
0	40	1.78	40	237	138	0.59	0	0.15		0.07	0.79	0.12	59	0
0	295	1.77	39	236	138	0.59		0.15	4.95	0.07	0.79	0.12	59	0
0	400			350	90									
0	401			351	130									
5	350	0		380	160		0							5
4	436	0.21	42	559	168	2.13	4	0.11	<.01	0.52	0.27	0.12	27	2
5	488	0.22	47	625	189	2.38	5	0.12	0	0.57	0.3	0.13	29	2
3	370	0.61	41	550	139	1.83		0.1	0.17	0.45	0.5	0.11	32	26
10	372	0.17	37	478	142	1.81	24	0.09	0.05	0.44	0.23	0.1	22	2
15	448	0.2	42	573	172	2.18	34	0.11	0.07	0.52	0.28	0.12	27	2
12	419	0.17	39	537	162	2.03	29	0.1	0.05	0.49	0.26	0.11	27	2
32	296	0.12	29	380	113	1.45	66	0.07	0.15	0.35	0.18	0.08	17	1
212	26	0.92	6	67	70	0.56	70	0.03	0.48	0.24	0.04	0.07	24	0
0	2	0.03	4	54	55	<.01	0	<.01	0	0.14	0.03	<.01	1	0

A Appendix

Table A–1

Food Composition

(Computer code number is for West Diet Analysis program)   (For purposes of calculations, use "0" for t, <1, <.1, <.01, etc.)

Computer Code Number	Food Description	Measure	Wt (g)	H₂O (%)	Ener (kcal)	Prot (g)	Carb (g)	Dietary Fiber (g)	Fat (g)	Fat Breakdown (g) Sat	Mono	Poly
	EGGS—Continued											
19508	Yolk	1 ea	17	52	55	3	1	0	5	1.6	2	0.7
	Cooked:											
19509	Fried in margarine	1 ea	46	69	92	6	<1	0	7	2	2.9	1.2
19510	Hard-cooked, shell removed	1 ea	50	75	78	6	1	0	5	1.6	2	0.7
19511	Hard-cooked, chopped	1 c	136	75	211	17	2	0	14	4.4	5.5	1.9
19517	Poached, no added salt	1 ea	50	76	74	6	<1		5	1.5	1.9	0.7
19516	Scrambled with milk & margarine	1 ea	61	73	101	7	1	0	7	2.2	2.9	1.3
	Substitute, liquid:											
19525	Egg substitute, liquid:	½ c	126	83	106	15	1	0	4	0.8	1.1	2
19581	Egg Beaters, Fleischmann's	½ c	122		60	12	2	0	0	0	0	0
19552	Egg substitute, liquid, prepared	½ c	105	80	107	12	2	0	6	1.1	2.1	2.1
	FATS AND OILS											
	Butter:											
8000	Tablespoon:	1 tbs	14	16	100	<1	<1	0	11	5.7	4.7	0.4
8025	Unsalted	1 tbs	14	18	100	<1	<1	0	11	7.2	2.9	0.4
8001	Pat (about 1 tsp)	1 ea	5	16	36	<1	<1	0	4	2	1.7	0.1
8142	Whipped	1 tsp	3	16	22	<1	<1	0	2	1.5	0.7	t
	Fats, cooking:											
8004	Beef fat/tallow	1 c	205	0	1849	0	0	0	205	102.1	85.7	8.2
8005	Chicken fat	1 c	205	0	1845	0	0	0	205	61.1	91.6	42.8
8007	Vegetable shortening	1 tbs	13	0	115	0	0	0	13	3.2	5.8	3.4
	Margarine:											
8041	Imitation (about 40% fat), soft	1 tbs	14	58	48	<1	<1	0	5	1.1	2.2	1.9
90234	Regular, hard (about 80% fat)	1 tbs	14	16	101	<1	<1	0	11	2.2	5	3.6
8043	Regular, soft (about 80% fat)	1 tbs	14	16	100	<1	<1	0	11	1.9	4	4.8
8485	Saffola, unsalted	1 tbs	14	20	100	0	0	0	11	2	3	4.5
8486	Saffola, reduced fat	1 tbs	14	37	60	0	0	0	8	1.3	2.7	4.4
	Spread:											
8044	Hard (about 60% fat)	1 tbs	14	37	76	<1	0	0	9	2	3.6	2.5
8045	Soft (about 60% fat)	1 tbs	14	37	76	<1	0	0	9	1.8	4.4	1.9
8602	Touch of Butter (47% fat)	1 tbs	14		60	0	0	0	7	1.5		
	Oils:											
8084	Canola	1 tbs	14	0	124	0	0	0	14	1	8.2	4.1
8009	Corn	1 tbs	14	0	124	0	0	0	14	1.8	3.4	8.2
8008	Olive	1 tbs	14	0	124	0	0	0	14	1.9	10.3	1.4
8361	Olive, extra virgin	1 tbs	14	0	126	0	0	0	14	2	10.8	1.3
8026	Peanut	1 tbs	14	0	124	0	0	0	14	2.4	6.5	4.5
8010	Safflower	1 tbs	14	0	124	0	0	0	14	0.9	2	10.4
8012	Soybean	1 tbs	14	0	124	0	0	0	14	2	3.3	8.1
8028	Soybean/cottonseed	1 tbs	14	0	124	0	0	0	14	2.5	4.1	6.7
8011	Sunflower	1 tbs	14	0	124	0	0	0	14	1.4	2.7	9.2
	Salad Dressings/Sandwich Spreads:											
	Store Brand:											
8013	Blue Cheese	1 tbs	15	32	76	1	1	0	8	1.5	1.8	4.2
	French:											
8015	Regular	1 tbs	16	37	73	<1	2		7	0.9	1.3	3.4
8014	Low calorie	1 tbs	16	54	37	<1	5	<1	2	0.2	0.9	0.8
	Italian:											
8020	Regular	1 tbs	15	56	44	<1	2	0	4	0.7	0.9	1.9
8016	Low calorie	1 tbs	15	85	11	<1	1	0	1	t	0.3	0.3
	Ranch:											
8428	Regular	1 tbs	15		80	0	<1	0	8	1.2		
8465	Low calorie	1 tbs	14		22	<1	1	<1	2	0.3		
8022	Russian	1 tbs	15	34	74	<1	2	0	8	1.1	1.8	4.4
8021	Mayo Type	1 tbs	15	40	58	<1	4	0	5	0.7	1.4	2.7
	Mayonnaise:											
8032	Imitation, low calorie	1 tbs	15	63	35	<1	2	0	3	0.5	0.7	1.6
8046	Regular (soybean)	1 tbs	14	15	100	<1	1	0	11	1.7	2.7	6
8148	Regular, low calorie, low sodium	1 tbs	14	63	32	<1	2	0	3	0.5	0.6	1.5
8141	Salad dressing, low calorie, oil free	1 tbs	15	88	4	<1	1	<1	<1	t	0	t

Appendix **A**

PAGE KEY: A–4 = Beverages   A–6 = Dairy   A–10 = Eggs   A–12 = Fat/Oil   A–14 = Fruit   A–20 = Bakery
A–26 = Grain   A–32 = Fish   A–32 = Meats   A–36 = Poultry   A–38 = Sausage   A–40 = Mixed/Fast   A–44 = Nuts/Seeds   A–46 = Sweets
A–48 = Vegetables/Legumes   A–58 = Vegetarian   A–62 = Misc   A–62 = Soups/Sauces   A–64 = Fast   A–78 = Convenience   A–80 = Baby foods

Chol (mg)	Calc (mg)	Iron (mg)	Magn (mg)	Pota (mg)	Sodi (mg)	Zinc (mg)	VT-A (RAE)	Thia (mg)	VT-E (mg α)	Ribo (mg)	Niac (mg)	V-B$_6$ (mg)	Fola (µg)	VT-C (mg)
210	22	0.46	1	19	8	0.39	65	0.03	0.44	0.09	<.01	0.06	25	0
210	27	0.91	6	68	94	0.55	91	0.03	0.56	0.24	0.04	0.07	23	0
212	25	0.6	5	63	62	0.52	84	0.03	0.52	0.26	0.03	0.06	22	0
577	68	1.62	14	171	169	1.43	230	0.09	1.4	0.7	0.09	0.16	60	0
211	26	0.92	6	66	147	0.55	70	0.03	0.48	0.24	0.04	0.07	24	0
215	43	0.73	7	84	171	0.61	87	0.03	0.52	0.27	0.05	0.07	18	0
1	67	2.65	11	416	223	1.64	23	0.14	0.34	0.38	0.14	<.01	19	0
0	40	2.16		170	250	1.2			1.61	1.7		0.16	64	0
1	83	1.85	11	337	201	1.25		0.09	0.83	0.29	0.12	0.01	11	0
30	3	<.01	0	3	81	0.01	96	<.01	0.32	<.01	<.01	<.01	<1	0
30	3	<.01	0	3	2	0.01	96	<.01	0.32	<.01	<.01	<.01	<1	0
11	1	<.01	0	1	29	<.01	34	<.01	0.12	<.01	<.01	<.01	<1	0
7	1	<.01	0	1	25	<.01	21	<.01	0.07	<.01	<.01	<.01	<1	0
223	0	0	0	0	0	0	0	0	5.54	0	0	0	0	0
174	0	0	0	0	0	0	0	0	5.54	0	0	0	0	0
0	0	0	0	0	0	0	0	0	0.1	0	0	0	0	0
0	3	0	0	4	134	0	115	<.01	0.33	<.01	<.01	<.01	<1	0
0	4	<.01	0	6	132	0	115	<.01	1.26	<.01	<.01	<.01	<1	0
0	4	0	0	5	151	0	102	<.01	0.98	<.01	<.01	<.01	<1	0
0	0			0									0	
0	0			115									0	
0	3	0	0	4	139	0	102	<.01	0.7	<.01	<.01	<.01	<1	0
0	3	0	0	4	139	0	102	<.01	0.7	<.01	<.01	<.01	<1	0
0	0	0	0	0	110				1.27					0
0	0	0	0	0	0	0	0	0	2.39	0	0	0	0	0
0	0	0	0	0	0	0	0	0	2	0	0	0	0	0
0	0	0.09	0	0	<1	0	0	0	2.01	0	0	0	0	0
									1.74					
0	0	<.01	0	0	0	<.01	0	0	2.2	0	0	0	0	0
0	0	0	0	0	0	0	0	0	4.77	0	0	0	0	0
0	0	<.01	0	0	0	0	0	0	1.29	0	0	0	0	0
0	0	0	0	0	0	0	0	0	1.69	0	0	0	0	0
0	0	0	0	0	0	0	0	0	5.75	0	0	0	0	0
3	12	0.03	0	6	164	0.04	10	<.01	0.9	0.02	0.02	<.01	4	0
0	4	0.13	1	11	134	0.05	4	<.01	0.8	<.01	0.03	0	0	0
0	2	0.14	1	17	129	0.03	4	<.01	0.05	<.01	0.07	<.01	<1	0
0	1	0.09	0	7	248	0.02	<1	<.01	0.75	<.01	0	<.01	0	0
1	1	0.1	1	13	205	0.03	<1	0	0.03	<.01	0	0.01	0	0
5	0	0			105		0							
3	12	0		16	120		0		0.7					0
3	3	0.09	0	24	130	0.06	2	<.01	0.6	<.01	0.09	<.01	2	1
4	2	0.03	0	1	107	0.03	10	<.01	0.31	<.01	<.01	<.01	1	0
4	0	0	0	2	75	0.02	0	0	0.3	0	0	0	0	0
5	3	0.07	0	5	80	0.02	12	0	0.73	0	<.01	0.08	1	0
3	0	0	0	1	15	0.02	1	0	0.53	<.01	0	0	<1	0
0	1	0.05	2	7	256	<.01	<1	<.01	<.01	<.01	<.01	<.01	<1	0

A   Appendix

Table A–1

Food Composition

(Computer code number is for West Diet Analysis program)   (For purposes of calculations, use "0" for t, <1, <.1, <.01, etc.)

Computer Code Number	Food Description	Measure	Wt (g)	H₂O (%)	Ener (kcal)	Prot (g)	Carb (g)	Dietary Fiber (g)	Fat (g)	Fat Breakdown (g) Sat	Mono	Poly
	FATS AND OILS—Continued											
8034	Salad dressing, from recipe, cooked	1 tbs	16	69	25	1	2	0	2	0.5	0.6	0.3
53122	Tartar Sauce, low calorie	1 tbs	14	63	31	<1	2	<1	3	0.4	0.6	1.4
	Thousand Island	1 tbs	16									
8024	Regular	1 tbs	16	47	59	<1	2		6	0.8	1.3	2.9
8023	Low Calorie	1 tbs	15	61	31	<1	3	<1	2	0.1	1	0.4
8035	Vinegar and oil	1 tbs	16	47	72	0	<1	0	8	1.5	2.4	3.9
	Kraft, Deliciously Right:											
8563	1000 Island	1 tbs	16		34	0	3	0	2	0.2		
8574	Cucumber ranch	1 tbs	16		31	0	1	0	3	0.5		
	Wishbone:											
8442	Creamy Italian, lite	1 tbs	15	72	26	<1	2		2	0.4	0.9	0.7
8413	Italian, lite	1 tbs	16		6	0	1		<1	0	0.2	0.1
8427	Ranch, lite	1 tbs	15		50	0	2	0	4	0.8		
	FRUITS and FRUIT JUICES											
	Apples:											
	Fresh, w/peel:											
3000	2¾" diam (about 3/lb w/cores)	1 ea	138	86	72	<1	19	3	<1	t	t	t
3001	3¼" diam (about 2/lb w/cores)	1 ea	212	86	110	1	29	5	<1	t	t	0.1
3004	Slices	1 c	110	87	53	<1	14	1	<1	t	t	t
3005	Dried, sulfured	10 ea	64	32	156	1	42	6	<1	t	t	t
3576	Cup	¼ c	40	30	107	<1	26	3	<1	0		
3008	Juice, bottled or canned	1 c	248	88	117	<1	29	<1	<1	t	t	t
3147	Applesauce, sweetened	1 c	255	80	194	<1	51	3	<1	t	t	0.1
3006	Applesauce, unsweetened	1 c	244	88	105	<1	28	3	<1	t	t	t
	Apricots:											
3157	Fresh, w/o pits (about 12 per lb w/pits)	3 ea	105	86	50	1	12		<1	t	0.2	t
	Canned (fruit and liquid):											
3011	Heavy syrup	1 c	240	78	199	1	52	4	<1	t	t	t
3633	Halves	3 ea	120	78	100	1	26	2	<1	t	t	t
3151	Juice pack	1 c	244	87	117	2	30	4	<1	t	t	t
3152	Halves	3 ea	108	87	52	1	13	2	<1	t	t	t
3013	Dried, halves	10 ea	35	31	84	1	22	3	<1	t	t	t
3580	Cup	¼ c	40	30	107	1	25	2	<1	0		
3217	Dried, cooked, unsweetened, w/liquid	1 c	250	76	212	3	55		<1	t	t	t
3015	Nectar, canned	1 c	251	85	141	1	36	2	<1	t	t	t
	Avocados, raw, edible part only:											
3210	California	1 ea	173	72	289	3	15	12	27	3.7	17	3.5
3212	Florida	1 ea	304	79	365	7	24		31	6	16.8	5.1
3017	Mashed, fresh, average	1 c	230	73	368	5	20	15	34	4.9	22.5	4.2
	Bananas:											
	Fresh, w/o peel:											
3020	Whole, 8¾ long (175g w/peel)	1 ea	118	75	105	1	27	3	<1	0.1	t	t
3021	Slices	1 c	150	75	134	2	34	4	<1	0.2	t	0.1
3023	Dehydrated slices	½ c	50	3	173	2	44	5	1	0.3	t	0.2
	Berries:											
3024	Blackberries, raw	1 c	144	88	62	2	14		1	t	t	0.4
	Blueberries:											
3029	Fresh	1 c	145	84	83	1	21	3	<1	t	t	0.2
3232	Frozen, sweetened	10 oz	284	77	230	1	62	6	<1	t	t	0.2
3231	Frozen, thawed	1 c	230	77	186	1	50	5	<1	t	t	0.1
	Cranberries:											
3042	Juice cocktail, vitamin C added	1 c	253	86	144	0	36	<1	<1	t	t	0.1
3276	Juice, low calorie	1 c	237	95	45	<1	11	0	<1	0	0	0
3223	Cranberry-apple juice, vitamin C added	1 c	245	82	174	<1	44	<1	<1	0	0	0
3040	Sauce, canned, strained	1 c	277	61	418	1	108	3	<1	t	t	0.2
3865	Dried	⅓ c	40		120	0	33	2	0	0	0	0
	Raspberries:											
3131	Fresh	1 c	123	86	64	1	15	8	1	t	t	0.5

PAGE KEY: A–4 = Beverages A–6 = Dairy A–10 = Eggs A–12 = Fat/Oil A–14 = Fruit A–20 = Bakery
A–26 = Grain A–32 = Fish A–32 = Meats A–36 = Poultry A–38 = Sausage A–40 = Mixed/Fast A–44 = Nuts/Seeds A–46 = Sweets
A–48 = Vegetables/Legumes A–58 = Vegetarian A–62 = Misc A–62 = Soups/Sauces A–64 = Fast A–78 = Convenience A–80 = Baby foods

Chol (mg)	Calc (mg)	Iron (mg)	Magn (mg)	Pota (mg)	Sodi (mg)	Zinc (mg)	VT-A (RAE)	Thia (mg)	VT-E (mg α)	Ribo (mg)	Niac (mg)	V-B$_6$ (mg)	Fola (µg)	VT-C (mg)
9	13	0.08	1	19	117	0.06	8	<.01	0.13	0.02	0.04	<.01	3	0
3	1	0.06	0	4	82	0.02	1	<.01	0.84	<.01	<.01	<.01	<1	0
4	3	0.19	1	17	138	0.04	2	0.23	0.18	<.01	0.07	0	0	0
0	2	0.14	1	30	125	0.03	2	<.01	0.15	<.01	0.07	0	0	0
0	0	0	0	1	<1	0	0	0	0.74	0	0	0	0	0
5	0	0		29	165		0							0
0	0	0		10	248		0							0
0	0	0			148		0	0	0.56	0	0			0
0	1	0			255			0	0.24	0	0			0
2	0	0			120		0							0
0	8	0.17	7	148	1	0.06	4	0.02	0.25	0.04	0.13	0.06	4	6
0	13	0.25	11	227	2	0.08	6	0.04	0.38	0.06	0.19	0.09	6	10
0	6	0.08	4	99	0	0.06	2	0.02	0.06	0.03	0.1	0.04	0	4
0	9	0.9	10	288	56	0.13	0	0	0.34	0.1	0.59	0.08	0	2
0	9	0.67		178	251		<1							1
0	17	0.92	7	295	7	0.07	<1	0.05	0.02	0.04	0.25	0.07	0	2
0	10	0.89	8	156	8	0.1	3	0.03	0.54	0.07	0.48	0.07	3	4
0	7	0.29	7	183	5	0.07	2	0.03	0.51	0.06	0.46	0.06	2	3
0	14	0.41	10	272	1	0.21	101	0.03	0.93	0.04	0.63	0.06	9	10
0	22	0.72	17	336	10	0.26	149	0.05	1.44	0.05	0.9	0.13	5	7
0	11	0.36	8	168	5	0.13	74	0.02	0.72	0.03	0.45	0.06	2	4
0	29	0.73	24	403	10	0.27	207	0.04	1.46	0.05	0.84	0.13	5	12
0	13	0.32	11	178	4	0.12	92	0.02	0.65	0.02	0.37	0.06	2	5
0	19	0.93	11	407	4	0.14	63	<.01	1.52	0.03	0.91	0.05	4	0
0	20	1.56		520	1		16							6
0	48	2.35	28	1028	10	0.35	160	0.01	3.82	0.06	2.29	0.13	8	1
0	18	0.95	13	286	8	0.23	166	0.02	0.78	0.04	0.65	0.06	3	2
0	22	1.06	50	877	14	1.18	12	0.13	3.41	0.25	3.31	0.5	107	15
0	30	0.52	73	1067	6	1.22	21	0.06	8.09	0.16	2.04	0.24	106	53
0	28	1.26	67	1116	16	1.47	16	0.15	4.76	0.3	4	0.59	133	23
0	6	0.31	32	422	1	0.18	4	0.04	0.12	0.09	0.78	0.43	24	10
0	8	0.39	40	537	2	0.22	4	0.05	0.15	0.11	1	0.55	30	13
0	11	0.57	54	746	2	0.3	6	0.09	0.2	0.12	1.4	0.22	7	4
0	42	0.89	29	233	1	0.76	16	0.03	1.68	0.04	0.93	0.04	36	30
0	9	0.41	9	112	1	0.23	4	0.05	0.83	0.06	0.61	0.08	9	14
0	17	1.11	6	170	3	0.17	6	0.06	1.48	0.15	0.72	0.17	20	3
0	14	0.9	5	138	2	0.14	5	0.05	1.2	0.12	0.58	0.14	16	2
0	8	0.38	5	46	5	0.18	1	0.02	0	0.02	0.09	0.05	0	90
0	21	0.09	5	59	7	0.05	<1	0	0.12	<.01	<.01	<.01	0	76
0	12	0.29	5	69	17	0.44	<1	0.01	0	0.05	0.15	0.05	0	78
0	11	0.61	8	72	80	0.14	6	0.04	2.3	0.06	0.28	0.04	3	6
0	0	0			0		0							0
0	31	0.85	27	186	1	0.52	2	0.04	1.07	0.05	0.74	0.07	26	32

Table A–1

Food Composition   (Computer code number is for West Diet Analysis program)   (For purposes of calculations, use "0" for t, <1, <.1, <.01, etc.)

Computer Code Number	Food Description	Measure	Wt (g)	H₂O (%)	Ener (kcal)	Prot (g)	Carb (g)	Dietary Fiber (g)	Fat (g)	Fat Breakdown (g)		
										Sat	Mono	Poly
	FRUITS and FRUIT JUICES—Continued											
71120	Frozen, sweetened	10 oz	284	73	293	2	74	12	<1	t	t	0.3
3235	Cup, thawed measure	1 c	250	73	258	2	65	11	<1	t	t	0.2
	Strawberries:											
3134	Fresh, whole, capped	1 c	144	91	46	1	11	3	<1	t	t	0.2
3236	Frozen, sliced, sweetened	1 c	255	73	245	1	66	5	<1	t	t	0.2
3663	Breadfruit	1 c	220	71	227	2	60	11	1	0.1	t	0.1
	Cherries:											
3035	Sour, red pitted, canned water pack	1 c	244	90	88	2	22	3	<1	t	t	t
3036	Sweet, red pitted, raw	10 ea	68	82	43	1	11	1	<1	t	t	t
3862	Dried	¼ c	40		120	2	26	3	0	0	0	0
	Dates:											
3044	Whole, without pits	10 ea	83	21	234	2	62	7	<1	t	t	t
3043	Chopped	1 c	178	21	502	4	134	14	1	t	t	t
3162	Figs, dried	10 ea	190	30	473	6	121	19	2	0.3	0.3	0.7
	Fruit cocktail, canned, not drained:											
3045	Heavy syrup pack	1 c	248	80	181	1	47	2	<1	t	t	t
3164	Juice pack	1 c	237	87	109	1	28	2	<1	t	t	t
	Grapefruit:											
	Raw 3¾" diam (half w/rind = 241g)											
3818	Pink/red, half fruit, edible part	1 ea	123	88	52	1	13	2	<1	t	t	t
3047	White, half fruit, edible part	1 ea	118	90	39	1	10	1	<1	t	t	t
3050	Canned sections with light syrup	1 c	254	84	152	1	39	1	<1	t	t	t
	Juice:											
3051	Fresh, white, raw	1 c	247	90	96	1	23	<1	<1	t	t	t
3052	Canned, unsweetened	1 c	247	90	94	1	22	<1	<1	t	t	t
3165	Sweetened	1 c	250	87	115	1	28	<1	<1	t	t	t
3053	Prepared from concentrate	1 c	247	89	101	1	24	<1	<1	t	t	t
	Grapes, European (adherent skin):											
	Fresh:											
3055	Thompson seedless	10 ea	50	81	34	<1	9	<1	<1	t	t	t
3056	Tokay/Emperor, seeded types	10 ea	50	81	34	<1	9	<1	<1	t	t	t
3064	Juice, prepared from frozen, vit C added	1 c	250	87	128	<1	32	<1	<1	t	t	t
3062	Low calorie	1 c	253	84	154	1	38	<1	<1	t	t	t
3636	Jackfruit, fresh, sliced	1 c	165	73	155	2	40	3	<1	0.1	t	0.1
3065	Kiwi fruit, raw, peeled (88g w/peel)	1 ea	76	83	46	1	11	2	<1	t	t	0.2
	Lemons:											
	Fresh:											
3066	Without peel and seeds (about 4/lb)	1 ea	58	89	17	1	5	2	<1	t	t	t
	Juice:											
254	Fresh:	1 c	244	91	61	1	21	1	0	0	0	0
3068	Tablespoon	1 tbs	15	91	4	<1	1	<1	0	0	0	0
3069	Canned or bottled	1 tbs	15	92	3	<1	1	<1	<1	t	t	t
258	Frozen, single strength, Unsweetened:	1 c	244	92	54	1	16	1	1	0.1	t	0.2
3070	Tablespoon	1 tbs	15	92	3	<1	1	<1	<1	t	t	t
	Lime juice:											
3072	Fresh	1 tbs	15	90	4	<1	1	<1	<1	t	t	t
3073	Canned or bottled, unsweetened	1 c	246	93	52	1	16	1	1	t	t	0.2
3758	Pomelos, raw	1 ea	609	89	231	5	59	6	<1			
3221	Mangos, raw, edible part (300g w/skin & seeds)	1 ea	207	82	135	1	35	4	1	0.1	0.2	0.1
	Melons:											
	Raw, without rind and contents:											
3076	Cantaloupe, 5" diam (2⅓ lb whole w/refuse), orange flesh	½ ea	276	90	94	2	23		1	0.1	t	0.2
3081	Honeydew, 6" diam (5⅟lb whole w/refuse), slice = ⅒ melon	1 pce	160	90	58	1	15	1	<1	t	t	t
3215	Nectarines, raw, w/o pits, 2" diam	1 ea	136	88	60	1	14	2	<1	t	0.1	0.2
	Oranges:											
	Fresh:											

PAGE KEY: A–4 = Beverages   A–6 = Dairy   A–10 = Eggs   A–12 = Fat/Oil   A–14 = Fruit   A–20 = Bakery
A–26 = Grain   A–32 = Fish   A–32 = Meats   A–36 = Poultry   A–38 = Sausage   A–40 = Mixed/Fast   A–44 = Nuts/Seeds   A–46 = Sweets
A–48 = Vegetables/Legumes   A–58 = Vegetarian   A–62 = Misc   A–62 = Soups/Sauces   A–64 = Fast   A–78 = Convenience   A–80 = Baby foods

Chol (mg)	Calc (mg)	Iron (mg)	Magn (mg)	Pota (mg)	Sodi (mg)	Zinc (mg)	VT-A (RAE)	Thia (mg)	VT-E (mg α)	Ribo (mg)	Niac (mg)	V-B6 (mg)	Fola (µg)	VT-C (mg)
0	43	1.85	37	324	3	0.51	9	0.05	2.04	0.13	0.65	0.1	74	47
0	38	1.62	32	285	2	0.45	8	0.05	1.8	0.11	0.57	0.08	65	41
0	23	0.6	19	220	1	0.2	1	0.03	0.42	0.03	0.56	0.07	35	85
0	28	1.5	18	250	8	0.15	3	0.04	0.59	0.13	1.02	0.08	38	106
0	37	1.19	55	1078	4	0.26	0	0.24	0.22	0.07	1.98	0.22	31	64
0	27	3.34	15	239	17	0.17	93	0.04	0.56	0.1	0.43	0.11	20	5
0	9	0.24	7	151	0	0.05	2	0.02	0.05	0.02	0.1	0.03	3	5
0	20	0.36			5		5							0
0	32	0.85	36	544	2	0.24	<1	0.04	0.04	0.05	1.06	0.14	16	0
0	69	1.82	77	1168	4	0.52	1	0.09	0.09	0.12	2.27	0.29	34	1
0	308	3.86	129	1292	19	1.04	1	0.16	0.66	0.16	1.18	0.2	17	2
0	15	0.72	12	218	15	0.2	25	0.04	0.99	0.05	0.93	0.12	7	5
0	19	0.5	17	225	9	0.21	36	0.03	0.95	0.04	0.96	0.12	7	6
0	27	0.1	11	166	0	0.09	71	0.05	0.16	0.04	0.25	0.07	16	38
0	14	0.07	11	175	0	0.08	2	0.04	0.15	0.02	0.32	0.05	12	39
0	36	1.02	25	328	5	0.2	0	0.1	0.23	0.05	0.62	0.05	23	54
0	22	0.49	30	400	2	0.12	5	0.1	0.54	0.05	0.49	0.11	25	94
0	17	0.49	25	378	2	0.22	1	0.1	0.1	0.05	0.57	0.05	25	72
0	20	0.9	25	405	5	0.15	1	0.1	0.1	0.06	0.8	0.05	25	67
0	20	0.35	27	336	2	0.12	1	0.1	0.1	0.05	0.54	0.11	10	83
0	5	0.18	4	96	1	0.04	2	0.03	0.1	0.04	0.09	0.04	1	5
0	5	0.18	4	96	1	0.04	2	0.03	0.1	0.04	0.09	0.04	1	5
0	10	0.25	10	52	5	0.1	1	0.04	0	0.06	0.31	0.1	2	60
0	23	0.61	25	334	8	0.13	1	0.07	0	0.09	0.66	0.16	8	0
0	56	0.99	61	500	5	0.69	25	0.05	0.25	0.18	0.66	0.18	23	11
0	26	0.24	13	237	2	0.11	3	0.02	1.11	0.02	0.26	0.05	19	70
0	15	0.35	5	80	1	0.03	1	0.02	0.09	0.01	0.06	0.05	6	31
0	17	0.07	15	303	2	0.12	2	0.07	0.37	0.02	0.24	0.12	32	112
0	1	<.01	1	19	<1	<.01	<1	<.01	0.02	<.01	0.02	<.01	2	7
0	2	0.02	1	15	3	<.01	<1	<.01	0.02	<.01	0.03	<.01	2	4
0	20	0.29	20	217	2	0.12	2	0.14	0.2	0.03	0.33	0.15	24	77
0	1	0.02	1	13	<1	<.01	<1	<.01	0.01	<.01	0.02	<.01	2	5
0	1	<.01	1	16	<1	<.01	<1	<.01	0.02	<.01	0.02	<.01	1	4
0	30	0.57	17	184	39	0.15	2	0.08	0.3	<.01	0.4	0.07	20	16
0	24	0.67	37	1315	6	0.49	2	0.21	0.55	0.16	1.34	0.22	158	371
0	21	0.27	19	323	4	0.08	79	0.12	2.32	0.12	1.21	0.28	29	57
0	25	0.58	33	737	44	0.5	466	0.11	0.14	0.05	2.03	0.2	58	101
10	10	0.27	16	365	29	0.14	5	0.06	0.03	0.02	0.67	0.14	30	29
0	8	0.38	12	273	0	0.23	23	0.05	1.05	0.04	1.53	0.03	7	7

A Appendix

Table A–1

Food Composition

(Computer code number is for West Diet Analysis program)   (For purposes of calculations, use "0" for t, <1, <.1, <.01, etc.)

Computer Code Number	Food Description	Measure	Wt (g)	H₂O (%)	Ener (kcal)	Prot (g)	Carb (g)	Dietary Fiber (g)	Fat (g)	Fat Breakdown (g) Sat	Mono	Poly
	FRUITS and FRUIT JUICES—Continued											
3082	Whole w/o peel & seeds, 2⅜" diam (180g w/peel & seeds)	1 ea	131	87	62	1	15	3	<1	t	t	t
3083	Sections, without membranes	1 c	180	87	85	2	21	4	<1	t	t	t
	Juice:											
3090	Fresh, all varieties	1 c	248	88	112	2	26	<1	<1	t	t	t
3093	Canned, unsweetened	1 c	249	89	105	1	25	<1	<1	t	t	t
3480	Calcium fortified	1 c	247		110	1	27	0	0	0	0	0
3092	Chilled	1 c	249	88	110	2	25	<1	1	t	0.1	0.2
3091	Prepared from concentrate	1 c	249	88	112	2	27	<1	<1	t	t	t
20004	Prepared from dry crystals	1 c	248	87	122	0	31	<1	0	0	0	0
3170	Orange and grapefruit juice, canned	1 c	247	89	106	1	25	<1	<1	t	t	t
	Papayas:											
	Fresh:											
3172	½"slices	1 c	140	89	55	1	14	3	<1	t	t	t
3171	Whole, 3" diam by 5⅛"	1 ea	304	89	119	2	30	5	<1	0.1	0.1	t
3095	Nectar, canned	1 c	250	85	142	<1	36	2	<1	0.1	0.1	t
	Peaches:											
	Fresh:											
3096	Whole, 2" diam	1 ea	98	89	38	1	9		<1	t	t	t
3097	Sliced	1 c	170	89	66	2	16		<1	t	0.1	0.1
	Canned, not drained:											
	Heavy syrup pack:											
3098	Cup	1 c	262	79	194	1	52	3	<1	t	t	0.1
3099	Half	1 ea	98	79	73	<1	20	1	<1	t	t	t
	Juice pack:											
3175	Cup	1 c	248	87	109	2	29	3	<1	t	t	t
3176	Half	1 ea	98	87	43	1	11	1	<1	t	t	t
	Dried:											
3100	Uncooked	10 ea	130	32	311	5	80	11	1	0.1	0.4	0.5
3214	Cooked, w/fruit & liquid	1 c	258	78	199	3	51	7	1	t	0.2	0.3
	Frozen, sweetened:											
3234	10-oz package,vitamin C added	1 ea	284	75	267	2	68	5	<1	t	0.1	0.2
57481	Cup, vitamin C added	1 c	250	75	235	2	60	4	<1	t	0.1	0.2
3101	Nectar, canned	1 c	249	86	134	1	35	1	<1	t	t	t
	Pears:											
	Fresh, with skin, cored:											
3103	Bartlett, 2½" diam (about 2½/lb)	1 ea	166	84	96	1	26	5	<1	t	t	t
3105	Bosc, 2⅛" diam (about 3/lb)	1 ea	139	84	81	1	21	4	<1	t	t	t
3106	D'Anjou, 3" diam (about 2/lb)	1 ea	209	84	121	1	32	6	<1	t	t	t
	Canned, fruit and liquid:											
	Heavy syrup pack:											
3107	Cup	1 c	266	80	197	1	51	4	<1	t	t	t
3108	Half	1 ea	76	80	56	<1	15	1	<1	t	t	t
	Juice pack:											
3179	Cup	1 c	248	86	124	1	32	4	<1	t	t	t
3180	Half	1 ea	76	86	38	<1	10	1	<1	t	t	t
3109	Dried halves	10 ea	175	27	458	3	122	13	1	t	0.2	0.3
3110	Nectar, canned	1 c	250	84	150	<1	39	2	<1	t	t	t
	Pineapple:											
3111	Fresh chunks, diced	1 c	155	86	74	1	20	2	<1	t	t	t
	Canned, not drained:											
	Heavy syrup pack:											
3114	Crushed, chunks, tidbits	½ c	127	79	99	<1	26	1	<1	t	t	t
3115	Slices	1 ea	49	79	38	<1	10	<1	<1	t	t	t
	Juice pack:											
3183	Crushed, chunks, tidbits	1 c	250	84	150	1	39	2	<1	t	t	t
3184	Slices	1 ea	47	84	28	<1	7	<1	<1	t	t	t
3120	Juice, canned, unsweetened:	1 c	250	86	140	1	34	<1	<1	t	t	t
	Plantains, yellow fleshed, without peel:											
3195	Raw slices (whole=179g w/o peel)	1 c	148	65	181	2	47	3	1	0.2	t	0.1
3196	Cooked, boiled, sliced	1 c	154	67	179	1	48	4	<1	0.1	t	t

PAGE KEY: A–4 = Beverages   A–6 = Dairy   A–10 = Eggs   A–12 = Fat/Oil   A–14 = Fruit   A–20 = Bakery
A–26 = Grain   A–32 = Fish   A–32 = Meats   A–36 = Poultry   A–38 = Sausage   A–40 = Mixed/Fast   A–44 = Nuts/Seeds   A–46 = Sweets
A–48 = Vegetables/Legumes   A–58 = Vegetarian   A–62 = Misc   A–62 = Soups/Sauces   A–64 = Fast   A–78 = Convenience   A–80 = Baby foods

Chol (mg)	Calc (mg)	Iron (mg)	Magn (mg)	Pota (mg)	Sodi (mg)	Zinc (mg)	VT-A (RAE)	Thia (mg)	VT-E (mg α)	Ribo (mg)	Niac (mg)	V-B$_6$ (mg)	Fola (µg)	VT-C (mg)
0	52	0.13	13	237	0	0.09	14	0.11	0.24	0.05	0.37	0.08	39	70
0	72	0.18	18	326	0	0.13	20	0.16	0.32	0.07	0.51	0.11	54	96
0	27	0.5	27	496	2	0.12	25	0.22	0.1	0.07	0.99	0.1	74	124
0	20	1.1	27	436	5	0.17	22	0.15	0.5	0.07	0.78	0.22	45	86
0	300	0		430	15		0	0	0		0	0	40	78
0	25	0.42	27	473	2	0.1	10	0.28	0.47	0.05	0.7	0.13	45	82
0	22	0.25	25	473	2	0.12	12	0.2	0.5	0.04	0.5	0.11	110	97
0	126	0.02	2	60	10	0.02	191	0	0	0.22	2.54	0.25	0	73
0	20	1.14	25	390	7	0.17	15	0.14	0.35	0.07	0.83	0.06	35	72
0	34	0.14	14	360	4	0.1	77	0.04	1.02	0.04	0.47	0.03	53	87
0	73	0.3	30	781	9	0.21	167	0.08	2.22	0.1	1.03	0.06	116	188
0	25	0.85	8	78	12	0.38	45	0.02	0.6	<.01	0.38	0.02	5	8
0	6	0.24	9	186	0	0.17	16	0.02	0.72	0.03	0.79	0.02	4	6
0	10	0.42	15	323	0	0.29	27	0.04	1.24	0.05	1.37	0.04	7	11
0	8	0.71	13	241	16	0.24	45	0.03	1.28	0.06	1.61	0.05	8	7
0	3	0.26	5	90	6	0.09	17	0.01	0.48	0.02	0.6	0.02	3	3
0	15	0.67	17	317	10	0.27	47	0.02	1.22	0.04	1.44	0.05	7	9
0	6	0.26	7	125	4	0.11	19	<.01	0.48	0.02	0.57	0.02	3	4
0	36	5.28	55	1295	9	0.74	140	<.01	0.25	0.28	5.69	0.09	0	6
0	23	3.38	34	826	5	0.46	26	0.01	0.15	0.05	3.92	0.1	0	10
0	9	1.05	14	369	17	0.14	40	0.04	1.76	0.1	1.85	0.05	9	268
0	8	0.92	12	325	15	0.12	35	0.03	1.55	0.09	1.63	0.04	8	235
0	12	0.47	10	100	17	0.2	32	<.01	0.72	0.03	0.72	0.02	2	13
0	15	0.28	12	198	2	0.17	2	0.02	0.2	0.04	0.26	0.05	12	7
0	13	0.24	10	165	1	0.14	1	0.02	0.17	0.03	0.22	0.04	10	6
0	19	0.36	15	249	2	0.21	2	0.03	0.25	0.05	0.33	0.06	15	9
0	13	0.59	11	173	13	0.21	0	0.03	0.21	0.06	0.64	0.04	3	3
0	4	0.17	3	49	4	0.06	0	<.01	0.06	0.02	0.18	0.01	1	1
0	22	0.72	17	238	10	0.22	1	0.03	0.2	0.03	0.5	0.03	2	4
0	7	0.22	5	73	3	0.07	<1	<.01	0.06	<.01	0.15	0.01	1	1
0	60	3.68	58	933	10	0.68	<1	0.01	0.1	0.25	2.4	0.13	0	12
0	12	0.65	8	32	10	0.18	<1	<.01	0.12	0.03	0.32	0.04	2	3
0	20	0.43	19	178	2	0.16	5	0.12	0.03	0.05	0.76	0.17	23	56
0	18	0.48	20	132	1	0.15	1	0.11	0.01	0.03	0.36	0.09	6	9
0	7	0.19	8	51	<1	0.06	<1	0.04	<.01	0.01	0.14	0.04	2	4
0	35	0.7	35	305	2	0.25	5	0.24	0.02	0.05	0.71	0.18	12	24
0	7	0.13	7	57	<1	0.05	1	0.04	<.01	<.01	0.13	0.03	2	4
0	42	0.65	32	335	2	0.28	1	0.14	0.05	0.06	0.64	0.24	58	27
0	4	0.89	55	739	6	0.21	83	0.08	0.21	0.08	1.02	0.44	33	27
0	3	0.89	49	716	8	0.2	69	0.07	0.2	0.08	1.16	0.37	40	17

A Appendix

Table A–1

Food Composition

(Computer code number is for West Diet Analysis program)   (For purposes of calculations, use "0" for t, <1, <.1, <.01, etc.)

Computer Code Number	Food Description	Measure	Wt (g)	H₂O (%)	Ener (kcal)	Prot (g)	Carb (g)	Dietary Fiber (g)	Fat (g)	Fat Breakdown (g)		
										Sat	Mono	Poly
	FRUITS and FRUIT JUICES—Continued											
	Plums:											
3121	Fresh, medium, 2⅛" diam	1 ea	66	87	30	<1	8	1	<1	t	t	t
	Canned, purple, not drained:											
	Heavy syrup pack:											
3124	Cup	1 c	258	76	230	1	60	2	<1	t	0.2	t
3125	Each	3 ea	138	76	123	<1	32	1	<1	t	t	t
	Juice Pack:											
3185	Cup	1 c	252	84	146	1	38	2	<1	t	t	t
3186	Each	3 ea	138	84	80	1	21	1	<1	t	t	t
3197	Pomegranate, fresh	1 ea	154	81	105	1	26	1	<1	t	t	t
	Prunes:											
	Dried, pitted:											
3126	Uncooked (10 = 97g w/pits, 84g w/o pits)	10 ea	84	31	202	2	54	6	<1	t	t	t
3127	Cooked, unsweetened, fruit & liq, (250 g w/pits)	1 c	248	70	265	2	70	8	<1	t	0.3	t
3128	Juice, bottled or canned	1 c	256	81	182	2	45	3	<1	t	t	t
	Raisins, seedless:											
3130	Cup, not pressed down	1 c	145	15	434	4	115	5	1	t	t	t
3764	One packet, ½ oz	½ oz	14	15	42	<1	11	1	<1	t	t	t
3133	Rhubarb, cooked, added sugar	1 c	240	68	278	1	75	5	<1	t	t	t
	Tangerines, without peel and seeds:											
3138	Fresh (2⅜" whole) 116g w/refuse	1 ea	84	88	37	1	9	2	<1	t	t	t
3237	Canned, light syrup, fruit and liquid	1 c	252	83	154	1	41	2	<1	t	t	t
3140	Juice, canned, sweetened	1 c	249	87	124	1	30	<1	<1	t	t	t
	Watermelon, raw, without rind and seeds:											
3143	Piece, 1/16th wedge	1 pce	286	91	86	2	22	1	<1	t	0.1	0.1
3142	Diced	1 c	152	91	46	1	11	1	<1	t	t	t
	BAKED GOODS: BREADS, CAKES, COOKIES, CRACKERS, PIES											
	Bagels:											
42100	Cinnamon raisin, 3½" diam.	1 ea	71	32	195	7	39	2	1	0.2	0.1	0.5
42620	Dunkin Donuts	1 ea	125		340	11	72	4	1	0		
42000	Plain, enriched, 3½" diam.	1 ea	71	33	195	7	38	2	1	0.2	t	0.5
42619	Dunkin Donuts	1 ea	125		330	12	68	3	1	0		
42596	Oat Bran	1 ea	110	33	280	12	59	4	1	0.2	0.3	0.5
42617	Whole Wheat	1 ea	110	28	291	12	62	10	2	0.3	0.2	0.6
	Biscuits:											
42001	From home recipe	1 ea	60	29	212	4	27	1	10	2.6	4.2	2.5
42002	From mix	1 ea	57	29	191	4	28	1	7	1.6	2.4	2.5
	Bread:											
42672	Cornbread, 2.5 x 2.5 x 1.5" piece	1 pce	65	49	152	4	23	2	5	1.6	2.4	0.5
42015	Croissants, 4½ x 4 x 1¾"	1 ea	57	23	231	5	26	1	12	6.6	3.1	0.6
42148	Croutons, seasoned	½ c	20	4	93	2	13	1	4	1	1.9	0.5
42004	Crumbs, dry, grated (see 364, 365 for soft crumbs)	1 c	108	7	427	14	78	5	6	1.3	1.1	2.2
42052	Boston brown, canned, 3¼" slice	1 pce	45	47	88	2	19	2	1	0.1	t	0.3
	Cracked wheat (¼ cracked-wheat & ¾ enriched wheat flour):											
42042	Slice (18 per loaf)	1 pce	25	36	65	2	12	1	1	0.2	0.5	0.2
	French/Vienna, enriched:											
42044	Slice, 4¾ x 4½"	1 pce	25	34	68	2	13	1	1	0.2	0.3	0.2
42043	French, slice, 5 x 2"	1 pce	25	34	68	2	13	1	1	0.2	0.3	0.2
	French toast: see Mixed Dishes, and Fast Foods, #691											
42476	Honey wheatberry	1 pce	38		100	3	18	2	2	0	0.5	1
	Italian, enriched:											
42046	Slice, 4½ x 3¼ x ¾"	1 pce	30	36	81	3	15	1	1	0.3	0.2	0.4
	Mixed grain, enriched:											
42047	Slice (18 per loaf)	1 pce	26	38	65	3	12	2	1	0.2	0.4	0.2
42048	Slice, toasted	1 pce	24	32	65	3	12	2	1	0.2	0.4	0.2

PAGE KEY: A–4 = Beverages   A–6 = Dairy   A–10 = Eggs   A–12 = Fat/Oil   A–14 = Fruit   A–20 = Bakery
A–26 = Grain   A–32 = Fish   A–32 = Meats   A–36 = Poultry   A–38 = Sausage   A–40 = Mixed/Fast   A–44 = Nuts/Seeds   A–46 = Sweets
A–48 = Vegetables/Legumes   A–58 = Vegetarian   A–62 = Misc   A–62 = Soups/Sauces   A–64 = Fast   A–78 = Convenience   A–80 = Baby foods

Chol (mg)	Calc (mg)	Iron (mg)	Magn (mg)	Pota (mg)	Sodi (mg)	Zinc (mg)	VT-A (RAE)	Thia (mg)	VT-E (mg α)	Ribo (mg)	Niac (mg)	V-B$_6$ (mg)	Fola (µg)	VT-C (mg)
0	4	0.11	5	104	0	0.07	11	0.02	0.17	0.02	0.28	0.02	3	6
0	23	2.17	13	235	49	0.18	34	0.04	0.46	0.1	0.75	0.07	8	1
0	12	1.16	7	126	26	0.1	18	0.02	0.25	0.05	0.4	0.04	4	1
0	25	0.86	20	388	3	0.28	126	0.06	0.45	0.15	1.19	0.07	8	7
0	14	0.47	11	213	1	0.15	69	0.03	0.25	0.08	0.65	0.04	4	4
0	5	0.46	5	399	5	0.18	8	0.05	0.92	0.05	0.46	0.16	9	9
0	36	0.78	34	615	2	0.37	33	0.04	0.36	0.16	1.58	0.17	3	1
0	47	1.02	45	796	2	0.47	42	0.06	0.47	0.25	1.79	0.54	0	7
0	31	3.02	36	707	10	0.54	<1	0.04	0.31	0.18	2.01	0.56	0	10
0	72	2.73	46	1086	16	0.32	0	0.15	0.17	0.18	1.11	0.25	7	3
0	7	0.26	4	105	2	0.03	0	0.01	0.02	0.02	0.11	0.02	1	0
0	348	0.5	29	230	2	0.19	10	0.04	0.65	0.06	0.48	0.05	12	8
0	12	0.08	10	132	1	0.2	29	0.09	0.13	0.02	0.13	0.06	17	26
0	18	0.93	20	197	15	0.6	106	0.13	0.25	0.11	1.12	0.11	13	50
0	45	0.5	20	443	2	0.07	32	0.15	0.37	0.05	0.25	0.08	12	55
0	20	0.69	29	320	3	0.29	80	0.09	0.14	0.06	0.51	0.13	9	23
0	11	0.36	15	170	2	0.15	43	0.05	0.08	0.03	0.27	0.07	5	12
0	13	2.7	20	105	229	0.8	15	0.27	0.22	0.2	2.19	0.04	79	0
0	40	3.6			470		0							4
0	53	2.53	21	72	379	0.62	0	0.38	0.21	0.22	3.24	0.04	75	0
0	20	4.5			690		0							4
0	13	3.39	34	126	558	0.99	1	0.36	0.36	0.37	3.26	0.05	108	0
0	32	3.52	116	379	592	2.52		0.34	0.99	0.28	5.75	0.3	66	0
2	141	1.74	11	73	348	0.32	14	0.21	0.78	0.19	1.77	0.02	37	0
2	105	1.17	14	107	544	0.35	15	0.2	0.23	0.2	1.72	0.04	30	0
22	71	0.82	13	105	356	0.38		0.12	0.64	0.16	0.93	0.05	6	0
38	21	1.16	9	67	424	0.43	117	0.22	0.48	0.14	1.25	0.03	50	0
1	19	0.56	8	36	248	0.19	1	0.1	0.08	0.08	0.93	0.02	21	0
0	198	5.22	46	212	791	1.57	0	1.04	0.09	0.44	7.16	0.13	116	0
0	32	0.94	28	143	284	0.22	11	<.01	0.14	0.05	0.5	0.04	5	0
0	11	0.7	13	44	134	0.31	0	0.09	0.15	0.06	0.92	0.08	15	0
0	19	0.63	7	28	152	0.22	0	0.13	0.08	0.08	1.19	0.01	37	0
0	19	0.63	7	28	152	0.22	0	0.13	0.08	0.08	1.19	0.01	37	0
0	20	0.72			200		0	0.12	0.24	0.07	0.8			0
0	23	0.88	8	33	175	0.26	0	0.14	0.09	0.09	1.31	0.01	57	0
0	24	0.9	14	53	127	0.33	0	0.11	0.09	0.09	1.13	0.09	31	0
0	24	0.9	14	53	127	0.33	0	0.08	0.08	0.08	1.02	0.08	28	0

Table A–1

Food Composition (Computer code number is for West Diet Analysis program) (For purposes of calculations, use "0" for t, <1, <.1, <.01, etc.)

Computer Code Number	Food Description	Measure	Wt (g)	H_2O (%)	Ener (kcal)	Prot (g)	Carb (g)	Dietary Fiber (g)	Fat (g)	Fat Breakdown (g) Sat	Mono	Poly
	BAKED GOODS: BREADS, CAKES, COOKIES, CRACKERS, PIES—Continued											
	Oatmeal, enriched:											
42049	Slice (18 per loaf)	1 pce	27	37	73	2	13	1	1	0.2	0.4	0.5
42050	Slice, toasted	1 pce	25	31	73	2	13	1	1	0.2	0.4	0.5
42007	Pita pocket bread, enr, 6" round	1 ea	60	32	165	5	33	1	1	t	t	0.3
	Pumpernickel (⅔ rye & ⅓ enr wheat flr):											
42006	Slice, 5 x 4 x ⅜"	1 pce	26	38	65	2	12	2	1	0.1	0.2	0.3
42054	Slice, toasted	1 pce	29	32	80	3	15	2	1	0.1	0.3	0.4
	Raisin, enriched:											
42051	Slice (18 per loaf)	1 pce	26	34	71	2	14	1	1	0.3	0.6	0.2
42055	Slice, toasted	1 pce	24	28	71	2	14	1	1	0.3	0.6	0.2
	Rye, light (⅓ rye & ⅔ enr wheat flr):											
42005	1-lb loaf	1 ea	454	37	1176	39	219	26	15	2.8	6	3.6
42005	Slice, 4¾ x 3¾ x 7⁄16"	1 pce	32	37	83	3	15	2	1	0.2	0.4	0.3
42056	Slice, toasted	1 pce	24	31	68	2	13	2	1	0.2	0.3	0.2
	Wheat (enr wheat & whole-wheat flour):											
42012	Slice (18 per loaf)	1 pce	25	37	65	2	12	1	1	0.2	0.4	0.2
42031	Slice, toasted	1 pce	23	32	65	2	12	1	1	0.2	0.4	0.2
	White, enriched:											
42138	Slice	1 pce	42	35	120	3	21	1	2	0.5	0.5	1.2
	Whole Wheat:											
42014	Slice (16 per loaf)	1 pce	28	38	69	3	13	2	1	0.3	0.5	0.3
42029	Slice, toasted	1 pce	25	30	69	3	13	2	1	0.3	0.5	0.3
	Bread stuffing, prepared from mix:											
42037	Dry type	1 c	200	65	356	6	43	6	17	3.5	7.6	5.2
	Cakes:											
	Prepared from mixes using enrich flour and veg shortening, w/frostings made from margarine:											
46004	Angel Food, 1⁄12 of cake	1 pce	28	33	72	2	16	<1	<1	t	t	0.1
46002	Boston cream pie, ⅛ of cake	1 pce	92	45	232	2	39	1	8	2.2	4.2	0.9
46005	Coffee Cake, ⅙ of cake	1 pce	56	30	178	3	30	1	5	1	2.2	1.8
46013	Devil's food, chocolate frosting, 1⁄16 of cake	1 pce	64	23	235	3	35	2	10	3.1	5.6	1.2
	Home recipes w/enrich flour:											
	Fruitcake, dark:											
46205	Piece, 1⁄32 of cake, ⅔" arc	1 pce	43	25	139	1	26	2	4	0.5	1.8	1.4
46015	Sheet, plain, made w/margarine, uncooked white frosting, ⅑ of cake	1 pce	64	22	239	2	38	<1	9	1.5	3.9	3.3
	Commercial:											
49004	Cheesecake, 1⁄12 of cake	1 pce	80	46	257	4	20	<1	18	7.9	6.9	1.3
46016	Pound cake, 1⁄17 of loaf, 2" slice	1 pce	28	25	109	2	14	<1	6	3.2	1.7	0.3
	Yellow, chocolate frosting, 2 layer:											
46012	Slice, 1⁄16 of cake	1 pce	64	22	243	2	35	1	11	3	6.1	1.4
	Snack:											
46011	Chocolate w/creme filling, Ding Dong	1 ea	50	20	188	2	30	<1	7	1.4	2.8	2.6
46008	Sponge cake w/creme filling, Twinkie	1 ea	43	20	157	1	27	<1	5	1.1	1.8	1.4
46001	Sponge cake, 1⁄12 of 12" cake	1 pce	38	30	110	2	23	<1	1	0.3	0.4	0.2
44032	Chex party mix	1 c	43	4	183	5	28	2	7	2.4	3.9	1.1
	Chips:											
44061	Bagel	5 pce	70	3	298	6	52	4	7	1.3	2.1	3.4
42396	Bagel, onion garlic, toasted	1 oz	28		181	5	30	3	7	1.6	4.9	0
43703	Potato chips	20 pce	28		148	2	15	1	10	3		
4022	Baked	11 pce	28		109	2	23	2	1	0		
	Cookies made with enriched flour:											
	Brownies with nuts:											
47000	Commercial w/frosting, 1½ x 1¾ x ⅞"	1 ea	61	14	247	3	39	1	10	2.6	5.5	1.4
47214	Fat free fudge, Entenmann's	1 pce	40		110	2	27	1	0	0	0	0
	Chocolate chip cookies:											
47001	Commercial, 2¼" diam	4 ea	60	12	275	2	35	2	15	4.4	7.8	2.1
47002	Home recipe, 2¼" diam	4 ea	64	6	312	4	37	2	18	5.2	6.6	5.4

Appendix A

Chol (mg)	Calc (mg)	Iron (mg)	Magn (mg)	Pota (mg)	Sodi (mg)	Zinc (mg)	VT-A (RAE)	Thia (mg)	VT-E (mg α)	Ribo (mg)	Niac (mg)	V-B$_6$ (mg)	Fola (μg)	VT-C (mg)
0	18	0.73	10	38	162	0.28	1	0.11	0.13	0.06	0.85	0.02	17	0
0	18	0.74	10	38	163	0.28	1	0.09	0.13	0.06	0.77	0.02	13	0
0	52	1.57	16	72	322	0.5	0	0.36	0.18	0.2	2.78	0.02	64	0
0	18	0.75	14	54	174	0.38	0	0.09	0.11	0.08	0.8	0.03	24	0
0	21	0.91	17	66	214	0.47	0	0.08	0.13	0.09	0.89	0.04	25	0
0	17	0.75	7	59	101	0.19	0	0.09	0.07	0.1	0.9	0.02	28	0
0	17	0.76	7	59	102	0.19	0	0.07	0.07	0.09	0.81	0.02	24	0
0	331	12.85	182	754	2996	5.18	2	1.97	1.5	1.52	17.27	0.34	499	2
0	23	0.91	13	53	211	0.36	<1	0.14	0.11	0.11	1.22	0.02	35	0
0	19	0.74	10	44	174	0.3	<1	0.09	0.09	0.08	0.9	0.02	25	0
0	26	0.83	12	50	132	0.26	0	0.1	0.07	0.07	1.03	0.02	23	0
0	26	0.83	12	50	132	0.26	0	0.08	0.07	0.06	0.93	0.02	19	0
1	24	1.25	8	61	151	0.27	9	0.17	0.36	0.16	1.51	0.02	38	0
0	20	0.92	24	71	148	0.54	<1	0.1	0.09	0.06	1.07	0.05	14	0
0	20	0.93	24	71	148	0.55	<1	0.08	0.08	0.05	0.97	0.05	10	0
0	64	2.18	24	148	1086	0.56	236	0.27	2.8	0.21	2.95	0.08	78	0
0	39	0.15	3	26	210	0.02	0	0.03	0.03	0.14	0.25	<.01	10	0
34	21	0.35	6	36	132	0.15	22	0.38	0.14	0.25	0.18	0.02	13	0
27	76	0.8	10	63	236	0.25	20	0.09	0.11	0.1	0.85	0.03	27	0
27	28	1.41	22	128	214	0.44	17	0.02	<.01	0.09	0.37	0.03	11	0
2	14	0.89	7	66	116	0.12	3	0.02	0.39	0.04	0.34	0.02	9	0
35	40	0.68	4	34	220	0.16	12	0.06	1.22	0.04	0.32	0.02	17	0
44	41	0.5	9	72	166	0.41	114	0.02	1.26	0.15	0.16	0.04	14	0
62	10	0.39	3	33	111	0.13	42	0.04	0.18	0.06	0.37	0.01	11	0
35	24	1.33	19	114	216	0.4	21	0.08	1.45	0.1	0.8	0.02	14	0
8	36	1.68	20	61	212	0.26	2	0.11	1.09	0.15	1.21	0.01	20	0
7	19	0.55	3	37	157	0.12	2	0.07	0.51	0.06	0.53	0.01	17	0
39	27	1.03	4	38	93	0.19	17	0.09	0.09	0.1	0.73	0.02	18	0
0	15	10.62	27	116	437	0.9	3	0.67	0.11	0.21	7.24	0.67	22	20
0	9	1.39	39	167	419	0.88		0.13	1.71	0.12	1.62	0.15	46	0
0	0	2.37			461		0	0.37	<.01	0.22	3.29			0
0	0	0			178		0							6
0	40	0.36			148		0							1
10	18	1.37	19	91	190	0.44	12	0.16	0.09	0.13	1.05	0.02	29	0
0	0	1.08		90	140		0		<.01					0
0	9	1.45	21	56	196	0.28	<1	0.07	1.74	0.12	0.97	0.1	23	0
20	25	1.57	35	143	231	0.6	92	0.12	1.84	0.11	0.87	0.05	21	0

A Appendix

Table A–1

Food Composition (Computer code number is for West Diet Analysis program) (For purposes of calculations, use "0" for t, <1, <.1, <.01, etc.)

Computer Code Number	Food Description	Measure	Wt (g)	H_2O (%)	Ener (kcal)	Prot (g)	Carb (g)	Dietary Fiber (g)	Fat (g)	Fat Breakdown (g) Sat	Mono	Poly
	BAKED GOODS: BREADS, CAKES, COOKIES, CRACKERS, PIES—Continued											
47013	From refrigerated dough, 2¼″ diam	4 ea	64	13	284	3	39	1	13	4.3	6.7	1.4
47012	Fig bars	4 ea	64	16	223	2	45	3	5	0.7	1.9	1.8
47376	Fruit bar, no fat	1 ea	28		90	2	21	0	0	0	0	0
47324	Fudge, fat free, Snackwell	1 ea	16	14	53	1	12	<1	<1	t	t	t
47154	Nabisco Newtons, fat free, all flavors	1 ea	23		69	1	16		0	0	0	0
47003	Oatmeal raisin, 2⅝″ diam	4 ea	60	6	261	4	41	2	10	1.9	4.1	3
47010	Peanut butter, home recipe, 2⅝″ diam	4 ea	80	6	380	7	47	2	19	3.6	8.7	5.8
47006	Sandwich-type, all	4 ea	40	2	189	2	28	1	8	1.5	3.4	2.9
47007	Shortbread, commercial, small	4 ea	32	4	161	2	21	1	8	2	4.3	1
47004	Sugar from refrigerated dough,2″ diam	4 ea	48	5	232	2	31	<1	11	2.8	6.2	1.4
47160	Vanilla sandwich, Snackwell's	2 ea	26	4	109	1	21	1	2	0.5	0.8	0.2
47008	Vanilla wafers	10 ea	40	5	176	2	29	1	6	1.5	2.6	1.6
	Crackers:											
43500	Cheese-enriched	10 ea	10	3	50	1	6	<1	3	0.9	1.2	0.2
43501	Cheese with peanut butter-enriched	4 ea	28	3	139	3	16	1	7	1.2	3.6	1.4
	Fat free-enriched:											
43596	Cracked pepper, Snackwell	1 ea	14	2	61	1	10	<1	2	0.3	0.6	0.2
43593	Wheat, Snackwell	7 ea	15	1	60	2	12	1	<1	0.1	0.1	0.1
43590	Whole wheat, herb seasoned	5 ea	14		50	2	11	2	0	0	0	0
43591	Whole wheat, onion	5 ea	14		50	2	11	2	0	0	0	0
43502	Graham-enriched	2 ea	14	4	59	1	11	<1	1	0.2	0.6	0.5
43509	Melba toast, plain-enriched	1 pce	5	5	20	1	4	<1	<1	t	t	t
44064	Rice cakes, unsalted-enriched	2 ea	18	6	70	1	15	1	1	0.1	0.2	0.2
43504	Rye wafer, whole grain	2 ea	22	5	73	2	18	5	<1	t	t	t
43506	Saltine-enriched	4 ea	12	4	52	1	9	<1	1	0.4	0.8	0.2
43586	Saltine, unsalted tops-enriched	2 ea	6		25	1	4	0	<1			
43543	Snack-type, round like Ritz-enriched	3 ea	9	4	45	1	5	<1	2	0.3	1	0.9
43584	Triscuits	7 ea	31	4	150	3	21		6	1	2	0.5
43558	Wasa extra crsip crispbread	1 pce	6	6	24	1	4	<1	<1	t	0.2	t
43508	Whole-wheat wafers	2 ea	8	3	35	1	5	1	1	0.3	0.5	0.5
	Pastry:											
	Danish:											
428	Round piece, plain, 4¼″ diam, 1″ high	1 ea	88	21	349	5	47	<1	17	3.5	10.6	1.6
45512	Ounce, plain	1 oz	28	21	111	2	15	<1	5	1.1	3.4	0.5
45513	Round piece with fruit	1 ea	94	29	335	5	45		16	3.3	10.1	1.6
42094	Pan dulce, sweet roll w/topping	1 ea	79	21	291	5	48	1	9	1.8	4	2.5
49009	Peach crisp, 3 x 3 piece	1 pce	139	73	165	1	30	2	5	0.8		
	Toaster:											
45504	Fortified (PopTarts)	1 ea	52	12	204	2	37	1	5	0.8	2.2	2
	Strudel:											
45647	Cream Cheese	1 ea	54		200	3	24	<1	10	3		
45648	French Toast	1 ea	54		200	3	24	<1	10	3		
	Doughnuts:											
45505	Cake type, plain, 3¼″ diam	1 ea	47	21	198	2	23	1	11	1.7	4.4	3.7
45698	Sugared, Dunkin Donuts	1 ea	67		310	4	28	1	20	4		
45506	Yeast-leavened, glazed, 3 ¾″diam	1 ea	60	25	242	4	27	1	14	3.5	7.7	1.7
45708	Dunkin Donuts	1 ea	46		160	3	23	1	7	2		
	Muffins:											
	English:											
42059	Plain, enriched	1 ea	57	42	134	4	26	2	1	0.1	0.2	0.5
42061	Toasted	1 ea	52	37	133	4	26	2	1	0.1	0.2	0.5
42082	Whole wheat	1 ea	66	46	134	6	27	4	1	0.2	0.3	0.6
44504	Cornmeal	1 ea	50	30	160	4	25	1	5	1.4	2.6	0.6
44614	Banana nut	1 ea	95		340	6	53	2	12	3		
44622	Blueberry	1 ea	95		310	5	51	2	10	2.5		
44616	Chocolate chip	1 ea	95		400	5	63	2	16	6		
	Granola Bars:											
23104	Soft	1 ea	28	6	124	2	19	1	5	2	1.1	1.5
23059	Hard	1 ea	25	4	118	3	16	1	5	0.6	1.1	3
47294	Fat free, all flavors	1 ea	42		140	2	35	3	0	0	0	0

PAGE KEY: A–4 = Beverages   A–6 = Dairy   A–10 = Eggs   A–12 = Fat/Oil   A–14 = Fruit   A–20 = Bakery
A–26 = Grain   A–32 = Fish   A–32 = Meats   A–36 = Poultry   A–38 = Sausage   A–40 = Mixed/Fast   A–44 = Nuts/Seeds   A–46 = Sweets
A–48 = Vegetables/Legumes   A–58 = Vegetarian   A–62 = Misc   A–62 = Soups/Sauces   A–64 = Fast   A–78 = Convenience   A–80 = Baby foods

Chol (mg)	Calc (mg)	Iron (mg)	Magn (mg)	Pota (mg)	Sodi (mg)	Zinc (mg)	VT-A (RAE)	Thia (mg)	VT-E (mg α)	Ribo (mg)	Niac (mg)	V-B$_6$ (mg)	Fola (µg)	VT-C (mg)
15	16	1.44	15	115	134	0.32	12	0.12	1.48	0.12	1.26	0.02	36	0
0	41	1.86	17	132	224	0.25	6	0.1	0.42	0.14	1.2	0.05	22	0
0	0	0.36			95		0		<.01					0
0	3	0.29	5	26	71	0.08		0.02	<.01	0.02	0.26	<.01		0
				77										
20	60	1.59	25	143	323	0.52	86	0.15	1.5	0.1	0.76	0.04	18	0
25	31	1.78	31	185	414	0.66	110	0.18	3.04	0.17	2.81	0.07	44	0
0	10	1.55	18	70	242	0.32	<1	0.03	0.63	0.07	0.83	<.01	20	0
6	11	0.88	5	32	146	0.17	6	0.11	0.11	0.11	1.07	0.03	22	0
15	43	0.88	4	78	225	0.13	6	0.09	0.1	0.06	1.16	0.01	34	0
0	17	0.61	5	28	95	0.16		0.05		0.07	0.69	0.01		0
20	19	0.95	6	39	125	0.14	3	0.11	0.09	0.13	1.24	0.03	24	0
1	15	0.48	4	14	100	0.11	3	0.06	<.01	0.04	0.47	0.06	15	0
0	14	0.76	16	61	199	0.29		0.15	0.66	0.08	1.63	0.04	26	0
0	24	0.51	4	16	117	0.11		0.04	0	0.05	0.75	<.01	11	0
0	28	0.58	7	43	169	0.21		0.04		0.07	0.73	0.02		0
0	0				80									2
0	0	0			80									2
0	3	0.52	4	19	85	0.11	<1	0.03	0.05	0.04	0.58	<.01	6	0
0	5	0.18	3	10	41	0.1	0	0.02	0.02	0.01	0.21	<.01	6	0
0	2	0.27	24	52	5	0.54	<1	0.01	0.02	0.03	1.41	0.03	4	0
0	9	1.31	27	109	175	0.62	<1	0.09	0.18	0.06	0.35	0.06	10	0
0	14	0.65	3	15	156	0.09	0	0.07	0.01	0.06	0.63	<.01	15	0
0		0.36	5		50				0.1					0
0	11	0.32	2	12	76	0.06	0	0.04	0.18	0.03	0.36	<.01	8	0
0		1.44		95	160									0
0	5	0.19	3	17	38	0.08		0.03	0.08	0.02	0.2	<.01	1	0
0	4	0.25	8	24	53	0.17	0	0.02	0.07	<.01	0.36	0.01	2	0
27	37	1.8	14	96	326	0.48	5	0.26	0.79	0.19	2.2	0.05	55	3
9	12	0.57	4	31	104	0.15	2	0.08	0.25	0.06	0.7	0.02	17	1
19	22	1.4	14	110	333	0.48	25	0.29	0.85	0.21	1.8	0.06	31	2
26	13	1.84	9	57	75	0.35		0.23	1.22	0.21	2.02	0.03	19	0
0	23	0.95	13	198	69	0.2		0.05	2.46	0.05	1.05	0.03	9	5
0	14	1.81	9	58	218	0.34	150	0.15	0.46	0.19	2.05	0.2	40	0
10	0	1.08			220		0							0
10	0	1.08			220		0							0
17	21	0.92	9	60	257	0.26	18	0.1	0.91	0.11	0.87	0.03	24	0
0	0	1.08			380		0							2
4	26	1.22	13	65	205	0.46	2	0.22	0.21	0.13	1.71	0.03	29	0
0	0	0.36			200		0							1
0	99	1.42	12	75	264	0.4	0	0.25	0.18	0.16	2.21	0.02	54	0
0	98	1.41	11	74	262	0.4	0	0.2	0.17	0.14	1.98	0.02	45	0
0	175	1.62	47	139	420	1.06	<1	0.2	0.27	0.09	2.25	0.11	32	0
31	38	0.97	10	66	398	0.32	20	0.12	0.75	0.14	1.05	0.05	28	0
35	40	1.8			210		0							1
35	40	1.44			190		0							0
35	40	1.8			190		0							0
0	29	0.72	21	91	78	0.42	0	0.08	0.34	0.05	0.14	0.03	7	0
0	15	0.74	24	84	74	0.51	2	0.07	0.33	0.03	0.4	0.02	6	0
0	0	3.6			5									0

A   Appendix

Table A–1

Food Composition

(Computer code number is for West Diet Analysis program) (For purposes of calculations, use "0" for t, <1, <.1, <.01, etc.)

Computer Code Number	Food Description	Measure	Wt (g)	H₂O (%)	Ener (kcal)	Prot (g)	Carb (g)	Dietary Fiber (g)	Fat (g)	Fat Breakdown (g)		
										Sat	Mono	Poly
	BAKED GOODS: BREADS, CAKES, COOKIES, CRACKERS, PIES—Continued											
	Pancakes, 4" diam:											
45001	Plain, from home recipe	1 ea	38	53	86	2	11	1	4	0.8	0.9	1.7
45002	Plain, from mix; egg, milk, oil added	1 ea	38	53	74	2	14	<1	1	0.2	0.3	0.3
	Piecrust, enriched flour, vegetable shortening, baked:											
45501	Home recipe, 9" shell	1 ea	180	10	949	12	86	3	62	15.5	27.3	16.4
45503	From mix, 1 pie shell	1 ea	160	11	802	11	81	3	49	12.3	27.7	6.2
48014	Pie, cherry, commercial fried	1 ea	128	38	404	4	55	3	21	3.1	9.5	6.9
44015	Pretzels, thin twists, 3¼ x 2¼ x ¼ "	10 pce	60	3	229	5	48	2	2	0.4	0.8	0.7
	Rolls & buns, enriched, commercial:											
42157	Cloverleaf rolls, 2½" diam, 2"high	1 ea	28	32	84	2	14	1	2	0.5	1	0.3
42021	Hot dog buns	1 ea	40	35	112	4	20	1	2	0.4	0.4	0.8
42020	Hamburger buns	1 ea	43	35	120	4	21	1	2	0.5	0.5	0.8
42022	Hard roll, white, 3¾" diam, 2"high	1 ea	57	31	167	6	30	1	2	0.3	0.6	1
42034	Submarine rolls/hoagies,11¼ x 3 x 2½"	1 ea	135	34	386	11	68	4	7	1.6	3.4	1.2
	Sports/fitness bar:											
	Clif Bars:											
4042	Chocolate brownie	1 ea	68	17	236	10	41	6	4	1		
62709	Chocolate chip	1 ea	68	15	238	10	42	5	4	1		
62711	Chocolate chip peanut	1 ea	68	15	241	12	39	5	5	1.1		
62716	Cookies and cream	1 ea	68	20	225	10	39	5	4	1.4		
62207	Forza energy bar	1 ea	70		231	10	45	4	1			
62205	Tiger sports bar	1 ea	65		260	11	33	2	9	1.9		
	Power bars:											
62275	Apple cinnamon	1 ea	65		230	10	45	3	2	0.5	1.5	0.5
62276	Banana	1 ea	65		230	9	45	3	2	0.5	1	0.5
62823	Chocolate harvest	1 ea	65		240	7	45	4	4	1		
62279	Malt nut	1 ea	65		230	10	45	3	2	0.5	1	1
62280	Mocha	1 ea	65		230	10	45	3	2	1	1	0.5
	Tortillas:											
42023	Corn, enriched, 6" diam	1 ea	26	44	58	1	12	1	1	t	0.2	0.3
42025	Flour, 10" diam	1 ea	72	27	234	6	40	2	5	1.3	2.7	0.8
42027	Taco shells	1 ea	14	6	66	1	9	1	3	0.5	1.3	1.2
	Waffles, 7" diam:											
45003	From home recipe	1 ea	75	42	218	6	25	1	11	2.1	2.6	5.1
45017	Whole grain, prepared from frozen	1 ea	39	43	105	4	13	1	4	1.2	1.8	1.1
	GRAIN PRODUCTS: CEREAL, FLOUR, GRAIN, PASTA and NOODLES, POPCORN											
	Grain:											
38070	Amaranth	1 c	195	10	729	28	129	30	13	3.2	2.8	5.6
	Barley:											
38002	Pearled, dry, uncooked	1 c	200	10	704	20	155	31	2	0.5	0.3	1.1
38003	Pearled, cooked	1 c	157	69	193	4	44	6	1	0.1	t	0.3
38072	Buckwheat, whole grain, dry	1 c	170	10	583	23	122	17	6	1.3	1.8	1.8
38027	Bulgar, dry, uncooked	1 c	140	9	479	17	106	26	2	0.3	0.2	0.8
38028	Bulgar, cooked	1 c	182	78	151	6	34	8	<1	t	t	0.2
38076	Couscous, cooked	1 c	157	73	176	6	36	2	<1	t	t	0.1
38329	Cracked wheat	1 c	120	10	407	16	87	14	2	0.4	0.3	0.9
38052	Millet, cooked	1 c	240	71	286	8	57	3	2	0.4	0.4	1.2
38064	Oat bran, dry	1/4 c	24	7	59	4	16	4	2	0.3	0.6	0.7
38079	Quinoa, dry	1 c	170	9	636	22	117	10	10	1	2.6	4
	Rice:											
38010	Brown, cooked	1 c	195	73	216	5	45	4	2	0.4	0.6	0.6
56998	Spanish, cooked	1 c	246		130	3	28	2	1	0		
	White, enriched, all types:											
38012	Regular/long grain, dry	1 c	185	12	675	13	148	2	1	0.3	0.4	0.3
38013	Regular/long grain, cooked	1 c	158	68	205	4	45	1	<1	0.1	0.1	0.1
38019	Instant, prepared without salt	1 c	165	76	162	3	35	1	<1	t	t	t
	Parboiled/converted:											
38015	Raw, dry	1 c	185	10	686	13	151	3	1	0.3	0.3	0.3

Appendix **A**

PAGE KEY: A–4 = Beverages   A–6 = Dairy   A–10 = Eggs   A–12 = Fat/Oil   A–14 = Fruit   A–20 = Bakery
A–26 = Grain   A–32 = Fish   A–32 = Meats   A–36 = Poultry   A–38 = Sausage   A–40 = Mixed/Fast   A–44 = Nuts/Seeds   A–46 = Sweets
A–48 = Vegetables/Legumes   A–58 = Vegetarian   A–62 = Misc   A–62 = Soups/Sauces   A–64 = Fast   A–78 = Convenience   A–80 = Baby foods

Chol (mg)	Calc (mg)	Iron (mg)	Magn (mg)	Pota (mg)	Sodi (mg)	Zinc (mg)	VT-A (RAE)	Thia (mg)	VT-E (mg α)	Ribo (mg)	Niac (mg)	V-B$_6$ (mg)	Fola (µg)	VT-C (mg)
22	83	0.68	6	50	167	0.21	21	0.08	0.36	0.11	0.6	0.02	14	0
5	48	0.59	8	66	239	0.15	4	0.08	0.32	0.08	0.65	0.03	14	0
0	18	5.2	25	121	976	0.79	0	0.7	0.56	0.5	5.95	0.04	121	0
0	96	3.44	24	99	1166	0.62	0	0.48	8.83	0.3	3.8	0.09	112	0
0	28	1.56	13	83	479	0.29	12	0.18	0.55	0.14	1.82	0.04	23	2
0	22	2.59	21	88	1029	0.51	0	0.28	0.21	0.37	3.15	0.07	103	0
0	33	0.88	6	37	146	0.22	<1	0.14	0.09	0.09	1.13	0.02	27	0
0	55	1.33	8	38	192	0.26	0	0.16	0.03	0.13	1.66	0.03	44	0
0	59	1.43	9	40	206	0.28	0	0.17	0.03	0.14	1.79	0.03	48	0
0	54	1.87	15	62	310	0.54	0	0.27	0.24	0.19	2.42	0.03	54	0
0	188	4.28	27	190	756	0.84	0	0.65	0.62	0.42	5.31	0.06	36	0
0	271	5.75	122	257	149	3.65	0	0.39	20.18	0.29	3.54	0.42	84	66
0	265	5.22	96	206	76	3.45	0	0.35	20.26	0.28	3.49	0.39	86	66
0	265	5.37	111	305	274	3.51	0	0.4	20.34	0.3	6.09	0.41	95	66
0	279	5.23	103	212	179	3.21	0	0.35	20.26	0.27	3.45	0.43	85	66
0	300	6.3	160	220	65	5.25		1.5	18.35	1.7	20	2	400	60
0	557	5.01	186		139			2.37		1.1	5.57	1.11		11
0	300	6.3	140	110	90	5.25	0	1.5	18.35	1.7	20	2	400	60
0	300	6.3	140	200	90	5.25	0	1.5	18.35	1.7	20	2	400	60
0	150	2.7	60		80	2.25	0	0.75	18.34	0.85	10	1	200	60
0	300	6.3	140	110	90	5.25	0	1.5	18.35	1.7	20	2	400	60
0	300	6.3	140	145	90	5.25	0	1.5	18.35	1.7	20	2	400	60
0	46	0.36	17	40	42	0.24	<1	0.03	0.02	0.02	0.39	0.06	26	0
0	90	2.38	19	94	344	0.51	0	0.38	0.4	0.21	2.57	0.04	75	0
0	22	0.35	15	25	51	0.2	5	0.03	0.42	<.01	0.19	0.05	1	0
52	191	1.73	14	119	383	0.51	49	0.2	1.72	0.26	1.55	0.04	34	0
37	102	0.81	16	90	132	0.44		0.08	0.55	0.13	0.77	0.04	7	0
0	298	14.8	519	714	41	6.2	0	0.16	2.01	0.41	2.51	0.43	96	8
0	58	5	158	560	18	4.26	2	0.38	0.04	0.23	9.21	0.52	46	0
0	17	2.09	35	146	5	1.29	1	0.13	0.02	0.1	3.24	0.18	25	0
0	31	3.74	393	782	2	4.08	0	0.17	1.75	0.72	11.93	0.36	51	0
0	49	3.44	230	574	24	2.7	1	0.32	0.08	0.16	7.16	0.48	38	0
0	18	1.75	58	124	9	1.04	<1	0.1	0.02	0.05	1.82	0.15	33	0
0	13	0.6	13	91	8	0.41	0	0.1	0.2	0.04	1.54	0.08	24	0
0	41	4.67	166	486	6	3.53	0	0.54	0.3	0.26	7.64	0.41	53	0
0	7	1.51	106	149	5	2.18	<1	0.25	0.05	0.2	3.19	0.26	46	0
0	14	1.3	56	136	1	0.75	0	0.28	0.24	0.05	0.22	0.04	12	0
0	102	15.72	357	1258	36	5.61	0	0.34	8.28	0.67	4.98	0.38	83	0
0	20	0.82	84	84	10	1.23	0	0.19	0.06	0.05	2.98	0.28	8	0
0	0	0			1340		0							0
0	52	7.97	46	213	9	2.02	0	1.07	0.2	0.09	7.76	0.3	427	0
0	16	1.9	19	55	2	0.77	0	0.26	0.06	0.02	2.33	0.15	92	0
0	13	1.04	8	7	5	0.4	0	0.12	0.02	0.08	1.45	0.02	116	0
0	111	6.59	57	222	9	1.78	0	1.1	0.06	0.13	6.72	0.65	475	0

A   Appendix

Table A–1

Food Composition

(Computer code number is for West Diet Analysis program)   (For purposes of calculations, use "0" for t, <1, <.1, <.01, etc.)

Computer Code Number	Food Description	Measure	Wt (g)	H₂O (%)	Ener (kcal)	Prot (g)	Carb (g)	Dietary Fiber (g)	Fat (g)	Fat Breakdown (g)		
										Sat	Mono	Poly
	GRAIN PRODUCTS: CEREAL, FLOUR, GRAIN, PASTA and NOODLES, POPCORN—Continued											
38016	Cooked	1 c	175	72	200	4	43	1	<1	0.1	0.1	0.1
38083	Sticky (Glutinous), cooked	1 c	174	77	169	4	37	2	<1	t	0.1	0.1
38021	Wild, cooked	1 c	164	74	166	7	35	3	1	t	t	0.3
38163	Rice and pasta (Rice-a-Roni), cooked	1 c	202	72	246	5	43	5	6	1.1	2.3	1.9
38034	Tapioca-pearl, dry	1 c	152	11	544	<1	135	1	<1	t	t	t
40002	Wheat, rolled, cooked	1 c	240	84	158	5	33	4	1	0.1	0.1	0.5
	Flour & Grain Fractions:											
38053	Buckwheat flour, dark	1 c	120	11	402	15	85	12	4	0.8	1.1	1.1
	Cornmeal:											
38059	Whole-ground, unbolted, dry	1 c	122	10	442	10	94	9	4	0.6	1.2	2
38004	Degermed, enriched, dry	1 c	138	12	505	12	107	10	2	0.3	0.6	1
38041	Degermed, enriched, baked	1 c	138	12	505	12	107	10	2	0.3	0.6	1
38056	Rye flour, medium	1 c	102	10	361	10	79	15	2	0.2	0.2	0.8
7502	Soy flour, low-fat	1 c	88	3	325	45	30	9	6	0.9	1.3	3.3
38024	Wheat bran, crude	1 c	58	10	125	9	37	25	2	0.4	0.4	1.3
38025	Wheat germ, raw	1 c	115	11	414	27	60	15	11	1.9	1.6	6.9
38026	Wheat germ, toasted	1 c	113	6	432	33	56	17	12	2.1	1.7	7.5
38055	Wheat germ, with brown sugar & honey	1 c	113	3	420	30	66	12	9	1.5	1.2	5.5
	Wheat flour:											
38030	All-purpose white flour, enriched	1 c	125	12	455	13	95	3	1	0.2	0.1	0.5
38033	Self-rising, enriched, unsifted	1 c	125	11	442	12	93	3	1	0.2	0.1	0.5
38032	Whole wheat, from hard wheats	1 c	120	10	407	16	87	15	2	0.4	0.3	0.9
	Breakfast Bars:											
47279	Store brand, fat free, all flavors	1 ea	38		110	2	26	3	0	0	0	0
	Snackwell:											
40255	Apple-cinnamon	1 ea	37	16	119	1	29	1	<1	t	t	0.1
40254	Blueberry	1 ea	37	16	121	1	29	1	<1	t	t	0.1
40253	Strawberry	1 ea	37	16	120	1	29	1	<1	t	t	0.1
	Breakfast cereals, hot, cooked: w/o salt added											
	Corn grits (hominy) enriched:											
38007	Regular/quick prep w/o salt, yellow:	1 c	242	85	143	3	31	1	<1	t	0.1	0.2
40089	Instant, prepared from packet, white	1 ea	137	82	93	2	21	1	<1	t	t	t
	Cream of wheat:											
40079	Regular, quick, instant	1 c	239	87	129	4	27	1	<1	t	t	0.3
40087	Mix and eat, plain, packet	1 ea	142	82	102	3	21	<1	<1	t	t	0.2
40006	Farina cereal, cooked w/o salt	1 c	233	88	112	3	24	1	<1	t	t	t
40014	Malt-O-Meal, cooked w/o salt	1 c	240	88	122	4	26	1	<1	0.1	t	t
40239	Maypo	1 c	216	83	153	5	29	4	2	0.4	0.6	0.5
	Oatmeal or rolled oats:											
40000	Cooked w/o salt, nonfortified	1 c	234	85	147	6	25	4	2	0.4	0.7	0.9
492	Plain, from packet, fortified	½ c	118	86	65	3	11	2	1	0.2	0.3	0.4
	Breakfast cereals, ready to eat:											
40095	All-Bran	1 c	62	3	161	8	46	20	2	0.3	0.5	1.3
40258	Alpha Bits	1 c	28	1	110	2	24	1	1	0.1	0.2	0.2
40098	Apple Jacks	1 c	33	3	129	1	30	1	1	0.1	0.2	0.3
40029	Bran Buds	1 c	90	3	225	6	72	39	2	0.4	0.4	1.1
40323	Bran Chex	1 c	49	2	156	5	39	8	1	0.2	0.3	0.7
40084	Honey BucWheat Crisp	1 c	38	5	147	4	31	3	1	0.2	0.2	0.5
40031	C.W. Post, with raisins	1 c	103	4	446	9	74	14	15	11	1.7	1.4
40032	Cap'n Crunch	1 c	37	2	148	2	31	1	2	0.6	0.4	0.3
40033	Cap'n Crunchberries	1 c	35	2	140	2	30	1	2	0.5	0.4	0.3
40034	Cap'n Crunch, peanut butter	1 c	35	2	146	2	28	1	3	0.7	1.4	0.8
40297	Cheerios	1 c	23	3	85	3	17	2	1	0.3	0.5	0.2
40102	Cocoa Krispies	1 c	41	3	156	1	36	1	1	0.8	0.2	t
40257	Cocoa Pebbles	1 c	32	3	127	1	28	1	1	1.2	0.1	t
40036	Corn Bran	1 c	36	2	121	2	31	6	1	0.3	0.3	0.4
40325	Corn Chex	1 c	28	2	104	2	24	1	<1	t	t	t
40195	Corn Flakes, Kellogg's	1 c	28	3	101	2	24	1	<1	t	t	t

Chol (mg)	Calc (mg)	Iron (mg)	Magn (mg)	Pota (mg)	Sodi (mg)	Zinc (mg)	VT-A (RAE)	Thia (mg)	VT-E (mg α)	Ribo (mg)	Niac (mg)	V-B$_6$ (mg)	Fola (µg)	VT-C (mg)
0	33	1.98	21	65	5	0.54	0	0.44	0.02	0.03	2.45	0.03	133	0
0	3	0.24	9	17	9	0.71	0	0.03	0.07	0.02	0.5	0.05	2	0
0	5	0.98	52	166	5	2.2	<1	0.09	0.39	0.14	2.11	0.22	43	0
2	16	1.9	24	85	1147	0.57	0	0.25	0.27	0.16	3.6	0.2	89	0
0	30	2.4	2	17	2	0.18	0	<.01	0	0	0	0.01	6	0
0	22	1.49	55	170	0	1.18	<1	0.17	0.58	0.13	2.13	0.17	34	0
0	49	4.87	301	692	13	3.74	0	0.5	0.38	0.23	7.38	0.7	65	0
0	7	4.21	155	350	43	2.22	13	0.47	0.51	0.25	4.43	0.37	30	0
0	7	5.7	55	224	4	0.99	15	0.99	0.21	0.56	6.95	0.35	322	0
0	7	5.7	55	224	4	0.99	28	0.74	0.23	0.48	6.6	0.27	52	0
0	24	2.16	76	347	3	2.03	0	0.29	0.81	0.12	1.76	0.27	19	0
0	165	5.27	202	2262	16	1.04	2	0.33	0.17	0.25	1.9	0.46	361	0
0	42	6.13	354	686	1	4.22	<1	0.3	0.86	0.33	7.88	0.76	46	0
0	45	7.2	275	1026	14	14.13	0	2.16	16.1	0.57	7.83	1.5	323	0
0	51	10.27	362	1070	5	18.84	6	1.89	18.07	0.93	6.32	1.11	398	7
0	56	9.1	307	1089	12	15.68	0	1.51	34.07	0.78	5.34	0.56	685	0
0	19	5.8	28	134	2	0.88	0	0.98	0.08	0.62	7.38	0.06	229	0
0	422	5.84	24	155	1588	0.78	0	0.84	0.06	0.52	7.29	0.06	245	0
0	41	4.66	166	486	6	3.52	1	0.54	0.98	0.26	7.64	0.41	53	0
0	20	0.72			25									1
0	17	5	6	68	103	3.88		0.39		0.44	5.2	0.52		0
0	14	4.83	5	44	107	3.85		0.39		0.44	5.2	0.52		0
0	14	4.82	6	47	102	3.83		0.39		0.44	5.2	0.52		2
0	7	1.45	12	51	5	0.17	5	0.2	0.05	0.13	1.75	0.05	80	0
0	8	7.96	10	38	288	0.18	<1	0.16	0.03	0.19	2.21	0.05	47	0
0	50	10.28	12	45	139	0.33	0	0.24	0.02	0	1.43	0.03	108	0
0	20	8.09	7	38	241	0.24	376	0.43	0.01	0.28	4.97	0.57	101	0
0	9	1.16	5	30	5	0.19	0	0.14	0.02	0.1	1.14	0.02	79	0
0	5	9.6	5	31	2	0.17	0	0.48	0.02	0.24	5.76	0.02	5	0
0	117	7.54	48	190	233	1.34	631	0.64	0.15	0.71	8.42	0.83	11	25
0	19	1.59	56	131	2	1.15	0	0.26	0.23	0.05	0.3	0.05	9	0
0	66	5.12	27	63	53	0.54	190	0.17	0.12	0.2	2.4	0.25	51	0
0	205	9.92	236	701	160	3.72	326	0.74	0.76	0.87	9.92	3.72	812	12
0	8	2.66	17	54	178	1.48		0.36	0.02	0.42	4.93	0.5	99	0
0	8	4.59	18	40	157	1.65	51	0.56	0.05	0.43	5.08	0.5	102	15
0	57	13.5	184	900	608	4.5	460	1.08	1.42	1.26	15.3	6.03	1210	18
0	29	13.99	69	216	345	6.48	5	0.64	0.56	0.26	8.62	0.88	173	26
0	54	10.86	43	142	361	0.68	914	0.9	8.99	1.03	12.06	1.88	11	36
0	50	16.38	74	261	161	1.64		1.34	0.72	1.54	18.13	1.85	364	0
0	6	7.07	21	74	277	5.87	3	0.58	0.34	0.66	7.83	0.78	576	0
0	7	6.65	20	72	245	5.54	2	0.55	0.26	0.63	7.38	0.74	545	0
0	4	6.42	24	83	260	5.35	3	0.53	0.42	0.61	7.27	0.71	545	0
0	77	6.21	31	74	209	2.88	115	0.29	0.08	0.33	3.84	0.38	153	5
0	53	6.15	16	66	251	1.97	202	0.49	0.25	0.57	6.56	0.66	135	20
0	4	1.99	12	47	173	1.65		0.41		0.47	5.52	0.55	110	0
0	26	11.1	19	75	309	5.5	3	0.18	0.24	0.62	7.34	0.73	533	0
0	93	8.4	8	23	269	3.5	141	0.35	0.05	0.4	4.68	0.47	187	6
0	2	8.4	3	25	203	0.08	150	0.36	0.04	0.43	5.01	0.5	102	6

A Appendix

Table A–1

Food Composition

(Computer code number is for West Diet Analysis program)   (For purposes of calculations, use "0" for t, <1, <.1, <.01, etc.)

Computer Code Number	Food Description	Measure	Wt (g)	H₂O (%)	Ener (kcal)	Prot (g)	Carb (g)	Dietary Fiber (g)	Fat (g)	Fat Breakdown (g)		
										Sat	Mono	Poly
	GRAIN PRODUCTS: CEREAL, FLOUR, GRAIN, PASTA and NOODLES, POPCORN—Continued											
40263	Corn Flakes, Post Toasties	1 c	28	4	101	2	24	1	<1	0	t	t
40206	Corn Pops	1 c	31	3	118	1	28	<1	<1	t	t	t
40205	Cracklin' Oat Bran	1 c	65	3	266	5	46	8	9	2.7	5.4	1.4
40040	Crispy Wheat `N Raisins	1 c	43	7	143	3	35	4	1	0.2	t	0.3
40041	Fortified Oat Flakes	1 c	48	3	180	8	36	1	1	0.2	0.3	0.4
40202	40% Bran Flakes, Kellogg's	1 c	39	3	124	4	31	7	1	0.2	0.2	0.4
40259	40% Bran Flakes, Post	1 c	47	4	150	4	38	8	1	0.2	0.2	0.7
40218	Froot Loops	1 c	32	3	126	1	28	1	1	0.5	0.1	0.2
40217	Frosted Flakes	1 c	41	3	150	1	37	1	<1	t	t	0.1
40043	Frosted Mini-Wheats	1 c	55	6	187	5	44	6	1	0.2	0.1	0.6
40105	Frosted Rice Krispies	1 c	35	3	133	1	31	<1	<1	t	0.1	0.1
40266	Fruity Pebbles	1 c	32	3	128	1	28	<1	1	0.3	0.6	0.4
40299	Golden Grahams	1 c	39	3	145	2	32	1	1	0.2	0.5	0.5
40048	Granola, homemade	½ c	61	5	299	9	32	5	15	2.8	4.7	6.5
40197	Granola, low fat, commercial	½ c	45	4	165	4	36	2	2	0.7	1.1	0.4
40277	Grape Nuts	½ c	55	4	197	6	45	5	1	0.2	0.2	0.6
40265	Grape Nuts Flakes	1 c	39	3	142	4	32	3	1	0.2	0.3	0.6
40238	Heartland Natural with raisins	1 c	110	5	468	11	76	6	16	4	4.2	6.2
40112	Honey & Nut Corn Flakes	1 c	37	2	150	3	31	1	2	0.3	0.8	0.6
40051	Honey Nut Cheerios	1 c	33	2	123	3	26	2	1	0.3	0.4	0.5
40052	HoneyBran	1 c	35	2	119	3	29	4	1	0.3	t	0.3
40264	HoneyComb	1 c	22	2	87	1	20	1	<1	0.1	0.1	0.2
40054	King Vitaman	1 c	21	2	81	1	18	1	1	0.2	0.2	0.2
40010	Kix	1 c	19	2	72	1	16	1	<1	t	0.1	0.1
40011	Life	1 c	44	4	165	4	34	3	2	0.4	0.7	0.6
40300	Lucky Charms	1 c	32	2	122	2	27	2	1	0.3	0.3	0.3
40124	Mueslix Five Grain	1 c	82	8	289	6	63	6	5	0.7	2	1.8
40008	Nature Valley Granola	1 c	113	4	510	12	74	7	20	2.6	13.3	3.8
40115	Nutri Grain Almond Raisin	1 c	40	6	147	3	31	3	2	t	1	1.2
40281	100% Bran	1 c	66	3	178	8	48	20	3	0.6	0.6	1.9
40063	100% Natural cereal, plain	1 c	104	2	473	11	69	8	20	8.6	4.6	2
40064	100% Natural with apples & cinnamon	1 c	104	2	477	11	70	7	20	15.5	1.8	1.3
40065	100% Natural with raisins & dates	1 c	110	4	496	12	72	7	20	13.6	3.7	1.7
40216	Product 19	1 c	30	3	100	2	25	1	<1	t	0.1	0.2
40066	Quisp	1 c	30	2	122	1	26	1	2	0.5	0.3	0.2
40209	Raisin Bran, Kellogg's	1 c	61	8	195	5	47	7	2	0.3	0.3	0.9
512	Raisin Bran, Post	1 c	59	9	187	5	46	8	1	0.2	0.2	0.7
40117	Raisin Squares	1 c	71	10	239	7	56	7	1	0.2	0.2	0.6
40333	Rice Chex	1 c	33	3	124	2	28	<1	<1	0.1	t	t
40017	Rice Krispies, Kellogg's	1 c	28	2	111	2	25	<1	<1	t	t	t
40018	Rice, puffed	1 c	14	4	54	1	12	<1	<1	t	t	t
40022	Shredded Wheat	1 c	43	5	154	5	35	4	1	0.1	0.1	0.4
40211	Special K	1 c	31	3	117	7	22	1	<1	0.1	0.1	0.2
40261	Super Golden Crisp	1 c	33	2	123	2	30	<1	<1	t	t	0.1
40068	Honey Smacks	1 c	36	2	139	2	32	1	1	0.1	0.2	0.3
40070	Tasteeos	1 c	24	2	94	3	19	3	1	0.2	0.2	0.2
40021	Total, wheat, with added calcium	1 c	40	3	130	3	30	3	1	0.2	0.2	0.4
40306	Trix	1 c	28	2	109	1	25	1	1	0.2	0.6	0.3
40335	Wheat Chex	1 c	46	2	159	5	37	5	1	0.2	0.1	0.4
40023	Wheat cereal, puffed, fortified	1 c	12	4	44	2	9	1	<1	t	t	0.1
40307	Wheaties	1 c	29	3	103	3	23	3	1	0.2	0.3	0.3
7508	Natto	1 c	175	55	371	31	25	9	19	2.8	4.3	10.9
	Pasta:											
	Cellophane Noodles:											
38146	Cooked	1 c	190	79	160	<1	39	<1	<1	t	t	t
7196	Dry	1 c	140	13	491	<1	121	1	<1	t	t	t
38048	Chow Mein, dry	1 c	45	1	237	4	26	2	14	2	3.5	7.8
	Cooked:											
38092	Fresh	2 oz	57	69	75	3	14	1	1	t	t	0.2
38118	Linguini/Rotini	1 c	140	66	197	7	40	2	1	0.1	0.1	0.4

Appendix A

PAGE KEY: A–4 = Beverages   A–6 = Dairy   A–10 = Eggs   A–12 = Fat/Oil   A–14 = Fruit   A–20 = Bakery
A–26 = Grain   A–32 = Fish   A–32 = Meats   A–36 = Poultry   A–38 = Sausage   A–40 = Mixed/Fast   A–44 = Nuts/Seeds   A–46 = Sweets
A–48 = Vegetables/Legumes   A–58 = Vegetarian   A–62 = Misc   A–62 = Soups/Sauces   A–64 = Fast   A–78 = Convenience   A–80 = Baby foods

Chol (mg)	Calc (mg)	Iron (mg)	Magn (mg)	Pota (mg)	Sodi (mg)	Zinc (mg)	VT-A (RAE)	Thia (mg)	VT-E (mg α)	Ribo (mg)	Niac (mg)	V-B$_6$ (mg)	Fola (µg)	VT-C (mg)
0	1	5.4	4	33	266	0.13		0.38	0.67	0.43	5	0.5	100	0
0	5	1.92	2	26	120	1.52	151	0.37	0.03	0.43	4.99	0.5	102	6
0	27	2.4	80	292	186	2.02	298	0.5	0.91	0.57	6.7	0.66	133	21
0	0	5.85	33	178	197	5.85	117	0.58	0.27	0.67	7.83	0.78	157	0
0	68	13.73	58	228	220	2.54		0.62	0.34	0.72	8.45	0.86	169	0
0	21	24.18	55	230	279	20.48	307	2.11	36.14	2.3	26.91	2.73	542	81
0	26	12.69	101	290	344	2.35		0.59		0.67	7.83	0.78	157	0
0	25	4.51	9	35	151	1.5	155	0.38	0.15	0.42	4.99	0.51	100	15
0	2	5.94	3	30	196	0.07	212	0.49	0.02	0.62	6.64	0.66	134	8
0	18	15.95	65	187	6	1.76	0	0.41	0.3	0.46	5.39	0.54	108	0
0	4	2.1	9	33	254	0.32	176	0.46	0.02	0.49	5.95	0.6	234	7
0	2	2.13	6	35	187	1.78		0.44		0.5	5.93	0.59	118	0
0	455	5.85	11	64	349	4.88	195	0.49	0.14	0.55	6.51	0.65	130	8
0	48	2.59	107	328	13	2.51	1	0.45	3.59	0.18	1.29	0.19	51	1
0	19	1.35	34	135	111	2.84	169	0.28	3.78	0.32	3.74	1.48	302	3
0	19	15.36	55	169	336	1.14		0.36		0.4	4.74	0.47	95	0
0	15	10.89	40	133	188	1.61		0.5	0.1	0.57	6.72	0.67	135	0
0	66	4.01	141	415	226	2.83	3	0.32	0.77	0.14	1.54	0.2	44	1
0	4	3.03	3	40	249	0.26	152	0.26	0.09	0.3	3.37	0.33	74	10
0	110	4.95	35	101	296	4.12	165	0.41	0.59	0.47	5.51	0.55	220	7
0	16	5.56	46	150	202	0.9	463	0.46	0.81	0.52	6.16	0.63	23	19
0	4	2.05	8	26	163	1.14		0.28		0.32	3.79	0.38	76	0
0	3	6.09	18	58	176	2.62	210	0.26	1.41	0.3	3.5	0.35	280	8
0	95	5.13	5	22	169	2.38	101	0.24	0.04	0.27	3.17	0.32	127	4
0	154	12.31	42	125	226	5.68	1	0.55	0.24	0.64	7.57	0.76	572	0
0	107	4.8	17	61	217	4	160	0.4	0.1	0.45	5.34	0.53	213	6
0	67	8.94	82	369	107	7.46	747	0.75	8.94	0.84	9.84	0.99	197	1
0	85	3.53	107	375	183	2.27	0	0.35	7.97	0.12	1.25	0.16	17	0
0	122	1	9	143	142	2.72	0	0.28	4	0.32	3.64	0.36	80	0
0	46	8.12	312	652	457	5.74	0	1.58	1.53	1.78	20.92	2.11	47	63
2	124	2.68	115	515	50	2.45	1	0.33	3.34	0.28	2.09	0.19	37	1
0	157	2.89	72	514	52	2		0.33	0.73	0.57	1.87	0.11	17	1
0	160	3.12	124	538	47	2.11		0.31	0.77	0.65	2.09	0.16	45	0
0	5	18.09	16	50	207	15.3	225	1.5	20.13	1.71	20.01	2.07	400	61
0	3	5.51	16	57	222	4.59	12	0.46	0.2	0.52	6.12	0.61	467	3
0	29	4.64	83	372	362	1.55	155	0.39	0.48	0.44	5.18	0.52	104	0
0	27	10.8	88	357	360	2.25		0.38		0.42	5	0.5	100	0
0	28	19.88	56	342	4	1.99	0	0.5	0.37	0.57	6.67	0.67	134	0
0	110	9.9	10	32	311	4.12	165	0.41	0.02	0.47	5.51	0.55	220	7
0	5	0.7	12	27	206	0.46	371	0.52	0	0.59	6.92	0.69	88	15
0	1	0.4	4	16	1	0.15	0	0.06	0.02	0.04	0.49	0	22	0
0	16	1.81	57	155	4	1.42	0	0.11	0.23	0.12	2.26	0.11	22	0
0	9	8.37	19	61	224	0.9	230	0.53	7.07	0.59	7.13	1.98	400	21
0	7	2.08	20	48	51	1.75		0.43	0.12	0.5	5.81	0.59	116	0
0	9	0.47	21	54	67	0.47	204	0.5	0.18	0.58	6.66	0.68	135	8
0	11	6.86	26	71	183	0.69	318	0.31	0.08	0.36	4.22	0.43	85	13
0	1333	24	32	119	256	20	200	2	26.84	2.27	26.68	2.67	533	80
0	93	4.2	3	16	181	3.5	140	0.35	0.56	0.4	4.68	0.47	93	6
0	92	13.34	37	172	410	3.68	138	0.34	0.33	0.39	4.6	0.46	368	6
0	3	0.53	16	44	1	0.37	0	0.08	0	0.05	0.63	0.02	18	0
0	0	7.83	31	107	210	7.25	145	0.72	0.18	0.82	9.66	0.97	193	6
0	380	15.05	201	1276	12	5.3	0	0.28	0.02	0.33	0	0.23	14	23
0	13	0.86	3	3	8	0.2	0	0.04	0.06	0	0.06	0.02	1	0
0	35	3.04	4	14	14	0.57	0	0.21	0.18	0	0.28	0.07	3	0
0	9	2.13	23	54	198	0.63	<1	0.26	1.57	0.19	2.68	0.05	40	0
19	3	0.65	10	14	3	0.32	3	0.12	0.09	0.09	0.57	0.02	36	0
0	10	1.96	25	43	1	0.74	0	0.29	0.08	0.14	2.34	0.05	108	0

A   Appendix

Table A–1

Food Composition

(Computer code number is for West Diet Analysis program)   (For purposes of calculations, use "0" for t, <1, <.1, <.01, etc.)

Computer Code Number	Food Description	Measure	Wt (g)	H$_2$O (%)	Ener (kcal)	Prot (g)	Carb (g)	Dietary Fiber (g)	Fat (g)	Fat Breakdown (g)		
										Sat	Mono	Poly
	GRAIN PRODUCTS: CEREAL, FLOUR, GRAIN, PASTA and NOODLES, POPCORN—Continued											
38047	Egg noodles, cooked, enriched	1 c	160	69	213	8	40	2	2	0.5	0.7	0.7
	Macaroni, cooked:											
38102	Enriched	1 c	140	66	197	7	40	2	1	0.1	0.1	0.4
38110	Whole wheat	1 c	140	67	174	7	37	4	1	0.1	0.1	0.3
38117	Vegetable, enriched	1 c	134	68	172	6	36	6	<1	t	t	t
	Spaghetti:											
38121	With salt, enriched	1 c	140	66	197	7	40	2	1	0.1	0.1	0.4
38060	Whole-wheat, cooked	1 c	140	67	174	7	37	6	1	0.1	0.1	0.3
38062	Spinach noodles, dry	3½ oz	100	8	372	13	75	11	2	0.2	0.2	0.6
	Popcorn:											
44012	Air popped, plain	1 c	8	4	31	1	6	1	<1	t	t	0.2
44013	Popped in vegetable oil/salted	1 c	11	3	55	1	6	1	3	0.5	0.9	1.5
44014	Sugar-syrup coated	1 c	35	3	151	1	28	2	4	1.3	1	1.6
	MEATS: FISH AND SHELLFISH											
	Fish:											
17029	Bass, baked or broiled	4 oz	113	69	165	27	0	0	5	1.1	2.1	1.5
17031	Bluefish, baked or broiled	4 oz	113	63	180	29	0	0	6	1.3	2.6	1.5
17126	Catfish, breaded/flour fried	4 oz	113	49	329	21	14	<1	20	4.5	9.2	5.2
	Cod:											
17037	Baked	4 oz	113	76	119	26	0	0	1	0.2	0.1	0.3
17000	Batter fried	4 oz	113	67	197	20	8	<1	9	1.8	3.6	3
17001	Poached, no added fat	4 oz	113	77	116	25	0	0	1	0.1	0.1	0.3
17002	Fish sticks, breaded pollock	2 ea	56	46	152	9	13	1	7	1.8	2.8	1.8
17068	Flounder/sole, baked	4 oz	113	73	132	27	0	0	2	0.4	0.3	0.7
17071	Grouper, baked or broiled	4 oz	113	73	133	28	0	0	1	0.3	0.3	0.5
17007	Haddock, breaded, fried	4 oz	113	60	247	23	10	<1	12	2.6	5.3	3.7
	Halibut:											
17291	Baked	4 oz	113	72	158	30	0	0	3	0.5	1.1	1.1
17259	Smoked	4 oz	113		203	34	0	0	4	0.6	1.2	1.5
17044	Raw	4 oz	113	78	124	24	0	0	3	0.4	0.8	0.8
17012	Herring, pickled	4 oz	113	55	296	16	11	0	20	2.7	13.5	1.9
	Ocean Perch:											
17093	Baked/Broiled	4 oz	113	73	137	27	0	0	2	0.4	0.9	0.6
17015	Breaded/Fried	4 oz	113	58	255	23	10	<1	13	2.7	5.8	3.9
19025	Octopus, raw	4 oz	113	80	93	17	2	0	1	0.3	0.2	0.3
17096	Pollock, baked, broiled, or poached	4 oz	113	74	128	27	0	0	1	0.3	0.2	0.6
17099	Salmon, broiled or baked	4 oz	113	62	244	31	0	0	12	2.2	6	2.7
17060	Sardines, Atlantic, canned, drained, 2=24g	4 oz	113	60	235	28	0	0	13	1.7	4.4	5.8
17022	Snapper, baked or broiled	4 oz	113	70	145	30	0	0	2	0.4	0.4	0.7
19047	Squid, fried in flour	4 oz	113	65	198	20	9	0	8	2.1	3.1	2.4
17080	Surimi	4 oz	113	76	112	17	8	0	1	0.2	0.2	0.5
17065	Swordfish, raw	4 oz	113	76	137	22	0	0	5	1.2	1.7	1
17066	Swordfish, baked or broiled	4 oz	113	69	175	29	0	0	6	1.6	2.2	1.3
17082	Trout, baked or broiled	4 oz	113	70	170	26	0	0	7	1.8	2	2.1
	Tuna:											
	Light, canned, drained solids:											
17025	Oil pack	1 c	145	60	287	42	0	0	12	2.2	4.3	4.2
17027	Water pack	1 c	154	75	179	39	0	0	1	0.4	0.2	0.5
17085	Bluefin, fresh	4 oz	113	68	163	26	0	0	6	1.4	1.8	1.6
	Shellfish:											
	Clams:											
19128	Raw meat only	1 ea	145	82	107	19	4	0	1	0.1	0.1	0.4
19002	Canned, drained	1 c	160	64	237	41	8	0	3	0.3	0.3	0.9
19000	Steamed, meat only	10 ea	95	64	141	24	5	0	2	0.2	0.2	0.5
	Crab, meat only:											
19033	Blue crab, cooked	1 c	118	77	120	24	0	0	2	0.3	0.3	0.8
19037	Imitation, from surimi	4 oz	113	74	115	14	12	0	1	0.3	0.2	0.8
19006	Lobster meat, cooked w/moist heat	1 c	145	76	142	30	2	0	1	0.2	0.2	0.1

PAGE KEY: A–4 = Beverages   A–6 = Dairy   A–10 = Eggs   A–12 = Fat/Oil   A–14 = Fruit   A–20 = Bakery
A–26 = Grain   A–32 = Fish   A–32 = Meats   A–36 = Poultry   A–38 = Sausage   A–40 = Mixed/Fast   A–44 = Nuts/Seeds   A–46 = Sweets
A–48 = Vegetables/Legumes   A–58 = Vegetarian   A–62 = Misc   A–62 = Soups/Sauces   A–64 = Fast   A–78 = Convenience   A–80 = Baby foods

Chol (mg)	Calc (mg)	Iron (mg)	Magn (mg)	Pota (mg)	Sodi (mg)	Zinc (mg)	VT-A (RAE)	Thia (mg)	VT-E (mg α)	Ribo (mg)	Niac (mg)	V-B$_6$ (mg)	Fola (µg)	VT-C (mg)
53	19	2.54	30	45	11	0.99	10	0.3	0.26	0.13	2.38	0.06	102	0
0	10	1.96	25	43	1	0.74	0	0.29	0.08	0.14	2.34	0.05	108	0
0	21	1.48	42	62	4	1.13	<1	0.15	0.42	0.06	0.99	0.11	7	0
0	15	0.66	25	42	8	0.59	7	0.15	0.12	0.08	1.44	0.03	87	0
0	10	1.96	25	43	140	0.74	0	0.29	0.08	0.14	2.34	0.05	108	0
0	21	1.48	42	62	4	1.13	<1	0.15	0.42	0.06	0.99	0.11	7	0
0	58	2.13	174	376	36	2.76	23	0.37	0.64	0.2	4.55	0.32	48	0
0	1	0.21	10	24	<1	0.28	1	0.02	0.02	0.02	0.16	0.02	2	0
0	1	0.31	12	25	97	0.29	1	0.01	0.55	0.01	0.17	0.02	2	0
2	15	0.61	12	38	72	0.2	1	0.02	0.42	0.02	0.77	<.01	2	0
98	116	2.16	43	515	102	0.94	40	0.1	0.84	0.1	1.72	0.16	19	2
86	10	0.7	47	539	87	1.18	156	0.08	0.71	0.11	8.19	0.52	2	0
91	62	1.88	36	391	240	1.17		0.46	2.87	0.21	3.78	0.22	17	1
62	16	0.55	47	276	88	0.66	16	0.1	0.92	0.09	2.84	0.32	9	1
56	33	0.81	28	437	104	0.57		0.08	1.49	0.11	2.58	0.37	10	2
52	10	0.37	31	484	90	0.56	9	0.02	0.33	0.05	2.45	0.45	7	3
63	11	0.41	14	146	326	0.37	17	0.07	0.29	0.1	1.19	0.03	24	0
77	20	0.38	66	389	119	0.71	15	0.09	0.75	0.13	2.46	0.27	10	0
53	24	1.29	42	537	60	0.58	56	0.09	0.71	<.01	0.43	0.4	11	0
88	70	2.02	49	377	194	0.63		0.11	1.93	0.12	4.94	0.31	16	0
46	68	1.21	121	651	78	0.6	61	0.08	1.23	0.1	8.05	0.45	16	0
59	87	1.56	154	833		0.78	86	0.11	1.11	0.14	10.83	0.64	22	0
36	53	0.95	94	508	61	0.47	53	0.07	0.96	0.08	6.61	0.39	14	0
15	87	1.38	9	78	983	0.6	292	0.04	1.93	0.16	3.73	0.19	2	0
61	155	1.33	44	396	108	0.69	16	0.15	1.84	0.15	2.75	0.31	11	1
71	151	1.88	39	335	201	0.75		0.18	2.87	0.2	2.98	0.24	13	1
54	60	5.99	34	396	260	1.9	51	0.03	1.36	0.05	2.37	0.41	18	6
108	7	0.32	82	437	131	0.68	28	0.08	0.89	0.09	1.86	0.08	5	0
98	8	0.62	35	424	75	0.58	71	0.24	1.42	0.19	7.54	0.25	6	0
160	432	3.3	44	449	571	1.48	36	0.09	2.31	0.26	5.93	0.19	14	0
53	45	0.27	42	590	64	0.5	40	0.06	0.71	<.01	0.39	0.52	7	2
294	44	1.14	43	315	346	1.97	12	0.06	2.09	0.52	2.94	0.07	16	5
34	10	0.29	49	127	162	0.37	23	0.02	0.71	0.02	0.25	0.03	2	0
44	5	0.92	31	325	102	1.3	41	0.04	0.56	0.11	10.94	0.37	2	1
56	7	1.18	38	417	130	1.66	46	0.05	0.71	0.13	13.32	0.43	2	1
78	97	0.43	35	506	63	0.58	17	0.17	0.57	0.11	6.52	0.39	21	2
26	19	2.02	45	300	513	1.3	33	0.06	1.26	0.17	17.98	0.16	7	0
46	17	2.36	42	365	521	1.19	26	0.05	0.51	0.11	20.45	0.54	6	0
43	9	1.15	56	285	44	0.68	740	0.27	1.13	0.28	9.78	0.51	2	0
49	67	20.27	13	455	81	1.99	130	0.12	0.45	0.31	2.56	0.09	23	19
107	147	44.74	29	1005	179	4.37	290	0.24	0.99	0.68	5.37	0.18	46	35
64	87	26.56	17	597	106	2.59	162	0.14	1.86	0.4	3.19	0.1	28	21
118	123	1.07	39	382	329	4.98	2	0.12	2.17	0.06	3.89	0.21	60	4
23	15	0.44	49	102	950	0.37	23	0.04	0.11	0.03	0.2	0.03	2	0
104	88	0.57	51	510	551	4.23	38	0.01	1.45	0.1	1.55	0.11	16	0

A Appendix

Table A–1

Food Composition   (Computer code number is for West Diet Analysis program)   (For purposes of calculations, use "0" for t, <1, <.1, <.01, etc.)

Computer Code Number	Food Description	Measure	Wt (g)	H₂O (%)	Ener (kcal)	Prot (g)	Carb (g)	Dietary Fiber (g)	Fat (g)	Fat Breakdown (g) Sat	Mono	Poly
	MEATS: FISH AND SHELLFISH—Continued											
	Oysters:											
	Raw:											
19026	Eastern	1 c	248	85	169	17	10	0	6	1.9	0.8	2.4
578	Pacific	1 c	248	82	201	23	12	0	6	1.3	0.9	2.2
	Cooked:											
19009	Eastern, breaded, fried, medium	5 ea	73	65	144	6	8	<1	9	2.3	3.4	2.4
19008	Western, simmered	5 ea	125	64	204	24	12	0	6	1.3	1	2.2
	Scallops:											
19030	Breaded, cooked from frozen	6 ea	93	58	200	17	9	<1	10	2.5	4.2	2.7
19046	Imitation, from surimi	4 oz	113	74	112	14	12	0	<1	t	t	0.2
	Shrimp:											
19012	Cooked, boiled, 2 large=11g	16 ea	88	77	87	18	0	0	1	0.3	0.2	0.4
19016	Canned, drained	1/2 c	64	73	77	15	1	0	1	0.2	0.2	0.5
19014	Fried, 2 large=15g,breaded	12 ea	90	53	218	19	10	<1	11	1.9	3.4	4.6
19032	Raw, large, about 7g each	14 ea	98	76	104	20	1	0	2	0.3	0.2	0.7
19039	Imitation, from surimi	4 oz	113	75	114	14	10	0	2	0.3	0.2	0.8
	MEATS: BEEF, LAMB, AND PORK											
	Beef:											
10008	Corned, canned	4 oz	113	58	282	31	0	0	17	7	6.7	0.7
10009	Dried, cured	1 oz	28	54	43	9	1	0	1	0.3	0.2	t
	Ground beef, broiled, patty:											
10030	Extra lean, about 16% fat	4 oz	113	54	299	32	0	0	18	7	7.8	0.7
10032	Lean, 21% fat	4 oz	113	53	316	32	0	0	20	7.8	8.7	0.7
10002	Rib, whole, rstd, choice, ¼" trim	4 oz	113	46	425	25	0	0	35	14.2	15.2	1.3
	Roast:											
10001	Blade, chuck, lean, brsd, choice, ¼" trim	4 oz	113	55	297	35	0	0	16	6.3	7	0.5
10016	Bottom round, brsd, ¼" trim	4 oz	113	52	311	32	0	0	19	7.2	8.3	0.7
10014	Bottom round, lean, brsd, choice, ¼" trim	4 oz	113	57	249	36	0	0	11	3.6	4.7	0.4
10000	Pot, chuck arm, brsd, choice, ¼" trim	4 oz	113	47	393	30	0	0	29	11.5	12.5	1.1
10017	Round eye, rstd, choice, ¼" trim	4 oz	113	59	272	30	0	0	16	6.2	6.8	0.6
10013	Round eye, lean, rstd, choice, ¼" trim	4 oz	113	65	198	33	0	0	6	2.3	2.7	0.2
	Steak:											
10056	Rib, lean, brld, ¼" trim	4 oz	113	58	250	32	0	0	13	5.1	5.3	0.4
10005	Top sirloin, lean, brld, choice, ¼" trim	4 oz	113	62	228	34	0	0	9	3.5	3.9	0.4
10006	T-bone, brld, choice, ¼" trim	4 oz	113	51	364	26	0	0	28	11	12.7	1
10007	T-bone, lean, brld, choice, ¼" trim	4 oz	113	61	232	30	0	0	11	4.1	5.6	0.3
	Variety meats:											
10010	Liver, fried	4 oz	113	62	198	30	6	0	5	1.7	0.7	0.7
10011	Tongue, cooked	4 oz	113	58	314	22	0	0	25	9.2	11.4	0.7
	Lamb:											
	Chop:											
13508	Shoulder arm, brsd, choice, ¼" trim	1 ea	70	44	242	21	0	0	17	6.9	7.1	1.2
13509	Shoulder arm, lean, brsd, ¼" trim	1 ea	55	49	153	20	0	0	8	2.8	3.4	0.5
13512	Loin, brld, choice, ¼" trim	1 ea	64	52	202	16	0	0	15	6.3	6.2	1.1
13513	Loin chop, lean, brld, choice, ¼" trim	1 ea	46	61	99	14	0	0	4	1.6	2	0.3
13517	Cutlet, avg of lean cuts, cooked	4 oz	113	54	330	28	0	0	23	9.9	9.8	1.7
	Leg:											
13500	Whole, rstd, choice,¼" trim	4 oz	113	57	292	29	0	0	19	7.8	7.9	1.3
13501	Whole, lean, rstd, choice, ¼" trim	4 oz	113	64	216	32	0	0	9	3.1	3.8	0.6
	Rib:											
13618	Rstd, choice, ¼" trim	4 oz	113	48	406	24	0	0	34	14.4	14.1	2.5
13511	Lean, rstd, choice, ¼" trim	4 oz	113	60	262	30	0	0	15	5.4	6.6	1
	Shoulder:											
13502	Whole, rstd, choice, ¼" trim	4 oz	113	56	312	25	0	0	23	9.5	9.2	1.8
13503	Whole, lean, rstd, choice, ¼" trim	4 oz	113	63	231	28	0	0	12	4.6	4.9	1.1

PAGE KEY: A–4 = Beverages A–6 = Dairy A–10 = Eggs A–12 = Fat/Oil A–14 = Fruit A–20 = Bakery
A–26 = Grain A–32 = Fish A–32 = Meats A–36 = Poultry A–38 = Sausage A–40 = Mixed/Fast A–44 = Nuts/Seeds A–46 = Sweets
A–48 = Vegetables/Legumes A–58 = Vegetarian A–62 = Misc A–62 = Soups/Sauces A–64 = Fast A–78 = Convenience A–80 = Baby foods

Chol (mg)	Calc (mg)	Iron (mg)	Magn (mg)	Pota (mg)	Sodi (mg)	Zinc (mg)	VT-A (RAE)	Thia (mg)	VT-E (mg α)	Ribo (mg)	Niac (mg)	V-B$_6$ (mg)	Fola (µg)	VT-C (mg)
131	112	16.52	117	387	523	225.21	74	0.25	2.11	0.24	3.42	0.15	25	9
124	20	12.67	55	417	263	41.22	201	0.17	2.11	0.58	4.98	0.12	25	20
59	45	5.07	42	178	304	63.6	66	0.11	1.66	0.15	1.2	0.05	23	3
125	20	11.5	55	378	265	41.55	182	0.16	1.06	0.55	4.52	0.11	19	16
57	39	0.76	55	310	432	0.99	21	0.04	1.77	0.1	1.4	0.13	34	2
25	9	0.35	49	116	898	0.37	23	0.01	0.12	0.02	0.35	0.03	2	0
172	34	2.72	30	160	197	1.37	60	0.03	1.21	0.03	2.28	0.11	4	2
111	38	1.75	26	134	108	0.81	12	0.02	0.6	0.02	1.76	0.07	1	1
159	60	1.13	36	202	310	1.24	51	0.12	1.35	0.12	2.76	0.09	16	1
149	51	2.36	36	181	145	1.09	53	0.03	1.08	0.03	2.5	0.1	3	2
41	21	0.68	49	101	797	0.37	23	0.03	0.12	0.04	0.19	0.03	2	0
97	14	2.35	16	154	1137	4.03	0	0.02	0.17	0.17	2.75	0.15	10	0
22	1	0.81	6	81	781	1.11	0	0.02	0	0.06	0.93	0.07	2	0
112	10	3.13	28	417	93	7.27	0	0.08	0.2	0.36	6.61	0.36	12	0
114	14	2.77	27	394	101	7.01	0	0.07	0.23	0.27	6.75	0.34	12	0
96	12	2.61	21	334	71	5.92	0	0.08	0.27	0.19	3.8	0.26	8	0
120	15	4.16	26	297	80	11.61	0	0.09	0.16	0.32	3.02	0.33	7	0
108	7	3.53	25	319	56	5.55	0	0.08	0.21	0.27	4.21	0.37	11	0
108	6	3.91	28	348	58	6.19	0	0.08	0.2	0.29	4.61	0.41	12	0
112	11	3.45	21	275	67	7.57	0	0.08	0.26	0.27	3.54	0.32	10	0
81	7	2.07	27	406	67	4.87	0	0.09	0.23	0.18	3.92	0.4	7	0
78	6	2.2	31	446	70	5.36	0	0.1	0.12	0.19	4.24	0.43	8	0
90	15	2.9	31	445	78	7.9	0	0.11	0.16	0.25	5.42	0.45	9	0
101	12	3.8	36	455	75	7.37	0	0.15	0.16	0.33	4.84	0.51	11	0
77	9	3.4	24	311	77	4.73	0	0.1	0.25	0.24	4.37	0.37	8	0
67	7	4.14	29	370	87	5.77	0	0.12	0.16	0.28	5.23	0.44	9	0
431	7	6.97	25	397	87	5.91	8751	0.2	0.52	3.87	19.75	1.16	294	1
149	6	2.95	17	208	73	46.22	0	0.02	0.34	0.33	3.94	0.18	8	1
84	18	1.67	18	214	50	4.26	0	0.05	0.1	0.18	4.66	0.08	13	0
67	14	1.48	16	186	42	4.01	0	0.04	0.1	0.15	3.48	0.07	12	0
64	13	1.16	15	209	49	2.23	0	0.06	0.08	0.16	4.54	0.08	12	0
44	9	0.92	13	173	39	1.9	0	0.05	0.07	0.13	3.15	0.07	11	0
110	12	2.26	25	340	77	4.67	0	0.12	0.15	0.32	7.48	0.16	19	0
105	12	2.24	27	354	75	4.97	0	0.11	0.17	0.31	7.45	0.17	23	0
101	9	2.4	29	382	77	5.58	0	0.12	0.2	0.33	7.16	0.19	26	0
110	25	1.81	23	306	82	3.94	0	0.1	0.11	0.24	7.63	0.12	17	0
99	24	2	26	356	92	5.05	0	0.1	0.17	0.26	6.96	0.17	25	0
104	23	2.23	26	284	75	5.91	0	0.1	0.16	0.27	6.95	0.15	24	0
98	21	2.41	28	299	77	6.83	0	0.1	0.2	0.29	6.51	0.17	28	0

A Appendix

Table A–1

Food Composition   (Computer code number is for West Diet Analysis program)   (For purposes of calculations, use "0" for t, <1, <.1, <.01, etc.)

Computer Code Number	Food Description	Measure	Wt (g)	H$_2$O (%)	Ener (kcal)	Prot (g)	Carb (g)	Dietary Fiber (g)	Fat (g)	Sat	Mono	Poly
	MEATS—Continued											
	Variety meats:											
13624	Brains, pan-fried	4 oz	113	76	164	14	0	0	11	2.9	2.1	1.2
13507	Sweetbreads, cooked	4 oz	113	60	264	26	0	0	17	7.7	6.2	0.8
13527	Tongue, cooked	4 oz	113	58	311	24	0	0	23	8.8	11.3	1.4
	Pork:											
	Cured:											
12000	Bacon, medium slices	3 pce	19	12	103	7	<1	0	8	2.6	3.5	0.9
12009	Breakfast strips, cooked	2 pce	23	27	106	7	<1	0	8	2.9	3.8	1.3
12002	Canadian-style bacon	2 pce	47	62	87	11	1	0	4	1.3	1.9	0.4
	Ham, roasted:											
12211	Reg, 11% fat	4 oz	113	65	201	26	0	0	10	3.5	5	1.6
12212	Extra lean, 5% fat	4 oz	113	68	164	24	2	0	6	2	3	0.6
12209	Extra lean, 4% fat, cnd	4 oz	113	69	154	24	1	0	6	1.8	2.8	0.5
	Chop:											
12029	Whole loin, brsd	1 ea	89	58	213	24	0	0	12	4.5	5.4	1
12033	Whole loin, lean, brsd	1 ea	80	61	163	23	0	0	7	2.7	3.3	0.6
12192	Center loin, w/bone, brld	1 ea	82	58	197	24	0	0	11	3.9	4.8	0.8
12025	Center loin, lean, brld	1 ea	74	61	149	22	0	0	6	2.2	2.7	0.4
12044	Center loin, pan fried	1 ea	78	53	216	23	0	0	13	4.7	5.5	1.5
12040	Blade loin, lean, pan fried	1 ea	63	59	152	16	0	0	10	3.3	3.9	1.2
	Leg:											
12016	Ham, whole, rstd	4 oz	113	55	308	30	0	0	20	7.3	8.9	1.9
12017	Ham, rump, lean, rstd	4 oz	113	61	233	35	0	0	9	3.2	4.3	0.9
	Rib:											
12050	Center loin, w/bone, rstd	4 oz	113	56	288	31	0	0	17	6.7	7.9	1.4
12055	Center loin, lean, w/bone, rstd	4 oz	113	59	252	32	0	0	13	4.9	5.9	1
	Shoulder:											
12003	Picnic, brsd	4 oz	113	48	372	32	0	0	26	9.6	11.7	2.6
12004	Picnic, lean, brsd	4 oz	113	54	280	36	0	0	14	4.7	6.5	1.3
12010	Spareribs, brsd	4 oz	113	40	449	33	0	0	34	12.6	15.2	3.1
14004	Rabbit, roasted (1 cup meat=140g)	4 oz	113	61	223	33	0	0	9	2.7	2.5	1.8
	Veal:											
11519	Short ribs, rstd	4 oz	113	60	258	27	0	0	16	6.1	6.1	1.1
11500	Liver, brsd	4 oz	113	60	217	32	4	0	7	2.2	1.3	1.2
14013	Venison (deer meat), roasted	4 oz	113	65	179	34	0	0	4	1.4	1	0.7
	Chicken:											
15016	Canned, boneless chicken	4 oz	113	69	186	25	0	0	9	2.5	3.6	2
	Fried, batter dipped:											
15013	Breast	1 ea	280	52	728	70	25	1	37	9.9	15.3	8.6
15030	Drumstick	1 ea	72	53	193	16	6	<1	11	3	4.6	2.7
15036	Thigh	1 ea	86	52	238	19	8	<1	14	3.8	5.8	3.4
15034	Wing	1 ea	49	46	159	10	5	<1	11	2.9	4.4	2.5
	Fried, flour coated:											
15003	Breast	1 ea	196	57	435	62	3	<1	17	4.8	6.9	3.8
15057	Breast, without skin	1 ea	172	60	322	58	1	0	8	2.2	3	1.8
15007	Drumstick	1 ea	49	57	120	13	1	<1	7	1.8	2.7	1.6
15009	Thigh	1 ea	62	54	162	17	2	<1	9	2.5	3.6	2.1
15011	Thigh, without skin	1 ea	52	59	113	15	1	0	5	1.4	2	1.3
15029	Wing	1 ea	32	49	103	8	1	<1	7	1.9	2.8	1.6
15902	Patty, breaded, cooked	1 ea	75	49	213	12	11	<1	13	4.1	6.4	1.7
	Roasted:											
15000	All types of meat	1 c	140	64	266	41	0	0	10	2.9	3.7	2.4
15027	Dark meat	1 c	140	63	287	38	0	0	14	3.7	5	3.2
15032	Light meat	1 c	140	65	242	43	0	0	6	1.8	2.2	1.4
15004	Breast, without skin	1 ea	172	65	284	53	0	0	6	1.7	2.1	1.3
15035	Drumstick, without skin	1 ea	44	65	76	12	0	0	2	0.7	0.8	0.6
15156	Leg, without skin	1 ea	95	65	181	26	0	0	8	2.2	2.9	1.9
15010	Thigh	1 ea	62	59	153	16	0	0	10	2.7	3.8	2.1
15012	Thigh, without skin	1 ea	52	63	109	13	0	0	6	1.6	2.2	1.3
15006	Stewed, all types	1 c	140	67	248	38	0	0	9	2.6	3.3	2.2

Chol (mg)	Calc (mg)	Iron (mg)	Magn (mg)	Pota (mg)	Sodi (mg)	Zinc (mg)	VT-A (RAE)	Thia (mg)	VT-E (mg α)	Ribo (mg)	Niac (mg)	V-B$_6$ (mg)	Fola (μg)	VT-C (mg)
2309	14	1.9	16	232	151	1.54	0	0.12	1.73	0.27	2.79	0.12	6	14
452	14	2.4	21	329	59	3.03	0	0.02	0.78	0.24	2.89	0.06	15	23
214	11	2.97	18	179	76	3.38	0	0.09	0.36	0.47	4.17	0.19	3	8
21	2	0.27	6	107	439	0.66	2	0.08	0.06	0.05	2.11	0.07	<1	0
24	3	0.45	6	107	483	0.85	0	0.17	0.06	0.08	1.75	0.08	1	0
27	5	0.39	10	183	727	0.8	0	0.39	0.16	0.09	3.25	0.21	2	0
67	9	1.51	25	462	1695	2.79	0	0.82	0.35	0.37	6.95	0.35	3	0
60	9	1.67	16	324	1359	3.25	0	0.85	0.28	0.23	4.55	0.45	3	0
34	7	1.04	24	393	1283	2.52	0	1.17	0.29	0.28	5.53	0.51	6	0
71	19	0.95	17	333	43	2.12	2	0.56	0.21	0.23	3.93	0.33	3	1
63	14	0.9	16	310	40	1.98	2	0.53	0.17	0.21	3.67	0.31	3	0
67	27	0.66	20	294	48	1.85	2	0.87	0.27	0.24	4.3	0.35	5	0
61	23	0.63	20	278	44	1.76	1	0.85	0.31	0.23	4.1	0.35	4	0
72	21	0.71	23	332	62	1.8	2	0.89	0.2	0.24	4.37	0.37	5	1
52	14	0.67	16	230	49	2.44	1	0.46	0.14	0.23	2.8	0.26	3	1
106	16	1.14	25	398	68	3.34	3	0.72	0.25	0.35	5.17	0.45	11	0
108	8	1.29	33	442	73	3.4	3	0.91	0.46	0.4	5.56	0.38	3	0
82	32	1.06	24	476	52	2.33	2	0.82	0.41	0.34	6.91	0.37	3	0
80	29	1.11	25	494	53	2.41	2	0.86	0.55	0.36	7.25	0.39	3	0
123	20	1.82	21	417	99	4.72	3	0.61	0.34	0.35	5.89	0.4	5	0
129	9	2.2	25	458	115	5.62	2	0.68	0.33	0.41	6.71	0.46	6	0
137	53	2.09	27	362	105	5.2	3	0.46	0.38	0.43	6.19	0.4	5	0
93	21	2.57	24	433	53	2.57	0	0.1	0.96	0.24	9.53	0.53	12	0
124	12	1.1	25	333	104	4.62	0	0.06	0.4	0.31	7.89	0.28	15	0
577	7	5.77	23	372	88	12.69	23894	0.21	0.77	3.23	14.86	1.04	374	1
127	8	5.05	27	379	61	3.11	0	0.2	0.28	0.68	7.58	0.42	5	0
70	16	1.79	14	156	568	1.59	38	0.02	0.29	0.15	7.15	0.4	5	2
238	56	3.5	67	563	770	2.66	56	0.32	2.97	0.41	29.46	1.2	42	0
62	12	0.97	14	134	194	1.68	19	0.08	0.88	0.15	3.67	0.19	13	0
80	15	1.25	18	165	248	1.75	25	0.1	1.05	0.2	4.91	0.22	16	0
39	10	0.63	8	68	157	0.68	17	0.05	0.52	0.07	2.58	0.15	9	0
174	31	2.33	59	508	149	2.16	29	0.16	1.12	0.26	26.93	1.14	12	0
157	28	1.96	53	475	136	1.86	12	0.14	0.72	0.22	25.43	1.1	7	0
44	6	0.66	11	112	44	1.42	12	0.04	0.41	0.11	2.96	0.17	5	0
60	9	0.92	16	147	55	1.56	18	0.06	0.52	0.15	4.31	0.2	7	0
53	7	0.76	14	135	49	1.45	11	0.05	0.3	0.13	3.7	0.2	5	0
26	5	0.4	6	57	25	0.56	12	0.02	0.18	0.04	2.14	0.13	2	0
45	12	0.94	15	184	399	0.78	11	0.07	1.46	0.1	5.04	0.23	8	0
125	21	1.69	35	340	120	2.94	22	0.1	0.38	0.25	12.84	0.66	8	0
130	21	1.86	32	336	130	3.92	31	0.1	0.38	0.32	9.17	0.5	11	0
119	21	1.48	38	346	108	1.72	13	0.09	0.38	0.16	17.39	0.84	6	0
146	26	1.79	50	440	127	1.72	10	0.12	0.46	0.2	23.58	1.03	7	0
41	5	0.57	11	108	42	1.4	8	0.03	0.12	0.1	2.67	0.17	4	0
89	11	1.24	23	230	86	2.72	18	0.07	0.26	0.22	6	0.35	8	0
58	7	0.83	14	138	52	1.46	30	0.04	0.17	0.13	3.95	0.19	4	0
49	6	0.68	12	124	46	1.34	10	0.04	0.14	0.12	3.39	0.18	4	0
116	20	1.64	29	252	98	2.79	21	0.07	0.38	0.23	8.56	0.36	8	0

Table A–1

Food Composition

(Computer code number is for West Diet Analysis program)   (For purposes of calculations, use "0" for t, <1, <.1, <.01, etc.)

Computer Code Number	Food Description	Measure	Wt (g)	H₂O (%)	Ener (kcal)	Prot (g)	Carb (g)	Dietary Fiber (g)	Fat (g)	Fat Breakdown (g) Sat	Mono	Poly
	MEATS—Continued											
	Variety meats:											
15025	Gizzards, simmered	1 c	145	68	212	44	0	0	4	1	0.8	0.5
15024	Hearts, simmered	1 c	145	65	268	38	<1	0	11	3.3	2.9	3.3
15215	Liver, simmered: Ounce	3 oz	85	67	142	21	1	0	6	1.8	1.2	1.1
	Duck:											
16295	Whole, w/skin, rstd, about 2.7 cups	½ ea	382	52	1287	73	0	0	108	36.9	49.3	13.9
14000	Whole, w/o skin, rstd, about 1.5 cups	½ ea	221	64	444	52	0	0	25	9.2	8.2	3.2
14003	Goose, whole, w/skin, rstd, about 5.5 cups	½ ea	774	52	2361	195	0	0	170	53.2	79.3	19.5
	Turkey:											
	Roasted, meat only:											
16002	Dark meat	4 oz	113	63	211	32	0	0	8	2.7	1.9	2.4
16158	Light meat	4 oz	113	66	177	34	0	0	4	1.2	0.6	1
16000	All types, chopped or diced	1 c	140	65	238	41	0	0	7	2.3	1.4	2
16003	Ground, cooked	4 oz	113	59	266	31	0	0	15	3.8	5.5	3.6
16010	With gravy, frozen package	3 oz	85	85	57	5	4	0	2	0.7	0.8	0.4
16307	Patty, breaded, fried	2 oz	57	50	161	8	9	<1	10	2.7	4.3	2.7
16308	Roasted, from frozen, seasoned	4 oz	113	68	175	24	3	0	7	2.1	1.4	1.9
16008	Roll, light meat	1 pce	28	72	41	5	<1	0	2	0.6	0.7	0.5
	Lunchmeat:											
	Turkey:											
	Breast:											
13259	Barbecued, Louis Rich	2 oz	56	72	57	11	2		<1	0.2	0.2	0.1
13108	Hickory smoked, Louis Rich	1 pce	80	73	80	16	2	0	1	0		
13110	Honey roasted, Louis Rich	1 pce	80	73	80	16	3	0	1	0.5		
13109	Oven roasted, Louis Rich	1 pce	80		70	16	0	0	1	0		
13114	Fat Free	1 pce	28	76	24	4	1	0	<1	t	t	t
13020	Pastrami	2 pce	57	71	80	10	1	0	4	1	1.2	0.9
13144	Salami	1 pce	28	72	41	4	<1	0	3	0.8	0.9	0.7
13014	Ham	2 pce	57	72	72	10	1	<1	3	0.9	1.1	0.8
	Bologna:											
13002	Beef	1 pce	23	54	72	2	1	0	6	2.6	2.8	0.2
13176	Beef, light, Oscar Mayer	1 pce	28	65	56	3	2	0	4	1.6	2	0.1
13006	Beef & pork	1 pce	28	52	85	4	2	0	7	2.7	2.9	0.3
13218	Healthy Favorites	1 pce	23		22	4	1	0	<1	0		
13032	Pork	1 pce	23	61	57	4	<1	0	5	1.6	2.2	0.5
13174	Regular, light, Oscar Mayer	1 pce	28	65	57	3	2	0	4	1.6	2	0.4
13007	Turkey	1 pce	28	65	59	3	1	<1	4	1.2	1.9	1.1
13149	Turkey, Louis Rich	1 pce	56	67	115	6	1	0	10	2.9	3.6	2.6
	Chicken:											
13222	Chicken breast, Healthy Favorites	4 pce	52		40	9	1	0	0	0	0	0
	Beef:											
13039	Corned beef loaf, jellied	1 pce	28	69	43	6	0	0	2	0.7	0.8	t
	Ham:											
13057	Chopped ham, packaged	2 pce	42	64	96	7	0	0	7	2.4	3.4	0.9
13220	Honey ham, Healthy Favorites	4 pce	52	73	55	9	2	0	1	0.4	0.6	0.1
13168	Oscar Mayer lower sodium ham	1 pce	21	73	23	3	1	0	1	0.3	0.4	t
13048	Mortadella lunchmeat	2 pce	30	52	93	5	1	0	8	2.9	3.4	0.9
13049	Olive loaf lunchmeat	2 pce	57	58	134	7	5	0	9	3.3	4.5	1.1
13051	Pickle & pimento loaf	2 pce	57	57	149	7	3	0	12	4.5	5.5	1.5
	Sausages:											
13035	Beerwurst/beer salami, beef	1 oz	28	57	77	4	1	<1	6	2.4	2.8	0.6
13031	Beerwurst/beer salami, pork	1 oz	28	61	67	4	1	0	5	1.8	2.5	0.7
13001	Berliner sausage	1 oz	28	61	64	4	1	0	5	1.7	2.2	0.4
13066	Braunschweiger sausage	2 pce	57	51	186	8	2	0	16	5.2	7.5	1.6
13036	Bratwurst-link	1 ea	70	51	226	10	2		19	7	9.3	2
13037	Cheesefurter/cheese smokie	2 ea	86	52	281	12	1	0	25	9	11.8	2.6
13070	Chorizo, pork & beef	1 ea	60	32	273	14	1	0	23	8.6	11	2.1
	Frankfurters:											
13008	Beef, large link, 8/package	1 ea	57	52	188	6	2	0	17	6.7	8.2	0.7
13010	Beef and pork, large link, 8/package	1 ea	45	56	137	5	1		12	4.8	6.2	1.2

Appendix **A**

PAGE KEY: A–4 = Beverages   A–6 = Dairy   A–10 = Eggs   A–12 = Fat/Oil   A–14 = Fruit   A–20 = Bakery
A–26 = Grain   A–32 = Fish   A–32 = Meats   A–36 = Poultry   A–38 = Sausage   A–40 = Mixed/Fast   A–44 = Nuts/Seeds   A–46 = Sweets
A–48 = Vegetables/Legumes   A–58 = Vegetarian   A–62 = Misc   A–62 = Soups/Sauces   A–64 = Fast   A–78 = Convenience   A–80 = Baby foods

Chol (mg)	Calc (mg)	Iron (mg)	Magn (mg)	Pota (mg)	Sodi (mg)	Zinc (mg)	VT-A (RAE)	Thia (mg)	VT-E (mg α)	Ribo (mg)	Niac (mg)	V-B₆ (mg)	Fola (μg)	VT-C (mg)
536	25	4.63	4	260	81	6.41	0	0.04	0.29	0.3	4.52	0.1	7	0
351	28	13.09	29	191	70	10.58	12	0.1	2.32	1.07	4.06	0.46	116	3
479	9	9.89	21	224	65	3.38	3384	0.25	0.7	1.69	9.39	0.64	491	24
321	42	10.31	61	779	225	7.11	241	0.66	2.67	1.03	18.43	0.69	23	0
197	27	5.97	44	557	144	5.75	51	0.57	1.55	1.04	11.27	0.55	22	0
704	101	21.9	170	2546	542	20.28	163	0.6	13.47	2.5	32.26	2.86	15	0
96	36	2.63	27	328	89	5.04	0	0.07	0.72	0.28	4.12	0.41	10	0
78	21	1.53	32	345	72	2.31	0	0.07	0.1	0.15	7.73	0.61	7	0
106	35	2.49	36	417	98	4.34	0	0.09	0.46	0.25	7.62	0.64	10	0
115	28	2.18	27	305	121	3.23	0	0.06	0.38	0.19	5.45	0.44	8	0
15	12	0.79	7	52	471	0.6	11	0.02	0.3	0.11	1.53	0.08	3	0
35	8	1.25	9	157	456	0.82	6	0.06	0.72	0.11	1.31	0.11	16	0
60	6	1.84	25	337	768	2.87	0	0.05	0.43	0.18	7.09	0.31	6	0
12	11	0.36	4	70	137	0.44	0	0.02	0.04	0.06	1.96	0.09	1	0
25	14	0.61	16		592	0.58	0							0
35	0	0.72			1060		0							0
35	0	0.72			940		0							0
35	0				910		0							0
9	3	0.31	8	57	334	0.24								0
31	5	0.95	8	148	596	1.23	0	0.03	0.13	0.14	2.01	0.15	3	0
21	11	0.35	6	60	281	0.65								0
41	5	1.33	13	164	635	1.48	4	0.02	0.36	0.08	1.21	0.12	4	0
13	7	0.25	3	40	248	2.09	3	<.01	0.08	0.02	0.58	0.04	2	3
12	4	0.34	4	44	322	0.53	0						4	0
17	24	0.34	5	88	206	0.64	7	0.06	0.1	0.05	0.71	0.08	2	0
8		0.18			255									
14	3	0.18	3	65	272	0.47	0	0.12	0.06	0.04	0.9	0.06	1	0
16	14	0.39	6	46	313	0.45	0	0.04		0.03	0.86	0.05	5	0
21	34	0.84	4	38	351	0.36	3	0.01	0.13	0.03	0.73	0.07	3	4
44	68	0.9	10	103	484	1.14	0	0.03		0.1	2.15	0.1		0
25		0.72			620									
13	3	0.57	3	28	267	1.15	0	0	0.05	0.03	0.49	0.03	2	0
21	3	0.35	7	134	576	0.81	0	0.27	0.1	0.09	1.63	0.15	<1	0
24	6	0.7	18	144	635	1.02	0							0
9	1	0.3	5	197	174	0.42	0							0
17	5	0.42	3	49	374	0.63	0	0.04	0.07	0.05	0.8	0.04	1	0
22	62	0.31	11	169	846	0.79	34	0.17	0.14	0.15	1.05	0.13	1	0
21	54	0.58	10	194	792	0.8	13	0.17	0.14	0.14	1.17	0.11	3	0
17	8	0.48	5	68	205	0.62	<1	0.07	0.05	0.05	0.83	0.06	1	0
17	2	0.21	4	71	347	0.48	0	0.16	0.06	0.05	0.91	0.1	1	0
13	3	0.32	4	79	363	0.69	0	0.11	0.06	0.06	0.87	0.06	1	0
103	5	6.38	6	113	661	1.6	2405	0.14	0.2	0.87	4.77	0.19	25	0
44	34	0.72	11	197	778	1.47	0	0.18	0.19	0.16	2.31	0.09	4	0
58	50	0.93	11	177	931	1.94	40	0.22	0	0.14	2.49	0.11	3	0
53	5	0.95	11	239	741	2.05		0.38	0.13	0.18	3.08	0.32	1	0
30	8	0.86	8	89	650	1.4	0	0.02	0.11	0.08	1.35	0.05	3	0
22	5	0.52	4	75	504	0.83	8	0.09	0.11	0.05	1.19	0.06	2	0

A Appendix

Table A–1

Food Composition

(Computer code number is for West Diet Analysis program)   (For purposes of calculations, use "0" for t, <1, <.1, <.01, etc.)

Computer Code Number	Food Description	Measure	Wt (g)	H₂O (%)	Ener (kcal)	Prot (g)	Carb (g)	Dietary Fiber (g)	Fat (g)	Fat Breakdown (g) Sat	Mono	Poly
	MEATS—Continued											
13012	Turkey frankfurter, 10/package	1 ea	45	63	102	6	1	0	8	2.7	2.5	2.2
13129	Turkey/chicken frank 8/pkg	1 ea	43	67	81	5	2	0	6	1.7	2.4	1.4
13043	Kielbasa sausage	1 pce	26	54	81	3	1	0	7	2.6	3.4	0.8
13044	Knockwurst sausage, link	1 ea	68	55	209	8	2	0	19	6.9	8.7	2
13021	Pepperoni	2 pce	11	31	51	2	<1	<1	4	1.8	2.1	0.3
13022	Polish	1 oz	28	53	91	4	<1	0	8	2.9	3.8	0.9
13024	Salami, pork and beef	2 pce	57	60	142	8	1	0	11	4.6	5.2	1.2
13026	Salami, pork and beef, dry	3 pce	30	35	125	7	1	0	10	3.7	5.1	1
13025	Salami, turkey	2 pce	57	54	126	8	11	<1	5	2	1.8	1.4
13029	Smoked link sausage, beef and pork	1 ea	68	54	218	8	2	0	20	6.6	8.3	2.7
13027	Smoked link sausage, pork	1 ea	68	39	265	15	1	0	22	7.7	10	2.6
13030	Summer sausage	2 pce	46	51	154	7	<1	0	14	5.5	6	0.6
13052	Turkey breakfast sausage	1 pce	28	60	64	6	0		5	1.6	1.8	1.2
13054	Vienna sausage, canned	2 ea	32	60	89	3	1	0	8	3	4	0.5
	Sandwich spreads:											
13034	Ham salad spread	2 tbsp	30	63	65	3	3	0	5	1.5	2.2	0.8
13056	Pork and beef	2 tbsp	30	60	70	2	4	<1	5	1.8	2.3	0.8
	MIXED DISHES											
15907	Almond Chicken	1 c	242	77	280	22	16	3	15	1.9	6.1	5.6
7084	Bean cake	1 ea	32	23	130	2	16	1	7	1	2.9	2.6
56124	Beef fajita	1 ea	223	65	399	23	36	3	18	5.5	7.6	3.5
56119	Beef flauta	1 ea	113	51	354	14	13	2	28	4.8	11.8	9.4
5513	Broccoli, batter fried	1 c	85	74	122	3	9	2	9	1.3	2.1	4.9
15903	Buffalo wings/spicy chicken wings	2 pce	32	53	98	8	<1	<1	7	1.8	2.7	1.8
56649	Cheeseburger deluxe	1 ea	219	52	563	28	38		33	15	12.6	2
56123	Chicken fajita	1 ea	223	65	363	20	44	5	12	2.3	5.5	3.1
56120	Chicken flauta	1 ea	113	55	330	13	12	2	26	4.2	10.7	9.3
50312	Chili con carne	½ c	127	77	128	12	11		4	1.7	1.7	0.3
15915	Chicken teriyaki, breast	1 ea	128	67	178	27	7	<1	4	0.9	1.1	0.9
45557	Chinese Pastry	1 oz	28	46	67	1	13	<1	2	0.2	0.5	0.8
56094	Chop suey with beef & pork	1 c	220	63	421	22	31		24	4.7	8.3	9.2
5461	Coleslaw	1 c	132	74	195	2	17		15	2.1	3.2	8.5
5366	Corn pudding	1 c	250	76	272	11	32	4	13	6.3	4.3	1.7
19539	Deviled egg (½ egg + filling)	1 ea	31	70	63	4	<1	0	5	1.2	1.7	1.5
	Egg Foo Yung Patty:											
56132	Meatless	1 ea	86	77	113	6	3	1	8	2	3.4	2.1
56290	With beef	1 ea	86	76	119	8	3	<1	8	2	2.9	2.2
56287	With chicken	1 ea	86	76	121	8	4	1	8	1.9	2.8	2.3
	Egg Roll:											
56110	Meatless	1 ea	64	70	101	3	10	1	6	1.2	2.9	1.3
57523	With Meat	1 ea	64	66	113	5	9	1	6	1.4	3	1.3
56003	Egg salad	1 c	183	57	584	17	3	0	56	10.5	17.4	23.9
56102	Falafel	1 ea	17	35	57	2	5		3	0.4	1.7	0.7
50182	Hot & Sour Soup (Chinese)	1 c	244	87	162	15	5	1	8	2.7	3.4	1.2
56659	Hamburger deluxe	1 ea	110	49	279	13	27		13	4.1	5.3	2.6
7081	Hummous/hummus	¼ c	62	65	110	3	12	2	5	0.7	3	1.3
16335	Kung Pao Chicken	1 c	162	54	431	29	11	2	31	5.2	13.9	9.7
	Lasagna:											
56108	With meat, homemade	1 pce	245	67	392	23	40		16	8	5.2	0.8
56071	Without meat, homemade	1 pce	218	69	306	16	40		10	5.6	2.5	0.6
56073	Chicken alfredo, w/broccoli	1 ea	340	75	389	24	41		14	6.7	5.5	0.8
57521	Lo mein, meatless	1 c	200	82	135	6	27	4	1	0.1	t	0.3
57522	Lo mein, with meat	1 c	200	72	283	20	21	3	14	2.6	4	6
56080	Moussaka (lamb & eggplant)	1 c	250	82	238	17	13	4	13	4.6		
5514	Mushrooms, batter fried	5 ea	70	63	156	2	11	1	12	1.5	3.6	6
56005	Potato salad with mayonnaise & eggs	½ c	125	76	179	3	14	2	10	1.8	3.1	4.7
56618	Pizza, combination, ½ of 12" round	1 pce	79	48	184	13	21		5	1.5	2.5	0.9
56619	Pizza, pepperoni, ½ of 12" round	1 pce	71	47	181	10	20		7	2.2	3.1	1.2
56098	Quiche Lorraine ⅛ of 8" quiche	1 pce	176	53	526	15	25	1	41	18.9	14.3	5.2
38067	Ramen noodles-cooked	1 c	227	86	154	3	20	1	7	1.7	1.2	3.3

Chol (mg)	Calc (mg)	Iron (mg)	Magn (mg)	Pota (mg)	Sodi (mg)	Zinc (mg)	VT-A (RAE)	Thia (mg)	VT-E (mg α)	Ribo (mg)	Niac (mg)	V-B$_6$ (mg)	Fola (µg)	VT-C (mg)
48	48	0.83	6	81	642	1.4	0	0.02	0.28	0.08	1.86	0.1	4	0
40	56	0.94	10	69	488	0.8								0
17	11	0.38	4	70	280	0.53	0	0.06	0.06	0.06	0.75	0.05	1	0
41	7	0.45	7	135	632	1.13	0	0.23	0.39	0.1	1.86	0.12	1	0
13	2	0.16	2	35	197	0.3	0	0.06	0.03	0.03	0.6	0.04	1	0
20	3	0.4	4	66	245	0.54	0	0.14	0.06	0.04	0.96	0.05	1	0
37	7	1.52	9	113	607	1.22	0	0.14	0.13	0.21	2.03	0.12	1	0
24	2	0.45	5	113	558	0.97	0	0.18	0.08	0.09	1.46	0.15	1	0
45	42	0.88	15	226	619	1.77	1	0.24	0.14	0.17	2.27	0.24	6	12
39	8	0.51	9	122	619	0.86	9	0.13	0.09	0.07	2	0.11	1	0
46	20	0.79	13	228	1020	1.92	0	0.48	0.17	0.17	3.08	0.24	3	1
34	6	1.17	6	125	571	1.18	0	0.07	0.1	0.15	1.98	0.12	1	0
23	5	0.51	6	75	188	0.96	0	0.03	0.14	0.08	1.4	0.08	1	0
17	3	0.28	2	32	305	0.51	0	0.03	0.07	0.03	0.52	0.04	1	0
11	2	0.18	3	45	274	0.33	0	0.13	0.52	0.04	0.63	0.04	<1	0
11	4	0.24	2	33	304	0.31	8	0.05	0.52	0.04	0.52	0.04	1	0
40	69	1.97	60	549	526	1.62		0.08	3.8	0.2	9.48	0.44	26	7
0	3	0.67	6	58	1	0.16	0	0.07	1.24	0.05	0.55	0.02	9	0
45	84	3.76	38	479	316	3.52	22	0.39	1.74	0.3	5.4	0.38	23	27
37	51	1.87	28	313	68	3.45		0.06	4.65	0.13	1.88	0.24	10	19
15	66	0.98	20	242	64	0.38		0.08	2.85	0.13	0.73	0.11	36	53
26	5	0.4	6	59	25	0.57		0.01	0.27	0.04	2.06	0.13	1	0
88	206	4.66	44	445	1108	4.6	140	0.39	1.18	0.46	7.38	0.28	81	8
39	101	3.32	48	534	343	1.65		0.43	1.71	0.33	6.12	0.38	42	37
35	50	0.95	27	269	71	1.13		0.05	4.36	0.09	3.1	0.22	8	18
67	34	2.6	23	347	505	1.79	42	0.06	0.81	0.57	1.24	0.17	23	1
82	27	1.71	35	309	1683	1.96		0.08	0.35	0.19	8.75	0.47	12	3
0	6	0.18	7	25	3	0.16	<1	0.02	0.26	<.01	0.27	0.04	1	0
43	39	4.19	54	519	950	3.48		0.36	1.8	0.37	5.73	0.39	44	20
7	45	0.96	12	236	356	0.26	48	0.05	5.28	0.04	0.11	0.15	51	11
250	100	1.4	38	402	138	1.25	135	1.03	0.52	0.32	2.47	0.3	62	7
122	15	0.35	3	37	50	0.3	50	0.02	0.61	0.15	0.02	0.05	13	0
185	31	1.04	12	117	317	0.7		0.04	1.22	0.26	0.43	0.09	30	5
166	25	1.01	11	139	131	1.01		0.05	1.06	0.23	0.65	0.13	22	3
167	27	0.82	11	136	132	0.76		0.05	1.1	0.23	0.89	0.12	22	3
30	14	0.81	9	97	274	0.25		0.08	0.85	0.11	0.8	0.05	13	3
37	15	0.83	10	124	274	0.46		0.16	0.8	0.13	1.28	0.09	10	2
581	74	1.81	13	181	464	1.45	263	0.09	7.66	0.66	0.09	0.46	61	0
0	9	0.58	14	99	50	0.26	<1	0.02	0.19	0.03	0.18	0.02	16	0
34	29	1.9	29	384	1011	1.51		0.27	0.15	0.25	5	0.2	13	1
26	63	2.63	22	227	504	2.06	4	0.23	0.82	0.2	3.68	0.12	52	2
0	30	0.97	18	107	150	0.68	<1	0.06	0.46	0.03	0.25	0.25	37	5
64	49	1.96	63	428	907	1.5		0.15	3.9	0.15	13.23	0.59	43	8
58	270	3.07	50	460	391	3.33		0.24	1.16	0.34	4.2	0.25	20	14
32	265	2.35	44	373	365	1.82		0.23	1.09	0.28	2.51	0.17	17	14
55	265	3.82	64	759	840	3.7		0.29	3.45	0.39	5.07	0.32	28	13
0	46	2.03	33	386	564	0.92	65	0.23	0.35	0.24	2.82	0.19	48	12
42	29	2.07	42	332	142	1.83		0.41	2.06	0.28	5.02	0.36	53	11
97	75	1.74	40	565	460	2.57		0.16	0.98	0.31	4.14	0.23	46	6
2	15	1.22	7	154	112	0.42		0.11	2.34	0.26	2.25	0.04	8	1
85	24	0.81	19	318	661	0.39	40	0.1	2.33	0.08	1.11	0.18	9	12
21	101	1.53	18	179	382	1.11	58	0.21		0.17	1.96	0.09	32	2
14	65	0.94	9	153	267	0.52	53	0.13		0.23	3.05	0.06	37	2
221	231	1.88	24	239	221	1.5		0.26	2.02	0.49	2.01	0.1	19	1
0	13	0.39	10	49	802	0.18	2	0.02	2.34	0.01	0.25	0.01	3	0

A   Appendix

Table A–1

Food Composition   (Computer code number is for West Diet Analysis program)   (For purposes of calculations, use "0" for t, <1, <.1, <.01, etc.)

Computer Code Number	Food Description	Measure	Wt (g)	H₂O (%)	Ener (kcal)	Prot (g)	Carb (g)	Dietary Fiber (g)	Fat (g)	Sat	Mono	Poly
	MIXED DISHES—Continued											
56302	Ravioli, meat	½ c	125	69	197	11	18	1	9	3	3.7	1
38145	Fried rice (meatless)	1 c	166	68	271	5	34	1	12	1.8	3.2	6.7
	Spaghetti (enriched) in tomato sauce:											
56097	With cheese, home recipe	1 c	250	77	260	9	37		9	2		
56100	With meatballs, home recipe	1 c	248	71	362	18	28	3	18	4.8		
56076	Spinach souffle	1 c	136	74	219	11	3		18	7.1	6.8	3.1
2995	Spring roll, vegetable	1 ea	63	49	158	4	20	1	7	0.9		
	Sushi:											
56313	Fish and vegetable	1 c	166	65	232	9	47	2	1	0.2	0.2	0.2
56314	Vegetable seaweed	1 c	166	71	194	4	43	1	<1	0.1	0.1	0.1
12900	Sweet & sour pork	1 c	226	77	231	15	25	2	8	2.1	3.2	2.3
15921	Sweet & sour chicken breast	1 ea	131	79	118	8	15	1	3	0.5	0.9	1.4
56916	Tabouli	1 c	160	77	199	3	16	4	15	2	10.8	1.4
	Thai dishes											
1984	Beef peanut satay, svg	1 ea	129	60	286	26	5	1	18	5.9		
1988	Drunkard noodles, svg	1 ea	366	80	344	5	51	4	14	1.8		
1994	Lemongrass vegetables, svg	1 ea	187	75	238	8	19	4	16	2.6		
1998	Peanut chicken, svg	1 ea	309	81	272	19	26	3	11	2.3		
2994	Spicy noodles, svg	1 ea	310	58	626	34	65	3	26	4.4		
2994	Sweet noodles, svg	1 ea	211	58	426	23	44	2	18	3		
56118	Three bean salad	1 c	150	81	140	4	15	5	8	1.1	1.7	4.4
56006	Waldorf salad	1 c	137	58	411	4	12	3	41	4.3		
56111	Wonton, meat filled	1 ea	19	45	55	3	5	<1	3	0.8	1.2	0.3
	FAST FOODS and SANDWICHES											
	(see end of this appendix for additional Fast Foods)											
	Burritos:											
66026	Beef & bean	1 ea	116	52	255	11	33	3	9	4.2	3.5	0.6
66025	Bean	1 ea	109	53	225	7	36	4	7	3.5	2.4	0.6
16255	Chicken con queso	1 ea	299		350	14	60	6	6	2.5		
	Cheeseburgers:											
66013	With bun, 4-oz patty	1 ea	166	51	417	21	35		21	8.7	7.8	2.7
66015	With bun, regular	1 ea	154	55	359	18	28		20	9.2	7.2	1.5
56668	Corndog	1 ea	175	47	460	17	56		19	5.2	9.1	3.5
15181	Chicken	1 ea	113		271	11	23	2	15			
66021	Enchilada	1 ea	163	63	319	10	29		19	10.6	6.3	0.8
66032	English muffin with egg, cheese, bacon	1 ea	146	57	308	18	28	2	13	5	5	1.7
	Fish sandwich:											
66010	Large, no cheese	1 ea	158	47	431	17	41	<1	23	5.2	7.7	8.2
66011	Regular, with cheese	1 ea	183	45	523	21	48	<1	29	8.1	8.9	9.4
	Hamburgers:											
56658	With bun, regular	1 ea	107	45	275	12	35	2	10	3.6	3.4	1
66006	With bun, 4-oz patty	1 ea	215	51	576	32	39		32	12	14.1	2.8
66004	Hotdog/frankfurter with bun	1 ea	98	54	242	10	18		15	5.1	6.9	1.7
	Lunchables:											
56938	Bologna & American cheese	1 ea	128		450	18	19	0	34	15		
56939	Ham & cheese	1 ea	128		320	22	19	0	17	8		
56930	Honey ham & Amer. w/choc pudding	1 ea	176		390	18	34	<1	20	9		
56931	Honey turkey & cheddar w/Jello	1 ea	163		320	17	27	<1	16	9		
56940	Pepperoni & American cheese	1 ea	128		480	20	19	0	36	17		
56936	Salami & American cheese	1 ea	128		430	18	18	0	32	15		
56937	Turkey & cheddar cheese	1 ea	128		360	20	20	1	22	11		
	SANDWICHES:											
	Avocado, chesse, tomato & lettuce:											
56021	On white bread, firm	1 ea	210	62	429	14	35	5	27	7.7		
56022	On part whole wheat	1 ea	201	63	402	14	30	6	27	7.8		
56023	On whole wheat	1 ea	214	63	424	15	33	8	28	8.2		
	Bacon, lettuce & tomato sandwich:											
56009	On white bread, soft	1 ea	124	52	318	10	29	2	18	4.1		
56011	On part whole wheat	1 ea	124	53	314	11	26	3	19	4.6		

Chol (mg)	Calc (mg)	Iron (mg)	Magn (mg)	Pota (mg)	Sodi (mg)	Zinc (mg)	VT-A (RAE)	Thia (mg)	VT-E (mg α)	Ribo (mg)	Niac (mg)	V-B$_6$ (mg)	Fola (µg)	VT-C (mg)
85	35	2.15	20	264	90	1.7		0.16	1.32	0.22	2.99	0.14	14	4
43	28	1.94	23	128	261	0.92		0.21	2.51	0.11	2.24	0.15	22	4
8	80	2.25	26	408		1.3		0.25	2.75	0.18	2.25	0.2	8	12
65	92	3.33	44	479	1133	3.4		0.25	2.46	0.3	4.38	0.29	68	16
184	230	1.35	38	201	763	1.29	267	0.09	1.22	0.3	0.48	0.12	80	
3	58	1.78	12	74	262	0.38		0.18	1.29	0.14	1.88	0.03	32	1
11	25	2.33	27	218	93	0.84		0.28	0.62	0.07	2.96	0.16	15	4
0	22	1.65	21	106	5	0.75	33	0.21	0.13	0.04	1.99	0.15	11	3
39	28	1.44	34	386	839	1.47		0.55	1.09	0.21	3.63	0.41	10	20
23	15	0.84	21	185	506	0.67		0.06	0.67	0.08	3.09	0.18	6	12
0	29	1.25	36	246	799	0.48	34	0.08	2.16	0.05	1.14	0.11	31	29
71	15	2.77	39	408	1173	5.81		0.1	1.23	0.24	4.75	0.41	16	3
0	37	1.77	37	524	1227	0.77		0.16	3.83	0.13	1.89	0.39	42	159
0	146	2.88	49	453	741	1.09		0.18	2.41	0.12	1.45	0.29	48	92
36	42	2.65	48	344	900	1.21		0.22	1.73	0.12	8.24	0.48	66	67
212	73	4.77	95	623	1660	2.31		0.26	1.97	0.32	9.47	0.55	90	20
144	49	3.25	65	424	1130	1.57		0.17	1.34	0.22	6.45	0.38	61	14
0	35	1.48	27	247	520	0.58		0.08	1.74	0.1	0.44	0.04	56	4
21	44	0.97	37	258	235	0.7		0.09	8.62	0.05	0.54	0.36	34	5
20	4	0.4	4	51	10	0.32		0.09	0.18	0.06	0.63	0.04	3	0
24	53	2.46	42	329	670	1.93	16	0.27	0.7	0.42	2.71	0.19	58	1
2	57	2.27	44	328	495	0.76	9	0.32	0.87	0.31	2.04	0.15	44	1
35	40	1.8			590									6
60	171	3.42	30	335	1051	3.49	71	0.35		0.28	8.05	0.18	61	2
52	182	2.65	26	229	976	2.62	82	0.32	1.34	0.23	6.38	0.15	65	2
79	102	6.18	18	262	973	1.31	60	0.28	0.7	0.7	4.16	0.09	103	0
53	60	2.71			738		0							0
44	324	1.32	51	240	784	2.51	99	0.08	1.47	0.42	1.91	0.39	65	1
250	161	2.6	25	212	777	1.66	188	0.53	0.6	0.48	3.55	0.16	73	2
55	84	2.61	33	340	615	1	33	0.33	0.87	0.22	3.4	0.11	85	3
68	185	3.5	37	353	939	1.17	130	0.46	1.83	0.42	4.23	0.11	92	3
30	127	2.74	24	254	539	2.27	4	0.29	0.43	0.24	3.95	0.12	52	2
103	92	5.55	45	527	742	5.8	2	0.34	1.61	0.41	6.73	0.37	84	1
44	24	2.31	13	143	670	1.98	0	0.24	0.27	0.27	3.65	0.05	48	0
85	300	2.7			1620									0
60	300	1.8			1770									
55	250	2.7			1540									
50	20	6			1360									
95	250	2.7			1840									
80	250	2.7			1740									
70	300	1.8			1650									
30	283	6.1	48	548	507	1.45		0.37	3.15	0.43	3.76	0.3	102	12
29	272	5.96	56	576	454	1.59		0.3	3.2	0.37	3.47	0.31	85	12
30	270	6.36	83	636	499	2.21		0.3	3.51	0.36	3.67	0.38	77	13
21	68	2.21	22	240	632	0.98		0.41	2.33	0.26	3.7	0.16	66	6
22	61	2.22	32	288	625	1.21		0.37	2.56	0.21	3.68	0.18	53	6

A Appendix

Table A–1

Food Composition

(Computer code number is for West Diet Analysis program)   (For purposes of calculations, use "0" for t, <1, <.1, <.01, etc.)

Computer Code Number	Food Description	Measure	Wt (g)	H₂O (%)	Ener (kcal)	Prot (g)	Carb (g)	Dietary Fiber (g)	Fat (g)	Sat	Mono	Poly
	MIXED DISHES—Continued											
56010	On whole wheat	1 ea	137	52	339	12	29	5	20	4.9		
	Cheese, grilled:											
56013	On white bread, soft	1 ea	119	37	399	17	30	1	23	11.9		
56015	On part whole wheat	1 ea	119	37	402	18	26	2	25	13.1		
56014	On whole wheat	1 ea	132	38	432	20	30	4	27	13.8		
56000	Chicken fillet	1 ea	182	47	515	24	39		29	8.5	10.4	8.4
	Chicken salad:											
56017	On white bread, soft	1 ea	110	41	366	10	31	2	22	2.7		
56019	On part whole wheat	1 ea	110	41	369	11	27	3	24	3.1		
56018	On whole wheat	1 ea	123	41	399	13	32	5	26	3.4		
56020	Corned beef & swiss on rye	1 ea	156	47	427	28	22	6	26	9.5		
	Egg salad:											
56025	On white bread, soft	1 ea	117	43	379	9	31	1	24	3.8		
56027	On part whole wheat	1 ea	116	44	378	10	27	2	26	4.3		
56026	On whole wheat	1 ea	130	44	410	11	31	5	28	4.6		
	Ham:											
56032	On rye bread	1 ea	150	52	345	22	29	4	15	3.2		
56029	On white bread, soft	1 ea	157	52	365	24	30	2	16	3.3		
56031	On part whole wheat	1 ea	156	54	355	25	26	2	17	3.6		
56030	On whole wheat	1 ea	169	53	378	27	29	4	18	3.9		
	Ham & cheese:											
56035	On white bread, soft	1 ea	157	48	423	24	31	2	23	8.1		
56037	On part whole wheat	1 ea	156	49	417	25	26	2	24	8.7		
56036	On whole wheat	1 ea	170	48	446	27	30	4	25	9.2		
56033	Ham & swiss on rye	1 ea	150	48	386	22	30	4	19	6.5		
	Ham salad:											
56064	On white bread, soft	1 ea	131	47	361	10	37	1	19	4.2		
56066	On part whole wheat	1 ea	131	48	358	11	33	2	21	4.7		
56065	On whole wheat	1 ea	144	48	383	12	37	4	22	5		
56038	Patty melt: Ground beef & cheese on rye	1 ea	182	46	561	37	22	6	37	12.7		
	Peanut butter & jelly:											
56040	On white bread, soft	1 ea	101	27	348	11	47	3	14	2.7		
56042	On part whole wheat	1 ea	101	27	339	12	45	6	15	3		
56041	On whole wheat	1 ea	114	27	398	13	51	5	17	3.6		
	Roast beef:											
66003	On a bun	1 ea	139	49	346	22	33		14	3.6	6.8	1.7
56044	On white bread, soft	1 ea	157	46	405	29	35	1	16	2.9		
56046	On part whole wheat	1 ea	156	47	398	30	30	2	17	3.2		
56045	On whole wheat	1 ea	169	47	423	32	34	4	18	3.4		
	Tuna salad:											
56048	On white bread, soft	1 ea	122	46	326	13	35	1	14	1.9		
56050	On part whole wheat	1 ea	122	47	322	14	32	2	16	2.2		
56049	On whole wheat	1 ea	135	47	347	16	36	4	17	2.4		
	Turkey:											
56052	On white bread, soft	1 ea	156	54	346	24	30	1	14	1.9		
56054	On part whole wheat	1 ea	155	55	336	25	25	2	15	2.1		
56053	On whole wheat	1 ea	169	54	360	27	29	4	16	2.3		
	Turkey ham:											
56103	On rye bread	1 ea	150	60	280	21	20	6	14	2.5		
56104	On white bread, soft	1 ea	156	55	331	21	30	2	14	2.5		
56106	On part whole wheat	1 ea	156	56	346	21	25	2	18	3.6		
56105	On whole wheat	1 ea	169	56	344	24	29	4	15	3		
57531	Taco	1 ea	171	58	369	21	27		21	11.4	6.6	1
	Tostada:											
66017	With refried beans	1 ea	144	66	223	10	27		10	5.4	3.1	0.7
66018	With beans & beef	1 ea	225	70	333	16	30		17	11.5	3.5	0.6
56062	With beans & chicken	1 ea	156	70	242	19	16	3	11	4.5		
	NUTS, SEEDS and PRODUCTS											
	Almonds:											
4571	Dry roasted, salted	1 c	138	3	824	30	27	16	73	5.6	46.4	17.5
4503	Slivered, packed, unsalted	1 c	108	5	624	23	21	13	55	4.2	34.7	13.2

Appendix **A**

PAGE KEY: A–4 = Beverages   A–6 = Dairy   A–10 = Eggs   A–12 = Fat/Oil   A–14 = Fruit   A–20 = Bakery
A–26 = Grain   A–32 = Fish   A–32 = Meats   A–36 = Poultry   A–38 = Sausage   A–40 = Mixed/Fast   A–44 = Nuts/Seeds   A–46 = Sweets
A–48 = Vegetables/Legumes   A–58 = Vegetarian   A–62 = Misc   A–62 = Soups/Sauces   A–64 = Fast   A–78 = Convenience   A–80 = Baby foods

Chol (mg)	Calc (mg)	Iron (mg)	Magn (mg)	Pota (mg)	Sodi (mg)	Zinc (mg)	VT-A (RAE)	Thia (mg)	VT-E (mg α)	Ribo (mg)	Niac (mg)	V-B$_6$ (mg)	Fola (µg)	VT-C (mg)
23	51	2.54	61	346	690	1.87		0.37	2.91	0.2	3.97	0.25	45	7
53	407	2	26	162	1155	2.03		0.29	1.01	0.4	2.37	0.08	60	0
57	431	2	38	208	1197	2.37		0.24	1.15	0.37	2.25	0.1	46	0
60	440	2.33	68	264	1293	3.13		0.24	1.44	0.36	2.46	0.16	37	0
60	60	4.68	35	353	957	1.87	31	0.33		0.24	6.81	0.2	100	9
30	76	2.21	20	146	483	0.79		0.3	5.48	0.24	4.09	0.27	63	1
33	70	2.25	32	194	468	1.04		0.25	6.08	0.2	4.15	0.31	49	1
35	59	2.6	63	252	528	1.76		0.25	6.65	0.18	4.47	0.38	39	1
83	267	3.03	28	232	1470	3.59		0.2	2.59	0.33	2.76	0.18	32	0
154	87	2.36	18	122	500	0.76		0.31	4.24	0.38	2.45	0.21	74	0
166	80	2.38	29	165	479	0.99		0.25	4.65	0.34	2.3	0.24	60	0
177	71	2.75	60	222	543	1.71		0.26	5.17	0.33	2.54	0.32	52	0
50	54	2.89	40	390	1245	2.76		0.84	2.36	0.42	6.42	0.37	54	0
54	76	3.1	32	384	1237	2.61		0.9	2.52	0.44	6.83	0.39	60	0
56	68	3.12	43	438	1252	2.92		0.88	2.72	0.4	6.91	0.42	44	0
59	57	3.46	72	500	1336	3.66		0.9	3.06	0.39	7.29	0.49	34	0
64	246	2.82	33	329	1366	2.71		0.7	2.58	0.46	5.36	0.31	61	0
67	248	2.82	44	379	1388	3.03		0.67	2.78	0.42	5.35	0.34	46	0
71	247	3.17	74	441	1487	3.8		0.69	3.14	0.41	5.7	0.41	36	0
56	241	2.68	43	370	1397	3		0.65	2.47	0.45	5.06	0.29	55	0
29	72	2.25	21	167	935	1.06		0.55	3.34	0.28	3.69	0.18	59	0
30	65	2.26	32	213	950	1.31		0.52	3.67	0.23	3.65	0.21	44	0
32	54	2.59	62	269	1029	2.02		0.53	4.08	0.21	3.92	0.28	34	0
113	222	4.17	39	391	714	7.04		0.25	3.52	0.46	6.14	0.37	37	0
1	76	2.29	52	240	429	1.05		0.29	2.55	0.23	5.44	0.15	79	2
0	54	2.41	88	320	418	1.82		0.22	2.9	0.15	5.43	0.22	53	2
0	80	2.66	75	335	465	1.5		0.28	3.24	0.21	6.41	0.2	76	2
51	54	4.23	31	316	792	3.39	11	0.38	0.19	0.31	5.87	0.26	57	2
43	76	4.13	30	436	1606	3.73		0.35	3.39	0.36	6.78	0.4	67	0
45	67	4.21	41	493	1643	4.11		0.29	3.63	0.32	6.86	0.44	51	0
47	57	4.6	70	557	1743	4.9		0.29	4.01	0.3	7.24	0.51	42	0
13	76	2.41	25	168	589	0.68		0.3	2.77	0.24	5.89	0.13	63	1
13	69	2.44	36	215	578	0.9		0.25	3.07	0.19	6.07	0.16	48	1
14	59	2.79	67	273	642	1.6		0.25	3.46	0.17	6.47	0.23	38	1
43	72	2.19	31	307	1589	1.33		0.31	3.28	0.29	9.29	0.42	60	0
45	63	2.15	42	356	1625	1.56		0.25	3.52	0.24	9.52	0.45	45	0
47	53	2.46	71	417	1736	2.25		0.25	3.92	0.22	10.07	0.53	35	0
55	51	4.04	27	343	1179	2.94		0.22	2.86	0.33	4.32	0.31	29	0
53	77	4.22	29	351	1252	2.85		0.32	2.83	0.41	5.29	0.29	62	0
57	68	4.22	39	395	1264	3.12		0.26	1.23	0.36	5.16	0.34	46	0
58	59	4.72	69	466	1361	3.95		0.26	3.41	0.36	5.63	0.39	37	0
56	221	2.41	70	474	802	3.93	108	0.15	1.88	0.44	3.21	0.24	68	2
30	210	1.89	59	403	543	1.9	45	0.1	1.15	0.33	1.32	0.16	43	1
74	189	2.45	68	490	871	3.17	101	0.09	1.8	0.5	2.86	0.25	86	4
55	146	1.57	41	263	386	1.93		0.09	0.66	0.15	4.26	0.31	25	5
0	367	6.22	395	1029	468	4.89	<1	0.1	35.88	1.19	5.31	0.17	46	0
0	268	4.64	297	786	1	3.63	<1	0.26	27.94	0.88	4.24	0.14	31	0

Table A–1

Food Composition

(Computer code number is for West Diet Analysis program)   (For purposes of calculations, use "0" for t, <1, <.1, <.01, etc.)

Computer Code Number	Food Description	Measure	Wt (g)	H₂O (%)	Ener (kcal)	Prot (g)	Carb (g)	Dietary Fiber (g)	Fat (g)	Fat Breakdown (g) Sat	Mono	Poly
	NUTS, SEEDS and PRODUCTS—Continued											
4502	Whole, dried, unsalted	1 oz	28	5	162	6	6	3	14	1.1	9	3.4
4534	Almond butter:	1 tbs	16	1	101	2	3	1	9	0.9	6.1	2
4572	Salted	1 tbs	16	1	101	2	3	1	9	0.9	6.1	2
4750	Brazil nuts, dry (about 7)	1 c	140	3	918	20	17	10	93	21.2	34.4	28.8
	Cashew nuts:											
4519	Dry roasted, salted	1 oz	28	2	161	4	9	1	13	2.6	7.6	2.2
4621	Dry roasted, unsalted	1 oz	28	2	161	4	9	1	13	2.6	7.6	2.2
4596	Oil roasted, salted	1 oz	28	2	163	5	8	1	13	2.4	7.3	2.4
4622	Oil roasted, unsalted:	1 c	130	3	754	22	39	4	62	11	33.7	11.1
4622	Ounce	1 oz	28	3	162	5	8	1	13	2.4	7.3	2.4
4537	Cashew butter, unsalted	1 tbs	16	3	94	3	4	<1	8	1.6	4.7	1.3
4662	Salted	1 tbs	16	3	94	3	4	<1	8	1.6	4.7	1.3
4538	Chestnuts, European, roasted, (1 cup = approx 17 kernels)	1 c	143	40	350	5	76	7	3	0.6	1.1	1.2
	Coconut, raw:											
4508	Piece 2 x 2 x ½"	1 pce	45	47	159	1	7	4	15	13.4	0.6	0.2
4507	Shredded/grated, unpacked	½ c	40	47	142	1	6	4	13	11.9	0.6	0.1
	Coconut, dried, shredded/grated:											
4510	Unsweetened	1 c	78	3	515	5	18	13	50	44.6	2.1	0.6
4511	Sweetened	1 c	93	13	466	3	44	4	33	29.3	1.4	0.4
4559	Coconut milk, canned	1 c	226	73	445	5	6	3	48	42.7	2	0.5
4514	Filberts/hazelnuts, chopped	1 oz	28	5	176	4	5	3	17	1.2	12.8	2.2
	Mixed Nuts:											
4592	Dry roasted, salted	1 c	137	2	814	24	35	12	70	9.5	43	14.8
4593	Oil roasted, salted	1 c	142	2	876	24	30	13	80	12.4	45	18.9
4533	Oil roasted, unsalted	1 c	142	2	876	24	30	14	80	12.4	45	18.9
4762	Peanuts, oil roasted, salted	1 oz	28	1	168	8	4	3	15	2.4	7.3	4.3
4578	Pecan halves, dried, unsalted	1 oz	28	4	193	3	4	3	20	1.7	11.4	6.1
4583	Dry roasted	¼ c	28	1	199	3	4	3	21	1.8	12.3	5.8
4554	Pine nuts/pinons, dried	1 oz	28	6	176	3	5	3	17	2.6	6.4	7.2
4520	Pistachios, dried, shelled	1 oz	28	4	156	6	8	3	12	1.5	6.5	3.8
4540	Dry roasted,salted,shelled	1 c	128	2	727	27	34	13	59	7.1	31	17.8
4522	Pumpkin kernels, dried, unsalted	1 oz	28	7	151	7	5	1	13	2.4	4	5.9
4625	Roasted, salted	1 c	227	7	1185	75	30	9	96	18.1	29.7	43.6
4524	Sesame seeds, hulled, dried	¼ c	38	5	225	8	6	5	21	2.9	7.9	9.1
8878	Soy nuts, BBQ	5 pce	28		119	12	9	4	4	1		
8877	Salted	5 pce	28		119	12	9	5	4	1		
	Sunflower seed kernels:											
4545	Dry	¼ c	36	5	205	8	7	4	18	1.9	3.4	11.8
4546	Oil roasted	¼ c	34	3	209	7	5	2	20	2	3.7	12.9
4532	Tahini (sesame butter)	1 tbs	15	3	91	3	3	1	8	1.2	3.2	3.7
44059	Trail Mix w/chocolate chips	1 c	146	7	707	21	66	8	47	8.9	19.8	16.5
4525	Black walnuts, chopped	1 oz	28	5	173	7	3	2	17	0.9	4.2	9.8
4556	English walnuts, chopped	1 oz	28	4	183	4	4	2	18	1.7	2.5	13.2
	SWEETENERS and SWEETS (see also Dairy (milk desserts) and Baked Goods)											
23000	Apple butter	2 tbs	36	56	62	<1	15	1	0	0	0	0
1124	Butterscotch topping	2 tbs	41	32	103	1	27	<1	<1	t	t	0
23069	Caramel topping	2 tbs	41	32	103	1	27	<1	<1	t	t	0
	Cake frosting, creamy vanilla:											
46009	Canned	2 tbs	39	15	164	0	26	<1	6	1.2	1.9	3.1
46018	From mix	2 tbs	39	12	161	<1	29	<1	5	0.7	1.5	1.1
	Candy:											
23405	Almond Joy candy bar	1 oz	28	8	134	1	17	1	8	4.9	1.5	0.3
23184	Butterscotch morsels	¼ c	43	8	243	0	27	0	12	10.6		
23015	Caramel, plain or chocolate	1 pce	10	8	38	<1	8	<1	1	0.7	t	t
90712	Chewing gum	1 pce	3	3	7	0	2		<1	t	t	t
25125	Sugarless	1 pce	3		6	0	2		0	0	0	0
	Chocolate:											
23016	Milk chocolate	1 oz	28	2	150	2	17	1	8	4	3.7	0.2

PAGE KEY: A–4 = Beverages   A–6 = Dairy   A–10 = Eggs   A–12 = Fat/Oil   A–14 = Fruit   A–20 = Bakery
A–26 = Grain   A–32 = Fish   A–32 = Meats   A–36 = Poultry   A–38 = Sausage   A–40 = Mixed/Fast   A–44 = Nuts/Seeds   A–46 = Sweets
A–48 = Vegetables/Legumes   A–58 = Vegetarian   A–62 = Misc   A–62 = Soups/Sauces   A–64 = Fast   A–78 = Convenience   A–80 = Baby foods

Chol (mg)	Calc (mg)	Iron (mg)	Magn (mg)	Pota (mg)	Sodi (mg)	Zinc (mg)	VT-A (RAE)	Thia (mg)	VT-E (mg α)	Ribo (mg)	Niac (mg)	V-B6 (mg)	Fola (μg)	VT-C (mg)
0	69	1.2	77	204	<1	0.94	<1	0.07	7.24	0.23	1.1	0.04	8	0
0	43	0.59	48	121	2	0.49	0	0.02	3.25	0.1	0.46	0.01	10	0
0	43	0.59	48	121	72	0.49	<1	0.02	4.16	0.1	0.46	0.01	10	0
0	224	3.4	526	923	4	5.68	0	0.86	8.02	0.05	0.41	0.14	31	1
0	13	1.68	73	158	179	1.57	0	0.06	0.26	0.06	0.39	0.07	19	0
0	13	1.68	73	158	4	1.57	0	0.06	0.26	0.06	0.39	0.07	19	0
0	12	1.69	76	177	86	1.5	0	0.1	0.26	0.06	0.49	0.09	7	0
0	56	7.86	355	822	17	6.96	0	0.47	1.2	0.28	2.26	0.42	32	0
0	12	1.69	76	177	4	1.5	0	0.1	0.26	0.06	0.49	0.09	7	0
0	7	0.8	41	87	2	0.83	0	0.05	0.25	0.03	0.26	0.04	11	0
0	7	0.8	41	87	98	0.83	0	0.05	0.15	0.03	0.26	0.04	11	0
0	41	1.3	47	847	3	0.82	1	0.35	0.72	0.25	1.92	0.71	100	37
0	6	1.09	14	160	9	0.5	0	0.03	0.11	<.01	0.24	0.02	12	1
0	6	0.97	13	142	8	0.44	0	0.03	0.1	<.01	0.22	0.02	10	1
0	20	2.59	70	424	29	1.57	0	0.05	0.34	0.08	0.47	0.23	7	1
0	14	1.79	46	313	244	1.69	0	0.03	0.36	0.02	0.44	0.25	7	1
0	41	7.46	104	497	29	1.27	0	0.05	1.47	0	1.44	0.06	32	2
0	32	1.32	46	190	0	0.69	<1	0.18	4.21	0.03	0.5	0.16	32	2
0	96	5.07	308	818	917	5.21	<1	0.27	14.99	0.27	6.44	0.41	68	1
0	153	4.56	334	825	926	7.21	<1	0.71	10.22	0.32	7.19	0.34	118	1
0	153	4.56	334	825	16	7.21	1	0.71	8.52	0.32	7.19	0.34	118	1
0	17	0.43	49	203	90	0.92	0	0.02	1.94	0.02	3.87	0.13	34	0
0	20	0.71	34	115	0	1.27	1	0.18	0.39	0.04	0.33	0.06	6	0
0	20	0.78	37	119	107	1.42	2	0.13	0.36	0.03	0.33	0.05	4	0
0	2	0.86	66	176	20	1.2	<1	0.35	0.98	0.06	1.22	0.03	16	1
0	30	1.16	34	287	<1	0.62	8	0.24	0.64	0.04	0.36	0.48	14	1
0	141	5.38	154	1334	518	2.94	17	1.08	2.47	0.2	1.82	1.63	64	3
0	12	4.19	150	226	5	2.09	5	0.06	0	0.09	0.49	0.06	16	1
0	98	33.91	1212	1830	1305	16.89	43	0.48	0	0.72	3.95	0.2	129	4
0	50	2.96	132	155	15	3.9	1	0.27	0.1	0.03	1.78	0.06	36	0
0	59	1.07			415		0							0
0	59	1.07			148		0							0
0	42	2.44	127	248	1	1.82	1	0.82	12.42	0.09	1.62	0.28	82	1
0	19	2.28	43	164	1	1.77	<1	0.11	13.25	0.1	1.4	0.27	80	0
0	21	0.95	53	69	<1	1.57	<1	0.24	0.34	0.02	0.85	0.02	15	0
6	159	4.95	235	946	177	4.58	3	0.6	15.62	0.33	6.43	0.38	95	2
0	17	0.87	56	146	1	0.94	1	0.02	0.5	0.04	0.13	0.16	9	0
0	27	0.81	44	123	1	0.87	<1	0.1	0.2	0.04	0.32	0.15	27	0
0	5	0.11	2	33	5	0.02	<1	<.01	<.01	<.01	0.02	0.01	<1	0
0	22	0.08	3	34	143	0.08	11	<.01	0	0.04	0.02	<.01	1	0
0	22	0.08	3	34	143	0.08	11	<.01	0	0.04	0.02	<.01	1	0
0	1	0.06	0	13	72	0.03	0	<.01	0.82	0.12	0.09	0	3	0
0	2	<.01	1	4	44	<.01	31	<.01	0.76	<.01	<.01	<.01	0	0
1	18	0.36	18	71	40	0.22		<.01	<.01	0.04	0.13	0.02		0
0	0	0		80	45		0	0.03		0.03	0.03			0
1	14	0.01	2	21	24	0.04	<1	<.01	0.28	0.03	0.02	<.01	<1	0
0	0	0	0	0	<1	0	0							0
				0	0									
6	53	0.66	18	104	22	0.56	14	0.03	0.57	0.08	0.11	0.01	3	0

A   Appendix

Table A–1

Food Composition

(Computer code number is for West Diet Analysis program)   (For purposes of calculations, use "0" for t, <1, <.1, <.01, etc.)

Computer Code Number	Food Description	Measure	Wt (g)	H₂O (%)	Ener (kcal)	Prot (g)	Carb (g)	Dietary Fiber (g)	Fat (g)	Fat Breakdown (g) Sat	Mono	Poly
	SWEETENERS and SWEETS (see also Dairy (milk desserts) and Baked Goods)—Continued											
23018	Milk chocolate with almonds	1 oz	28	2	147	3	15	2	10	4.8	3.8	0.6
23058	Milk chocolate with rice cereal	1 oz	28	2	139	2	18	1	7	4.4	2.4	0.2
23012	Semisweet chocolate chips	1 c	168	1	805	7	106	10	50	29.8	16.7	1.6
23057	Sweet Dark chocolate (candy bar)	1 ea	41	1	218	2	24	3	13	7.9	2.1	0.2
23024	Fondant candy, uncoated (mints, candy corn, other)	1 pce	16	7	60	0	15	0	<1	0	0	0
23409	Gumdrops	1 c	182	1	721	0	180	<1	0	0	0	0
23031	Hard candy-all flavors	1 pce	6	1	24	0	6	0	<1	0	0	0
23033	Jellybeans	10 pce	11	6	41	0	10	<1	<1	0	0	0
23025	Fudge, chocolate	1 pce	17	10	70	<1	13	<1	2	1	0.5	t
23046	M&M's plain chocolate candy	10 pce	7	2	34	<1	5	<1	1	0.9	0.2	t
23048	M&M's peanut chocolate candy	10 pce	20	2	103	2	12	1	5	2.1	2.2	0.8
23037	MARS almond bar	1 ea	50	4	234	4	31	1	12	3.6	5.3	2
23038	MILKY WAY candy bar	1 ea	60	6	254	3	43	1	10	4.7	3.6	0.4
23021	Milk chocolate-coated peanuts	1 c	149	2	773	20	74	7	50	21.8	19.3	6.5
23081	Peanut brittle, recipe	1 c	147	1	711	11	104	4	28	6.1	11.9	6.7
23036	Skor English toffee candy bar	1 ea	39	2	209	1	24	1	13	7.3	3.6	0.5
1131	Snickers candy bar (2.2oz)	1 ea	62	5	297	5	37	2	15	5.6	6.5	3
23086	Fruit Roll-Up (small)	1 ea	14	10	52	<1	12	<1	<1	t	0.2	t
23174	Fruit juice bar (2.5 fl oz)	1 ea	77	78	63	1	16	1	<1	t	0	t
23052	Gelatin dessert/Jello, prepared	½ c	135	84	84	2	19	0	0	0	0	0
23093	SugarFree	½ c	117	95	23	1	5	0	0	0	0	0
25001	Honey	1 tbs	21	17	64	<1	17	<1	0	0	0	0
23003	Jellies:	1 tbs	19	30	51	<1	13	<1	<1	t	t	t
23004	Packet	1 ea	14	30	37	<1	10	<1	<1	t	t	t
23005	Marmalade	1 tbs	20	33	49	<1	13	<1	0	0	0	0
23007	Marshmallows	1 ea	7	16	22	<1	6	<1	<1	t	t	t
23071	Marshmallow creme topping	2 tbs	38	20	122	<1	30	<1	<1	t	t	t
25003	Molasses	2 tbs	41	22	119	0	31	0	<1	t	t	t
23050	Popsicle/ice pops	1 ea	128	80	92	0	24	0	0	0	0	0
23171	Rice crispie bar	1 ea	28	13	107	1	20	<1	3	0.6	1.3	0.8
	Sugars:											
25005	Brown sugar	1 c	220	2	829	0	214	0	0	0	0	0
25006	White sugar, granulated	1 tbs	12	0	46	0	12	0	0	0	0	0
25007	Packet	1 ea	6	0	23	0	6	0	0	0	0	0
	Sweeteners:											
25038	Equal, packet	1 ea	1	12	4	<1	1	0	<1	t	t	t
25208	Sweet 'N Low, packet	1 ea	1		4	0	1	0	0	0	0	0
	Syrups											
23014	Chocolate, hot fudge type	2 tbs	43	22	150	2	27	1	4	1.7	1.7	0.1
23056	Thin type	2 tbs	38	29	93	1	25	1	<1	0.3	0.2	t
23042	Pancake table syrup (corn and maple)	2 tbs	40	38	94	0	25	<1	0	0	0	0
	VEGETABLES AND LEGUMES											
	Amaranth leaves:											
5375	Raw, chopped	1 c	28	92	6	1	1	<1	<1	t	t	t
5376	Raw, each	1 ea	14	92	3	<1	1	<1	<1	t	t	t
5377	Cooked	1 c	132	91	28	3	5	2	<1	t	t	0.1
6033	Arugula, raw, chopped	½ c	10	92	2	<1	<1	<1	<1	t	t	t
5000	Artichokes, cooked globe (300 g with refuse)	1 ea	120	84	60	4	13	6	<1	t	t	t
	Artichoke hearts:											
5192	Cooked from frozen	1 c	168	86	76	5	15	8	1	0.2	t	0.4
5191	Marinated	1 c	130		116	5	14	5	7	0		
7962	In water	½ c	100		38	2	6	0	0	0	0	0
	Asparagus, green:											
5003	Cooked from fresh, cuts and tips	½ c	90	93	20	2	4	2	<1	t	t	0.1
5004	Spears, ½" diam at base	4 ea	60	93	13	1	2	1	<1	t	t	t
5005	Cooked from frozen, cuts and tips	½ c	90	94	16	3	2	1	<1	t	t	0.2
5006	Spears, ½" diam at base	4 ea	60	94	11	2	1	1	<1	t	t	0.1
5007	Canned, spears, ½" diam at base	4 pce	72	94	14	2	2		<1	0.1	t	0.2

Appendix **A**

PAGE KEY: A–4 = Beverages   A–6 = Dairy   A–10 = Eggs   A–12 = Fat/Oil   A–14 = Fruit   A–20 = Bakery
A–26 = Grain   A–32 = Fish   A–32 = Meats   A–36 = Poultry   A–38 = Sausage   A–40 = Mixed/Fast   A–44 = Nuts/Seeds   A–46 = Sweets
A–48 = Vegetables/Legumes   A–58 = Vegetarian   A–62 = Misc   A–62 = Soups/Sauces   A–64 = Fast   A–78 = Convenience   A–80 = Baby foods

Chol (mg)	Calc (mg)	Iron (mg)	Magn (mg)	Pota (mg)	Sodi (mg)	Zinc (mg)	VT-A (RAE)	Thia (mg)	VT-E (mg α)	Ribo (mg)	Niac (mg)	V-B$_6$ (mg)	Fola (µg)	VT-C (mg)
5	63	0.46	25	124	21	0.38	12	0.02	1.26	0.12	0.21	0.01	4	0
5	48	0.21	14	96	41	0.31	17	0.02	0.56	0.08	0.13	0.02	4	0
0	54	5.26	193	613	18	2.72	0	0.09	0.39	0.15	0.72	0.06	5	0
2	12	0.87	13	206	2	<.01		0	0.08	<.01	0	0	0	0
0	0	<.01	0	1	3	0	0	<.01	0	<.01	<.01	<.01	0	0
0	5	0.73	2	9	80	0	0	0.01	0	0.02	0.02	<.01	0	0
0	0	0.02	0	0	2	<.01	0	<.01	0	<.01	<.01	<.01	0	0
0	0	0.01	0	4	6	<.01	0	<.01	0	<.01	<.01	<.01	0	0
2	8	0.3	6	22	8	0.19	7	<.01	0.03	0.01	0.03	<.01	1	0
1	7	0.08	2	14	4	0.08	2	<.01	0.08	0.01	0.01	<.01	<1	0
2	20	0.23	15	69	10	0.48	5	0.02	0.51	0.03	0.82	0.02	8	0
8	84	0.55	36	162	85	0.56	8	0.02	3.88	0.16	0.47	0.03	4	0
8	78	0.46	20	145	144	0.43	11	0.02	0.75	0.13	0.21	0.03	4	1
13	155	1.95	143	748	61	3.56	51	0.17	5.16	0.26	6.33	0.31	12	0
18	40	1.79	62	247	654	1.28	57	0.2	3.76	0.06	3.89	0.12	68	0
21	51	0.22	4	60	124	0.07		<.01	0.02	0.04	0.05	0.01	1	0
8	58	0.47	35	185	165	0.97	26	0.08	0.68	0.12	0.98	0.04	19	0
0	4	0.14	3	41	44	0.03	1	0.01	0.08	<.01	0.01	0.04	1	17
0	4	0.15	3	41	3	0.04	1	<.01	0.1	0.01	0.12	0.02	5	7
0	4	0.03	1	1	101	0.01	0	0	0	<.01	<.01	0	1	0
0	4	0.01	1	1	56	0	0	0	0	0	0	0	0	0
0	1	0.09	0	11	1	0.05	0	0	0	<.01	0.03	<.01	<1	0
0	1	0.04	1	10	6	<.01	<1	<.01	0	<.01	<.01	<.01	<1	0
0	1	0.03	1	8	4	<.01	<1	<.01	0	<.01	<.01	<.01	<1	0
0	8	0.03	0	7	11	<.01	1	<.01	0.01	<.01	0.01	<.01	2	1
0	0	0.02	0	0	6	<.01	0	<.01	0	<.01	<.01	<.01	<1	0
0	1	0.08	1	2	30	0.02	<1	<.01	0	<.01	0.03	<.01	<1	0
0	84	1.94	99	600	15	0.12	0	0.02	0	<.01	0.38	0.27	0	0
0	0	0	1	5	15	0.03	0	0	0	0	0	0	0	0
0	2	0.51	4	12	123	0.15			0.1	0.41	0.11	1.31	0.13	28	4
0	187	4.2	64	761	86	0.4	0	0.02	0	0.02	0.18	0.06	2	0
0	0	<.01	0	0	0	0	0	0	0	<.01	0	0	0	0
0	0	<.01	0	0	0	0	0	0	0	<.01	0	0	0	0
0	0	<.01	0	0	<1	0	0	0	0	0	0	0	0	0
0	0	0			0	0	0							0
1	43	0.68	28	194	149	0.36	2	0.03	1.07	0.12	0.16	0.03	2	0
0	5	5.15	25	183	58	0.28		<.01	0.01	0.31	12.76	<.01	2	0
0	1	0.01	1	6	33	0.03	0	<.01	0	<.01	<.01	<.01	0	0
0	60	0.65	15	171	6	0.25	0	<.01	0.22	0.04	0.18	0.05	24	12
0	30	0.32	8	86	3	0.13	0	<.01	0.11	0.02	0.09	0.03	12	6
0	276	2.98	73	846	28	1.16	183	0.03	0.66	0.18	0.74	0.23	75	54
0	16	0.15	5	37	3	0.05	12	<.01	0.04	<.01	0.03	<.01	10	2
0	54	1.55	72	425	114	0.59	11	0.08	0.23	0.08	1.2	0.13	61	12
0	35	0.94	52	444	89	0.6	13	0.1	0.27	0.27	1.54	0.15	200	8
0	0	0			488		0							46
0	0	1.35		0	250		6							4
0	21	0.82	13	202	13	0.54	45	0.15	0.34	0.13	0.98	0.07	134	7
0	14	0.55	8	134	8	0.36	30	0.1	0.23	0.08	0.65	0.05	89	5
0	16	0.5	9	155	3	0.37	36	0.06	1.08	0.09	0.93	0.02	122	22
0	11	0.34	6	103	2	0.25	24	0.04	0.72	0.06	0.62	0.01	81	15
0	12	1.32	7	124	207	0.29	30	0.04	0.22	0.07	0.69	0.08	69	13

Table A–1

Food Composition

(Computer code number is for West Diet Analysis program)   (For purposes of calculations, use "0" for t, <1, <.1, <.01, etc.)

Computer Code Number	Food Description	Measure	Wt (g)	H₂O (%)	Ener (kcal)	Prot (g)	Carb (g)	Dietary Fiber (g)	Fat (g)	Fat Breakdown (g) Sat	Mono	Poly
	VEGETABLES AND LEGUMES—Continued											
	Bamboo shoots:											
5401	Canned, drained slices	1 c	131	94	25	2	4		1	0.1	t	0.2
6736	Raw slices	1 c	151	91	41	4	8	3	<1	0.1	t	0.2
5249	Cooked slices	1 c	120	96	14	2	2	1	<1	t	t	0.1
	Beans (see also alphabetical listing this section):											
7034	Adzuki beans, cooked	½ c	115	66	147	9	28	8	<1	t		
7012	Black beans, cooked	½ c	86	66	114	8	20	7	<1	0.1	t	0.2
	Canned beans (white/navy):											
7004	With pork and tomato sauce	½ c	127	73	124	7	25	6	1	0.5	0.6	0.2
7023	With sweet sauce	½ c	130	71	144	7	27	7	2	0.7	0.8	0.2
56101	With frankfurters	½ c	130	69	185	9	20	9	9	3.1	3.7	1.1
	Lima beans:											
5247	Fordhooks, cooked from frozen	½ c	85	73	88	5	16	5	<1	t	t	0.1
5019	Baby, cooked from frozen	½ c	90	72	94	6	18	5	<1	t	t	0.1
7010	Cooked from dry, drained	½ c	94	70	108	7	20	7	<1	t	t	0.2
7352	Red Mexican, cooked f/dry	½ c	112	70	127	8	24	9	<1	t	t	0.2
	Snap bean/green string beans:											
5011	Cooked from fresh	½ c	63	89	22	1	5	2	<1	t	t	t
5013	Cooked from frozen	½ c	68	91	19	1	4	2	<1	t	t	t
5015	Canned, drained	½ c	68	93	14	1	3	1	<1	t	t	t
5194	Snap bean, yellow, cooked f/fresh	½ c	63	89	22	1	5	2	<1	t	t	t
	Bean sprouts (mung):											
5020	Raw	½ c	52	90	16	2	3	1	<1	t	t	t
5246	Cooked, stir fried	½ c	62	84	31	3	7	1	<1	t	t	t
5021	Cooked, boiled, drained	½ c	62	93	13	1	3	<1	<1	t	t	t
5197	Canned, drained	½ c	63	96	8	1	1	1	<1	t	t	t
	Beets:											
5022	Cooked from fresh, sliced or diced	½ c	85	87	37	1	8	2	<1	t	t	t
5023	Whole beets, 2" diam	2 ea	100	87	44	2	10	2	<1	t	t	t
5024	Canned, sliced or diced	½ c	79	91	24	1	6	1	<1	t	t	t
5310	Pickled slices	½ c	114	82	74	1	19	3	<1	t	t	t
5025	Beet greens, cooked, drained	½ c	72	89	19	2	4	2	<1	t	t	t
	Broccoli:											
5027	Broccoli, raw, spear	1 ea	31	89	11	1	2	1	<1	t	t	t
5029	Cooked from fresh, spears	1 ea	180	89	63	4	13	6	1	0.1	t	0.3
5028	Chopped	½ c	78	89	27	2	6	3	<1	t	t	0.1
5234	Cooked from frozen, spear, small piece	½ c	92	91	26	3	5	3	<1	t	t	t
5030	Chopped	½ c	92	91	26	3	5	3	<1	t	t	t
5679	Broccoflower-steamed	½ c	78	90	25	2	5	2	<1	t	t	t
5033	Brussels sprouts, cooked from fresh	½ c	78	89	28	2	6	2	<1	t	t	0.2
5035	Cooked from frozen	½ c	78	87	33	3	6	3	<1	t	t	0.2
5036	Cabbage, common, raw, chopped	1 c	70	92	17	1	4		<1	t	t	t
5038	Cooked, drained	1 c	150	94	33	2	7		1	t	t	0.3
	Cabbage, Chinese:											
5041	Bok Choy, raw, shredded	1 c	70	95	9	1	2	1	<1	t	t	t
5237	Cooked, drained	1 c	170	96	20	3	3		<1	t	t	0.1
5535	Kim chee style	1 c	150	92	31	2	6	2	<1	t	t	0.1
5040	Pe Tsai, raw, chopped	1 c	76	94	12	1	2		1	<1	t	t
5235	Cooked	1 c	119	95	17	2	3	2	<1	t	t	t
6766	Cabbage, red, raw, chopped	1 c	89	90	28	1	7	2	<1	t	t	t
5238	Cooked, drained	1 c	150	91	44	2	10	4	<1	t	t	t
5043	Cabbage, savoy, raw, chopped	1 c	70	91	19	1	4	2	<1	t	t	t
5044	Cooked	1 c	145	92	35	3	8	4	<1	t	t	t
5511	Capers	1 ea	5		0	<1	<1		<1			
	Carrots:											
5045	Raw, whole, 7½ x 1⅛"	1 ea	72	88	30	1	7	2	<1	t	t	t
5046	Grated	½ c	55	88	23	1	5	2	<1	t	t	t
5047	Cooked from raw	½ c	78	90	27	1	6	2	<1	t	t	t
5358	From frozen	½ c	73	90	27	<1	6	2	<1	t	t	0.2
5199	Canned	½ c	73	93	18	<1	4	1	<1	t	t	t

PAGE KEY: A–4 = Beverages   A–6 = Dairy   A–10 = Eggs   A–12 = Fat/Oil   A–14 = Fruit   A–20 = Bakery
A–26 = Grain   A–32 = Fish   A–32 = Meats   A–36 = Poultry   A–38 = Sausage   A–40 = Mixed/Fast   A–44 = Nuts/Seeds   A–46 = Sweets
A–48 = Vegetables/Legumes   A–58 = Vegetarian   A–62 = Misc   A–62 = Soups/Sauces   A–64 = Fast   A–78 = Convenience   A–80 = Baby foods

Chol (mg)	Calc (mg)	Iron (mg)	Magn (mg)	Pota (mg)	Sodi (mg)	Zinc (mg)	VT-A (RAE)	Thia (mg)	VT-E (mg α)	Ribo (mg)	Niac (mg)	V-B6 (mg)	Fola (µg)	VT-C (mg)
0	10	0.42	5	105	9	0.85	1	0.03	0.83	0.03	0.18	0.18	4	1
0	20	0.76	5	805	6	1.66	2	0.23	1.51	0.11	0.91	0.36	11	6
0	14	0.29	4	640	5	0.56	0	0.02	0.8	0.06	0.36	0.12	2	0
0	32	2.3	60	612	9	2.04	<1	0.13	0.12	0.07	0.82	0.11	139	0
0	23	1.81	60	305	1	0.96	<1	0.21	0.07	0.05	0.43	0.06	128	0
9	71	4.17	44	381	559	7.44	5	0.07	0.69	0.06	0.63	0.09	29	4
9	79	2.16	44	346	437	1.95	1	0.06	0.7	0.08	0.46	0.11	48	4
8	62	2.25	36	306	559	2.43	5	0.08	0.6	0.07	1.17	0.06	39	3
0	26	1.55	36	258	59	0.63	8	0.06	0.25	0.05	0.91	0.1	18	11
0	25	1.76	50	370	26	0.5	7	0.06	0.58	0.05	0.69	0.1	14	5
0	16	2.25	40	478	2	0.89	0	0.15	0.17	0.05	0.4	0.15	78	0
0	42	1.87	48	371	6	0.87	<1	0.13	0.08	0.07	0.38	0.12	94	2
0	28	0.41	11	92	1	0.16	22	0.05	0.28	0.06	0.39	0.04	21	6
0	33	0.6	16	86	6	0.33	19	0.02	0.24	0.06	0.26	0.04	16	3
0	18	0.61	9	74	178	0.2	15	0.01	0.19	0.04	0.14	0.03	22	3
0	29	0.81	16	188	2	0.23	3	0.05	0.28	0.06	0.39	0.04	21	6
0	7	0.47	11	77	3	0.21	1	0.04	0.05	0.06	0.39	0.05	32	7
0	8	1.18	20	136	6	0.56	1	0.09	<.01	0.11	0.74	0.08	43	10
0	7	0.4	9	63	6	0.29	1	0.03	0.04	0.06	0.51	0.03	18	7
0	9	0.27	6	17	88	0.18	<1	0.02	0.03	0.04	0.14	0.02	6	0
0	14	0.67	20	259	65	0.3	2	0.02	0.03	0.03	0.28	0.06	68	3
0	16	0.79	23	305	77	0.35	2	0.03	0.04	0.04	0.33	0.07	80	4
0	12	1.44	13	117	153	0.17	1	<.01	0.02	0.03	0.12	0.05	24	3
0	13	0.47	17	169	301	0.3	1	0.01	0.15	0.05	0.29	0.06	31	3
0	82	1.37	49	654	174	0.36	276	0.08	1.3	0.21	0.36	0.1	10	18
0	15	0.23	7	98	10	0.13	10	0.02	0.24	0.04	0.2	0.05	20	28
0	72	1.21	38	527	74	0.81	176	0.11	2.61	0.22	1	0.36	194	117
0	31	0.52	16	229	32	0.35	76	0.05	1.13	0.1	0.43	0.16	84	51
0	47	0.56	18	166	22	0.28	52	0.05	1.21	0.07	0.42	0.12	28	37
0	30	0.56	12	131	10	0.26	52	0.05	1.21	0.07	0.42	0.12	52	37
0	25	0.55	16	251	18	0.39	3	0.06	0.23	0.07	0.59	0.14	38	49
0	28	0.94	16	247	16	0.26	30	0.08	0.34	0.06	0.47	0.14	47	48
0	20	0.37	14	226	12	0.19	36	0.08	0.4	0.09	0.42	0.23	79	36
0	33	0.41	10	172	13	0.13	6	0.04	0.1	0.03	0.21	0.07	30	23
0	46	0.26	12	146	12	0.14	10	0.09	0.18	0.08	0.42	0.17	30	30
0	74	0.56	13	176	46	0.13	156	0.03	0.06	0.05	0.35	0.14	46	32
0	158	1.77	19	631	58	0.29	360	0.05	0.15	0.11	0.73	0.28	70	44
0	145	1.28	27	375	995	0.36	213	0.07	0.24	0.1	0.75	0.34	88	80
0	59	0.24	10	181	7	0.17	12	0.03	0.09	0.04	0.3	0.18	60	21
0	38	0.36	12	268	11	0.21	57	0.05	0.14	0.05	0.6	0.21	63	19
0	40	0.71	14	216	24	0.2	50	0.06	0.1	0.06	0.37	0.19	16	51
0	63	0.99	26	393	12	0.38	3	0.11	0.18	0.09	0.57	0.34	36	16
0	24	0.28	20	161	20	0.19	35	0.05	0.12	0.02	0.21	0.13	56	22
0	44	0.55	35	267	35	0.33	64	0.07	0.15	0.03	0.03	0.22	67	25
0	2	0.05			105		1							0
0	24	0.22	9	230	50	0.17	433	0.05	0.48	0.04	0.71	0.1	14	4
0	18	0.16	7	176	38	0.13	331	0.04	0.36	0.03	0.54	0.08	10	3
0	23	0.27	8	183	45	0.16	671	0.05	0.8	0.03	0.5	0.12	11	3
0	26	0.39	8	140	43	0.26	607	0.02	0.74	0.03	0.3	0.06	8	2
0	18	0.47	6	131	177	0.19	407	0.01	0.54	0.02	0.4	0.08	7	2

A Appendix

Table A–1

Food Composition

(Computer code number is for West Diet Analysis program)   (For purposes of calculations, use "0" for t, <1, <.1, <.01, etc.)

Computer Code Number	Food Description	Measure	Wt (g)	H$_2$O (%)	Ener (kcal)	Prot (g)	Carb (g)	Dietary Fiber (g)	Fat (g)	Fat Breakdown (g) Sat	Mono	Poly
	VEGETABLES AND LEGUMES—Continued											
5226	Juice, canned	1 c	236	89	94	2	22	2	<1	t	t	0.2
5625	Cassava, cooked	1 c	137	59	221	2	53	2	<1	0.1	0.1	t
5049	Cauliflower, flowerets, raw:	½ c	50	92	12	1	3	1	<1	t	t	t
5051	Cooked from fresh, drained	½ c	62	93	14	1	3	2	<1	t	t	0.1
5053	From frozen, drained	½ c	90	94	17	1	3	2	<1	t	t	t
5055	Celery, raw, large outer stalk, 8 x 1½" (root end)	1 ea	40	95	6	<1	1		<1	t	t	t
5054	Diced	1 c	120	95	17	1	4		<1	t	t	t
5200	Celeriac/celery root, cooked	1 c	155	92	42	1	9	2	<1	t	t	0.2
5057	Chard, swiss, raw, chopped	1 c	36	93	7	1	1	1	<1	t	t	t
5059	Cooked	1 c	175	93	35	3	7	4	<1	t	t	t
5413	Chayote fruit, raw	1 ea	203	95	35	2	8	3	<1	t	t	0.1
5414	Cooked	1 c	160	93	38	1	8	4	1	0.1	t	0.3
	Chickpeas (see Garbanzo Beans #854)											
5061	Collards, cooked from raw	½ c	95	92	25	2	5	3	<1	t	t	0.2
5062	From frozen	½ c	85	88	31	3	6	2	<1	t	t	0.2
	Corn, yellow:											
5364	Cooked from frozen, on cob, 3½" long	1 ea	63	73	59	2	14	2	<1	t	0.1	0.2
5065	Kernels, cooked from frozen	½ c	82	77	66	2	16	2	1	t	0.2	0.3
5068	Canned, cream style	½ c	128	79	92	2	23	2	1	t	0.2	0.3
5067	Whole kernel, vacuum pack	½ c	105	77	83	3	20	2	1	t	0.2	0.2
	Cowpeas (see Black-eyed peas #814-816)											
5610	Cucumber, kim chee style	1 c	150	91	32	2	7	2	<1	t	t	t
5241	Dandelion greens, raw	1 c	55	86	25	1	5	2	<1	t	t	0.2
5242	Chopped, cooked, drained	1 c	105	90	35	2	7	3	1	0.2	t	0.3
5072	Eggplant, cooked	1 c	99	90	35	1	9	2	<1	t	t	t
5202	Endive, fresh, chopped	1 c	50	94	8	1	2		2	<1	t	t
856	Escarole/curly endive-chopped	1 c	50	94	8	1	2	2	<1	t	t	t
7001	Garbanzo beans (Chickpeas), cooked	1 c	164	60	269	15	45	12	4	0.4	1	1.9
7961	Grape leaf, raw:	1 ea	3	73	3	<1	1	<1	<1	t	t	t
7914	Cup	1 c	14	73	13	1	2	2	<1	t	t	0.1
7021	Great northern beans, cooked	1 c	177	69	209	15	37	12	1	0.2	t	0.3
5077	Jerusalem artichoke, raw slices	1 c	150	78	114	3	26	2	<1	0	t	t
5224	Jicama	1 c	120	90	46	1	11		<1	t	t	t
5075	Kale, cooked from raw	1 c	130	91	36	2	7	3	1	t	t	0.3
5076	From frozen	1 c	130	90	39	4	7	3	1	t	t	0.3
7292	Kidney beans, canned	1 c	256	77	218	13	40	16	1	0.1	t	0.5
5078	Kohlrabi, raw slices	1 c	135	91	36	2	8	5	<1	t	t	t
5079	Cooked	1 c	165	90	48	3	11	2	<1	t	t	t
5205	Leeks, raw, chopped	1 c	89	83	54	1	13	2	<1	t	t	0.1
5203	Cooked, chopped	1 c	104	91	32	1	8	1	<1	t	t	0.1
7006	Lentils, cooked from dry	1 c	198	70	230	18	40	16	1	0.1	0.1	0.3
5390	Lentils, sprouted, stir fried	1 c	124	69	125	11	26	5	1	t	0.1	0.2
5389	Raw	1 c	77	67	82	7	17	3	<1	t	t	0.2
	Lettuce:											
5082	Butterhead/Boston, head, 5" diameter	¼ ea	41	96	5	1	1	<1	<1	t	t	t
5081	Leaves, inner or outer	4 ea	30	96	4	<1	1	<1	<1	t	t	t
5083	Iceberg/cripshead, chopped	1 c	55	96	6	<1	1		<1	t	t	t
5085	Head, 6" diameter	1 ea	539	96	54	4	11		1	t	t	0.3
866	Wedge, ¼ head	1 ea	135	96	14	1	3		<1	t	t	t
5086	Looseleaf, chopped	½ c	28	95	4	<1	1	<1	<1	t	t	t
5088	Romaine, chopped	½ c	28	95	5	<1	1	1	<1	t	t	t
5089	Romaine, inner leaf	3 pce	30	95	5	<1	1	1	<1	t	t	t
5528	Luffa, cooked (Chinese okra)	1 c	178	90	57	3	13	4	<1	t	t	t
6778	Manioc, raw	1 c	206	60	330	3	78	4	1	0.2	0.2	t
	Mushrooms:											
5090	Raw, sliced	½ c	35	92	8	1	1	<1	<1	t	t	t
5092	Cooked from fresh, pieces	½ c	78	91	22	2	4	2	<1	t	t	0.1
1962	Stir fried, shitake slices	½ c	73	83	40	1	10	2	<1	t	t	t
5094	Canned, drained	½ c	78	91	20	1	4	2	<1	t	t	t
5612	Mushroom caps, pickled	8 ea	47	92	11	1	2	<1	<1	t	t	t

Chol (mg)	Calc (mg)	Iron (mg)	Magn (mg)	Pota (mg)	Sodi (mg)	Zinc (mg)	VT-A (RAE)	Thia (mg)	VT-E (mg α)	Ribo (mg)	Niac (mg)	V-B$_6$ (mg)	Fola (µg)	VT-C (mg)
0	57	1.09	33	689	68	0.42	2256	0.22	2.74	0.13	0.91	0.51	9	20
0	21	0.35	28	338	18	0.45	1	0.1	0.26	0.06	1.06	0.11	24	19
0	11	0.22	8	152	15	0.14	<1	0.03	0.04	0.03	0.26	0.11	28	23
0	10	0.2	6	88	9	0.11	1	0.03	0.04	0.03	0.25	0.11	27	27
0	15	0.37	8	125	16	0.12	<1	0.03	0.05	0.05	0.28	0.08	37	28
0	16	0.08	4	104	32	0.05	9	<.01	0.11	0.02	0.13	0.03	14	1
0	48	0.24	13	312	96	0.16	26	0.03	0.32	0.07	0.38	0.09	43	4
0	40	0.67	19	268	95	0.31	0	0.04	0.31	0.06	0.66	0.16	5	6
0	18	0.65	29	136	77	0.13	110	0.01	0.68	0.03	0.14	0.04	5	11
0	102	3.96	150	961	313	0.58	536	0.06	3.31	0.15	0.63	0.15	16	32
0	35	0.69	24	254	4	1.5	0	0.05	0.24	0.06	0.95	0.15	189	16
0	21	0.35	19	277	2	0.5	3	0.04	0.19	0.06	0.67	0.19	29	13
0	133	1.1	19	110	15	0.22	386	0.04	0.84	0.1	0.55	0.12	88	17
0	178	0.95	26	213	42	0.23	489	0.04	1.06	0.1	0.54	0.1	65	22
0	2	0.38	18	158	3	0.4	8	0.11	0.05	0.04	0.96	0.14	20	3
0	2	0.39	23	191	1	0.52	8	0.02	0.06	0.05	1.08	0.08	29	3
0	4	0.49	22	172	365	0.68	5	0.03	0.09	0.07	1.23	0.08	58	6
0	5	0.44	24	195	286	0.48	4	0.04	0.04	0.08	1.23	0.06	51	9
0	14	7.23	12	176	1532	0.76	25	0.04	0.24	0.04	0.69	0.16	34	5
0	103	1.7	20	218	42	0.23	136	0.1	2.63	0.14	0.44	0.14	15	19
0	147	1.89	25	244	46	0.29	521	0.14	3.57	0.18	0.54	0.17	14	19
0	6	0.25	11	122	1	0.12	2	0.08	0.41	0.02	0.59	0.09	14	1
0	26	0.42	8	157	11	0.4	54	0.04	0.22	0.04	0.2	<.01	71	3
0	26	0.42	8	157	11	0.4	54	0.04	0.22	0.04	0.2	<.01	71	3
0	80	4.74	79	477	11	2.51	2	0.19	0.57	0.1	0.86	0.23	282	2
0	11	0.08	3	8	<1	0.02	41	<.01	0.06	0.01	0.07	0.01	2	0
0	51	0.37	13	38	1	0.09	193	<.01	0.28	0.05	0.33	0.06	12	2
0	120	3.77	88	692	4	1.56	<1	0.28	0.53	0.1	1.21	0.21	181	2
0	21	5.1	26	644	6	0.18	2	0.3	0.28	0.09	1.95	0.12	20	6
0	14	0.72	14	180	5	0.19	1	0.02	0.55	0.03	0.24	0.05	14	24
0	94	1.17	23	296	30	0.31	885	0.07	1.1	0.09	0.65	0.18	17	53
0	179	1.22	23	417	20	0.23	956	0.06	1.2	0.15	0.87	0.11	18	33
0	61	3.23	72	658	873	1.41	0	0.27	1.54	0.23	1.17	0.06	131	3
0	32	0.54	26	472	27	0.04	3	0.07	0.65	0.03	0.54	0.2	22	84
0	41	0.66	31	561	35	0.51	3	0.07	0.86	0.03	0.64	0.25	20	89
0	53	1.87	25	160	18	0.11	74	0.05	0.82	0.03	0.36	0.21	57	11
0	31	1.14	15	90	10	0.06	2	0.03	0.63	0.02	0.21	0.12	25	4
0	38	6.59	71	731	4	2.51	1	0.33	0.22	0.14	2.1	0.35	358	3
0	17	3.84	43	352	12	1.98	2	0.27	0.11	0.11	1.49	0.2	83	16
0	19	2.47	28	248	8	1.16	2	0.18	0.07	0.1	0.87	0.15	77	13
0	14	0.51	5	98	2	0.08	68	0.02	0.07	0.03	0.15	0.03	30	2
0	10	0.37	4	71	2	0.06	50	0.02	0.05	0.02	0.11	0.02	22	1
0	11	0.19	4	84	5	0.09	9	0.02	0.02	0.01	0.07	0.03	31	2
0	108	1.89	43	819	49	0.86	86	0.2	0.16	0.11	0.67	0.25	302	21
0	27	0.47	11	205	12	0.22	22	0.05	0.04	0.03	0.17	0.06	76	5
0	10	0.24	4	54	8	0.05	104	0.02	0.08	0.02	0.1	0.03	11	5
0	9	0.27	4	69	2	0.06	81	0.02	0.04	0.02	0.09	0.02	38	7
0	10	0.29	4	74	2	0.07	87	0.02	0.04	0.02	0.09	0.02	41	7
0	112	0.8	101	573	9	0.98	52	0.23	1.23	0.1	1.55	0.33	81	29
0	33	0.56	43	558	29	0.7	2	0.18	0.39	0.1	1.76	0.18	56	42
0	1	0.18	3	110	1	0.18	0	0.03	<.01	0.15	1.35	0.04	6	1
0	5	1.36	9	278	2	0.68	0	0.06	<.01	0.23	3.48	0.07	14	3
0	2	0.32	10	85	3	0.97	0	0.03	0.01	0.12	1.1	0.12	15	0
0	9	0.62	12	101	332	0.56	0	0.07	<.01	0.02	1.24	0.05	9	0
0	2	0.51	5	140	2	0.28	0	0.03	0.05	0.16	1.42	0.03	6	1

A   Appendix

Table A–1

Food Composition

(Computer code number is for West Diet Analysis program) (For purposes of calculations, use "0" for t, <1, <.1, <.01, etc.)

Computer Code Number	Food Description	Measure	Wt (g)	H₂O (%)	Ener (kcal)	Prot (g)	Carb (g)	Dietary Fiber (g)	Fat (g)	Fat Breakdown (g)		
										Sat	Mono	Poly
	VEGETABLES AND LEGUMES—Continued											
5096	Mustard greens, cooked from raw	½ c	70	94	10	2	1	1	<1	t	t	t
5097	From frozen	½ c	75	94	14	2	2		<1	t	t	t
7022	Navy beans, cooked from dry	1 c	182	63	258	16	48	12	1	0.3	t	0.4
	Okra, cooked:											
5098	From fresh pods	8 ea	85	93	19	2	4		<1	t	t	t
5100	From frozen slices	1 c	184	91	52	4	11	5	1	0.1	t	0.1
5644	Batter fried from fresh	1 c	92	67	175	2	14	2	13	1.7	3.1	7.1
	Onions, yellow/white:											
7499	Raw, chopped	½ c	80	89	34	1	8	1	<1	t	t	t
7808	Raw, sliced	½ c	58	89	24	1	6	1	<1	t	t	t
7812	Cooked, drained, chopped	½ c	105	88	46	1	11	1	<1	t	t	t
5113	Dehydrated flakes	¼ c	14	4	49	1	12	1	<1	t	t	t
5530	Onions, pearl, cooked	½ c	93	88	41	1	9	1	<1	t	t	t
5114	Spring/green onions, bulb and top, chopped	½ c	50	90	16	1	4	1	<1	t	t	t
5190	Onion rings, breaded, heated from frozen	2 ea	20	28	81	1	8	<1	5	1.7	2.2	1
5522	Palm hearts, cooked slices	1 c	146	70	150	4	39	2	<1	t	t	0.1
26012	Parsley, raw, chopped	½ c	30	88	11	1	2	1	<1	t	t	t
5212	Parsnips, sliced, cooked	½ c	78	80	55	1	13	3	<1	t	t	t
	Peas:											
7018	Black-eyed, cooked from dry, drained	½ c	86	70	100	7	18	6	<1	0.1	t	0.2
5213	From fresh, drained	½ c	82	75	80	3	17	4	<1	t	t	0.1
5115	From frozen, drained	½ c	85	66	112	7	20	5	1	0.1	t	0.2
5122	Edible pod peas, cooked	½ c	80	89	34	3	6	2	<1	t	t	t
5119	Green, canned, drained:	½ c	85	82	59	4	11	3	<1	t	t	0.1
5267	Unsalted	½ c	124	86	66	4	12	4	<1	t	t	0.2
5118	Green, cooked from frozen	½ c	80	80	62	4	11	4	<1	t	t	0.1
6836	Snow peas, raw	½ c	49	89	21	1	4	1	<1	t	t	t
5121	each	10 ea	34	89	14	1	3	1	<1	t	t	t
7020	Split, green, cooked from dry	½ c	98	69	116	8	21	8	<1	t	t	0.2
5123	Peas & carrots, cooked from frozen	½ c	80	86	38	2	8	2	<1	t	t	0.2
5281	Canned w/liquid	½ c	128	88	49	3	11	3	<1	t	t	0.2
	Peppers, hot:											
5063	Hot green chili, canned	½ c	68	92	14	1	3	1	<1	t	t	t
5400	Raw	1 ea	45	88	18	1	4	1	<1	t	t	t
5288	Hot red chili, raw, diced	1 tbs	9	88	4	<1	1	<1	<1	t	t	t
5293	Jalapeno, chopped, canned	½ c	68	89	18	1	3	2	1	t	t	0.3
	Peppers, sweet:											
6846	Green, whole, raw	1 ea	119	94	24	1	6	2	<1	t	t	t
5126	Cooked, chopped	½ c	68	92	19	1	5	1	<1	t	t	t
5128	Red, raw, chopped	½ c	75	92	20	1	5		<1	t	t	0.1
5294	Raw, each	1 ea	74	92	19	1	4		<1	t	t	0.1
5278	Cooked, chopped	½ c	68	92	19	1	5	1	<1	t	t	t
5441	Yellow, raw, whole	1 ea	186	92	50	2	12	2	<1	t	t	0.2
5442	Strips	10 pce	52	92	14	1	3	<1	<1	t	t	t
7013	Pinto beans, cooked from dry	½ c	85	64	119	8	21	7	1	t	t	0.2
5229	Poi - two finger	½ c	120	72	134	<1	33	<1	<1	t	t	t
	Potatoes:											
	Baked in oven, 4¾" x 2⅓" diam											
5129	Flesh only	1 ea	156	75	145	3	34	2	<1	t	t	t
5339	Skin only	1 ea	58	47	115	2	27	5	<1	t	t	t
	Baked in microwave, 4¾" x 2⅓" dm:											
5340	With skin	1 ea	202	72	212	5	49	5	<1	t	t	t
5345	Flesh only	1 ea	156	74	156	3	36	2	<1	t	t	t
5350	Skin only	1 ea	58	64	77	3	17	3	<1	t	t	t
	Boiled, about 2½" diam:											
5133	Peeled after boiling	1 ea	136	77	118	3	27	2	<1	t	t	t
5135	Peeled before boiling	1 ea	135	77	116	2	27	2	<1	t	t	t
5139	French fried, strips, oven heated	10 ea	50	35	166	2	20	2	9	3	5.7	0.7
5140	Hashed browns from frozen	1 c	156	56	340	5	44	3	18	7	8	2.1

Chol (mg)	Calc (mg)	Iron (mg)	Magn (mg)	Pota (mg)	Sodi (mg)	Zinc (mg)	VT-A (RAE)	Thia (mg)	VT-E (mg α)	Ribo (mg)	Niac (mg)	V-B$_6$ (mg)	Fola (μg)	VT-C (mg)
0	52	0.49	10	141	11	0.08	221	0.03	0.85	0.04	0.3	0.07	51	18
0	76	0.84	10	104	19	0.15	266	0.03	1.01	0.04	0.19	0.08	52	10
0	127	4.51	107	670	2	1.93	<1	0.37	0.73	0.11	0.97	0.3	255	2
0	65	0.24	31	115	5	0.37	12	0.11	0.23	0.05	0.74	0.16	39	14
0	177	1.23	94	431	6	1.14	31	0.18	0.59	0.23	1.44	0.09	269	22
2	61	1.26	36	190	122	0.5		0.18	3.04	0.14	1.44	0.12	38	10
0	18	0.15	8	115	2	0.13	<1	0.04	0.02	0.02	0.07	0.12	15	5
0	13	0.11	6	84	2	0.09	<1	0.03	0.01	0.01	0.05	0.09	11	4
0	23	0.25	12	174	3	0.22	<1	0.04	0.02	0.02	0.17	0.14	16	5
0	36	0.22	13	227	3	0.26	<1	0.07	0.03	0.01	0.14	0.22	23	10
0	20	0.22	10	154	3	0.2	0	0.04	0.12	0.02	0.15	0.12	14	5
0	36	0.74	10	138	8	0.2	25	0.03	0.28	0.04	0.26	0.03	32	9
0	6	0.34	4	26	75	0.08	2	0.06	0.14	0.03	0.72	0.02	13	0
0	26	2.47	15	2637	20	5.45	5	0.07	0.73	0.25	1.25	1.06	30	10
0	41	1.86	15	166	17	0.32	126	0.03	0.22	0.03	0.39	0.03	46	40
0	29	0.45	23	286	8	0.2	0	0.06	0.78	0.04	0.56	0.07	45	10
0	21	2.16	46	239	3	1.11	1	0.17	0.24	0.05	0.43	0.09	179	0
0	105	0.92	43	343	3	0.84	33	0.08	0.18	0.12	1.15	0.05	104	2
0	20	1.8	42	319	4	1.21	3	0.22	0.26	0.05	0.62	0.08	120	2
0	34	1.58	21	192	3	0.3	43	0.1	0.31	0.06	0.43	0.12	23	38
0	17	0.81	14	147	214	0.6	23	0.1	0.03	0.07	0.62	0.05	37	8
0	22	1.26	21	124	11	0.87	89	0.14	0.02	0.09	1.04	0.08	36	12
0	19	1.22	18	88	58	0.54	84	0.23	0.02	0.08	1.18	0.09	47	8
0	21	1.02	12	98	2	0.13	26	0.07	0.19	0.04	0.29	0.08	21	29
0	15	0.71	8	68	1	0.09	18	0.05	0.13	0.03	0.2	0.05	14	20
0	14	1.26	35	355	2	0.98	<1	0.19	0.03	0.05	0.87	0.05	64	0
0	18	0.75	13	126	54	0.36	374	0.18	0.42	0.05	0.92	0.07	21	6
0	29	0.96	18	128	333	0.74	370	0.09	0.24	0.07	0.74	0.11	23	8
0	5	0.34	10	127	798	0.12	24	0.01	0.47	0.03	0.54	0.1	7	46
0	8	0.54	11	153	3	0.14	27	0.04	0.31	0.04	0.43	0.13	10	109
0	1	0.09	2	29	1	0.02	4	<.01	0.06	<.01	0.11	0.05	2	13
0	16	1.28	10	131	1136	0.23	58	0.03	0.47	0.03	0.27	0.13	10	7
0	12	0.4	12	208	4	0.15	21	0.07	0.44	0.03	0.57	0.27	13	96
0	6	0.31	7	113	1	0.08	10	0.04	0.36	0.02	0.32	0.16	11	51
0	5	0.32	9	158	2	0.19	118	0.04	1.18	0.06	0.73	0.22	14	142
0	5	0.32	9	156	1	0.18	116	0.04	1.17	0.06	0.72	0.22	13	141
0	6	0.31	7	113	1	0.08	187	0.04	1.12	0.02	0.32	0.16	11	116
0	20	0.86	22	394	4	0.32	19	0.05	1.28	0.05	1.66	0.31	48	341
0	6	0.24	6	110	1	0.09	5	0.01	0.36	0.01	0.46	0.09	14	95
0	36	1.77	35	246	9	0.86	0	0.08	0.8	0.06	0.17	0.08	146	1
0	19	1.06	29	220	14	0.26	4	0.16	2.76	0.05	1.32	0.33	25	5
0	8	0.55	39	610	8	0.45	0	0.16	0.06	0.03	2.18	0.47	14	20
0	20	4.08	25	332	12	0.28	1	0.07	0.02	0.06	1.78	0.36	13	8
0	22	2.5	55	903	16	0.73	0	0.24	0.1	0.06	3.46	0.69	24	31
0	8	0.64	39	641	11	0.51	0	0.2	0.06	0.04	2.54	0.5	19	24
0	27	3.45	21	377	9	0.3	0	0.04	0.02	0.04	1.29	0.29	10	9
0	7	0.42	30	515	5	0.41	<1	0.14	0.01	0.03	1.96	0.41	14	18
0	11	0.42	27	443	7	0.36	<1	0.13	0.01	0.03	1.77	0.36	12	10
0	6	0.83	12	270	306	0.2	0	0.04	0.25	0.02	1.33	0.11	11	3
0	23	2.36	27	680	53	0.5	0	0.17	0.3	0.03	3.78	0.2	11	10

Table A–1

Food Composition   (Computer code number is for West Diet Analysis program)   (For purposes of calculations, use "0" for t, <1, <.1, <.01, etc.)

Computer Code Number	Food Description	Measure	Wt (g)	H₂O (%)	Ener (kcal)	Prot (g)	Carb (g)	Dietary Fiber (g)	Fat (g)	Fat Breakdown (g) Sat	Mono	Poly
	VEGETABLES AND LEGUMES—Continued											
	Mashed:											
5137	Home recipe with whole milk	½ c	105	79	87	2	18	2	1	0.3	0.1	t
5272	Home recipe with milk and marg	½ c	105	75	119	2	18	2	4	1	1.8	1.3
5138	Prepared from flakes with milk and marg	½ c	110	76	124	2	17	3	6	1.6	2.5	1.7
	Potato products, prepared:											
5276	Au gratin from dry mix	½ c	123	79	114	3	16	1	5	3.2	1.4	0.2
5275	From home recipe,using butter	½ c	122	74	161	6	14	2	9	4.3	3.2	1.3
5271	Scalloped from dry mix	½ c	122	79	113	3	16	1	5	3.2	1.5	0.2
5270	From home recipe,using butter	½ c	123	81	106	4	13	2	5	1.7	1.7	0.9
	Potato Salad (see Mixed Dishes #715)											
5265	Potato Puffs, cooked from frozen	½ c	64	53	142	2	20	2	7	3.3	2.8	0.5
5396	Pumpkin, cooked from fresh, mashed	½ c	123	94	25	1	6	1	<1	t	t	t
5142	Canned	½ c	123	90	42	1	10	4	<1	0.2	t	t
5451	Radicchio, raw, shredded	½ c	20	93	5	<1	1	<1	<1	t	t	t
5452	Leaf	10 ea	80	93	18	1	4	1	<1	t	t	t
5143	Red radishes	10 ea	45	95	7	<1	2		<1	t	t	t
1793	Daikon radishes (Chinese) raw	½ c	44	95	8	<1	2	1	<1	t	t	t
7024	Refried beans, canned	½ c	126	76	118	7	20	7	2	0.6	0.7	0.2
7225	Rutabaga, cooked cubes	½ c	85	89	33	1	7	2	<1	t	t	t
5145	Sauerkraut, canned with liquid	½ c	118	93	22	1	5	3	<1	t	t	t
6857	Seaweed, kelp, raw	½ c	40	82	17	1	4	1	<1	t	t	t
5260	Seaweed, spirulina, dried	½ c	8	5	23	5	2	<1	1	0.2	t	0.2
5427	Shallots, raw, chopped	1 tbs	10	80	7	<1	2	<1	<1	t	t	t
5666	Snow Peas, stir fried	½ c	83	89	35	2	6	2	<1	t	t	t
7015	Soybeans, cooked from dry	½ c	86	63	149	14	9	5	8	1.1	1.7	4.4
7063	dry roasted	½ c	86	1	388	34	28	7	19	2.7	4.1	10.5
	Soybean products:											
7503	Miso	½ c	138	41	284	16	39	7	8	1.2	1.9	4.7
	Soy milk (see Dairy)											
	Tofu (soybean curd):											
	Silken:											
7540	Extra firm	½ c	126	88	69	9	3	<1	2	0.4	0.4	1.3
7542	Firm	½ c	126	87	78	9	3	<1	3	0.5	0.7	1.9
7500	Regular	½ c	124	87	76	8	2	<1	5	0.7	1	2.6
7721	Block	1 pce	116	85	88	9	2	<1	6	0.8	1.2	3.1
7720	Firm	1 pce	81	70	117	13	3	2	7	1	1.6	4
7519	Dried, frozen	1 pce	17	6	82	8	2	1	5	0.7	1.1	2.9
90039	Fried	1 pce	13	51	35	2	1	1	3	0.4	0.6	1.5
7541	Soft	½ c	124	89	68	6	4	<1	3	0.4	0.6	1.9
7518	Prepared with nigari	½ c	126	84	97	10	4	1	6	0.8	1.2	3.2
	Spinach:											
5146	Raw, chopped	1/2 c	15	91	3	<1	1	<1	<1	t	t	t
5147	Cooked, from fresh, drained	½ c	90	91	21	3	3	2	<1	t	t	t
5148	From frozen (leaf)	½ c	95	89	30	4	5	4	<1	t	0	0.2
5149	Canned, unsalted	½ c	107	92	25	3	4	3	1	t	t	0.2
	Spinach souffle (see Mixed Dishes)											
	Squash, summer varieties,cooked w/skin:											
5152	Varieties averaged	½ c	90	94	18	1	4	1	<1	t	t	0.1
5322	Crookneck	½ c	90	94	18	1	4	1	<1	t	t	0.1
5327	Zucchini	½ c	90	95	14	1	4	1	<1	t	t	t
	Squash, winter varieties, cooked:											
5303	Average of all varieties, baked, cubes	1 c	205	89	76	2	18	6	1	0.2	t	0.4
5314	Acorn, baked, mashed	½ c	123	83	69	1	18	5	<1	t	t	t
5316	Boiled, mashed	½ c	122	90	41	1	11	3	<1	t	t	t
	Butternut squash:											
5317	Baked, mashed	½ c	103	88	41	1	11	3	<1	t	t	t
5274	Cooked from frozen	½ c	120	88	47	1	12	3	<1	t	t	t
5453	Hubbard, baked, mashed	½ c	120	85	60	3	13	3	1	0.2	t	0.3
5454	Boiled, mashed	½ c	118	91	35	2	8	3	<1	t	t	0.2
5455	Spaghetti, baked or boiled	½ c	77	92	21	1	5	1	<1	t	t	t

A Appendix

PAGE KEY: A–4 = Beverages   A–6 = Dairy   A–10 = Eggs   A–12 = Fat/Oil   A–14 = Fruit   A–20 = Bakery
A–26 = Grain   A–32 = Fish   A–32 = Meats   A–36 = Poultry   A–38 = Sausage   A–40 = Mixed/Fast   A–44 = Nuts/Seeds   A–46 = Sweets
A–48 = Vegetables/Legumes   A–58 = Vegetarian   A–62 = Misc   A–62 = Soups/Sauces   A–64 = Fast   A–78 = Convenience   A–80 = Baby foods

Chol (mg)	Calc (mg)	Iron (mg)	Magn (mg)	Pota (mg)	Sodi (mg)	Zinc (mg)	VT-A (RAE)	Thia (mg)	VT-E (mg α)	Ribo (mg)	Niac (mg)	V-B$_6$ (mg)	Fola (µg)	VT-C (mg)
2	23	0.28	19	311	317	0.3	4	0.09	0.02	0.05	1.18	0.24	8	7
1	21	0.27	20	342	350	0.32	43	0.1	0.44	0.05	1.23	0.26	9	11
4	54	0.24	20	256	365	0.2	51	0.12	0.77	0.06	0.74	<.01	8	11
18	102	0.39	18	269	540	0.3	64	0.02	1.48	0.1	1.15	0.05	9	4
18	145	0.78	24	483	528	0.84	78	0.08	0.64	0.14	1.21	0.21	13	12
13	44	0.46	17	248	416	0.3	43	0.02	0.18	0.07	1.26	0.05	12	4
7	70	0.7	23	465	412	0.49	41	0.08	0.4	0.11	1.3	0.22	14	13
0	19	1	12	243	477	0.19	<1	0.13	0.03	0.05	1.38	0.15	11	4
0	18	0.7	11	283	1	0.28	308	0.04	0.98	0.1	0.51	0.05	11	6
0	32	1.71	28	253	6	0.21	957	0.03	1.3	0.07	0.45	0.07	15	5
0	4	0.11	3	60	4	0.12	<1	<.01	0.45	<.01	0.05	0.01	12	2
0	15	0.46	10	242	18	0.5	1	0.01	1.81	0.02	0.2	0.05	48	6
0	11	0.15	4	105	18	0.13	<1	<.01	0	0.02	0.11	0.03	11	7
0	12	0.18	7	100	9	0.07	0	<.01	0	<.01	0.09	0.02	12	10
10	44	2.09	42	336	377	1.47	0	0.03	0	0.02	0.4	0.18	14	8
0	41	0.45	20	277	17	0.3	<1	0.07	0.27	0.03	0.61	0.09	13	16
0	35	1.73	15	201	780	0.22	1	0.02	0.12	0.03	0.17	0.15	28	17
0	67	1.14	48	36	93	0.49	2	0.02	0.35	0.06	0.19	<.01	72	1
0	10	2.28	16	109	84	0.16	2	0.19	0.4	0.29	1.03	0.03	8	1
0	4	0.12	2	33	1	0.04	6	<.01	<.01	<.01	0.02	0.03	3	1
0	36	1.73	20	166	3	0.22	5	0.11	0.32	0.06	0.47	0.13	28	42
0	88	4.42	74	443	1	0.99	<1	0.13	0.3	0.25	0.34	0.2	46	1
0	120	3.4	196	1173	2	4.1	0	0.37	3.96	0.65	0.91	0.19	176	4
0	91	3.78	58	226	5033	4.58	6	0.13	0.01	0.34	1.19	0.3	46	0
0	39	1.5	34	194	79	0.76	0	0.1	0.18	0.04	0.3	0.01		0
0	40	1.3	34	244	45	0.77	0	0.13	0.24	0.05	0.31	0.01		0
0	138	1.38	33	149	10	0.79	<1	0.06	0.01	0.05	0.66	0.06	55	0
0	406	6.22	35	140	8	0.93	5	0.09	0.01	0.06	0.23	0.05	17	0
0	553	2.15	47	192	11	1.27	6	0.13	0.02	0.08	0.31	0.07	23	0
0	62	1.65	10	3	1	0.83	4	0.08	0.01	0.05	0.2	0.05	16	0
0	125	0.63	12	19	2	0.26	0	0.02	<.01	<.01	0.01	0.01	4	0
0	38	1.02	36	223	6	0.64	0	0.12	0.25	0.05	0.37	0.01		0
0	204	1.83	58	222	10	1.27	0	0.03	0.13	0.01	0.08	42	0
0	15	0.41	12	84	12	0.08	70	0.01	0.3	0.03	0.11	0.03	29	4
0	122	3.21	78	419	63	0.68	472	0.09	1.87	0.21	0.44	0.22	131	9
0	145	1.86	78	287	92	0.47	573	0.07	3.36	0.17	0.42	0.13	115	2
0	136	2.46	81	370	29	0.49	524	0.02	2.08	0.15	0.42	0.11	105	15
0	24	0.32	22	173	1	0.35	10	0.04	0.13	0.04	0.46	0.06	18	5
0	20	0.33	14	153	0	0.2	7	0.04	0.11	0.04	0.46	0.08	18	5
0	12	0.32	20	228	3	0.16	50	0.04	0.11	0.04	0.39	0.07	15	4
0	45	0.9	27	896	2	0.45	535	0.03	0.25	0.14	1.01	0.33	41	20
0	54	1.14	53	538	5	0.21	26	0.21	0.15	0.02	1.08	0.24	23	13
0	32	0.68	32	321	4	0.13	50	0.12	0.15	<.01	0.65	0.14	13	8
0	42	0.62	30	293	4	0.13	575	0.07	1.33	0.02	1	0.13	20	16
0	23	0.7	11	160	2	0.14	200	0.06	0.16	0.05	0.56	0.08	19	4
0	20	0.56	26	430	10	0.18	362	0.09	0.14	0.06	0.67	0.21	19	11
0	12	0.33	15	253	6	0.12	236	0.05	0.14	0.03	0.39	0.12	12	8
0	16	0.26	8	90	14	0.15	5	0.03	0.09	0.02	0.62	0.08	6	3

A Appendix

Table A–1

Food Composition

(Computer code number is for West Diet Analysis program)   (For purposes of calculations, use "0" for t, <1, <.1, <.01, etc.)

Computer Code Number	Food Description	Measure	Wt (g)	H₂O (%)	Ener (kcal)	Prot (g)	Carb (g)	Dietary Fiber (g)	Fat (g)	Sat	Mono	Poly
	VEGETABLES AND LEGUMES—Continued											
5154	Succotash, cooked from frozen	½ c	85	74	79	4	17	3	1	0.1	0.1	0.4
	Sweet potatoes:											
5155	Baked in skin, peeled, 5 x 2" diam	1 ea	114	76	103	2	24	4	<1	t	t	t
5159	Boiled without skin, 5 x 2" diam	1 ea	151	80	115	2	27	4	<1	t	0	t
5166	Candied pieces, 2½ x 2"	1 pce	105	67	144	1	29	3	3	1.4	0.7	0.2
5162	Canned, solid pack	½ c	128	74	129	3	30	2	<1	t	t	0.1
5164	Vacuum pack, mashed	½ c	127	76	116	2	27	2	<1	t	t	0.1
5543	Taro shoots, cooked slices	1 c	140	95	20	1	4	1	<1	t	t	t
5544	Taro, tahitian, cooked slices	1 c	137	86	60	6	9	1	1	0.2	t	0.4
5445	Tomatillos, raw, each	1 ea	34	92	11	<1	2	1	<1	t	t	0.1
5444	Chopped	1 c	132	92	42	1	8	3	1	0.2	0.2	0.6
	Tomatoes:											
5169	Raw, whole, 2⅗" diam	1 ea	123	94	22	1	5	1	<1	t	t	0.1
6492	Whole, small	1 ea	62	94	11	1	2	1	<1	t	t	t
5170	Chopped	1 c	180	94	32	2	7	2	<1	t	t	0.2
5178	Cooked from raw	1 c	240	94	43	2	10	2	<1	t	t	0.1
5179	Canned, solids and liquid:	1 c	240	94	41	2	9	2	<1	t	t	0.1
6992	Unsalted	1 c	240	94	46	2	10	2	<1	t	t	0.1
5446	Sundried:	1 c	54	15	139	8	30	7	2	0.2	0.3	0.6
5447	Pieces	10 pce	20	15	52	3	11	2	1	t	t	0.2
5448	Oil pack, drained	10 pce	30	54	64	2	7	2	4	0.6	2.6	0.6
5188	Tomato juice, canned:	1 c	243	94	41	2	10	1	<1	t	t	t
5397	Unsalted	1 c	243	94	41	2	10	1	<1	t	t	t
	Tomato products, canned:											
5181	Paste-no added salt	1 c	262	74	215	11	50	12	1	0.3	0.2	0.4
5225	Puree-no added salt	1 c	250	88	95	4	22	5	1	t	t	0.2
5180	Sauce-with salt	1 c	245	89	78	3	18	4	1	t	t	0.2
5183	Turnips, cubes, cooked from fresh	1 c	156	94	34	1	8	3	<1	t	t	t
5185	Turnip greens, cooked from fresh, leaves and stems	1 c	144	93	29	2	6	5	<1	t	t	0.1
5186	From frozen, chopped	1 c	164	90	48	5	8	6	1	0.2	t	0.3
20080	Vegetable juice cocktail, canned	1 c	242	94	46	2	11	2	<1	t	t	t
5305	Vegetables, mixed, canned, drained	½ c	81	87	40	2	8	2	<1	t	t	t
5187	Frozen, cooked, drained	½ c	91	83	59	3	12	4	<1	t	t	t
5386	Water chestnuts, Chinese, raw	½ c	62	73	60	1	15	2	<1	t	t	t
5387	Canned, slices	½ c	70	86	35	1	9	2	<1	t	t	t
5388	Whole	4 ea	28	86	14	<1	3	1	<1	t	t	t
5222	Watercress, fresh, chopped	½ c	17	95	2	<1	<1	<1	<1	t	t	t
	VEGETARIAN FOODS:											
7509	Bacon strips, meatless	3 ea	15	49	46	2	1	<1	4	0.7	1.1	2.3
7038	Baked beans, canned	1/2 c	127	73	118	6	26	6	1	0.1	t	0.2
7526	Bakon Crumbles	1/4 c	7	8	31	2	2	1	2	0		
7723	Chicken, fillet	3 oz	85		90	15	8	4	2	0		
7557	Chili w/meat substitute	½ c	107	65	141	19	15	5	2	0.3	0.6	0.9
	Frankfurters:											
7550	Meatless	1 ea	51	58	102	10	4	2	5	0.8	1.2	2.6
8127	Deli	1 ea	76	72	80	16	4	1	0	0	0	0
8835	Tofu wiener	1 ea	38		45	9	2	0	<1	0		
8839	Jumbo	1 ea	76		100	16	7	2	2	0		
8840	Hot and spicy	1 ea	52		70	13	3	2	1	0		
	Garden Burger patties:											
7504	Regular	1 ea	71		110	6	16	3	3	1.5		
8811	Mushroom	1 ea	71		120	6	18	4	2	1		
8813	Roasted vegetable	1 ea	71		120	6	18	4	2	1		
7652	Veggie medley	1 ea	71		90	5	18	3	0	0	0	0
7505	Garden Sausage, patty	1 ea	71		83	8	3	3	6	0		
57433	Nuteena	1 ea	55	58	162	6	6	2	13	5.2	5.8	1.7
7556	Pot pie, meatless	1 ea	227	60	510	14	41	5	32	8.6	12.4	9.6
8831	Soyburger mix	⅓ c	55	73	60	10	4	3	0	0	0	0

Chol (mg)	Calc (mg)	Iron (mg)	Magn (mg)	Pota (mg)	Sodi (mg)	Zinc (mg)	VT-A (RAE)	Thia (mg)	VT-E (mg α)	Ribo (mg)	Niac (mg)	V-B$_6$ (mg)	Fola (µg)	VT-C (mg)
0	13	0.76	20	225	38	0.38	8	0.06	0.15	0.06	1.11	0.08	28	5
0	43	0.79	31	542	41	0.36	1096	1.65	0.81	0.12	1.7	0.33	7	22
0	41	1.09	27	347	41	0.3	1190	0.08	1.42	0.07	0.81	0.25	9	19
8	27	1.19	12	198	74	0.16	219	0.02	3.99	0.04	0.41	0.04	12	7
0	38	1.7	31	269	96	0.27	968	0.03	0.35	0.12	1.22	0.3	14	7
0	28	1.13	28	396	67	0.23	507	0.05	1.27	0.07	0.94	0.24	22	34
0	20	0.57	11	482	3	0.76	4	0.05	1.4	0.07	1.13	0.16	4	26
0	204	2.14	70	854	74	0.14	121	0.06	3.7	0.27	0.66	0.16	10	52
0	2	0.21	7	91	<1	0.07	2	0.01	0.13	0.01	0.63	0.02	2	4
0	9	0.82	26	354	1	0.29	8	0.06	0.5	0.05	2.44	0.07	9	15
0	12	0.33	14	292	6	0.21	52	0.05	0.66	0.02	0.73	0.1	18	16
0	6	0.17	7	147	3	0.11	26	0.02	0.33	0.01	0.37	0.05	9	8
0	18	0.49	20	427	9	0.31	76	0.07	0.97	0.03	1.07	0.14	27	23
0	26	1.63	22	523	26	0.34	58	0.09	1.34	0.05	1.28	0.19	31	55
0	74	2.33	26	451	307	0.34	14	0.11	1.7	0.11	1.76	0.22	19	22
0	72	1.32	29	545	24	0.38	17	0.11	1.92	0.07	1.76	0.22	19	34
0	59	4.91	105	1851	1131	1.07	24	0.29	<.01	0.26	4.89	0.18	37	21
0	22	1.82	39	685	419	0.4	9	0.11	<.01	0.1	1.81	0.07	14	8
0	14	0.8	24	470	80	0.23	19	0.06	0.16	0.11	1.09	0.1	7	31
0	24	1.04	27	556	654	0.36	56	0.11	0.78	0.08	1.64	0.27	49	44
0	24	1.04	27	556	24	0.36	56	0.11	0.78	0.08	1.64	0.27	49	44
5	94	7.81	110	2657	257	1.65	199	0.16	11.27	0.4	8.06	0.57	31	57
5	45	4.45	58	1098	70	0.9	65	0.06	4.92	0.2	3.66	0.32	28	26
0	32	2.5	39	811	1284	0.49	42	0.06	5.1	0.16	2.39	0.24	22	17
0	51	0.28	14	276	25	0.19	0	0.04	0.03	0.04	0.47	0.1	14	18
0	197	1.15	32	292	42	0.2	549	0.06	2.71	0.1	0.59	0.26	170	39
0	249	3.18	43	367	25	0.67	882	0.09	4.36	0.12	0.77	0.11	64	36
0	27	1.02	27	467	653	0.48	189	0.1	12.1	0.07	1.76	0.34	51	67
0	22	0.85	13	236	121	0.33	471	0.04	0.28	0.04	0.47	0.06	19	4
0	23	0.75	20	154	32	0.45	195	0.06	0.4	0.11	0.77	0.07	17	3
0	7	0.04	14	362	9	0.31	0	0.09	0.74	0.12	0.62	0.2	10	2
0	3	0.61	4	83	6	0.27	0	<.01	0.35	0.02	0.25	0.11	4	1
0	1	0.24	1	33	2	0.11	0	<.01	0.14	<.01	0.1	0.04	2	0
0	20	0.03	4	56	7	0.02	40	0.02	0.17	0.02	0.03	0.02	2	7
0	3	0.36	3	26	220	0.06	1	0.66	1.03	0.07	1.13	0.07	6	0
0	64	0.37	41	376	504	1.78	6	0.19	0.67	0.08	0.54	0.17	30	4
0	16	0.28			163		0							0
0	80	1.8			170		0							1
0	54	4.38	36	366	355	1.27	37	0.12	1.28	0.07	1.22	0.15	82	6
0	17	0.92	9	76	219	0.61	0	0.56	0.98	0.61	8.16	0.5	40	0
0	40	0.72			590		0							1
0	20	2.16	8	90	240	0.6	0	0.15						0
0	20	4.5	16	170	480	3.75	0	0.38						0
0	20	3.6	16	135	400	3	38	0.3						0
20	60	0			560		0							0
20	100	0.36			370									0
15	100	0.36			330		0							0
0	40	0			280									0
0	33	1.19			198		0							0
0	9	0.27		166	119	0.46	0	0.1		0.35	1.04	0.45		0
20	68	2.96	32	378	486	1.08		0.82	4.23	0.44	5.17	0.41	58	10
0	40	2.7		240	270	3	0	0.22		0.14	3	0.2		0

A   Appendix

A–59

Table A–1

Food Composition

(Computer code number is for West Diet Analysis program)   (For purposes of calculations, use "0" for t, <1, <.1, <.01, etc.)

Computer Code Number	Food Description	Measure	Wt (g)	H₂O (%)	Ener (kcal)	Prot (g)	Carb (g)	Dietary Fiber (g)	Fat (g)	Fat Breakdown (g)		
										Sat	Mono	Poly
	VEGETARIAN FOODS:—Continued											
7751	Griller patty	1 ea	113		150	21	11	6	4	0		
7562	Patty, with cheese	1 ea	135	51	308	20	30	4	12	3.6	3.6	3.6
7517	Soy protein isolate	1 oz	28	5	95	23	2	2	1	0.1	0.2	0.5
7564	Tempeh	1 c	166	60	320	31	16		18	3.7	5	6.4
7670	Vegan burger, patty	1 ea	78	71	83	13	7	4	<1	t	0.3	0.1
8842	Veggie slices, soy	1 pce	15		19	3	1	<1	<1	0		
8830	Veggie ground soy	⅓ c	55	73	60	10	4	3	0	0	0	0
7902	Vegetarian burger mix	½ c	47		170	8	30	5	3	0		
7511	Vegetarian sausage link	1 ea	25	50	64	5	2	1	5	0.7	1.1	2.3
7512	Patty	1 ea	38	50	98	7	4	1	7	1.1	1.7	3.5
	Vegetarian foods, Green Giant harvest burger											
7673	Italian, patty	1 ea	90		140	17	8	5	4	1.5	0.5	0.5
7674	Original, patty	1 ea	90	65	138	18	7	6	4	1	2.1	0.3
6658	Mix, frozen	⅔ c	54		90	15	7	3	0	0	0	0
7675	Southwestern, patty	1 ea	90		140	16	9	5	4	1.5	0	0.5
	Vegetarian Foods, Loma Linda											
7727	Chik nuggets, frozen	5 pce	85	47	245	12	13	5	16	2.5	4	8.8
7767	Fakin Bakin, bacon bits	1 tsp	3		12	1	1	0	1	0		
7860	Smokey tempeh strip	1 pce	19		27	3	2	<1	1	0.2		
7744	Franks, big, canned	1 ea	51	58	118	12	2		7	0.8	1.5	3.7
7997	Gimme Lean! Meatless ground beef	2 oz	57		70	9	8	1	0	0	0	0
7998	Meatless sausage	2 oz	57		70	9	8	1	0	0	0	0
8159	Italian sausage, link	1 ea	40	68	60	5	5	0	2	1		
7747	Linketts, canned	1 ea	35	60	72	7	3	2	4	0.7	1.2	2.5
57434	Redi-burger, patty	1 ea	85	59	172	16	7	5	10	1.5	2.4	5.8
7755	Swiss stake w/gravy, canned	1 pce	92	71	120	9	8	4	6	0.8	1.5	3.3
57435	Vege-Burger, patty	1 ea	55	71	66	11	3	2	2	0.4	0.6	0.5
7999	Light	3 oz	85	64	130	16	12	2	1	0		
	Vegetarian foods, Morningstar Farms:											
7766	Better-n-eggs	¼ c	57	88	26	5	<1	0	<1	t	0.1	0.1
57436	Breakfast links	2 pce	45	68	64	9	2	1	2	0.4	0.5	1.1
7752	Breakfast strips	2 pce	16	43	56	2	2	1	4	0.7	0.9	2.6
477	Buffalo wings	5 ea	85	50	204	13	18	3	9	1.2	2.2	4.8
7725	Burger crumbles, svg	1 ea	55	60	116	11	3	3	6	1.6	2.3	2.5
7726	Burger, spicy black bean	1 ea	78	60	115	12	15	5	1	0.2	0.2	0.4
62541	Chik nuggets	4 pce	86	53	183	14	17	4	7	0.6	1.1	2.7
7665	Chik pattie	1 ea	71	56	148	10	14	3	6	0.9	1.6	3.6
7724	Frank, deli	1 ea	57	52	141	13	5	3	8	1.1	2.5	4.2
7722	Garden vege pattie	1 ea	67	60	119	11	10	4	4	0.5	1.1	2.2
7746	Grillers	1 ea	64	56	139	15	5	2	6	1.1	1.5	3.1
62545	Griller Prime	1 ea	71	56	160	17	4	2	8	1	3.6	3.8
7789	Ground burger	½ c	55	71	62	10	4	1	<1	t	t	0.3
62359	Sausage breakfast patty	1 ea	38	54	79	10	4	2	3	0.5	0.7	1.3
	Vegetarian foods, Worthington:											
7634	Beef style, meatless, frzn	3 pce	55	58	113	9	4	3	7	1.2	2.7	2.6
7732	Burger, meatless, patty	¼ c	55	71	60	9	2	1	2	0.3	0.5	1.1
7610	Choplets, slices, canned	2 pce	92	72	93	17	3	2	2	0.9	0.3	0.3
7608	Corned beef style, meatless, frzn	4 pce	57	55	138	10	5	2	9	1.9	4.1	3.1
7607	Country stew, canned	1 c	240	81	208	13	20	5	9	1.6	2.3	4.8
7612	Numete, slices, canned	1 pce	55	58	132	6	5	3	10	2.4	4.4	2.7
7613	Prime stakes, slices, canned	1 pce	92	71	136	9	4	4	9	1.4	2.9	4.9
7617	Protose, slices, canned	1 pce	55	53	131	13	5	3	7	1	3	2.4
7606	Roast, dinner, meatless, frzn	1 ea	85	63	180	12	5	3	13	2.2	5	5.2
7619	Saucette links, canned	1 pce	38	62	86	6	2	1	6	1.1	1.6	3.8
7620	Savory slices, canned	1 pce	28	66	48	3	2	1	3	1.2	1.3	0.6
7735	Stakelets, frzn	1 pce	71	58	145	12	6	2	8	1.4	2.7	3.9
7625	Turkee slices, canned	1 pce	33	64	68	5	1	1	5	0.8	1.9	2.1
7782	Vegan meatless patty	1 ea	71		90	10	12	2	0	0	0	0
7765	Vita Burger chunks	¼ c	21	6	75	10	6	4	1	0.3	0.2	0.6

Chol (mg)	Calc (mg)	Iron (mg)	Magn (mg)	Pota (mg)	Sodi (mg)	Zinc (mg)	VT-A (RAE)	Thia (mg)	VT-E (mg α)	Ribo (mg)	Niac (mg)	V-B$_6$ (mg)	Fola (µg)	VT-C (mg)
0					475									
9	158	2.94	27	242	922	1.91		0.8	1.58	0.59	8.32	0.84	65	1
0	50	4.06	11	23	281	1.13	0	0.05	0	0.03	0.4	0.03	49	0
0	184	4.48	134	684	15	1.89	0	0.13	0.03	0.59	4.38	0.36	40	0
0	80	2.66	15	398	351	0.69		0.23	<.01	0.51	3.77	0.18	225	0
0	10	0.87	4	31	104	0.73	2	0.07						0
0	40	2.7		240	270	3	0	0.22		0.14	3	0.2		0
0	60	1.8			320									2
0	16	0.93	9	58	222	0.36	0	0.59	0.52	0.1	2.8	0.21	6	0
0	24	1.41	14	88	337	0.55	0	0.89	0.8	0.15	4.25	0.31	10	0
0	80	2.7			370	6.75	0	0.3		0.14	4	0.3		0
0	102	3.85	70	432	411	8.07		0.32	1.56	0.2	6.3	0.39	22	0
0	100	1.8			370	4.5	0	0.22		0.1	3	0.16		0
0	80	2.7			370	6.75	0	0.3		0.14	4	0.3		0
2	40	1.4		153	709	0.43	0	0.67		0.3	2.89	0.45		0
0	0	0			30		0							0
0	30	0.3			77									0
0	10	0.99		61	224	1.2		0.28		0.68	5.78	0.67		
0	40	1.81			241		0							2
0	40	1.81			292		0							2
0	20	1.08			160		0							2
1	4	0.39		29	160	0.46	0	0.13		0.22	0.64	0.29		0
1	12	1.06		121	455	1.1	0	0.14		0.3	1.9	0.51		0
2	24	0.31		225	433	0.41	0	1.25		0.65	5.41	1		0
0	8	0.5		30	114	0.58	0	0.2		0.25	0.78	0.31		0
0	110	3.96			410		0							0
1	24	0.83		60	98	0.6		0.05		0.36	0	0.11		0
1	9	1.77		46	355	0.35	0	5.43		0.15	2.5	0.38		0
0	3	0.33		16	228	0.06	0	0.67		0.05	0.75	0.08		0
2	39	2.75		299	628	0.65		0.31		0.26	2.82	0.37		0
0	40	3.2	1	89	238	0.82		4.96	0.35	0.18	1.49	0.27		0
1	56	1.84	44	269	499	0.93		8.06	0.36	0.14	0	0.21		0
1	37	1.95		297	632	0.83	0	1.28		0.29	2.64	0.24		0
1	24	2.31		174	514	0.58	0	1.39		0.23	2.26	0.18		0
1	22	0.77	5	63	545	0.48		0.18	1.59	0.03	0	0.01		0
1	48	1.21	29	180	382	0.58	134	6.47	0.55	0.1	0	0	59	0
2	22	2.5		122	269	0.67	0	11.81		0.2	4.07	0.48		0
1	25	1.43		159	365	0.87	0	0.23		0.15				0
0	21	0.62		135	262	0.72	0	0.35		0.15	6.39	0.36		0
1	18	1.92	1	102	259	0.37		5.38	0.3	0.13	1.84	0.19		0
0	4	2.63		44	624	0.22	0	0.89		0.34	6.46	0.56		0
0	4	1.73		25	269	0.38	0	0.13		0.1	1.96	0.24		0
0	6	0.37		40	500	0.65	0	0.05		0.06	0	0.06		0
1	6	1.17		58	524	0.26	0	10.61		0.07	1.36	0.3		0
2	51	5.09		270	826	1.03	108	1.85		0.29	4.22	0.86		0
0	10	1.12		155	272	0.56	0	0.08		0.06	0.54	0.2		0
2	12	0.38		82	445	0.38	0	0.12		0.13	1.98	0.38		0
0	1	1.84		50	283	0.7	0	0.18		0.13	1.34	0.24		0
2	36	2.87		38	566	0.64	0	2.13		0.26	6.02	0.6		0
1	9	1.15		25	205	0.26	0	0.59		0.08	0.1	0.13		0
0	0	0.47		14	179	0.08	0	0.08		0.06	0.48	0.1		0
2	49	0.99		95	484	0.5	0	1.51		0.12	3.1	0.26		0
1	3	0.48		16	203	0.11	0	1.13		0.05	0.39	0.09		0
0	40	4.5			230		0							0
0	27	1.83		499	353	4.79	0	1.23		0.17	3.06	0.12		0

A Appendix

Table A–1

Food Composition

(Computer code number is for West Diet Analysis program)   (For purposes of calculations, use "0" for t, <1, <.1, <.01, etc.)

Computer Code Number	Food Description	Measure	Wt (g)	H₂O (%)	Ener (kcal)	Prot (g)	Carb (g)	Dietary Fiber (g)	Fat (g)	Fat Breakdown (g)		
										Sat	Mono	Poly
	MISCELLANEOUS											
28006	Baking powder, low sodium	1 tsp	5	6	5	<1	2	<1	<1	t	t	t
28003	Baking soda	1 tsp	5	0	0	0	0	0	0	0	0	0
26001	Basil, dried	1 tbsp	5	6	13	1	3	2	<1	t	t	0.1
38063	Carob flour	1 c	103	4	229	5	92	41	1	t	0.2	0.2
27000	Catsup	1 tbsp	15	70	14	<1	4	<1	<1	t	t	t
26027	Cayenne/red pepper	1 tbsp	5	8	16	1	3	1	1	0.2	0.1	0.4
26040	Celery seed	1 tsp	2	6	8	<1	1	<1	1	t	0.3	t
26002	Chili powder	1 tbsp	8	8	25	1	4	3	1	0.2	0.3	0.6
23010	Chocolate, baking, unsweetened	1 oz	28	1	140	4	8	5	15	9.1	4.5	0.4
	(For other chocolate items, see Sweeteners & Sweets)											
26038	Cilantro/Coriander, fresh	1 tbsp	1	92	0	<1	<1	<1	<1	t	t	t
26003	Cinnamon	1 tsp	2	10	5	<1	2	1	<1	t	t	t
30197	Cornstarch	1 tbsp	8	8	30	<1	7	<1	<1	t	t	t
26004	Curry powder	1 tsp	2	10	6	<1	1	1	<1	t	0.1	t
26021	Dill weed, dried	1 tbsp	3	7	8	1	2	<1	<1	t	t	t
26005	Garlic cloves	1 ea	3	59	4	<1	1	<1	<1	t	t	t
26007	Garlic powder	1 tsp	3	6	10	1	2	<1	<1	t	t	t
23009	Gelatin, dry, unsweetened: Envelope	1 ea	7	13	23	6	0	0	<1	t	t	t
26043	Ginger root, slices, raw	2 pce	5	79	4	<1	1	<1	<1	t	t	t
27004	Horseradish, prepared	1 tbsp	15	85	7	<1	2	<1	<1	t	t	t
7081	Hummous/hummus	1 c	246	65	435	12	49	10	21	2.8	12.1	5.1
27072	Mustard, country dijon	1 tsp	5		5	1	1	0	0	0	0	0
6313	Mustard, gai choy chinese	1 tbsp	16	97	1	<1	<1	<1	0	0	0	0
	Miso (see Vegetables and Legumes, Soybean products)											
27009	Olives, ripe, pitted	5 ea	22	80	25	<1	1	1	2	0.3	1.7	0.2
26008	Onion powder	1 tsp	2	5	7	<1	2	<1	<1	t	t	t
26009	Oregano, ground	1 tsp	2	7	6	<1	1	1	<1	t	t	0.1
26010	Paprika	1 tsp	2	10	6	<1	1	1	<1	t	t	0.2
26011	Parsley, freeze dried	¼ c	1	2	3	<1	<1	<1	<1	t	t	t
26016	Pepper, black	1 tsp	2	11	5	<1	1	1	<1	t	t	t
27012	Pickles, dill, medium, 3¾ x 1¼" diam	1 ea	65	92	12	<1	3		<1	t	t	t
27016	Sweet, medium	1 ea	35	65	41	<1	11	<1	<1	t	t	t
	Popcorn (see Grain Products)											
44006	Potato chips:	10 pce	20	2	107	1	11	1	7	2.2	2	2.4
44076	Unsalted	1 oz	28	2	150	2	15	1	10	3.1	2.8	3.4
26031	Sage, ground	1 tsp	1	8	3	<1	1	<1	<1	t	t	t
27020	Salsa, from recipe	1 tbsp	15	95	3	<1	1	<1	<1	t	t	t
26014	Salt	1 tsp	6	0	0	0	0	0	0	0	0	0
	Salt Substitutes:											
26090	Morton, salt substitute	1 tsp	6	0	0	0	<1		0	0	0	0
26048	Morton, light salt	1 tsp	6	0	0	0	<1		<1			
26098	Seasoned salt, no MSG	1 tsp	5	5	0	0	0	0	0	0	0	0
27007	Vinegar, cider	½ c	120	94	17	0	7	0	0	0	0	0
27148	Balsamic	1 tbsp	15	64	21	0	5	0	0	0	0	0
27154	Malt	1 tbsp	15	90	5	0	1	0	0	0	0	0
27158	Tarragon	1 tbsp	15	95	3	0	<1	0	0	0	0	0
27156	White wine	1 tbsp	15	89	5	0	2	0	0	0	0	0
28001	Yeast, baker's, dry, active, package	1 ea	7	8	21	3	3	1	<1	t	0.2	t
	SOUPS, SAUCES, AND GRAVIES											
	SOUPS, canned:											
	Unprepared, condensed:											
50650	Cream of celery	1 c	251	85	181	3	18	2	11	2.8	2.6	5
50654	Cream of chicken	1 c	251	82	233	7	19	1	15	4.2	6.6	3
50666	Cream of mushroom	1 c	251	81	259	4	19	1	19	5.1	3.6	8.9
50668	Onion	1 c	246	86	113	8	16	2	3	0.5	1.5	1.3
	Prepared w/equal volume of whole milk:											
50008	Clam chowder, New England	1 c	248	85	164	9	17	1	7	3	2.3	1.1
50015	Cream of celery	1 c	248	86	164	6	15	1	10	3.9	2.5	2.7

Appendix A

PAGE KEY: A–4 = Beverages   A–6 = Dairy   A–10 = Eggs   A–12 = Fat/Oil   A–14 = Fruit   A–20 = Bakery
A–26 = Grain   A–32 = Fish   A–32 = Meats   A–36 = Poultry   A–38 = Sausage   A–40 = Mixed/Fast   A–44 = Nuts/Seeds   A–46 = Sweets
A–48 = Vegetables/Legumes   A–58 = Vegetarian   A–62 = Misc   A–62 = Soups/Sauces   A–64 = Fast   A–78 = Convenience   A–80 = Baby foods

Chol (mg)	Calc (mg)	Iron (mg)	Magn (mg)	Pota (mg)	Sodi (mg)	Zinc (mg)	VT-A (RAE)	Thia (mg)	VT-E (mg α)	Ribo (mg)	Niac (mg)	V-B$_6$ (mg)	Fola (µg)	VT-C (mg)
0	217	0.41	1	505	4	0.04	0	0	0	0	0	0	0	0
0	0	0	0	0	1368	0	0	0	0	0	0	0	0	0
0	106	2.1	21	172	2	0.29	23	<.01	0.37	0.02	0.35	0.12	14	3
0	358	3.03	56	852	36	0.95	1	0.05	0.65	0.47	1.95	0.38	30	0
0	3	0.08	3	57	167	0.04	7	<.01	0.22	0.07	0.23	0.02	2	2
0	7	0.39	8	101	2	0.12	104	0.02	1.49	0.05	0.44	0.12	5	4
0	35	0.9	9	28	3	0.14	<1	<.01	0.02	<.01	0.06	0.02	<1	0
0	22	1.14	14	153	81	0.22	119	0.03	2.32	0.06	0.63	0.29	8	5
0	28	4.87	92	232	7	2.7	0	0.04	0.11	0.03	0.38	<.01	8	0
0	1	0.02	0	5	<1	<.01	3	<.01	0.02	<.01	0.01	<.01	1	0
0	25	0.76	1	10	1	0.04	<1	<.01	0.02	<.01	0.03	<.01	1	1
0	0	0.04	0	0	1	<.01	0	0	0	0	0	0	0	0
0	10	0.59	5	31	1	0.08	1	<.01	0.44	<.01	0.07	0.02	3	0
0	54	1.46	14	99	6	0.1	9	0.01		<.01	0.08	0.05		2
0	5	0.05	1	12	1	0.03	0	<.01	<.01	<.01	0.02	0.04	<1	1
0	2	0.08	2	33	1	0.08	0	0.01	0.02	<.01	0.02	0.09	<1	1
0	4	0.08	2	1	14	<.01	0	<.01	0	0.02	<.01	<.01	2	0
0	1	0.03	2	21	1	0.02	0	<.01	0.01	<.01	0.04	<.01	1	0
0	8	0.06	4	37	47	0.12	<1	<.01	<.01	<.01	0.06	0.01	9	4
0	121	3.86	71	426	595	2.68	1	0.22	1.84	0.13	0.98	0.98	145	19
0				10	120									
5	0.07			1		14							4	
0	19	0.73	1	2	192	0.05	4	<.01	0.36	0	<.01	<.01	0	0
0	7	0.05	2	19	1	0.05	0	<.01	<.01	<.01	0.01	0.02	3	0
0	32	0.88	5	33	<1	0.09	7	<.01	0.38	<.01	0.12	0.02	5	1
0	4	0.47	4	47	1	0.08	53	0.01	0.6	0.03	0.31	0.08	2	1
0	2	0.54	4	63	4	0.06	32	0.01	0.06	0.02	0.1	0.01	2	1
0	9	0.58	4	25	1	0.03	<1	<.01	0.01	<.01	0.02	<.01	<1	0
0	6	0.34	7	75	833	0.09	6	<.01	0.06	0.02	0.04	<.01	1	1
0	1	0.21	1	11	329	0.03	3	<.01	0.03	0.01	0.06	<.01	<1	0
0	5	0.33	13	255	119	0.22	0	0.03	1.82	0.04	0.77	0.13	9	6
0	7	0.46	19	357	2	0.31	0	0.05	2.55	0.06	1.07	0.18	13	9
0	17	0.28	4	11	<1	0.05	3	<.01	0.07	<.01	0.06	0.03	3	0
0	1	0.05	1	23	1	0.02	3	<.01	0.04	<.01	0.06	0.01	2	2
0	1	0.02	0	0	2325	<.01	0	0	0	0	0	0	0	0
33		0	3018	1										
2		4	1518	1182										
0					1583									
0	7	0.72	26	120	1	0	0	0	0	0	0	0	0	0
2	0.15		10	3			0.15			0.15	0.15		0	
2	0.15		14	4			0.15			0.15	0.15		2	
0	0.15		2	1			0.15			0.15	0.15		0	
1	0.15		12	1			0.15			0.15	0.15		0	
0	4	1.16	7	140	4	0.45	0	0.17	0	0.38	2.78	0.11	164	0
28	80	1.25	13	246	1900	0.3	55	0.06	3.49	0.1	0.67	0.03	5	1
20	68	1.2	5	176	1973	1.25	108	0.06	1.36	0.12	1.64	0.03	3	0
3	65	1.05	10	168	1737	1.18	20	0.06	2.03	0.17	1.62	0.03	8	2
0	54	1.35	5	138	2116	1.23	7	0.07	0.54	0.05	1.21	0.1	30	2
22	186	1.49	22	300	992	0.79	57	0.07	0.45	0.24	1.03	0.13	10	3
32	186	0.69	22	310	1009	0.2	114	0.07	0.97	0.25	0.44	0.06	7	1

Table A–1

Food Composition

(Computer code number is for West Diet Analysis program)   (For purposes of calculations, use "0" for t, <1, <.1, <.01, etc.)

Computer Code Number	Food Description	Measure	Wt (g)	H₂O (%)	Ener (kcal)	Prot (g)	Carb (g)	Dietary Fiber (g)	Fat (g)	Fat Breakdown (g)		
										Sat	Mono	Poly
	SOUPS, SAUCES, AND GRAVIES—Continued											
50006	Cream of chicken	1 c	248	85	191	7	15	<1	11	4.6	4.5	1.6
50011	Cream of mushroom	1 c	248	85	203	6	15	<1	14	5.1	3	4.6
50026	Cream of potato	1 c	248	87	149	6	17	<1	6	3.8	1.7	0.6
50024	Oyster stew	1 c	245	89	135	6	10	0	8	5	2.1	0.3
50012	Tomato	1 c	248	85	161	6	22	3	6	2.9	1.6	1.1
	Prepared with equal volume of water:											
50000	Bean with bacon	1 c	253	84	172	8	23	9	6	1.5	2.2	1.8
50003	Beef noodle	1 c	244	92	83	5	9	1	3	1.1	1.2	0.5
50005	Chicken noodle	1 c	241	92	75	4	9	1	2	0.7	1.1	0.6
50020	Chicken rice	1 c	241	94	60	4	7	1	2	0.5	0.9	0.4
50007	Chili beef	1 c	250	85	170	7	21	10	7	3.4	2.8	0.3
50021	Clam chowder, Manhattan	1 c	244	92	78	2	12	1	2	0.4	0.4	1.3
50018	Cream of chicken	1 c	244	91	117	3	9	<1	7	2.1	3.3	1.5
50049	Cream of mushroom	1 c	244	90	129	2	9	<1	9	2.4	1.7	4.2
50009	Minestrone	1 c	241	91	82	4	11	1	3	0.6	0.7	1.1
50022	Onion	1 c	241	93	58	4	8	1	2	0.3	0.7	0.7
50025	Split pea & ham	1 c	253	82	190	10	28	2	4	1.8	1.8	0.6
50028	Tomato	1 c	244	90	85	2	17	<1	2	0.4	0.4	1
50014	Vegetable beef	1 c	244	92	78	6	10	<1	2	0.9	0.8	0.1
50013	Vegetarian vegetable	1 c	241	92	72	2	12	<1	2	0.3	0.8	0.7
50052	Ready to serve, Chunky chicken soup	1 c	251	84	178	13	17	2	7	2	3	1.4
	SOUPS, dehydrated:											
	Prepared with water:											
50032	Beef broth/bouillon	1 c	244	97	20	1	2	0	1	0.3	0.3	t
50034	Chicken broth	1 c	244	97	22	1	1	0	1	0.3	0.4	0.4
50037	Chicken noodle	1 c	252	94	58	2	9	<1	1	0.3	0.5	0.4
50036	Cream of chicken	1 c	261	91	107	2	13	<1	5	3.4	1.2	0.4
50040	Onion	1 c	246	96	27	1	5	1	1	0.1	0.3	t
50041	Split pea	1 c	255	87	125	7	21	3	1	0.4	0.7	0.3
50043	Tomato vegetable	1 c	253	94	56	2	10	1	1	0.4	0.3	t
	Unprepared, dry products:											
50051	Beef bouillon, packet	1 ea	6	3	14	1	1	0	1	0.3	0.2	t
50054	Onion soup, packet	1 ea	39	4	115	5	21	4	2	0.5	1.4	0.3
	SAUCES											
53097	From home recipe, lowfat cheese sauce	¼ c	61	74	81	6	4	<1	5	1.8	1.8	0.8
	Ready to serve:											
53396	Alfredo sauce, reduced fat	¼ c	69		144	5	9	0	9	6.2		
53000	Barbeque sauce	1 tbsp	16	81	12	<1	2	<1	<1	t	0.1	0.1
53355	Creole sauce	¼ c	62	89	25	1	4	1	1	t	0.2	0.3
53392	Pesto sauce	2 tbsp	16		83	2	1	<1	8	1.5		
53002	Soy sauce	1 tbsp	16	71	8	1	1	<1	<1	t	t	t
53349	Szechuan sauce	1 tbsp	16	71	21	<1	3	<1	1	0.1	0.3	0.4
53004	Teriyaki sauce	1 tbsp	18	68	15	1	3	<1	0	0	0	0
53011	Spaghetti with meat sauce	1 c	250	85	178	7	19	4	8	1.8	3.3	1.8
	GRAVIES											
53023	Beef	1 c	233	87	123	9	11	1	5	2.7	2.2	0.2
53022	Chicken	1 c	238	85	188	5	13	1	14	3.4	6.1	3.6
53026	Mushroom	1 c	238	89	119	3	13	1	6	1	2.8	2.4
	FAST FOOD RESTAURANTS											
	ARBY'S											
	Roast beef sandwiches:											
56336	Regular	1 ea	155		326	21	35	2	14	6.9		
56337	Junior	1 ea	89		200	11	23	1	8	3.4		
56338	Super	1 ea	254		467	23	50	3	22	8.3		
69056	Beef 'n cheddar	1 ea	194		451	23	42	2	23	8.8		
56341	Chicken breast sandwich	1 ea	204		539	24	46	2	29	4.9		
56342	Ham'n cheese sandwich	1 ea	169		338	23	35	1	13	4.5		
69048	Italian sub sandwich	1 ea	297		742	28	47	3	50	14.3		
56343	Turkey sandwich, deluxe	1 ea	218		292	26	37	3	6	0.6	2.5	2.6
69044	Turkey sub sandwich	1 ea	277		570	24	46	2	33	8.1		

Appendix A

PAGE KEY: A–4 = Beverages A–6 = Dairy A–10 = Eggs A–12 = Fat/Oil A–14 = Fruit A–20 = Bakery
A–26 = Grain A–32 = Fish A–32 = Meats A–36 = Poultry A–38 = Sausage A–40 = Mixed/Fast A–44 = Nuts/Seeds A–46 = Sweets
A–48 = Vegetables/Legumes A–58 = Vegetarian A–62 = Misc A–62 = Soups/Sauces A–64 = Fast A–78 = Convenience A–80 = Baby foods

Chol (mg)	Calc (mg)	Iron (mg)	Magn (mg)	Pota (mg)	Sodi (mg)	Zinc (mg)	VT-A (RAE)	Thia (mg)	VT-E (mg α)	Ribo (mg)	Niac (mg)	V-B$_6$ (mg)	Fola (µg)	VT-C (mg)
27	181	0.67	17	273	1047	0.67	179	0.07	0.25	0.26	0.92	0.07	7	1
20	179	0.6	20	270	918	0.64	35	0.08	1.24	0.28	0.91	0.06	10	2
22	166	0.55	17	322	1061	0.67	52	0.08	0.1	0.24	0.64	0.09	10	1
32	167	1.05	20	235	1041	10.34	56	0.07	0.49	0.23	0.34	0.06	10	4
17	159	1.81	22	449	744	0.3	64	0.13	1.24	0.25	1.52	0.16	17	68
3	81	2.05	46	402	951	1.04	46	0.09	0.76	0.03	0.57	0.04	33	2
5	15	1.1	5	100	952	1.54	7	0.07	0.68	0.06	1.07	0.04	20	0
7	17	0.77	5	55	1106	0.39	36	0.05	0.1	0.06	1.39	0.03	22	0
7	17	0.75	0	101	815	0.27	22	0.02	0.1	0.02	1.13	0.02	0	0
12	42	2.12	30	525	1035	1.4	75	0.06	1.52	0.08	1.07	0.16	18	4
2	27	1.63	12	188	578	0.98	56	0.03	0.34	0.04	0.82	0.1	10	4
10	34	0.61	2	88	986	0.63	163	0.03	0.2	0.06	0.82	0.02	2	0
2	46	0.51	5	100	881	0.59	15	0.05	0.95	0.09	0.72	0.01	5	1
2	34	0.92	7	313	911	0.75	118	0.05	0.07	0.04	0.94	0.1	36	1
0	27	0.67	2	67	1053	0.6	<1	0.03	0.19	0.02	0.6	0.05	14	1
8	23	2.28	48	400	1007	1.32	23	0.15	0.15	0.08	1.47	0.07	3	2
0	12	1.76	7	264	695	0.24	29	0.09	2.32	0.05	1.42	0.11	15	66
5	17	1.12	5	173	791	1.54	95	0.04	0.37	0.05	1.03	0.08	10	2
0	22	1.08	7	210	822	0.46	116	0.05	0.41	0.05	0.92	0.06	10	1
30	25	1.73	8	176	889	1	68	0.09	0.33	0.17	4.42	0.05	5	1
0	10	0.02	7	37	1362	0.07	0	<.01	0.1	0.02	0.36	0	0	0
0	15	0.07	5	24	1484	0	2	<.01	0.07	0.03	0.2	0	2	0
10	5	0.5	8	33	577	0.2	3	0.2	0.13	0.08	1.09	0.03	18	0
3	76	0.26	5	214	1185	1.57	21	0.1	0.6	0.2	2.61	0.05	5	1
0	12	0.15	5	64	849	0.05	0	0.03	0	0.06	0.48	0	2	0
3	20	0.94	43	224	1148	0.56	3	0.21	0.03	0.14	1.26	0.05	41	0
0	8	0.63	20	104	1146	0.18	10	0.06	0.35	0.05	0.79	0.05	10	6
1	4	0.06	3	27	1019	0	0	<.01	0.13	0.01	0.27	0.01	2	0
2	55	0.58	25	260	3493	0.23	<1	0.11	0.03	0.24	1.99	0.04	6	1
9	167	0.24	10	101	307	0.74		0.03	0.51	0.14	0.16	0.03	4	0
31	103	0	8	93	618			0		0.11	0		0	0
0	3	0.14	3	28	130	0.03	<1	<.01	<.01	<.01	0.14	0.01	1	1
0	35	0.31	9	187	339	0.1		0.03	0.61	0.02	0.53	0.07	9	0
4	65	0.09		26	137			0		0.02	0		2	0
0	3	0.32	5	29	914	0.06	0	<.01	0	0.02	0.54	0.03	3	0
0	2	0.12	2	13	218	0.02		<.01	0.07	<.01	0.1	<.01	1	0
0	4	0.31	11	40	690	0.02		<.01	0	0.01	0.23	0.02	4	0
15	54	2.1	43	742	982	1.27		0.13	2.98	0.12	3.46	0.31	25	19
7	14	1.63	5	189	1305	2.33	2	0.07	0.05	0.08	1.54	0.02	5	0
5	48	1.12	5	259	1373	1.9	2	0.04	0.31	0.1	1.05	0.02	5	0
0	17	1.57	5	252	1357	1.67	0	0.08	0.19	0.15	1.6	0.05	29	0
44	59	3.55	16	422	879	3.75	0							0
28	41	1.86	8	201	483	1.5		0.18		0.26	6.6	0.1	7	
47	62	3.73	25	533	1099	3.73		0.39		0.58	12.4	0.3	21	1
49	98	3.53		321	1146	2.94		0.42		0.63	9.8			1
88	78	1.77	30	330	1138	0.15		0.22		0.54	8.99	0.38	18	4
89	149	2.68	31	380	1441	0.89		0.82		0.37	7.75	0.31	26	1
114	238	2.57		565	2323			0.91		0.49	8.19			2
45	90	2.02	33	394	1157	1.69		0.09		0.46	17.22	0.58	22	1
91	181	0.33		500	1964			13.2		0.54	18.8			2

Table A–1

Food Composition

(Computer code number is for West Diet Analysis program)   (For purposes of calculations, use "0" for t, <1, <.1, <.01, etc.)

Computer Code Number	Food Description	Measure	Wt (g)	H₂O (%)	Ener (kcal)	Prot (g)	Carb (g)	Dietary Fiber (g)	Fat (g)	Fat Breakdown (g) Sat	Mono	Poly
	FAST FOOD RESTAURANTS—Continued											
	Milkshakes:											
2124	Chocolate	1 ea	340	71	411	9	72	0	14	6.9		
2125	Jamocha	1 ea	326	72	386	8	67	0	12	5.7		
2126	Vanilla	1 ea	312	72	369	8	65	0	12	5.5		
6431	Salad, roast chicken	1 ea	400		152	19	14	6	2	0		
	Source: Arby's											
	BOSTON MARKET											
	Chicken											
15247	Half chicken with skin, svg	1 ea	277		590	70	4	0	33	10		
15246	Dark meat, svg	1 ea	95		190	22	1	0	10	3		
15249	Dark meat with skin, svg	1 ea	125		320	30	2	0	21	6		
15244	White meat, svg	1 ea	140		170	33	2	0	4	1		
15245	White meat, with skin, svg	1 ea	152	64	280	40	2	0	12	3.5		
57530	Chicken pot pie	1 ea	425		780	32	61	4	46	13		
15248	Turkey breast, svg	1 ea	142		170	36	1	0	1	0.5		
	Sandwiches											
69084	Ham and turkey club	1 ea	266		430	29	64	4	6	2		
69081	Ham	1 ea	266		440	25	66	4	8	2.5		
69076	Chicken	1 ea	281	64	430	34	62	4	4	1		
69075	Chicken with cheese and sauce	1 ea	352		750	41	72	5	33	12		
69079	Turkey	1 ea	266		400	45	61	4	4	1		
69078	Turkey with cheese and sauce	1 ea	337		710	45	68	4	28	10		
69083	Turkey and ham with cheese and sauce	1 ea	379		890	47	79	4	43	20		
69074	Meatloaf	1 ea	351		690	40	86	6	21	7		
69082	Meatloaf with cheese	1 ea	383		860	46	95	6	33	16		
	Side dishes											
7393	BBQ baked beans	¾ c	201		270	8	48	12	5	2		
7387	New Potatoes	¾ c	131		130	3	25	2	2	0		
7390	Mashed potatoes	⅔ c	161		190	3	24	1	9	6		
28257	Red beans and rice	1 c	227		260	8	45	4	5	0		
7388	Whole kernel corn	¾ c	146		180	5	30	2	4	0.5		
7386	Steamed vegetables	⅔ c	105		35	2	7	3	<1	0		
52111	Caesar side salad	4 oz	113		199	7	7	1	17	4.5		
52103	Chunky chicken salad	¾ c	158		370	28	3	1	27	4.5		
52108	Mediterranean pasta salad	¾ c	129		170	4	16	2	10	2.5		
	Source: Boston Chicken, Inc.											
	BURGER KING											
56346	Croissant sandwich, egg, sausage & cheese	1 ea	176	45	575	22	30	1	41	15		
	Sandwiches:											
56354	Whopper	1 ea	270		660	28	51	4	38	11.7		
56355	Whopper with cheese	1 ea	294		757	33	53	4	46	16.5		
57002	BK broiler chicken sandwich	1 ea	248		529	29	50	3	24	4.8		
56352	Cheeseburger	1 ea	138		375	20	31	2	18	9.1		
56360	Chicken sandwich	1 ea	229		675	26	54	3	40	8.2		
56356	Double beef	1 ea	351		915	48	53	4	57	19.9		
56357	Double beef & cheese	1 ea	375	52	1012	53	55	4	64	24.8		
56353	Double cheeseburger with bacon	1 ea	218		649	40	34	2	39	19.1		
56351	Hamburger	1 ea	126		328	18	31	2	14	6.1		
56362	Ocean catch fish fillet	1 ea	255		688	23	65	4	37	13.6		
15158	Chicken tenders	1 ea	88		251	16	15	1	14	3.4		
6141	French fries (salted), svg	1 ea	116		357	4	46	4	18	5		
42429	French toast sticks, svg	1 ea	141		491	8	58	3	25	5.7		

PAGE KEY: A–4 = Beverages   A–6 = Dairy   A–10 = Eggs   A–12 = Fat/Oil   A–14 = Fruit   A–20 = Bakery
A–26 = Grain   A–32 = Fish   A–32 = Meats   A–36 = Poultry   A–38 = Sausage   A–40 = Mixed/Fast   A–44 = Nuts/Seeds   A–46 = Sweets
A–48 = Vegetables/Legumes   A–58 = Vegetarian   A–62 = Misc   A–62 = Soups/Sauces   A–64 = Fast   A–78 = Convenience   A–80 = Baby foods

Chol (mg)	Calc (mg)	Iron (mg)	Magn (mg)	Pota (mg)	Sodi (mg)	Zinc (mg)	VT-A (RAE)	Thia (mg)	VT-E (mg α)	Ribo (mg)	Niac (mg)	V-B$_6$ (mg)	Fola (µg)	VT-C (mg)
39	428	0.62	48	410	317	1.5		0.12		0.68	0.8	0.14	14	2
37	411	0.59	36	525	320	1.48		0.12		0.68	0.8	0.14	14	2
35	393	0.85	36	686	283	1.49		0.12		0.68	4	0.14	37	2
38	76	1.71		877	667			0.31		0.54	5.6			0
290	0	2.7			1010		0							0
115	0	1.08			440		0							0
155	0	1.8			500		0							0
85	0	0.72			480		0							0
175	0	1.08			510		0							0
135	40	3.6			1480									4
100	20	1.8			850		0							0
55	80	2.7			1330									9
45	80	1.8			1450									9
65	100	1.8			910									9
135	400	2.7			1860									9
60	100	2.7			1070									9
110	500	2.7			1390									9
150	800	4.5			2310									9
120	100	4.5			1610									15
165	400	4.5			2270									15
0	100	3.6			540									6
0	0	0.72			150		0							12
25	60	0.36			570									6
5	60	2.7			1050									12
0	0	0.36			170		10							5
0	20	0.36			35		200							21
15	199	1.08			448									12
120	20	0.72			800									4
10	60	1.08			490									9
219	173	3.11			1173									0
78	97	5.24			913									9
102	243	5.24			1349									9
101	58	3.46			1067									6
56	152	2.74			761									0
72	82	2.76			1360									0
149	149	7.16			1014									9
169	298	7.14			1448									9
128	266	4.79			1244									0
46	82	2.77			543									0
48	78	3.49			1163									0
34	0	0.41			606		0							0
0	20	0.71			684		0							15
0	76	2.27			554		0							0

A

Appendix

A-67

Table A–1

Food Composition

(Computer code number is for West Diet Analysis program)　　　(For purposes of calculations, use "0" for t, <1, <.1, <.01, etc.)

Computer Code Number	Food Description	Measure	Wt (g)	H₂O (%)	Ener (kcal)	Prot (g)	Carb (g)	Dietary Fiber (g)	Fat (g)	Fat Breakdown (g) Sat	Mono	Poly
	FAST FOOD RESTAURANTS—Continued											
56363	Onion rings, svg	1 ea	124		436	5	55	4	22	5.5		
2127	Milk shakes, chocolate	1 ea	284		315	9	57	3	6	3.6		
2129	Vanilla	1 ea	284		315	9	57	1	6	3.6		
48134	Fried apple pie	1 ea	113		340	2	52	1	14	3		
	Source: Burger King Corporation											
	CARL'S JR											
	Sandwiches:											
91413	Carl's catch	1 ea	201	51	510	18	50	1	27	7		
91407	Charbroiled BBQ Chicken	1 ea	199		280	25	37	2	3	1		
91411	Bacon swiss chicken	1 ea	291		720	32	66	3	36	10		
91410	Ranch crispy chicken	1 ea	266		730	29	77	4	34	7.1		
91406	Hamburger	1 ea	134		330	18	34	1	13	5		
91402	Famous star hamburger	1 ea	254		580	25	49	2	32	9		
91403	Super star hamburger	1 ea	345		790	42	49	2	46	14		
91404	Western bacon cheeseburger	1 ea	225		650	32	63	2	30	12		
91405	Western bacon double cheeseburger	1 ea	308		900	51	64	2	49	21		
91424	Charbroiled chicken salad to go	1 ea	350		200	25	12	3	7	3.5		
	Side dishes:											
91437	Bacon, 2 slices, svg	1 ea	9	15	50	3	0	0	4	1.5		
91418	Criss-cut french fries, svg	1 ea	139		410	5	43	4	24	5		
91414	French fries, svg	1 ea	92	36	290	5	37	3	14	3		
91415	Onion rings, svg	1 ea	127		430	7	53	3	21	5		
	Source: Carl Karcher Enterprises, Inc.											
	CHICK-FIL-A											
	Sandwiches:											
69153	Chargrilled chicken	1 ea	150		290	26	30	1	7	2.1		
69154	Chicken salad club	1 ea	232		413	34	36	2	15	5.7		
	Salads:											
52139	Carrot and raisin	1 ea	76		109	1	18	2	4	0.8		
69181	Chargrilled chicken caesar	1 ea	241		240	31	6	2	10	6		
52134	Chicken garden, charbroiled	1 ea	397	86	257	33	11	4	9	4.3		
52135	Chick-n-strips	1 ea	451		487	43	27	4	23	7.2		
52138	Cole slaw	1 ea	79		158	1	11	2	13	1.9		
69184	Chargrilled chicken cool wrap	1 ea	240		390	31	53	3	7	3		
69183	Chicken caesar cool wrap	1 ea	227		460	38	51	3	11	6		
69188	Chicken breast fillet, breaded	1 ea	105		230	23	10	0	11	2.5		
15263	Chicken nuggets, svg	1 ea	110	54	253	25	12	1	12	2.4		
15262	Chicken-n- strips, svg	1 ea	119	54	275	28	13	0	12	2.8		
50885	Hearty breast of chicken soup, svg	1 ea	215	88	100	9	13	1	2	0		
7973	Waffle potato fries, svg	1 ea	85	34	280	3	37	5	14	5		
46489	Cheesecake, svg	1 ea	88		322	6	28	2	20	11.4		
49134	Fudge nut brownie, svg	1 ea	74	13	330	4	45	2	15	3.5		
20601	Icedream, svg	1 ea	127		151	4	26	0	4	1.9		
48214	Lemon pie, svg	1 ea	99	39	280	6	45	3	9	3.1		
20602	Lemonade, svg	1 ea	255		170	0	41	0	<1	0		
	Source: Chick-Fil-A											
	DAIRY QUEEN											
	Ice cream cones:											
2144	Small vanilla	1 ea	142		230	6	38	0	7	4.5		
2143	Regular vanilla	1 ea	213		355	9	57	0	10	6.5		
2142	Large vanilla	1 ea	253		410	10	65	0	12	8		
2136	Chocolate dipped	1 ea	220		490	8	59	1	24	13		

PAGE KEY: A–4 = Beverages   A–6 = Dairy   A–10 = Eggs   A–12 = Fat/Oil   A–14 = Fruit   A–20 = Bakery
A–26 = Grain   A–32 = Fish   A–32 = Meats   A–36 = Poultry   A–38 = Sausage   A–40 = Mixed/Fast   A–44 = Nuts/Seeds   A–46 = Sweets
A–48 = Vegetables/Legumes   A–58 = Vegetarian   A–62 = Misc   A–62 = Soups/Sauces   A–64 = Fast   A–78 = Convenience   A–80 = Baby foods

Chol (mg)	Calc (mg)	Iron (mg)	Magn (mg)	Pota (mg)	Sodi (mg)	Zinc (mg)	VT-A (RAE)	Thia (mg)	VT-E (mg α)	Ribo (mg)	Niac (mg)	V-B$_6$ (mg)	Fola (μg)	VT-C (mg)
0	136	0			627	0								0
25	250	1.29			193									0
18	286	0			243									4
0	0	1.44			470									0
80	150	1.8			1030									2
60	80	2.7			830									5
75	250	3.6			1610									6
59	177	4.24			1436									6
45	60	3.6			480	0								2
70	100	4.5			910									6
130	100	5.4			910									9
80	200	4.5			1430									1
155	300	6.3			1770									1
75	200	1.8			440									27
10	0	0			140	0								0
0	20	1.8			950	0								12
0	0	1.08			170	0								21
0	20	0.36			700	0								0
62	83	1.86			1034	0								0
92	172	3.1			1573									4
0	17	0.3			75									3
85	350	0.36			1170									0
100	214	0.51			1042									9
122	215	1.55			974									9
15	30	0.27			135									20
70	200	3.6			1120									5
85	400	3.6			1540									0
60	40	1.08			990									0
68	39	1.05			1061	0								0
77	44	1.19			628	0								0
20	40	0			940	0								0
15	20	0			105	0								21
85	57	0			255									0
20	20	1.8			210									0
14	94	0.34			75									0
96	131	0			193									4
0	0	0.36			10	0								15
20	200	1.08			115									1
32	269	1.94			172									3
40	350	1.8			200									2
30	250	1.8			190									2

Table A–1

Food Composition (Computer code number is for West Diet Analysis program) (For purposes of calculations, use "0" for t, <1, <.1, <.01, etc.)

Computer Code Number	Food Description	Measure	Wt (g)	H₂O (%)	Ener (kcal)	Prot (g)	Carb (g)	Dietary Fiber (g)	Fat (g)	Fat Breakdown (g) Sat	Mono	Poly
	FAST FOOD RESTAURANTS—Continued											
2154	Chocolate sundae	1 ea	234		400	8	71	0	10	6		
2131	Banana split	1 ea	369		510	8	96	3	12	8		
2151	Peanut buster parfait	1 ea	305		730	16	99	2	31	17		
2133	Buster bar	1 ea	149		450	10	41	2	28	12		
2135	Dilly bar	1 ea	85	55	210	3	21	0	13	7		
	Milkshakes:											
2152	Large	1 ea	461	72	600	13	101	<1	16	10	2	2
2145	Malted	1 ea	418	68	610	13	106	<1				
2351	Misty slush, small	1 ea	454		220	0	56	0	0	0	0	0
2359	Starkiss	1 ea	85		80	0	21	0	0	0	0	0
	Sandwiches:											
56372	Cheeseburger, double	1 ea	219		540	35	30	2	31	16		
56371	Single	1 ea	152	55	340	20	29	2	17	8		
56379	Chicken	1 ea	191		466	18	45	4	24	4.2		
69029	Chicken fillet, grilled	1 ea	184		310	24	30	3	10	2.5		
56381	Fish fillet sandwich	1 ea	170	58	370	16	39	2	16	3.5		
56382	With cheese	1 ea	184	56	420	19	40	2	21	6		
56368	Hamburger, single	1 ea	138	56	290	17	29	2	12	5		
56369	Hamburger, double	1 ea	212		440	30	29	2	22	10		
	Hotdogs:											
56374	Regular	1 ea	99	55	240	9	19	1	14	5		
56375	With cheese	1 ea	113		290	12	20	1	18	8	8	2
56376	With chili	1 ea	128		297	13	20	2	19	8.1		
6143	French fries, small	1 ea	112	41	347	4	42	3	18	3.5		
56383	Onion rings	1 ea	113		320	5	39	3	16	4		
	Source: International Dairy Queen											
	HARDEES'S											
	Sandwiches:											
69061	Frisco burger hamburger	1 ea	242		793	36	41	2	54	15.3		
56422	Chicken sandwich	1 ea	187		444	23	39	2	17	4.6		
6146	French fries, svg	1 ea	96		289	3	38	0	14	1.7		
	Source: Hardees Food Systems											
	IN-N-OUT BURGERS											
81110	Hamburger with spread	1 ea	243		390	16	39	3	19	5		
81111	Hamburger w/mustard and ketchup	1 ea	243		310	16	41	3	10	4		
81113	Cheeseburger with spread	1 ea	268		480	22	39	3	27	10		
81114	Cheeseburger w/mustard and ketchup	1 ea	268		400	22	41	3	18	9		
81115	Cheeseburger, protein style, without bun	1 ea	300		330	18	11	2	25	9		
81116	Double cheeseburger with spread	1 ea	328		670	37	40	3	41	18		
81117	Double cheeseburger with mustard and ketchup	1 ea	328		590	37	42	3	32	17		
81118	Double cheeseburger, protein style, without bun	1 ea	361		520	33	11	2	39	17		
	Source: In-N-Out Burgers											
	JACK IN THE BOX											
	Breakfast items:											
56430	Breakfast jack sandwich	1 ea	126		280	17	28	1	12	5		
56431	Sausage crescent	1 ea	181		660	20	37	0	48	15		
56432	Supreme crescent	1 ea	164		530	21	37	0	34	10		
	Sandwiches:											
69032	Bacon cheeseburger	1 ea	274		760	39	39	2	50	17		

PAGE KEY: A–4 = Beverages   A–6 = Dairy   A–10 = Eggs   A–12 = Fat/Oil   A–14 = Fruit   A–20 = Bakery
A–26 = Grain   A–32 = Fish   A–32 = Meats   A–36 = Poultry   A–38 = Sausage   A–40 = Mixed/Fast   A–44 = Nuts/Seeds   A–46 = Sweets
A–48 = Vegetables/Legumes   A–58 = Vegetarian   A–62 = Misc   A–62 = Soups/Sauces   A–64 = Fast   A–78 = Convenience   A–80 = Baby foods

Chol (mg)	Calc (mg)	Iron (mg)	Magn (mg)	Pota (mg)	Sodi (mg)	Zinc (mg)	VT-A (RAE)	Thia (mg)	VT-E (mg α)	Ribo (mg)	Niac (mg)	V-B₆ (mg)	Fola (µg)	VT-C (mg)
30	250	1.44			210									0
30	250	1.8			180									15
35	300	1.8			400									1
15	150	1.08			280									0
10	100	0.36			75									0
50	450	1.44		660	260			0.15		0.68	0.8			0
45				570										
0	0	0			20	0								0
0	0	0			10	0								0
115	250	4.5			1130									4
55	150	3.6			850									4
28	75	4.19			1016									2
50	200	2.7			1040	0								0
45	40	1.8		280	630	0		0.3		0.22	3			0
60	100	1.8		290	850			0.3		0.26	5			0
45	60	2.7			630									4
90	60	4.5			680									6
25	60	1.8			730									4
40	150	1.8		180	950			0.22		0.17	2			4
41	135	1.62			983									3
0	20	1.07			872	0								4
0	20	1.44			180	0								0
111					1192									
61					1081									
0					331									
40	40	3.6			640									15
35	40	3.6			720									15
60	200	3.6			1000									15
55	200	3.6			1080									15
60	200	1.08			720									18
120	350	5.4			1430									15
115	350	5.4			1510									15
120	350	1.08			1160									18
190	150	3.6		120	750									10
240	100	1.8		160	860									0
225	100	1.8		165	1060									4
135	250	4.5		530	1570			0.27		0.54	9.96	0.44		9

A

Appendix

Table A–1

Food Composition

(Computer code number is for West Diet Analysis program) (For purposes of calculations, use "0" for t, <1, <.1, <.01, etc.)

Computer Code Number	Food Description	Measure	Wt (g)	H₂O (%)	Ener (kcal)	Prot (g)	Carb (g)	Dietary Fiber (g)	Fat (g)	Sat	Mono	Poly
	FAST FOOD RESTAURANTS—Continued											
1655	Chicken sandwich	1 ea	164		400	15	38	3	21	3		
56443	Chicken supreme	1 ea	305		830	33	66	3	49	7		
56366	Double cheeseburger	1 ea	158		440	24	31	2	24	11		
69033	Grilled sourdough burger	1 ea	233		690	34	37	2	45	15		
56436	Jumbo jack burger	1 ea	271		550	27	43	2	30	10		
56437	Jumbo jack burger with cheese	1 ea	296		640	31	44	2	38	15		
56377	Tacos, regular	1 ea	90	66	170	7	12	2	10	3.5		
56378	Super	1 ea	138	63	270	12	19	4	17	6		
57014	Chicken teriyaki bowl	1 ea	502		670	26	128	3	4	1		
6150	French fries	1 ea	113	40	350	4	46	3	16	4		
6149	Hash browns	1 ea	57		170	1	14	1	12	2		
2164	Milkshake, strawberry	1 ea	382		640	10	85	0	28	15		
48135	Apple turnover	1 ea	107	40	340	4	41	2	18	4		
	Source: Jack in the Box Restaurant, Inc											
	KENTUCKY FRIED CHICKEN											
	Original Recipe chicken:											
15163	Side breast	1 ea	153		400	29	16	1	24	6	14.4	3.6
15165	Drumstick	1 ea	61	56	140	13	4	0	9	2	5.3	1.7
15166	Thigh	1 ea	91	54	250	16	6	1	18	4.5	10.2	3.4
15167	Wing	1 ea	47	48	140	9	5	0	10	2.5	5.8	1.7
	Hot & spicy chicken:											
15185	Center breast	1 ea	180	48	505	38	23	1	29	8		
15183	Thigh	1 ea	107		355	19	13	1	26	7		
15187	Wing	1 ea	55	37	210	10	9	1	15	4		
	Extra Crispy Recipe chicken:											
15169	Center breast	1 ea	168	48	470	39	17	1	28	8	16.7	3.3
15170	Drumstick	1 ea	67	48	195	15	7	1	12	3	7.4	1.6
15171	Thigh	1 ea	118		380	21	14	1	27	7	15.8	4.2
15172	Wing	1 ea	55	35	220	10	10	1	15	4	8.9	2.1
7139	Baked beans	½ c	167		203	6	35	6	3	1.1	1.4	0.7
42331	Buttermilk biscuit	1 ea	56		180	4	20	1	10	2.5	5.4	2.1
56451	Coleslaw, svg	1 ea	142		232	2	26	3	14	2	3.5	7
6152	Corn-on-the-cob	1 ea	162		150	5	35	2	2	0	0.6	0.9
15177	Chicken, hot wings, svg	1 ea	135		471	27	18	2	33	8		
56681	Macaroni & cheese, svg	1 ea	153		180	7	21	2	8	3		
56453	Mashed potatoes & gravy, svg	1 ea	136	82	120	1	17	2	6	1	3.6	1.4
56454	Potato salad	½ c	160		230	4	23	3	14	1.8	4.5	7.8
6188	Potato wedges, svg	1 ea	135		278	5	28	5	13	4		
	Source: Kentucky Fried Chicken Corp											
	McDONALD'S											
	Sandwiches:											
69010	Big Mac	1 ea	216		590	24	47	3	34	11		
69013	Filet-o-fish	1 ea	156	43	470	15	45	1	26	5		
69011	Quarter-pounder	1 ea	171		428	23	37	2	21	8		
69012	Quarter-pounder with cheese	1 ea	199		527	28	38	2	30	12.9		
15174	Chicken McNuggets	4 pce	71		187	10	13	1	11	2.5		
	Sauces (packet):											
27070	Hot mustard	1 ea	28		60	1	7	1	4	0		
53176	Barbecue	1 ea	28		45	0	10	0	0	0	0	0
53177	Sweet & sour	1 ea	28		50	0	11	0	0			
	Low-fat (frozen yogurt) milk shakes:											
2167	Chocolate	1 ea	295		360	11	60	1	9	6		
2168	Strawberry	1 ea	294	72	360	11	60	0	9	6		
2169	Vanilla	1 ea	293		360	11	59	0	9	6		
	Low-fat (frozen yogurt) sundaes:											

PAGE KEY: A–4 = Beverages A–6 = Dairy A–10 = Eggs A–12 = Fat/Oil A–14 = Fruit A–20 = Bakery
A–26 = Grain A–32 = Fish A–32 = Meats A–36 = Poultry A–38 = Sausage A–40 = Mixed/Fast A–44 = Nuts/Seeds A–46 = Sweets
A–48 = Vegetables/Legumes A–58 = Vegetarian A–62 = Misc A–62 = Soups/Sauces A–64 = Fast A–78 = Convenience A–80 = Baby foods

Chol (mg)	Calc (mg)	Iron (mg)	Magn (mg)	Pota (mg)	Sodi (mg)	Zinc (mg)	VT-A (RAE)	Thia (mg)	VT-E (mg α)	Ribo (mg)	Niac (mg)	V-B$_6$ (mg)	Fola (µg)	VT-C (mg)
40	100	2.7		200	770									5
65	200	3.6		250	2140									9
80	250	4.5		290	1100									1
105	200	4.5		480	1180									9
75	150	4.5		490	880									9
105	250	4.5		530	1340									9
15	100	1.08	40	235	390	1.38								0
30	200	1.44	49	365	630	1.8								2
15	100	4.5		620	1730									24
0	10	0.72		590	710		0							6
0	10	0.18		100	250		0							0
85	350	0		620	300									0
0	10	1.8		85	510									10
135	40	1.08			1116									1
75	20	0.72			422									1
95	20	0.72			747									1
55	20	0.36			414									1
162	60	1.08			1170									1
126	20	0.72			630									1
55	20	0.72			350									1
160	20	1.08			874									1
77	20	0.72			375									1
118	20	1.08			625									1
55	20	0.36			415									1
5	86	1.93			814									1
0	20	1.08			560									1
8	30	0.36			285									34
0	20	0.36			20		5							4
150	40	1.44			1230									1
10	150	0.36			860									1
1	20	0.36			440									1
15	20	2.7			540									1
5	20	1.79			744		5							1
85	300	4.5	46	455	1090	4.8		0.49	1.01	0.44	6.07	0.25	49	4
50	200	1.8	34	286	890	0.76			1.64					0
70	199	4.47	34	405	835	4.66		0.39	0.36	0.32	6.78	0.24	27	2
95	348	4.48			1303				0.81		6.78		33	2
35	9	0.7	16	199	355	0.65	0		0.93		4.87			0
5	7	0.72		27	240			<.01		<.01	0.14			0
0	3	0		45	250		0	<.01		<.01	0.15			4
0	2	0.14		7	140			0		<.01	0.07			0
40	350	0.72		543	250									1
40	350	0.72		542	180									6
40	350	0.36		533	250									1

A Appendix

Table A–1

Food Composition   (Computer code number is for West Diet Analysis program)   (For purposes of calculations, use "0" for t, <1, <.1, <.01, etc.)

Computer Code Number	Food Description	Measure	Wt (g)	H₂O (%)	Ener (kcal)	Prot (g)	Carb (g)	Dietary Fiber (g)	Fat (g)	Fat Breakdown (g)		
										Sat	Mono	Poly
	FAST FOOD RESTAURANTS—Continued											
2170	Hot caramel	1 ea	182		360	7	61	0	10	6		
2171	Hot fudge	1 ea	179	59	340	8	52	1	12	9		
2172	Strawberry	1 ea	178		290	7	50		7	5		
2166	Vanilla	1 ea	90		150	4	23	0	4	3		
47147	Cookies, McDonaldland	1 ea	42	13	169	2	28	1	6	1.5		
47146	Chocolaty chip	1 ea	35		175	2	23	1	9	5		
42333	Muffin, apple bran, fat-free	1 ea	114		300	6	61	3	3	0.5		
48136	Pie, apple	1 ea	77		260	3	34	1	13	3.5		
	Breakfast items:											
42064	English muffin with spread	1 ea	63	33	189	5	30	2	6	2.4	1.5	1.3
69005	Egg McMuffin	1 ea	137		292	17	27	1	12	4.5		
45069	Hotcakes with marg & syrup	1 ea	228		600	9	104	0	17	3		
19579	Scrambled eggs	1 ea	102		160	13	1	0	11	3.5		
12230	Pork sausage	1 ea	43		170	6	0	0	16	5		
69006	Sausage McMuffin	1 ea	112		360	13	26	1	23	8		
69007	Sausage McMuffin with egg	1 ea	163	52	443	19	27	1	28	10.1		
42332	Biscuit with biscuit spread	1 ea	84	33	292	5	37	1	13	3		
69003	Biscuit with sausage	1 ea	127	36	465	11	34	1	32	9.1		
69004	Biscuit with sausage & egg	1 ea	178	49	538	18	34	1	36	11		
69002	Biscuit with bacon, egg, cheese	1 ea	157	44	496	21	32	1	32	10.3		
56479	Garden salad	1 ea	149		100	7	4	2	6	3		
	Source: McDonald's Corporation											
	PIZZA HUT											
	Pan pizza:											
56481	Cheese	2 pce	216		569	24	55	4	27	11.8		
56482	Pepperoni	2 pce	208		549	22	55	4	27	9.8		
56483	Supreme	2 pce	273		657	27	60	6	35	12.3		
56484	Super supreme	2 pce	286		680	28	60	6	36	12		
	Thin 'n crispy pizza:											
56485	Cheese	2 pce	174		409	20	45	4	18	10.2		
56486	Pepperoni	2 pce	168		394	19	44	4	19	8.3		
56487	Supreme	2 pce	232		496	24	46	4	26	11.9		
56488	Super supreme	2 pce	247		532	25	44	4	28	11.4		
	Hand tossed pizza:											
56489	Cheese	2 pce	216		489	24	57	4	20	10.2		
56490	Pepperoni	2 pce	208		502	23	50	4	23	10.8		
56492	Super supreme	2 pce	286		599	29	64	7	27	11.1		
56493	Pepperoni personal pan pizza	1 ea	255		615	26	69	5	28	10.9		
	Source: Pizza Hut.											
	SUBWAY											
	Deli style sandwich:											
69102	Ham	1 ea	171		253	13	42	4	5	1.8		
69103	Roast beef	1 ea	180		262	15	42	4	5	2.4		
69107	Tuna with light mayo	1 ea	178		350	14	38	3	17	4.8		
69101	Turkey	1 ea	180		262	15	43	4	4	1.8		
	Sandwiches, 6 inch:											
69117	Club on white bread	1 ea	246		309	23	44	4	6	1.9		
69113	Cold cut trio on white bread	1 ea	246		421	20	45	4	20	6.7		
69115	Ham on white bread	1 ea	232		290	18	46	4	5	1.5		
69139	Italian B.M.T. on white bread	1 ea	246		476	23	46	4	24	8.9		
69129	Meatball on white bread	1 ea	260		480	22	48	5	24	9.1		
69127	Melt with turkey, ham, bacon, cheese, on white bread	1 ea	251		397	23	46	4	16	5.8		
69121	Roast Beef on white bread	1 ea	232		303	20	47	4	5	2.1		

A
Appendix

PAGE KEY: A–4 = Beverages   A–6 = Dairy   A–10 = Eggs   A–12 = Fat/Oil   A–14 = Fruit   A–20 = Bakery
A–26 = Grain   A–32 = Fish   A–32 = Meats   A–36 = Poultry   A–38 = Sausage   A–40 = Mixed/Fast   A–44 = Nuts/Seeds   A–46 = Sweets
A–48 = Vegetables/Legumes   A–58 = Vegetarian   A–62 = Misc   A–62 = Soups/Sauces   A–64 = Fast   A–78 = Convenience   A–80 = Baby foods

Chol (mg)	Calc (mg)	Iron (mg)	Magn (mg)	Pota (mg)	Sodi (mg)	Zinc (mg)	VT-A (RAE)	Thia (mg)	VT-E (mg α)	Ribo (mg)	Niac (mg)	V-B$_6$ (mg)	Fola (μg)	VT-C (mg)
35	250				180									1
30	250	0.72			170								1	
30	200	0.36			95									1
20	100	0.36			75									1
0	15	1.33	8	46	184	0.29	0		0.74		1.5			0
25	12	0.9	15	89	106	0.25		0.09	0.58	0.1	0.92			0
0	100	1.44	20	117	380	0.5	0	0.22	0	0.22	2.01	0.04	8	1
0	20	1.08	7	63	200	0.21		0.18	1.38	0.11	1.42	0.03	8	24
13	103	1.59	13	69	386	0.42	32	0.25	0.13	0.32	2.61	0.04	57	1
237	201	2.72	24	199	796	1.56		0.49	0.85	0.44	3.32	0.15	33	1
20	100	4.5	28	292	770	0.55			1.23				<1	0
425	40	1.08	10	126	170	1.06	150	0.07	0.92	0.51	0.06	0.12	44	0
35	7	0.36	7	102	290	0.78	0	0.18	0.26	0.06	1.7	0.09		0
45	200	1.8	22	191	740	1.51			0.66				16	0
257	252	2.72	26	251	895	2.06			1.11				30	0
0	49	2.19	10	116	779	0.33			0.89		2.46		5	0
40	45	2.7	16	221	1055	1.08		0.51	1.13	0.31	4.19	0.13	5	0
269	88	2.97	21	283	1110	1.68			1.59		4.14		28	0
258	155	2.79	21	253	1456	1.69			1.54		3.43		31	0
75	150	1.08			120	75								15
20	393	3.53			1159									5
29	196	3.53			1197									5
41	308	3.69			1375									12
50	300	3.6			1560									12
20	409	2.95			1208									5
31	207	2.99			1265									5
40	297	3.57			1408									18
48	285	3.42			1596									17
20	408	2.93			1325									5
36	359	3.23			1417									43
44	333	3.99			1618									13
30	298	4.46			1419									6
12	72	4.34			927									14
18	72	6.44			787									14
26	159	3.81			879									13
18	72	4.29			870									14
34	58	5.21			1254									20
53	144	5.17			1608									23
25	60	3.6			1270									21
55	149	3.57			1885									24
50	136	4.89			1232									24
44	145	3.49			1677									23
21	63	6.58			951									22

A   Appendix

Table A–1

Food Composition

(Computer code number is for West Diet Analysis program)   (For purposes of calculations, use "0" for t, <1, <.1, <.01, etc.)

Computer Code Number	Food Description	Measure	Wt (g)	H₂O (%)	Ener (kcal)	Prot (g)	Carb (g)	Dietary Fiber (g)	Fat (g)	Fat Breakdown (g) Sat	Mono	Poly
	FAST FOOD RESTAURANTS—Continued											
69125	Roasted chicken breast on white bread	1 ea	246		334	24	49	5	5	2.1		
69147	Seafood and crab with light mayo on white bread	1 ea	246		396	15	50	5	15	4.3		
69119	Steak and cheese on white bread	1 ea	257		392	24	48	5	14	5		
69143	Tuna with light mayo on white bread	1 ea	246		434	19	44	4	21	5.8		
69111	Turkey on white bread	1 ea	232		293	19	48	4	5	1.6		
69137	Turkey and ham on white bread	1 ea	232		290	20	46	4	5	1.5		
69109	Veggie delite on white bread	1 ea	175		242	9	46	4	3	1.1		
	Salads:											
52124	B.M.T., classic italian	1 ea	331		302	17	12	3	20	8.6		
52115	Club	1 ea	331		154	17	12	3	4	1.5		
52120	Cold cut trio	1 ea	330		240	15	11	3	16	6.3		
52123	Ham	1 ea	316		120	12	12	3	3	1.1		
52129	Meatball	1 ea	345		319	17	17	4	20	9		
52131	Melt	1 ea	336		211	18	12	3	11	4.8		
52126	Roast beef	1 ea	316		131	13	11	3	3	1.6		
52119	Roasted chicken breast	1 ea	331		153	17	13	3	3	1.1		
52116	Seafood and crab with light mayo	1 ea	331		211	9	18	4	12	3.7		
52130	Steak and cheese	1 ea	342		195	18	13	4	9	3.8		
52118	Tuna with light mayo	1 ea	331		253	14	11	3	17	4.2		
52113	Veggie delite	1 ea	260		56	2	10	3	1	0		
	Cookies:											
47658	With M&M's	1 ea	48	5	235	2	32	1	11	4.3		
47659	Chocolate chunk	1 ea	48		235	2	32	1	11	4.3		
47656	Oatmeal raisin	1 ea	48		213	3	32	2	9	2.7		
47657	Peanut butter	1 ea	48	5	235	4	28	1	13	4.3		
47660	Sugar	1 ea	48		245	2	30	0	13	4.3		
47661	White chip macadamia	1 ea	48		235	2	30	1	12	4.3		
	Source: Subway International											
	TACO BELL											
	Burritos:											
56519	Bean with red sauce	1 ea	198		370	14	55	8	10	3.5		
56688	Chicken burrito	1 ea	171		283	14	34	3	10	4.1		
56522	Supreme with red sauce	1 ea	255		452	19	52	7	19	8.2		
56691	7 layer burrito	1 ea	283		530	18	67	10	22	8		
	Tacos:											
56525	Soft taco	1 ea	90		191	9	19	2	9	4.1		
56526	Soft taco supreme	1 ea	142		276	12	23	3	15	7.4		
56689	Soft taco, chicken	1 ea	121		232	17	23	1	7	3.1		
56693	Soft taco, steak	1 ea	128		282	12	21	1	17	4.5		
56528	Tostada with red sauce	1 ea	177		260	11	30	7	10	4.2		
56531	Mexican pizza	1 ea	220		560	21	47	7	32	11.2		
56537	Taco salad with salsa	1 ea	539		799	31	74	13	42	15.2		
56533	Nachos, regular	1 ea	99	40	320	5	33	2	19	4.5		
56534	Nachos, bellgrande	1 ea	312		790	20	81	12	44	13.2		
56536	Pintos & cheese with red sauce	1 ea	120	69	169	9	19	6	7	3.3		
53186	Taco sauce, packet	1 ea	11		4	0	<1	0	0	0	0	0
45585	Cinnamon twists	1 ea	28		128	1	22	0	4	0.8		
	Source: Taco Bell Corporation											
	WENDY'S											
56571	Cheeseburger, bacon	1 ea	166		382	20	34	2	19	7		
69059	Chicken sandwich, grilled	1 ea	189		302	24	36	2	7	1.5		
	Baked potatoes:											
6167	Plain	1 ea	284		310	7	72	7	0	0	0	0

PAGE KEY: A–4 = Beverages   A–6 = Dairy   A–10 = Eggs   A–12 = Fat/Oil   A–14 = Fruit   A–20 = Bakery
A–26 = Grain   A–32 = Fish   A–32 = Meats   A–36 = Poultry   A–38 = Sausage   A–40 = Mixed/Fast   A–44 = Nuts/Seeds   A–46 = Sweets
A–48 = Vegetables/Legumes   A–58 = Vegetarian   A–62 = Misc   A–62 = Soups/Sauces   A–64 = Fast   A–78 = Convenience   A–80 = Baby foods

Chol (mg)	Calc (mg)	Iron (mg)	Magn (mg)	Pota (mg)	Sodi (mg)	Zinc (mg)	VT-A (RAE)	Thia (mg)	VT-E (mg α)	Ribo (mg)	Niac (mg)	V-B$_6$ (mg)	Fola (µg)	VT-C (mg)
47	63	5.63			1042									22
24	145	3.47			1235									23
35	151	8.13			1215									24
39	14	3.47			1148									23
21	63	3.76			1055									22
25	60	3.6			1220									21
0	63	3.8			538									22
59	108	1.16			1714									32
36	41	18.5			1141									31
57	157	1.88			1431									31
27	44	1.18			1170									33
55	100	1.79			1047									36
48	106	1.14			1490									32
22	44	1.97			787									33
49	44	1.18			874									33
26	105	1.14			1023									32
38	109	3.91			966									33
42	105	1.14			928									32
0	45	1.21			346									33
16	0	1.15			112	0								0
11	0	1.15			112	0								0
16	0	1.15			192	0								0
11	0	1.15			213									0
16	0	1.15			144	0								0
16	0	1.15			171									0
10	200	2.7			1200									5
31	138	1.86			876									6
41	206	2.78			1368									9
25	300	3.6			1360									5
23	91	1.64			564									2
42	159	1.91			668									5
37	122	1.32			672									1
30	101	1.45			655									4
16	156	1.5			739									5
46	356	3.67			1049									6
66	405	6.37			1689									21
5	80	0.72			530	0								0
35	203	2.74			1317									6
14	141	1.01			656									3
0	0	0			102									0
0	0	0.29			120	0								0
55	151	3.62		322	895									9
55	80	2.71		432	744									9
0	20	3.6		1190	25		0							36

A Appendix

Table A-1

Food Composition

(Computer code number is for West Diet Analysis program)   (For purposes of calculations, use "0" for t, <1, <.1, <.01, etc.)

Computer Code Number	Food Description	Measure	Wt (g)	H₂O (%)	Ener (kcal)	Prot (g)	Carb (g)	Dietary Fiber (g)	Fat (g)	Fat Breakdown (g) Sat	Mono	Poly
	FAST FOOD RESTAURANTS—Continued											
56579	With bacon & cheese	1 ea	380		580	18	79	7	22	6		
56580	With broccoli & cheese	1 ea	411		480	9	81	9	14	3		
56582	With sour cream & chives	1 ea	439		521	10	103	10	8	5.6		
50311	Chili	1 ea	227		200	17	21	5	6	2.5		
2176	Frosty dairy dessert	1 ea	227		330	8	56	0	8	5		
	Source: Wendy's International											
	CONVENIENCE FOODS & MEALS											
	HAAGEN DAZS											
70642	Ice cream bar, vanilla almond	1 ea	87		304	5	21	1	22	11.5		
70645	Lemon sorbet	½ c	113		120	0	31	<1	0	0	0	0
70646	Raspberry	½ c	105		120	0	30	2	0	0	0	0
	Source: Pillsbury											
	HEALTHY CHOICE											
	Entrees:											
18825	Fish, lemon pepper	1 ea	303		320	14	50	5	7	2		
81039	Lasagna	1 ea	383		420	26	59	6	9	3		
11119	Meatloaf, traditional	1 ea	340	78	316	15	52	6	5	2.5	1.9	0.6
	Low-Fat ice cream:											
2184	Brownie	½ c	71		120	3	22	1	2	1	0.3	0.7
2185	Chocolate chip	½ c	71		120	3	21	1	2	1	1	0
2105	Cookie & cream	½ c	71		120	3	21	1	2	1	1	0
2123	Rocky road	½ c	71		140	3	28	1	2	1	1	0
	Source: ConAgra Frozen Foods, Omaha, NE											
	HEALTH VALLEY											
	Soups, fat-free:											
50355	Beef broth, no salt added	1 c	240		18	5	0	0	0	0	0	0
50366	Beef broth, w/salt	1 c	240		30	5	2	0	0	0	0	0
50363	Black bean & vegetable	1 c	240		110	11	24	12	0	0	0	0
50364	Chicken broth	1 c	240		30	6	0	0	0	0	0	0
50365	14 garden vegetable	1 c	240		80	6	17	4	0	0	0	0
50362	Lentil & carrot	1 c	240		90	10	25	14	0	0	0	0
50361	Split pea & carrot	1 c	240		110	8	17	4	0	0	0	0
50360	Tomato vegetable	1 c	240		80	6	17	5	0	0	0	0
	Source: Health Valley											
	LA CHOY											
83016	Egg rolls, mini, chicken, svg	1 ea	106		108	3	13	1	5	1.3		
83013	Egg rolls, mini, shrimp, svg	1 ea	106		98	3	14	1	3	0.8		
	Source: Beatrice/Hunt-Wesson											
	LEAN CUISINE											
	Dinners:											
56901	Baked cheese ravioli	1 ea	241		260	12	38	4	7	3.5	1.5	0.5
15964	Chicken chow mein	1 ea	255		240	14	37	3	3	1	1.5	0.5
56740	Lasagna	1 ea	291		293	19	37	4	8	4.4	2	1
56702	Macaroni & cheese	1 ea	255		261	13	38	2	6	3.6	1.3	0.4
56732	Spaghetti w/meatballs	1 ea	269	75	299	18	40	5	8	2.1	2.7	1.3
56734	French bread sausage pizza	1 ea	170		210	8	24	1	9	3.5		
	Source: Stouffer's Foods Corp, Solon OH											

Appendix A

PAGE KEY: A–4 = Beverages   A–6 = Dairy   A–10 = Eggs   A–12 = Fat/Oil   A–14 = Fruit   A–20 = Bakery
A–26 = Grain   A–32 = Fish   A–32 = Meats   A–36 = Poultry   A–38 = Sausage   A–40 = Mixed/Fast   A–44 = Nuts/Seeds   A–46 = Sweets
A–48 = Vegetables/Legumes   A–58 = Vegetarian   A–62 = Misc   A–62 = Soups/Sauces   A–64 = Fast   A–78 = Convenience   A–80 = Baby foods

Chol (mg)	Calc (mg)	Iron (mg)	Magn (mg)	Pota (mg)	Sodi (mg)	Zinc (mg)	VT-A (RAE)	Thia (mg)	VT-E (mg α)	Ribo (mg)	Niac (mg)	V-B₆ (mg)	Fola (µg)	VT-C (mg)
40	200	3.6		1410	950									42
5	200	4.5		1400	510									72
21	84	5.07		1731	56									51
35	80	1.8		470	870									2
35	300	1.08		590	200									0
74	123	0.59		180	66					0.15				0
0	0	0		30	5	0								4
0	0	0		56	0	0								2
30	20	1.08			480									30
35	150	3.6		500	580			0.3		0.26	2			6
37	48	2.24			459									55
5	100	0		268	55									0
5	100	0		240	50									0
5	100			254	90			0.03		0.15				0
5	100	0		168	60			0.03		0.15				0
0				196	74						0.98			
0	0	0			160	0								5
0	40	3.6			280									9
0	20	1.8			170	0								1
0	40	1.8			250									15
0	60	5.4			220									2
0	40	5.4			230									9
0	40	5.4			240									9
8	10	0.56			335									0
5	10	0.56			377									0
35	150	1.44	42	450	590	1.5		0.06		0.26	1.2	0.2	48	5
35	40	0.72	30	300	590	1.1		0.15		0.17	5			0
29	244	1.41	44	596	576	2.9		0.15		0.25	3	0.32		6
18	180	0.65		423	567		0	0.12		0.25	1.2			0
5	94	2.37		539	465									6
10	100	2.7	39	165	630	2.2		0.45		0.51	0.05	0.07		1

A Appendix

Table A–1

Food Composition (Computer code number is for West Diet Analysis program) (For purposes of calculations, use "0" for t, <1, <.1, <.01, etc.)

Computer Code Number	Food Description	Measure	Wt (g)	H₂O (%)	Ener (kcal)	Prot (g)	Carb (g)	Dietary Fiber (g)	Fat (g)	Fat Breakdown (g)		
										Sat	Mono	Poly
	CONVENIENCE FOODS & MEALS—Continued											
	TASTE ADVENTURE SOUPS											
50325	Black bean	1 c	242		807	51	148	36	4	0.9	0.3	1.5
50324	Curry lentil	1 c	241		795	66	135	71	3	0.4	0.5	1.2
50326	Lentil chili	1 c	242		411	24	75	15	2			
50323	Split pea	1 c	244		807	58	143	60	3	0.4	0.6	1.2
	Source: Taste Adventure Soups											
	BABY FOODS											
60465	Apple juice	½ c	125	88	59	0	15	<1	<1	t	t	t
60475	Applesauce, strained	1 tbsp	16	87	8	<1	2	<1	<1	t	t	t
60502	Carrots, strained	1 tbsp	14	92	4	<1	1	<1	<1	t	t	t
60563	Cereal, mixed, milk added	1 tbsp	15	75	17	1	2	<1	1	0.3		
60622	Cereal, rice, milk added	1 tbsp	15	75	17	1	3	<1	1	0.3		
60515	Chicken and noodles, strained	1 tbsp	16	86	11	<1	1	<1	<1	t	0.1	t
60603	Peas, strained	1 tbsp	15	88	6	1	1	<1	<1	t	t	t
60634	Teething biscuits	1 ea	11	6	43	1	8	<1	<1	0.2	0.2	t

PAGE KEY: A–4 = Beverages   A–6 = Dairy   A–10 = Eggs   A–12 = Fat/Oil   A–14 = Fruit   A–20 = Bakery
A–26 = Grain   A–32 = Fish   A–32 = Meats   A–36 = Poultry   A–38 = Sausage   A–40 = Mixed/Fast   A–44 = Nuts/Seeds   A–46 = Sweets
A–48 = Vegetables/Legumes   A–58 = Vegetarian   A–62 = Misc   A–62 = Soups/Sauces   A–64 = Fast   A–78 = Convenience   A–80 = Baby foods

Chol (mg)	Calc (mg)	Iron (mg)	Magn (mg)	Pota (mg)	Sodi (mg)	Zinc (mg)	VT-A (RAE)	Thia (mg)	VT-E (mg α)	Ribo (mg)	Niac (mg)	V-B$_6$ (mg)	Fola (μg)	VT-C (mg)
0	296	12.18	405	3521	1978	8.67	39	2.12	0.12	0.48	4.8	0.84	1043	1
0	140	22.04	256	2170	2182	8.51	48	1.12	0.72	0.6	6.33	1.25	1005	16
			1476	1016										
0	140	10.65	272	2324	1729	7.14		1.71	2.75	0.51	6.84	0.42	646	5
0	5	0.71	4	114	4	0.04	1	<.01	0.75	0.02	0.1	0.04	0	72
0	1	0.04	0	11	<1	<.01	<1	<.01	0.03	<.01	<.01	<.01	<1	6
0	3	0.05	1	27	5	0.02	80	<.01	0.15	<.01	0.06	0.01	2	1
2	33	1.56	4	30	7	0.11	4	0.06		0.09	0.87	<.01	2	0
2	36	1.83	7	28	7	0.1	3	0.07		0.08	0.78	0.02	1	0
3	4	0.1	2	22	4	0.09	18	<.01	0.03	<.01	0.12	0.01	2	0
0	3	0.14	2	17	1	0.05	3	0.01	0.01	<.01	0.15	0.01	4	1
0	29	0.39	4	36	40	0.1	3	0.03	0.03	0.06	0.48	0.01	5	1

A Appendix

Reliable Sources of Nutrition Information

Many sources of nutrition information are available to consumers, but the quality of the information they provide varies widely. All of the sources listed here provide scientifically based information.

Expert Advice

Registered dietitians (hospitals and the yellow pages) Public health nutritionists (public health departments) College nutrition instructors/professors (colleges and universities) Extension Service home economists (state and county U.S. Department of Agriculture Extension Service offices) Consumer affairs staff of the Food and Drug Administration (national, regional, and state FDA offices)

You can find hundreds of toll-free telephone numbers for health information through the following Website: **http://www.healthfinder.gov/**. After you are connected, search the term toll-free numbers.

U.S. Government

- Federal Trade Commission (FTC)
  Public Reference Branch
  (202) 326-2222
  www.ftc.gov

- Food and Drug Administration (FDA)
  Office of Consumer Affairs, HFE 1
  Room 16-85
  5600 Fishers Lane
  Rockville, MD 20857
  (301) 443-1544
  www.fda.gov

- FDA Consumer Information Line
  (301) 827-4420

- FDA Office of Food Labeling, HFS 150
  Washington, DC 20204
  (202) 205-4561; fax (202) 205-4564
  www.cfsan.fda.gov

- FDA Office of Plant and Dairy Foods
  and Beverages
  HFS 300
  200 C Street SW
  Washington, DC 20204
  (202) 205-4064; fax (202) 205-4422

- FDA Office of Special Nutritionals,
  HFS 450
  200 C Street SW
  Washington, DC 20204
  (202) 205-4168; fax (202) 205-5295

- Food and Nutrition Information Center
  National Agricultural Library, Room 304
  10301 Baltimore Avenue
  Beltsville, MD 20705-2351
  (301) 504-5719; fax (301) 504-6409
  www.nal.usda.gov/fnic

- Food Research Action Center (FRAC)
  1875 Connecticut Avenue NW, Suite 540
  Washington, DC 20009
  (202) 986-2200; fax (202) 986-2525

- Superintendent of Documents
  U.S. Government Printing Office
  Washington, DC 20402
  (202) 512-1071
  www.access.gpo.gov/su_docs

- U.S. Department of Agriculture (USDA)
  14th Street SW and Independence
  Avenue
  Washington, DC 20250
  (202) 720-2791
  www.usda.gov/fcs

- USDA Center for Nutrition Policy and
  Promotion
  1120 20th Street NW, Suite 200
  North Lobby
  Washington, DC 20036
  (202) 208-2417
  www.usda.gov/fcs/cnpp.htm

- USDA Food Safety and Inspection
  Service
  Food Safety Education Office,
  Room 1180-S
  Washington, DC 20250
  (202) 690-0351
  www.usda.gov/fsis

- U.S. Department of Education (DOE)
  Accreditation Agency Evaluation Branch
  7th and D Street SW
  ROB 3, Room 3915
  Washington, DC 20202-5244
  (202) 708-7417

- U.S. Department of Health and Human
  Services
  200 Independence Avenue SW
  Washington, DC 20201
  (202) 619-0257
  www.os.dhhs.gov

- U.S. Environmental Protection
  Agency (EPA)
  401 Main Street SW
  Washington, DC 20460
  (202) 260-2090
  www.epa.gov

- U.S. Public Health Service
  Assistant Secretary of Health
  Humphrey Building, Room 725-H
  200 Independence Avenue SW
  Washington, DC 20201
  (202) 690-7694

Health Canada
Headquarters

- Health Canada
  A.L. 0900C2
  Ottawa, Canada
  K1A 0K9
  Telephone: (613) 957-2991
  TTY: 1-800-267-1245
  http://www.hc-sc.gc.ca/

Regional Headquarters

- British Columbia/Yukon
  Suite 405, Winch Building
  757 West Hastings Street
  Vancouver, BC
  V6C 1A1
  Tel: (604) 666-2083
  Fax: (604) 666-2258

- Alberta/NWT
  Suite 710, Canada Place
  9700 Jasper Avenue
  Edmonton, AB
  T5J 4C3
  Tel: (780) 495-2651
  Fax: (780) 495-3285

- Manitoba/Saskatchewan
  391 York Avenue, Suite 425
  Winnipeg, MB
  R3C 0P4
  Tel: (204) 983-2508
  Fax: (204) 983-3972

- Ontario/Nunavut
  25 St. Clair Avenue East, 4th Floor
  Toronto, ON
  M4T 1M2
  Tel: (416) 973-4389
  Toll free: 1-866-999-7612
  Fax: (416) 973-1423

- Quebec
  Room 218, Complexe Guy-Favreau
  East Tower
  200 René Lévesque Blvd. West
  Montreal, QC
  H2Z 1X4
  Tel: (514) 283-2306
  Fax: (514) 283-6739

- Atlantic
  Suite 1525, 15th Floor, Maritime Centre
  1505 Barrington Street
  Halifax, NS B3J 3Y6
  Tel: (902) 426-2700
  Fax: (902) 426-9689

International Agencies

- Food and Agriculture Organization of
  the United Nations (FAO)
  Liaison Office for North America
  2175 K Street, Suite 300
  Washington, DC 20437
  (202) 653-2400
  www.fao.org

- International Food Information Council
  Foundation
  1100 Connecticut Avenue NW, Suite 430
  Washington, DC 20036
  (202) 296-6540
  ificinfo.health.org

- UNICEF
  3 United Nations Plaza
  New York, NY 10017
  (212) 326-7000
  www.unicef.com

- World Health Organization (WHO)
  Regional Office
  525 23rd Street NW
  Washington, DC 20037
  (202) 974-3000
  www.who.org

Professional Nutrition Organizations

- American Dietetic Association (ADA)
  216 West Jackson Boulevard, Suite 800
  Chicago, IL 60606-6995
  (800) 877-1600; (312) 899-0040
  www.eatright.org

- ADA, The Nutrition Hotline
  (800) 366-1655

- American Society for Clinical Nutrition
  9650 Rockville Pike
  Bethesda, MD 20814-3998
  (301) 530-7110; fax (301) 571-1863
  www.faseb.org/ascn

- Dietitians of Canada
  480 University Avenue, Suite 604
  Toronto, Ontario M5G 1V2, Canada
  (416) 596-0857; fax (416) 596-0603
  www.dietitians.ca

- Human Nutrition Institute (INACG)
  1126 Sixteenth Street NW
  Washington, DC 20036
  (202) 659-0789
  www.ilsi.org

- National Academy of Sciences/
  National Research Council (NAS/NRC)
  2101 Constitution Avenue, NW
  Washington, DC 20418
  (202) 334-2000
  www.nas.edu

- National Institute of Nutrition
  265 Carling Avenue, Suite 302
  Ottawa, Ontario K1S 2E1
  (613) 235-3355; fax (613) 235-7032
  www.nin.ca

- Society for Nutrition Education
  7101 Wisconsin Avenue, Suite 901
  Bethesda, MD 20814-4805
  (301) 656-4938

Aging

- Administration on Aging
  330 Independence Avenue SW
  Washington, DC 20201
  (202) 619-0724
  www.aoa.dhhs.gov

- American Association of Retired Persons
  (AARP)
  601 E Street NW
  Washington, DC 20049
  (202) 434-2277
  www.aarp.org

- National Aging Information Center
  330 Independence Avenue SW
  Washington, DC 20201
  (202) 619-7501
  www.aoa.dhhs.gov/naic

- National Institute on Aging
  Public Information Office
  31 Center Drive, MSC 2292
  Bethesda, MD 20892
  (301) 496-1752
  www.nih.gov/nia

Alcohol and Drug Abuse

- Al-Anon Family Group Headquarters, Inc.
  1600 Corporate Landing Parkway
  Virginia Beach, VA 23454-5617
  (800) 356-9996
  www.al-anon.alateen.org

- Alateen
  1600 Corporate Landing Parkway
  Virginia Beach, VA 23454-5617
  (800) 356-9996
  www.al-anon.alateen.org

- Alcoholics Anonymous (AA)
  General Service Office
  475 Riverside Drive
  New York, NY 10115
  (212) 870-3400
  www.aa.org

- Narcotics Anonymous (NA)
  P.O. Box 9999
  Van Nuys, CA 91409
  (818) 773-9999; fax (818) 700-0700
  www.wsoinc.com

- National Clearinghouse for Alcohol and
  Drug Information (NCADI)
  P.O. Box 2345
  Rockville, MD 20847-2345
  (800) 729-6686
  www.health.org

- National Council on Alcoholism and
  Drug Dependence (NCADD)
  12 West 21st Street
  New York, NY 10010
  (800) NCA-CALL or (800) 622-2255
  (212) 206-6770; fax (212) 645-1690
  www.ncadd.org

- U.S. Center for Substance Abuse
  Prevention
  1010 Wayne Avenue, Suite 850
  Silver Spring, MD 20910
  (301) 459-1591 ext. 244;
  fax (301) 495-2919
  www.covesoft.com/csap.html

Consumer Organizations

- Center for Science in the Public Interest
  (CSPI)
  1875 Connecticut Avenue NW, Suite 300
  Washington, DC 20009-5728
  (202) 332-9110; fax (202) 265-4954
  www.cspinet.org

B Appendix

- Choice in Dying, Inc.
  1035 30th Street NW
  Washington, DC 20007
  (202) 338-9790; fax (202) 338-0242
  www.choices.org

- Consumer Information Center
  Pueblo, CO 81009
  (888) 8 PUEBLO or (888) 878-3256
  www.pueblo.gsa.gov

- Consumers Union of US Inc.
  101 Truman Avenue
  Yonkers, NY 10703-1057
  (914) 378-2000
  www.consunion.org

- National Council Against Health Fraud,
  Inc. (NCAHF)
  P.O. Box 1276
  Loma Linda, CA 92354
  (909) 824-4690
  www.ncahf.org

Fitness

- American College of Sports Medicine
  P.O. Box 1440
  Indianapolis, IN 46206-1440
  (317) 637-9200
  _www.acsm.org/sportsmed

- American Council on Exercise (ACE)
  5820 Oberlin Drive, Suite 102
  San Diego, CA 92121
  (800) 529-8227
  www.acefitness.org

- President's Council on Physical Fitness
  and Sports
  Humphrey Building, Room 738
  200 Independence Avenue SW
  Washington, DC 20201
  (202) 690-9000; fax (202) 690-5211
  _www.indiana.edu/~preschal

- Shape Up America!
  6707 Democracy Boulevard, Suite 306
  Bethesda, MD 20817
  (301) 493-5368
  www.shapeup.org

- Sport Medicine and Science Council of
  Canada
  1600 James Naismith Drive, Suite 314
  Gloucester, Ontario K1B 5N4, Canada
  (613) 748-5671; fax (613) 748-5729
  www.smscc.ca

Food Safety

- Alliance for Food & Fiber
  Food Safety Hotline
  (800) 266-0200

- FDA Center for Food Safety and Applied
  Nutrition
  200 C Street SW
  Washington, DC 20204
  (800) FDA-4010 or (800) 332-4010
  vm.cfsan.fda.gov

- National Lead Information Center
  (800) LEAD-FYI or (800) 532-3394
  (800) 424-LEAD or (800) 424-5323

- National Pesticide Telecommunications
  Network (NPTN)
  Oregon State University
  333 Weniger Hall
  Corvallis, OR 97331-6502
  (541) 737-6091
  _www.ace.orst.edu/info/nptn

- USDA Meat and Poultry Hotline
  (800) 535-4555

- U.S. EPA Safe Drinking Water Hotline
  (800) 426-4791

Health and Disease

- Alzheimer's Disease Education and
  Referral Center
  P. O. Box 8250
  Silver Spring, MD 20907-8250
  (800) 438-4380
  www.alzheimers.org

- Alzheimer's Disease Information and
  Referral Service
  919 North Michigan Avenue, Suite 1000
  Chicago, IL 60611
  (800) 272-3900
  www.alz.org

- American Academy of Allergy, Asthma,
  and Immunology
  611 East Wells Street
  Milwaukee, WI 53202
  (414) 272-6071; fax (414) 276-3349
  www.aaaai.org

- American Cancer Society
  National Home Office
  1599 Clifton Road NE
  Atlanta, GA 30329-4251
  (800) ACS-2345 or (800) 227-2345
  www.cancer.org

- American Council on Science and
  Health
  1995 Broadway, 2nd Floor
  New York, NY 10023-5860
  (212) 362-7044; fax (212) 362-4919
  www.acsh.org

- American Dental Association
  211 East Chicago Avenue
  Chicago, IL 60611
  (312) 440-2800
  www.ada.org

- American Diabetes Association
  1660 Duke Street
  Alexandria, VA 22314
  (800) 232-3472 or (703) 549-1500
  www.diabetes.org

- American Heart Association
  Box BHG, National Center
  7320 Greenville Avenue
  Dallas, TX 75231
  (800) 275-0448 or (214) 373-6300
  www.amhrt.org

- American Institute for Cancer Research
  1759 R Street NW
  Washington, DC 20009
  (800) 843-8114 or (202) 328-7744;
  fax (202) 328-7226
  www.aicr.org

- American Medical Association
  515 North State Street
  Chicago, IL 60610
  (312) 464-5000
  www.ama-assn.org

- American Public Health Association
  (APHA)
  1015 Fifteenth Street NW, Suite 300
  Washington, DC 20005
  (202) 789-5600
  www.apha.org

- American Red Cross
  National Headquarters
  8111 Gatehouse Road
  Falls Church, VA 22042
  (703) 206-7180
  www.redcross.org

- Canadian Diabetes Association
  15 Toronto Street, Suite 800
  Toronto, ON M5C 2E3
  (800) BANTING or (800) 226-8464
  (416) 363-3373
  www.diabetes.ca

- Canadian Public Health Association
  400-1565 Carling Avenue
  Ottawa, Ontario K1Z 8R1
  (613) 725-3769; fax (613) 725-9826
  www.cpha.ca

- Centers for Disease Control and
  Prevention (CDC)
  1600 Clifton Road NE
  Atlanta, GA 30333
  (404) 639-3311
  www.cdc.gov

- The Food Allergy Network
  10400 Eaton Place, Suite 107
  Fairfax, VA 22030-2208
  (800) 929-4040 or (703) 691-3179
  www.foodallergy.org

Appendix **B**

- Internet Health Resources
  www.ihr.com
- National AIDS Hotline (CDC)
  (800) 342-AIDS (English)
  (800) 344-SIDA (Spanish)
  (800) 2437-TTY (Deaf)
  (900) 820-2437
- National Cancer Institute
  Office of Cancer Communications
  Building 31, Room 10824
  Bethesda, MD 20892
  (800) 4-CANCER or (800) 422-6237
  www.nci.nih.gov
- National Diabetes Information
  Clearinghouse
  1 Information Way
  Bethesda, MD 20892-3560
  (301) 654-3327
  www.niddk.nih.gov
- National Digestive Disease Information
  Clearinghouse (NDDIC)
  2 Information Way
  Bethesda, MD 20892-3570
  (301) 654-3810
  www.niddk.nih.gov
- National Health Information Center
  (NHIC)
  Office of Disease Prevention and Health
  Promotion
  (800) 336-4797
  nhic-nt.health.org
- National Heart, Lung, and Blood
  Institute
  Information Center
  P.O. Box 30105
  Bethesda, MD 20824-0105
  (301) 251-1222
  _www.nhlbi.nih.gov/nhlbi/nhlbi.htm
- National Institute of Allergy and
  Infectious Diseases
  Office of Communications
  Building 31, Room 7A50
  31 Center Drive, MSC2520
  Bethesda, MD 20892-2520
  (301) 496-5717
  www.niaid.nih.gov
- National Institute of Dental Research
  (NIDR)
  National Institute of Health
  Bethesda, MD 20892-2190
  (301) 496-4261
  www.nidr.nih.gov
- National Institutes of Health (NIH)
  9000 Rockville Pike
  Bethesda, MD 20892
  (301) 496-2433
  www.nih.gov

- National Osteoporosis Foundation
  1150 17th Street NW, Suite 500
  Washington, DC 20036
  (202) 223-2226
  www.nof.org
- Office of Disease Prevention and Health
  Promotion
  odphp.osophs.dhhs.gov
- Office on Smoking and Health (OSH)
  **_www.americanheart.org/heart.org/
  Heart_and_stroke_A_Z_Guide/osh.html**

Infancy and Childhood

- American Academy of Pediatrics
  141 Northwest Point Boulevard
  Elk Grove Village, IL 60007-1098
  (847) 228-5005
  www.aap.org
- Association of Birth Defect Children, Inc.
  930 Woodcock Road, Suite 225
  Orlando, FL 32803
  (407) 245-7035
  www.birthdefects.org
- Canadian Paediatric Society
  100-2204 Walkley Road
  Ottawa, ON K1G 4G8
  (613) 526-9397; fax (613) 526-3332
  www.cps.ca
- National Center for Education in
  Maternal & Child Health
  2000 15th Street North, Suite 701
  Arlington, VA 22201-2617
  (703) 524-7802
  www.ncemch.org

Pregnancy and Lactation

- American College of Obstetricians and
  Gynecologists Resource Center
  409 12th Street SW
  Washington, DC 20024-2188
  (202) 638-5577
  www.acog.org
- La Leche International, Inc.
  1400 N. Meacham Road
  Schaumburg, IL 60173
  (847) 519-7730
  www.lalecheleague.org
- March of Dimes Birth Defects Foundation
  1275 Mamaroneck Avenue
  White Plains, NY 10605
  (914) 428-7100
  www.sunkist.com

World Hunger

- Bread for the World
  1100 Wayne Avenue, Suite 1000
  Silver Spring, MD 20910
  (301) 608-2400
  www.bread.org

- Center on Hunger, Poverty and
  Nutrition Policy
  Tufts University School of Nutrition
  11 Curtis Avenue
  Medford, MA 02155
  (617) 627-3956
- Freedom from Hunger
  P.O. Box 2000
  1644 DaVinci Court
  Davis, CA 95617
  (530) 758-6200
  www.freefromhunger.org
- Oxfam America
  26 West Street
  Boston, MA 02111
  (617) 482-1211
  www.oxfamamerica.org
- SEEDS Magazine
  P.O. Box 6170
  Waco, TX 76706
  (254) 755-7745
  _www.helwys.com/seedhome.htm
- Worldwatch Institute
  1776 Massachusetts Avenue NW,
  Suite 800
  Washington, DC 20036
  (202) 452-1999
  www.worldwatch.org

Scientific Literature

Nutrition Journals

American Journal of Clinical Nutrition

British Journal of Nutrition

Human Nutrition, Applied Nutrition

Journal of the American College of Nutrition

Journal of the American Dietetic Association

Journal of the Canadian Dietetic Association

Journal of Food Composition and Analysis

Journal of Nutrition

Journal of Nutrition Education

Nutrition Abstracts and Reviews

Nutrition and Metabolism

Nutrition Reports International

Nutrition Research

Nutrition Reviews

Nutrition Today

Other Journals

American Journal of Epidemiology

American Journal of Nursing

American Journal of Public Health

Annals of Internal Medicine

Annals of Surgery

B Appendix

Canadian Journal of Public Health

Caries Research

Food Technology

Gastroenterology

International Journal of Obesity

Journal of the American Dental Association

Journal of the American Medical Association

Journal of Clinical Investigation

Journal of Food Science

Journal of Home Economics

Journal of Pediatrics

Lancet

New England Journal of Medicine

Nutrition Today

Pediatrics

Science

The Scientist

The U.S. Food Exchange System

The U.S. exchange system divides the foods suitable for use in planning a healthy diet into six lists—the starch/bread, meat/meat alternate, vegetable, fruit, milk, and fat lists.[a] These lists are shown in Tables C.1 through C.6. Following these lists are three other sets of foods: free foods, combination foods, and foods for occasional use (Tables C.7, C.8, and C.9).

[a]The Exchange Lists are the basis of a meal planning system designed by a committee of the American Diabetes Association and The American Dietetic Association. While designed primarily for people with diabetes and others who must follow special diets, the Exchange Lists are based on principles of good nutrition that apply to everyone. © 1989 American Diabetes Association, Inc., The American Dietetic Association.

Table C.1

The U.S.. Exchange System: Starch/Bread List

15 g carbohydrate, 3 g protein, trace fat, 80 cal

Amount	Food	Amount	Food
Cereals/Grains/Pasta		**Bread**	
½ c	Bran cereals, concentrated ⌇	½ (1 oz)	Bagels
½ c	Bran cereals, flaked ⌇	2 (⅔ oz)	Bread sticks, crisp, 4" × ½"
½ c	Bulgur, cooked	1 c	Croutons, low-fat
½ c	Cooked cereals	½	English muffins
2½ tbs	Cornmeal, dry	½ (1 oz)	Frankfurter or hamburger buns
3 tbs	Grapenuts	½ loaf	Pita, 6" across
½ c	Grits, cooked	1 (1 oz)	Plain rolls, small
¾ c	Other ready-to-eat unsweetened cereals	1 slice (1 oz)	Raisin, unfrosted
½ c	Pasta, cooked	1 slice (1 oz)	Rye, pumpernickel ⌇
1½ c	Puffed cereals	1	Tortillas, 6" across
⅓ c	Rice, white or brown, cooked	1 slice (1 oz)	White (including French, Italian)
½ c	Shredded wheat	1 slice (1 oz)	Whole-wheat
3 tbs	Wheat germ ⌇		
Dried Beans/Peas/Lentils		**Crackers/Snacks**	
¼ c	Baked beans ⌇	8	Animal crackers
⅓ c	Beans and peas, cooked, such as kidney, white, split, black-eyed ⌇	3	Graham crackers, 2½" square
		¾ oz	Matzoth
⅓ c	Lentils, cooked ⌇	5 slices	Melba toast
		24	Oyster crackers
Starchy Vegetables		3 c	Popcorn, popped, no fat added
½ c	Corn ⌇	¾ oz	Pretzels
1 cob	Corn on the cob, 6" long ⌇	4	Rye crisp, 2" × 3½"
½ c	Lima beans ⌇	6	Saltine-type crackers
½ c	Peas, green, canned or frozen ⌇	2 to 4 (¾ oz)	Whole-wheat crackers, no fat added (crisp breads)
½ c	Plantains ⌇		
1 small (3 oz)	Potatoes, baked	**Starch Foods Prepared with Fat**	
½ c	Potatoes, mashed	(Count as 1 starch/bread serving, plus 1 fat serving.)	
¾ c	Squash, winter (acorn, butternut)	1	Biscuits, 2½" across
⅓ c	Yams, sweet potatoes, plain	½ c	Chow mein noodles
		1 (2 oz)	Cornbread, 2" cube
		6	Crackers, round butter type
		10 (1½ oz)	French fries, 2" to 3½" long
		1	Muffins, plain, small
		2	Pancakes, 4" across
		¼ c	Stuffing, bread, prepared
		2	Taco shells, 6" across
		1	Waffles, 4½" square
		4 to 6 (1 oz)	Whole-wheat crackers, fat added

⌇ 3 grams or more dietary fiber per serving. Average fiber contents of whole-grain products is 2 grams per serving. For starch foods not on this list, the general rule is that ½ cup cereal, grain, or pasta is 1 serving: 1 ounce of a bread product is 1 serving.

Table C.2

U.S. Exchange System: Meat/Meat Alternate Lists

Lean meat = 7 g protein, 3 g fat, 55 cal; medium-fat meat = 7 g protein, 5 g fat, 75 cal; high-fat meat = 7 g protein, 8 g fat, 100 cal.

Category	Amount	Food	Category	Amount	Food
Lean Meat and Alternates			**Medium-Fat Meat and Alternates** *(continued)*		
Beef	1 oz	USDA Good or Choice grades of lean beef, such as round, sirloin, and flank steak; tenderloin; chipped beef 🖉	Veal	1 oz	Cutlet, ground or cubed, unbreaded
Pork	1 oz	Leak pork, such as fresh ham; canned, cured, or boiled ham 🖉, Canadian bacon 🖉, tenderloin	Poultry	1 oz	Chicken (with skin), domestic duck or goose (well-drained of fat), ground turkey
Veal	1 oz	All cuts are lean except veal cutlets (ground or cubed); examples of lean veal: chops and roasts	Fish	¼ c	Tuna 🖉, canned in oil and drained
Poultry	1 oz	Chicken, turkey, Cornish hen (without skin)		¼ c	Salmon 🖉, canned
Fish	1 oz	All fresh and frozen fish	Cheese		Skim or part-skim milk cheeses, such as:
	2 oz	Crab, lobster, scallops, shrimp, clams (fresh or canned in water 🖉)		¼ c	Ricotta
	6 medium	Oysters		1 oz	Mozzarella
	¼ c	Tuna 🖉, canned in water		1 oz	Diet cheeses 🖉 (with 56 to 80 cal/oz)
	1 oz	Herring, uncreamed or smoked	Other	1 oz	86% fat-free lunch meat 🖉
	2 medium	Sardines, canned		1	Eggs (high in cholesterol, limit to 3 per week)
Wild game	1 oz	Venison, rabbit, squirrel		¼ c	Egg substitutes with 56 to 80 cal per ¼ c
	1 oz	Pheasant, duck, goose (without skin)		4 oz	Tofu, 2½" × 2¾" × 1"
Cheese	¼ c	Any cottage cheese		1 oz	Liver, hearts, kidneys, sweetbreads (high in cholesterol)
	2 tbs	Grated parmesan	**High-Fat Meat and Alternates**[a]		
	1 oz	Diet cheeses 🖉 (with less than 55 cal/oz)	Beef	1 oz	Most USDA Prime cuts of beef, such as ribs, corned beef 🖉
Other	1 oz	95% fat-free lunch meats	Pork	1 oz	Spareribs, ground pork, pork sausages 🖉 (patties or links)
	3 whites	Egg whites	Lamb	1 oz	Patties, ground lamb
	¼ c	Egg substitutes with less than 55 cal per ¼ c	Fish	1 oz	Any fried fish product
Medium-Fat Meat and Alternates			Cheese	1 oz	All regular cheeses 🖉, such as American, blue, Cheddar, Monterey, Swiss
Beef	1 oz	Most beef products fall into this category; examples: all ground beef, roasts (rib, chuck, rump), steak (cubed, porterhouse, T-bone), meatloaf	Other	1 oz	Lunch meats 🖉, such as bologna, salami, pimento loaf
				1 oz	Sausage 🖉, such as Polish, Italian
Pork	1 oz	Most pork products fall into this category; examples: chops, loin roast, Boston butt, cutlets		1 oz	Knockwurst, smoked
				1 oz	Bratwurst 🖉
Lamb	1 oz	Most lamb products fall into this category; examples: chops, leg, roast		1 (10/lb)	Frankfurters 🖉 (turkey or chicken)
				1 tbs	Peanut butter (contains unsaturated fat)
			Count as 1 high-fat meat plus 1 fat exchange:		
			1 frank	(10/lb)	Frankfurters 🖉 (beef, pork, or combination)

🖉 400 milligrams or more sodium per exchange. Meats contribute no fiber to the diet.

[a] These items are high in saturated fat, cholesterol, and calories and should be used no more than three times per week.

C Appendix

Table C.3
U.S. Exchange System: Vegetable List

5 g carbohydrate, 2 g protein, 25 cal
All portion sizes, except as otherwise noted, are ½ c of any cooked vegetable or vegetable juice, 1 c of any raw vegetable.

Artichokes, ½ medium	Cabbage, cooked	Leeks	Spinach, cooked
Asparagus	Carrots	Mushrooms, cooked	Summer squash (crookneck)
Bean sprouts	Cauliflower	Okra	Tomatoes, 1 large
Beans (green, wax, Italian)	Eggplant	Onions	Tomato/vegetable juice ✎
Beets	Green peppers	Pea pods	Turnips
Broccoli	Greens (collard, mustard,	Rutabagas	Water chestnuts
Brussels sprouts	turnip)	Sauerkraut ✎	Zucchini, cooked
	Kohlrabi		

Starchy vegetables such as corn, peas, and potatoes are found on the Starch/Bread List.
For free vegetables, see the Free Food List (Table D.7).

✎ 400 milligrams or more sodium per serving. Most vegetable servings contain 2 to 3 grams dietary fiber.

Table C.4
U.S. Exchange System: Fruit List

15 g carbohydrate, 60 cal
All portion sizes, unless otherwise noted, are ½ c fresh fruit or fruit juice, ¼ c dried fruit.

Amount	Food	Amount	Food
Fresh, Frozen, and Unsweetened Canned Fruit		**Dried Fruit**	
1	Apples, raw, 2" across	4 rings	Apples 🖋
½ c	Applesauce, unsweetened	7 halves	Apricots 🖋
4	Apricots, medium, raw	2½ medium	Dates
½ c (4 halves)	Apricots, canned	1½	Figs 🖋
½	Bananas, 9" long	3 medium	Prunes 🖋
¾ c	Blackberries, raw 🖋	2 tbs	Raisins
¾ c	Blueberries, raw 🖋		
⅓	Cantaloupe, 5" across	**Fruit Juice**	
1 c	Cantaloupe, cubes	½ c	Apple juice/cider
12	Cherries, large, raw	⅓ c	Cranberry juice cocktail
½ c	Cherries, canned	⅓ c	Grape juice
2	Figs, raw, 2" across	½ c	Grapefruit juice
½ c	Fruit cocktail, canned	½ c	Orange juice
½	Grapefruit, medium	½ c	Pineapple juice
¾ c	Grapefruit, segments	⅓ c	Prune juice
15	Grapes, small		
⅛	Honeydew melon, medium		
1 c	Honeydew melon, cubes		
1	Kiwis, large		
¾ c	Mandarin oranges		
½	Mangoes, small		
1	Nectarines, 1½" across 🖋		
1	Oranges, 2½" across		
1 c	Papayas		
1 (¾ c)	Peaches, 2¾" across		
½ c (2 halves)	Peaches, canned		
½ large or 1 small	Pears		
½ c (2 halves)	Pears, canned		
2	Persimmons, medium, native		
¾ c	Pineapple, raw		
⅓ c	Pineapple, canned		
2	Plums, raw, 2" across		
½	Pomegranates 🖋		
1 c	Raspberries, raw 🖋		
1¼ c	Strawberries, raw, whole		
2	Tangerines, 2½" across		
1¼ c	Watermelon, cubes		

🖋 3 grams or more dietary fiber per serving. Average fiber contents of fresh, frozen, and dry fruits: 2 grams per serving.

C Appendix

Table C.5
U.S. Exchange System: Milk List

Nonfat and very low-fat milk = 12 g carbohydrate, 8 g protein, trace fat, 90 cal; low-fat milk = 12 g carbohydrate, 8 g protein, = g fat, 120 cal; whole milk = 12 g carbohydrate, 8 g protein, 8 g fat, 150 cal.

Amount	Food	Amount	Food
Nonfat and Very Low-Fat Milk		**Low-Fat Milk**	
1 c	Nonfat milk	1 c fluid	2% milk
1 c	½% milk	8 oz	Plain low-fat yogurt, with added nonfat milk solids
1 c	1% milk		
⅓ c	Dry nonfat milk	**Whole Milk**	
½ c	Evaporated nonfat milk	1 c	Whole milk
1 c	Low-fat buttermilk	½ c	Evaporated whole milk
8 oz	Plain nonfat yogurt	8 oz	Whole plain yogurt

Table C.6
U.S. Exchange System: Fat List

5 g fat, 45 cal

Amount	Food	Amount	Food
Unsaturated Fats		**Saturated Fats**	
⅛ medium	Avocados	1 slice	Bacon[a]
1 tsp	Margarine	1 tsp	Butter
1 tbs	Margarine, diet[a]	½ oz	Chitterlings
1 tsp	Mayonnaise	2 tbs	Coconut, shredded
1 tbs	Mayonnaise, reduced calorie[a]	2 tbs	Coffee whitener, liquid
	Nuts and seeds:	4 tsp	Coffee whitener, powder
6 whole	Almonds, dry roasted	1 tbs	Cream (heavy, whipping)
1 tbs	Cashews, dry roasted	2 tbs	Cream (light, coffee, table)
20 small or 10 large	Peanuts	2 tbs	Cream (sour)
2 whole	Pecans	1 tbs	Cream cheese
2 tsp	Pumpkin seeds	¼ oz	Salt pork[a]
1 tbs	Other nuts	Two tablespoons of low-calorie salad dressing is a free food.	
1 tbs	Seeds, pine nuts, sunflower seeds (without shells)		
2 whole	Walnuts		
1 tsp	Oil (corn, cottonseed, safflower, soybean, sunflower, olive, peanut)		
10 small or 5 large	Olives[a]		
1 tbs	Salad dressing, all varieties[a]		
2 tsp	Salad dressing, mayonnaise type		
1 tbs	Salad dressing, mayonnaise type, reduced calorie		
2 tbs	Salad dressing, reduced calorie		

[a]If more than one or two servings are eaten, these foods provide 400 milligrams or more sodium.

400 milligrams or more sodium per serving.

Table C.7
U.S. Exchange System: Free Foods

A free food is any food or drink that contains less than 20 cal/serving. People with diabetes are advised to eat as much as they want of those items that have no serving size specified. They may eat two or three servings per day of those items that have a specific serving size. It is suggested that they spread the servings out through the day.

Amount	Food	Amount	Food
Drinks	Bouillon, low-sodium	**Condiments**	
	Bouillon or broth without fat	1 tbs	Catsup
	Carbonated drinks, sugar-free		Horseradish
	Carbonated water		Mustard
	Club soda		Pickles, dill, unsweetened
1 tbs	Cocoa powder, unsweetened	2 tbs	Salad dressing, low-calorie
	Coffee/tea	1 tbs	Taco sauce
	Drink mixes, sugar-free		Vinegar
	Tonic water, sugar-free	**Seasonings**	Basil, fresh
Nonstick Pan Spray			Celery seeds
Fruit			Chili powder
½ c	Cranberries, unsweetened		Chives
½ c	Rhubarb, unsweetened		Cinnamon
Vegetables (raw, 1 c)	Cabbage		Curry
	Celery		Dill
	Chinese cabbage		Flavoring extracts (almond, butter, lemon, peppermint, vanilla, walnut, etc.)
	Cucumbers		
	Green onions		Garlic
	Hot peppers		Garlic powder
	Mushrooms		Herbs
	Radishes		Hot pepper sauce
	Zucchini		Lemon
Salad Greens	Endive		Lemon juice
	Escarole		Lemon pepper
	Lettuce		Lime
	Romaine		Lime juice
	Spinach		Mint
Sweet Substitutes	Candy, hard, sugar-free		Onion powder
	Gelatin, sugar-free		Oregano
	Gum, sugar-free		Paprika
2 tsp	Jam/jelly, sugar-free		Pepper
1 to 2 tbs	Pancake syrup, sugar-free		Pimento
	Sugar substitutes (saccharin, aspartame, acesulfame-K)		Soy sauce
			Soy sauce, low-sodium ("lite")
2 tbs	Whipped topping		Spices
		¼ c	Wine, used in cooking
			Worcestershire sauce

3 grams or more dietary fiber per serving.

400 milligrams or more sodium per serving.

Table C.8
U.S. Exchange Combination Foods

Much of the food we eat is mixed together in various combinations. These combination foods do not fit into only one exchange list. It can be quite hard to tell what is in a certain casserole dish or baked food item. This is a list of average values for some typical combination foods. This list will help you fit these foods into your meal plan. Ask your dietitian for information about any other foods you'd like to eat. The *American Diabetes Association/American Dietetic Association Family Cookbooks* and the *American Diabetes Associates Holiday Cookbook* have many recipes and further information about many foods, including combination foods. Check your library or local bookstore.

Food	Amount	Exchanges	Food	Amount	Exchanges
Casseroles, homemade	1 c (8 oz)	2 starch, 2 medium-fat meat, 1 fat	Spaghetti and meatballs, canned	1 c (8 oz)	2 starch, 1 medium-fat meat, 1 fat
Cheese pizza, thin crust	¼ of 15 oz or ¼ of 10"	2 starch, 1 medium-fat meat, 1 fat	Sugar-free pudding, made with nonfat milk	½ c	1 starch
Chili with beans, commercial	1 c (8 oz)	2 starch, 2 medium-fat meat, 2 fat	**If beans are used as a meat substitute:**		
Chow mein without noodles or rice	2 c (16 oz)	1 starch, 2 vegetable, 2 lean meat	Dried beans, peas, lentils	1 c (cooked)	2 starch, 1 lean meat
Macaroni and cheese	1 c (8 oz)	2 starch, 1 medium-fat meat, 2 fat			
Soups:					
Bean	1 c (8 oz)	1 starch, 1 vegetable, 1 lean meat			
Chunky, all varieties	10¾-oz can	1 starch, 1 vegetable, 1 medium-fat meat			
Cream, made with water	1 c (8 oz)	1 starch, 1 fat			
Vegetable or broth	1 c (8 oz)	1 starch			

🌾 400 milligrams or more sodium per serving.   🌾 3 grams or more dietary fiber per serving.

Table C.9
U.S. Exchange Foods for Occasional Use

The following list includes average exchange values for some foods high in sugar and fat. People are advised to use them only occasionally and in moderate amounts.

Food	Amount	Exchanges	Food	Amount	Exchanges
Angel food cake	1/12 cake	2 starch	Granola bars	1 small	1 starch, 1 fat
Cake, no icing	1/12 cake or a 3" square	2 starch, 2 fat	Ice cream, any flavor	½ c	1 starch, 2 fat
			Ice milk, any flavor	½ c	1 starch, 1 fat
Cookies	2 small, 1¾" across	1 starch, 1 fat	Sherbet, any flavor	¼ c	1 starch
Frozen fruit yogurt	⅓ c	1 starch	Snack chips, all varieties	1 oz	1 starch, 2 fat
Gingersnaps	3	1 starch	Vanilla wafers	6 small	1 starch, 1 fat
Granola	¼ c	1 starch, 1 fat			

🌾 If more than one serving is eaten, these foods have 400 milligrams or more sodium.

Table of Intentional Food Additives

Table D.1

A guide to Intentional Food Additives

Abbreviation	Type of Additive	Uses
AA	Anticaking agents	Keeps dry powders and crystals from clumping together (e.g., salt, powdered sugar).
C	Colors	Synthetic (laboratory-made) vegetable and fruit concentrates and other substances used to color foods (e.g., soft drinks, frosting).
E	Emulsifiers	Used to make oil and water mix (e.g., salad dressings, sauces).
EX	Extenders	"Fillers" such as fruit pulp and texturized protein (e.g., fruit drinks, hamburger).
FA	Flavoring agents	Used to add particular flavors to food (e.g., pudding, rye bread).
N	Nutrients	Used to add vitamins or minerals to foods (e.g., breakfast cereals, skim milk).
P	Preservatives	Used to keep food from spoiling (e.g., breads, breakfast cereals).
S	Sweeteners	Used to sweeten foods. Some are "artificial," such as aspartame and saccharin, and some are extracted from plants, such as sugar cane and sugar beets (e.g., soft drinks, catsup).
T	Texturizers	Used to improve the texture of food by stabilizing moisture content, dryness, volume, tenderness, or hardness (e.g., cakes, breads).
TH	Thickeners	Used to improve the consistency of foods (e.g., low-fat salad dressings, low-calorie jams).

Common Food Additives and Their Primary Function

Alginates (T)
Alpha tocopherol (vitamin E) (N)
Alpha tocopheryl acetate (vitamin E) (N)
Ascorbic acid (vitamin C) (P)
Aspartame (NutraSweet™) (S)
Baking powder (T)
Beet juice (C)
Beet sugar (S)
Beta-carotene (N)
BHA (P)
BHT (P)
Calcium carbonate (calcium) (N)
Calcium pantothenate (pantothenic acid, a B vitamin) (N)
Calcium propionate (P)
Calcium silicate (AA)
Cane sugar (S)

Carotene (C)
Carrageenan (E, TH)
Cellulose gum (TH)
Chromium chloride (N)
Citric acid (FA, P)
Corn syrup (S)
Cupric oxide (copper) (N)
Cyanocobalamin (vitamin B_{12}) (N)
Dextrin (TH)
Dextrose (S)
Dibasic calcium phosphate (calcium phosphorus) (N)
Diglycerides (E)
EDTA (P)
Extracts (FA)
FD&C blue no. 1 (C)
FD&C red no. 3 (use is being phased out) (C)

continued

D Appendix

Table D.1
A guide to Intentional Food
Additives, continued

Common Food Additives and Their Primary Function	
FD&C yellow no. 5 (C)	Salt (FA, P)
Ferrous fumarate (iron) (N)	Silicondioxide (AA)
Ferrous sulfate (iron) (N)	Sodium aluminum phosphate (T)
Fructose (S)	Sodium ascorbite (vitamin C) (N)
Fruit pulp (EX)	Sodium benzoate (P)
Gelatin (TH)	Sodium bicarbonate (baking soda) (T)
Glycerol (T)	Sodium bisulfite (P)
Glyceryl abietate (E, T)	Sodium chloride (P, FA)
Guar gum (T, TH)	Sodium citrate (P)
High-fructose corn syrup (S)	Sodium erythorbate (P)
Honey (S)	Sodium hexametaphosphate (P, T)
Hydrolyzed protein (EX)	Sodium metabisulfite (P)
Lecithin (E)	Sodium molydate (molybdenum) (N)
Magnesium oxide (magnesium) N	Sodium nitrate (P)
Maltodextrin (S, TH)	Sodium nitrite (P)
Manganese sulfate (manganese) (N)	Sodium propionate (P)
Modified food starch (TH)	Sodium saccharin (S)
Monoglycerides (E)	Sodium selenate (selenium) (N)
Monosodium glutamate (MSG) (FA)	Sodium stearolyn-2-lactylate (P, E)
Natural flavorings (FA)	Sodium sulfite (P)
Niacinamide (niacin, vitamin B_3) (N)	Sorbitan monostearate (E)
Nitrates (P)	Sorbitol (S, T)
Nitrites (P)	Sorghum (S, P)
Paprika (C)	Starches (TH)
Pectin (TH)	Sucrose (S)
Phosphoric acid (P)	Sugar (P)
Phytonadione (vitamin K) (N)	Sulfur dioxide (P)
Polysorbates (P)	Sweeteners (FA)
Potassium benzoate (P)	Texturized protein (EX)
Potassium bicarbonate (T)	Thiamin hydrochloride (thiamin, B_1) (N)
Potassium chloride (N)	Thiamin mononitrate (thiamin, B_1) (N)
Potassium iodite (iodine) (N)	Tocopherols (P, N)
Potassium metabisulfite (P)	Turmeric (C)
Potassium sorbate (P)	Vitamin A palmitate (vitamin A) (N)
Propyl gallate (P)	Vitamin C (ascorbic acid) (N, P)
Propyleneglycol (T)	Vitamin E (N, P)
Pyridoxine hydrochloride (vitamin B_6) (N)	Xanthan gum (T)
Reduced iron (iron) (N)	Xylitol (S)
Retinol (vitamin A) (N)	Yeast (T)
Riboflavin (vitamin B_2) (N)	Zinc oxide (zinc) (N)
Saccharin (S)	

Cells

This appendix presents an overview of the basic structure and functions of *cells* in the human body. Cell structure and function are central to discussions of nutrition for it is within cells that nutrients are utilized to sustain life and health. The life-sustaining processes that take place within each of the more than one hundred trillion cells in the human body are maintained by the nutrients we consume in our diet. Chemical reactions within the cells produce energy from carbohydrates, proteins, and fats; cause proteins to be broken down or built up; and in thousands of other ways keep us in a state of health.

A cell is a basic unit of life. Any substance that does not consist of one or more cells cannot be alive. Cells are the "building blocks" of tissues (such as muscles and bones), organs (such as the kidneys and liver), and systems (the respiratory and digestive systems, for example). Normal cell health and functioning are maintained when a state of nutritional and environmental utopia exists within and around the cells. A disruption in the availability of nutrients or the presence of harmful substances in the cell's environment can initiate disorders that eventually affect our health or growth. Health problems in general begin with disruptions in the normal activity of cells.

The types and amounts of food and supplements people consume affect the cells' environment and their ability to function normally. Excessive or inadequate supplies of nutrients and other chemical substances disrupt cell functions and result in health problems. Humans remain in a state of health as long as their cells do.

➡ **cell**

The basic unit of life, of which all living things are composed. Every cell is surrounded by a membrane and contains cytoplasm, within which are organelles and a nucleus; the cell nucleus contains chromosomes.

Cell Structures and Functions

A generalized diagram of a human cell is shown in Illustration E.1. Although all cells have some structures and functions in common, the specific functions performed, and the structures that support those functions, can vary a good deal from cell to cell. Cells lining the esophagus, stomach, and intestines, for example, are specialized to produce and secrete mucus that helps food pass through the digestive tract. Red blood cells are specially formed to transport oxygen and carbon dioxide.

Every cell is surrounded by a cell membrane that helps move nutrients into and out of the cell. Inside the cell membrane lies the cytoplasm, a fluid material that contains many organelles, tubes, and particles. Among the organelles in the cytoplasm are ribosomes, mitochondria, and lysosomes. Each of these "little organs" is encased in a cell membrane and performs specific functions. Ribosomes assemble amino acids into proteins following the instructions of DNA and its messenger RNA. The mitochondria are made of intricately folded membranes that bear thousands of highly organized sets of enzymes on their surfaces. These enzymes are actively involved in the production of energy and are found in particularly dense quantities in muscle cells. The lysosomes are like packets of enzymes. The enzymes are used to break down old cell particles that are being recycled and to destroy substances that are harmful to the body.

Cytoplasm also contains a highly organized system of membranes called the endoplasmic reticulum. When these membranes are dotted with ribosomes, they are called "rough endoplasmic reticulum." When ribosomes are absent, the endoplasmic reticulum is referred to as "smooth." Some membranes within the cytoplasm form tubes that collect certain types of cellular material and transport it out of the cell. These membranous tubes are called "Golgi apparatus." The rough and smooth endoplasmic reticulum are continuous with the Golgi apparatus, so secretions produced throughout the cell can be collected and transported to its exterior.

Within each cell is a nucleus covered by a two-layer membrane. The nucleus contains chromosomes and the genetic material DNA. DNA encodes all of the instructions a cell needs to conduct protein synthesis and to replicate life.

All cells within the body are part of a complex communication system that uses hormones, electrical impulses, and other chemical messengers to link each cell to the others. No cell is an island that operates independently from the others.

Illustration E.1
Generalized structure of a human cell.

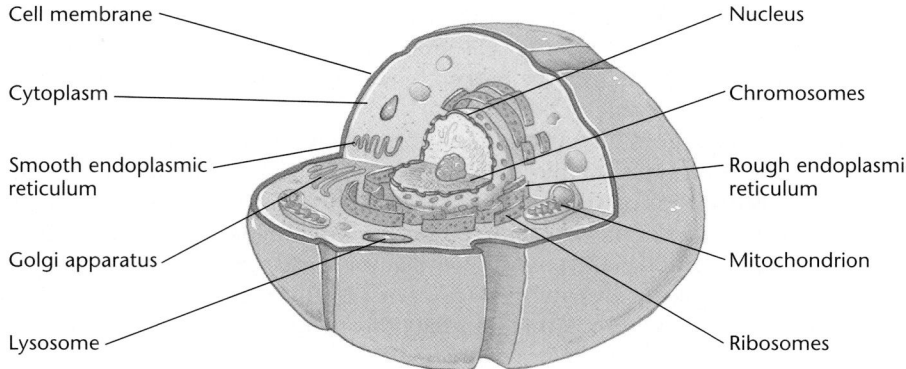

cell membrane, Nucleus, Cytoplasm, Chromosomes, Smooth endoplasmic reticulum, Rough endoplasmic reticulum, Golgi apparatus, Mitochondrion, Lysosome, Ribosomes

cell membrane: the membrane that surrounds the cell and encloses its contents; made primarily of lipid and protein.

chromosomes: a set of structures within the nucleus of every cell that contain the cell's genetic material, DNA, associated with other materials (primarily proteins).

cytoplasm (SIGH-toe-plazm): the cell contents, except for the nucleus.
cyto = cell
plasm = a form

Golgi (GOAL-gee) **apparatus:** a set of membranes within the cell where secretory materials are packaged for export.

lysosomes: cellular organelles; membrane-enclosed sacs of degradative enzymes.
lysis = dissolution

mitochondria (my-toe-KON-dree-uh; *singular* **mitochondrion**): the cellular organelles responsible for producing ATP aerobically; made of membranes (lipid and protein) with enzymes mounted on them.
mitos = thread (referring to their slender shape)

chondros = cartilage (referring to their external appearance)

nucleus: a major membrane-enclosed body within every cell, which contains the cell's genetic material, DNA, embedded in chromosomes.
nucleus = a kernel

organelles: membrane-bound subcellular structures such as ribosomes, mitochondria, and lysosomes.
organelle = little organ

ribosomes: protein-making organelles in cells; composed of RNA and protein.
ribo = containing the sugar ribose
some = body

rough endoplasmic reticulum (en-doh-PLAZ-mic reh-TIC-you-lum): intracellular membranes dotted with ribosomes, where protein synthesis takes place.
endo = inside
plasm = the cytoplasm

smooth endoplasmic reticulum: smooth intracellular membranes bearing no ribosomes.

WHO: Nutrition Recommendations Canada: Choice System and Guidelines

This appendix first presents nutrition recommendations from the World Health Organization (WHO) and then provides details for Canadians on Canada's *Food Guide to Healthy Eating* and on the exchange system (called the choice system).

Nutrition Recommendations from WHO

The World Health Organization (WHO) has assessed the relationships between diet and the development of chronic diseases. Its recommendations include:

- Total energy: sufficient to support normal growth, physical activity, and healthy body weight (body mass index = 20 to 22).
- Total fat: 15 to 30 percent of total energy.
- Saturated fat: less than 10 percent of total energy.
- Total carbohydrate: 55 to 75 percent of total energy.
- Added sugars: less than 10 percent of total energy.
- Protein: 10 to 15 percent of total energy.
- Salt: less than 5 grams/day, preferably iodized.
- Fruit and vegetables: at least 400 grams (almost 1 pound) daily.
- Physical activity: one hour per day of moderate intensity on most days of the week.

Canada's *Food Guide to Healthy Eating*

Figure F-1 presents the 1992 Canada's *Food Guide to Healthy Eating,* which interprets Canada's *Guidelines for Healthy Eating* (see Table 2-1 on p. 42) for consumers and recommends a range of servings to consume daily from each of the four food groups. The following publications, which are available from Health Canada, through its website, explain how to use the *Guide: Using the Food Guide; Food Guide Facts: Background for Educators and Communicators; Canada's Food Guide to Healthy Eating—Focus on Preschoolers: Background for Educators and Communicators;* and *Canada's Food Guide to Healthy Eating—Focus on Children Six to Twelve Years: Background for Educators and Communicators.* Figure F-2 presents Canada's Physical Activity Guide.

Canada's *Guidelines for Healthy Eating* and Canada's *Food Guide to Healthy Eating* are being reviewed for consistency with the new Dietary Reference Intakes. Check the website for the Health Canada Office of Nutrition Policy and Promotion, **www.hc-sc.gc.ca/hpfb-dgpsa/onpp-bppn/**, for the status of the review.

F Appendix

FIGURE F-1 Canada's *Food Guide to Healthy Eating*

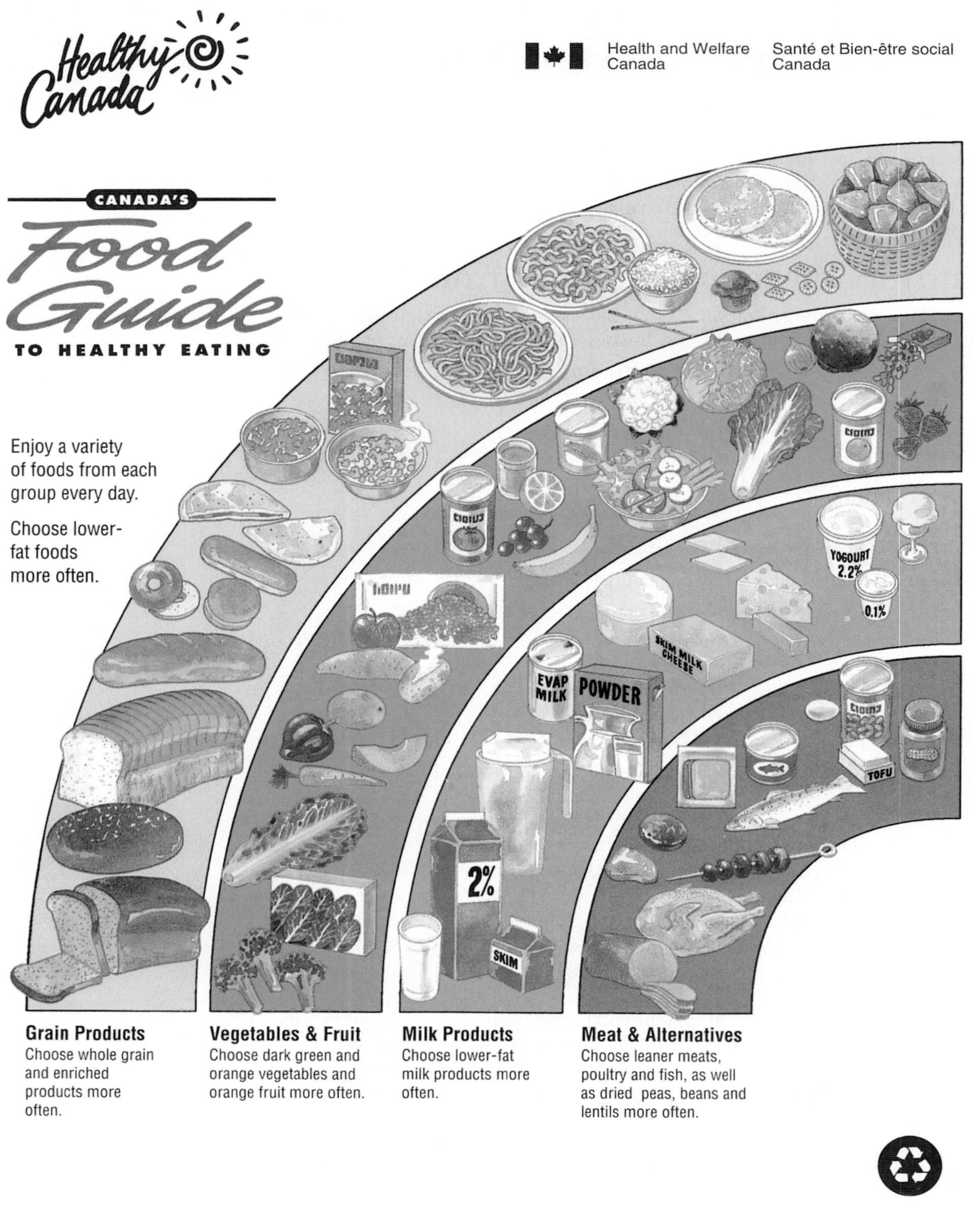
Healthy Canada
Health and Welfare Canada
Santé et Bien-être social Canada
CANADA'S
Food Guide
TO HEALTHY EATING
Enjoy a variety of foods from each group every day.
Choose lower-fat foods more often.
YOGOURT 2.2%
0.1%
SKIM MILK CHEESE
EVAP MILK
POWDER
2%
SKIM
TOFU
Grain Products
Choose whole grain and enriched products more often.
Vegetables & Fruit
Choose dark green and orange vegetables and orange fruit more often.
Milk Products
Choose lower-fat milk products more often.
Meat & Alternatives
Choose leaner meats, poultry and fish, as well as dried peas, beans and lentils more often.

Appendix F

FIGURE F-1 Canada's *Food Guide to Healthy Eating*—continued

CANADA'S Food Guide TO HEALTHY EATING FOR PEOPLE FOUR YEARS AND OVER

Different People Need Different Amounts of Food

The amount of food you need every day from the 4 food groups and other foods depends on your age, body size, activity level, whether you are male or female and if you are pregnant or breast-feeding. That's why the Food Guide gives a lower and higher number of servings for each food group. For example, young children can choose the lower number of servings, while male teenagers can go to the higher number. Most other people can choose servings somewhere in between.

Grain Products
5–12 SERVINGS PER DAY

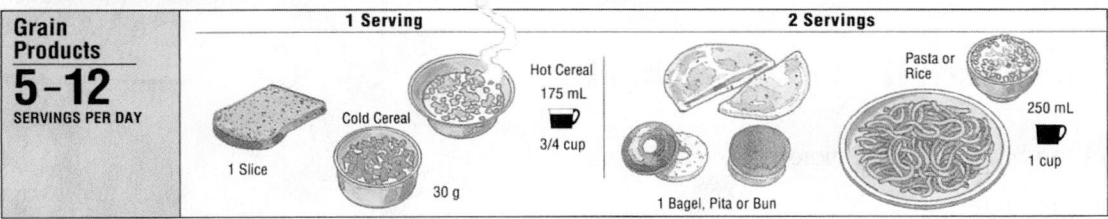

Vegetables & Fruit
5–10 SERVINGS PER DAY

Milk Products
SERVINGS PER DAY
Children 4–9 years: 2–3
Youth 10–16 years: 3–4
Adults: 2–4
Pregnant & Breast-feeding Women: 3–4

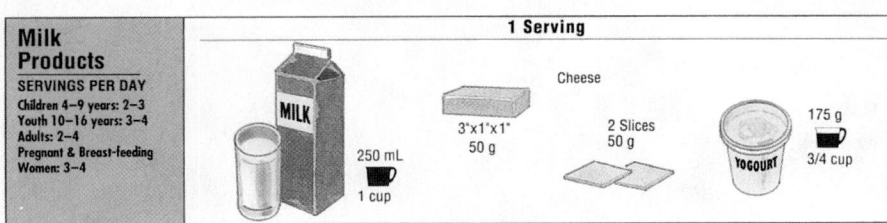

Other Foods

Taste and enjoyment can also come from other foods and beverages that are not part of the 4 food groups. Some of these foods are higher in fat or Calories, so use these foods in moderation.

Meat & Alternatives
2–3 SERVINGS PER DAY

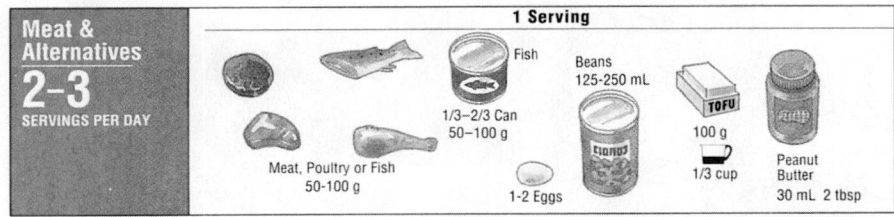

Enjoy eating well, being active and feeling good about yourself. That's VITALIT

© Minister of Supply and Services Canada 1992   Cat. No. H39-252/1992E   No changes permitted. Reprint permission not required.
ISBN 0-662-19648-1

F Appendix

FIGURE F-2 Canada's Physical Activity Guide

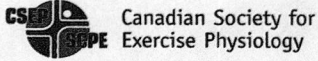

CANADA'S
Physical Activity Guide
to Healthy Active Living
Physical activity improves health.
Every little bit counts, but more is even better – everyone can do it!
Get active your way – build physical activity into your daily life...
• at home
• at school
• at work
• at play
• on the way
...that's active living!
TAKE THE STAIRS
Increase
Endurance Activities
Increase
Flexibility Activities
Increase
Strength Activities
Reduce
Sitting for long periods
Health Canada
Santé Canada
CSEP SCPE
Canadian Society for Exercise Physiology

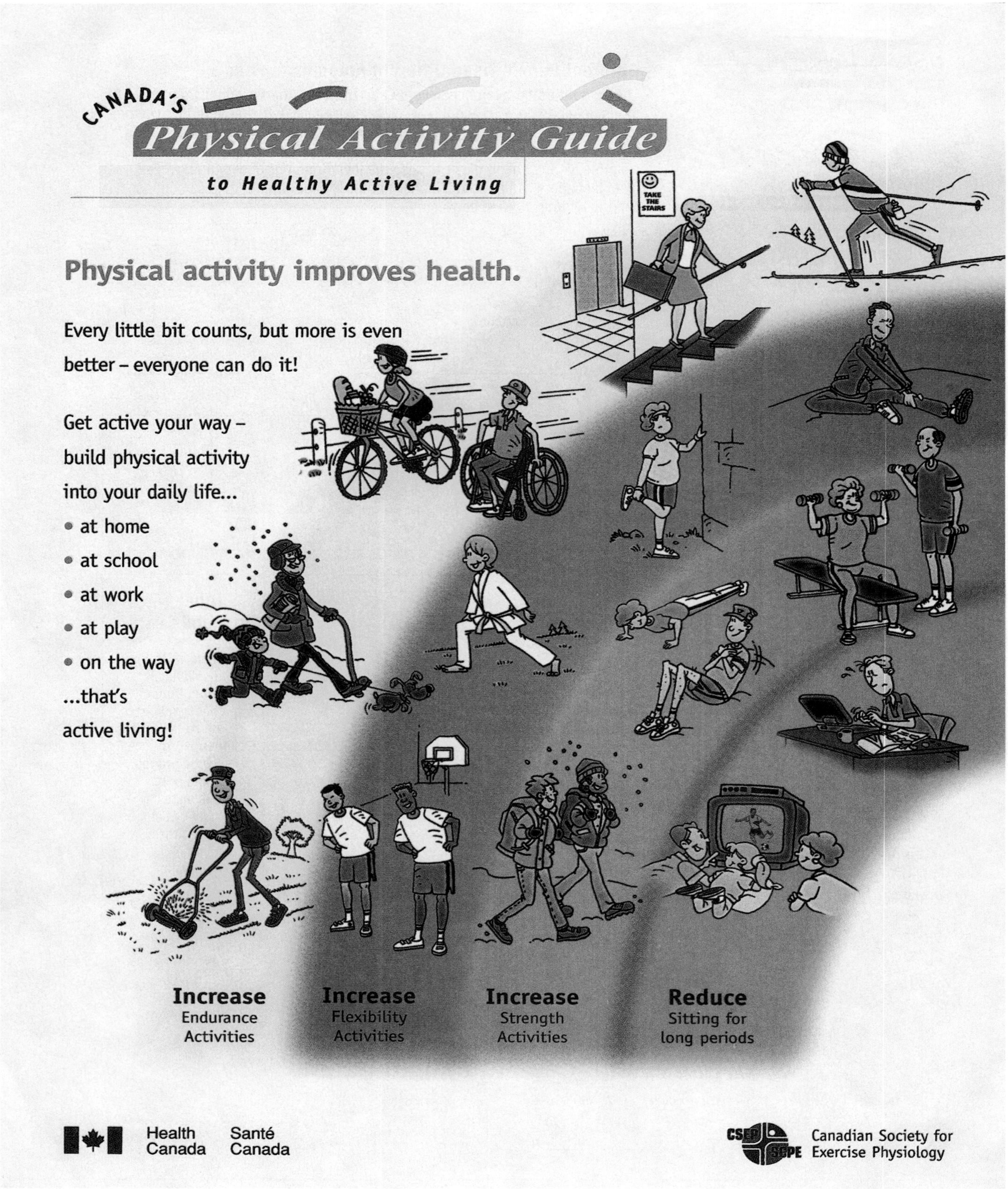

FIGURE F-2   Canada's Physical Activity Guide—continued

Choose a variety of activities from these three groups:

Endurance

4-7 days a week
Continuous activities for your heart, lungs and circulatory system.

Flexibility

4-7 days a week
Gentle reaching, bending and stretching activities to keep your muscles relaxed and joints mobile.

Strength

2-4 days a week
Activities against resistance to strengthen muscles and bones and improve posture.

Starting slowly is very safe for most people. Not sure? Consult your health professional.

For a copy of the *Guide Handbook* and more information: **1-888-334-9769**, or **www.paguide.com**

Eating well is also important. Follow *Canada's Food Guide to Healthy Eating* to make wise food choices.

Get Active Your Way, Every Day—For Life!

Scientists say accumulate 60 minutes of physical activity every day to stay healthy or improve your health. As you progress to moderate activities you can cut down to 30 minutes, 4 days a week. Add-up your activities in periods of at least 10 minutes each. Start slowly... and build up.

Time needed depends on effort

Very Light Effort	Light Effort *60 minutes*	Moderate Effort *30-60 minutes*	Vigorous Effort *20-30 minutes*	Maximum Effort
• Strolling	• Light walking	• Brisk walking	• Aerobics	• Sprinting
• Dusting	• Volleyball	• Biking	• Jogging	• Racing
	• Easy gardening	• Raking leaves	• Hockey	
	• Stretching	• Swimming	• Basketball	
		• Dancing	• Fast swimming	
		• Water aerobics	• Fast dancing	

Range needed to stay healthy

You Can Do It – Getting started is easier than you think

Physical activity doesn't have to be very hard. Build physical activities into your daily routine.

- Walk whenever you can – get off the bus early, use the stairs instead of the elevator.
- Reduce inactivity for long periods, like watching TV.
- Get up from the couch and stretch and bend for a few minutes every hour.
- Play actively with your kids.
- Choose to walk, wheel or cycle for short trips.

- Start with a 10 minute walk – gradually increase the time.
- Find out about walking and cycling paths nearby and use them.
- Observe a physical activity class to see if you want to try it.
- Try one class to start, you don't have to make a long-term commitment.
- Do the activities you are doing now, more often.

Benefits of regular activity:

- better health
- improved fitness
- better posture and balance
- better self-esteem
- weight control
- stronger muscles and bones
- feeling more energetic
- relaxation and reduced stress
- continued independent living in later life

Health risks of inactivity:

- premature death
- heart disease
- obesity
- high blood pressure
- adult-onset diabetes
- osteoporosis
- stroke
- depression
- colon cancer

No changes permitted. Permission to photocopy this document in its entirety not required.
Cat. No. H39-429/1998-1E ISBN 0-662-86627-7

F Appendix

Canada's Choice System for Meal Planning

The *Good Health Eating Guide* is the Canadian choice system of meal planning.[1] It contains several features similar to those of the U.S. exchange system including the following:

- Foods are divided into lists according to carbohydrate, protein, and fat content.
- Foods are interchangeable within a group.
- Most foods are eaten in measured amounts.
- An energy value is given for each food group.

Tables F-1 through F-8 present the Canadian choice system.

---

[1]The tables for the Canadian choice system are adapted from the *Good Health Eating Guide Resource*, copyright 1994, with permission of the Canadian Diabetes Association.

TABLE F-1 Canadian Choice System: Starch Foods

1 starch choice = 15 g carbohydrate (starch), 2 g protein, 290 kJ (68 kcal)

Food	Measure	Mass (Weight)
Breads		
Bagels	½	30 g
Bread crumbs	50 mL (¼ c)	30 g
Bread cubes	250 mL (1 c)	30 g
Bread sticks	2	20 g
Brewis, cooked	50 mL (¼ c)	45 g
Chapati	1	20 g
Cookies, plain	2	20 g
English muffins, crumpets	½	30 g
Flour	40 mL (2½ tbs)	20 g
Hamburger buns	½	30 g
Hot dog buns	½	30 g
Kaiser rolls	½	30 g
Matzo, 15 cm	1	20 g
Melba toast, rectangular	4	15 g
Melba toast, rounds	7	15 g
Pita, 20 cm (8") diameter	¼	30 g
Pita, 15 cm (6") diameter	½	30 g
Plain rolls	1 small	30 g
Pretzels	7	20 g
Raisin bread	1 slice	30 g
Rice cakes	2	30 g
Roti	1	20 g
Rusks	2	20 g
Rye, coarse or pumpernickel	½ slice	30 g
Soda crackers	6	20 g
Tortillas, corn (taco shell)	1	30 g
Tortilla, flour	1	30 g
White (French and Italian)	1 slice	25 g
Whole-wheat, cracked-wheat, rye, white enriched	1 slice	30 g
Cereals		
Bran flakes, 100% bran	125 mL (½ c)	30 g
Cooked cereals, cooked	125 mL (½ c)	125 g
Dry	30 mL (2 tbs)	20 g

(continued on the next page)

TABLE F-1   Canadian Choice System: Starch Foods—continued

1 starch choice = 15 g carbohydrate (starch), 2 g protein, 290 kJ (68 kcal)

Food	Measure	Mass (Weight)
Cornmeal, cooked	125 mL (½ c)	125 g
Dry	30 mL (2 tbs)	20 g
Ready-to-eat unsweetened cereals	125 mL (½ c)	20 g
Shredded wheat biscuits, rectangular or round	1	20 g
Shredded wheat, bite size	125 mL (½ c)	20 g
Wheat germ	75 mL (⅓ c)	30 g
Cornflakes	175 mL (⅔ c)	20 g
Rice Krispies	175 mL (⅔ c)	20 g
Cheerios	200 mL (¾ c)	20 g
Muffets	1	20 g
Puffed rice	300 mL (1¼ c)	15 g
Puffed wheat	425 mL (1⅔ c)	20 g

Grains

Food	Measure	Mass (Weight)
Barley, cooked	125 mL (½ c)	120 g
Dry	30 mL (2 tbs)	20 g
Bulgur, kasha, cooked, moist	125 mL (½ c)	70 g
Cooked, crumbly	75 mL (⅓ c)	40 g
Dry	30 mL (2 tbs)	20 g
Rice, cooked, brown & white (short & long grain)	125 mL (½ c)	70 g
Rice, cooked, wild	75 mL (⅓ c)	70 g
Tapioca, pearl and granulated, quick cooking, dry	30 mL (2 tbs)	15 g
Couscous, cooked moist	125 mL (½ c)	70 g
Dry	30 mL (tbs)	20 g
Quinoa, cooked moist	125 mL (½ c)	70 g
Dry	30 mL (2 tbs)	20 g

Pastas

Food	Measure	Mass (Weight)
Macaroni, cooked	125 mL (½ c)	70 g
Noodles, cooked	125 mL (½ c)	80 g
Spaghetti, cooked	125 mL (½ c)	70 g

Starchy Vegetables

Food	Measure	Mass (Weight)
Beans and peas, dried, cooked	125 mL (½ c)	80 g
Breadfruit	1 slice	75 g
Corn, canned, whole kernel	125 mL (½ c)	85 g
Corn on the cob	½ medium cob	140 g
Cornstarch	30 mL (2 tbs)	15 g
Plantains	⅓ small	50 g
Popcorn, air-popped, unbuttered	750 mL (3 c)	20 g
Potatoes, whole (with or without skin)	½ medium	95 g
Yams, sweet potatoes (with or without skin)	½	75 g

Food	Choices per Serving	Measure	Mass (Weight)
Note: Food items found in this category provide more than 1 starch choice:			
Bran flakes	1 starch + ½ sugar	150 mL (⅔ c)	24 g
Croissant, small	1 starch + 1½ fats	1 small	35 g
Large	1 starch + 1½ fats	½ large	30 g
Corn, canned creamed	1 starch + ½ fruits and vegetables	12 mL (½ c)	113 g
Potato chips	1 starch + 2 fats	15 chips	30 g
Tortilla chips (nachos)	1 starch + 1½ fats	13 chips	20 g
Corn chips	1 starch + 2 fats	30 chips	30 g

(continued on the next page)

F Appendix

TABLE F-1 | Canadian Choice System: Starch Foods—continued

1 starch choice = 15 g carbohydrate (starch), 2 g protein, 290 kJ (68 kcal)

Food	Choices per Serving	Measure	Mass (Weight)
Cheese twists	1 starch + 1½ fats	30 chips	30 g
Cheese puffs	1 starch + 2 fats	27 chips	30 g
Tea biscuit	1 starch + 2 fats	1	30 g
Pancakes, homemade using 50 mL (¼ c) batter (6″ diameter)	1½ starches + 1 fat	1 medium	50 g
Potatoes, french fried (homemade or frozen)	1 starch + 1 fat	10 regular size	35 g
Soup, canned* (prepared with equal volume of water)	1 starch	250 mL (1 c)	260 g
Waffles, packaged	1 starch + 1 fat	1	35 g

*Soup can vary according to brand and type. Check the label for Food Choice Values and Symbols or the core nutrient listing.

TABLE F-2 | Canadian Choice System: Fruits and Vegetables

1 fruits and vegetables choice = 10 g carbohydrate, 1 g protein, 190 kJ (44 kcal)

Food	Measure	Mass (Weight)
Fruits (fresh, frozen, without sugar, canned in water)		
Apples, raw (with or without skin)	½ medium	75 g
Sauce unsweetened	125 mL (½ c)	120 g
Sweetened	see *Combined Food Choices (Table I-8)*	
Apple butter	20 mL (4 tsp)	20 g
Apricots, raw	2 medium	115 g
Canned, in water	4 halves, plus 30 mL (2 tbs) liquid	110 g
Bake-apples (cloudberries), raw	125 mL (½ c)	120 g
Bananas, with peel	½ small	75 g
Peeled	½ small	50 g
Berries (blackberries, blueberries, boysenberries, huckleberries, loganberries, raspberries)		
Raw	125 mL (½ c)	70 g
Canned, in water	125 mL (½ c), plus 30 mL (2 tbs) liquid	100 g
Cantaloupe, raw, with rind	¼ wedge	240 g
Cubed or diced	250 mL (1 c)	160 g
Cherries, raw, with pits	10	75 g
Raw, without pits	10	70 g
Canned, in water, with pits	75 mL (⅓ c), plus 30 mL (2 tbs) liquid	90 g
Canned, in water, without pits	75 mL (⅓ c), plus 30 mL (2 tbs) liquid	85 g
Crabapples, raw	1 small	55 g
Cranberries, raw	250 mL (1 c)	100 g
Figs, raw	1 medium	50 g
Canned, in water	3 medium, plus 30 mL (2 tbs) liquid	100 g
Foxberries, raw	250 mL (1 c)	100 g
Fruit cocktail, canned, in water	125 mL (½ c), plus 30 mL (2 tbs) liquid	120 g
Fruit, mixed, cut-up	125 mL (½ c)	120 g
Gooseberries, raw	250 mL (1 c)	150 g
Canned, in water	250 mL (1 c), plus 30 mL (2 tbs) liquid	230 g
Grapefruit, raw, with rind	½ small	185 g
Raw, sectioned	125 mL (½ c)	100 g
Canned, in water	125 mL (½ c), plus 30 mL (2 tbs) liquid	120 g

(continued on the next page)

TABLE F-2 Canadian Choice System: Fruits and Vegetables—continued

1 fruits and vegetables choice = 10 g carbohydrate, 1 g protein, 190 kJ (44 kcal)

Food	Measure	Mass (Weight)
Grapes, raw, slip skin	125 mL (½ c)	75 g
Raw, seedless	125 mL (½ c)	75 g
Canned, in water	75 mL (⅓ c), plus 30 mL (2 tbs) liquid	115 g
Guavas, raw	½	50 g
Honeydew melon, raw, with rind	½	225 g
Cubed or diced	250 mL (1 c)	170 g
Kiwis, raw, with skin	2	155 g
Kumquats, raw	3	60 g
Loquats, raw	8	130 g
Lychee fruit, raw	8	120 g
Mandarin oranges, raw, with rind	1	135 g
Raw, sectioned	125 mL (½ c)	100 g
Canned, in water	125 mL (½ c), plus 30 mL (2 tbs) liquid	100 g
Mangoes, raw, without skin and seed	⅓	65 g
Diced	75 mL (⅓ c)	65 g
Nectarines	½ medium	75 g
Oranges, raw, with rind	1 small	130 g
Raw, sectioned	125 mL (½ c)	95 g
Papayas, raw, with skin and seeds	¼ medium	150 g
Raw, without skin and seeds	¼ medium	100 g
Cubed or diced	125 mL (½ c)	100 g
Peaches, raw, with seed and skin	1 large	100 g
Raw, sliced or diced	125 mL (½ c)	100 g
Canned in water, halves or slices	125 mL (½ c), plus 30 mL (2 tbs) liquid	120 g
Pears, raw, with skin and core	½	90 g
Raw, without skin and core	½	85 g
Canned, in water, halves	1 half plus 30 mL (2 tbs) liquid	90 g
Persimmons, raw, native	1	30 g
Raw, Japanese	¼	50 g
Pineapple, raw	1 slice	75 g
Raw, diced	125 mL (½ c)	75 g
Canned, in juice, diced	75 mL (⅓ c), plus 15 mL (1 tbs) liquid	55 g
Canned, in juice, sliced	1 slice, plus 15 mL (1 tbs) liquid	55 g
Canned, in water, diced	125 mL (½ c), plus 30 mL (2 tbs) liquid	100 g
Canned, in water, sliced	2 slices, plus 15 mL (1 tbs) liquid	100 g
Plums, raw	2 small	60 g
Damson	6	65 g
Japanese	1	70 g
Canned, in apple juice	2, plus 30 mL (2 tbs) liquid	70 g
Canned, in water	3, plus 30 mL (2 tbs) liquid	100 g
Pomegranates, raw	½	140 g
Strawberries, raw	250 mL (1 c)	150 g
Frozen/canned, in water	250 mL (1 c), plus 30 mL (2 tbs) liquid	240 g
Rhubarb	250 mL (1 c)	150 g
Tangelos, raw	1	205 g
Tangerines, raw	1 medium	115 g
Raw, sectioned	125 mL (½ c)	100 g

(continued on the next page)

TABLE F-2 — Canadian Choice System: Fruits and Vegetables—continued

1 fruits and vegetables choice = 10 g carbohydrate, 1 g protein, 190 kJ (44 kcal)

Food	Measure	Mass (Weight)
Watermelon, raw, with rind	1 wedge	310 g
Cubed or diced	250 mL (1 c)	160 g
Dried Fruit		
Apples	5 pieces	15 g
Apricots	4 halves	15 g
Banana flakes	30 mL (2 tbs)	15 g
Currants	30 mL (2 tbs)	15 g
Dates, without pits	2	15 g
Peaches	½	15 g
Pears	½	15 g
Prunes, raw, with pits	2	15 g
Raw, without pits	2	10 g
Stewed, no liquid	2	20 g
Stewed, with liquid	2, plus 15 mL (1 tbs) liquid	35 g
Raisins	30 mL (2 tbs)	15 g
Juices (no sugar added or unsweetened)		
Apricot, grape, guava, mango, prune	50 mL (¼ c)	55 g
Apple, carrot, papaya, pear, pineapple, pomegranate	75 mL (⅓ c)	80 g
Cranberry	see *Sugars (Table I-4)*	
Clamato	see *Sugars (Table I-4)*	
Grapefruit, loganberry, orange, raspberry, tangelo, tangerine	125 mL (½ c)	130 g
Tomato, tomato-based mixed vegetables	250 mL (1 c)	255 g
Vegetables (fresh, frozen, or canned)		
Artichokes, French, globe	2 small	50 g
Beets, diced or sliced	125 mL (½ c)	85 g
Carrots, diced, cooked or uncooked	125 mL (½ c)	75 g
Chestnuts, fresh	5	20 g
Parsnips, mashed	125 mL (½ c)	80 g
Peas, fresh or frozen	125 mL (½ c)	80 g
Canned	75 mL (⅓ c)	55 g
Pumpkin, mashed	125 mL (½ c)	45 g
Rutabagas, mashed	125 mL (½ c)	85 g
Sauerkraut	250 mL (1 c)	235 g
Snow peas	250 mL (1 c)	135 g
Squash, yellow or winter, mashed	125 mL (½ c)	115 g
Succotash	75 mL (⅓ c)	55 g
Tomatoes, canned	250 mL (1 c)	240 g
Tomato paste	50 mL (¼ c)	55 g
Tomato sauce*	75 mL (⅓ c)	100 g
Turnips, mashed	125 mL (½ c)	115 g
Vegetables, mixed	125 mL (½ c)	90 g
Water chestnuts	8 medium	50 g

*Tomato sauce varies according to brand name. Check the label or discuss with your dietitian.

TABLE F-3 — Canadian Choice System: Milk

Type of Milk	Carbohydrate (g)	Protein (g)	Fat (g)	Energy
Nonfat (0%)	6	4	0	170 kJ (40 kcal)
1%	6	4	1	206 kJ (49 kcal)
2%	6	4	2	244 kJ (58 kcal)
Whole (4%)	6	4	4	319 kJ (76 kcal)

Food	Measure	Mass (Weight)
Buttermilk (higher in salt)	125 mL (½ c)	125 g
Evaporated milk	50 mL (¼ c)	50 g
Milk	125 mL (½ c)	125 g
Powdered milk, regular	30 mL (2 tbs)	15 g
Instant	50 mL (¼ c)	15 g
Plain yogurt	125 mL (½ c)	125 g

Food	Choices per Serving	Measure	Mass (Weight)
Note: Food items found in this category provide more than 1 milk choice:			
Milk shake	1 milk + 3 sugars + ½ protein	250 mL (1 c)	300 g
Chocolate milk, 2%	2 milks 2% + 1 sugar	250 mL (1 c)	300 g
Frozen yogurt	1 milk + 1 sugar	125 mL (½ c)	125 g

TABLE F-4   Canadian Choice System: Sugars

1 sugar choice = 10 g carbohydrate (sugar), 167 kJ (40 kcal)

Food	Measure	Mass (Weight)
Beverages		
Condensed milk	15 mL (1 tbs)	
Flavoured fruit crystals*	75 mL (⅓ c)	
Iced tea mixes*	75 mL (⅓ c)	
Regular soft drinks	125 mL (½ c)	
Sweet drink mixes*	75 mL (⅓ c)	
Tonic water	125 mL (½ c)	
*These beverages have been made with water.		
Miscellaneous		
Bubble gum (large square)	1 piece	5 g
Cranberry cocktail	75 mL (⅓ c)	80 g
Cranberry cocktail, light	350 mL (1⅓ c)	260 g
Cranberry sauce	30 mL (2 tbs)	
Hard candy mints	2	5 g
Honey, molasses, corn & cane syrup	10 mL (2 tsp)	15 g
Jelly bean	4	10 g
Licorice	1 short stick	10 g
Marshmallows	2 large	15 g
Popsicle	1 stick (½ popsicle)	
Powdered gelatin mix		
(Jello®) (reconstituted)	50 mL (¼ c)	
Regular jam, jelly, marmalade	15 mL (1 tbs)	
Sugar, white, brown, icing, maple	10 mL (2 tsp)	10 g
Sweet pickles	2 small	100 g
Sweet relish	30 mL (2 tbs)	

Food	Choices per Serving	Measures	Mass (Weight)
The following food items provide more than 1 sugar choice:			
Brownie	1 sugar + 1 fat	1	20 g
Clamato juice	1½ sugars	175 mL (⅔ c)	
Fruit salad, light syrup	1 sugar + 1 fruits & vegetables	125 mL (½ c)	130 g
Aero® bar	2½ sugars + 2½ fats	1 bar	43 g
Smarties®	4½ sugars + 2 fats	1 box	60 g
Sherbet	3 sugars + ½ fat	125 mL (½ c)	95 g

TABLE F-5 Canadian Choice System: Protein Foods

1 protein choice = 7 g protein, 3 g fat, 230 kJ (55 kcal)

Food	Measure	Mass (Weight)
Cheese		
Low-fat cheese, about 7% milk fat	1 slice	30 g
Cottage cheese, 2% milkfat or less	50 mL (¼ c)	55 g
Ricotta, about 7% milkfat	50 mL (¼ c)	60 g
Fish		
Anchovies	see *Extras (Table I-7)*	
Canned, drained (e.g., mackerel, salmon, tuna packed in water)	(⅓ of 6.5 oz can)	30 g
Cod tongues, cheeks	75 mL (⅓ c)	50 g
Fillet or steak (e.g., Boston blue, cod, flounder, haddock, halibut, mackerel, orange roughy, perch, pickerel, pike, salmon, shad, snapper, sole, swordfish, trout, tuna, whitefish)	1 piece	30 g
Herring	⅓ fish	30 g
Sardines, smelts	2 medium or 3 small	30 g
Squid, octopus	50 mL (¼ c)	40 g
Shellfish		
Clams, mussels, oysters, scallops, snails	3 medium	30 g
Crab, lobster, flaked	50 mL (¼ c)	30 g
Shrimp, fresh	5 large	30 g
Frozen	10 medium	30 g
Canned	18 small	30 g
Dry pack	50 mL (¼ c)	30 g
Meat and Poultry (e.g., beef, chicken, goat, ham, lamb, pork, turkey, veal, wild game)		
Back, peameal bacon	3 slices, thin	30 g
Chop	½ chop, with bone	40 g
Minced or ground, lean or extra-lean	30 mL (2 tbs)	30 g
Sliced, lean	1 slice	30 g
Steak, lean	1 piece	30 g
Organ Meats		
Hearts, liver	1 slice	30 g
Kidneys, sweetbreads, chopped	50 mL (¼ c)	30 g
Tongue	1 slice	30 g
Tripe	5 pieces	60 g
Soyabean		
Bean curd or tofu	½ block	70 g
Eggs		
In shell, raw or cooked	1 medium	50 g
Without shell, cooked or poached in water	1 medium	45 g
Scrambled	50 mL (¼ c)	55 g

Note: The following choices provide more than 1 protein exchange:

Food	Choices per Serving	Measures	Mass (Weight)
Cheese			
Cheeses	1 protein + 1 fat	1 piece	25 g
Cheese, coarsely grated (e.g., cheddar)	1 protein + 1 fat	50 mL (¼ c)	25 g

(continued on the next page)

F — Appendix

TABLE F-5   Canadian Choice System: Protein Foods—continued

1 protein choice = 7 g protein, 3 g fat, 230 kJ (55 kcal)
Note: The following choices provide more than 1 protein exchange:

Food	Choices per Serving	Measures	Mass (Weight)
Cheese, dry, finely grated (e.g., parmesan)	1 protein + 1 fat	45 mL	15 g
Cheese, ricotta, high fat	1 protein + 1 fat	50 mL (¼ c)	55 g
Fish			
Eel	1 protein + 1 fat	1 slice	50 g
Meat			
Bologna	1 protein + 1 fat	1 slice	20 g
Canned lunch meats	1 protein + 1 fat	1 slice	20 g
Corned beef, canned	1 protein + 1 fat	1 slice	25 g
Corned beef, fresh	1 protein + 1 fat	1 slice	25 g
Ground beef, medium-fat	1 protein + 1 fat	30 mL (2 tbs)	25 g
Meat spreads, canned	1 protein + 1 fat	45 mL	35 g
Mutton chop	1 protein + 1 fat	½ chop, with bone	35 g
Paté	see *Fats and Oils (Table I-6)*		
Sausages, garlic, Polish or knockwurst	1 protein + 1 fat	1 slice	50 g
Sausages, pork, links	1 protein + 1 fat	1 link	25 g
Spareribs or shortribs, with bone	1 protein + 1 fat	1 large	65 g
Stewing beef	1 protein + 1 fat	1 cube	25 g
Summer sausage or salami	1 protein + 1 fat	1 slice	40 g
Wieners, hot dog	1 protein + 1 fat	½ medium	25 g
Miscellaneous			
Blood pudding	1 protein + 1 fat	1 slice	25 g
Peanut butter	1 protein + 1 fat	15 mL (1 tbs)	15 g

Appendix **F**

TABLE F-6 Canadian Choice System: Fats and Oils

1 fat choice = 5 g fat, 190 kJ (45 kcal)

Food	Measure	Mass (Weight)	Food	Measure	Mass (Weight)
Avocado*	⅛	30 g	Nuts (continued):		
Bacon, side, crisp*	1 slice	5 g	Sesame seeds	15 mL (1 tbs)	10 g
Butter*	5 mL (1 tsp)	5 g	Sunflower seeds		
Cheese spread	15 mL (1 tbs)	15 g	Shelled	15 mL (1 tbs)	10 g
Coconut, fresh*	45 mL (3 tbs)	15 g	In shell	45 mL (3 tbs)	15 g
Coconut, dried*	15 mL (1 tbs)	10 g	Walnuts	4 halves	10 g
Cream, Half and half			Oil, cooking and salad	5 mL (1 tsp)	5 g
(cereal), 10%*	30 mL (2 tbs)	30 g	Olives, green	10	45 g
Light (coffee), 20%*	15 mL (1 tbs)	15 g	Ripe black	7	57 g
Whipping, 32 to 37%*	15 mL (1 tbs)	15 g	Pâté, liverwurst,	15 mL (1 tbs)	15 g
Cream cheese*	15 mL (1 tbs)	15 g	meat spreads		
Gravy*	30 mL (2 tbs)	30 g	Salad dressing: blue,	10 mL (2 tsp)	10 g
Lard*	5 mL (1 tsp)	5 g	French, Italian,		
Margarine	5 mL (1 tsp)	5 g	mayonnaise,		
Nuts, shelled:			Thousand Island	5 mL (1 tsp)	5 g
Almonds	8	5 g	Salad dressing,	30 mL (2 tbs)	30 g
Brazil nuts	2	10 g	low-kcalorie		
Cashews	5	10 g	Salt pork, raw	5 mL (1 tsp)	5 g
Filberts, hazelnuts	5	10 g	or cooked*		
Macadamia	3	5 g	Sesame oil	5 mL (1 tsp)	5 g
Peanuts	10	10g	Sour cream		
Pecans	5 halves	5 g	12% milkfat	30 mL (2 tbs)	30 g
Pignolias, pine nuts	25 mL (5 tsp)	10 g	7% milkfat	60 mL (4 tbs)	60 g
Pistachios, shelled	20	10 g	Shortening*	5 mL (1 tsp)	
Pistachios, in shell	20	20 g			
Pumpkin and squash seeds	20 mL (4 tsp)	10 g			

*These items contain higher amounts of saturated fat.

Appendix F

TABLE F-7   Canadian Choice System: Extras

Extras have no more than 2.5 g carbohydrate, 60 kJ (14 kcal).

Vegetables 125 mL (½ c)

Artichokes
Asparagus
Bamboo shoots
Bean sprouts, mung or soya
Beans, string, green, or yellow
Bitter melon (balsam pear)
Bok choy
Broccoli
Brussels sprouts
Cabbage
Cauliflower
Celery
Chard
Cucumbers
Eggplant
Endive
Fiddleheads
Greens: beet, collard, dandelion, mustard, turnip, etc.
Kale
Kohlrabi
Leeks
Lettuce
Mushrooms
Okra
Onions, green or mature
Parsley
Peppers, green, yellow or red
Radishes
Rapini
Rhubarb
Sauerkraut
Shallots
Spinach
Sprouts: alfalfa, radish, etc.
Tomato wedges
Watercress
Zucchini

Free Foods (may be used without measuring)

Artificial sweetener, such as cyclamate or aspartame	Lime juice or lime wedges
	Marjoram, cinnamon, etc.
Baking powder, baking soda	Mineral water
Bouillon from cube, powder, or liquid	Mustard
	Parsley
Bouillon or clear broth	Pimentos
Chowchow, unsweetened	Salt, pepper, thyme
Coffee, clear	Soda water, club soda
Consommé	Soya sauce
Dulse	Sugar-free Crystal Drink
Flavorings and extracts	Sugar-free Jelly Powder
Garlic	Sugar-free soft drinks
Gelatin, unsweetened	Tea, clear
Ginger root	Vinegar
Herbal teas, unsweetened	Water
Horseradish, uncreamed	Worcestershire sauce
Lemon juice or lemon wedges	

Condiments

Food	Measure
Anchovies	2 fillets
Barbecue sauce	15 mL (1 tbs)
Bran, natural	30 mL (2 tbs)
Brewer's yeast	5 mL (1 tsp)
Carob powder	5 mL (1 tsp)
Catsup	5 mL (1 tsp)
Chili sauce	5 mL (1 tsp)
Cocoa powder	5 mL (1 tsp)
Cranberry sauce, unsweetened	15 mL (1 tbs)
Dietetic fruit spreads	5 mL (1 tsp)
Maraschino cherries	1
Nondairy coffee whitener	5 mL (1 tsp)
Nuts, chopped pieces	5 mL (1 tsp)
Pickles	
unsweetened dill	2
sour mixed	11
Sugar substitutes, granular	5 mL (1 tsp)
Whipped toppings	15 mL (1 tbs)

TABLE F-8 Canadian Choice System: Combined Food Choices

Food	Choices per Serving	Measure	Mass (Weight)
Angel food cake	½ starch + 2½ sugars	1/12 cake	50 g
Apple crisp	½ starch + 1½ fruits & vegetables + 1 sugar + 1–2 fats	125 mL (½ c)	
Applesauce, sweetened	1 fruits & vegetables + 1 sugar	125 mL (½ c)	
Beans and pork in tomato sauce	1 starch + ½ fruits & vegetables + ½ sugar + 1 protein	125 mL (½ c)	135 g
Beef burrito	2 starches + 3 proteins + 3 fats		110 g
Brownie	1 sugar + 1 fat	1	20 g
Cabbage rolls*	1 starch + 2 proteins	3	310 g
Caesar salad	2–4 fats	20 mL dressing (4 tsp)	
Cheesecake	½ starch + 2 sugars + ½ protein + 5 fats	1 piece	80 g
Chicken fingers	1 starch + 2 proteins + 2 fats	6 small	100 g
Chicken and snow pea Oriental	2 starches + ½ fruits & vegetables + 3 proteins + 1 fat	500 mL (2 c)	
Chili	1½ starches + ½ fruits & vegetables + 3½ protein	300 mL (1¼ c)	325 g
Chips			
Potato chips	1 starch + 2 fats	15 chips	30 g
Corn chips	1 starch + 2 fats	30 chips	30 g
Tortilla chips	1 starch + 1½ fats	13 chips	
Cheese twist	1 starch + 1½ fats	30 chips	30 g
Chocolate bar			
Aero®	2½ sugars + 2½ fats	bar	43 g
Smarties®	4½ sugars + 2 fats	package	60 g
Chocolate cake (without icing)	1 starch + 2 sugars + 3 fats	1/10 of a 8" pan	
Chocolate devil's food cake (without icing)	2 starches + 2 sugars + 3 fats	1/12 of a 9" pan	
Chocolate milk	2 milks 2% + 1 sugar	250 mL (1 c)	300 g
Clubhouse (triple-decker) sandwich	3 starches + 3 proteins + 4 fats		
Cookies			
Chocolate chip	½ starch + ½ sugar + 1½ fats	2	22 g
Oatmeal	1 starch + 1 sugar + 1 fat	2	40 g
Donut (chocolate glazed)	1 starch + 1½ sugars + 2 fats	1	65 g
Egg roll	1 starch + ½ protein + 1 fat		75 g
Four bean salad	1 starch + ½ protein + 1 fat	125 mL (½ c)	
French toast	1 starch + ½ protein + 2 fats	1 slice	65 g
Fruit in heavy syrup	1 fruits & vegetables + 1½ sugars	125 mL (½ c)	
Granola bar	½ starch + 1 sugar + 1–2 fats		30 g
Granola cereal	1 starch + 1 sugar + 2 fats	125 mL (½ c)	45 g
Hamburger	2 starches + 3 proteins + 2 fats	junior size	

*If eaten with sauce, add ½ fruits & vegetables exchange.

(continued on the next page)

F Appendix

TABLE F-8 — Canadian Choice System: Combined Food Choices—continued

Food	Choices per Serving	Measure	Mass (Weight)
Ice cream and cone, plain flavour			
Ice cream	½ milk + 2–3 sugars + 1–2 fats		100 g
Cone	½ sugar		4 g
Lasagna			
Regular cheese	1 starch + 1 fruits & vegetables + 3 proteins + 2 fats	3″ × 4″ piece	
Low-fat cheese	1 starch + 1 fruits & vegetables + 3 proteins	3″ × 4″ piece	
Legumes			
Dried beans (kidney, navy, pinto, fava, chick peas)	2 starches + 1 protein	250 mL (1 c)	180 g
Dried peas	2 starches + 1 protein	250 mL (1 c)	210 g
Lentils	2 starches + 1 protein	250 mL (1 c)	210 g
Macaroni and cheese	2 starches + 2 proteins + 2 fats	250 mL (1 c)	210 g
Minestrone soup	1½ starches + ½ fruits & vegetables + ½ fat	250 mL (1 c)	
Muffin	1 starch + ½ sugar + 1 fat	1 small	45 g
Nuts (dry or roasted without any oil added):			
Almonds, dried sliced	½ protein + 2 fats	50 mL (¼ c)	22 g
Brazil nuts, dried unblanched	½ protein + 2½ fats	5 large	23 g
Cashew nuts, dry roasted	½ starch + ½ protein + 2 fats	50 mL (¼ c)	28 g
Filbert hazelnut, dry	½ protein + 3½ fats	50 mL (¼ c)	30 g
Macadamia nuts, dried	½ protein + 4 fats	50 mL (¼ c)	28 g
Peanuts, raw	1 protein + 2 fats	50 mL (¼ c)	30 g
Pecans, dry roasted	½ fruits & vegetables + 3 fats	50 mL (¼ c)	22 g
Pine nuts, pignolia dried	1 protein + 3 fats	50 mL (¼ c)	34 g
Pistachio nuts, dried	½ fruits & vegetables + ½ protein + 2½ fats	50 mL (¼ c)	27 g
Pumpkin seeds, roasted	2 proteins + 2½ fats	50 mL (¼ c)	47 g
Sesame seeds, whole dried	½ fruits & vegetables + ½ protein + 2½ fats	50 mL (¼ c)	30 g
Sunflower kernel, dried	½ protein + 1½ fats	50 mL (¼ c)	17 g
Walnuts, dried chopped	½ protein + 3 fats	50 mL (¼ c)	26 g
Perogies	2 starches + 1 protein + 1 fat	3	
Pie, fruit	1 starch + 1 fruits & vegetables + 2 sugars + 3 fats	1 piece	120 g
Pizza, cheese	1 starch + 1 protein + 1 fat	1 slice (⅛ of a 12″)	50 g
Pork stir fry	½ to 1 fruits & vegetables + 3 proteins	200 mL (¾ c)	
Potato salad	1 starch + 1 fat	125 mL (½ c)	130 g
Potatoes, scalloped	2 starches + 1 milk + 1–2 fats	200 mL (¾ c)	210 g
Pudding, bread or rice	1 starch + 1 sugar + 1 fat	125 mL (½ c)	
Pudding, vanilla	1 milk + 2 sugars	125 mL (½ c)	
Raisin bran cereal	1 starch + ½ fruits & vegetables + ½ sugar	175 mL (⅔ c)	40 g
Rice krispie squares	½ starch + 1½ sugars + ½ fat	1 square	30 g
Shepherd's pie	2 starches + 1 fruits & vegetables + 3 proteins	325 mL (1⅓ c)	
Sherbet, orange	3 sugars + ½ fat	125 mL (½ c)	
Spaghetti and meat sauce	2 starches + 1 fruits & vegetables + 2 proteins + 3 fats	250 mL (1 c)	
Stew	2 starches + 2 fruits & vegetables + 3 proteins + ½ fat	200 mL (¾ c)	
Sundae	4 sugars + 3 fats	125 mL (½ c)	
Tuna casserole	1 starch + 2 proteins + ½ fat	125 mL (½ c)	
Yogurt, fruit bottom	1 fruits & vegetables + 1 milk + 1 sugar	125 mL (½ c)	125 g
Yogurt, frozen	1 milk + 1 sugar	125 mL (½ c)	125 g

Glossary

Note to the reader: If you have a customized book, this glossary may include some entries that will not be found in your book.

absorption
The process by which nutrients and other substances are transferred from the digestive system into body fluids for transport throughout the body.

adequate diet
A diet consisting of foods that together supply sufficient protein, vitamins, and minerals and enough calories to meet a person's need for energy.

Adequate Intakes (AIs)
Provisional RDAs developed when there is insufficient evidence to support a specific level of intake.

aerobic fitness
A state of respiratory and circulatory health as measured by the ability to deliver oxygen to muscles, and the capacity of muscles to use the oxygen for physical activity.

age-related macular degeneration
Eye damage caused by oxidation of the macula, the central portion of the eye that allows you to see details clearly. It is the leading cause of blindness in U.S. adults over the age of 65. Antioxidants provided by the carotenoids in dark green, leafy vegetables such as kale, collard greens, spinach, and Swiss chard may help prevent macular degeneration.

alcohol sugars
Simple sugars containing an alcohol group in their molecular structure. The most common are xylitol, mannitol, and sorbitol.

alcoholism
An illness characterized by a dependence on alcohol and by a level of alcohol intake that interferes with health, family and social relations, and job performance.

amylophagia (am-e-low-phag-ah)
Laundry starch or cornstarch eating.

anaphylactic shock (an-ah-fa-lac-tic)
Reduced oxygen supply to the heart and other tissues due to the body's reaction to an allergen in food. Symptoms of anaphylactic shock include paleness, a flush discoloration of the lips and fingertips, a "faceless" expression, weak and rapid pulse, and difficulty breathing.

anorexia nervosa
An eating disorder characterized by extreme weight loss, poor body image, and irrational fears of weight gain and obesity.

antibodies
Blood proteins that help the body fight particular diseases. They help the body develop an immunity, or resistance, to many diseases.

antioxidants
Chemical substances that prevent or repair damage to cells caused by exposure to oxidizing agents such as environmental pollutants, smoke, ozone, and oxygen. Oxidation reactions are also a normal part of cellular processes. Beta-carotene, vitamin E, and vitamin C function as antioxidants.

appetite
The desire to eat; a pleasant sensation that is aroused by thoughts of the taste and enjoyment of food.

association
The finding that one condition is correlated with, or related to another condition, such as a disease or disorder. For example, diets low in vegetables are associated with breast cancer. Associations do *not* prove that one condition (such as a diet low in vegetables) *causes* an event (such as breast cancer). They indicate that a statistically significant relationship between a condition and an event exists.

atherosclerosis
"Hardening of the arteries" due to a build-up of plaque.

ATP, ADP Adenosine triphosphate (ah-den-o-scene tri-phos-fate) and adenosine diphosphate. Molecules containing a form of phosphorous that can trap energy obtained from the macronutrients. ADP becomes ATP when it traps energy, and returns to being ADP when it releases energy for muscular and other work.

balanced diet
A diet that provides neither too much nor too little of nutrients and other components of food such as fat and fiber.

basal metabolism
Energy used to support body processes such as growth, health, tissue repair and maintenance, and other functions. Assessed while at rest, basal metabolism includes energy the body expends for breathing, the pumping of the heart, the maintenance of body temperature, and other life-sustaining, ongoing functions.

bile
A yellowish-brown or green fluid produced by the liver, stored in the gallbladder, and secreted into the small intestine. It acts like a detergent, breaking down globs of fat entering the small intestine to droplets, making the fats more accessible to the action of lipase.

binge eating
The consumption of a large amount of food in a small amount of time.

binge-eating disorder
An eating disorder characterized by periodic binge eating, which normally is not followed by vomiting or the use of laxatives. People must experience eating binges twice a week on average over a period of 6 months to qualify for the diagnosis.

bioavailability
The amount of a nutrient consumed that is available for absorption and use by the body.

biotechnology
As applied to food products, the process of modifying the composition of foods by biologically altering their genetic makeup. Also called *genetic engineering* of foods. The food products produced are sometimes referred to as "GM" and GMOs (genetically modified organisms).

body mass index (BMI)
An indicator of the appropriateness of a person's weight for their height. It is calculated by dividing weight in kilograms by height in meters. It can also be calculated using the method shown in Table 9.2.

bulimia nervosa
An eating disorder characterized by recurrent episodes of rapid, uncontrolled eating of large amounts of food in a short period of time. Episodes of binge eating are often followed by purging.

calorie (calor = heat)
A unit of measure used to express the amount of energy produced by foods in the form of heat. The calorie used in nutrition is the large "Calorie," or the "kilocalorie" (kcal). It equals the amount of energy needed to raise the temperature of 1 kilogram of water (about 4 cups) from 15 to 16°C (59 to 61°F). The term *kilocalorie*, or "calorie" as used in this text, is gradually being replaced by the "kilojoule" (kJ) in the United States; 1 kcal = 4.2 kJ.

cancer
A group of diseases in which abnormal cells grow out of control and can spread throughout the body. Cancer is not contagious and has many causes.

carbohydrates
Chemical substances in foods that consist of a simple sugar molecule or multiples of them in various forms.

cardiovascular disease
Disorders related to plaque build-up in arteries of the heart, brain, and other organs and tissues.

cataracts
Complete or partial clouding over the lens of the eye.

cause and effect
A finding that demonstrates that a condition causes a particular event. For example, vitamin C deficiency causes the deficiency disease scurvy.

celiac disease
A disease characterized by inflammation of the small intestine lining resulting from a genetically based intolerance to gluten. The inflammation produces diarrhea, fatty stools, weight loss, and vitamin and mineral deficiencies. (Also called *celiac sprue* and *gluten-sensitive enteropathy.*)

cholesterol
A fat-soluble, colorless liquid found in many animal products but not in plants. High cholesterol intake is related to the development of heart disease.

chronic diseases
Slow-developing, long-lasting diseases that are not contagious (e.g., heart disease, cancer, diabetes). They can be treated but not always cured.

circulatory system
The heart, arteries, capillaries, and veins responsible for circulating blood throughout the body.

cirrhosis of the liver
Degeneration of the liver, usually caused by excessive alcohol intake over a number of years.

clinical trial
A study design in which one group of randomly assigned subjects (or subjects selected by the "luck of the draw") receives an active treatment and another group receives an inactive treatment, or "sugar pill," called the placebo.

coenzymes
Chemical substances, including many vitamins, that activate specific enzymes. Activated enzymes increase the rate at which reactions take place in the body, such as the breakdown of fats or carbohydrates in the small intestine and the conversion of glucose and fatty acids into energy within cells.

cofactors
Individual minerals required for the activity of certain proteins. For example: • iron is needed for hemoglobin's function in oxygen and carbon dioxide transport, • zinc is needed to activate or is a structural component of over 200 enzymes, and • magnesium activates over 300 enzymes involved in the formation of energy and proteins.

colostrum
The milk produced during the first few days after delivery. It contains more antibodies, protein, and certain minerals than the mature milk that is produced later. It is thicker than mature milk and has a yellowish color.

complementary protein sources
Plant sources of protein that together provide sufficient quantities of the nine essential amino acids.

complete proteins
Proteins that contain all of the essential amino acids in amounts needed to support growth and tissue maintenance.

complex carbohydrates
The form of carbohydrate found in starchy vegetables, grains, and dried beans and in many types of dietary fiber. The most common form of starch is made of long chains of interconnected glucose units.

control group
Subjects in a study who do not receive the active treatment or who do *not* have the condition under investigation. Control periods, or times when subjects are not receiving the treatment, are sometimes used instead of a control group.

critical period
A specific interval of time during which cells of a tissue or organ are genetically programmed to multiply. If the supply of nutrients needed for cell multiplication is not available during the specific time interval, the growth and development of the tissue or organ are permanently impaired.

cruciferous vegetables
Sulfur-containing vegetables whose outer leaves form a cross (or crucifix). Vegetables in this family include broccoli, cabbage, cauliflower, brussels sprouts, mustard and collard greens, kale, bok choy, kohlrabi, rutabaga, turnips, broccoflower, and watercress.

Daily Values (DVs)
Scientifically agreed-upon daily dietary intake standards for fat, saturated fat, cholesterol, carbohydrate, dietary fiber, and protein intake compatible with health. DVs are intended for use on nutrition labels only and are listed in the Nutrition Facts panel. "% Daily Value" on Nutrition Facts panels is calculated as the percentage of each DV supplied by a serving of the labeled food.

development
Processes involved in enhancing functional capabilities. For example, the brain grows, but the ability to reason develops.

diabetes
A disease characterized by abnormal utilization of carbohydrates by the body and elevated blood glucose levels. There are three main types of diabetes: type 1, type 2, and gestational diabetes. The word *diabetes* in this unit refers to type 2, which is by far the most common form of diabetes.

diarrhea
The presence of three or more liquid stools in a 24-hour period.

dietary fiber
Naturally occurring, intact forms of nondigestible carbohydrates in plants and "woody" plant cell walls. Oat and wheat bran, and raffinose in dried beans, are examples of this type of fiber.

dietary supplements
Any products intended to supplement the diet, including vitamin and mineral supplements; proteins, enzymes, and amino acids; fish oils and fatty acids; hormones and hormone precursors; and herbs and other plant extracts. Such products must be labeled "Dietary Supplement."

dietary thermogenesis
Thermogenesis means "the production of heat." Dietary thermogenesis is the energy expended during food ingestion, the digestion of food, and the absorption and utilization of nutrients. Some of the energy escapes as heat. It accounts for approximately 10% of the body's total energy need. Also called *diet-induced thermogenesis* and *thermic effect of foods or feeding.*

digestion
The mechanical and chemical processes whereby ingested food is converted into substances that can be absorbed by the intestinal tract and utilized by the body.

disaccharides (di = two, saccharide = sugar)
Simple sugars consisting of two molecules of monosaccharides linked together. Sucrose, maltose, and lactose are disaccharides.

DNA (deoxyribonucleic acid)
Genetic material contained in cells that initiates and directs the production of proteins in the body.

double blind
A study in which neither the subjects participating in the research nor the scientists performing the research know which subjects are receiving the treatment and which are getting the placebo. Both subjects and investigators are "blind" to the treatment administered.

double-blind, placebo-controlled food challenge
A test used to determine the presence of a food allergy or other adverse reaction to a food. In this test, neither the patient nor the care provider knows whether a suspected offending food or a placebo is being tested.

duodenal (do-odd-en-all) and stomach ulcers
Open sores in the lining of the duodenum (the uppermost part of the small intestine) or the stomach.

edema
Swelling due to an accumulation of fluid in body tissues.

electrolytes
Minerals such as sodium and potassium that carry a charge when in solution. Many electrolytes help the body maintain an appropriate amount of fluid.

empty-calorie foods
Foods that provide an excess of calories in relation to nutrients. Soft drinks, candy, sugar, alcohol, and fats are considered empty-calorie foods.

enrichment
The replacement of thiamin, riboflavin, niacin, and iron lost when grains are refined.

environmental trigger
An environmental factor, such as inactivity, a high-fat diet, or a high sodium intake, that causes a genetic tendency toward a disorder to be expressed.

enzymes
Protein substances that speed up chemical reactions. Enzymes are found throughout the body but are present in particularly large amounts in the digestive system.

epidemiological studies
Research that seeks to identify conditions related to particular events within a population. This type of research does *not* identify cause-and-effect relationships. For example, much of the information known about diet and cancer is based on epidemiological studies that have found that diets low in vegetables and fruits are associated with the development of heart disease.

ergogenic aids (ergo = work; genic = producing)
Substances that increase the capacity for muscular work.

essential amino acids
Amino acids that cannot be synthesized in adequate amounts by humans and therefore must be obtained from the diet.

They are sometimes referred to as "indispensable amino acids."

essential fatty acids
Components of fats (linoleic acid—pronounced lynn-oh-lay-ick and alpha-linolenic acid—lynn-oh-len-ick) required in the diet.

essential hypertension
Hypertension of no known cause; also called primary or idiopathic hypertension, it accounts for 90–95% of all cases of hypertension.

essential nutrients
Substances required for normal growth and health that the body cannot generally produce, or produce in sufficient amounts; they must be obtained in the diet.

experimental group
Subjects in a study who receive the treatment being tested or have the condition that is being investigated.

fermentation
The process by which carbohydrates are converted to ethanol by the action of the enzymes in yeast.

fetus
A baby in the womb from the eighth week of pregnancy until birth. (Before then, it is referred to as an embryo.)

flatulence (flat-u-lens)
Presence of excess gas in the stomach and intestines.

food additives
Any substances added to food that become part of the food or affect the characteristics of the food. The term applies to substances added both intentionally and unintentionally to food.

food allergen
A substance in food (almost always a protein) that is identified as harmful by the body and elicits an allergic reaction from the immune system.

food allergy
Adverse reaction to a normally harmless substance in food that involves the body's immune system. (Also called food hypersensitivity.)

food insecurity
Limited or uncertain availability of safe, nutritious foods.

food intolerance
Adverse reaction to a normally harmless substance in food that does not involve the body's immune system.

food security
Access at all times to a sufficient supply of safe, nutritious foods.

foodborne illnesses
An illness related to consumption of foods or beverages containing disease-causing bacteria, viruses, parasites, toxins, or other contaminants.

fortification
The addition of one or more vitamins and/or minerals to a food product.

free radicals
Chemical substances (usually oxygen or hydrogen) that are missing an electron. The absence of the electron makes the chemical substances reactive and prone to oxidizing nearby atoms or molecules by stealing an electron from them.

fruitarian
A form of vegetarian diet in which fruits are the major ingredient. Such diets provide inadequate amounts of a variety of nutrients.

functional fiber
Specific types of nondigestible carbohydrates and connective tissues that have beneficial effects on health. Two examples of functional fibers are psyllium and pectin.

functional foods
Generally taken to mean foods, fortified foods, and enhanced food products that may benefit health beyond the effects of essential nutrients they contain.

geophagia (ge-oh-phag-ah)
Clay or dirt eating.

gestational diabetes
Diabetes first discovered during pregnancy.

glycemic index (GI)
A measure of the extent to which blood glucose level is raised by a 50-gram portion of a carbohydrate-containing food compared to 50 grams of glucose or white bread.

glycemic load (GL)
A measure of the extent to which blood glucose level is raised by a given amount of a carbohydrate-containing food. GL is calculated by multiplying a food's GI by its carbohydrate content.

glycogen
The body's storage form of glucose. Glycogen is stored in the liver and muscles.

growth
A process characterized by increases in cell number and size.

heart disease
One of a number of disorders that result when circulation of blood to parts of the heart is inadequate. Also called coronary heart disease. ("Coronary" refers to the blood vessels at the top of the heart. They look somewhat like a crown.)

heartburn
A condition that results when acidic stomach contents are released into the esophagus, usually causing a burning sensation.

hemoglobin
The iron-containing protein in red blood cells.

hemorrhoids (hem-or-oids)
Swelling of veins in the anus or rectum.

histamine (hiss-tah-mean)
A substance released in allergic reactions. It causes dilation of blood vessels, itching, hives, and a drop in blood pressure and stimulates the release of stomach acids and other fluids. Antihistamines neutralize the effects of histamine and are used in the treatment of some cases of allergies.

homocysteine
A compound produced when the amino acid methionine is converted to another amino acid, cysteine. High blood levels of homocysteine increase the risk of hardening of the arteries, heart attack, and stroke.

hunger
Unpleasant physical and psychological sensations (weakness, stomach pains, irritability) that lead people to acquire and ingest food.

hydrogenation
The addition of hydrogen to unsaturated fatty acids.

hypertension
High blood pressure. It is defined as blood pressure exerted inside blood vessel walls that typically exceeds 140/90 millimeters of mercury.

hypoglycemia
A disorder resulting from abnormally low blood glucose levels. Symptoms of hypoglycemia include irritability, nervousness, weakness, sweating, and hunger. These symptoms are relieved by consuming glucose or foods that provide carbohydrate.

hypothesis
A statement made prior to initiating a study of the relationship sought to be proved (found to be true) by the research.

immune system
Body tissues that provide protection against bacteria, viruses, and other substances identified by cells as harmful.

incomplete proteins
Proteins that are deficient in one or more essential amino acids.

initiation
The start of the cancer process; it begins with the alteration of DNA within cells.

insulin resistance
A condition in which cell membrane have reduced sensitivity to insulin so that more insulin than normal is required to transport a given amount of glucose into cells. It is characterized by elevated levels of serum insulin, glucose, and triglycerides, and increased blood pressure.

iron deficiency
A disorder that results from a depletion of iron stores in the body. It is characterized by weakness, fatigue, short attention span, poor appetite, increased susceptibility to infection, and irritability.

iron-deficiency anemia
A condition that results when the content of hemoglobin in red blood cells is reduced due to a lack of iron. It is characterized by the signs of iron deficiency plus paleness, exhaustion, and a rapid heart rate.

irritable bowel syndrome (IBS)
A disorder of bowel function characterized by chronic or episodic gas, abdominal pain, diarrhea or constipation, or both.

kwashiorkor (kwa-she-or-kor)
A deficiency disease primarily caused by a lack of complete protein in the diet. It usually occurs after children are taken off breast milk and given solid foods containing low-quality protein.

lactose intolerance
The term for gastrointestinal symptoms (flatulence, bloating, abdominal pain, diarrhea, and "rumbling in the bowel") resulting from the consumption of more lactose than can be digested with available lactase.

lactose intolerance
The term for gastrointestinal symptoms (flatulence, bloating, abdominal pain, diarrhea, and "rumbling in the bowel") resulting from the consumption of more lactose than can be digested with available lactase.

lactose maldigestion
A disorder characterized by reduced digestion of lactose due to the low availability of the enzyme lactase.

life expectancy
The average length of life of people of a given age.

lipids
Compounds that are insoluble in water and soluble in fat. Triglycerides, saturated and unsaturated fats, and essential fatty acids are examples of lipids, or "fats."

lymphatic system
A network of vessels that absorb some of the products of digestion and transport them to the heart, where they are mixed with the substances contained in blood.

macronutrients
The group name for the energy-yielding nutrients of carbohydrate, protein, and fat. They are called macronutrients because we need relatively large amounts of them in our daily diet.

malnutrition
Poor nutrition resulting from an excess or lack of calories or nutrients.

marasmus
A condition of severe body wasting due to deficiencies of both protein and calories. Also called protein-energy malnutrition and protein-calorie malnutrition.

maximal oxygen consumption
The highest amount of oxygen that can be delivered to, and utilized by, muscles for physical activity. Also called VO_2 max and maximal volume of oxygen.

meta-analysis
An analysis of data from multiple studies. Results are based on larger samples than the individual studies and are therefore more reliable. Differences in methods and subjects among the studies may bias the results of meta-analyses.

metabolic syndrome
A constellation of metabolic abnormalities that increase the risk of heart disease and cancer. Metabolic syndrome is characterized by insulin resistance, abdominal obesity, high blood pressure and triglycerides levels, low levels of HDL cholesterol, and impaired glucose tolerance. It is also called *Syndrome X* and *insulin resistance syndrome*.

metabolism
The chemical changes that take place in the body. The conversion of glucose to energy or to body fat is an example of a metabolic process.

minerals
In the context of nutrition, minerals are specific, single atoms that perform particular functions in the body. There are 15 essential minerals—or minerals required in the diet.

monosaccharides
(mono = one, saccharide = sugar): Simple sugars consisting of one sugar molecule. Glucose, fructose, and galactose are monosaccharides.

myoglobin
The iron-containing protein in muscle cells.

neural tube defects
Malformations of the spinal cord and brain. They are among the most common and severe fetal malformations, occurring in approximately 1 in every 1000 pregnancies. Neural tube defects include

spina bifida (spinal cord fluid protrudes through a gap in the spinal cord; shown in Illustration 29.6), anencephaly (absence of the brain or spinal cord), and encephalocele (protrusion of the brain through the skull).

nonessential amino acids
Amino acids that can be readily produced by humans from components of the diet. Also referred to as "dispensable amino acids."

nonessential nutrients
Nutrients required for normal growth and health that the body can manufacture in sufficient quantities from other components of the diet. We do not require a dietary source of nonessential nutrients.

nutrient-dense foods
Foods that contain relatively high amounts of nutrients compared to their calorie value. Broccoli, collards, bread, cantaloupe, and lean meats are examples of nutrient-dense foods.

nutrients
Chemical substances found in food that are used by the body for growth and health. The six categories of nutrients are carbohydrates, proteins, fats, vitamins, minerals, and water.

nutrition
The study of foods, their nutrients and other chemical constituents, and the effects of food constituents on health.

obesity
A condition characterized by excess body fat.

osteoporosis (osteo = bones; poro = porous, osis = abnormal condition)
A condition in which bones become fragile and susceptible to fracture due to a loss of calcium and other minerals.

overweight
A high weight-for-height.

pagophagia (pa-go-phag-ah)
Ice eating.

peer review
Evaluation of the scientific merit of research or scientific reports by experts in the area under review. Studies published in scientific journals have gone through peer review prior to being accepted for publication.

% Daily Value (%DV)
Scientifically agreed-upon standards of daily intake of nutrients from the diet developed for use on nutrition labels. The "% Daily Values" listed in nutrition labels represent the percentages of the standards obtained from one serving of the food product.

phenylketonuria (feen-ol-key-tone-u-re-ah), PKU
A rare genetic disorder related to the lack of the enzyme phenylalanine hydroxylase. Lack of this enzyme causes the essential amino acid phenylalanine to build up in blood.

physical fitness
The health of the body as measured by muscular strength, endurance, and flexibility in the conduct of physical activity.

phytochemicals (phyto = plant)
Chemical substances in plants, some of which perform important functions in the human body. Phytochemicals give plants color and flavor, participate in processes that enable plants to grow, and protect plants against insects and diseases.

pica (pike-eh)
The regular consumption of nonfood substances such as clay or laundry starch.

placebo
A "sugar pill," an imitation treatment given to subjects in research.

placebo effect
Changes in health or perceived health that result from expectations that a "treatment" will produce an effect on health.

plant stanols or sterols
Substances in corn, wheat, oats, rye, olives, wood, and some other plants that are similar in structure to cholesterol but that are not absorbed by the body. They decrease cholesterol absorption.

plaque (dental)
A soft, sticky, white material on teeth; formed by bacteria.

plaque (arterial)
Deposits of cholesterol, other fats, calcium, and cell materials in the lining of the inner wall of arteries.

plumbism
Lead (primarily from old paint flakes) eating.

polysaccharides (poly = many, saccharide = sugar)
Carbohydrates containing many molecules of sugar linked together. Starch, glycogen, and dietary fiber are the three major types of polysaccharides.

prebiotics
"Intestinal fertilizer." Certain fiber-like forms of nondigestible carbohydrates that support the growth of beneficial bacteria in the gut.

precursor
In nutrition, a nutrient that can be converted into another nutrient. (Also called provitamin.) Beta-carotene is a precursor of vitamin A.

prediabetes
A condition in which blood glucose levels are higher than normal but not high enough for the diagnosis of diabetes. It is characterized by impaired glucose tolerance, or fasting blood glucose levels between 110 and 126 mg/dl.

probiotics
Non-harmful bacteria and some yeasts that help colonize the intestinal tract with beneficial microorganisms and that sometimes replace colonies of harmful microorganisms. Most common probiotic strains are Lactobacilli and Bifidobacteria.

progression
The uncontrolled growth of abnormal cells.

promotion
The period in cancer development when the number of cells with altered DNA increases.

prostate
A gland located above the testicles in males. The prostate secretes a fluid that surrounds sperm.

protein
Chemical substance in foods made up of chains of amino acids.

purging
The use of self-induced vomiting, laxatives, or diuretics (water pills) to prevent weight gain.

Recommended Dietary Allowances (RDAs)
Intake levels of essential nutrients that meet the nutritional needs of practically all healthy people while decreasing the risk of certain chronic diseases.

remodeling
The breakdown and buildup of bone tissue.

restrained eating
The purposeful restriction of food intake below desired amounts in order to control body weight.

salt sensitivity
A genetically determined condition in which a person's blood pressure rises when high amounts of salt or sodium are consumed. Such individuals are sometimes identified by blood pressure increases of 10% or more when switched from a low-salt (1–3 grams) to a high-salt (12–15 grams) diet.

satiety
A feeling of fullness or of having had enough to eat.

saturated fats
The type of fat that tends to raise blood cholesterol levels and the risk for heart disease. They are solid at room temperature and are found primarily in animal products such as meat, butter, and cheese.

serotonin (pronounced sare-uh-tone-in)
A neurotransmitter, or chemical messenger, for nerve cell activities that excite or inhibit various behaviors and body functions. It plays a role in mood, appetite regulation, food intake, respiration, pain transmission, blood vessel constriction, and other body processes.

simple sugars
Carbohydrates that consist of a glucose, fructose, or galactose molecule; or a combination of glucose and either fructose or galactose. High-fructose corn syrup and alcohol sugars are also considered simple sugars. Simple sugars are often referred to as "sugars."

single-gene defects
Disorders resulting from one abnormal gene. Also called "inborn errors of metabolism." Over 800 single-gene defects have been cataloged, and most are very rare.

starch
Complex carbohydrates made up of complex chains of glucose molecules. Starch is the primary storage form of carbohydrate in plants. The vast majority of carbohydrate in our diet consists of starch, monosaccharides, and disaccharides.

statistically significant
Research findings that likely represent a true or actual result and not one due to chance.

stroke
The event that occurs when a blood vessel in the brain suddenly ruptures or becomes blocked, cutting off blood supply to a portion of the brain. Stroke is often associated with "hardening of the arteries" in the brain. (Also called a *cerebral vascular accident*.)

tooth decay
The disintegration of teeth due to acids produced by bacteria in the mouth that feed on sugar. Also called dental caries or cavities.

total fiber
The sum of functional and dietary fiber.

trans fats
A type of unsaturated fat present in hydrogenated oils, margarine, shortening, pastries, and some cooking oils that increase the risk of heart disease. Fats containing fatty acids in the trans form are generally referred to as trans fats.

trimester
One-third of the normal duration of pregnancy. The first trimester is 0 to 13 weeks, the second is 13 to 26 weeks, and the third is 26 to 40 weeks.

tryptophan (pronounced trip-tuh-fan)
An essential amino acid that is used to form the chemical messenger serotonin (among other functions). Tryptophan is generally present in lower amounts in food protein than most other essential amino acids. It can be produced in the body from niacin, a B vitamin.

type 1 diabetes
A disease characterized by high blood glucose levels resulting from destruction of the insulin-producing cells of the pancreas. This type of diabetes was called juvenile-onset diabetes and insulin-dependent diabetes in the past, and its official medical name is type 1 diabetes mellitus.

type 2 diabetes
A disease characterized by high blood glucose levels due to the body's inability to use insulin normally, or to produce enough insulin. This type of diabetes was called adult-onset diabetes and non-insulin-dependent diabetes in the past, and its official medical name is type 2 diabetes mellitus.

underweight
Usually defined as a low weight-for-height. May also represent a deficit of body fat.

unsaturated fats
The type of fat that tends to lower blood cholesterol level and the risk of heart disease. They are liquid at room temperature and found in foods such as nuts, seeds, fish, shellfish, and vegetable oils.

vitamins
Chemical substances that perform specific functions in the body.

water balance
The ratio of the amount of water outside cells to the amount inside cells; this balance is needed for normal cell functioning.

zoochemicals
Chemical substances in animal foods, some of which perform important functions in the body.

Index

Note to the reader: If you have a customized book, this index may include some entries that will not be found in your book.

Daily Values for Food Labels

The Daily Values are standard values developed by the Food and Drug Administration (FDA) for use on food labels. The values are based on 2000 kcalories a day for adults and children over 4 years old.

Proteins, Vitamins, and Minerals

Nutrient	Amount
Protein[a]	50 g
Thiamin	1.5 mg
Riboflavin	1.7 mg
Niacin	20 mg NE
Biotin	300 µg
Pantothenic acid	10 mg
Vitamin B_6	2 mg
Folate	400 µg
Vitamin B_{12}	6 µg
Vitamin C	60 mg
Vitamin A	5000 IU[b]
Vitamin D	400 IU[b]
Vitamin E	30 IU[b]
Vitamin K	80 µg
Calcium	1000 mg
Iron	18 mg
Zinc	15 mg
Iodine	150 µg
Copper	2 mg
Chromium	120 µg
Selenium	70 µg
Molybdenum	75µg
Manganese	2 mg
Chloride	3400 mg
Magnesium	400 mg
Phosphorus	1000 mg

[a] The Daily Values for protein vary for different groups of people: pregnant women, 60 g; nursing mothers, 65 g; infants under 1 year, 14 g; children 1 to 4 years, 16 g.
[b] Equivalent values for nutrients expressed as IU are: vitamin A, 1500 RAE (assumes a mixture of 40% retinol and 60% beta-carotene); vitamin D, 10 µg; vitamin E, 20 mg.

Nutrients and Food Components

Food Component	Amount	Calculation Factors
Fat	65 g	30% of kcalories
Saturated fat	20 g	10% of kcalories
Cholesterol	300 mg	Same regardless of kcalories
Carbohydrate (total)	300 g	60% of kcalories
Fiber	25 g	11.5 g per 1000 kcalories
Protein	50 g	10% of kcalories
Sodium	2400 mg	Same regardless of kcalories
Potassium	3500 mg	Same regardless of kcalories

GLOSSARY OF NUTRIENT MEASURES

kcal: kcalories; a unit by which energy is measured (Chapter 1 provides more details).

g: grams; a unit of weight equivalent to about 0.03 ounces.

mg: milligrams; one-thousandth of a gram.

µg: micrograms; one-millionth of a gram.

IU: international units; an old measure of vitamin activity determined by biological methods (as opposed to new measures that are determined by direct chemical analyses). Many fortified foods and supplements use IU on their labels.
- For vitamin A, 1 IU = 0.3 µg retinol, 3.6 µg β-carotene, or 7.2 µg other vitamin A carotenoids.
- For vitamin D, 1 IU = 0.025 µg cholecalciferol.
- For vitamin E, 1 IU = 0.67 natural α-tocopherol (other conversion factors are used for different forms of vitamin E).

mg NE: milligrams niacin equivalents; a measure of niacin activity (Chapter 10 provides more details).
- 1 NE = 1 mg niacin.
  = 60 mg tryptophan (an amino acid).

µg DFE: micrograms dietary folate equivalents; a measure of folate activity (Chapter 10 provides more details).
- 1 µg DFE = 1 µg food folate.
  = 0.6 µg fortified food or supplement folate.
  = 0.5 µg supplement folate taken on an empty stomach.

µg RAE: micrograms retinol activity equivalents; a measure of vitamin A activity (Chapter 11 provides more details).
- 1 µg RAE = 1 µg retinol.
  = 12 µg β-carotene.
  = 24 µg other vitamin A carotenoids.

mmol: millimoles; one-thousanth of a mole, the molecular weight of a substance. To convert mmol to mg, multiply by the atomic weight of the substance.
- For sodium, mmol × 23 = mg Na.
- For chloride, mmol × 35.5 = mg Cl.
- For sodium chloride, mmol × 58.5 = mg NaCl.